Graphing Concepts

Odd Functions
A function f is an odd function if $f(-x) = -f(x)$ for all x in the domain of f. The graph of an odd function is symmetric with respect to the origin.

Even Functions
A function is an even function if $f(-x) = f(x)$ for all x in the domain of f. The graph of an even function is symmetric with respect to the y-axis.

Vertical and Horizontal Translations
If f is a function and c is a positive constant, then the graph of

- $y = f(x) + c$ is the graph of $y = f(x)$ shifted up vertically c units.
- $y = f(x) - c$ is the graph of $y = f(x)$ shifted down vertically c units.
- $y = f(x + c)$ is the graph of $y = f(x)$ shifted left horizontally c units.
- $y = f(x - c)$ is the graph of $y = f(x)$ shifted right horizontally c units.

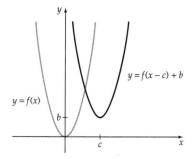

Reflections
If f is a function then the graph of

- $y = -f(x)$ is the graph of $y = f(x)$ reflected across the x-axis.
- $y = f(-x)$ is the graph of $y = f(x)$ reflected across the y-axis.

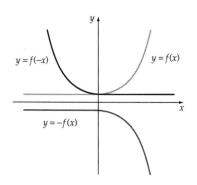

Vertical Shrinking and Stretching
- If $c > 0$ and the graph of $y = f(x)$ contains the point (x, y), then the graph of $y = c \cdot f(x)$ contains the point (x, cy).
- If $c > 1$, the graph of $y = cf(x)$ is obtained by stretching the graph of $y = f(x)$ away from the x-axis by a factor of c.
- If $0 < c < 1$, the graph of $y = cf(x)$ is obtained by shrinking the graph of $y = f(x)$ toward the x-axis by a factor of c.

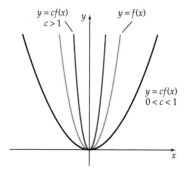

Horizontal Shrinking and Stretching
- If $a > 0$ and the graph of $y = f(x)$ contains the point (x, y), then the graph of $y = f(ax)$ contains the point $\left(\dfrac{1}{a}x, y \right)$.
- If $a > 1$, the graph of $y = f(ax)$ is a *horizontal shrinking* of the graph of $y = f(x)$.
- If $0 < a < 1$, the graph of $y = f(ax)$ is a *horizontal stretching* of the graph of $y = f(x)$.

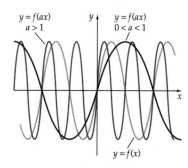

COLLEGE
ALGEBRA

COLLEGE
ALGEBRA

THIRD EDITION

Richard N. Aufmann | Vernon C. Barker | Richard D. Nation

Palomar College

HOUGHTON MIFFLIN COMPANY   Boston   New York

Sponsoring Editor: *Maureen O'Connor*
Associate Editor: *Dawn Nuttall*
Senior Project Editor: *Cynthia Harvey*
Associate Project Editor: *Tamela Ambush*
Senior Production/Design Coordinator: *Carol Merrigan*
Manufacturing Manager: *Florence Cadran*
Marketing Manager: *Charles Cavaliere*

Cover concept and design: *Stoltze Design*
Photographer: *Masao Ota/Photonica*

PHOTO CREDITS

Chapter 1: p. 1, (top) The Granger Collection, (middle) Michael Dwyer/Stock Boston, (bottom) Francesco Regineto/The Image Bank. **Chapter 2:** p. 69, (top) Turner and Devries/The Image Bank, (middle) David Young Wolff/Tony Stone Images, (bottom) Grant Faint/The Image Bank. **Chapter 3:** p. 131, (top) Bill Gallery/Stock Boston, (middle) G & V Chapman/The Image Bank, (bottom) Bill Gallery/Stock Boston. **Chapter 4:** p. 233, (top) Benn Mitchell/The Image Bank, (middle) Benn Mitchell/The Image Bank, (bottom) John P. Kelly/The Image Bank. **Chapter 5:** p. 289, (top) Tony Craddock/Tony Stone Images, (middle) Frank Siteman/Stock Boston, (bottom) Edward Miller/Stock Boston, (in text) Coris/Bettmann. **Chapter 6:** p. 345, (top) Hitoshi Ikemata/The Image Bank, (middle) John Livzey/Tony Stone Images, (bottom) Ken Whitmore/Tony Stone Images. **Chapter 7:** p. 381, (top) Michael Grecco/Stock Boston, (middle) Michael Melford/The Image Bank, (bottom) Marc Romanelli/The Image Bank. **Chapter 8:** p. 435, (top) Max Dannebaum/The Image Bank, (middle) Tony Stone Images, (bottom) The Image Bank. **Chapter 9:** p. 491, (top) Stephen Johnson/Tony Stone Images, (middle) Stephen Johnson/Tony Stone Images, (bottom) Stephen Johnson/Tony Stone Images.

Printed in the U.S.A.

Library of Congress Catalog Card Number: 96-76856

ISBN:
Text: 0-395-78644-4
Instructor's Annotated Edition: 0-395-81530-4

3456789-VH-00   99   98

CONTENTS

9 SEQUENCES, SERIES, AND PROBABILITY   491

PREFACE

Mathematics education continues to evolve at an ever-increasing rate. There is greater emphasis on *doing* mathematics rather than merely duplicating mathematics through extensive drill. Students are urged to investigate concepts, apply those concepts, and then present their findings. Technological advances in graphing calculators, computers, and software make it possible to explore the interdependence of mathematics and its application.

This third edition of *College Algebra* continues to build on the success of the second edition, enhancing the features of that edition and including new features that demonstrate the dynamic link between math concepts and math models. The special features include

- Topics for Discussion

- Projects

- Exploring Concepts with Technology

Topics for Discussion precede the exercise set in each section of the text. These topics can serve as group discussion or writing assignments.

Projects included at the end of each exercise set are designed to encourage students to research and write about mathematics and its applications. In the *Instructor's Resource Manual* there are additional projects that may be assigned. To ensure that projects are contemporary and in adequate supply, we have written additional projects that can be found on the Internet. Also included there are more modeling exercises arranged by subject area. These problems are updated once a semester and can be found at http://www.hmco.com.

A special end-of-chapter feature, **Exploring Concepts with Technology,** extends ideas introduced in the text by using technology to investigate applications or mathematical topics. These explorations can serve as group projects, class discussions, or extra-credit assignments.

 Technology is introduced very naturally in the text to illustrate or enhance a concept. Our intention is to demonstrate appropriate technology when necessary to support a concept. We try to foster the idea that the concept motivates the use of technology, not to introduce technology for technology's sake. The technology icon (shown at this beginning of this paragraph) is used throughout the text to identify discussions of specific techniques for using graphing utilities, as well as examples and exercises that call for the use of technology. Optional exercises requiring the use of a graphing utility are designed to develop in students an appreciation for both the power and the limitations of technology. These exercises are supplemented by our new *Graphing Workbook*, which offers over 600 additional problems that may be worked using graphing utilities.

Despite the changes to this edition, we have retained our basic goal: to provide a comprehensive and mathematically sound treatment of the topics considered essential for a college algebra and trigonometry course. To help students master concepts, we have tried to maintain a balance among theory, application, modeling, and drill. Carefully developed mathematics is complemented by abundant, creative applications that are contemporary and represent a wide range of disciplines. Many application problems are accompanied by a diagram that helps the student visualize the mathematics involved.

The Features in Detail

Interactive Presentation   *College Algebra* is written in a style that encourages the student to interact with the textbook. At various places throughout the text, for example, we pose a question to the student about the material being read. This question encourages the reader to pause and think about the current discussion and to frame an answer to the question. To make sure the student does not miss important information, the answer to the question is provided as a footnote on the same page.

Each section contains a variety of worked-out examples. Each example is given a descriptive title so the student can see at a glance what type of problem it illustrates. Many examples are accompanied by annotations that help the student follow the logic of the solution as it moves from step to step. After the worked-out example, we give the number of an exercise in that section's exercise set for the student to work. Such exercises are printed in red in the exercise set, and their *complete solutions* can be found in the Solutions to Selected Exercises section at the end of the text.

Worked-out examples include a descriptive title and annotations that help students follow the solution.

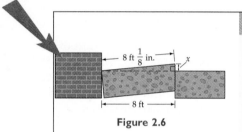

Figure 2.6

EXAMPLE 7   *Solve a Construction Application*

Concrete slabs often crack and buckle if proper expansion joints are not installed. Suppose a concrete slab expands as a result of an increase in temperature, as shown in **Figure 2.6**. Determine the height x, to the nearest inch, to which the concrete will rise as a consequence of this expansion.

Solution

Use the Pythagorean Theorem.

$$\left(8 \text{ feet} + \frac{1}{8} \text{ inch}\right)^2 = x^2 + (8 \text{ feet})^2$$

$$(96.125)^2 = x^2 + (96)^2 \qquad \bullet \textbf{ Change units to inches.}$$

$$(96.125)^2 - (96)^2 = x^2$$

$$\sqrt{(96.125)^2 - (96)^2} = x \qquad \bullet \textbf{ Only the positive root is}$$
$$\qquad\qquad\qquad\qquad\qquad\qquad \textbf{taken because } x > 0.$$

$$4.9 \approx x$$

Thus, to the nearest inch, the concrete will rise 5 inches.

TRY EXERCISE 68, EXERCISE SET 2.3

68. CONSTRUCTION A concrete slab cracks and expands as a result of an increase in temperature, as shown in the following figure. Determine the height x, to the nearest inch, to which the concrete will rise as a consequence of this expansion.

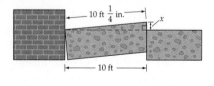

Students are directed to try an exercise similar to the example.

68. $\left(10 \text{ feet} + \frac{1}{4} \text{ inch}\right)^2 = (10 \text{ feet})^2 + x^2$

$\sqrt{(120.25)^2 - (120)^2} = x$ • **Change feet to inches.**

$7.75 = x$

To the nearest inch, the concrete will rise 8 inches.

A complete solution to the exercise is given in the Solutions to Selected Exercises section at the end of the text.

Extensive Exercise Sets The exercise sets of *College Algebra* were carefully developed to provide the student with a variety of exercises. The exercises range from drill and practice to interesting challenges and were chosen to illustrate many facets of topics discussed in the text. Each exercise set emphasizes skill building, skill maintenance, and (as appropriate) applications. Included in each exercise set are Supplemental Exercises that incorporate material from previous chapters, offer extensions of topics, require data analysis, and present challenge problems or problems in the "prove or disprove" format. Four types of exercises are identified by icons:

 Writing exercises  Graphing Utility exercises

 Group exercises | Data Analysis exercises

Applications One way to stimulate student interest in mathematics is through applications. The applications in *College Algebra* have been taken from many disciplines, including agriculture, architecture, biology, business, chemistry, earth science, economics, engineering, medicine, and physics. Besides providing motivation to study mathematics, the applications help students develop good problem-solving skills.

Projects One of our goals in writing this text has been to involve the student with the text. As mentioned earlier, we do this through various pedagogical features, such as pausing in the course of developing a concept to ask the student to answer a pertinent question. *Projects* are another feature designed to engage the student in mathematics, this time through writing. The projects at the end of each section's exercise set provide guidelines for further investigations. Some projects ask the student to solve a more complex application problem. Others ask the student to write a proof of some statement. Still others invite the student to chronicle the procedure used to solve a problem and to suggest extensions to that problem. Many of these projects are ideal candidates for small-group assignments.

Exploring Concepts with Technology Calculators and computers have expanded the limits of the types of problems that students can solve. To take advantage of the new technologies, we have incorporated in each chapter some optional extensions of ideas presented in that chapter. These problems are not so much conceptually difficult as they are computationally messy. For each of these problems, we encourage the student to use a calculator or computer to investigate solutions. As the student progresses through a solution, we challenge the student to think about the pitfalls of computational solutions.

Changes for the Third Edition

Thanks to the helpful suggestions of our users and reviewers, we have made some changes to this edition that strengthen the text. We sincerely appreciate all suggestions we receive.

    Essays and Projects from the previous edition have been renamed *Projects* and are now found at the end of every section rather than at the end of each chapter. We have included additional projects in the *Instructor's Resource Manual.* These projects and others can be found on the Internet at http://www.hmco.com. Supplemental application problems can also be found through this home page. These problems are updated once each semester.

Graphing technology is introduced and used as a natural outgrowth of content. Rather than present technology for its own sake, we introduce technology as it is needed and always to illustrate a concept. In this way, we hope to encourage students to see technology as an aid to better *understanding* rather than just a tool for obtaining answers. All graphing technology material is optional.

Topics for Discussion precede the exercise set in each section. These questions, which are conceptual rather than computational, can be used for group discussions or as writing exercises.

We have included three types of margin notes. *Instructor Notes* are printed in the Instructor's Annotated Edition only. *Take Note* and *Point of Interest* are margin notes that are printed in both the Instructor's Annotated Edition and the student text. *Point of Interest* notes contain interesting sidelights about mathematics, its history, or its application. Each *Take Note* alerts the student to a point that requires special attention.

Besides adding many contemporary application problems, we have made the following organizational and topical changes.

- Chapter 1 has been reorganized so that integer and rational exponents appear in one section.

- In Chapter 2 we have added compound inequalities to Section 2.5. This provides better motivation for absolute value inequalities in Section 2.6.

- There are a number of changes to Chapter 3. Section 3.3 now includes finding linear models for a data set. The introduction to the difference quotient is motivated by an example to help students understand why this quotient is important. In Section 3.8, we added many graphs to present visual representations of direct and inverse variation.

- Chapter 4 now has approximation of zeros integrated throughout the chapter. Therefore, the "Approximation of Zeros" section has been eliminated.

- The number and variety of applications in Chapter 5 have been increased.

- Chapter 8 now includes an application to motivate addition and scalar multiplication of matrices. We have included more application problems and some problems on matrices as transformations.

- Chapter 9 now includes the binomial probability formula and exercises that cover its use.

- A Glossary that defines key terms has been added at the end of the text.

Supplements for the Instructor

College Algebra has an unusually comprehensive set of teaching aids for the instructor.

Instructor's Annotated Edition   This is an exact replica of the student text except that annotations for the instructor are liberally distributed throughout. These annotations, labeled "Instructor Notes," are printed in blue in the margin. These notes include teaching tips, warnings about common errors, suggestions for the use of graphing utilities, and historical vignettes. In addition, a suggested homework assignment is given at the beginning of each exercise set.

Solutions Manual   The *Solutions Manual* contains worked-out solutions for all end-of-section, supplemental, and end-of-chapter exercises.

Projects on the Internet   Projects like those at the end of each exercise set in the text can also be found on the Internet at http://www.hmco.com. Contemporary and motivating, these projects are updated every semester to reflect the relevance of mathematics to world events occurring right now. Furthermore, additional modeling exercises are added to take advantage of the latest web technology. These modeling exercises enable users to see the results immediately when they interact with the data and graphics.

Instructor's Resource Manual with Chapter Tests   The *Instructor's Resource Manual* contains the printed testing program, which is the first of three sources of testing material available to the user. Six printed tests (in two formats: free response and multiple choice) are provided for each chapter. The *Instructor's Resource Manual* also includes additional projects that can be assigned as group activities or for extra credit. Moreover, there are suggestions for course sequencing, suggestions for incorporating graphing utilities, and outlines for solutions of the projects.

Computerized Test Generator with On-Line Testing   The Computerized Test Generator is the second source of testing material. The database contains more than 3600 test items. These questions are unique to the test generator and do not repeat items provided in the *Instructor's Resource Manual* testing program. The Test Generator is designed to produce an unlimited number of tests for each chapter of the text, including cumulative tests and final exams. It is available for the IBM PC and compatible computers and for the Macintosh. DOS and Windows versions also offer algorithms that can produce an unlimited number of some types of test questions and provide new **on-line testing** and **gradebook** functions.

Printed Test Bank   The Printed Test Bank, the third component of the testing material, is a printout of all items in the Computerized Test Generator. Instructors employing the Test Generator can use the Test Bank to select specific items from the database. Instructors who do not have access to a computer can use the Test Bank to select items to include on tests prepared by hand.

Supplements for the Student

In addition to the *Student Study Guide,* the *Graphing Workbook,* the Math Assistant software, and the Houghton Mifflin Video Library, two computerized study aids accompany this text: The Review Tutor covers prerequisite material, and the Computer Tutor covers material in the text.

Student Study Guide with Internet Guide   The *Student Study Guide* contains complete solutions to all odd-numbered exercises in the text as well as study tips and a practice test for each chapter. The *Internet Guide* explains the basics of using the Internet—such as how to gain access, how to search, and how to download information—and also provides a list of useful and interesting math sites.

Graphing Workbook   The *Graphing Workbook* offers over 600 exercises that can be solved using a graphing utility. These exercises are designed to extend and explore such concepts as approximating roots of equations, translating graphs, and solving inequalities. Students may complete the exercises individually or in small groups.

Math Assistant   This software is instructional and allows students to practice a skill, such as finding the inverse of a matrix, as well as to perform numerical calculations. In addition, the software includes a function grapher that graphs elementary functions and polar equations. The Math Assistant is available for the IBM PC, compatible microcomputers, and the Macintosh.

Houghton Mifflin Video Library   These review videos contain 32 segments that cover the essential topics in this series. The videos, professionally produced specifically for the text, offer a valuable resource for further instruction and review.

Review Tutor   The Review Tutor is a self-paced, interactive computer tutorial covering all necessary prerequisite material, such as would be found in an intermediate algebra course, that the student will need to know to progress successfully through this course. It is algorithmically based and includes color, animation, and free response. The algorithmic feature allows the Tutor to provide an unlimited number of practice problems. This tutorial is available for the Macintosh and for IBM PC and compatible computers running Windows.

Computer Tutor   The *Computer Tutor,* an interactive computer tutorial covering every college algebra topic, has been completely revised. It is now an algorithmically based tutor that includes color and animation. The algorithmic feature allows the Tutor to provide an unlimited number of practice problems. A management system is also available to help instructors track student progress. The *Computer Tutor* can be used in several ways: (1) to cover material the student missed because of absence from class, (2) to reinforce instruction on a concept the student has not yet mastered, or (3) to review material in preparation for examinations. This tutorial is available for the Macintosh and for the IBM PC and compatible computers running Windows.

Acknowledgments

The authors would like to thank the people who have reviewed this manuscript and provided many valuable suggestions:

Art Bukowski, *University of Alaska at Anchorage*
Richard Eells, *Roxbury Community College, MA*
Sudhir Goel, *Valdosta State University, GA*
JoAnne Kennedy, *LaGuardia Community College, NY*
Michael Longfritz, *Rensselaer Polytechnic Institute, NY*
Frank Mauz, *Honolulu Community College*
Susan Pfeifer, *Butler County Community College, KS*
Beverly Reed, *Kent State University, OH*
Mary Rice, *University of Nevada–Las Vegas*
Helen Salzberg, *Rhode Island College*
Ming Xue, *Massachusetts Institute of Technology*

LIST OF APPLICATIONS

Lowest cost of car rentals, 110, 118, 130

Lowest cost of checking accounts, 118

Lowest cost of video rentals, 118

Monthly car payments, 297

Monthly condominium fees, 129

Nutritional content of diets, 469

Pool maintenance, 8, 98

Price of battery and calculator, 129

Price of book, 87

Price of computer, 86

Price of magazine subscription, 129

Price of yacht, 87

Rate of cooling of a can of soda, 320

Strategies for playing Monopoly, 381

Temperature of dessert in freezer, 282

EARTH SCIENCE

Carbon dating of a volcanic eruption, 334

Crop yield after successive plantings, 338

Crop yield as function of fertilizer, 160

Distance to horizon, 107

Richter scale measure of an earthquake, 310–311, 313, 338

Soil chemistry, 469

ECOLOGY AND GENETICS

Biological diversity, 306

Birth rates, 340

Capture-recapture models, 457–458

Chromosome breakage, 533

Depletion of aluminum resources, 321–322

Depletion of coal resources, 321, 322

Depletion of oil resources, 313

Logistic model of population growth, 331–333

Malthusian model of population growth, 331

Population growth of cities and towns, 334

Population of bison, 336

Population of mosquitoes, 209

Population of squirrels in nature preserve, 336

Population of walrus colony, 336

Predator-prey interactions, 458

Rate of oil leakage, 338

Verhulst population model, 127

FINANCE

Compound interest, 28, 325–326, 333, 343

Continuous compounding, 326–327, 333–334, 344

Doubling a sum of money, 327, 333

Future value of an ordinary annuity, 510–511, 512

Investment in stocks and bonds, 390

Present value of an ordinary annuity, 62

Rates of U.S. Treasury securities, 312

Retirement planning, 323

Simple interest investment in two accounts, 81–82, 86, 129–130

Tripling a sum of money, 333–334

GEOMETRY

Altitude of a triangle, 97

Area of a rectangle as a function of the length, 159

Area of a snowflake, 491

Area of a triangle in terms of the three sides, 161

Areas of geometric figures and factoring, 53

Box made from a square piece of material, 155–156, 159

Cube equivalent in volume to two spheres, 106

Curve fitting: circle, 402, 434

Curve fitting: plane, 434

Curve fitting: quadratic, 400–401, 402, 434

Cylinder inscribed in a cone, 159

Diagonal of a square, 97

Diagonals of a polygon, 98

Diameter of the base of a right circular cone, 106, 129

Distance from homeplate to second base, 97

Fencing a rectangular area, 95, 97

Length and width of a rectangle, 79–80, 85, 128, 129

Lengths of the sides of a triangle, 85

Maximum area of a rectangle, 185, 250

Maximum volume of a box, 247, 250

Minimum surface area of a cylinder, 279, 282

Perimeter of an ellipse, 365–366

Perimeter of a snowflake, 513

Radius of a circle circumscribed about a triangle, 107

Radius of a circle inscribed in a triangle, 107

Radius of a cone, 106

Rectangle inscribed in an ellipse, 365

Tangent to a parabola, 354

Transformation matrices, 459, 460–461

Volume variation of a cone, 159, 223

LINEAR PROGRAMMING

Automotive engine reconditioning, 429

Diets from two food groups, 429

Industrial solvents, 426–427, 429

Manufacture of answering machines, 429

Manufacture of sporting goods, 429

Maximizing a farmer's profit, 429, 434

Minimum cost of ice cream, 429

Nutrition of farm animals, 425–426, 430

MATHEMATICS

Attractors, 99

Cardano's solution of the cubic, 267

Chaos, 126–127

Codes, 435, 470–471

Combinations, 40, 42

Continued fractions, 62

Dirichlet function, 198

INT function, 153–154

Mandelbrot set, 63–64

Napierian logarithms, 313–314

Newton's approximation of a square root, 498

Simpson's Rule, 187

Stirling's formula, 337, 498

Tower of Hanoi, 520

Triangle inequality, 13

Zeller's Congruence, 161

MEDICINE

Amount of anesthetic in a patient, 321

Bacterial growth, 297

Concentration of medication in blood, 207–208

Healing rate of a skin wound, 343

Kidney stone treatment, 345

Linear model of AIDS cases, 173

Maximum exercise heart rate, 76–77

FUNDAMENTAL CONCEPTS

Georg Cantor

Railroad tracks stretch forward an infinite point on the horizon.

The "infinity" of space.

How Large Is Infinity?

The German mathematician Georg Cantor (1845–1918) developed the idea of the cardinality of a set. The cardinality of a finite set is the number of elements in the set. For example, the set $\{5, 7, 11\}$ has a cardinality of 3. The set of natural numbers $\{1, 2, 3, 4, 5, 6, 7, \ldots\}$ is an infinite set. Cantor denoted its cardinality by the symbol $\aleph_0$, which is read "aleph null."

The set of whole numbers consists of all the elements of the set of natural numbers and the number 0. The following display shows a one-to-one correspondence between the set of natural numbers and the set of whole numbers.

$$\{1, \quad 2, \quad 3, \quad 4, \quad 5, \quad \ldots, \quad n, \quad \ldots\}$$
$$\updownarrow \quad \updownarrow \quad \updownarrow \quad \updownarrow \quad \updownarrow \qquad \updownarrow$$
$$\{0, \quad 1, \quad 2, \quad 3, \quad 4, \quad \ldots, \quad n-1, \quad \ldots\}$$

Cantor reasoned that because of this one-to-one correspondence, the set of natural numbers and the set of whole numbers both have the same cardinality, namely $\aleph_0$.

Cantor was also able to show that the set of irrational numbers has a cardinality that is different from $\aleph_0$. The idea that some infinite sets have more elements than other infinite sets was not readily accepted. Previous to Cantor's work, the philosopher Voltaire (1694–1778) expressed the following opinion:

> We admit, in geometry, not only infinite magnitudes, that is to say magnitudes greater than any assignable magnitude, but infinite magnitudes infinitely greater, the one than the other. This astonishes our dimension of brains, which is only about six inches long, five broad, and six in depth, in the largest of heads.

1.1   THE REAL NUMBER SYSTEM

Human beings share the desire to organize and classify. Ancient astronomers classified stars into groups called constellations. Modern astronomers continue to classify stars by such characteristics as color, mass, size, temperature, and distance from earth. In mathematics it is useful to place numbers with similar characteristics into **sets.** The following sets of numbers are used extensively in the study of algebra:

Integers	$\{\ldots,-3,-2,-1,0,1,2,3,\ldots\}$
Rational numbers	{all terminating or repeating decimals}
Irrational numbers	{all nonterminating, nonrepeating decimals}
Real numbers	{all rational or irrational numbers}

If a decimal terminates or repeats a block of digits, then the number is a rational number. Rational numbers can also be written in the form p/q, where p and q are integers and $q \neq 0$. For example,

$$\frac{3}{4} = 0.75 \quad \text{and} \quad \frac{5}{11} = 0.\overline{45}$$

are rational numbers. The bar over the 45 means that the block repeats without end; that is, $0.\overline{45} = 0.454545\ldots$.

In its decimal form, an irrational number neither terminates nor repeats. For example, $0.272272227\ldots$ is a nonterminating, nonrepeating decimal and thus is an irrational number. One of the best-known irrational numbers is pi, denoted by the Greek symbol π. The number π is defined as the ratio of the circumference of a circle to its diameter. Often in applications, the rational number 3.14 or the rational number 22/7 is used as an approximation of the irrational number π.

Every real number is either a rational number or an irrational number. If a real number is written in decimal form, it is a terminating decimal, a repeating decimal, or a nonterminating and nonrepeating decimal.

Each member of a set is called an **element** of the set. Set A is a **subset** of set B if every element of set A is also an element of set B. The set of **negative integers** $\{-1,-2,-3,-4,-5,\ldots\}$ is a subset of the set of integers. The set of **positive integers** $\{1,2,3,4,5,\ldots\}$ (also known as the set of **natural numbers**) is also a subset of the integers. **Figure 1.1** illustrates the subset relationships among the sets defined above.

Prime numbers and *composite numbers* play an important role in almost every branch of mathematics. A **prime number** is a positive integer greater than 1 that has no positive-integer factors other than itself and 1. The 10 smallest prime numbers are 2, 3, 5, 7, 11, 13, 17, 19, 23, and 29. Each of these numbers has only itself and 1 as factors.

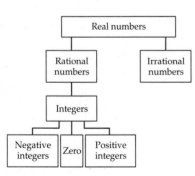

Figure 1.1

A **composite number** is a positive integer greater than 1 that is not a prime number. For example, 10 is a composite number because 10 has both 2 and 5 as factors. The 10 smallest composite numbers are 4, 6, 8, 9, 10, 12, 14, 15, 16, and 18.

| EXAMPLE 1 | *Classify Real Numbers* |

Determine which of the following numbers are

a. integers   **b.** rational numbers   **c.** irrational numbers

d. real numbers   **e.** prime numbers   **f.** composite numbers

$$-0.2, \quad 0, \quad 0.\overline{3}, \quad \pi, \quad 6, \quad 7, \quad 41, \quad 51, \quad 0.71771777177771\ldots$$

Solution

a. Integers: 0, 6, 7, 41, 51

b. Rational numbers: $-0.2, 0, 0.\overline{3}, 6, 7, 41, 51$

c. Irrational numbers: $0.71771777177771\ldots, \pi$

d. Real numbers: $-0.2, 0, 0.\overline{3}, \pi, 6, 7, 41, 51, 0.71771777177771\ldots$

e. Prime numbers: 7, 41

f. Composite numbers: 6, 51

TRY EXERCISE 2, EXERCISE SET 1.1

POINT OF INTEREST

Prime numbers are used to encrypt messages on the Internet via a coding system called **PGP (Pretty Good Privacy)**. This is part of an encryption scheme called **Public Key Cryptography.**

Sets are often written using **set-builder notation,** which makes use of a variable and a characteristic property that the elements of the set alone possess. This notation is especially useful to describe infinite sets. The set-builder notation

$$\{x^2 \mid x \text{ is an integer}\}$$

is read as "the set of all elements x^2 such that x is an integer." This is the infinite set of **perfect squares:** $\{0, 1, 4, 9, 16, 25, 36, 49, \ldots\}$.

The **empty set** or **null set** is a set without any elements. The set of numbers that are both prime and also composite is an example of the null set. The null set is denoted by the symbol $\varnothing$.

Just as addition and subtraction are operations performed on real numbers, there are operations performed on sets. Two of these set operations are called *intersection* and *union*. The **intersection** of sets A and B, denoted by $A \cap B$, is the set of all elements belonging to both set A and set B. The **union** of sets A and B, denoted by $A \cup B$, is the set of all elements belonging to set A, to set B, or to both.

| EXAMPLE 2 | *Find the Intersection and the Union of Two Sets* |

Find each intersection or union, given $A = \{0, 1, 4, 6, 9\}$, $B = \{1, 3, 5, 7, 9\}$ and $P = \{x \mid x \text{ is a prime number} < 10\}$.

a. $A \cap B$ **b.** $A \cap P$ **c.** $A \cup B$ **d.** $A \cup P$

Solution

a. $A \cap B = \{0, 1, 4, 6, 9\} \cap \{1, 3, 5, 7, 9\}$
$= \{1, 9\}$ • **Only 1 and 9 belong to both sets.**

b. First determine that $P = \{2, 3, 5, 7\}$. Therefore,
$A \cap P = \{0, 1, 4, 6, 9\} \cap \{2, 3, 5, 7\}$
$= \varnothing$ • **There are no common elements.**

c. $A \cup B = \{0, 1, 4, 6, 9\} \cup \{1, 3, 5, 7, 9\}$
$= \{0, 1, 3, 4, 5, 6, 7, 9\}$ • **List the elements of the first set. Include elements from the second set that are not already listed.**

d. $A \cup P = \{0, 1, 4, 6, 9\} \cup \{2, 3, 5, 7\} = \{0, 1, 2, 3, 4, 5, 6, 7, 9\}$

TRY EXERCISE 4, EXERCISE SET 1.1

PROPERTIES OF REAL NUMBERS

Addition, multiplication, subtraction, and *division* are the operations of arithmetic. **Addition** of the two real numbers a and b is designated by $a + b$. If $a + b = c$, then c is the **sum** and the real numbers a and b are called the **terms.**

Multiplication of the real numbers a and b is designated by ab or $a \cdot b$. If $ab = c$, then c is the **product** and the real numbers a and b are called **factors** of c.

The number $-b$ is referred to as the **additive inverse** of b. **Subtraction** of the real numbers a and b is designated by $a - b$ and is defined as the sum of a and the additive inverse of b. That is,

$$a - b = a + (-b)$$

If $a - b = c$, then c is called the **difference** of a and b.

The **multiplicative inverse** or **reciprocal** of the nonzero number b is $1/b$. The **division** of a and b, designated by $a \div b$ with $b \neq 0$, is defined as the product of a and the reciprocal of b. That is,

$$a \div b = a\left(\frac{1}{b}\right) \quad \text{provided } b \neq 0$$

If $a \div b = c$, then c is called the **quotient** of a and b.

The notation $a \div b$ is often represented by the fractional notation a/b or $\frac{a}{b}$.

The real number a is the **numerator,** and the nonzero real number b is the **denominator** of the fraction.

PROPERTIES OF REAL NUMBERS

Let a, b, and c be real numbers.

	Addition Properties	Multiplication Properties
Closure	$a + b$ is a unique real number.	ab is a unique real number.
Commutative	$a + b = b + a$	$ab = ba$
Associative	$(a + b) + c = a + (b + c)$	$(ab)c = a(bc)$
Identity	There exists a unique real number 0 such that $a + 0 = 0 + a = a$.	There exists a unique real number 1 such that $a \cdot 1 = 1 \cdot a = a$.
Inverse	For each real number a, there is a unique real number $-a$ such that $a + (-a) = (-a) + a = 0$.	For each *nonzero* real number a, there is a unique real number $1/a$ such that $a \cdot \dfrac{1}{a} = \dfrac{1}{a} \cdot a = 1$.
Distributive		$a(b + c) = ab + ac$

We can identify which property of real numbers has been used to rewrite expressions by closely comparing the expressions and noting any changes.

EXAMPLE 3   *Identify Properties of Real Numbers*

Identify the property of real numbers illustrated in each statement.

a. $(2a)b = 2(ab)$ **b.** $\left(\dfrac{1}{5}\right)11$ is a real number.

c. $4(x + 3) = 4x + 12$ **d.** $(a + 5b) + 7c = (5b + a) + 7c$

e. $\left(\dfrac{1}{2} \cdot 2\right)a = 1 \cdot a$ **f.** $1 \cdot a = a$

Solution

a. Associative property of multiplication

b. Closure property of multiplication of real numbers

c. Distributive property

d. Commutative property of addition

e. Inverse property of multiplication

f. Identity property of multiplication

TRY EXERCISE 16, EXERCISE SET 1.1

An **equation** is a statement of equality between two numbers or two expressions. There are four basic properties of equality that relate to equations.

PROPERTIES OF EQUALITY

Let a, b, and c be real numbers.

Reflexive	$a = a$
Symmetric	If $a = b$, then $b = a$.
Transitive	If $a = b$ and $b = c$, then $a = c$.
Substitution	If $a = b$, then a may be replaced by b in any expression that involves a.

EXAMPLE 4   *Identify Properties of Equality*

Identify the property of equality illustrated in each statement.

a. If $3a + b = c$, then $c = 3a + b$.   **b.** $5(x + y) = 5(x + y)$

c. If $4a - 1 = 7b$ and $7b = 5c + 2$, then $4a - 1 = 5c + 2$.

d. If $a = 5$ and $b(a + c) = 72$, then $b(5 + c) = 72$.

Solution

a. Symmetric   **b.** Reflexive   **c.** Transitive   **d.** Substitution

TRY EXERCISE 18, EXERCISE SET 1.1

The following properties of fractions will be used throughout this text.

PROPERTIES OF FRACTIONS

For all fractions a/b and c/d, where $b \neq 0$ and $d \neq 0$:

Equality	$\dfrac{a}{b} = \dfrac{c}{d}$ if and only if $ad = bc$
Equivalent fractions	$\dfrac{a}{b} = \dfrac{ac}{bc}, \quad c \neq 0$
Addition	$\dfrac{a}{b} + \dfrac{c}{b} = \dfrac{a + c}{b}$
Subtraction	$\dfrac{a}{b} - \dfrac{c}{b} = \dfrac{a - c}{b}$
Multiplication	$\dfrac{a}{b} \cdot \dfrac{c}{d} = \dfrac{ac}{bd}$
Division	$\dfrac{a}{b} \div \dfrac{c}{d} = \dfrac{a}{b} \cdot \dfrac{d}{c} = \dfrac{ad}{bc}, \quad c \neq 0$
Sign	$-\dfrac{a}{b} = \dfrac{-a}{b} = \dfrac{a}{-b}$

The equality property of fractions contains the terminology "if and only if," which implies each of the following:

$$\text{If } \frac{a}{b} = \frac{c}{d}, \qquad \text{then } ad = bc.$$

$$\text{If } ad = bc, \qquad \text{then } \frac{a}{b} = \frac{c}{d}.$$

The number zero has many special properties. The following division properties of zero play an important role in this text.

DIVISION PROPERTIES OF ZERO

1. For $a \neq 0$, $\dfrac{0}{a} = 0$.   (Zero divided by any nonzero number is zero.)

2. $\dfrac{a}{0}$ is undefined.   (Division by zero is undefined.)

The properties of fractions can be used to find the sum, difference, product, or quotient of fractions.

EXAMPLE 5   *Compute with Fractions*

Use the properties of fractions to perform the indicated operations. Assume that $a \neq 0$.

a. $\dfrac{2a}{3} - \dfrac{a}{5}$ **b.** $\dfrac{2a}{5} \cdot \dfrac{3a}{4}$ **c.** $\dfrac{5a}{6} \div \dfrac{3a}{4}$ **d.** $\dfrac{0}{3a}$

Solution

a. Rewrite each fraction as an equivalent fraction with a common denominator of 15 by multiplying both the numerator and the denominator of 2a/3 by 5 and by multiplying both the numerator and the denominator of a/5 by 3.

$$\frac{2a}{3} - \frac{a}{5} = \frac{2a(5)}{3(5)} - \frac{a(3)}{5(3)} = \frac{10a}{15} - \frac{3a}{15} = \frac{10a - 3a}{15} = \frac{7a}{15}$$

b. $\dfrac{2a}{5} \cdot \dfrac{3a}{4} = \dfrac{(2a)(3a)}{(5)(4)} = \dfrac{6a^2}{20} = \dfrac{3a^2}{10}$

c. $\dfrac{5a}{6} \div \dfrac{3a}{4} = \dfrac{5a}{6} \cdot \dfrac{4}{3a} = \dfrac{20a}{18a} = \dfrac{10}{9}$

d. $\dfrac{0}{3a} = 0$ • Zero divided by any nonzero number is zero.

TRY EXERCISE 30, EXERCISE SET 1.1

TOPICS FOR DISCUSSION

1. Archimedes determined that $223/71 < \pi < 22/7$. Is it possible to find an exact expression for π of the form a/b, where a and b are integers?

2. Is the intersection of two infinite sets always an infinite set? Is the union of two infinite sets always an infinite set?

3. Explain why division by zero is not allowed.

4. Explain the similarities and differences between rational and irrational numbers.

EXERCISE SET 1.1

In Exercises 1 and 2, determine which of the numbers are a. **integers,** b. **rational numbers,** c. **irrational numbers,** d. **real numbers,** e. **prime numbers,** f. **composite numbers.**

1. $-3 \quad 4 \quad \dfrac{1}{5} \quad 11 \quad 3.14 \quad 57 \quad 0.252252225\ldots$

2. $5.\overline{17} \quad -4.25 \quad \dfrac{1}{4} \quad \pi \quad 21 \quad 53 \quad 0.45454545\ldots$

In Exercises 3 to 14, use $A = \{0, 1, 2, 3, 4\}$, $B = \{1, 3, 5, 11\}$, $C = \{1, 3, 6, 10\}$, **and** $D = \{0, 2, 4, 6, 8, 10\}$ **to find the indicated intersection or union.**

3. $A \cap B$
4. $A \cap C$
5. $B \cap C$
6. $B \cap D$
7. $A \cap D$
8. $C \cap D$
9. $A \cup B$
10. $A \cup C$
11. $B \cup C$
12. $B \cup D$
13. $A \cup D$
14. $C \cup D$

In Exercises 15 to 28, identify the property of real numbers or the property of equality that is illustrated.

15. $3 + (2 + 5) = (3 + 2) + 5$
16. $6 + (2 + 7) = 6 + (7 + 2)$
17. $1 \cdot a = a$
18. If $a + b = 2$, then $2 = a + b$.
19. $a(bx) = a(bx)$
20. If $x + 2y = 7$ and $7 = y$, then $x + 2(7) = 7$.
21. If $x = 2(y + z)$ and $2(y + z) = 5w$, then $x = 5w$.
22. $p(q + r) = pq + pr$
23. $m + (-m) = 0$
24. $t\left(\dfrac{1}{t}\right) = 1$

25. $7(a + b) = 7(b + a)$
26. $8(gh + 5) = 8(hg + 5)$
27. If $x + 2y = 7$ and $w = 7$, then $x + 2y = w$.
28. $5[x + (y + z)] = 5x + 5(y + z)$

In Exercises 29 to 38, use the properties of fractions to perform the indicated operations. State each answer in lowest terms. Assume a **is a nonzero real number.**

29. $\dfrac{2a}{7} - \dfrac{5a}{7}$
30. $\dfrac{2a}{5} + \dfrac{3a}{7}$
31. $\dfrac{-3a}{5} + \dfrac{a}{4}$
32. $\dfrac{7}{8}a - \dfrac{13}{5}a$
33. $\dfrac{-5}{7} \cdot \dfrac{2}{3}$
34. $\dfrac{7}{11} \cdot \dfrac{-22}{21}$
35. $\dfrac{12a}{5} \div \dfrac{-2a}{3}$
36. $\dfrac{2}{5} \div 3\dfrac{2}{3}$
37. $\dfrac{2a}{3} - \dfrac{4a}{5}$
38. $\dfrac{1}{2a} - \dfrac{3}{a}$

39. **POOL MAINTENANCE** One pipe can fill a pool in 11 hours. A second pipe can fill the same pool in 15 hours. Assume the first pipe fills $1/11$ of the pool every hour and the second pipe fills $1/15$ of the pool every hour.

   a. Find the amount of the pool the two pipes together fill in 3 hours.

   b. Find the amount of the pool they fill together in x hours.

40. **FOCAL LENGTH OF A MIRROR** The relationship between the distance of an object d_0 from a curved mirror, the distance of its image d_i from the mirror, and the focal

length f of the mirror is given by the **mirror equation:**

$$\frac{1}{f} = \frac{1}{d_0} + \frac{1}{d_i}$$

What is the focal length[1] f of a mirror for which $d_0 = 25$ centimeters and $d_i = -5$ centimeters?

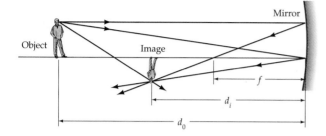

Object

Image

Mirror

41. State the multiplicative inverse of $7\frac{3}{8}$.

42. State the multiplicative inverse of $-4\frac{2}{5}$.

43. Show by an example that the operation of subtraction of real numbers is not a commutative operation.

44. Show by an example that the operation of division of nonzero real numbers is not a commutative operation.

45. Show by an example that the operation of subtraction of real numbers is not an associative operation.

46. Show by an example that the operation of division of real numbers is not an associative operation.

In Exercises 47 to 56, classify each statement as true or false.

47. $a/0$ is the multiplicative inverse of $0/a$.

48. $(-1/\pi)$ is the multiplicative inverse of $-\pi$.

49. If $p = q + \dfrac{t}{2}$ and $q + \dfrac{t}{2} = \dfrac{1}{2}s$, then $\dfrac{1}{2}s = p$.

50. If $a - b = 7$, then $7 = b - a$.

51. The sum of two composite numbers is a composite number.

52. All integers are natural numbers.

53. Every real number is either a rational or an irrational number.

54. Every rational number is either even or odd.

55. 1 is the only positive integer that is not prime and not composite.

56. All repeating decimals are rational numbers.

---

[1] For convex mirrors, both the focal length f and the image distance d_i are *negative* quantities.

57. Use a calculator to write each of the following rational numbers as a decimal. If the number is represented by a nonterminating decimal, then use a *bar* over the repeating portion of the decimal.

a. $\dfrac{8}{11}$ **b.** $\dfrac{33}{40}$ **c.** $\dfrac{2}{7}$ **d.** $\dfrac{5}{37}$

58. Use a calculator to determine whether 3.14 or 22/7 is a closer approximation to π.

59. Use a calculator to complete the following table.

x	0.1	0.01	0.001	0.0000001
$\dfrac{\sqrt{x+9}-3}{x}$				

Now make a guess as to the number the fraction seems to be approaching as x assumes the values of real numbers that are closer and closer to zero.

60. Use a calculator to complete the following table.

x	0.1	0.01	0.001	0.0000001
$\dfrac{\dfrac{1}{2} - \dfrac{1}{x+2}}{x}$				

Now make a guess as to the number the fraction seems to be approaching as x assumes the values of real numbers that are closer and closer to zero.

SUPPLEMENTAL EXERCISES

In Exercises 61 to 64, list the elements of the set.

61. $A = \{x \mid x$ is a composite number less than 11$\}$

62. $B = \{x \mid x$ is an even prime number$\}$

63. $C = \{x \mid 50 < x < 60$ and x is a prime number$\}$

64. $D = \{x \mid x$ is the smallest odd composite number$\}$

65. Which of the properties of real numbers are satisfied by the set of positive integers?

66. Which of the properties of real numbers are satisfied by the set of integers?

67. Which of the properties of real numbers are satisfied by the set of rational numbers?

68. Which of the properties of real numbers are satisfied by the set of irrational numbers?

69. GOLDBACH'S CONJECTURE In 1742 Christian Goldbach conjectured that every even number greater than 2 can be written as the sum of two prime numbers. Many mathematicians have tried to prove or disprove this conjecture without succeeding. Show that Goldbach's conjecture is true for the following even numbers.

a. 12   **b.** 30

70. TWIN PRIMES If the natural numbers n and $n + 2$ are both prime numbers, then they are said to be twin primes. For example, 11 and 13 are twin primes. It is not known whether the set of twin primes is an infinite set or a finite set. List all the twin primes less than 50.

PROJECTS

1. 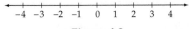 **NUMBER THEORY** *Theorem:* If a number of the form 111...1 is a prime number, then the number of 1's is a prime number. For instance, the numbers

$$11 \quad \text{and} \quad 1111111111111111111$$

are prime numbers, and the number of 1's in each number is a prime number (2 in the first number and 19 in the second number).

a. The number $111 = 3 \cdot 37$, so 111 is not a prime number. Explain why this does not contradict the above theorem.

b. What is the converse of a theorem? State the converse of the theorem above.

c. If a theorem is true, is the converse of a theorem also true? Explain your answer.

2. 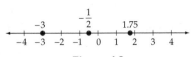 **PERFECT SQUARES** Explain why a perfect-square integer must have an odd number of distinct natural-number divisors.

SECTION 1.2 INTERVALS, ABSOLUTE VALUE, AND DISTANCE

The real numbers can be represented geometrically by a **coordinate axis** called a **real number line. Figure 1.2** shows a portion of a real number line. The number associated with a particular point on a real number line is called the **coordinate** of the point. It is customary to label those points whose coordinates are integers. The point corresponding to zero is called the **origin**, denoted 0. Numbers to the right of the origin are **positive real numbers;** numbers to the left of the origin are **negative real numbers.**

A real number line provides a picture of the real numbers. That is, each real number corresponds to one and only one point on the real number line, and each point on a real number line corresponds to one and only one real number. This type of correspondence is referred to as a **one-to-one correspondence.** The real numbers -3, $-1/2$, and 1.75 are graphed in **Figure 1.3.**

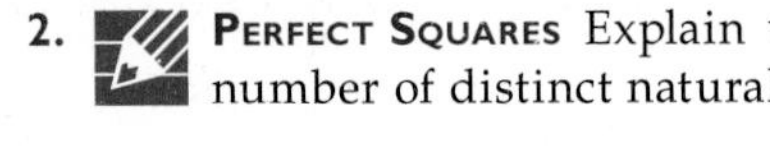

Certain order relationships exist between real numbers. For example, if a and b are real numbers, then

a **equals** b (denoted by $a = b$) if $a - b = 0$.

a is **greater than** b (denoted by $a > b$) if $a - b$ is positive.

a is **less than** b (denoted by $a < b$) if $b - a$ is positive.

Figure 1.2

Figure 1.3

On a horizontal number line, the notation

$a = b$ implies that the point with coordinate a is the same point as the point with coordinate b.

$a > b$ implies that the point with coordinate a is to the right of the point with coordinate b.

$a < b$ implies that the point with coordinate a is to the left of the point with coordinate b.

The **inequality** symbols $<$ and $>$ are sometimes combined with the equality symbol in the following manner:

$a \geq b$ This is read "a is greater than or equal to b," which means $a > b$ or $a = b$.

$a \leq b$ This is read "a is less than or equal to b," which means $a < b$ or $a = b$.

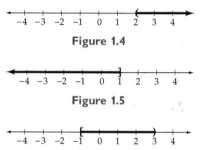

Figure 1.4

Figure 1.5

Figure 1.6

Inequalities can be used to represent subsets of real numbers. For example, the inequality $x > 2$ represents all real numbers greater than 2; **Figure 1.4** shows its graph. The parenthesis at 2 means that 2 is not part of the graph.

The inequality $x \leq 1$ represents all real numbers less than or equal to 1; **Figure 1.5** shows its graph. The bracket at 1 means that 1 is part of the graph.

The inequality $-1 \leq x < 3$ represents all real numbers between -1 and 3, including -1 but not including 3. **Figure 1.6** shows its graph.

INTERVAL NOTATION

Subsets of real numbers can also be represented by a compact form of notation called **interval notation**. For example, $[-1, 3)$ is the interval notation for the subset of real numbers in **Figure 1.6.**

In general, the interval notation

(a, b) represents all real numbers between a and b, not including a and not including b. This is an **open interval**.

$[a, b]$ represents all real numbers between a and b, including a and including b. This is a **closed interval**.

$(a, b]$ represents all real numbers between a and b, not including a but including b. This is a **half-open interval**.

$[a, b)$ represents all real numbers between a and b, including a but not including b. This is a **half-open interval**.

Figure 1.7 shows the four subsets of real numbers that are associated with the four interval notations (a, b), $[a, b]$, $(a, b]$, and $[a, b)$.

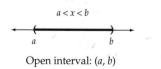
Open interval: (a, b)

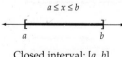
Closed interval: $[a, b]$

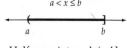
Half-open interval: $(a, b]$

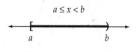
Half-open interval: $[a, b)$

Figure 1.7
Finite intervals

Subsets of the real numbers whose graphs extend forever in one or both directions can be represented by interval notation using the **infinity symbol** ∞ or the **negative infinity symbol** $-\infty$.

As **Figure 1.8** shows, the interval notation

$(-\infty, a)$ represents all real numbers less than a.

(b, ∞) represents all real numbers greater than b.

$(-\infty, a]$ represents all real numbers less than or equal to a.

$[b, \infty)$ represents all real numbers greater than or equal to b.

$(-\infty, \infty)$ represents all real numbers.

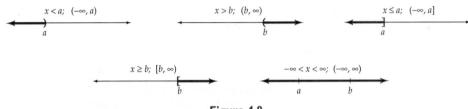

Figure 1.8
Infinite intervals

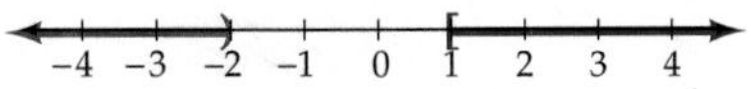

Some graphs consist of more than one interval of the real number line. **Figure 1.9** is a graph of the interval $(-\infty, -2)$, along with the interval $[1, \infty)$.

The word *or* is used to denote the union of two sets. The word *and* is used to denote intersection. Thus the graph in **Figure 1.9** is denoted by the inequality notation

$$x < -2 \quad \text{or} \quad x \geq 1$$

To represent this graph using interval notation, use the union symbol $\cup$ and write $(-\infty, -2) \cup [1, \infty)$.

Figure 1.9

EXAMPLE 1   *Graph Intervals and Inequalities*

Graph the following. Also write **a.** and **b.** using interval notation, and write **c.** and **d.** using inequality notation.

a. $-2 \leq x < 3$ **b.** $x \geq -3$ **c.** $[-4, -2] \cup [0, \infty)$ **d.** $(-\infty, 2)$

Solution

a. 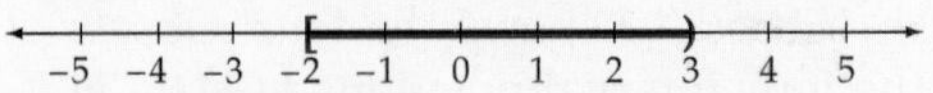 $[-2, 3)$

b. 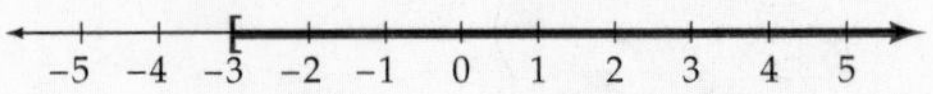 $[-3, \infty)$

c.  $-4 \leq x \leq -2 \quad \text{or} \quad x \geq 0$

d. 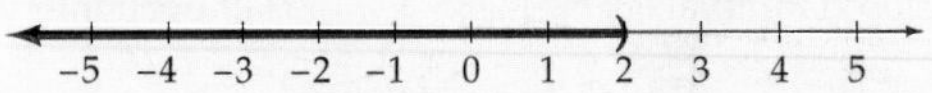 $x < 2$

TRY EXERCISE 16, EXERCISE SET 1.2

ABSOLUTE VALUE

The *absolute value* of the real number a, denoted $|a|$, is the distance between a and 0 on the number line. For example, $|2| = 2$ and $|-2| = 2$. In general, if $a \geq 0$, then $|a| = a$; however, if $a < 0$, then $|a| = -a$ because $-a$ is positive when $a < 0$. This leads us to the following definition.

TAKE NOTE

The second part of the definition of absolute value states that if $a < 0$, then $|a| = -a$. For instance, if $a = -4$, then

$$|a| = |-4| = -(-4) = 4$$

DEFINITION OF ABSOLUTE VALUE

The **absolute value** of the real number a is defined by

$$|a| = \begin{cases} a & \text{if } a \geq 0 \\ -a & \text{if } a < 0 \end{cases}$$

Rule

The following theorems can be derived by using the definition of absolute value.

TAKE NOTE

Note the term *nonnegative* that is used at the right. Nonnegative means greater than *or equal to* zero. Positive means greater than zero.

ABSOLUTE VALUE THEOREMS

For all real numbers a and b,

Nonnegative	$|a| \geq 0$	
Product	$|ab| = |a||b|$	*Rules*
Quotient	$\left|\dfrac{a}{b}\right| = \dfrac{|a|}{|b|}, \quad b \neq 0$	
Triangle inequality	$|a + b| \leq |a| + |b|$	
Difference	$|a - b| = |b - a|$	

The definition of absolute value and the absolute value theorems can be used to write some expressions without absolute value symbols. For instance, because $1 - \pi < 0$,

$$|1 - \pi| = -(1 - \pi) = \pi - 1$$

More complicated expressions can also be simplified by using these theorems. For example, given $-1 < x < 1$,

$$|x + 3| - |x - 2| = (x + 3) - [-(x - 2)]$$
$$= (x + 3) + (x - 2)$$
$$= 2x + 1$$

• $-1 < x < 1$. Thus $|x - 2| = -(x - 2)$.

EXAMPLE 2   *Evaluate Absolute Value Expressions*

Write $\left|\dfrac{2x}{|x| + |x - 2|}\right|$, given $0 < x < 2$, without absolute value symbols.

Continued ▶

Solution

Use the quotient theorem to write the expression as a quotient of absolute values.

$$\left| \frac{2x}{|x| + |x - 2|} \right| = \frac{|2x|}{||x| + |x - 2||}$$

Because $0 < x < 2$, $|2x| = 2x$, $|x| = x$, and $|x - 2| = -x + 2$. Substituting yields

$$\frac{|2x|}{||x| + |x - 2||} = \frac{2x}{|x + (-x + 2)|} = \frac{2x}{|2|} = \frac{2x}{2} = x$$

> **TRY EXERCISE 56, EXERCISE SET 1.2**

The definition of *distance* between any two points on a real number line makes use of absolute value.

DISTANCE BETWEEN POINTS ON A REAL NUMBER LINE

For any real numbers a and b, the **distance** between the graph of a and the graph of b is denoted by $d(a, b)$, where

$$d(a, b) = |a - b|$$

EXAMPLE 3   *Find the Distance Between Points*

Find the distance between the points whose coordinates are given.

a. $5, -2$ **b.** $-\pi, -2$

Solution

a. $d(5, -2) = |5 - (-2)| = |5 + 2| = |7| = 7$

b. $d(-\pi, -2) = |-\pi - (-2)| = |-\pi + 2|$ • $-\pi + 2 < 0$. Thus
$$= -(-\pi + 2) = \pi - 2 \qquad |-\pi + 2| = -(-\pi + 2).$$

> **TRY EXERCISE 64, EXERCISE SET 1.2**

Absolute value notation and the notion of distance can also be used to describe intervals.

EXAMPLE 4   *Use Absolute Value Notation*

Express "the distance between a real number x and 7 is less than 2" using absolute value notation.

Solution

The distance between x and 7 is $|x - 7|$. To express that this distance is less than 2, we write $|x - 7| < 2$. See **Figure 1.10**.

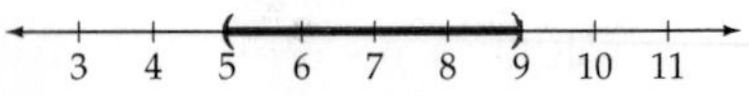

Figure 1.10

> **TRY EXERCISE 80, EXERCISE SET 1.2**

TOPICS FOR DISCUSSION

1. Explain the similarities and differences between open intervals and closed intervals.

2. Discuss why it is *not* correct to write intervals of real numbers such as $(a, \infty]$ or $[-\infty, \infty]$.

3. Discuss the correctness of the statement "If x is a real number, then $|x| = x$."

4. What is an order relation? Can all things be ordered? For instance, can colors such as blue, brown, aqua, purple, and yellow be put in order? Are the letters of the alphabet ordered?

EXERCISE SET 1.2

In Exercises 1 and 2, graph each number on a real number line.

1. $-4; -2; \dfrac{7}{4}; 2.5$

2. $-3.5; 0; 3; \dfrac{9}{4}$

In Exercises 3 to 14, replace the ☐ with the appropriate symbol ($<$, $=$, or $>$).

3. $\dfrac{5}{2} \square 4$

4. $-\dfrac{3}{2} \square -3$

5. $\dfrac{2}{3} \square 0.6666$

6. $\dfrac{1}{5} \square 0.2$

7. $1.75 \square 2.23$

8. $1.25 \square 1.3$

9. $0.\overline{36} \square \dfrac{4}{11}$

10. $0.4 \square \dfrac{4}{9}$

11. $\dfrac{10}{5} \square 2$

12. $\dfrac{0}{2} \square -\dfrac{0}{5}$

13. $\pi \square 3.14159$

14. $\dfrac{22}{7} \square \pi$

In Exercises 15 to 26, graph each inequality and write the inequality using interval notation.

15. $3 < x < 5$

16. $-2 \le x < 1$

17. $x < 3$

18. $x \ge 4$

19. $x \ge 0$ and $x < 3$

20. $x > -4$ and $x \le 4$

21. $x < -3$ or $x \ge 2$

22. $x \le 2$ or $x > 3$

23. $x > 3$ and $x < 4$

24. $x > -5$ or $x < 1$

25. $x \le 3$ and $x > -1$

26. $x < 5$ and $x \le 2$

In Exercises 27 to 38, graph each interval and write each interval as an inequality.

27. $[-4, 1]$

28. $[-2, 3)$

29. $(1, 5)$

30. $(1, 4]$

31. $[2.5, \infty)$

32. $(-\infty, 3]$

33. $(-\infty, 2)$

34. (π, ∞)

35. $(-\infty, 2] \cup (3, \infty)$

36. $(-\infty, 1) \cup (4, \infty)$

37. $(-\infty, 3) \cup (3, \infty)$

38. $(-\infty, 1) \cup [2, \infty)$

In Exercises 39 to 46, use the given notation or graph to supply the notation or graph that is marked with a question mark.

	Inequality Notation	Interval Notation	Graph
39.	$x \le 3$	?	?
40.	?	$(-2, \infty)$	?
41.	?	?	(graph: open interval from -1 to 2, marks at $-2\ -1\ 0\ 1\ 2$)
42.	$-3 \le x < -1$	?	?
43.	?	$[1, 4]$	?
44.	?	?	(graph: open ray, marks at $-3\ -2\ -1\ 0\ 1$)
45.	?	$[-2, \pi)$	?
46.	$x < 2$ or $x \ge 4$	?	?

In Exercises 47 to 60, write each expression without absolute value symbols.

47. $|4|$

48. $|-8|$

49. $|-27.4|$

50. $|3| - |-7|$

51. $-|-3| - |8|$

52. $|4| \, |-8|$

53. $|y^2 + 10|$

54. $|x^2 + 1|$

55. $|-1 - \pi|$

56. $|x + 6| + |x - 2|$, given $0 < x < 1$

57. $|x - 4| + |x + 5|$, given $2 < x < 3$

58. $|x + 1| + |x - 3|$, given $x > 5$

59. $\left| \dfrac{x + 7}{|x| + |x - 1|} \right|$, given $0 < x < 1$

60. $\left| \dfrac{x + 3}{\left|x - \dfrac{1}{2}\right| + \left|x + \dfrac{1}{2}\right|} \right|$, given $0 < x < 0.2$

In Exercises 61 to 72, find the distance between the points whose coordinates are given.

61. 8, 1

62. −2, −7

63. −3, 5

64. −5, 8

65. 16, −34

66. −108, 22

67. −38, −5

68. π, 3

69. −π, 3

70. $\frac{1}{7}$, −$\frac{1}{2}$

71. $\frac{1}{3}$, $\frac{3}{4}$

72. 0, −8

In Exercises 73 to 80, use absolute value notation to describe the given expression.

73. Distance between a and 2

74. Distance between b and −7

75. $d(m, n)$

76. $d(p, -8)$

77. The distance between a and 4 is less than z.

78. The distance between z and 5 is greater than 4.

79. The distance between x and −2 is less than 7.

80. The distance between y and −3 is greater than 6.

In Exercises 81 to 84, write interval notation for the given expression.

81. x is a real number and $x \neq 3$.

82. x is a real number whose square is nonnegative.

83. x is a real number whose absolute value is less than 3.

84. x is a real number whose absolute value is greater than 2.

In Exercises 85 to 87, determine whether each statement is true or false.

85. $|x|$ is a positive number.

86. $|-y| = y$

87. If $m < 0$, then $|m| = -m$.

88. For any two different real numbers x and y, the smaller of the two numbers is given by

$$\frac{1}{2}(x + y - |x - y|)$$

Verify the statement given at the end of the preceding column for

a. $x = 5$ and $y = 8$

b. $x = -2$ and $y = 7$

c. $x = -4$ and $y = -7$

89. Prove that the expression in Exercise 88 yields the smaller of the numbers x and y. *Hint:* Evaluate the expression for the two cases

$$x > y \quad \text{and} \quad x < y$$

90. The inequality $|a + b| \leq |a| + |b|$ is called the triangle inequality. For what values of a and b does

$$|a + b| = |a| + |b|?$$

SUPPLEMENTAL EXERCISES

In Exercises 91 to 94, use inequalities to describe the given statement.

91. The interest I is not greater than $120.

92. The rent R will be at least $650 a month.

93. The property has an area A that is at least 2 acres but less than 3 acres.

94. The distance D is greater than 7 miles, and it is not more than 8 miles.

In Exercises 95 to 102, use absolute value notation to describe the given statement.

95. x is closer to 2 than it is to 6.

96. x is closer to a than it is to b.

97. x is farther from 3 than it is from −7.

98. x is farther from 0 than it is from 5.

99. x is more than 2 units from 4 but less than 7 units from 4.

100. x is more than b units from a but less than c units from a.

101. x is within δ units of a.

102. x is not equal to a, but it is within δ units of a.

103. Prove the product theorem: $|ab| = |a|\,|b|$

104. Prove the quotient theorem: $\left|\dfrac{a}{b}\right| = \dfrac{|a|}{|b|}$

PROJECTS

1. INFINITE SETS Explain how Georg Cantor (see page 1) was able to prove that the set of irrational numbers has a cardinality that is larger than the set of rational numbers. One source of information is *From Zero to Infinity* by Constance Reid (New York: Thomas Y. Crowell, 1964).

1.3 INTEGER AND RATIONAL NUMBER EXPONENTS

A compact method of writing $5 \cdot 5 \cdot 5 \cdot 5$ is 5^4. The expression 5^4 is written in **exponential notation.** Similarly, we can write

$$\frac{2x}{3} \cdot \frac{2x}{3} \cdot \frac{2x}{3} \quad \text{as} \quad \left(\frac{2x}{3}\right)^3$$

Exponential notation can be used to express the product of any expression that is used repeatedly as a factor.

DEFINITION OF NATURAL NUMBER EXPONENTS

If b is any real number and n is any natural number, then

$$b^n = \underbrace{b \cdot b \cdot b \cdots b}_{n \text{ factors of } b}$$

In the expression b^n, b is the **base,** n is the **exponent,** and b^n is the *n*th **power of *b.***

For instance,

$$(-5)^4 = (-5)(-5)(-5)(-5) = 625$$
$$-5^4 = (5 \cdot 5 \cdot 5 \cdot 5) = -625$$

Note the difference between $(-5)^4 = 625$ and $-5^4 = -625$. The parentheses in $(-5)^4$ indicate that the base is -5; however, the expression -5^4 means $-(5^4)$. This time the base is 5.

DEFINITION OF b^0

For any nonzero real number b, $b^0 = 1$.

Any nonzero real number raised to the zero power equals 1. For example,

$$7^0 = 1 \qquad \left(\frac{1}{2}\right)^0 = 1 \qquad (-3)^0 = 1 \qquad \pi^0 = 1 \qquad (a^2 + 1)^0 = 1$$

DEFINITION OF b^{-n}

If $b \neq 0$ and n is any natural number, then $b^{-n} = \dfrac{1}{b^n}$ and $\dfrac{1}{b^{-n}} = b^n$.

Here are some examples of this definition.

$$3^{-2} = \frac{1}{3^2} = \frac{1}{9} \qquad \frac{1}{4^{-3}} = 4^3 = 64 \qquad \frac{5^{-2}}{7^{-1}} = \frac{7}{5^2} = \frac{7}{25}$$

RESTRICTION AGREEMENT

The expressions 0^0, 0^n where n is a negative integer, and $x/0$ are all undefined expressions. Therefore, all values of variables in this text are restricted to avoid any one of these expressions.

For instance, in the expression

$$\frac{x^0 y^{-3}}{z - 4}$$

we assume that $x \neq 0$, $y \neq 0$, and $z \neq 4$.

Simplifying exponential expressions requires use of the following properties of exponents.

PROPERTIES OF EXPONENTS

If m, n, and p are integers and a and b are real numbers, then

Product $b^m \cdot b^n = b^{m+n}$

Quotient $\dfrac{b^m}{b^n} = b^{m-n}, \quad b \neq 0$

Power $(b^m)^n = b^{mn} \qquad (a^m b^n)^p = a^{mp} b^{np}$

$\left(\dfrac{a^m}{b^n}\right)^p = \dfrac{a^{mp}}{b^{np}}, \quad b \neq 0$

Exponential expressions such as a^{b^c} can be confusing. The generally accepted meaning of a^{b^c} is $a^{(b^c)}$. However, some graphing calculators do not evaluate exponential expressions in this way. Enter 2^3^4 in a graphing calculator. If the result is approximately 2.42×10^{24}, then the calculator evaluated $2^{(3^4)}$. If the result is 4096, then the calculator evaluated $(2^3)^4$. To ensure that you calculate the value you intend, we strongly urge you to use parentheses. For instance, entering 2^(3^4) will produce 2.42×10^{24} and entering (2^3)^4 will produce 4096.

To simplify an expression involving exponents, write the expression in a form in which *each base appears at most once* and *no powers of powers or negative exponents appear.*

EXAMPLE 1   *Simplify Exponential Expressions*

Simplify.   **a.** $\left(\dfrac{2abc^2}{5a^2b}\right)^3$ **b.** $\dfrac{x^n y^{2n}}{x^{n-1} y^n}$

Solution

a. $\left(\dfrac{2abc^2}{5a^2b}\right)^3 = \left(\dfrac{2c^2}{5a}\right)^3$ • **The quotient property**

$\qquad = \dfrac{8c^6}{125a^3}$ • **A power property**

b. $\dfrac{x^n y^{2n}}{x^{n-1} y^n} = x^{n-(n-1)} y^{2n-n}$ • **The quotient property**

$\qquad = xy^n$

TRY EXERCISE 24, EXERCISE SET 1.3

SCIENTIFIC NOTATION

The exponent theorems provide a compact method of writing very large or very small numbers. The method is called *scientific notation*. A number written in **scientific notation** has the form $a \cdot 10^n$, where n is an integer and $1 \leq a < 10$. The following procedure is used to change a number from its decimal form to scientific notation.

For numbers greater than 10, move the decimal point to the position to the right of the first digit. The exponent n will equal the number of places the decimal point has been moved. For example,

$$7{,}430{,}000 = 7.43 \times 10^6$$

6 places

For numbers less than 1, move the decimal point to the right of the first nonzero digit. The exponent n will be negative, and its absolute value will equal the number of places the decimal point has been moved. For example,

$$0.00000078 = 7.8 \times 10^{-7}$$

7 places

To change a number from scientific notation to its decimal form, reverse the procedure. That is, if the exponent is positive, move the decimal point to the right the same number of places as the exponent. For example,

$$3.5 \times 10^5 = 350{,}000$$

5 places

If the exponent is negative, move the decimal point to the left the same number of places as the absolute value of the exponent. For example,

$$2.51 \times 10^{-8} = 0.0000000251$$

8 places

Most scientific calculators display very large and very small numbers in scientific notation. The number $450{,}000^2$ is displayed as $\boxed{2.025 \quad 11}$. This means $450{,}000^2 = 2.025 \times 10^{11}$.

RATIONAL EXPONENTS AND RADICALS

To this point, the expression b^n has been defined for real numbers b and integers n. Now we wish to extend the definition of exponents to include rational numbers so that expressions such as $2^{1/2}$ will be meaningful. Not just any definition will do. We want a definition of rational exponents for which the properties of integer exponents are true. The following example shows the direction we can take to accomplish our goal.

If the product property for exponential expressions is to hold for rational exponents, then for rational numbers p and q, $b^p b^q = b^{p+q}$. For example,

$$9^{1/2} \cdot 9^{1/2} \quad \text{must equal} \quad 9^{1/2+1/2} = 9^1 = 9$$

Thus $9^{1/2}$ must be a square root of 9. That is, $9^{1/2} = 3$.

The example suggests that $b^{1/n}$ can be defined in terms of roots according to the following definition.

DEFINITION OF $b^{1/n}$

If n is an even positive integer and $b \geq 0$, then $b^{1/n}$ is the nonnegative real number such that $(b^{1/n})^n = b$.

If n is an odd positive integer, then $b^{1/n}$ is the real number such that $(b^{1/n})^n = b$.

As examples,

- $25^{1/2} = 5$ because $5^2 = 25$.

- $(-64)^{1/3} = -4$ because $(-4)^3 = -64$.

- $16^{1/2} = 4$ because $4^2 = 16$.

- $-16^{1/2} = -(16^{1/2}) = -4$.

- $(-16)^{1/2}$ is not a real number.

- $(-32)^{1/5} = -2$ because $(-2)^5 = -32$.

If n is an even positive integer and $b < 0$, then $b^{1/n}$ is a *complex number*. Complex numbers are discussed in Section 1.4.

To define expressions such as $8^{2/3}$, we will extend our definition of exponents even further. Because we want the power property $(b^p)^q = b^{pq}$ to be true for rational exponents also, we must have $(b^{1/n})^m = b^{m/n}$. With this in mind, we make the following definition.

DEFINITION OF $b^{m/n}$

For all positive integers m and n such that m/n is in simplest form, and for all real numbers b for which $b^{1/n}$ is a real number,

$$b^{m/n} = (b^{1/n})^m = (b^m)^{1/n}$$

Because $b^{m/n}$ is defined as $(b^{1/n})^m$ and also as $(b^m)^{1/n}$, we can evaluate expressions such as $8^{4/3}$ in more than one way. For example, $8^{4/3}$ can be evaluated in either of the following ways:

$$8^{4/3} = (8^{1/3})^4 = 2^4 = 16$$

$$8^{4/3} = (8^4)^{1/3} = 4096^{1/3} = 16$$

Of the two methods, the $b^{m/n} = (b^{1/n})^m$ method is usually easier to apply, provided you can evaluate $b^{1/n}$.

Some graphing calculators do not evaluate $b^{m/n}$ when $b < 0$. Here is something you can try with your calculator. Evaluate $8^{1/3}$, $8^{2/3}$, $(-8)^{1/3}$, and $(-8)^{2/3}$. A typical display for the first three expressions is shown in **Figure 1.11.**

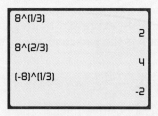

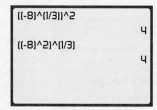

Figure 1.11   Figure 1.12

Each result in **Figure 1.11** is correct. The problem occurs with the fourth expression, $(-8)^{2/3}$. Try entering **(-8)^(2/3)**. The answer should be 4, but some calculators display an error message for this expression. You can still use your calculator to evaluate this expression, but you must use parentheses. Two possible expressions you can enter are shown in **Figure 1.12.**

The following exponent properties were stated earlier, but they are restated here to remind you that they have now been extended to apply to rational exponents.

PROPERTIES OF RATIONAL EXPONENTS

If p, q, and r represent rational numbers and a and b are positive real numbers, then

Product $\qquad b^p \cdot b^q = b^{p+q}$

Quotient $\qquad \dfrac{b^p}{b^q} = b^{p-q}$

Power $\qquad (b^p)^q = b^{pq} \qquad (a^p b^q)^r = a^{pr} b^{qr} \qquad \left(\dfrac{a^p}{b^q}\right)^r = \dfrac{a^{pr}}{b^{qr}} \qquad b^{-p} = \dfrac{1}{b^p}$

Recall that an exponential expression is in simplest form when no powers of powers or negative exponents appear and each base occurs at most once.

EXAMPLE 2   Simplify Exponential Expressions

Simplify: $\left(\dfrac{x^2 y^3}{x^{-3} y^5}\right)^{1/2}$ (Assume $x > 0, y > 0$.)

Solution

$$\left(\frac{x^2 y^3}{x^{-3} y^5}\right)^{1/2} = (x^5 y^{-2})^{1/2} = \frac{x^{5/2}}{y}$$

TRY EXERCISE 36, EXERCISE SET 1.3

Radicals, expressed by the notation $\sqrt[n]{b}$, are also used to denote roots. The number b is the **radicand,** and the positive integer n is the **index** of the radical.

DEFINITION OF $\sqrt[n]{b}$

If n is a positive integer and b is a real number such that $b^{1/n}$ is a real number, then $\sqrt[n]{b} = b^{1/n}$.

If the index n equals 2, then the radical $\sqrt[2]{b}$ is written as simply $\sqrt{b}$, and it is referred to as the **principal square root of b** or simply the **square root of b.**

The symbol $\sqrt{b}$ is reserved to represent the nonnegative square root of b. To represent the negative square root of b, write $-\sqrt{b}$. For example, $\sqrt{25} = 5$, whereas $-\sqrt{25} = -5$.

DEFINITION OF $(\sqrt[n]{b})^m$

For all positive integers n, all integers m, and all real numbers b such that $\sqrt[n]{b}$ is a real number, $(\sqrt[n]{b})^m = \sqrt[n]{b^m} = b^{m/n}$.

The equations

$$b^{m/n} = \sqrt[n]{b^m} \qquad \text{and} \qquad b^{m/n} = (\sqrt[n]{b})^m$$

can be used to write exponential expressions such as $b^{m/n}$ in radical form. Use the denominator n as the index of the radical and the numerator m as the power of the radicand or as the power of the radical. For example,

$$(5xy)^{2/3} = (\sqrt[3]{5xy})^2 = \sqrt[3]{25x^2y^2}$$

• Use the denominator 3 as the index of the radical and the numerator 2 as the power of the radical.

The equations

$$b^{m/n} = \sqrt[n]{b^m} \qquad \text{and} \qquad b^{m/n} = (\sqrt[n]{b})^m$$

can also be used to write radical expressions in exponential form. For example,

$$\sqrt{(2ab)^3} = (2ab)^{3/2}$$

• Use the index 2 as the denominator of the power and the exponent 3 as the numerator of the power.

The definition of $\sqrt[n]{b^m}$ can often be used to evaluate radical expressions. For instance,

$$(\sqrt[3]{8})^4 = 8^{4/3} = (8^{1/3})^4 = 2^4 = 16$$

Care must be exercised when simplifying even roots (square roots, fourth roots, sixth roots,...) of variable expressions. Consider $\sqrt{x^2}$ when $x = 5$ and when $x = -5$.

Case 1 If $x = 5$, then $\sqrt{x^2} = \sqrt{5^2} = \sqrt{25} = 5 = x$.

Case 2 If $x = -5$, then $\sqrt{x^2} = \sqrt{(-5)^2} = \sqrt{25} = 5 = -x$.

These two cases suggest that

$$\sqrt{x^2} = \begin{cases} x, & \text{if } x \geq 0 \\ -x, & \text{if } x < 0 \end{cases}$$

Recalling the definition of absolute value, we can write this more compactly as $\sqrt{x^2} = |x|$.

Simplifying odd roots of a variable expression does not require using the absolute value symbol. Consider $\sqrt[3]{x^3}$ when $x = 5$ and when $x = -5$.

Case 1 If $x = 5$, then $\sqrt[3]{x^3} = \sqrt[3]{5^3} = \sqrt[3]{125} = 5 = x$.

Case 2 If $x = -5$, then $\sqrt[3]{x^3} = \sqrt[3]{(-5)^3} = \sqrt[3]{-125} = -5 = x$.

Thus $\sqrt[3]{x^3} = x$.

Although we have illustrated this principle only for square roots and cube roots, the same reasoning can be applied to other cases. The general result is given below.

PROPERTIES OF $\sqrt[n]{b^n}$

If n is an even natural number and b is a real number, then

$$\sqrt[n]{b^n} = |b|$$

If n is an odd natural number and b is a real number, then

$$\sqrt[n]{b^n} = b$$

Here are some examples of these properties.

$$\sqrt[4]{16z^4} = 2|z| \qquad \sqrt[5]{32a^5} = 2a$$

Because radicals are defined in terms of rational powers, the properties of radicals are similar to those of exponential expressions.

PROPERTIES OF RADICALS

If m and n are natural numbers and a and b are nonnegative real numbers, then

Product $\sqrt[n]{a} \cdot \sqrt[n]{b} = \sqrt[n]{ab}$

Quotient $\dfrac{\sqrt[n]{a}}{\sqrt[n]{b}} = \sqrt[n]{\dfrac{a}{b}}$

Index $\sqrt[m]{\sqrt[n]{a}} = \sqrt[mn]{a}$

A radical is in **simplest form** if it meets all of the following criteria.

1. The radicand contains only powers less than the index. ($\sqrt{x^5}$ does not satisfy this requirement because 5, the exponent, is greater than 2, the index.)

2. The index of the radical is as small as possible. ($\sqrt[9]{x^3}$ does not satisfy this requirement because $\sqrt[9]{x^3} = x^{3/9} = x^{1/3} = \sqrt[3]{x}$.)

3. The denominator has been rationalized. That is, no radicals appear in the denominator. ($1/\sqrt{2}$ does not satisfy this requirement.)

4. No fractions appear under the radical sign. ($\sqrt[4]{2/x^3}$ does not satisfy this requirement.)

Radical expressions are simplified by using the properties of radicals. Here are some examples.

EXAMPLE 3   *Simplify Radical Expressions*

Simplify.   **a.** $\sqrt[4]{32x^3y^4}$ **b.** $\sqrt[3]{162x^4y^6}$

Solution

a. $\sqrt[4]{32x^3y^4} = \sqrt[4]{2^5x^3y^4} = \sqrt[4]{(2^4y^4) \cdot (2x^3)}$ • Factor and group factors that can be written as a power of the index.

$= \sqrt[4]{2^4y^4} \cdot \sqrt[4]{2x^3}$ • Use the product property of radicals.

$= 2|y|\sqrt[4]{2x^3}$ • Recall that for n even, $\sqrt[n]{b^n} = |b|$.

b. $\sqrt[3]{162x^4y^6} = \sqrt[3]{(2 \cdot 3^4)x^4y^6}$ • Factor and group factors that can be written as a power of the index.

$= \sqrt[3]{(3xy^2)^3 \cdot (2 \cdot 3x)}$

$= \sqrt[3]{(3xy^2)^3} \cdot \sqrt[3]{6x}$ • Use the product property of radicals.

$= 3xy^2\sqrt[3]{6x}$ • Recall that for n odd, $\sqrt[n]{b^n} = b$.

TRY EXERCISE 76, EXERCISE SET 1.3

Like radicals have the same radicand and the same index. For instance,

$$3\sqrt[3]{5xy^2} \quad \text{and} \quad -4\sqrt[3]{5xy^2}$$

are like radicals. Addition and subtraction of like radicals are accomplished by using the distributive property. For example,

$$4\sqrt{3x} - 9\sqrt{3x} = (4 - 9)\sqrt{3x} = -5\sqrt{3x}$$
$$2\sqrt[3]{y^2} + 4\sqrt[3]{y^2} - \sqrt[3]{y^2} = (2 + 4 - 1)\sqrt[3]{y^2} = 5\sqrt[3]{y^2}$$

The sum $2\sqrt{3} + 6\sqrt{5}$ cannot be simplified further because the radicands are not the same. The sum $3\sqrt[3]{x} + 5\sqrt[4]{x}$ cannot be simplified because the indices are not the same.

Sometimes it is possible to simplify radical expressions that do not appear to be like radicals by simplifying each radical expression.

EXAMPLE 4   *Combine Radical Expressions*

Perform the indicated operation: $5\sqrt[3]{16} + 2\sqrt[3]{128}$

Solution

$$5\sqrt[3]{16} + 2\sqrt[3]{128} = 5\sqrt[3]{2^4} + 2\sqrt[3]{2^7} = 5\sqrt[3]{2^3}\sqrt[3]{2} + 2\sqrt[3]{2^6}\sqrt[3]{2}$$
$$= 5 \cdot 2\sqrt[3]{2} + 2 \cdot 2^2\sqrt[3]{2} = 10\sqrt[3]{2} + 8\sqrt[3]{2}$$
$$= 18\sqrt[3]{2}$$

TRY EXERCISE 84, EXERCISE SET 1.3

Multiplication of radical expressions is accomplished by using the distributive property. For instance,

$$\sqrt{5}(\sqrt{20} - 3\sqrt{15}) = \sqrt{5}(\sqrt{20}) - \sqrt{5}(3\sqrt{15}) \qquad \bullet \textbf{ Use the distributive property.}$$

$$= \sqrt{100} - 3\sqrt{75} \qquad \bullet \textbf{ Multiply the radicals.}$$
$$= 10 - 3 \cdot 5\sqrt{3} \qquad \bullet \textbf{ Simplify.}$$
$$= 10 - 15\sqrt{3}$$

The product of more complicated radical expressions may require repeated use of the distributive property.

EXAMPLE 5   *Multiply Radical Expressions*

Perform the indicated operation: $(\sqrt{3} + 5)(\sqrt{3} - 2)$

Solution

$$(\sqrt{3} + 5)(\sqrt{3} - 2)$$
$$= (\sqrt{3} + 5)\sqrt{3} - (\sqrt{3} + 5)2 \qquad \bullet \textbf{ Use the distributive property.}$$
$$= (\sqrt{3}\sqrt{3} + 5\sqrt{3}) - (2\sqrt{3} + 2 \cdot 5) \qquad \bullet \textbf{ Use the distributive property.}$$
$$= 3 + 5\sqrt{3} - 2\sqrt{3} - 10$$
$$= -7 + 3\sqrt{3}$$

TRY EXERCISE 92, EXERCISE SET 1.3

To **rationalize the denominator** of a fraction means to write it in an equivalent form that does not involve any radicals in its denominator.

EXAMPLE 6   *Rationalize the Denominator*

Rationalize the denominator.   **a.** $\dfrac{5}{\sqrt[3]{a}}$ **b.** $\sqrt{\dfrac{3}{32y}}$

Solution

a. $\dfrac{5}{\sqrt[3]{a}} = \dfrac{5}{\sqrt[3]{a}} \cdot \dfrac{\sqrt[3]{a^2}}{\sqrt[3]{a^2}} = \dfrac{5\sqrt[3]{a^2}}{\sqrt[3]{a^3}} = \dfrac{5\sqrt[3]{a^2}}{a}$ • Use $\sqrt[3]{a} \cdot \sqrt[3]{a^2} = \sqrt[3]{a^3} = a.$

b. $\sqrt{\dfrac{3}{32y}} = \dfrac{\sqrt{3}}{\sqrt{32y}} = \dfrac{\sqrt{3}}{4\sqrt{2y}} = \dfrac{\sqrt{3}}{4\sqrt{2y}} \cdot \dfrac{\sqrt{2y}}{\sqrt{2y}} = \dfrac{\sqrt{6y}}{8y}$

> **TRY EXERCISE 104, EXERCISE SET 1.3**

To rationalize the denominator of a fractional expression such as

$$\frac{1}{\sqrt{m} + \sqrt{n}}$$

we make use of the conjugate of $\sqrt{m} + \sqrt{n}$, which is $\sqrt{m} - \sqrt{n}$. The product of these conjugate pairs does not involve a radical.

$$(\sqrt{m} + \sqrt{n})(\sqrt{m} - \sqrt{n}) = m - n$$

In Example 7 we use the conjugate of the denominator to rationalize the denominator.

EXAMPLE 7   *Rationalize the Denominator*

Rationalize the denominator.   **a.** $\dfrac{2}{\sqrt{3} + \sqrt{2}}$ **b.** $\dfrac{a + \sqrt{5}}{a - \sqrt{5}}$

Solution

a. $\dfrac{2}{\sqrt{3} + \sqrt{2}} = \dfrac{2}{\sqrt{3} + \sqrt{2}} \cdot \dfrac{\sqrt{3} - \sqrt{2}}{\sqrt{3} - \sqrt{2}} = \dfrac{2\sqrt{3} - 2\sqrt{2}}{3 - 2} = 2\sqrt{3} - 2\sqrt{2}$

b. $\dfrac{a + \sqrt{5}}{a - \sqrt{5}} = \dfrac{a + \sqrt{5}}{a - \sqrt{5}} \cdot \dfrac{a + \sqrt{5}}{a + \sqrt{5}} = \dfrac{a^2 + 2a\sqrt{5} + 5}{a^2 - 5}$

> **TRY EXERCISE 110, EXERCISE SET 1.3**

TOPICS FOR DISCUSSION

1. Given that a is a real number, discuss when the expression $a^{p/q}$ represents a real number.

2. The expressions $-a^n$ and $(-a)^n$ do not always represent the same number. Discuss the situations in which the two expressions are equal and those in which they are not equal.

3. Most calculators will automatically convert a number to scientific notation whenever the calculation produces a number that cannot be represented exactly. However, there are limits as to how large or small a number can be and still have a calculator representation. Discuss the limits of your calculator and the practical effects they would have on the most demanding calculations. For instance, can you find the distance (in miles) to the Orion nebula, which is 1600 light-years away? Can your calculator represent the mass of a quark (a subatomic particle)?

4. If you enter the expression for $\sqrt{5}$ on your calculator, the calculator will respond with 2.236067977 or some number close to that. Is this the exact value of $\sqrt{5}$? Is it possible to find the exact decimal value of $\sqrt{5}$ with a calculator? with a computer?

5. Values of variables can be stored in graphing calculators and then used for calculations. Store in A the value 5 and in B the value 8. Now use your calculator to find 1/AB and 1/A*B. Are the results the same? Discuss how your calculator evaluates these two expressions. Discuss how your calculator evaluates 1/AB^-1. Which of these operations is consistent with the Order of Operations Agreement you learned to apply in previous classes?

EXERCISE SET 1.3

In Exercises 1 to 22, evaluate each expression.

1. -4^4

2. $(-4)^4$

3. $\left(\dfrac{2^2 \cdot 3^{-5}}{2^{-3} \cdot 5^4}\right)^0$

4. -7^0

5. $\left(\dfrac{4}{9}\right)^{-2}$

6. $\left(\dfrac{5^{-3} \cdot 7}{3^{-2}}\right)^{-1}$

7. $4^{3/2}$

8. $16^{3/2}$

9. $-9^{1/2}$

10. $-25^{1/2}$

11. $-64^{2/3}$

12. $-125^{2/3}$

13. $(-64)^{2/3}$

14. $(-125)^{2/3}$

15. $9^{-1/2}$

16. $16^{-1/2}$

17. $27^{-2/3}$

18. $4^{-3/2}$

19. $\left(\dfrac{9}{16}\right)^{1/2}$

20. $\left(\dfrac{4}{25}\right)^{3/2}$

21. $\left(\dfrac{8}{27}\right)^{-2/3}$

22. $\left(\dfrac{4}{9}\right)^{-3/2}$

31. $(2ab^{-3})^2(-2a^{-1}b^2)^2$

32. $\left[\left(\dfrac{b^{-3}}{a^2}\right)^2\left(\dfrac{a^{-2}}{ab}\right)^{-1}\right]^0$

33. $(81x^4y^{12})^{1/4}$

34. $(625a^8b^4)^{1/4}$

35. $\dfrac{a^{3/4}b^{1/2}}{a^{1/4}b^{1/5}}$

36. $\dfrac{x^{1/3}y^{5/6}}{x^{3/2}y^{1/6}}$

37. $a^{1/3}(a^{5/3} + 7a^{2/3})$

38. $m^{3/4}(m^{1/4} - 8m^{5/4})$

39. $(p^{1/2} + q^{1/2})(p^{1/2} - q^{1/2})$

40. $(c + d^{1/3})(c - d^{1/3})$

41. $\left(\dfrac{m^2n^4}{m^{-2}n}\right)^{1/2}$

42. $\left(\dfrac{r^3s^{-2}}{rs^4}\right)^{1/2}$

43. $\dfrac{x^{n+1/2} \cdot x^{-n}}{x^{1/2}}$

44. $\dfrac{r^{n/2} \cdot r^{2n}}{r^{-n}}$

45. $\dfrac{r^{1/n}}{r^{1/m}}$

46. $\dfrac{s^{2/n}}{s^{-n/2}}$

In Exercises 23 to 46, simplify each exponential expression.

23. $(2x^2y^3)(3x^5y)$

24. $\left(\dfrac{2ab^2c^3}{5ab^2}\right)^3$

25. $\dfrac{(3xy^{-3})^2}{(2xy)^{-2}}$

26. $(2x^{-3}y^0)(3^{-1}xy)^2$

27. $\left(\dfrac{3x}{y}\right)^{-1}$

28. $(x^2y^{-3})^{-2}$

29. $a^{-1} + b^{-1}$

30. $\dfrac{4a^2(bc)^{-1}}{(-2)^2a^3b^{-2}c}$

In Exercises 47 to 50, write each number in scientific notation.

47. 21,000,000

48. 163,000,000

49. 0.00095

50. 0.0000000821

In Exercises 51 to 54, change each number from scientific notation to decimal notation.

51. 6.5×10^3

52. 6.86×10^{-9}

53. 2.17×10^{-4}

54. 3.75×10^0

In Exercises 55 to 60, write each exponential expression in radical form.

55. $(3x)^{1/2}$ **56.** $(6y)^{1/3}$ **57.** $5(xy)^{1/4}$

58. $2a(bc)^{1/5}$ **59.** $(5w)^{2/3}$ **60.** $(a + b)^{3/4}$

In Exercises 61 to 66, write each radical in exponential form.

61. $\sqrt[3]{17k}$ **62.** $4\sqrt{3m}$ **63.** $\sqrt[5]{a^2}$

64. $3\sqrt[4]{5n}$ **65.** $\sqrt{\dfrac{7a}{3}}$ **66.** $\sqrt[3]{\dfrac{5b^2}{7}}$

In Exercises 67 to 78, simplify each radical.

67. $\sqrt{45}$ **68.** $\sqrt{75}$ **69.** $\sqrt[3]{24}$

70. $\sqrt[3]{135}$ **71.** $\sqrt[3]{-81}$ **72.** $\sqrt[3]{-250}$

73. $-\sqrt[3]{32}$ **74.** $-\sqrt[3]{243}$ **75.** $\sqrt{24x^3y^2}$

76. $\sqrt{18x^2y^5}$ **77.** $-\sqrt[3]{16a^3y^7}$ **78.** $-\sqrt[3]{54c^2d^5}$

In Exercises 79 to 86, simplify each radical and then combine like radicals.

79. $7\sqrt{3} - \sqrt{3}$ **80.** $8\sqrt{11} - \sqrt{11}$

81. $\sqrt{8} - 5\sqrt{2}$ **82.** $\sqrt{27} + 4\sqrt{3}$

83. $2\sqrt[3]{2} - \sqrt[3]{16}$ **84.** $5\sqrt[3]{3} + 2\sqrt[3]{81}$

85. $3x\sqrt[3]{8x^3y^4} + 4y\sqrt[3]{64x^6y}$ **86.** $4\sqrt{a^5b} - a^2\sqrt{ab}$

In Exercises 87 to 96, find the indicated product of the radical expressions. Express each result in simplest form.

87. $(\sqrt{5} + 8)(\sqrt{5} + 3)$ **88.** $(\sqrt{7} + 4)(\sqrt{7} - 1)$

89. $(\sqrt{2x} + 3)(\sqrt{2x} - 3)$ **90.** $(7 - \sqrt{3a})(7 + \sqrt{3a})$

91. $(5\sqrt{2y} + \sqrt{3z})^2$ **92.** $(3\sqrt{5y} - 4)^2$

93. $(\sqrt{x - 3} + 5)^2$ **94.** $(\sqrt{x + 7} - 3)^2$

95. $(\sqrt{2x + 5} + 7)^2$ **96.** $(\sqrt{9x - 2} + 11)^2$

In Exercises 97 to 112, simplify each expression by rationalizing the denominator. Write the result in simplest form.

97. $\dfrac{2}{\sqrt{2}}$ **98.** $\dfrac{3x}{\sqrt{3}}$ **99.** $\sqrt{\dfrac{5}{18}}$ **100.** $\sqrt{\dfrac{7}{40}}$

101. $\dfrac{3}{\sqrt[3]{2}}$ **102.** $\dfrac{2}{\sqrt[3]{4}}$ **103.** $\dfrac{4}{\sqrt[3]{8x^2}}$ **104.** $\dfrac{2}{\sqrt[4]{4y}}$

105. $\sqrt{\dfrac{10}{18}}$ **106.** $\sqrt{\dfrac{14}{40}}$ **107.** $\sqrt{\dfrac{2x}{27y}}$ **108.** $\sqrt{\dfrac{4c}{50d}}$

109. $\dfrac{3}{\sqrt{5} + \sqrt{x}}$ **110.** $\dfrac{5}{\sqrt{y} - \sqrt{3}}$

111. $\dfrac{\sqrt{7}}{2 - \sqrt{7}}$ **112.** $\dfrac{6\sqrt{6}}{5 + \sqrt{6}}$

113. **COLOR MONITORS** The number of colors that a computer with a color monitor can display is sometimes indicated in *bits* (a bit is a *binary digit*). For example, a computer with a 4-bit color interface can display $2^4 = 16$ colors. Determine how many colors each of the following can display.

a. A computer with an 8-bit interface.

b. A computer with a 16-bit interface.

114. **PHYSIOLOGY** It has been estimated that the human eye can detect 36,000 different colors. How many of these colors (to the nearest 1000) would go undetected by a human using a computer with a 24-bit color interface? (See Exercise 113.)

115. **ASTRONOMY** Pluto is 5.91×10^{12} meters from the Sun. The speed of light is 3.00×10^8 meters per second. Find the time it takes light from the Sun to reach Pluto.

116. **ASTRONOMY** The Earth's mean distance from the Sun is 9.3×10^7 miles. This distance is called the *astronomical unit* (AU). Jupiter is 5.2 AU from the Sun. Find the distance in miles from the Sun to Jupiter.

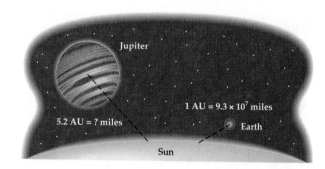

117. **FINANCE** A principal P invested at an annual interest rate r compounded n times per year yields a balance A given by the formula

$$A = P\left(1 + \frac{r}{n}\right)^n$$

Find the balance after 1 year when \$4500 is deposited in an account with an annual interest rate of 8% that is compounded monthly.

118. **FINANCE** You plan to save 1¢ the first day of a month, 2¢ the second day, and 4¢ the third day and to continue this pattern of saving twice what you saved on the previous day for every day in a month that has 30 days.

a. How much money will you need to save on the 30th day?

b. How much money will you have after 30 days?

(*Hint:* Note that after 2 days you will have saved $2^2 - 1 = 3$¢ and that after 3 days you will have saved $2^3 - 1 = 7$¢.)

119. OPTICS The percent P of light that will pass through a frosted glass is given by the equation $P = 10^{-kd}$, where d is the thickness of the glass in centimeters and k is a constant that depends on the glass. Find, to the nearest percent, the amount of light that will pass through frosted glass for which

a. $k = 0.15$, $d = 0.6$ centimeters.

b. $k = 0.15$, $d = 1.2$ centimeters.

120. FOOD SCIENCE The number of hours h needed to cook a pot roast that weighs p pounds can be approximated by using the formula $h = 0.9p^{0.6}$.

a. Find the time (to the nearest hundredth of an hour) required to cook a 12-pound pot roast.

b. If pot roast A weighs twice as much as pot roast B, then roast A should be cooked for a period of time that is how many times longer than the time required for roast B to cook?

SUPPLEMENTAL EXERCISES

In Exercises 121 to 124, write each expression as an equivalent expression in which the variables x and y occur only once.

121. $\dfrac{x^n y^{n+2}}{x^{n-3} y}$

122. $\left(\dfrac{x^n y^{2n}}{y^{3-n}}\right)^{-2}$

123. $\left(\dfrac{x^{3n} y^{2n}}{x^{-2n} y^{3n+1}}\right)^{-1}$

124. $\left(\dfrac{x^{4-n} y^{n+4}}{xy^{n-4}}\right)^{2}$

In Exercises 125 to 128, find the value of p for which the statement is true.

125. $a^{2/5} a^p = a^2$

126. $b^{-3/4} b^{2p} = b^3$

127. $\dfrac{x^{-3/4}}{x^{3p}} = x^4$

128. $(x^4 x^{2p})^{1/2} = x$

129. Which is larger, $3^{(3^3)}$ or $(3^3)^3$?

130. If $2^x = y$, then find 2^{x-3} in terms of y.

131. Prove: $\sqrt{a^2 + b^2} \neq a + b$.
(*Hint:* Find a counter-example.)

132. When does $\sqrt[3]{a^3 + b^3} = a + b$?
(*Hint:* Cube each side of the equation.)

In Exercises 133 to 138, rationalize the numerator.

133. $\dfrac{\sqrt{4+h} - 2}{h}$

134. $\dfrac{\sqrt{9+h} - 3}{h}$

135. $\dfrac{\sqrt{a+h} - \sqrt{a}}{h}$

136. $\dfrac{\sqrt{2x+2h} - \sqrt{2x}}{h}$

137. $\sqrt{n^2 + 1} - n \left(Hint: \sqrt{n^2+1} - n = \dfrac{\sqrt{n^2+1} - n}{1} \right)$

138. $\sqrt{n^2 + n} - n \left(Hint: \sqrt{n^2+n} - n = \dfrac{\sqrt{n^2+n} - n}{1} \right)$

PROJECTS

1. **RELATIVITY THEORY** A moving object has energy, called kinetic energy, by virtue of its motion. As mentioned earlier in this chapter, the theory of relativity uses

$$K.E._r = mc^2 \left[\frac{1}{\sqrt{1 - \dfrac{v^2}{c^2}}} - 1 \right]$$

as the formula for kinetic energy. When the speed of an object is much less than the speed of light (3.0×10^8 meters per second) the formula

$$K.E._n = \frac{1}{2} mv^2$$

is used. In each formula, v is the velocity of the object in meters per second, m is its mass in kilograms, and c is the speed of light given above. Calculate the percent error (in **a.** through **e.**) for each of the given velocities. The formula for percent error is

$$\% \text{ error} = \frac{|K.E._r - K.E._n|}{K.E._r} \times 100$$

a. $v = 30$ meters per second (speeding car on an expressway)

b. $v = 240$ meters per second (speed of a commercial jet)

c. $v = 3.0 \times 10^7$ meters per second (10% of the speed of light)

d. $v = 1.5 \times 10^8$ meters per second (50% of the speed of light)

e. $v = 2.7 \times 10^8$ meters per second (90% of the speed of light)

f. Use your answers from **a.** through **e.** to give a reason why the formula for kinetic energy given by $K.E_n$ is adequate for most of our common experiences involving motion (walking, running, bicycle, car, plane).

g. According to relativity theory, the mass, m, of an object changes as its velocity according to

$$m = \frac{m_0}{\sqrt{1 - \dfrac{v^2}{c_2}}}$$

where m_0 is the rest mass of the object. The approximate rest mass of an electron is 9.11×10^{-31} kilogram. What is the percent change, from its rest mass, in the mass of an electron that is traveling at $0.99c$ (99% of the speed of light)?

h. According to the theory of relativity, a particle (such as an electron or a space craft) cannot exceed the speed of light. Explain why the equation for $K.E_r$ suggests that conclusion.

1.4 COMPLEX NUMBERS

There is no real number whose square is a negative number. For example, there is no real number x such that $x^2 = -1$. In the seventeenth century a new number, called an *imaginary number,* was defined. The square of an **imaginary number** is a negative real number. The letter i was chosen to represent an imaginary number whose square is -1.

POINT OF INTEREST

Electrical engineers use i to represent the current in a circuit. To avoid confusion between i for current and the imaginary unit i, these engineers use j as the imaginary unit.

DEFINITION OF i

The number i, called the **imaginary unit,** is the number such that

$$i^2 = -1$$

Many of the solutions to equations in the remainder of this text will involve radicals such as $\sqrt{-a}$, where a is a positive real number. The expression $\sqrt{-a}$, with $a > 0$, is defined as follows:

DEFINITION OF $\sqrt{-a}$

For any positive real number a,

$$\sqrt{-a} = i\sqrt{a}$$

This definition with $a = 1$ implies that $\sqrt{-1} = i$. It is often used to write the square root of a negative real number as the product of the imaginary unit i and a positive real number. For example,

$$\sqrt{-4} = i\sqrt{4} = 2i \qquad \text{and} \qquad \sqrt{-7} = i\sqrt{7}$$

DEFINITION OF A COMPLEX NUMBER

If a and b are real numbers and i is the imaginary unit, then $a + bi$ is called a **complex number**. The real number a is called the **real part** and the real number b is called the **imaginary part** of the complex number.

> **TAKE NOTE**
>
> **Even though b is a real number, it is called the imaginary part of the complex number $a + bi$. For example, the complex number $3 + 8i$ has the real number 8 as its imaginary part.**

The real numbers are a subset of the complex numbers. This can be observed by letting $b = 0$. Then $a + bi = a + 0i = a$, which is a real number. It can be shown that the associative, commutative, distributive, and identity properties also apply to complex numbers. Any number that can be written in the form $0 + bi = bi$, where b is a nonzero real number, is an **imaginary number** (or a pure imaginary number). For example, i, $3i$, and $-0.5i$ are all imaginary numbers.

A complex number is in **standard form** when it is written in the form $a + bi$.

EXAMPLE 1   *Write Complex Numbers in Standard Form*

Write each complex number in standard form.

a. $3 + \sqrt{-4}$ **b.** $\sqrt{-37} - 3$

Solution

> **TAKE NOTE**
>
> **The expression $\sqrt{a}\, i$ is often written $i\sqrt{a}$ so that it is not mistaken for $\sqrt{ai}$.**

Use the definition $\sqrt{-a} = i\sqrt{a}$.

a. $3 + \sqrt{-4} = 3 + i\sqrt{4} = 3 + 2i$ • $a + bi$ form with $a = 3$ and $b = 2$.

b. $\sqrt{-37} - 3 = i\sqrt{37} - 3 = -3 + i\sqrt{37}$

TRY EXERCISE 2, EXERCISE SET 1.4

DEFINITION OF ADDITION AND SUBTRACTION OF COMPLEX NUMBERS

If $a + bi$ and $c + di$ are complex numbers, then

Addition $(a + bi) + (c + di) = (a + c) + (b + d)i$

Subtraction $(a + bi) - (c + di) = (a - c) + (b - d)i$

To add two complex numbers, add their real parts to produce the real part of the sum and add their imaginary parts to produce the imaginary part of the sum.

EXAMPLE 2   Add or Subtract Complex Numbers

Perform the indicated operation.

a. $(4 + 2i) + (3 + 7i)$ **b.** $i - (3 - 4i)$

Solution

a. $(4 + 2i) + (3 + 7i) = (4 + 3) + (2 + 7)i = 7 + 9i$

b. $i - (3 - 4i) = (0 + 1i) - (3 - 4i)$ • $i = 0 + 1i$
$= (0 - 3) + [1 - (-4)]i = -3 + 5i$

TRY EXERCISE 16, EXERCISE SET 1.4

DEFINITION OF MULTIPLICATION OF COMPLEX NUMBERS

If $a + bi$ and $c + di$ are complex numbers, then

$$(a + bi)(c + di) = (ac - bd) + (ad + bc)i$$

Because every complex number can be written as a sum of two terms, it is natural to perform multiplication on complex numbers in a manner consistent with the operation of multiplication defined on binomials and the definition $i^2 = -1$. Thus to multiply complex numbers, it is not necessary to memorize the definition of multiplication.

EXAMPLE 3   Multiply Complex Numbers

Simplify: $(3 + 5i)(2 - 4i)$

Solution

$(3 + 5i)(2 - 4i) = 6 - 12i + 10i - 20i^2$
$= 6 - 12i + 10i - 20(-1)$ • **Substitute −1 for i^2.**
$= 6 - 12i + 10i + 20$ • **Simplify.**
$= 26 - 2i$

TRY EXERCISE 24, EXERCISE SET 1.4

The complex numbers $a + bi$ and $a - bi$ are called **complex conjugates** or **conjugates** of each other. The conjugate of the complex number z is denoted by $\bar{z}$. For example,

$$\overline{3 + 2i} = 3 - 2i \quad \text{and} \quad \overline{7 - 11i} = 7 + 11i$$

Consider the product of a complex number and its conjugate. For instance,

$(3 + 2i)(3 - 2i) = 9 - 6i + 6i - 4i^2$ • **3 + 2i and 3 − 2i are conjugates.**
$= 9 - 4(-1)$
$= 13$

Note that the product is a *real* number. This is always true.

PRODUCT OF CONJUGATES

The product of a complex number and its conjugate is a real number. That is, if $a + bi$ and $a - bi$ are complex conjugates, then

$$(a + bi)(a - bi) = a^2 + b^2$$

The fact that the product of a complex number and its conjugate is a real number can be used to find the quotient of two complex numbers. For instance, to find the quotient $\dfrac{a + bi}{c + di}$, multiply the numerator and denominator by the conjugate of the denominator.

EXAMPLE 4   *Divide Complex Numbers*

Find the quotient: $\dfrac{3 + 2i}{5 - i}$

Solution

$$\dfrac{3 + 2i}{5 - i} = \dfrac{(3 + 2i)(5 + i)}{(5 - i)(5 + i)}$$

• **Multiply numerator and denominator by $5 + i$, which is the conjugate of the denominator.**

$$= \dfrac{15 + 3i + 10i + 2i^2}{25 + 1}$$

$$= \dfrac{13 + 13i}{26} = \dfrac{1}{2} + \dfrac{1}{2}i$$

• **Write in standard form.**

TRY EXERCISE 34, EXERCISE SET 1.4

The following powers of i illustrate a pattern:

$$i^1 = i \qquad\qquad i^5 = i^4 \cdot i = (1)i = i$$

$$i^2 = -1 \qquad\qquad i^6 = i^4 \cdot i^2 = (1)(-1) = -1$$

$$i^3 = i^2 \cdot i = (-1)i = -i \qquad\qquad i^7 = i^4 \cdot i^3 = (1)(-i) = -i$$

$$i^4 = i^2 \cdot i^2 = (-1)(-1) = 1 \qquad i^8 = (i^4)^2 = 1^2 = 1$$

Because $i^4 = 1$, $(i^4)^n = 1$ for any integer n. Thus it is possible to evaluate powers of i by factoring out powers of i^4, as shown in the following example:

$$i^{25} = (i^4)^6(i) = 1^6(i) = i$$

The following theorem can be used to evaluate powers of i. Essentially, it makes use of division to eliminate powers of i^4.

POWERS OF i

If n is a positive integer, then $i^n = i^r$, where r is the remainder of the division of n by 4.

EXAMPLE 5   *Evaluate Powers of i*

Evaluate: i^{543}

Solution

Use the theorem on powers of i.

$$i^{543} = i^3 = -i \qquad \bullet \text{ Remainder of } 543 \div 4 \text{ is 3.}$$

Try Exercise 54, Exercise Set 1.4

Caution   To compute $\sqrt{a}\,\sqrt{b}$ when both a and b are negative numbers, write each radical in terms of i before multiplying. For example,

Correct method $\qquad \sqrt{-1}\,\sqrt{-1} = i \cdot i = i^2 = -1$

Incorrect method $\qquad \sqrt{-1}\,\sqrt{-1} = \sqrt{(-1)(-1)} = \sqrt{1} = 1$

EXAMPLE 6   *Simplify Products Involving Radicals with Negative Radicands*

Simplify.

a. $\sqrt{-16}\,\sqrt{-25}$ **b.** $\sqrt{-9}\,\sqrt{-7}$ **c.** $(2 + \sqrt{-5})(2 - \sqrt{-5})$

Solution

a. $\sqrt{-16}\,\sqrt{-25} = (4i)(5i) = 20i^2 = -20$

b. $\sqrt{-9}\,\sqrt{-7} = (3i)(i\sqrt{7}) = 3i^2\sqrt{7} = -3\sqrt{7}$

c. $(2 + \sqrt{-5})(2 - \sqrt{-5}) = (2 + i\sqrt{5})(2 - i\sqrt{5}) = 4 + 5 = 9$

Try Exercise 68, Exercise Set 1.4

Topics for Discussion

1. Define the term *complex number*. Is a complex number different from an imaginary number? Is it different from a real number? Are there real numbers that are not complex numbers? Are there complex numbers that are not real numbers?

2. Division of real numbers is defined in terms of multiplication. For instance, 15/3 means the number that, when multiplied by 3, is 15. Discuss whether division of complex numbers satisfies that same property.

3. If a and b are real numbers and $ab = 0$, then $a = 0$ or $b = 0$. Discuss whether the same is true for complex numbers. That is, if u and v are complex numbers and $uv = 0$, is one of the complex numbers equal to zero?

4. The concept of number may seem obvious, but this concept has been refined since the time of the first written records of history. The idea of a

fraction such as 2/3 was apparent to Egyptians 5000 years ago. However, these same people did not recognize the fraction 3/4 and thought of this number as the sum of 1/2 and 1/4. Negative numbers did not gain acceptance until the fourteenth century, and complex numbers did not achieve acceptance until the sixteenth century. Why do you think a complete concept of number did not occur immediately? Why did it evolve over time?

5. Although $2/4 = 1/2$, it is not necessarily true that $a^{2/4} = a^{1/2}$. Discuss what restrictions must be placed on a to guarantee the equality of $a^{2/4}$ and $a^{1/2}$.

EXERCISE SET 1.4

In Exercises 1 to 10, write the complex number in standard form.

1. $2 + \sqrt{-9}$

2. $3 + \sqrt{-25}$

3. $4 - \sqrt{-121}$

4. $5 - \sqrt{-144}$

5. $8 + \sqrt{-3}$

6. $9 - \sqrt{-75}$

7. $\sqrt{-16} + 7$

8. $\sqrt{-49} + 3$

9. $\sqrt{-81}$

10. $-\sqrt{-100}$

In Exercises 11 to 28, simplify and write the complex number in standard form.

11. $(2 + 5i) + (3 + 7i)$

12. $(1 - 3i) + (6 + 2i)$

13. $(-5 - i) + (9 - 2i)$

14. $5 + (3 - 2i)$

15. $(8 - 6i) - (10 - i)$

16. $(-3 + i) - (-8 + 2i)$

17. $(7 - 3i) - (-5 - i)$

18. $7 - (3 - 2i)$

19. $8i - (2 - 3i)$

20. $(4i - 5) - 2$

21. $3(2 + 7i) + 5(2 - i)$

22. $8(4 - i) - (4 - 3i)$

23. $(2 + 3i)(4 - 5i)$

24. $(5 - 3i)(-2 - 4i)$

25. $(5 + 7i)(5 - 7i)$

26. $(-3 - 5i)(-3 + 5i)$

27. $(8i + 11)(-7 + 5i)$

28. $(9 - 12i)(15i + 7)$

In Exercises 29 to 48, write each expression as a complex number in standard form.

29. $\dfrac{4 + i}{3 + 5i}$

30. $\dfrac{5 - i}{4 + 5i}$

31. $\dfrac{1}{7 - 3i}$

32. $\dfrac{1}{-8 + i}$

33. $\dfrac{3 + 2i}{3 - 2i}$

34. $\dfrac{5 - 7i}{5 + 7i}$

35. $\dfrac{2i}{11 + i}$

36. $\dfrac{3i}{5 - 2i}$

37. $\dfrac{6 + i}{i}$

38. $\dfrac{5 - i}{-i}$

39. $(3 - 5i)^2$

40. $(-5 + 7i)^2$

41. $(1 - i) - 2(4 + i)^2$

42. $(4 - i) - 5(2 + 3i)^2$

43. $(1 - i)^3$

44. $(2 + i)^3$

45. $(2i)(8i)$

46. $(-5)(7i)$

47. $(5i)^2(-3i)$

48. $(-6i)(-5i)^2$

In Exercises 49 to 64, simplify and write the complex number as i, $-i$, 1, or -1.

49. i^3

50. $-i^3$

51. i^5

52. $-i^5$

53. i^{10}

54. i^{28}

55. $-i^{40}$

56. i^{40}

57. i^{223}

58. i^{553}

59. i^{2001}

60. i^{5000}

61. i^{5042}

62. i^0

63. i^{-1}

64. $i^{10,000}$

In Exercises 65 to 72, simplify each product.

65. $\sqrt{-1}\,\sqrt{-4}$

66. $\sqrt{-16}\,\sqrt{-49}$

67. $\sqrt{-64}\,\sqrt{-5}$

68. $\sqrt{-3}\,\sqrt{-121}$

69. $(3 + \sqrt{-2})(3 - \sqrt{-2})$

70. $(4 + \sqrt{-81})(4 - \sqrt{-81})$

71. $(5 + \sqrt{-16})^2$

72. $(3 - \sqrt{-144})^2$

In Exercises 73 to 80, evaluate

$$\frac{-b \pm \sqrt{b^2 - 4ac}}{2a}$$

for the given values of a, b, and c. Write your final answer as a complex number in standard form.

73. $a = 3, b = -3, c = 3$

74. $a = 1, b = -3, c = 10$

75. $a = 2, b = 4, c = 4$

76. $a = 4, b = -4, c = 2$

77. $a = 2, b = 6, c = 6$

78. $a = 6, b = -5, c = 5$

79. $a = 2, b = 1, c = 3$

80. $a = 3, b = 2, c = 4$

The *absolute value of the complex number* **$a + bi$ is denoted by $|a + bi|$ and defined as the real number $\sqrt{a^2 + b^2}$. In Exercises 81 to 88, find the indicated absolute value of each complex number.**

81. $|3 + 4i|$

82. $|5 + 12i|$

83. $|2 - 5i|$

84. $|4 - 4i|$

85. $|7 - 4i|$

86. $|11 - 2i|$

87. $|-3i|$

88. $|18i|$

SUPPLEMENTAL EXERCISES

In Exercises 89 to 92, use the complex number $z = a + bi$ and its conjugate $\bar{z} = a - bi$ to establish each result.

89. Prove that the absolute value of a complex number and the absolute value of its conjugate are equal.

90. Prove that the difference of a complex number and its conjugate is a pure imaginary number.

91. Prove that the conjugate of the sum of two complex numbers equals the sum of the conjugates of the two numbers.

92. Prove that the conjugate of the product of two complex numbers equals the product of the conjugates of the two numbers.

93. Show that if $x = 1 + i\sqrt{3}$, then $x^2 - 2x + 4 = 0$.

94. Show that if $x = 1 - i\sqrt{3}$, then $x^2 - 2x + 4 = 0$.

95. A set T is closed under the operation of addition if the sum of any two elements of T is also an element of T. Is $T = \{1, -1, i, -i\}$ closed under the operation of addition?

96. A set T is closed under the operation of multiplication if the product of any two elements of T is also an element of T. Is $T = \{1, -1, i, -i\}$ closed under the operation of multiplication?

97. Simplify:
$$[(3 + \sqrt{5}) + (7 - \sqrt{3})i][(3 + \sqrt{5}) - (7 - \sqrt{3})i]$$

98. Simplify: $[2 - (3 - \sqrt{5})i][2 + (3 - \sqrt{5})i]$

99. Simplify: $\left(\dfrac{-1}{2} + \dfrac{\sqrt{3}}{2}i\right)^3$

100. Simplify $(a + bi)^3$, where a and b are real numbers.

101. Simplify: $i + i^2 + i^3 + i^4 + \cdots + i^{28}$

102. Simplify: $i + i^2 + i^3 + i^4 + \cdots + i^{100}$

103. When we say $\sqrt[3]{8} = 2$, we are really stating that the *real* cube root of 8 is 2. There are, however, two other cube roots of 8, both of which are complex numbers. Show that
$$(-1 + i\sqrt{3})^3 = 8 \quad \text{and} \quad (-1 - i\sqrt{3})^3 = 8$$
Explain how this shows that $-1 + i\sqrt{3}$ and $-1 - i\sqrt{3}$ are cube roots of 8.

104. It is possible, although we will not show the technique here, to take the square root of a complex number. Verify that
$$\sqrt{i} = \frac{\sqrt{2}}{2}(1 + i)$$
by showing that
$$\left[\frac{\sqrt{2}}{2}(1 + i)\right]^2 = i$$

PROJECTS

1. **QUATERNIONS** William Rowan Hamilton (1805–1865) created a system of mathematics called the quaternions. It took Hamilton 13 years to determine how quaternions should be multiplied. The solution to the problem came to him as he walked across a bridge in Ireland. He was so excited that he took out a pocket knife and etched his discovery on the railing of the wooden bridge. Write an essay on Hamilton. Include in your essay the equations Hamilton carved on the bridge. Give some examples to show that you can add and multiply quaternions. Determine whether quaternions satisfy the commutative property of multiplication.

Reproduced by kind permission of An Post.

SECTION 1.5   POLYNOMIALS

A **monomial** is a constant, a variable, or a product of a constant and one or more variables, with the variables having only nonnegative integer exponents. The constant is called the **numerical coefficient** or simply the **coefficient** of the monomial. The **degree of a monomial** is the sum of the exponents of the variables. For example, $-5xy^2$ is a monomial with coefficient -5 and degree 3.

The algebraic expression $3x^{-2}$ is not a monomial because it cannot be written as a product of a constant and a variable with a *nonnegative* integer exponent.

A sum of a finite number of monomials is called a **polynomial.** Each monomial is called a **term** of the polynomial. The **degree of a polynomial** is the largest degree of the terms in the polynomial.

Terms that have exactly the same variables raised to the same powers are called **like terms.** For example, $14x^2$ and $-31x^2$ are like terms; however, $2x^3y$ and $7xy$ are not like terms because x^3y and xy are not identical.

A polynomial is said to be simplified if all its like terms have been combined. For example, the simplified form of $4x^2 + 3x + 5x$ is $4x^2 + 8x$. A simplified polynomial that has two terms is a **binomial,** and a simplified polynomial that has three terms is a **trinomial.** For example, $4x + 7$ is a binomial, and $2x^3 - 7x^2 + 11$ is a trinomial.

A nonzero constant, such as 5, is called a **constant polynomial.** It has degree zero because $5 = 5x^0$. The number 0 is defined to be a polynomial with no degree.

GENERAL FORM OF A POLYNOMIAL

The **general form of a polynomial** of degree n in the variable x is

$$a_n x^n + a_{n-1} x^{n-1} + \cdots + a_2 x^2 + a_1 x + a_0$$

where $a_n \neq 0$ and n is a nonnegative integer. The coefficient a_n is the **leading coefficient,** and a_0 is the **constant term.**

If a polynomial in the variable x is written with decreasing powers of x, then it is in **standard form.** For example, the polynomial

$$3x^2 - 4x^3 + 7x^4 - 1$$

is written in standard form as

$$7x^4 - 4x^3 + 3x^2 - 1$$

The following table shows the leading coefficient, degree, terms, and coefficients of the given polynomials.

Polynomial	Leading Coefficient	Degree	Terms	Coefficients
$9x^2 - x + 5$	9	2	$9x^2, -x, 5$	$9, -1, 5$
$11 - 2x$	-2	1	$-2x, 11$	$-2, 11$
$x^3 + 5x - 3$	1	3	$x^3, 5x, -3$	$1, 5, -3$

To add polynomials, we combine like terms.

EXAMPLE 1 *Add Polynomials*

Simplify: $(3x^2 + 7x - 5) + (4x^2 - 2x + 1)$

Solution

$$(3x^2 + 7x - 5) + (4x^2 - 2x + 1) = (3x^2 + 4x^2) + (7x - 2x) + [(-5) + 1]$$
$$= 7x^2 + 5x - 4$$

TRY EXERCISE 24, EXERCISE SET 1.5

The **additive inverse of the polynomial** $3x - 7$ is

$$-(3x - 7) = -3x + 7$$

To subtract a polynomial, we add its additive inverse. For example,

$$(2x - 5) - (3x - 7) = (2x - 5) + (-3x + 7)$$
$$= [2x + (-3x)] + [(-5) + 7]$$
$$= -x + 2$$

The distributive property is used to find the product of polynomials. For instance, to find the product of $(3x - 4)$ and $(2x^2 + 5x + 1)$, we treat $3x - 4$ as a *single* quantity and *distribute it* over the trinomial $2x^2 + 5x + 1$, as shown in Example 2.

EXAMPLE 2   *Multiply Polynomials*

Simplify: $(3x - 4)(2x^2 + 5x + 1)$

Solution

$(3x - 4)(2x^2 + 5x + 1)$
$= (3x - 4)(2x^2) + (3x - 4)(5x) + (3x - 4)(1)$
$= (3x)(2x^2) - 4(2x^2) + (3x)(5x) - 4(5x) + (3x)(1) - 4(1)$
$= 6x^3 - 8x^2 + 15x^2 - 20x + 3x - 4$
$= 6x^3 + 7x^2 - 17x - 4$

TRY EXERCISE 32, EXERCISE SET 1.5

In the following calculation, a vertical format has been used to find the product of $(x^2 + 6x - 7)$ and $(5x - 2)$. Note that like terms are arranged in the same vertical column.

$$
\begin{array}{r}
x^2 + 6x - 7 \\
5x - 2 \\
\hline
-2x^2 - 12x + 14 \\
5x^3 + 30x^2 - 35x \\
\hline
5x^3 + 28x^2 - 47x + 14
\end{array}
$$

If the terms of the binomials $(a + b)$ and $(c + d)$ are labeled as shown below, then the product of the two binomials can be computed mentally by the **FOIL method.**

In the following illustration, we find the product of $(7x - 2)$ and $(5x + 4)$ by the FOIL method.

$$\begin{array}{cccc} \text{First} & \text{Outer} & \text{Inner} & \text{Last} \end{array}$$
$$(7x - 2)(5x + 4) = (7x)(5x) + (7x)(4) + (-2)(5x) + (-2)(4)$$
$$= 35x^2 + 28x - 10x - 8$$
$$= 35x^2 + 18x - 8$$

Certain products occur so frequently in algebra that they deserve special attention.

Special Product Formulas

Special Form	Formula(s)
(Sum)(Difference)	$(x + y)(x - y) = x^2 - y^2$
(Binomial)^2	$(x + y)^2 = x^2 + 2xy + y^2$ $(x - y)^2 = x^2 - 2xy + y^2$

The variables x and y in these special product formulas can be replaced by other algebraic expressions, as shown in Example 3.

EXAMPLE 3   *Use the Special Product Formulas*

Find each special product.   **a.** $(7x + 10)(7x - 10)$ **b.** $(2y^2 + 11z)^2$

Solution

a. $(7x + 10)(7x - 10) = (7x)^2 - (10)^2 = 49x^2 - 100$

b. $(2y^2 + 11z)^2 = (2y^2)^2 + 2[(2y^2)(11z)] + (11z)^2 = 4y^4 + 44y^2z + 121z^2$

TRY EXERCISE 60, EXERCISE SET 1.5

Many application problems require you to *evaluate polynomials*. To **evaluate a polynomial**, substitute the given value(s) for the variable(s) and then perform the indicated operations using the **Order of Operations Agreement.**

THE ORDER OF OPERATIONS AGREEMENT

If grouping symbols are present, evaluate by performing the operations within the grouping symbols, innermost grouping symbol first, while observing the order given in steps 1 to 3.

1.   First, evaluate each power.

2.   Next, do all multiplications and divisions, working from left to right.

3.   Last, do all additions and subtractions, working from left to right.

EXAMPLE 4   *Evaluate a Polynomial*

Evaluate the polynomial $2x^3 - 6x^2 + 7$ for $x = -4$.

Solution

$$
\begin{aligned}
2x^3 - 6x^2 + 7 &= 2(-4)^3 - 6(-4)^2 + 7 \\
&= 2(-64) - 6(16) + 7 \\
&= -128 - 96 + 7 \\
&= -217
\end{aligned}
$$

- Substitute -4 for x.
- Evaluate the powers.
- Perform the multiplications.
- Perform the additions and subtractions.

TRY EXERCISE 72, EXERCISE SET 1.5

EXAMPLE 5   *Solve an Application*

The number of singles tennis matches that can be played between n tennis players is given by the polynomial $\dfrac{1}{2}n^2 - \dfrac{1}{2}n$. Find the number of singles tennis matches that can be played among 4 tennis players.

Solution

$$
\frac{1}{2}n^2 - \frac{1}{2}n = \frac{1}{2}(4)^2 - \frac{1}{2}(4)
$$

- Substitute 4 for n.

$$
= \frac{1}{2}(16) - \frac{1}{2}(4) = 8 - 2 = 6
$$

Therefore, 4 tennis players can play a total of 6 singles matches. See **Figure 1.13**.

TRY EXERCISE 82, EXERCISE SET 1.5

Figure 1.13
4 tennis players can play a total of 6 singles matches.

EXAMPLE 6   *Solve an Application*

A scientist determines that the average time in seconds that it takes a particular computer to determine whether an n-digit natural number is prime or composite is given by

$$0.002n^2 + 0.002n + 0.009, \quad 20 \le n \le 40$$

The average time in seconds that it takes the computer to factor an n-digit number is given by

$$0.00032(1.7)^n, \quad 20 \le n \le 40$$

Estimate the average time it takes the computer to

a.   determine whether a 30-digit number is a prime or a composite.

b.   factor a 30-digit number.

Solution

a. $0.002n^2 + 0.002n + 0.009 = 0.002(30)^2 + 0.002(30) + 0.009$

$$\approx 1.8 + 0.06 + 0.009 = 1.869 \approx 2 \text{ seconds}$$

b. $0.00032(1.7)^n = 0.00032(1.7)^{30}$

$$\approx 0.00032(8{,}193{,}465.726)$$

$$\approx 2600 \text{ seconds}$$

TRY EXERCISE 84, EXERCISE SET 1.5

TOPICS FOR DISCUSSION

1. Discuss the definition of the term *polynomial.* Give some examples of expressions that are polynomials and some examples of expressions that are not polynomials.

2. Suppose that P and Q are both polynomials of degree n. Discuss the degrees of $P + Q$, $P - Q$, PQ, $P + P$, and $P - P$.

3. Suppose that you evaluate a polynomial P of degree n for larger and larger values of x (for instance, when $x = 1, 2, 3, 4, \ldots$). Discuss whether the value of the polynomial would eventually (for very large values of x) continually increase, decrease, or fluctuate between increasing and decreasing.

4. Discuss the similarities and differences among monomials, binomials, trinomials, and polynomials.

EXERCISE SET 1.5

In Exercises 1 to 10, match the descriptions, labeled A, B, C, ..., J, with the appropriate examples.

A. $x^3y + xy$ **B.** $7x^2 + 5x - 11$

C. $\dfrac{1}{2}x^2 + xy + y^2$ **D.** $4xy$

E. $8x^3 - 1$ **F.** $3 - 4x^2$

G. 8 **H.** $3x^5 - 4x^2 + 7x - 11$

I. $8x^4 - \sqrt{5}x^3 + 7$ **J.** 0

1. A monomial of degree 2.

2. A binomial of degree 3.

3. A polynomial of degree 5.

4. A binomial with leading coefficient of -4.

5. A zero-degree polynomial.

6. A fourth-degree polynomial that has a third-degree term.

7. A trinomial with integer coefficients.

8. A trinomial in x and y.

9. A polynomial with no degree.

10. A fourth-degree binomial.

In Exercises 11 to 16, for each polynomial determine its *a.* **standard form,** *b.* **degree,** *c.* **coefficients,** *d.* **leading coefficient,** *e.* **terms.**

11. $2x + x^2 - 7$ **12.** $-3x^2 - 11 - 12x^4$

13. $x^3 - 1$ **14.** $4x^2 - 2x + 7$

15. $2x^4 + 3x^3 + 5 + 4x^2$ **16.** $3x^2 - 5x^3 + 7x - 1$

In Exercises 17 to 22, determine the degree of the given polynomial.

17. $3xy^2 - 2xy + 7x$ **18.** $x^3 + 3x^2y + 3xy^2 + y^3$

19. $4x^2y^2 - 5x^3y^2 + 17xy^3$ **20.** $-9x^5y + 10xy^4 - 11x^2y^2$

21. xy **22.** $5x^2y - y^4 + 6xy$

In Exercises 23 to 34, perform the indicated operations and simplify if possible by combining like terms. Write the result in standard form.

23. $(3x^2 + 4x + 5) + (2x^2 + 7x - 2)$

24. $(5y^2 - 7y + 3) + (2y^2 + 8y + 1)$

25. $(4w^3 - 2w + 7) + (5w^3 + 8w^2 - 1)$

26. $(5x^4 - 3x^2 + 9) + (3x^3 - 2x^2 - 7x + 3)$

27. $(r^2 - 2r - 5) - (3r^2 - 5r + 7)$

28. $(7s^2 - 4s + 11) - (-2s^2 + 11s - 9)$

29. $(u^3 - 3u^2 - 4u + 8) - (u^3 - 2u + 4)$

30. $(5v^4 - 3v^2 + 9) - (6v^4 + 11v^2 - 10)$

31. $(4x - 5)(2x^2 + 7x - 8)$

32. $(5x - 7)(3x^2 - 8x - 5)$

33. $(3x^2 - 2x + 5)(2x^2 - 5x + 2)$

34. $(2y^3 - 3y + 4)(2y^2 - 5y + 7)$

In Exercises 35 to 52, use the FOIL method to find the indicated product.

35. $(2x + 4)(5x + 1)$

36. $(5x - 3)(2x + 7)$

37. $(y + 2)(y + 1)$

38. $(y + 5)(y + 3)$

39. $(4z - 3)(z - 4)$

40. $(5z - 6)(z - 1)$

41. $(a + 6)(a - 3)$

42. $(a - 10)(a + 4)$

43. $(b - 4)(b + 6)$

44. $(b + 5)(b - 2)$

45. $(5x - 11y)(2x - 7y)$

46. $(3a - 5b)(4a - 7b)$

47. $(9x + 5y)(2x + 5y)$

48. $(3x - 7z)(5x - 7z)$

49. $(6w - 11x)(2w - 3x)$

50. $(4m + 5n)(2m - 5n)$

51. $(3p + 5q)(2p - 7q)$

52. $(2r - 11s)(5r + 8s)$

In Exercises 53 to 58, perform the indicated operations and simplify.

53. $(4d - 1)^2 - (2d - 3)^2$

54. $(5c - 8)^2 - (2c - 5)^2$

55. $(r + s)(r^2 - rs + s^2)$

56. $(r - s)(r^2 + rs + s^2)$

57. $(3c - 2)(4c + 1)(5c - 2)$

58. $(4d - 5)(2d - 1)(3d - 4)$

In Exercises 59 to 68, use the special product formulas to perform the indicated operation.

59. $(3x + 5)(3x - 5)$

60. $(4x^2 - 3y)(4x^2 + 3y)$

61. $(3x^2 - y)^2$

62. $(6x + 7y)^2$

63. $(4w + z)^2$

64. $(3x - 5y^2)^2$

65. $[(x - 2) + y]^2$

66. $[(x + 3) - y]^2$

67. $[(x + 5) + y][(x + 5) - y]$

68. $[(x - 2y) + 7][(x - 2y) - 7]$

In Exercises 69 to 76, evaluate the given polynomial for the indicated value of the variable.

69. $x^2 + 7x - 1$, for $x = 3$

70. $x^2 - 8x + 2$, for $x = 4$

71. $-x^2 + 5x - 3$, for $x = -2$

72. $-x^2 - 5x + 4$, for $x = -5$

73. $3x^3 - 2x^2 - x + 3$, for $x = -1$

74. $5x^3 - x^2 + 5x - 3$, for $x = -1$

75. $1 - x^5$, for $x = -2$

76. $1 - x^3 - x^5$, for $x = 2$

In Exercises 77 to 79, evaluate the given polynomial for the indicated values of the variable.

77. $4x^2 - 5x - 4$ for: **a.** $x = 4.3$ **b.** $x = 4.4$

78. $8x^3 - 2x^2 + 6.4x - 7.1$ for: **a.** $x = 1.2$ **b.** $x = 1.3$

79. $x^3 - 2x^2 - 5x + 11$ for: **a.** $x = 0.001$ **b.** $x = 0.0001$

80. **HIGHWAY ENGINEERING** On an expressway, the recommended *safe distance* between cars in feet is given by $0.015v^2 + v + 10$, where v is the speed of the car in miles per hour. Find the safe distance when

a. $v = 30$ mph   **b.** $v = 55$ mph

81. Find the number of chess matches that can be played between the members of a group of 150 people. Use the formula from Example 5.

82. The number of committees consisting of exactly 3 people that can be formed from a group of n people is given by the polynomial

$$\frac{1}{6}n^3 - \frac{1}{2}n^2 + \frac{1}{3}n$$

Find the number of committees consisting of exactly 3 people that can be formed from a group of 21 people.

83. **COMPUTER SCIENCE** If n is a positive integer, then $n!$, which is read "n factorial," is given by

$$n(n - 1)(n - 2) \cdots 2 \cdot 1$$

For example, $4! = 4 \cdot 3 \cdot 2 \cdot 1 = 24$. A statistician determines that each time a statistical package is booted up on a particular computer, the time in seconds required to compute $n!$ is given by the polynomial

$$1.9 \times 10^{-6}n^2 - 3.9 \times 10^{-3}n$$

where $1000 \le n \le 10{,}000$. Using this polynomial, estimate the time it takes the computer to calculate 4000! and 8000!. Assume the statistical package is booted up before each calculation.

84. COMPUTER SCIENCE A computer scientist determines that the time in seconds it takes a particular computer to calculate n digits of π is given by the polynomial

$$4.3 \times 10^{-6}n^2 - 2.1 \times 10^{-4}n$$

where $1000 \le n \le 10{,}000$. Estimate the time it takes the computer to calculate π to

a. 1000 digits **b.** 5000 digits **c.** 10,000 digits

SUPPLEMENTAL EXERCISES

The following special product formulas can be used to find the cube of a binomial.

$$(x + y)^3 = x^3 + 3x^2y + 3xy^2 + y^3$$
$$(x - y)^3 = x^3 - 3x^2y + 3xy^2 - y^3$$

In Exercises 85 to 90, make use of the above special product formulas to find the indicated products.

85. $(a + b)^3$ **86.** $(a - b)^3$ **87.** $(x - 1)^3$

88. $(y + 2)^3$ **89.** $(2x - 3y)^3$ **90.** $(3x + 5y)^3$

PROJECTS

1. ODD NUMBERS Every odd number can be written in the form $2n - 1$, and every even number can be expressed as $2n$, where n is a natural number. Explain, by writing a few paragraphs and giving the supporting mathematics, why the product of two odd numbers is an odd number, the product of two even numbers is an even number, and the product of an even number and an odd number is an even number.

2. PRIME NUMBERS Fermat's Little Theorem states, "If n is a prime number and a is *any* natural number, then $a^n - a$ is divisible by n." For instance, for $n = 11$ and $a = 14$, $\dfrac{14^{11} - 14}{11} = 368{,}142{,}288{,}150$. The important aspect of this theorem is that no matter what natural number is chosen for a, $a^{11} - a$ is evenly divisible by 11.

Knowing whether a number is prime plays a central role in the security of computer systems. A restatement (called the *contrapositive*) of Fermat's Little Theorem as "If n is a number and a is some number for which $a^n - a$ is *not* divisible by n, then n is *not* a prime number" is used to determine when a number is *not* prime. For example, if $n = 14$, then $\dfrac{2^{14} - 2}{14} = \dfrac{8191}{7}$, and thus there is some number ($a = 2$) for which $2^{14} - 2$ is not evenly divisible by 14. Therefore, 14 is not prime.

a. Explain the meaning of the *contrapositive* (used above) of a theorem. Use your explanation to write the contrapositive of "If two triangles are congruent, then they are similar."

b. $7^{14} - 7$ is divisible by 14. Explain why this does not contradict the fact that 14 is not a prime.

c. Explain the meaning of the *converse* of a theorem. State the converse of Fermat's Little Theorem.

d. The number 561 has the property that $a^{561} - a$ is divisible by 561 for all natural numbers a. Can you use Fermat's Little Theorem to conclude that 561 is a prime number? Explain.

e. Suppose that $a^n - a$ is divisible by n for all values of a. Can you conclude that n is a prime number? Explain your answer.

f. Find a definition of a Carmichael number. What do Carmichael numbers have to do with **e.**?

1.6 FACTORING

Writing a polynomial as a product of polynomials of lower degree is called **factoring**. Factoring is an important procedure that is often used to simplify fractional expressions and to solve equations.

In this section we consider only the factorization of polynomials that have integer coefficients. Also, we are concerned only with **factoring over the integers.** That is, we search only for polynomial factors that have integer coefficients.

The first step in any factorization of a polynomial is to use the distributive property to factor out the **greatest common factor (GCF)** of the terms of the polynomial. Given two or more exponential expressions with the same prime number base or the same variable base, the GCF is the exponential expression with the smallest exponent. For example,

$$2^3 \text{ is the GCF of } 2^3, 2^5, \text{ and } 2^8, \quad \text{and} \quad a \text{ is the GCF of } a^4 \text{ and } a.$$

The GCF of two or more monomials is the product of the GCFs of all of the *common* bases. For example, to find the GCF of $27a^3b^4$ and $18b^3c$, factor the coefficients into prime factors and then write each common base with its smallest exponent.

$$27a^3b^4 = 3^3 \cdot a^3 \cdot b^4$$

$$18b^3c = 2 \cdot 3^2 \cdot b^3 \cdot c$$

The only common bases are 3 and b. The product of these common bases with their smallest exponents is 3^2b^3. The GCF of $27a^3b^4$ and $18b^3c$ is $9b^3$.

EXAMPLE 1   *Factor Out the Greatest Common Factor*

Factor out the GCF.

a. $10x^3 + 6x$ **b.** $15x^{2n} + 9x^{n+1} - 3x^n$ (where n is a positive integer)

c. $(m + 5)(x + 3) + (m + 5)(x - 10)$

Solution

a. $\begin{aligned} 10x^3 + 6x &= (2x)(5x^2) + (2x)(3) \\ &= 2x(5x^2 + 3) \end{aligned}$ • The GCF is $2x$.
    • Factor out the GCF.

b. $15x^{2n} + 9x^{n+1} - 3x^n$
    $\begin{aligned} &= (3x^n)(5x^n) + (3x^n)(3x) - (3x^n)(1) \\ &= 3x^n(5x^n + 3x - 1) \end{aligned}$ • The GCF is $3x^n$.
    • Factor out the GCF.

c. $\begin{aligned} (m + 5)&(x + 3) + (m + 5)(x - 10) \\ &= (m + 5)[(x + 3) + (x - 10)] \\ &= (m + 5)(2x - 7) \end{aligned}$ • Use the distributive property to factor out $(m + 5)$.
    • Simplify.

TRY EXERCISE 6, EXERCISE SET 1.6

Some polynomials can be **factored by grouping.** Pairs of terms that have a common factor are first grouped together. The process makes repeated use of the distributive property, as shown in the following factorization of $6y^3 - 21y^2 - 4y + 14$.

$$6y^3 - 21y^2 - 4y + 14$$

$$= (6y^3 - 21y^2) - (4y - 14) \qquad \bullet \text{ Group the first two terms and the last two terms.}$$

$$= 3y^2(2y - 7) - 2(2y - 7) \qquad \bullet \text{ Factor out the GCF from each of the groups.}$$

$$= (2y - 7)(3y^2 - 2) \qquad \bullet \text{ Factor out the common binomial factor.}$$

Some trinomials of the form $x^2 + bx + c$ can be factored by a trial procedure. This method makes use of the FOIL method in reverse. For example, consider the following products:

$$(x + 3)(x + 5) = x^2 + 5x + 3x + (3)(5) \qquad = x^2 + 8x + 15$$

$$(x - 2)(x - 7) = x^2 - 7x - 2x + (-2)(-7) = x^2 - 9x + 14$$

$$(x + 4)(x - 9) = x^2 - 9x + 4x + (4)(-9) \qquad = x^2 - 5x - 36$$

The coefficient of *x* is the sum of the constant terms of the binomials.

The constant term of the trinomial is the product of the constant terms of the binomials.

POINTS TO REMEMBER TO FACTOR $x^2 + bx + c$

1. The constant term c of the trinomial is the product of the constant terms of the binomials.

2. The coefficient b in the trinomial is the sum of the constant terms of the binomials.

3. If the constant term c of the trinomial is positive, the constant terms of the binomials have the same sign as the coefficient b of the trinomial.

4. If the constant term c of the trinomial is negative, the constant terms of the binomials have opposite signs.

EXAMPLE 2 *Factor a Trinomial of the Form $x^2 + bx + c$*

Factor: $x^2 + 7x - 18$

Solution

We must find two binomials whose first terms have a product of x^2 and whose last terms have a product of -18; also, the sum of the product of

Continued • ➤

the outer terms and the product of the inner terms must be $7x$. Begin by listing the possible integer factorizations of -18.

Factors of -18	Sum of the Factors
$1 \cdot (-18)$	$1 + (-18) = -17$
$(-1) \cdot 18$	$(-1) + 18 = 17$
$2 \cdot (-9)$	$2 + (-9) = -7$
$(-2) \cdot 9$	$(-2) + 9 = 7$

• **Stop. This is the desired sum.**

Thus -2 and 9 are the numbers whose sum is 7 and whose product is -18. Therefore,

$$x^2 + 7x - 18 = (x - 2)(x + 9)$$

The FOIL method can be used to verify that the factorization is correct.

TRY EXERCISE 18, EXERCISE SET 1.6

The trial method can sometimes be used to factor trinomials of the form $ax^2 + bx + c$, which do not have a leading coefficient of 1. We use the factors of a and c to form trial binomial factors. Factoring trinomials of this type may require testing many factors. To reduce the number of trial factors, make use of the following points.

POINTS TO REMEMBER TO FACTOR $ax^2 + bx + c, \quad a > 0$

1. If the constant term of the trinomial is positive, the constant terms of the binomials have the same sign as the coefficient b in the trinomial.

2. If the constant term of the trinomial is negative, the constant terms of the binomials have opposite signs.

3. If the terms of the trinomial do not have a common factor, then neither binomial will have a common factor.

EXAMPLE 3   *Factor a Trinomial of the Form $ax^2 + bx + c$*

Factor: $6x^2 - 11x + 4$

Solution

Because the constant term of the trinomial is positive and the coefficient of the x term is negative, the constant terms of the binomials will both be negative. This time we find factors of the first term as well as factors of the constant term.

Factors of $6x^2$	Factors of 4 (both negative)
$x, 6x$	$-1, -4$
$2x, 3x$	$-2, -2$

Use these factors to write trial factors. Use the FOIL method to see whether any of the trial factors produce the correct middle term. If the terms of a trinomial do not have a common factor, then a binomial factor cannot have a common factor (point 3). Such trial factors need not be checked.

Trial Factors	Middle Term
$(x - 1)(6x - 4)$	Common factor
$(x - 4)(6x - 1)$	$-1x - 24x = -25x$
$(x - 2)(6x - 2)$	Common factor
$(2x - 1)(3x - 4)$	$-8x - 3x = -11x$

- **6x and 4 have a common factor.**

- **6x and 2 have a common factor.**
- **This is the correct middle term.**

Thus $6x^2 - 11x + 4 = (2x - 1)(3x - 4)$.

TRY EXERCISE 22, EXERCISE SET 1.6

Sometimes it is impossible to factor a polynomial into the product of two polynomials having integer coefficients. Such polynomials are said to be **nonfactorable over the integers.** For example, $x^2 + 3x + 7$ is nonfactorable over the integers because there are no integers whose product is 7 and whose sum or difference is 3.

If you have difficulty factoring a trinomial, you may wish to use the following theorem. It will indicate whether the trinomial is factorable over the integers.

FACTORIZATION THEOREM

The trinomial $ax^2 + bx + c$, with integer coefficients a, b, and c, can be factored as the product of two binomials with integer coefficients if and only if $b^2 - 4ac$ is a perfect square.

EXAMPLE 4   *Apply the Factorization Theorem*

Determine whether each trinomial is factorable over the integers.

a. $4x^2 + 8x - 7$ **b.** $6x^2 - 5x - 4$

Solution

a. The coefficients of $4x^2 + 8x - 7$ are $a = 4$, $b = 8$, and $c = -7$. Applying the factorization theorem yields

$$b^2 - 4ac = 8^2 - 4(4)(-7) = 176$$

Because 176 is not a perfect square, the trinomial is nonfactorable over the integers.

Continued •➤

b. The coefficients of $6x^2 - 5x - 4$ are $a = 6, b = -5$, and $c = -4$. Thus

$$b^2 - 4ac = (-5)^2 - 4(6)(-4) = 121$$

Because 121 is a perfect square, the trinomial is factorable over the integers. Using the methods we have developed, we find

$$6x^2 - 5x - 4 = (3x - 4)(2x + 1)$$

TRY EXERCISE 30, EXERCISE SET 1.6

Some polynomials of degree greater than 2 can be factored by the trial procedure. Consider $2x^6 + 9x^3 + 9$. Because all the signs of the trinomial are positive, the coefficients of all the terms in the binomial factors must be positive.

Factors of $2x^6$	Factors of 9 (both positive)
$x^3, 2x^3$	1, 9 3, 3

The factors $(x^3 + 3)$ and $(2x^3 + 3)$ are the only trial factors whose product has the correct middle term, $9x^3$. Thus $2x^6 + 9x^3 + 9 = (x^3 + 3)(2x^3 + 3)$.

Some polynomials can be factored by making use of the following factoring formulas.

Factoring Formulas

Special Form	Formula(s)
Difference of two squares	$x^2 - y^2 = (x + y)(x - y)$
Perfect-square trinomials	$x^2 + 2xy + y^2 = (x + y)^2$ $x^2 - 2xy + y^2 = (x - y)^2$
Sum of cubes	$x^3 + y^3 = (x + y)(x^2 - xy + y^2)$
Difference of cubes	$x^3 - y^3 = (x - y)(x^2 + xy + y^2)$

The monomial a^2 is a square of a, and a is called a **square root** of a^2. The factoring formula

$$x^2 - y^2 = (x + y)(x - y)$$

indicates that the **difference of two squares** can be written as the product of the sum and the difference of the square roots of the squares.

TAKE NOTE

The polynomial $x^2 + y^2$ is the *sum* of two squares. You may be tempted to factor it in a manner similar to the method used on the *difference* of two squares; however, $x^2 + y^2$ is nonfactorable over the integers.

EXAMPLE 5   *Factor the Difference of Squares*

Factor: $49x^2 - 144$

Solution

$49x^2 - 144 = (7x)^2 - (12)^2$ • Recognize the difference-of-squares form.

$= (7x + 12)(7x - 12)$ • The binomial factors are the sum and the difference of the square roots of the squares.

TRY EXERCISE 38, EXERCISE SET 1.6

A **perfect-square trinomial** is a trinomial that is the square of a binomial. For example, $x^2 + 6x + 9$ is a perfect-square trinomial because

$$(x + 3)^2 = x^2 + 6x + 9$$

Every perfect-square trinomial can be factored by the trial method, but it generally is faster to factor perfect-square trinomials by using the factoring formulas.

EXAMPLE 6 *Factor a Perfect-Square Trinomial*

Factor: $16m^2 - 40mn + 25n^2$

Solution

$16m^2 - 40mn + 25n^2$
$= (4m)^2 - 2(4m)(5n) + (5n)^2$ • **Recognize the perfect-square**
$= (4m - 5n)^2$ **trinomial form.**

TRY EXERCISE 48, EXERCISE SET 1.6

The product of the same three factors is called a **cube**. For example, $8a^3$ is a cube because $8a^3 = (2a)^3$. The **cube root** of a cube is one of the three equal factors. To factor the sum or the difference of two cubes, use the appropriate factoring formula. It helps to use the following patterns, which involve the signs of the terms.

In the factorization of the sum or difference of two cubes, the terms of the binomial factor are the cube roots of the cubes. For example,

$$8a^3 - 27b^3 = (2a)^3 - (3b)^3 = (2a - 3b)(4a^2 + 6ab + 9b^2)$$

EXAMPLE 7 *Factor the Sum or Difference of Cubes*

Factor. **a.** $8a^3 + b^3$ **b.** $a^3 - 64$

Solution

---

a. $8a^3 + b^3 = (2a)^3 + b^3$ • **Recognize the sum-of-cubes form.**

$= (2a + b)(4a^2 - 2ab + b^2)$ • **Factor.**

b. $a^3 - 64 = a^3 - 4^3$ • **Recognize the difference-of-cubes form.**

$= (a - 4)(a^2 + 4a + 16)$ • **Factor.**

TRY EXERCISE 54, EXERCISE SET 1.6

Here is a general factoring strategy for polynomials:

GENERAL FACTORING STRATEGY

1. Factor out the GCF of all terms.

2. Try to factor a binomial as

    a. the difference of two squares.

    b. the sum or difference of two cubes.

3. Try to factor a trinomial

    a. as a perfect-square trinomial.

    b. using the trial method.

4. Try to factor a polynomial with more than three terms by grouping.

5. After each factorization, examine the new factors to see whether they can be factored.

EXAMPLE 8   *Factor Using the General Factoring Strategy*

Completely factor: $x^6 + 7x^3 - 8$

Solution

Factor $x^6 + 7x^3 - 8$ as the product of two binomials.

$$x^6 + 7x^3 - 8 = (x^3 + 8)(x^3 - 1)$$

Now factor $x^3 + 8$, which is the sum of two cubes, and factor $x^3 - 1$, which is the difference of two cubes.

$$x^6 + 7x^3 - 8 = (x + 2)(x^2 - 2x + 4)(x - 1)(x^2 + x + 1)$$

TRY EXERCISE 64, EXERCISE SET 1.6

When you factor by grouping, some experimentation may be necessary to find a grouping that is of the form of one of the special factoring formulas.

EXAMPLE 9   *Factor by Grouping*

Factor by grouping.   **a.** $a^2 + 10ab + 25b^2 - c^2$ **b.** $p^2 + p - q - q^2$

Solution

a. $a^2 + 10ab + 25b^2 - c^2$

    $= (a^2 + 10ab + 25b^2) - c^2$ • **Group the terms of the perfect-square trinomial.**

    $= (a + 5b)^2 - c^2$ • **Factor the trinomial.**

    $= [(a + 5b) + c][(a + 5b) - c]$ • **Factor the difference of squares.**

    $= (a + 5b + c)(a + 5b - c)$ • **Simplify.**

b. $p^2 + p - q - q^2$

$\quad = p^2 - q^2 + p - q$ • **Rearrange the terms.**

$\quad = (p^2 - q^2) + (p - q)$ • **Regroup.**

$\quad = (p + q)(p - q) + (p - q)$ • **Factor the difference of squares.**

$\quad = (p - q)(p + q + 1)$ • **Factor out the common factor $(p - q)$.**

TRY EXERCISE 74, EXERCISE SET 1.6

TOPICS FOR DISCUSSION

1. Discuss the meaning of the phrase *nonfactorable over the integers*.

2. You know that if $ab = 0$, then $a = 0$ or $b = 0$. Suppose a polynomial is written in factored form and then set equal to zero. For instance, suppose

$$x^2 - 2x - 15 = (x - 5)(x + 3) = 0$$

Discuss what implications this has for the values of x. Do not only answer this question for the polynomial above, but also discuss the implications for the values of x for any polynomial written as a product of linear factors and then set equal to zero.

3. Let P be a polynomial of degree n. Discuss the number of possible distinct linear polynomials that can be a factor of P.

4. A method of evaluating polynomials, sometimes called Horner's method, involves factoring a polynomial in a certain manner. For instance,

$$4x^3 - 2x^2 + 5x - 3 = [(4x - 2)x + 5]x - 3$$
$$5x^4 - 2x^3 + 4x^2 + x - 6 = \{[(5x - 2)x + 4]x + 1\}x - 6$$

To evaluate the polynomial, the factored form is evaluated. Discuss the advantages and disadvantages of using this method to evaluate a polynomial.

5. Recall that if n is a natural number, $n! = n(n - 1)(n - 2)\cdots 3 \cdot 2 \cdot 1$. Explain why none of the following consecutive integers is a prime number.

$$5! + 2 \qquad 5! + 3 \qquad 5! + 4 \qquad 5! + 5$$

How many numbers are in the following list of consecutive integers? How many of those numbers are prime numbers?

$$k! + 2, \, k! + 3, \, k! + 4, \, k! + 5, \ldots, k! + k$$

Explain why this result means that there are arbitrarily long sequences of consecutive natural numbers that do not contain a prime number.

EXERCISE SET 1.6

In Exercises 1 to 8, factor out the GCF from each polynomial.

1. $5x + 20$

2. $8x^2 + 12x - 40$

3. $-15x^2 - 12x$

4. $-6y^2 - 54y$

5. $10x^2y + 6xy - 14xy^2$

6. $6a^3b^2 - 12a^2b + 72ab^3$

7. $(x - 3)(a + b) + (x - 3)(a + 3b)$

8. $(x - 4)(2a - b) + (x + 4)(2a - b)$

In Exercises 9 to 14, factor by grouping in pairs.

9. $3x^3 + x^2 + 6x + 2$

10. $18w^3 + 15w^2 + 12w + 10$

11. $ax^2 - ax + bx - b$

12. $a^2y^2 - ay^3 + ac - cy$

13. $6w^3 + 4w^2 - 15w - 10$

14. $10z^3 - 15z^2 - 4z + 6$

In Exercises 15 to 28, factor each trinomial.

15. $x^2 + 7x + 12$

16. $x^2 + 9x + 20$

17. $a^2 - 10a - 24$

18. $b^2 + 12b - 28$

19. $6x^2 + 25x + 4$

20. $8a^2 - 26a + 15$

21. $51x^2 - 5x - 4$

22. $57y^2 + y - 6$

23. $6x^2 + xy - 40y^2$

24. $8x^2 + 10xy - 25y^2$

25. $x^4 + 6x^2 + 5$

26. $x^4 + 11x^2 + 18$

27. $6x^4 + 23x^2 + 15$

28. $9x^4 + 10x^2 + 1$

In Exercises 29 to 34, use the factorization theorem to determine whether each trinomial is factorable over the integers.

29. $8x^2 + 26x + 15$

30. $16x^2 + 8x - 35$

31. $4x^2 - 5x + 6$

32. $6x^2 + 8x - 3$

33. $6x^2 - 14x + 5$

34. $10x^2 - 4x - 5$

In Exercises 35 to 44, factor each difference of squares.

35. $x^2 - 9$

36. $x^2 - 64$

37. $4a^2 - 49$

38. $81b^2 - 16c^2$

39. $1 - 100x^2$

40. $1 - 121y^2$

41. $x^4 - 9$

42. $y^4 - 196$

43. $(x + 5)^2 - 4$

44. $(x - 3)^2 - 16$

In Exercises 45 to 52, factor each perfect-square trinomial.

45. $x^2 + 10x + 25$

46. $y^2 + 6y + 9$

47. $a^2 - 14a + 49$

48. $b^2 - 24b + 144$

49. $4x^2 + 12x + 9$

50. $25y^2 + 40y + 16$

51. $z^4 + 4z^2w^2 + 4w^4$

52. $9x^4 - 30x^2y^2 + 25y^4$

In Exercises 53 to 60, factor each sum or difference of cubes.

53. $x^3 - 8$

54. $b^3 + 64$

55. $8x^3 - 27y^3$

56. $64u^3 - 27v^3$

57. $8 - x^6$

58. $1 + y^{12}$

59. $(x - 2)^3 - 1$

60. $(y + 3)^3 + 8$

In Exercises 61 to 80, use the general factoring strategy to completely factor each polynomial. If the polynomial does not factor, then state that it is nonfactorable over the integers.

61. $18x^2 - 2$

62. $4bx^3 + 32b$

63. $16x^4 - 1$

64. $81y^4 - 16$

65. $12ax^2 - 23axy + 10ay^2$

66. $6ax^2 - 19axy - 20ay^2$

67. $3bx^3 + 4bx^2 - 3bx - 4b$

68. $2x^6 - 2$

69. $72bx^2 + 24bxy + 2by^2$

70. $64y^3 - 16y^2z + yz^2$

71. $(w - 5)^3 + 8$

72. $5xy + 20y - 15x - 60$

73. $x^2 + 6xy + 9y^2 - 1$

74. $4y^2 - 4yz + z^2 - 9$

75. $8x^2 + 3x - 4$

76. $16x^2 + 81$

77. $5x(2x - 5)^2 - (2x - 5)^3$

78. $6x(3x + 1)^3 - (3x + 1)^4$

79. $4x^2 + 2x - y - y^2$

80. $a^2 + a + b - b^2$

SUPPLEMENTAL EXERCISES

In Exercises 81 and 82, find all positive values of k such that the trinomial is a perfect-square trinomial.

81. $x^2 + kx + 16$

82. $36x^2 + kxy + 100$

In Exercises 83 and 84, find k such that the trinomial is a perfect-square trinomial.

83. $x^2 + 16x + k$

84. $x^2 - 14xy + ky^2$

In Exercises 85 and 86, use the general strategy to completely factor each polynomial. In each exercise n represents a positive integer.

85. $x^{4n} - 1$

86. $x^{4n} - 2x^{2n} + 1$

In Exercises 87 to 90, write, in its factored form, the area of the shaded portion of each geometric figure.

87.

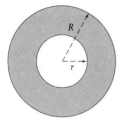

88.

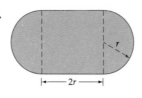

89.

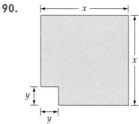

90.

The product $(a + bi)(a - bi) = a^2 + b^2$ can be used to factor the sum of two squares over the set of complex numbers. For example,

$$x^2 + 25 = (x + 5i)(x - 5i)$$

In Exercises 91 to 94, factor each polynomial over the set of complex numbers.

91. $x^2 + 9$

92. $y^2 + 121$

93. $4x^2 + 81$

94. $144y^2 + 625$

In Exercises 95 to 98, evaluate each polynomial for the given value of x.

95. $x^2 + 36; x = 6i$

96. $x^2 + 100; x = -10i$

97. $x^2 - 6x + 10; x = 3 - i$

98. $x^2 + 10x + 29; x = -5 + 2i$

PROJECTS

1. **GEOMETRY** The ancient Greeks used geometric figures and the concept of area to illustrate many algebraic concepts. The factoring formula $x^2 - y^2 = (x + y)(x - y)$ can be illustrated by the figure at the left below.

   a. Which regions are represented by $(x + y)(x - y)$?

   b. Which regions are represented by $x^2 - y^2$?

   c. Explain why the area of the regions listed in **a.** must equal the area of the regions listed in **b.**

2. **GEOMETRY** What algebraic formula does the geometric figure in the middle below illustrate?

3. **GEOMETRY** Show how the figure at the right below can be used to illustrate the factoring formula for the difference of two cubes.

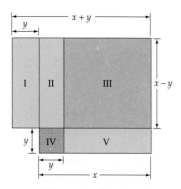

Figure for Project 1

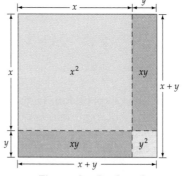

Figure for Project 2

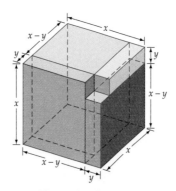

Figure for Project 3

1.7 RATIONAL EXPRESSIONS

A **rational expression** is a fraction in which the numerator and denominator are polynomials. For example,

$$\frac{3}{x+1} \quad \text{and} \quad \frac{x^2 - 4x - 21}{x^2 - 9}$$

are rational expressions.

The **domain of a rational expression** is the set of all real numbers that can be used as replacements for the variable. Any value of the variable that causes division by zero is excluded from the domain of the rational expression. For example, the domain of

$$\frac{7x}{x^2 - 5x} \quad x \neq 0, \, x \neq 5$$

is the set of all real numbers except 0 and 5. Both 0 and 5 are excluded values because the denominator $x^2 - 5x$ equals zero when $x = 0$ and also when $x = 5$. Sometimes the excluded values are specified to the right of a rational expression, as shown here. However, a rational expression is meaningful only for those real numbers that are not excluded values, regardless of whether the excluded values are specifically stated.

Rational expressions have properties similar to the properties of rational numbers.

PROPERTIES OF RATIONAL EXPRESSIONS

For all rational expressions P/Q and R/S where $Q \neq 0$ and $S \neq 0$,

Equality $\qquad\qquad \dfrac{P}{Q} = \dfrac{R}{S}$ if and only if $PS = QR$

Equivalent expressions $\qquad \dfrac{P}{Q} = \dfrac{PR}{QR}, \quad R \neq 0$

Sign $\qquad\qquad\qquad -\dfrac{P}{Q} = \dfrac{-P}{Q} = \dfrac{P}{-Q}$

To **simplify a rational expression,** factor the numerator and the denominator. Then use the equivalent expressions property to eliminate factors common to both the numerator and the denominator. A rational expression is *simplified* when 1 is the only common polynomial factor of both the numerator and the denominator.

EXAMPLE 1   *Simplify a Rational Expression*

Simplify: $\dfrac{7 + 20x - 3x^2}{2x^2 - 11x - 21}$

Solution

$$\frac{7 + 20x - 3x^2}{2x^2 - 11x - 21} = \frac{(7 - x)(1 + 3x)}{(x - 7)(2x + 3)} \qquad \bullet \text{ Factor.}$$

$$= \frac{-(x - 7)(1 + 3x)}{(x - 7)(2x + 3)} \qquad \bullet \text{ Use } (7 - x) = -(x - 7).$$

$$= \frac{-\cancel{(x - 7)}(1 + 3x)}{\cancel{(x - 7)}(2x + 3)}$$

$$= \frac{-(1 + 3x)}{2x + 3} = -\frac{3x + 1}{2x + 3} \quad x \neq 7, x \neq -\frac{3}{2}$$

TRY EXERCISE 2, EXERCISE SET 1.7

Caution A rational expression like $(x + 3)/3$ does not simplify to $x + 1$ because

$$\frac{x + 3}{3} = \frac{x}{3} + \frac{3}{3} = \frac{x}{3} + 1$$

Rational expressions can be simplified by dividing nonzero *factors* common to the numerator and the denominator, but not terms.

Arithmetic operations are defined on rational expressions just as they are on rational numbers.

ARITHMETIC OPERATIONS DEFINED ON RATIONAL EXPRESSIONS

For all rational expressions P/Q, R/Q, and R/S where $Q \neq 0$ and $S \neq 0$,

Addition $\qquad \dfrac{P}{Q} + \dfrac{R}{Q} = \dfrac{P + R}{Q}$

Subtraction $\qquad \dfrac{P}{Q} - \dfrac{R}{Q} = \dfrac{P - R}{Q}$

Multiplication $\qquad \dfrac{P}{Q} \cdot \dfrac{R}{S} = \dfrac{PR}{QS}$

Division $\qquad \dfrac{P}{Q} \div \dfrac{R}{S} = \dfrac{P}{Q} \cdot \dfrac{S}{R} = \dfrac{PS}{QR} \quad R \neq 0$

Factoring and the equivalent expressions property of rational expressions are used in the multiplication and division of rational expressions.

EXAMPLE 2 *Divide a Rational Expression*

Simplify: $\dfrac{x^2 + 6x + 9}{x^3 + 27} \div \dfrac{x^2 + 7x + 12}{x^3 - 3x^2 + 9x}$

Solution

$\dfrac{x^2 + 6x + 9}{x^3 + 27} \div \dfrac{x^2 + 7x + 12}{x^3 - 3x^2 + 9x}$

$= \dfrac{(x + 3)^2}{(x + 3)(x^2 - 3x + 9)} \div \dfrac{(x + 4)(x + 3)}{x(x^2 - 3x + 9)}$ • **Factor.**

$= \dfrac{(x + 3)^2}{(x + 3)(x^2 - 3x + 9)} \cdot \dfrac{x(x^2 - 3x + 9)}{(x + 4)(x + 3)}$ • **Multiply by the reciprocal.**

$= \dfrac{\cancel{(x + 3)^2}\, x\, \cancel{(x^2 - 3x + 9)}}{\cancel{(x + 3)}\, \cancel{(x^2 - 3x + 9)}\,(x + 4)\cancel{(x + 3)}}$ • **Simplify.**

$= \dfrac{x}{x + 4}$

TRY EXERCISE 16, EXERCISE SET 1.7

Addition of rational expressions with a **common denominator** is accomplished by writing the sum of the numerators over the common denominator. For example,

$$\frac{5x}{18} + \frac{x}{18} = \frac{5x + x}{18} = \frac{6x}{18} = \frac{x}{3}$$

If the rational expressions do not have a common denominator, then they can be written as equivalent rational expressions that have a common denominator by multiplying the numerator and denominator of each of the rational expressions by the required polynomials. The following procedure can be used to determine the least common denominator (LCD) of rational expressions. It is similar to the process used to find the LCD of rational numbers.

DETERMINING THE LCD OF RATIONAL EXPRESSIONS

1. Factor each denominator completely and express repeated factors using exponential notation.

2. Identify the largest power of each factor in any single factorization. The LCD is the product of each factor raised to its largest power.

For example,

$$\frac{1}{x + 3} \qquad \text{and} \qquad \frac{5}{2x - 1}$$

have an LCD of $(x + 3)(2x - 1)$. The rational expressions

$$\frac{5x}{(x + 5)(x - 7)^3} \quad \text{and} \quad \frac{7}{x(x + 5)^2(x - 7)}$$

have an LCD of $x(x + 5)^2(x - 7)^3$.

EXAMPLE 3 *Add and Subtract Rational Expressions*

Perform the indicated operation and then simplify if possible.

a. $\dfrac{5x}{48} + \dfrac{x}{15}$ b. $\dfrac{x}{x^2 - 4} - \dfrac{2x - 1}{x^2 - 3x - 10}$

Solution

a. Determine the prime factorization of the denominators.

$$48 = 2^4 \cdot 3 \quad \text{and} \quad 15 = 3 \cdot 5$$

The desired common denominator is the product of each of the prime factors raised to its largest power. Thus the common denominator is $2^4 \cdot 3 \cdot 5 = 240$. Write each rational expression as an equivalent rational expression with a denominator of 240.

$$\frac{5x}{48} + \frac{x}{15} = \frac{5x \cdot 5}{48 \cdot 5} + \frac{x \cdot 16}{15 \cdot 16} = \frac{25x}{240} + \frac{16x}{240} = \frac{41x}{240}$$

b. Factor each denominator to determine the LCD of the rational expressions.

$$x^2 - 4 = (x + 2)(x - 2)$$
$$x^2 - 3x - 10 = (x + 2)(x - 5)$$

The LCD is $(x + 2)(x - 2)(x - 5)$. Forming equivalent rational expressions that have the LCD, we have

$$\frac{x}{x^2 - 4} - \frac{2x - 1}{x^2 - 3x - 10}$$

$$= \frac{x(x - 5)}{(x + 2)(x - 2)(x - 5)} - \frac{(2x - 1)(x - 2)}{(x + 2)(x - 5)(x - 2)}$$

$$= \frac{x^2 - 5x - (2x^2 - 5x + 2)}{(x + 2)(x - 2)(x - 5)} = \frac{x^2 - 5x - 2x^2 + 5x - 2}{(x + 2)(x - 2)(x - 5)}$$

$$= \frac{-x^2 - 2}{(x + 2)(x - 2)(x - 5)} = -\frac{x^2 + 2}{(x + 2)(x - 2)(x - 5)}$$

TRY EXERCISE 30, EXERCISE SET 1.7

COMPLEX FRACTIONS

A **complex fraction** is a fraction whose numerator or denominator contains one or more fractions. Complex fractions can be simplified by using one of the following two methods.

METHODS FOR SIMPLIFYING COMPLEX FRACTIONS

Method 1: Multiply by the LCD
1. Determine the LCD of all the fractions in the complex fraction.

2. Multiply both the numerator and the denominator of the complex fraction by the LCD.

3. If possible, simplify the resulting rational expression.

Method 2: Multiply by the reciprocal of the denominator
1. Simplify the numerator to a single fraction and the denominator to a single fraction.

2. Multiply the numerator by the reciprocal of the denominator.

3. If possible, simplify the resulting rational expression.

EXAMPLE 4   *Simplify Complex Fractions*

Simplify: $\dfrac{\dfrac{2}{x-2}+\dfrac{1}{x}}{\dfrac{3x}{x-5}-\dfrac{2}{x-5}}$

Solution

First simplify the numerator to a single fraction and then simplify the denominator to a single fraction.

$$\frac{\dfrac{2}{x-2}+\dfrac{1}{x}}{\dfrac{3x}{x-5}-\dfrac{2}{x-5}}=\frac{\dfrac{2\cdot x}{(x-2)\cdot x}+\dfrac{1\cdot(x-2)}{x\cdot(x-2)}}{\dfrac{3x-2}{x-5}}$$

• Simplify numerator and denominator.

$$=\frac{\dfrac{2x+(x-2)}{x(x-2)}}{\dfrac{3x-2}{x-5}}=\frac{\dfrac{3x-2}{x(x-2)}}{\dfrac{3x-2}{x-5}}$$

$$=\frac{\cancel{3x-2}}{x(x-2)}\cdot\frac{x-5}{\cancel{3x-2}}$$

• Multiply by the reciprocal of the denominator.

$$=\frac{x-5}{x(x-2)}$$

TRY EXERCISE 42, EXERCISE SET 1.7

EXAMPLE 5 *Simplify a Fraction*

Simplify the fraction $\dfrac{c^{-1}}{a^{-1} + b^{-1}}$.

Solution

The fraction written without negative exponents becomes

$$\frac{c^{-1}}{a^{-1} + b^{-1}} = \frac{\dfrac{1}{c}}{\dfrac{1}{a} + \dfrac{1}{b}} \qquad \bullet \text{ Using } x^{-n} = \frac{1}{x^n}.$$

$$= \frac{\dfrac{1}{c} \cdot abc}{\left(\dfrac{1}{a} + \dfrac{1}{b}\right) abc} \qquad \bullet \text{ Multiply the numerator and the denominator by } abc, \text{ which is the LCD.}$$

$$= \frac{ab}{bc + ac}$$

> **TAKE NOTE**
>
> It is a mistake to write
>
> $$\frac{c^{-1}}{a^{-1} + b^{-1}} \quad \text{as} \quad \frac{a + b}{c}$$
>
> because a^{-1} and b^{-1} are *terms* and cannot be treated as *factors*.

TRY EXERCISE 60, EXERCISE SET 1.7

EXAMPLE 6 *Solve an Application*

The *average speed* for a round trip is given by the complex fraction

$$\frac{2}{\dfrac{1}{v_1} + \dfrac{1}{v_2}}$$

where v_1 is the average speed on the way to your destination and v_2 is the average speed on your return trip. Find the average speed for a round trip if $v_1 = 50$ mph and $v_2 = 40$ mph.

Solution

Evaluate the complex fraction with $v_1 = 50$ and $v_2 = 40$.

$$\frac{2}{\dfrac{1}{v_1} + \dfrac{1}{v_2}} = \frac{2}{\dfrac{1}{50} + \dfrac{1}{40}} = \frac{2}{\dfrac{1 \cdot 4}{50 \cdot 4} + \dfrac{1 \cdot 5}{40 \cdot 5}} \qquad \bullet \text{ Substitute and simplify the denominator}$$

$$= \frac{2}{\dfrac{4}{200} + \dfrac{5}{200}} = \frac{2}{\dfrac{9}{200}}$$

$$= 2 \cdot \frac{200}{9} = \frac{400}{9} = 44\frac{4}{9}$$

The average speed of the round trip is $44\frac{4}{9}$ mph.

TRY EXERCISE 64, EXERCISE SET 1.7

QUESTION   In Example 6, why is the average speed of the round trip *not* the average of v_1 and v_2?

TOPICS FOR DISCUSSION

1. Discuss the meaning the phrase *rational expression*. Is a rational expression the same as a fraction? If not, give some examples of a fraction that is not a rational expression.

2. What is the domain of a rational expression?

3. Explain why the following is *not* correct.

$$\frac{2x^2 + 5}{x^2} = 2 + 5 = 7$$

4. Consider the rational expression $\dfrac{x^2 - 3x - 10}{x^2 + x - 30}$. By simplifying this expression, we have

$$\frac{x^2 - 3x - 10}{x^2 + x - 30} = \frac{(x - 5)(x + 2)}{(x - 5)(x + 6)} = \frac{x + 2}{x + 6}$$

Does this really mean that $\dfrac{x^2 - 3x - 10}{x^2 + x - 30} = \dfrac{x + 2}{x + 6}$ for every value of x? If not, for what values of x are the two expressions equal?

EXERCISE SET 1.7

In Exercises 1 to 10, simplify each rational expression.

1. $\dfrac{x^2 - x - 20}{3x - 15}$

2. $\dfrac{2x^2 - 5x - 12}{2x^2 + 5x + 3}$

3. $\dfrac{x^3 - 9x}{x^3 + x^2 - 6x}$

4. $\dfrac{x^3 + 125}{2x^3 - 50x}$

5. $\dfrac{a^3 + 8}{a^2 - 4}$

6. $\dfrac{y^3 - 27}{-y^2 + 11y - 24}$

7. $\dfrac{x^2 + 3x - 40}{-x^2 + 3x + 10}$

8. $\dfrac{2x^3 - 6x^2 + 5x - 15}{9 - x^2}$

9. $\dfrac{4y^3 - 8y^2 + 7y - 14}{-y^2 - 5y + 14}$

10. $\dfrac{x^3 - x^2 + x}{x^3 + 1}$

In Exercises 11 to 40, simplify each expression.

11. $\left(-\dfrac{4a}{3b^2}\right)\left(\dfrac{6b}{a^4}\right)$

12. $\left(\dfrac{12x^2y}{5z^4}\right)\left(-\dfrac{25x^2z^3}{15y^2}\right)$

13. $\left(\dfrac{6p^2}{5q^2}\right)^{-1}\left(\dfrac{2p}{3q^2}\right)^2$

14. $\left(\dfrac{4r^2s}{3t^3}\right)^{-1}\left(\dfrac{6rs^3}{5t^2}\right)$

15. $\dfrac{x^2 + x}{2x + 3} \cdot \dfrac{3x^2 + 19x + 28}{x^2 + 5x + 4}$

16. $\dfrac{x^2 - 16}{x^2 + 7x + 12} \cdot \dfrac{x^2 - 4x - 21}{x^2 - 4x}$

17. $\dfrac{3x - 15}{2x^2 - 50} \cdot \dfrac{2x^2 + 16x + 30}{6x + 9}$

18. $\dfrac{y^3 - 8}{y^2 + y - 6} \cdot \dfrac{y^2 + 3y}{y^3 + 2y^2 + 4y}$

19. $\dfrac{12y^2 + 28y + 15}{6y^2 + 35y + 25} \div \dfrac{2y^2 - y - 3}{3y^2 + 11y - 20}$

20. $\dfrac{z^2 - 81}{z^2 - 16} \div \dfrac{z^2 - z - 20}{z^2 + 5z - 36}$

ANSWER   Because you were traveling slower on the return trip, the return trip took longer than the time spent going to your destination. More time was spent traveling at the slower speed. Thus the average speed is less than the average of v_1 and v_2.

21. $\dfrac{a^2 + 9}{a^2 - 64} \div \dfrac{a^3 - 3a^2 + 9a - 27}{a^2 + 5a - 24}$

22. $\dfrac{6x^2 + 13xy + 6y^2}{4x^2 - 9y^2} \div \dfrac{3x^2 - xy - 2y^2}{2x^2 + xy - 3y^2}$

23. $\dfrac{p + 5}{r} + \dfrac{2p - 7}{r}$

24. $\dfrac{2s + 5t}{4t} + \dfrac{-2s + 3t}{4t}$

25. $\dfrac{x}{x - 5} + \dfrac{7x}{x + 3}$

26. $\dfrac{2x}{3x + 1} + \dfrac{5x}{x - 7}$

27. $\dfrac{5y - 7}{y + 4} - \dfrac{2y - 3}{y + 4}$

28. $\dfrac{6x - 5}{x - 3} - \dfrac{3x - 8}{x - 3}$

29. $\dfrac{4z}{2z - 3} + \dfrac{5z}{z - 5}$

30. $\dfrac{3y - 1}{3y + 1} - \dfrac{2y - 5}{y - 3}$

31. $\dfrac{x}{x^2 - 9} - \dfrac{3x - 1}{x^2 + 7x + 12}$

32. $\dfrac{m - n}{m^2 - mn - 6n^2} + \dfrac{3m - 5n}{m^2 + mn - 2n^2}$

33. $\dfrac{1}{x} + \dfrac{2}{3x - 1} \cdot \dfrac{3x^2 + 11x - 4}{x - 5}$

34. $\dfrac{2}{y} - \dfrac{3}{y + 1} \cdot \dfrac{y^2 - 1}{y + 4}$

35. $\dfrac{q + 1}{q - 3} - \dfrac{2q}{q - 3} \div \dfrac{q + 5}{q - 3}$

36. $\dfrac{p}{p + 5} + \dfrac{p}{p - 4} \div \dfrac{p + 2}{p^2 - p - 12}$

37. $\dfrac{1}{x^2 + 7x + 12} + \dfrac{1}{x^2 - 9} + \dfrac{1}{x^2 - 16}$

38. $\dfrac{2}{a^2 - 3a + 2} + \dfrac{3}{a^2 - 1} - \dfrac{5}{a^2 + 3a - 10}$

39. $\left(1 + \dfrac{2}{x}\right)\left(3 - \dfrac{1}{x}\right)$

40. $\left(4 - \dfrac{1}{z}\right)\left(4 + \dfrac{2}{z}\right)$

In Exercises 41 to 58, simplify each complex fraction.

41. $\dfrac{4 + \dfrac{1}{x}}{1 - \dfrac{1}{x}}$

42. $\dfrac{3 - \dfrac{2}{a}}{5 + \dfrac{3}{a}}$

43. $\dfrac{\dfrac{x}{y} - 2}{y - x}$

44. $\dfrac{3 + \dfrac{2}{x - 3}}{4 + \dfrac{1}{2 + \dfrac{1}{x}}}$

45. $\dfrac{5 - \dfrac{1}{x + 2}}{1 + \dfrac{3}{1 + \dfrac{3}{x}}}$

46. $\dfrac{\dfrac{1}{(x + h)^2} - 1}{h}$

47. $\dfrac{1 + \dfrac{1}{b - 2}}{1 - \dfrac{1}{b + 3}}$

48. $r - \dfrac{r}{r + \dfrac{1}{3}}$

49. $\dfrac{1 - \dfrac{1}{x^2}}{1 + \dfrac{1}{x}}$

50. $\dfrac{1}{\dfrac{1}{a} + \dfrac{1}{b}}$

51. $2 - \dfrac{m}{1 - \dfrac{1 - m}{-m}}$

52. $\dfrac{\dfrac{x + h + 1}{x + h} - \dfrac{x}{x + 1}}{h}$

53. $\dfrac{\dfrac{1}{x} - \dfrac{x - 4}{x + 1}}{\dfrac{x}{x + 1}}$

54. $\dfrac{\dfrac{2}{y} - \dfrac{3y - 2}{y - 1}}{\dfrac{y}{y - 1}}$

55. $\dfrac{\dfrac{1}{x + 3} - \dfrac{2}{x - 1}}{\dfrac{x}{x - 1} + \dfrac{3}{x + 3}}$

56. $\dfrac{\dfrac{x + 2}{x^2 - 1} + \dfrac{1}{x + 1}}{\dfrac{x}{2x^2 - x - 1} + \dfrac{1}{x - 1}}$

57. $\dfrac{\dfrac{x^2 + 3x - 10}{x^2 + x - 6}}{\dfrac{x^2 - x - 30}{2x^2 - 15x + 18}}$

58. $\dfrac{\dfrac{2y^2 + 11y + 15}{y^2 - 4y - 21}}{\dfrac{6y^2 + 11y - 10}{3y^2 - 23y + 14}}$

In Exercises 59 to 62, simplify each algebraic fraction. Write all answers with positive exponents.

59. $\dfrac{a^{-1} + b^{-1}}{a - b}$

60. $\dfrac{e^{-2} - f^{-1}}{ef}$

61. $\dfrac{a^{-1}b - ab^{-1}}{a^2 + b^2}$

62. $(a + b^{-2})^{-1}$

63. **AVERAGE SPEED** According to Example 6, the average speed for a round trip in which the average speed on the way to your destination was v_1 and the average speed on your return was v_2 is given by the complex fraction

$$\dfrac{2}{\dfrac{1}{v_1} + \dfrac{1}{v_2}}$$

a. Find the average speed for a round trip by helicopter with $v_1 = 180$ mph and $v_2 = 110$ mph.

b. Simplify the complex fraction.

64. **RELATIVITY THEORY** Using Einstein's theory of relativity, the "sum" of the two speeds v_1 and v_2 is given by the complex fraction

$$\dfrac{v_1 + v_2}{1 + \dfrac{v_1 v_2}{c^2}}$$

where c is the speed of light.

a. Evaluate this expression with $v_1 = 1.2 \times 10^8$ mph, $v_2 = 2.4 \times 10^8$ mph, and $c = 6.7 \times 10^8$ mph.

b. Simplify the complex fraction.

65. Find the rational expression in simplest form that represents the sum of the reciprocals of the consecutive integers x and $x + 1$.

66. Find the rational expression in simplest form that represents the positive difference between the reciprocals of the consecutive even integers x and $x + 2$.

67. Find the rational expression in simplest form that represents the sum of the reciprocals of the consecutive even integers $x - 2$, x, and $x + 2$.

68. Find the rational expression in simplest form that represents the sum of the reciprocals of the squares of the consecutive even integers $x - 2$, x, and $x + 2$.

SUPPLEMENTAL EXERCISES

In Exercises 69 to 72, simplify each algebraic fraction.

69. $\dfrac{(x + 5) - x(x + 5)^{-1}}{x + 5}$

70. $\dfrac{(y + 2) + y^2(y + 2)^{-1}}{y + 2}$

71. $\dfrac{x^{-1} - 4y}{(x^{-1} - 2y)(x^{-1} + 2y)}$

72. $\dfrac{x + y}{x - y} \cdot \dfrac{x^{-1} - y^{-1}}{x^{-1} + y^{-1}}$

73. **FINANCE** The **present value** of an ordinary annuity is given by

$$R\left[\frac{1 - \dfrac{1}{(1 + i)^n}}{i}\right]$$

where n is the number of payments of R dollars each invested at an interest rate of i per conversion period. Simplify the complex fraction.

74. **ELECTRICITY** The total resistance of the three resistances R_1, R_2, and R_3 in parallel is given by

$$\frac{1}{\dfrac{1}{R_1} + \dfrac{1}{R_2} + \dfrac{1}{R_3}}$$

Simplify the complex fraction.

PROJECTS

1. **CONTINUED FRACTIONS** The complex fraction shown at the right is called a **continued fraction.** The three dots in $\dfrac{1}{1 + \cdots}$ indicate that the pattern continues in the same manner. A **convergent** of a complex fraction is an approximation of the continued fraction that is found by stopping the process at some point.

$$\cfrac{1}{1 + \cfrac{1}{1 + \cfrac{1}{1 + \cfrac{1}{1 + \cdots}}}}$$

a. Calculate the convergent $C_2 = \cfrac{1}{1 + \cfrac{1}{1 + 1}}$.

b. Calculate the convergent $C_3 = \cfrac{1}{1 + \cfrac{1}{1 + \cfrac{1}{1 + 1}}}$.

c. Calculate the convergent $C_5 = \cfrac{1}{1 + \cfrac{1}{1 + \cfrac{1}{1 + \cfrac{1}{1 + 1}}}}$.

d. Show that $C_5 \approx \dfrac{-1 + \sqrt{5}}{2}$. Using some techniques from more advanced math courses, it can be shown that the convergents of the continued fraction become closer and closer to $\dfrac{-1 + \sqrt{5}}{2}$.

2. **REPRESENTATION OF π** There are a few continued-fraction representations for π. Find two of these representations. Compute the value of π accurate to 4 decimal places using a convergent from each of the continued fractions you found.

EXPLORING CONCEPTS WITH TECHNOLOGY

The Mandelbrot Replacement Procedure

The following procedure is called the **Mandelbrot replacement procedure.**

MANDELBROT REPLACEMENT PROCEDURE

Pick a complex number s.

1. Square s and add the result to s.

2. Square the last result and add it to s.

3. Repeat step 2.

The number s is referred to as the seed of the procedure. The number s is a seed in the sense that each seed produces a different sequence of numbers. Some seeds produce sequences that grow without bound. Some seeds produce sequences that grow toward some constant. Still other seeds yield sequences that are cyclic. Consider the following illustrations.

- Let the seed $s = 1$.

$$1^2 + 1 = 2, \qquad 2^2 + 1 = 5, \qquad 5^2 + 1 = 26, \qquad 26^2 + 1 = 677$$

As the replacement procedure continues, we get larger and larger numbers.

- Let the seed $s = -1$.

$$(-1)^2 + (-1) = 0, \qquad 0^2 + (-1) = -1, \qquad (-1)^2 + (-1) = 0, \ldots$$

As the replacement procedure continues, the results *cycle:* $0, -1, 0, -1, 0, \ldots$.

- Let the seed $s = 0.25$.

$$(0.25)^2 + 0.25 = 0.3125, \qquad (0.3125)^2 + 0.25 = 0.34765625,$$
$$(0.34765625)^2 + 0.25 \approx 0.3708648682, \ldots$$

1. Use your calculator to continue the Mandelbrot replacement procedure. What number do you have after

   a. 25 applications of step 2?

   b. 50 applications of step 2?

   c. 75 applications of step 2?

   d. What constant do you think the sequence of numbers is approaching?

- Let the seed $s = i$.

$$i^2 + i = -1 + i, \qquad (-1 + i)^2 + i = -i, \ldots$$

2. **a.** What is the next number produced by the Mandelbrot replacement procedure?

b. What happens as the procedure is continued?

The Mandelbrot replacement procedure can be used to determine special kinds of numbers called **attractors**. The attractors produced by the Mandelbrot replacement procedure are an essential part of the Mandelbrot set, which is shown in black in **Figure 1.14.**

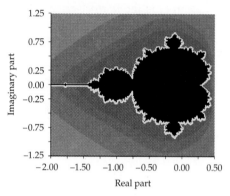

Figure 1.14

CHAPTER 1 REVIEW

1.1 The Real Number System

- The following sets of numbers are used extensively in the study of algebra:

Integers	$\{\ldots, -3, -2, -1, 0, 1, 2, 3, \ldots\}$
Rational numbers	{all terminating or repeating decimals}
Irrational numbers	{all nonterminating, nonrepeating decimals}
Real numbers	{all rational or irrational numbers}

1.2 Intervals, Absolute Value, and Distance

- The absolute value of the real number a is defined by

$$|a| = \begin{cases} a & \text{if } a \geq 0 \\ -a & \text{if } a < 0 \end{cases}$$

- For any real numbers a and b, the distance between the graph of a and the graph of b is denoted by $d(a, b)$, where $d(a, b) = |a - b|$.

1.3 Integer and Rational Number Exponents

- If b is any real number and n is any natural number, then

$$b^n = \underbrace{b \cdot b \cdot b \cdot \cdots \cdot b}_{n \text{ factors of } b}$$

- For any nonzero real number b, $b^0 = 1$.

- If $b \neq 0$ and n is any natural number, then $b^{-n} = \dfrac{1}{b^n}$ and $\dfrac{1}{b^{-n}} = b^n$.

- **Properties of Rational Exponents**
  If p, q, and r represent rational numbers, and a and b are positive real numbers, then

  Product $\quad b^p \cdot b^q = b^{p+q}$

  Quotient $\quad \dfrac{b^p}{b^q} = b^{p-q}$

  Power $\quad (b^p)^q = b^{pq} \qquad (a^p b^q)^r = a^{pr} b^{qr}$

  $\qquad\qquad \left(\dfrac{a^p}{b^q}\right)^r = \dfrac{a^{pr}}{b^{qr}} \qquad b^{-p} = \dfrac{1}{b^p}$

- **Properties of Radicals**
  If m and n are natural numbers and a and b are nonnegative real numbers, then

  Product $\quad \sqrt[n]{a} \cdot \sqrt[n]{b} = \sqrt[n]{ab}$

  Quotient $\quad \dfrac{\sqrt[n]{a}}{\sqrt[n]{b}} = \sqrt[n]{\dfrac{a}{b}}$

  Index $\quad \sqrt[m]{\sqrt[n]{b}} = \sqrt[mn]{b}$

1.4 Complex Numbers

- **Definition of i**
  The number i, called the imaginary unit, is the number such that $i^2 = -1$.

- **Definition of a Complex Number**
  If a and b are real numbers and i is the imaginary unit, then $a + bi$ is called a complex number. The complex numbers $a + bi$ and $a - bi$ are called complex conjugates or conjugates of each other.

- **Powers of i**
  If n is a positive integer, then $i^n = i^r$, where r is the remainder of the division of n by 4.

1.5 Polynomials

- A polynomial is an expression of the form

  $$a_n x^n + a_{n-1} x^{n-1} + \cdots + a_2 x^2 + a_1 x + a_0$$

- Special product formulas are as follows:

Special Form	Formula(s)
(Sum)(Difference)	$(x + y)(x - y) = x^2 - y^2$
(Binomial)^2	$(x + y)^2 = x^2 + 2xy + y^2$ $(x - y)^2 = x^2 - 2xy + y^2$

1.6 Factoring

- Factoring formulas are as follows:

Special Form	Formula(s)
Difference of two squares	$x^2 - y^2 = (x + y)(x - y)$
Perfect-square trinomials	$x^2 + 2xy + y^2 = (x + y)^2$ $x^2 - 2xy + y^2 = (x - y)^2$
Sum of cubes	$x^3 + y^3 = (x + y)(x^2 - xy + y^2)$
Difference of cubes	$x^3 - y^3 = (x - y)(x^2 + xy + y^2)$

- To factor a polynomial, use the general factoring strategy.

1.7 Rational Expressions

- A rational expression is a fraction in which the numerator and denominator are polynomials. The properties of rational expressions are used to simplify a rational expression and to find the sum, difference, product, and quotient of two rational expressions.

- Complex fractions can be simplified in either of the following ways:

  Method 1: Multiply both the numerator and the denominator by the LCD of all the fractions in the complex fraction.

  Method 2: Simplify the numerator to a single fraction and the denominator to a single fraction. Multiply the numerator by the reciprocal of the denominator.

CHAPTER 1 TRUE/FALSE EXERCISES

In Exercises 1 to 10, answer true or false. If the statement is false, give an example to show that the statement is false.

1. If a and b are real numbers, then $|a - b| = |b - a|$.

2. If a is a real number, then $a^2 \geq a$.

3. The set of rational numbers is closed under the operation of addition.

4. The set of irrational numbers is closed under the operation of addition.

5. Let $x \oplus y$ denote the average of the two real numbers x and y. That is,
   $$x \oplus y = \frac{x + y}{2}$$
   The operation $\oplus$ is an associative operation because $(x \oplus y) \oplus z = x \oplus (y \oplus z)$ for all real numbers x, y, and z.

6. Using interval notation, we write the inequality $x > a$ as $[a, \infty)$.

7. If n is a real number, then $\sqrt{n^2} = n$.

8. Every real number is a complex number.

9. The sum of a complex number z and its conjugate $\bar{z}$ is a real number.

10. The product of a complex number z and its conjugate $\bar{z}$ is a real number.

CHAPTER 1 REVIEW EXERCISES

In Exercises 1 to 4, classify each number as one or more of the following: integer, rational number, irrational number, real number, prime number, composite number.

1. 3   **2.** $\sqrt{7}$ **3.** $-\dfrac{1}{2}$ **4.** $0.\overline{5}$

In Exercises 5 and 6, use $A = \{1, 5, 7\}$ and $B = \{2, 3, 5, 11\}$ to find the indicated intersection or union.

5. $A \cup B$ **6.** $A \cap B$

In Exercises 7 to 14, identify the real number property or the property of equality that is illustrated.

7. $5(x + 3) = 5x + 15$

8. $a(3 + b) = a(b + 3)$

9. $(6c)d = 6(cd)$

10. $\sqrt{2} + 3$ is a real number.

11. $7 + 0 = 7$

12. $1x = x$

13. If $7 = x$, then $x = 7$.

14. If $3x + 4 = y$, and $y = 5z$, then $3x + 4 = 5z$.

In Exercises 15 and 16, graph each inequality and write the inequality using interval notation.

15. $-4 < x \leq 2$ **16.** $x \leq -1$ or $x > 3$

In Exercises 17 and 18, graph each interval and write each interval as an inequality.

17. $[-3, 2)$ **18.** $(-1, \infty)$

In Exercises 19 to 22, write each real number without absolute value symbols.

19. $|7|$ **20.** $|2 - \pi|$ **21.** $|4 - \pi|$ **22.** $|-11|$

In Exercises 23 and 24, find the distance on the real number line between the points whose coordinates are given.

23. $-3, 14$ **24.** $\sqrt{5}, -\sqrt{2}$

In Exercises 25 and 26, evaluate each expression.

25. $-5^2 + (-11)$ **26.** $\dfrac{(2^2 \cdot 3^{-2})^2}{3^{-1} \cdot 2^3}$

In Exercises 27 and 28, simplify each expression.

27. $(3x^2y)(2x^3y)^2$ **28.** $\left(\dfrac{2a^2b^3c^{-2}}{3ab^{-1}}\right)^2$

In Exercises 29 and 30, write each number in scientific notation.

29. 620,000   **30.** 0.0000017

In Exercises 31 and 32, change each number from scientific notation to decimal form.

31. 3.5×10^4 **32.** 4.31×10^{-7}

In Exercises 33 to 36, perform the indicated operation and express each result as a polynomial in standard form.

33. $(2a^2 + 3a - 7) + (-3a^2 - 5a + 6)$

34. $(5b^2 - 11) - (3b^2 - 8b - 3)$

35. $(2x^2 + 3x - 5)(3x^2 - 2x + 4)$

36. $(3y - 5)^3$

In Exercises 37 to 40, completely factor each polynomial.

37. $3x^2 + 30x + 75$ **38.** $25x^2 - 30xy + 9y^2$

39. $20a^2 - 4b^2$ **40.** $16a^3 + 250$

In Exercises 41 and 42, simplify each rational expression.

41. $\dfrac{6x^2 - 19x + 10}{2x^2 + 3x - 20}$ **42.** $\dfrac{4x^3 - 25x}{8x^4 + 125x}$

In Exercises 43 to 46, perform the indicated operation and simplify if possible.

43. $\dfrac{10x^2 + 13x - 3}{6x^2 - 13x - 5} \cdot \dfrac{6x^2 + 5x + 1}{10x^2 + 3x - 1}$

44. $\dfrac{15x^2 + 11x - 12}{25x^2 - 9} \div \dfrac{3x^2 + 13x + 12}{10x^2 + 11x + 3}$

45. $\dfrac{x}{x^2 - 9} + \dfrac{2x}{x^2 + x - 12}$

46. $\dfrac{3x}{x^2 + 7x + 12} - \dfrac{x}{2x^2 + 5x - 3}$

In Exercises 47 and 48, simplify each complex fraction.

47. $\dfrac{2 + \dfrac{1}{x - 5}}{3 - \dfrac{2}{x - 5}}$

48. $\dfrac{1}{2 + \dfrac{3}{1 + \dfrac{4}{x}}}$

In Exercises 49 and 50, evaluate each exponential expression.

49. $25^{1/2}$

50. $-27^{2/3}$

In Exercises 51 to 54, simplify each expression.

51. $x^{2/3} \cdot x^{3/4}$

52. $\left(\dfrac{8x^{5/4}}{x^{1/2}}\right)^{2/3}$

53. $\left(\dfrac{x^2 y}{x^{1/2} y^{-3}}\right)^{1/2}$

54. $(x^{1/2} - y^{1/2})(x^{1/2} + y^{1/2})$

In Exercises 55 to 64, simplify each radical expression. Assume the variables are positive real numbers.

55. $\sqrt{48a^2 b^7}$

56. $\sqrt{12a^3 b}$

57. $\sqrt{72x^2 y}$

58. $\sqrt{18x^3 y^5}$

59. $\sqrt{\dfrac{54xy^3}{10x}}$

60. $-\sqrt{\dfrac{24xyz^3}{15z^6}}$

61. $\dfrac{7x}{\sqrt[3]{2x^2}}$

62. $\dfrac{5y}{\sqrt[3]{9y}}$

63. $\sqrt[3]{-135x^2 y^7}$

64. $\sqrt[3]{-250xy^6}$

In Exercises 65 and 66, write the complex number in standard form and give its conjugate.

65. $3 - \sqrt{-64}$

66. $\sqrt{-4} + 6$

In Exercises 67 to 70, simplify and write the complex number in standard form.

67. $(3 + 7i) + (2 - 5i)$

68. $(6 - 8i) - (9 - 11i)$

69. $(5 + 3i)(2 - 5i)$

70. $\dfrac{4 + i}{7 - 2i}$

In Exercises 71 to 74, simplify and write each complex number as i, $-i$, 1, or -1.

71. i^{20}

72. i^{57}

73. $\dfrac{1}{i^{28}}$

74. i^{-200}

CHAPTER 1 TEST

1. For real numbers a, b, and c, identify the property that is illustrated by $(a + b)c = ac + bc$.

2. Given $A = \{0, 2, 4, 6, 8\}$ and $B = \{1, 3, 5, 7, 9\}$, find $A \cup B$.

3. Write $|-3| - |-6|$ without absolute value symbols.

4. Find the distance between the points -12 and -5 on the number line.

5. Simplify: $(-2x^0 y^{-2})^2 (-3x^2 y^{-1})^{-2}$

6. Simplify: $\dfrac{(2a^{-1}bc^{-2})^2}{(3^{-1}b)(2^{-1}ac^{-2})^3}$

7. Write 0.00137 in scientific notation.

8. Simplify: $(x - 2y)(x^2 - 2x + y)$

9. Evaluate the polynomial $3y^3 - 2y^2 - y + 2$ for $y = -3$.

10. Factor: $7x^2 + 34x - 5$

11. Factor: $3ax - 12bx - 2a + 8b$

12. Factor: $16x^4 - 2xy^3$

13. Factor: $x^4 + x^3 y - x - y$

14. Simplify: $\dfrac{x^4 - 2x^3 - x + 2}{x^3 - x^2 - x + 1}$

15. Simplify: $\dfrac{x}{x^2 + x - 6} - \dfrac{2}{x^2 - 5x + 6}$

16. Simplify: $\dfrac{2x^2 + 3x - 2}{x^2 - 3x} \div \dfrac{2x^2 - 7x + 3}{x^3 - 3x^2}$

17. Simplify: $\dfrac{3}{a + b} \cdot \dfrac{a^2 - b^2}{2a - b} - \dfrac{5}{a}$

18. Simplify: $x - \dfrac{x}{x + \dfrac{1}{2}}$

19. Simplify: $\dfrac{x^{1/3}y^{-3/4}}{x^{-1/2}y^{3/2}}$

20. Simplify: $3x\sqrt[3]{81xy^4} - 2y\sqrt[3]{3x^4y}$

21. Simplify: $\dfrac{x}{\sqrt[4]{2x^3}}$

22. Simplify: $\dfrac{3}{\sqrt{x}+2}$

23. Simplify $3(2-3i)-4(1-5i)$ and write the complex number in standard form.

24. Simplify $(2-3i)(4+i)$ and write the complex number in standard form.

25. Write $\dfrac{2i}{-2+3i}$ as a complex number in standard form.

CHAPTER

EQUATIONS AND INEQUALITIES

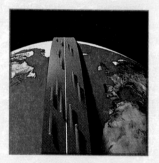

The *Information Superhighway* puts the world at your fingertips.

The Internet can be used to find information on any subject, for example, travel.

The **Global Village:** Internet accessibility fosters learning across cultures.

Mathematics on the World Wide Web

Mathematics is a dynamic subject. New results are discovered daily. To keep up with some of the latest advances, many mathematicians and many students make use of the Internet. If you have access to the Internet, check out some of the following sources, which are listed on the World Wide Web.

Mathematical Resources on the Web
"http://www.math.ufl.edu/math/math-web.html#HomePages"

Mathematics: UofA Software
"http://www.math.arizona.edu/software/uasft.html"

Yahoo—Science:Mathematics
"http://www.yahoo.com/Science/Mathematics"

Fractal pictures and animations
"http://www.cnam.fr/fractals.html"

Mathbrowser Home Page
"http://www.mathsoft.com/browser/index.html"

Calculus&Mathematica Home Page
"http://www-cm.math.uiuc.edu/"

Home Page of Edward Dunne
"http://www.math.okstate.edu/~dunne/"

Texas Instruments
"http://www.ti.com/calc"

2.1 LINEAR EQUATIONS

SOLVE BY PRODUCING EQUIVALENT EQUATIONS

An **equation** is a statement about the equality of two expressions. If either of the expressions contains a variable, the equation may be a true statement for some values of the variable and a false statement for other values. For example, the equation $2x + 1 = 7$ is a true statement for $x = 3$, but it is false for any number except 3. The number 3 is said to **satisfy** the equation $2x + 1 = 7$, because substituting 3 for x produces $2(3) + 1 = 7$, which is a true statement.

To **solve** an equation means to find all values of the variable that satisfy the equation. The values that satisfy an equation are called **solutions** or **roots** of the equation. For instance, 2 and 3 are both solutions of $x^2 - 5x + 6 = 0$.

Equivalent equations are equations that have exactly the same solution(s). The process of solving an equation involving the variable x is often accomplished by producing a sequence of equivalent equations until we produce an equation or equations of the form

$$x = \text{a constant}$$

To produce these equivalent equations that lead us to the solution(s), we often perform one or more of the following procedures.

PROCEDURES THAT PRODUCE EQUIVALENT EQUATIONS

1. Simplification of an expression on either side of the equation by such procedures as (i) combining like terms and (ii) applying the properties explained in Chapter 1, such as the commutative, associative, and distributive properties.

   $2x + 3 + 5x = -11$ and $7x + 3 = -11$ are equivalent equations.

2. Addition or subtraction of the same quantity on both sides of an equation.

   $3x - 7 = 2$ and $3x = 9$ are equivalent equations.

3. Multiplication or division by the same nonzero quantity on both sides of an equation.

   $\dfrac{5}{6}x = 10$ and $x = 12$ are equivalent equations.

Many applications can be modeled by *linear equations*.

DEFINITION OF A LINEAR EQUATION

A **linear equation** in the single variable x is an equation that can be written in the form

$$ax + b = 0$$

where a and b are real numbers, with $a \neq 0$.

Linear equations are generally solved by applying the procedures that produce equivalent equations.

EXAMPLE 1   *Solve a Linear Equation*

Solve: $\dfrac{3}{4}x - 6 = 0$

Solution

$$\frac{3}{4}x - 6 = 0$$

$$\frac{3}{4}x - 6 + 6 = 0 + 6 \qquad \bullet \textbf{ Add 6 to each side.}$$

$$\frac{3}{4}x = 6$$

$$\left(\frac{4}{3}\right)\left(\frac{3}{4}x\right) = \left(\frac{4}{3}\right)(6) \qquad \bullet \textbf{ Multiply each side by } \frac{4}{3}.$$

$$x = 8$$

Because 8 satisfies the original equation (see the *Take Note*), 8 is the solution.

TRY EXERCISE 2, EXERCISE SET 2.1

If an equation involves fractions, it is helpful to multiply each side of the equation by the LCD of all the denominators to produce an equivalent equation that does not contain fractions.

EXAMPLE 2   *Solve by Clearing Fractions*

Solve: $\dfrac{2}{3}x + 10 - \dfrac{x}{5} = \dfrac{36}{5}$

Continued • ➤

Solution

$$\frac{2}{3}x + 10 - \frac{x}{5} = \frac{36}{5}$$

$$15\left(\frac{2}{3}x + 10 - \frac{x}{5}\right) = 15\left(\frac{36}{5}\right)$$ • Multiply each side of the equation by **15**, the **LCD** of the denominators.

$$10x + 150 - 3x = 108$$ • Simplify.

$$7x + 150 = 108$$

$$7x + 150 - 150 = 108 - 150$$ • Subtract **150** from each side.

$$7x = -42$$

$$\frac{7x}{7} = \frac{-42}{7}$$ • Divide each side by **7**.

$$x = -6$$ • Check as before.

TRY EXERCISE 12, EXERCISE SET 2.1

EXAMPLE 3   *Solve an Equation by Applying Properties*

Solve: $(x + 2)(5x + 1) = 5x(x + 1)$

Solution

$$(x + 2)(5x + 1) = 5x(x + 1)$$

$$5x^2 + 11x + 2 = 5x^2 + 5x$$ • Simplify each product.

$$11x + 2 = 5x$$ • Subtract $5x^2$ from each side.

$$6x + 2 = 0$$ • Subtract $5x$ from each side.

$$6x = -2$$ • Subtract **2** from each side.

$$x = -\frac{1}{3}$$ • Divide each side of the equation by **6**.

TRY EXERCISE 20, EXERCISE SET 2.1

CONTRADICTIONS, CONDITIONAL EQUATIONS, AND IDENTITIES

An equation that has no solutions is called a **contradiction**. The equation $x = x + 1$ is a contradiction. No number is equal to itself increased by 1.

An equation that is true for some values of the variable but not true for other values of the variable is called a **conditional equation**. For example, $x + 2 = 8$ is a conditional equation because it is true for $x = 6$ and false for any number not equal to 6.

An **identity** is an equation that is true for *every* real number for which all terms of the equation are defined. Examples of identities include the equations $x + x = 2x$ and $(x + 3)^2 = x^2 + 6x + 9$.

EXAMPLE 4	*Verify an Identity*

Verify the identity $\dfrac{3(x^3 - 8)}{x - 2} = 3x^2 + 6x + 12, \ x \neq 2.$

Solution

Simplify the left side of the equation.

$$\dfrac{3(x^3 - 8)}{x - 2} = \dfrac{3(x - 2)(x^2 + 2x + 4)}{x - 2}$$

• Factor the difference of cubes and simplify.

$$= 3(x^2 + 2x + 4)$$
$$= 3x^2 + 6x + 12$$

Because we have shown that it is possible to write the left side of the equation exactly as the right side is written, we have verified the identity.

TRY EXERCISE 30, EXERCISE SET 2.1

Multiplying each side of an equation by the same *nonzero* number always yields an equivalent equation. If each side of an equation is multiplied by an expression that involves a variable, then we restrict the variable so that the expression is not equal to zero. Example 5b illustrates the fact that you may produce incorrect results if you fail to restrict the variable.

EXAMPLE 5	*Solve Equations That Have Restrictions*

Solve each equation.   **a.** $\dfrac{x}{x - 3} = \dfrac{9}{x - 3} - 5$ **b.** $1 + \dfrac{x}{x - 5} = \dfrac{5}{x - 5}$

Solution

TAKE NOTE

When we multiply both sides of an equation by $x - a$, we assume that $x \neq a$.

a. First, note that the denominator $x - 3$ would equal zero if x were 3. To produce a simpler equivalent equation, multiply each side by $x - 3$, with the restriction that $x \neq 3$.

$$(x - 3)\left(\dfrac{x}{x - 3}\right) = (x - 3)\left(\dfrac{9}{x - 3} - 5\right)$$

$$x = (x - 3)\left(\dfrac{9}{x - 3}\right) - (x - 3)5$$

$$x = 9 - 5x + 15$$

$$6x = 24$$

$$x = 4$$

Substituting 4 for x in the original equation establishes that 4 is the solution.

Continued • ▶

Determine the incorrect step in the following "proof" that 2 = 1.

$a = b$	Given.
$a^2 = ab$	Multiply by a.
$a^2 - b^2 = ab - b^2$	Subtract b^2.
$(a + b)(a - b) = b(a - b)$	Factor.
$\dfrac{(a + b)(a - b)}{(a - b)} = \dfrac{b(a - b)}{(a - b)}$	Divide by $a - b$.
$a + b = b$	Simplify.
$b + b = b$	Substitute b for a.
$2b = b$	Simplify.
$2 = 1$	Divide by b.

b. To produce a simpler equivalent equation, multiply each side of the equation by $x - 5$, with the restriction that $x \neq 5$.

$$(x - 5)\left(1 + \frac{x}{x - 5}\right) = (x - 5)\left(\frac{5}{x - 5}\right)$$

$$(x - 5)1 + (x - 5)\left(\frac{x}{x - 5}\right) = 5$$

$$x - 5 + x = 5$$

$$2x = 10$$

$$x = 5$$

Although we have obtained 5 as a proposed solution, 5 is *not* a solution of the original equation because it contradicts our restriction $x \neq 5$. Substitution of 5 for x in the original equation results in denominators of 0. In this case the original equation has no solutions.

TRY EXERCISE 38, EXERCISE SET 2.1

APPLICATIONS

Linear equations can often be used to model real-world data.

EXAMPLE 6	*An Application Concerning Median Annual Earnings*

The graph shown in **Figure 2.1** reveals that although many college graduates enter the job market through low-wage positions, the average college graduate starts to earn more than the average student who holds only a high school diploma within a few years of graduation.

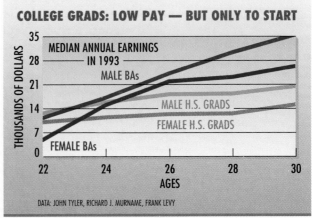

Reprinted from the September 4, 1995 issue of *Business Week* by special permission. © by McGraw-Hill, Inc.

Figure 2.1

The median annual earnings for male college graduates can be modeled by the following linear equation.

$$\text{Earnings} = 2875a - 51{,}250$$

where the earnings are in dollars per year and a is the age of the worker, with $22 \leq a \leq 30$.

   Determine the age (to the nearest year) at which the average male college graduate can expect annual earnings of $30,000.

Solution

Replace "Earnings" with 30,000 and solve for a.

$$30{,}000 = 2875a - 51{,}250$$

$$81{,}250 = 2875a \qquad \text{• Add 51,250 to each side.}$$

$$a = \frac{81{,}250}{2875} \qquad \text{• Divide both sides by 2875.}$$

$$a \approx 28 \qquad \text{• Round to the nearest year.}$$

According to the model, the average male college graduate can expect his annual earnings to reach $30,000 at age $a = 28$.

Try Exercise 58, Exercise Set 2.1

Topics for Discussion

1. A student doubles each side of the equation $\frac{1}{2}x + 3 = 4$ to produce the equation $x + 3 = 8$. Has the student produced an equivalent equation? Explain.

2. Are the equations $2(x - 3) = 14$ and $2(x - 3 + 3) = 14 + 3$ equivalent equations? Explain.

3. If $P = Q$, is it also true that $Q = P$?

4. Are the equations $x^2 = 9$ and $x = 3$ equivalent equations? Explain.

5. A classmate claims that equations of the form $\frac{a}{b} = \frac{c}{d}$ can be cleared of denominators by *cross-multiplying* instead of multiplying both sides of the equation by the LCD (least common denominator) of the denominators as shown in Example 2. Do you agree? Explain.

Exercise Set 2.1

In Exercises 1 to 28, solve and check each equation.

1. $2x + 10 = 40$

2. $-3y + 20 = 2$

3. $5x + 2 = 2x - 10$

4. $4x - 11 = 7x + 20$

5. $2(x - 3) - 5 = 4(x - 5)$

6. $5(x - 4) - 7 = -2(x - 3)$

7. $4(2r - 17) + 5(3r - 8) = 0$

8. $6(5s - 11) - 12(2s + 5) = 0$

9. $\dfrac{3}{4}x + \dfrac{1}{2} = \dfrac{2}{3}$

10. $\dfrac{x}{4} - 5 = \dfrac{1}{2}$

11. $\dfrac{2}{3}x - 5 = \dfrac{1}{2}x - 3$

12. $\dfrac{1}{2}x + 7 - \dfrac{1}{4}x = \dfrac{19}{2}$

13. $0.2x + 0.4 = 3.6$

14. $0.04x - 0.2 = 0.07$

15. $x + 0.08(60) = 0.20(60 + x)$

16. $6(t + 1.5) = 12t$

17. $\dfrac{3}{5}(n + 5) - \dfrac{3}{4}(n - 11) = 0$

18. $-\dfrac{5}{7}(p + 11) + \dfrac{2}{5}(2p - 5) = 0$

19. $3(x + 5)(x - 1) = (3x + 4)(x - 2)$

20. $5(x + 4)(x - 4) = (x - 3)(5x + 4)$

21. $5[x - (4x - 5)] = 3 - 2x$

22. $6[3y - 2(y - 1)] - 2 + 7y = 0$

23. $\dfrac{40 - 3x}{5} = \dfrac{6x + 7}{8}$

24. $\dfrac{12 + x}{-4} = \dfrac{5x - 7}{3} + 2$

25. $0.08x + 0.12(4000 - x) = 432$

26. $0.075y + 0.06(10{,}000 - y) = 727.50$

27. $0.115x + 0.0975(8000 - x) = 823.75$

28. $0.145x + 0.109(4000) = 0.12(4000 + x)$

In Exercises 29 to 36, determine whether the equation is an identity, a conditional equation, or a contradiction.

29. $-3(x - 5) = -3x + 15$

30. $2x + \dfrac{1}{3} = \dfrac{6x + 1}{3}$

31. $2y + 7 = 3(y - 1)$

32. $x^2 + 10x = x(x + 10)$

33. $\dfrac{4y + 7}{4} = y + 7$

34. $(x + 3)^2 = x^2 + 9$

35. $2x + 5 = x + 9 + x$

36. $(x - 3)(x + 4) = x^2 + 4x - 11$

In Exercises 37 to 56, solve and check each equation.

37. $\dfrac{3}{x + 2} = \dfrac{5}{2x - 7}$

38. $\dfrac{4}{y + 2} = \dfrac{7}{y - 4}$

39. $\dfrac{30}{10 + x} = \dfrac{20}{10 - x}$

40. $\dfrac{6}{8 + x} = \dfrac{4}{8 - x}$

41. $\dfrac{3x}{x + 4} = 2 - \dfrac{12}{x + 4}$

42. $\dfrac{8}{2m + 1} - \dfrac{1}{m - 2} = \dfrac{5}{2m + 1}$

43. $2 + \dfrac{9}{r - 3} = \dfrac{3r}{r - 3}$

44. $\dfrac{t}{t - 4} + 3 = \dfrac{4}{t - 4}$

45. $\dfrac{5}{x - 3} - \dfrac{3}{x - 2} = \dfrac{4}{x - 3}$

46. $\dfrac{4}{x - 1} + \dfrac{7}{x + 7} = \dfrac{5}{x - 1}$

47. $\dfrac{2x + 5}{3x - 1} = 1$

48. $\dfrac{4x - 1}{3x + 2} = \dfrac{5}{6}$

49. $(y + 3)^2 = (y + 4)^2 + 1$

50. $(z - 7)^2 = (z - 2)^2 + 9$

51. $\dfrac{x}{x - 3} = \dfrac{x + 4}{x + 2}$

52. $\dfrac{x}{x - 5} = \dfrac{x + 7}{x + 1}$

53. $\dfrac{x + 3}{x + 5} = \dfrac{x - 3}{x - 4}$

54. $\dfrac{x - 6}{x + 4} = \dfrac{x - 1}{x + 2}$

55. $\dfrac{4x - 3}{2x} = \dfrac{2x - 4}{x - 2}$

56. $\dfrac{x + 3}{x + 1} = \dfrac{x + 6}{x + 4}$

57. **MEDIAN ANNUAL EARNINGS** Assuming that the earnings equation

$$\text{Earnings} = 2875a - 51{,}250$$

from Example 6 is valid for $a \geq 30$, determine the age (to the nearest year) at which the average male college graduate can expect to earn an annual income of $45,000.

58. **MEDIAN ANNUAL EARNINGS** The earnings equation

$$\text{Earnings} = 990a - 9000$$

can be used to model the median annual earnings of male high school graduates (see Example 6). Determine the age (to the nearest year) at which the average male high school graduate can expect to earn an annual income of $19,000.

To benefit from an aerobic exercise program, many experts recommend that you exercise three to five times a week for 20 minutes to an hour. It is also important that your heart rate be in the *training zone*, which is defined by the following linear equations, where a is your age in years and the heart rate is in beats per minute.[1]

Maximum exercise heart rate = $0.85(220 - a)$

Minimum exercise heart rate = $0.65(220 - a)$

59. **MAXIMUM EXERCISE HEART RATE** Find the maximum exercise heart rate and the minimum exercise heart rate for a person who is 25 years of age. (Round to the nearest beat per minute.)

[1] "The Heart of the Matter," *American Health*, September 1995.

60. MAXIMUM EXERCISE HEART RATE How old is a person who has a maximum exercise heart rate of 153 beats per minute?

SUPPLEMENTAL EXERCISES

In Exercises 61 to 64, determine whether the given pair of equations are equivalent.

61. $3x - 11 = -5$, $\dfrac{3x - 11}{x - 2} = \dfrac{-5}{x - 2}$

62. $3x - 9 = x - 3$, $\dfrac{3x - 9}{x - 3} = \dfrac{x - 3}{x - 3}$

63. $\dfrac{1}{t} = \dfrac{1}{a} + \dfrac{1}{b}$, $t = \dfrac{ab}{a + b}$, where t is a variable and a and b are nonzero constants, $a \neq -b$.

64. $\dfrac{2}{x} = \dfrac{1}{x - 1}$, $2(x - 1) = x$

65. Let a, b, and c be real constants. Show that an equation of the form $ax + b = c$ has $x = \dfrac{c - b}{a}$ $(a \neq 0)$ as its solution.

66. Let a, b, c, and d be real constants. Show that an equation of the form $ax + b = cx + d$ has $x = \dfrac{d - b}{a - c}$ $(a - c \neq 0)$ as its solution.

In Exercises 67 to 72, solve each equation for x.

67. $\sqrt{7}x - 3 = 7$

68. $\sqrt{8}x + 2 = 14$

69. $\sqrt{3}x - 5 = \sqrt{27}x + 2$

70. $\sqrt{20}x + 14 = \sqrt{5}x - 8$

71. $a^2x - b = b^2x + a$ (assume $a \neq \pm b$)

72. $a^3x - a^2 + ab = b^2 - b^3x$ (assume $a \neq -b$)

PROJECTS

1. PERFECT GAMES In baseball, a **perfect game** is a game in which one of the teams gives up no hits, no walks, and no errors. Statistics show that a batter will get on base roughly 30% of the time. Thus the probability that a pitcher will retire the batter is 70%, or 0.7 as a decimal. The probability that a pitcher will retire two batters in a row is $0.7^2 = 0.49$. The probability is 0.7^{27} that a pitcher will retire 27 batters in succession and thus pitch a perfect game.[2]

a. Explain why the linear equation

$$p = 2(0.7^{27})x$$

provides a good estimate of the number of perfect games p we can expect after x games are completed.

b. Check a major league baseball almanac to determine how many perfect games have been played in the last 40 years and how many games have been played in the last 40 years.

c. Use the linear equation in **a.** to estimate how many perfect games we should expect to have been pitched over the last 40 years of major league baseball. How does this result compare with the actual result found in **b.**?

[2] *A Mathematician Reads the Newspaper* by John Allen Paulos, New York: BasicBooks, A Division of HarperCollins Publishers, Inc., 1995.

SECTION

2.2 FORMULAS AND APPLICATIONS

A **formula** is an equation that expresses known relationships between two or more variables. Table 2.1 lists several formulas from geometry that are used in this text. The variable P represents perimeter, C represents circumference of a circle, A represents area, S represents surface area of an enclosed solid, and V represents volume.

Table 2.1	Formulas from Geometry			
Rectangle	Square	Triangle	Circle	Parallelogram
$P = 2l + 2w$ $A = lw$	$P = 4s$ $A = s^2$	$P = a + b + c$ $A = \frac{1}{2}bh$	$C = \pi d = 2\pi r$ $A = \pi r^2$	$P = 2b + 2s$ $A = bh$
Rectangular Solid	Right Circular Cone	Sphere	Right Circular Cylinder	Frustum of a Cone
$S = 2(wh + lw + hl)$ $V = lwh$	$S = \pi r\sqrt{r^2 + h^2} + \pi r^2$ $V = \frac{1}{3}\pi r^2 h$	$S = 4\pi r^2$ $V = \frac{4}{3}\pi r^3$	$S = 2\pi rh + 2\pi r^2$ $V = \pi r^2 h$	$S = \pi(R + r)\sqrt{h^2 + (R - r)^2} + \pi r^2 + \pi R^2$ $V = \frac{1}{3}\pi h(r^2 + rh + R^2)$

It is often necessary to solve a formula for a specified variable. Begin the process by isolating all terms that contain the specified variable on one side of the equation and all terms that do not contain the specified variable on the other side.

TAKE NOTE

- In Example 1a, the solution $l = \dfrac{P - 2w}{2}$ can also be written as $l = \dfrac{P}{2} - w$.
- In Example 1b, the restriction $x - z \neq 0$ is necessary to ensure that each side is divided by a *nonzero* expression.

EXAMPLE 1 *Solve a Formula for a Specified Variable*

a. Solve $2l + 2w = P$ for l.   **b.** Solve $xy - z = yz$ for y.

Solution

a. $2l + 2w = P$

$\quad\quad 2l = P - 2w$ • Subtract $2w$ from each side to isolate the $2l$ term.

$\quad\quad l = \dfrac{P - 2w}{2}$ • Divide each side by 2.

b. To solve for y, first isolate the terms that involve the variable y on the left side of the equation.

$$xy - z = yz$$

$$xy - yz - z = 0$$ • **Subtract yz from each side so that all terms that contain y are on the same side of the equation.**

$$xy - yz = z$$ • **Add z to each side to isolate the terms that contain y.**

$$y(x - z) = z$$ • **Factor y from each term on the left side of the equation.**

$$y = \frac{z}{x - z}$$ • **Divide each side of the equation by x − z, x − z ≠ 0.**

> **TRY EXERCISE 4, EXERCISE SET 2.2**

APPLICATIONS

People with good problem-solving skills generally work application problems by applying specific techniques in a series of small steps.

> **GUIDELINES FOR SOLVING APPLICATION PROBLEMS**
>
> **1.** Read the problem carefully. If necessary, reread the problem several times.
>
> **2.** When appropriate, draw a sketch and label parts of the drawing with the specific information given in the problem.
>
> **3.** Determine the unknown quantities, and label them with variables. Write down any equation that relates the variables.
>
> **4.** Use the information from step 3, along with a known formula or some additional information given in the problem, to write an equation.
>
> **5.** Solve the equation obtained in step 4, and check to see whether these results satisfy all the conditions of the original problem.

EXAMPLE 2 *Solve an Application*

The length of a rectangular garden is 2 feet longer than three times its width. If the perimeter of the garden is 92 feet, find the width and the length of the garden.

Continued •➤

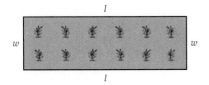

Figure 2.2

Solution

1. Read the problem carefully.

2. Draw a rectangle as shown in **Figure 2.2.**

3. Label the length of the rectangle l and the width of the rectangle w. The problem states that the length l is 2 feet greater than three times the width w. Thus l and w are related by the equation

$$l = 3w + 2$$

4. Because the problem involves the length, width, and perimeter of a rectangle, we use the geometric formula $2l + 2w = P$. To write an equation that involves only constants and a single variable (say, w), substitute 92 for P and $3w + 2$ for l.

$$2l + 2w = P$$
$$2(3w + 2) + 2w = 92$$

5. Solve for the unknown w.

$$6w + 4 + 2w = 92$$
$$8w + 4 = 92$$
$$8w = 88$$
$$w = 11$$

Because the length l is two more than three times the width,

$$l = 3(11) + 2 = 35$$

A check verifies that 35 is two more than three times 11. Also, twice the length (70) plus twice the width (22) gives the perimeter (92). The width of the rectangle is 11 feet, and its length is 35 feet.

> **TRY EXERCISE 22, EXERCISE SET 2.2**

Many *uniform motion* problems can be solved by using the formula $d = rt$, where d is the distance traveled, r is the rate of speed, and t is the time.

> **EXAMPLE 3** *Solve a Uniform Motion Problem*

A runner runs a course at a constant speed of 6 mph. One hour after the runner begins, a cyclist starts on the same course at a constant speed of 15 mph. How long after the runner starts does the cyclist overtake the runner?

Solution

If we represent the time the runner has spent on the course by t, then the time the cyclist takes to overtake the runner is $t - 1$. The following table organizes the information and helps us determine how to write the distances each person travels.

	rate r	$\cdot$	time t	$=$	distance d
Runner	6	$\cdot$	t	$=$	$6t$
Cyclist	15	$\cdot$	$t-1$	$=$	$15(t-1)$

Figure 2.3 indicates that the runner and the cyclist cover the same distance. Thus

$$6t = 15(t-1)$$
$$6t = 15t - 15$$
$$-9t = -15$$
$$t = 1\frac{2}{3}$$

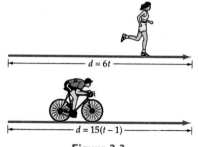

$d = 6t$

$d = 15(t-1)$

Figure 2.3

A check will verify that the cyclist does overtake the runner $1\frac{2}{3}$ hours after the runner starts.

TRY EXERCISE 30, EXERCISE SET 2.2

Many business applications can be solved by using the equation

Profit = revenue − cost

EXAMPLE 4   *Solve a Business Application*

It costs a tennis shoe manufacturer $26.55 to produce a pair of tennis shoes that sells for $49.95. How many pairs of tennis shoes must the manufacturer sell to make a profit of $14,274.00?

Solution

The *profit* is equal to the *revenue* minus the *cost*. If x equals the number of pairs of tennis shoes to be sold, then the revenue will be $49.95x$ and the cost will be $26.55x$. Therefore,

$$\text{Profit} = \text{revenue} - \text{cost}$$
$$14{,}274.00 = 49.95x - 26.55x$$
$$14{,}274.00 = 23.40x$$
$$610 = x$$

The manufacturer must sell 610 pairs of tennis shoes to make the desired profit.

TRY EXERCISE 36, EXERCISE SET 2.2

Simple interest problems can be solved by using the formula $I = Prt$, where I is the interest, P is the principal, r is the simple interest rate per period, and t is the number of periods.

EXAMPLE 5 *Solve a Simple Interest Problem*

An accountant invests part of a $6000 bonus in a 5% simple interest account and the remainder of the money at 8.5% simple interest. Together the investments earn $370 per year. Find the amount invested at each rate.

Solution

Let x be the amount invested at 5%. The remainder of the money is $6000 − x$, which will be the amount invested at 8.5%. Using $I = Prt$, with $t = 1$ year, yields

$$\text{Interest at 5\%} = x \cdot 0.05 = 0.05x$$

$$\text{Interest at 8.5\%} = (6000 − x) \cdot (0.085) = 510 − 0.085x$$

The interest earned on the two accounts equals $370.

$$0.05x + (510 − 0.085x) = 370$$
$$−0.035x + 510 = 370$$
$$−0.035x = −140$$
$$x = 4000$$

Therefore, the accountant invested $4000 at 5% and the remaining $2000 at 8.5%. Check as before.

TRY EXERCISE 40, EXERCISE SET 2.2

Percent mixture problems involve combining solutions or alloys that have different concentrations of a common substance. Percent mixture problems can be solved by using the formula $pA = Q$, where p is the percent of concentration, A is the amount of the solution or alloy, and Q is the quantity of a substance in the solution or alloy. For example, in 4 liters of a 25% acid solution, p is the percent of acid (25%), A is the amount of solution (4 liters), and Q is the amount of acid in the solution, which equals $(0.25) \cdot (4)$ liters $= 1$ liter.

EXAMPLE 6 *Solve a Percent Mixture Problem*

A chemist mixes an 11% hydrochloric acid with a 6% hydrochloric acid solution. How many milliliters (ml) of each solution should the chemist use to make a 600-milliliter solution that is 8% hydrochloric acid?

Solution

Let x be the number of milliliters of the 11% solution. Because the final solution will have a total of 600 milliliters of fluid, $600 − x$ is the number of milliliters of the 6% solution. See **Figure 2.4.**

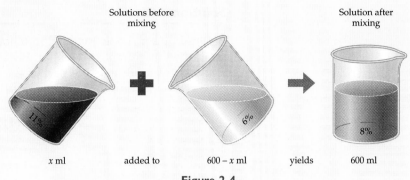

Figure 2.4

Because all the hydrochloric acid in the final solution comes from either the 11% solution or the 6% solution, the number of milliliters of hydrochloric acid in the 11% solution added to the number of milliliters of hydrochloric acid in the 6% solution must equal the number of milliliters of hydrochloric acid in the 8% solution.

$$\begin{pmatrix} \text{ml of acid in} \\ \text{11\% solution} \end{pmatrix} + \begin{pmatrix} \text{ml of acid in} \\ \text{6\% solution} \end{pmatrix} = \begin{pmatrix} \text{ml of acid in} \\ \text{8\% solution} \end{pmatrix}$$

$$0.11x + 0.06(600 - x) = 0.08(600)$$

$$0.11x + 36 - 0.06x = 48$$

$$0.05x + 36 = 48$$

$$0.05x = 12$$

$$x = 240$$

Therefore, the chemist should use 240 milliliters of the 11% solution and 360 milliliters of the 6% solution to make a 600-milliliter solution that is 8% hydrochloric acid.

TRY EXERCISE 44, EXERCISE SET 2.2

To solve a *work problem*, use the equation

Rate of work × time worked = part of task completed

For example, if a painter can paint a wall in 15 minutes, then the painter can paint 1/15 of the wall in 1 minute. The painter's *rate of work* is 1/15 of the wall each minute. In general, if a task can be completed in x minutes, then the rate of work is $1/x$ of the task each minute.

EXAMPLE 7 *Solve a Work Problem*

Pump A can fill a pool in 6 hours and pump B can fill the same pool in 3 hours. How long will it take to fill the pool if both pumps are used?

Continued •➤

Solution

Because pump A fills the pool in 6 hours, 1/6 represents the part of the pool filled by pump A in 1 hour. Because pump B fills the pool in 3 hours, 1/3 represents the part of the pool filled by pump B in 1 hour.
   Let $t = $ the number of hours to fill the pool together. Then

$$t \cdot \frac{1}{6} = \frac{t}{6} \qquad \bullet \text{ Part of the pool filled by pump } A$$

$$t \cdot \frac{1}{3} = \frac{t}{3} \qquad \bullet \text{ Part of the pool filled by pump } B$$

$$\begin{pmatrix} \text{Part filled} \\ \text{by pump } A \end{pmatrix} + \begin{pmatrix} \text{Part filled} \\ \text{by pump } B \end{pmatrix} = \begin{pmatrix} 1 \text{ filled} \\ \text{pool} \end{pmatrix}$$

$$\frac{t}{6} \quad + \quad \frac{t}{3} \quad = \quad 1$$

Multiplying each side of the equation by 6 produces

$$t + 2t = 6$$
$$3t = 6$$
$$t = 2$$

Check: Pump A fills 2/6, or 1/3, of the pool in 2 hours and pump B fills 2/3 of the pool in 2 hours, so 2 hours is the time required to fill the pool if both pumps are used.

TRY EXERCISE 54, EXERCISE SET 2.2

TOPICS FOR DISCUSSION

1. A student solves the formula $A = P + Prt$ for the variable P. The student's answer is $P = A - Prt$. Is this a correct response? Explain.

2. A student takes reciprocals of each term to write the formula

$$\frac{1}{f} = \frac{1}{d_0} + \frac{1}{d_i}$$

   as $f = d_0 + d_i$. Did this technique produce a valid formula? Explain.

3. In the formula $S = a_1/(1 - r)$, what restrictions are placed on the variable r?

4. A tutor states that the formula $A = \frac{1}{2}bh$ can also be expressed as $A = \frac{bh}{2}$. Do you agree?

5. A tutor claims that a runner who runs the length of a track at 8 yards per second and then jogs back at 2 yards per second will take the same amount of time as a second runner who runs the length of the same track and back at a rate of 5 yards per second. Do you agree?

EXERCISE SET 2.2

In Exercises 1 to 18, solve the formula for the specified variable.

1. $V = \dfrac{1}{3}\pi r^2 h$; h (geometry)

2. $P = S - Sdt$; t (business)

3. $I = Prt$; t (business)

4. $A = P + Prt$; P (business)

5. $F = \dfrac{Gm_1 m_2}{d^2}$; m_1 (physics)

6. $A = \dfrac{1}{2}h(b_1 + b_2)$; b_1 (geometry)

7. $s = v_0 t - 16t^2$; v_0 (physics)

8. $\dfrac{1}{f} = \dfrac{1}{d_o} + \dfrac{1}{d_i}$; f (astronomy)

9. $Q_w = m_w c_w (T_f - T_w)$; T_w (physics)

10. $T\Delta t = Iw_f - Iw_i$; I (physics)

11. $a_n = a_1 + (n - 1)d$; d (mathematics)

12. $y - y_1 = m(x - x_1)$; x (mathematics)

13. $S = \dfrac{a_1}{1 - r}$; r (mathematics)

14. $\dfrac{P_1 V_1}{T_1} = \dfrac{P_2 V_2}{T_2}$; V_2 (chemistry)

15. $\dfrac{w_1}{w_2} = \dfrac{f_2 - f}{f - f_1}$; f_1 (hydrostatics)

16. $v = \dfrac{v_1 + v_2}{1 + \dfrac{v_1 v_2}{c^2}}$; v_1 (physics)

17. $f_{LC} = f_v \dfrac{v + v_{LC}}{v}$; v_{LC} (physics)

18. $F_1 d_1 + F_2 d_2 = F_3 d_3 + F_4 d_4$; F_3 (physics)

In Exercises 19 to 58, solve by using the Guidelines for Solving Application Problems (see page 79).

19. One-fifth of a number plus one-fourth of the number is five less than one-half the number. What is the number?

20. The numerator of a fraction is 4 less than the denominator. If the numerator is increased by 14 and the denominator is decreased by 10, the resulting number is 5. What is the original fraction?

21. GEOMETRY The length of a rectangle is 3 feet less than twice the width of the rectangle. If the perimeter of the rectangle is 174 feet, find the width and the length.

22. GEOMETRY The width of a rectangle is 1 meter more than half the length of the rectangle. If the perimeter of the rectangle is 110 meters, find the width and the length.

23. GEOMETRY A triangle has a perimeter of 84 centimeters. Each of the two longer sides of the triangle is three times as long as the shortest side. Find the length of each side of the triangle.

24. GEOMETRY A triangle has a perimeter of 161 miles. Each of the two smaller sides of the triangle is two-thirds the length of the longest side. Find the length of each side of the triangle.

25. CONSECUTIVE NATURAL NUMBERS Find two consecutive natural numbers whose sum is 1745.

26. CONSECUTIVE ODD INTEGERS Find three consecutive odd integers whose sum is 2001.

27. CONSECUTIVE EVEN INTEGERS The difference of the squares of two consecutive positive even integers is 76. Find the integers.

28. CONSECUTIVE INTEGERS The product of two consecutive integers is 90 less than the product of the next two integers. Find the four integers.

29. UNIFORM MOTION Running at an average rate of 6 meters per second, a sprinter ran to the end of a track and then jogged back to the starting point at an average rate of 2 meters per second. The total time for the sprint and the jog back was 2 minutes 40 seconds. Find the length of the track.

30. UNIFORM MOTION A motorboat left a harbor and traveled to an island at an average rate of 15 knots. The average speed on the return trip was 10 knots. If the total trip took 7.5 hours, how far is the harbor from the island?

31. UNIFORM MOTION A plane leaves an airport traveling at an average speed of 240 kilometers per hour. How long will it take a second plane traveling the same route at an average speed of 600 kilometers per hour to catch up with the first plane if it leaves 3 hours later?

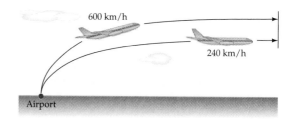

32. UNIFORM MOTION A plane leaves Chicago headed for Los Angeles at 540 mph. One hour later, a second plane leaves Los Angeles headed for Chicago at 660 mph. If the air route from Chicago to Los Angeles is 1800 miles, how long will it take for the planes to pass by each other? How far from Chicago will they be at that time?

33. FINDING AN AVERAGE A student has test scores of 80, 82, 94, and 71. What score does the student need on the next test to produce an average score of 85?

34. FINDING AN AVERAGE A student has test scores of 90, 74, 82, and 90. The next examination is the final examination, which will count as two tests. What score does the student need on the final examination to produce an average score of 85?

35. BUSINESS It costs a manufacturer of sunglasses $8.95 to produce sunglasses that sell for $29.99. How many sunglasses must the manufacturer sell to make a profit of $17,884?

36. BUSINESS It costs a restaurant owner 18 cents per glass for orange juice, which is sold for 75 cents per glass. How many glasses of orange juice must the restaurant owner sell to make a profit of $2337?

37. BUSINESS The price of a computer fell 20% this year. If the computer now costs $750, how much did it cost last year?

38. BUSINESS The price of a magazine subscription rose 4% this year. If the subscription now costs $26, how much did it cost last year?

39. INVESTMENT An investment adviser invested $14,000 in two accounts. One investment earned 8% annual simple interest, and the other investment earned 6.5%

annual simple interest. The amount of interest earned for 1 year was $1024. How much was invested in each account?

40. INVESTMENT A total of $7500 is deposited into two simple interest accounts. On one account the annual simple interest rate is 5%, and on the second account the annual simple interest rate is 7%. The amount of interest earned for 1 year was $405. How much was invested in each account?

41. INVESTMENT An investment of $2500 is made at an annual simple interest rate of 5.5%. How much additional money must be invested at an annual simple interest rate of 8% so that the total interest earned is 7% of the total investment?

42. INVESTMENT An investment of $4600 is made at an annual simple interest rate of 6.8%. How much additional money must be invested at an annual simple interest rate of 9% so that the total interest earned is 8% of the total investment?

43. METALLURGY How many grams of pure silver must a silversmith mix with a 45% silver alloy to produce 200 grams of a 50% alloy?

44. CHEMISTRY How many liters of a 40% sulfuric acid solution should be mixed with 4 liters of a 24% sulfuric acid solution to produce a 30% solution?

45. NURSING How many liters of water should be evaporated from 160 liters of a 12% saline solution so that the solution that remains is a 20% saline solution?

46. AUTOMOTIVE A radiator contains 6 liters of a 25% antifreeze solution. How much should be drained and replaced with pure antifreeze to produce a 33% antifreeze solution?

47. COMMERCE A ballet performance brought in $61,800 on the sale of 3000 tickets. If the tickets sold for $14 and $25, how many of each were sold?

48. COMMERCE A vending machine contains $41.25. The machine contains 255 coins, which consist only of nickels, dimes, and quarters. If the machine contains twice as many dimes as nickels, how many of each type of coin does the machine contain?

49. COMMERCE A coffee shop decides to blend a coffee that sells for $12 per pound with a coffee that sells for $9 per pound to produce a blend that will sell for $10 per pound. How much of each should be used to yield 20 pounds of the new blend?

50. DETERMINE NUMBER OF COINS A bag contains 42 coins, with a total weight of 246 grams. If the bag contains only gold coins that weigh 8 grams each and silver coins that weigh 5 grams each, how many gold and how many silver coins are in the bag?

51. METALLURGY How much pure gold should be melted with 15 grams of 14-karat gold to produce 18-karat gold? *Hint:* A karat is a measure of the purity of gold in an alloy. Pure gold measures 24 karats. An alloy that measures x karats is $x/24$ gold. For example, 18-karat gold is $18/24 = 3/4$ gold.

52. METALLURGY How much 14-karat gold should be melted with 4 ounces of pure gold to produce 18-karat gold? (*Hint:* See Exercise 51.)

53. INSTALL ELECTRICAL WIRES An electrician can install the electric wires in a house in 14 hours. A second electrician requires 18 hours. How long would it take both electricians, working together, to install the wires?

54. PRINT A REPORT Printer A can print a report in 3 hours. Printer B can print the same report in 4 hours. How long would it take both printers, working together, to print the report?

55. BUILD A FENCE A worker can build a fence in 8 hours. With the help of an assistant, the fence can be built in 5 hours. How long would it take the assistant to build the fence alone?

56. REPAIR A ROOF A roofer and an assistant can repair a roof together in 6 hours. The assistant can complete the repair alone in 14 hours. If both the roofer and the assistant work together for 2 hours and then the assistant is left alone to finish the job, how much longer will the assistant need to finish the repairs?

57. DETERMINE INDIVIDUAL PRICES A book and a bookmark together sell for $10.10. If the price of the book is $10.00 more than the price of the bookmark, find the price of the book and the price of the bookmark.

58. SHARE AN EXPENSE Three people decide to share the cost of a yacht. By bringing in an additional partner, they can reduce the cost for each by $4000. What is the total cost of the yacht?

SUPPLEMENTAL EXERCISES

The *Archimedean law of the lever* **states that for a lever to be in a state of balance with respect to a point called the fulcrum, the sum of the downward forces times their respective distances from the fulcrum on one side of the fulcrum must equal the sum of the downward forces times their respective distances from the fulcrum on the other side of the fulcrum. The accompanying figure shows this relationship.**

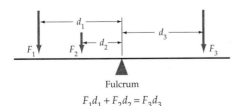

Fulcrum

$$F_1 d_1 + F_2 d_2 = F_3 d_3$$

59. LOCATE THE FULCRUM A 100-pound person 8 feet from the fulcrum and a 40-pound person 5 feet from the fulcrum balance with a 160-pound person on a teeter-totter. How far from the fulcrum is the 160-pound person?

60. LOCATE THE FULCRUM A lever 21 feet long has a force of 117 pounds applied to one end of the lever and a force of 156 pounds applied to the other end. Where should the fulcrum be located to produce a state of balance?

61. DETERMINE A FORCE How much force applied 5 feet from the fulcrum is needed to lift a 400-pound weight that is located on the other side, 0.5 foot from the fulcrum?

62. DETERMINE A FORCE Two workers need to lift a 1440-pound rock. They use a 6-foot steel bar with the fulcrum 1 foot from the rock, as the accompanying figure shows. One worker applies 180 pounds of force to the other end of the lever. How much force will the second worker need to apply 1 foot from that end to lift the rock?

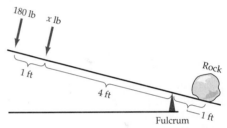

63. SPEED OF SOUND IN AIR Two seconds after firing a rifle at a target, the shooter hears the impact of the bullet. Sound travels at 1100 feet per second and the bullet at 1865 feet per second. Determine the distance to the target (to the nearest foot).

64. SPEED OF SOUND IN WATER Sound travels through sea water 4.62 times faster than through air. The sound of an exploding mine on the surface of the water and partially submerged reaches a ship through the water 4 seconds before it reaches the ship through the air. How far is the ship from the explosion? Use 1100 feet per second as the speed of sound through the air.

65. AGE OF DIOPHANTUS The work of the ancient Greek mathematician Diophantus had great influence on later European number theorists. Nothing is known about his personal life except for the information given in the following epigram. "Diophantus passed 1/6 of his life in childhood, 1/12 in youth, and 1/7 more as a bachelor. Five years after his marriage was born a son who died four years before his father, at 1/2 his father's (final) age." How old was Diophantus when he died?

66. EQUIVALENT TEMPERATURES The relationship between the Fahrenheit temperature (F) and the Celsius temperature (C) is given by the formula

$$F = \frac{9}{5}C + 32$$

At what temperature will a Fahrenheit thermometer and a Celsius thermometer read the same?

PROJECTS

1. **A WORK PROBLEM AND ITS EXTENSIONS** If a pump can fill a pool in A hours and a second pump can fill the same pool in B hours, then the total time T in hours to fill the pool with both pumps working together is given by

$$T = \frac{AB}{A + B}$$

   a. Verify this formula for T.

   b. Consider the case where a pool is to be filled by three pumps. One can fill the pool in A hours, a second in B hours, and a third in C hours. Derive a formula in terms of A, B, and C for the total time T needed to fill the pool.

   c. Consider the case where a pool is to be filled by n pumps. One pump can fill the pool in A_1 hours, a second in A_2 hours, a third in A_3 hours, ..., and the nth pump can fill the pool in A_n hours. Write a formula in terms of $A_1, A_2, A_3, \ldots, A_n$ for the total time T needed to fill the pool.

   The chart at the right is called an *alignment chart* or a *nomogram*. If you know any two of the values A, B, and T, then you can use the alignment chart to determine the unknown value. For example, the straight line segment that connects 3 on the A-axis with 6 on the B-axis crosses the T-axis at 2. Thus the total time required for a pump that takes 3 hours to fill the pool and a pump that takes 6 hours to fill the pool is 2 hours when they work together.

   d. Consider the case where a pool is to be filled by three pumps. One can fill the pool in $A = 6$ hours, a second in $B = 8$ hours, and a third in $C = 12$ hours. Write a few sentences explaining how you could make use of the alignment chart at the right to show that it takes about 2.7 hours for the three pumps to fill the pool when they work together.

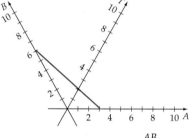

Alignment Chart for $T = \dfrac{AB}{A + B}$

2. **RESISTANCE OF PARALLEL CIRCUITS** The alignment chart shown in Project 1 can be used to solve some problems in electronics that concern the total resistance of a *parallel* circuit. Read an electronics text, and write a paragraph or two that explain this problem and how it is related to the problem of filling a pool with two pumps.

2.3 QUADRATIC EQUATIONS

A **quadratic equation** in x is an equation that can be written in the **standard quadratic form** $ax^2 + bx + c = 0, a \neq 0$.

Several methods can be used to solve quadratic equations. If the quadratic polynomial $ax^2 + bx + c$ can be factored over the integers, then the equation can be solved by factoring and using the **zero product property.**

ZERO PRODUCT PROPERTY

If A and B are algebraic expressions, then

$$AB = 0 \quad \text{if and only if} \quad A = 0 \text{ or } B = 0$$

This property states that when the product of two factors equals zero, then at least one of the factors is zero.

EXAMPLE 1   *Solve by Using the Zero Product Property*

Solve each quadratic equation.

a. $3x^2 + 10x = 8$ **b.** $x^2 + 10x + 25 = 0$

Solution

a.
$$3x^2 + 10x = 8$$
$$3x^2 + 10x - 8 = 0$$ • **Write in standard quadratic form.**

$$(3x - 2)(x + 4) = 0$$ • **Factor.**

$3x - 2 = 0$ or $x + 4 = 0$ • **Apply the zero product property.**

$3x = 2$ or $x = -4$

$x = \dfrac{2}{3}$ or $x = -4$ • **Check as before.**

The solutions of $3x^2 + 10x - 8 = 0$ are -4 and $\dfrac{2}{3}$.

b.
$$x^2 + 10x + 25 = 0$$
$$(x + 5)^2 = 0$$ • **Factor.**

$x + 5 = 0$ or $x + 5 = 0$ • **Apply the zero product property.**

$x = -5$ or $x = -5$ • **Check as before.**

The only solution of $x^2 + 10x + 25 = 0$ is -5.

TRY EXERCISE 4, EXERCISE SET 2.3

In Example 1b, the solution or root -5 is called a **double root** of the equation because the application of the zero product property produced the two identical equations $x + 5 = 0$, both of which have a root of -5.

SOLVING QUADRATIC EQUATIONS BY TAKING SQUARE ROOTS

The quadratic equation $x^2 = c$ can be solved by factoring and applying the zero product property to yield the roots $\sqrt{c}$ and $-\sqrt{c}$.

$$x^2 = c$$
$$x^2 - c = 0$$
$$(x + \sqrt{c})(x - \sqrt{c}) = 0$$
$$x + \sqrt{c} = 0 \qquad \text{or} \qquad x - \sqrt{c} = 0$$
$$x = -\sqrt{c} \qquad \text{or} \qquad x = \sqrt{c}$$

This result is known as the square root theorem, which we will use to solve quadratic equations that can be written in the form $A^2 = B$.

> **THE SQUARE ROOT THEOREM**
>
> If A and B are algebraic expressions such that
> $$A^2 = B, \quad \text{then} \quad A = \pm\sqrt{B}$$

EXAMPLE 2 *Solve by Using the Square Root Theorem*

Use the square root theorem to solve each quadratic equation.

a. $(x + 1)^2 = 49$ **b.** $(x - 3)^2 = -28$

Solution

a. $(x + 1)^2 = 49$

$x + 1 = \pm\sqrt{49}$ • **Apply the square root theorem.**

$x + 1 = \pm 7$ • **Solve for x.**

$x = -1 \pm 7$

Thus $x = -1 - 7 = -8$ or $x = -1 + 7 = 6$.

The solutions of $(x + 1)^2 = 49$ are -8 and 6.

b. $(x - 3)^2 = -28$

$x - 3 = \pm\sqrt{-28}$ • **Apply the square root theorem.**

$x - 3 = \pm 2i\sqrt{7}$ • **Solve for x.**

$x = 3 \pm 2i\sqrt{7}$

The solutions of $(x - 3)^2 = -28$ are $3 - 2i\sqrt{7}$ and $3 + 2i\sqrt{7}$.

TRY EXERCISE 18, EXERCISE SET 2.3

SOLVING QUADRATIC EQUATIONS BY COMPLETING THE SQUARE

Consider the following binomial squares and their perfect-square trinomial products.

Square of a Binomial		Perfect-Square Trinomial
$(x + 6)^2$	$=$	$x^2 + 12x + 36$
$(x - 3)^2$	$=$	$x^2 - 6x + 9$

In each perfect-square trinomial, the coefficient of x^2 is 1, and the constant term of the perfect-square trinomial is the square of half the coefficient of its x term.

$$x^2 + 12x + 36, \quad \left(\frac{1}{2} \cdot 12\right)^2 = 36$$

$$x^2 - 6x + 9, \quad \left(\frac{1}{2}(-6)\right)^2 = 9$$

Adding, to a binomial of the form $x^2 + bx$, the constant that makes that binomial a perfect-square trinomial is called **completing the square**. For example, to complete the square of $x^2 + 8x$, add

$$\left(\frac{1}{2} \cdot 8\right)^2 = 16$$

to produce the perfect-square trinomial $x^2 + 8x + 16$.

Completing the square is a powerful method because it can be used to solve any quadratic equation.

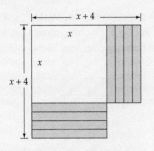
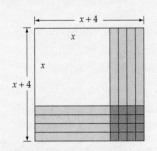
EXAMPLE 3   *Solve by Completing the Square*

Solve: $x^2 = 2x - 6$

Solution

$$
\begin{aligned}
x^2 &= 2x - 6 \\
x^2 - 2x &= -6 \qquad \bullet \text{ Isolate the constant term.} \\
x^2 - 2x + 1 &= -6 + 1 \qquad \bullet \text{ Complete the square.} \\
(x - 1)^2 &= -5 \qquad \bullet \text{ Factor and simplify.} \\
x - 1 &= \pm\sqrt{-5} \qquad \bullet \text{ Apply the square root theorem.} \\
x &= 1 \pm i\sqrt{5} \qquad \bullet \text{ Solve for } x.
\end{aligned}
$$

The solutions are $1 - i\sqrt{5}$ and $1 + i\sqrt{5}$.

TRY EXERCISE 34, EXERCISE SET 2.3

Completing the square by adding the square of half the coefficient of the x term requires that the coefficient of the x^2 term be 1. If the coefficient of the x^2 term is not 1, multiply each term on each side of the equation by the reciprocal of the coefficient of x^2.

EXAMPLE 4   *Solve by Completing the Square*

Solve: $2x^2 + 8x - 15 = 0$

Solution

$$
\begin{aligned}
2x^2 + 8x - 15 &= 0 \\
2x^2 + 8x &= 15 \qquad \bullet \text{ Isolate the constant term.} \\
\frac{1}{2}(2x^2 + 8x) &= \frac{1}{2}(15) \qquad \bullet \text{ Multiply both sides of the equation} \\
& \qquad\qquad\qquad \text{ by the reciprocal of the leading} \\
x^2 + 4x &= \frac{15}{2} \qquad\qquad \text{ coefficient.} \\
x^2 + 4x + 4 &= \frac{15}{2} + 4 \qquad \bullet \text{ Add the square of half the } x \\
& \qquad\qquad\qquad \text{ coefficient to both sides.}
\end{aligned}
$$

Continued • ➤

$$(x + 2)^2 = \frac{23}{2}$$ • Factor and simplify.

$$x + 2 = \pm\sqrt{\frac{23}{2}}$$ • Apply the square root theorem.

$$x = -2 \pm \frac{\sqrt{46}}{2}$$ • Add -2 to each side of the equation, and rationalize the denominator.

$$x = \frac{-4 \pm \sqrt{46}}{2}$$

The solutions are $\dfrac{-4 - \sqrt{46}}{2}$ and $\dfrac{-4 + \sqrt{46}}{2}$.

TRY EXERCISE 38, EXERCISE SET 2.3

SOLVE QUADRATIC EQUATIONS BY USING THE QUADRATIC FORMULA

Completing the square on $ax^2 + bx + c = 0$ ($a \neq 0$) produces a formula for x in terms of the coefficients a, b, and c. The formula is known as the **quadratic formula**, and applying it is another way to solve quadratic equations.

POINT OF INTEREST

There is a general procedure to solve "by radicals" the general cubic

$$ax^3 + bx^2 + cx + d = 0$$

and the general quartic

$$ax^4 + bx^3 + cx^2 + dx + e = 0$$

However, it has been proved that there are no general procedures that can solve "by radicals" general equations of degree 5 or larger.

THE QUADRATIC FORMULA

If $ax^2 + bx + c = 0$, $a \neq 0$, then

$$x = \frac{-b \pm \sqrt{b^2 - 4ac}}{2a}$$

Proof We assume a is a positive real number. If a were a negative real number, then we could multiply each side of the equation by -1 to make it positive.

$$ax^2 + bx + c = 0 \quad (a \neq 0)$$ • Given.

$$ax^2 + bx = -c$$ • Isolate the constant term.

$$x^2 + \frac{b}{a}x = -\frac{c}{a}$$ • Multiply each term on each side of the equation by $1/a$.

$$x^2 + \frac{b}{a}x + \left(\frac{b}{2a}\right)^2 = \left(\frac{b}{2a}\right)^2 - \frac{c}{a}$$ • Complete the square.

$$\left(x + \frac{b}{2a}\right)^2 = \frac{b^2}{4a^2} - \frac{c}{a}$$ • Factor the left side. Simplify the powers on the right side.

$$\left(x + \frac{b}{2a}\right)^2 = \frac{b^2}{4a^2} - \frac{4a}{4a} \cdot \frac{c}{a}$$ • Use a common denominator to simplify the right side.

$$x + \frac{b}{2a} = \pm\sqrt{\frac{b^2 - 4ac}{4a^2}}$$ • Apply the square root theorem.

$$x + \frac{b}{2a} = \pm\frac{\sqrt{b^2 - 4ac}}{2a} \qquad \bullet \textbf{ Because } a > 0, \sqrt{4a^2} = 2a.$$

$$x = -\frac{b}{2a} \pm \frac{\sqrt{b^2 - 4ac}}{2a} \qquad \bullet \textbf{ Add } -\frac{b}{2a} \textbf{ to each side.}$$

$$x = \frac{-b \pm \sqrt{b^2 - 4ac}}{2a} \qquad\qquad\qquad\qquad \blacklozenge$$

As a general rule, you should first try to solve quadratic equations by factoring. If the factoring process proves difficult, then solve either by using the quadratic formula or by completing the square.

EXAMPLE 5   *Solve by Using the Quadratic Formula*

Solve: $4x^2 - 8x + 1 = 0$

Solution

The coefficients are $a = 4$, $b = -8$, and $c = 1$.

$$x = \frac{-(-8) \pm \sqrt{(-8)^2 - 4(4)(1)}}{2(4)} = \frac{8 \pm \sqrt{48}}{8} = \frac{8 \pm 4\sqrt{3}}{8} = \frac{2 \pm \sqrt{3}}{2}$$

The solutions are $\dfrac{2 - \sqrt{3}}{2}$ and $\dfrac{2 + \sqrt{3}}{2}$.

TRY EXERCISE 48, EXERCISE SET 2.3

THE DISCRIMINANT

In the quadratic formula

$$x = \frac{-b \pm \sqrt{b^2 - 4ac}}{2a}$$

the expression $b^2 - 4ac$ is called the **discriminant** of the quadratic formula. If $b^2 - 4ac \geq 0$, then $\sqrt{b^2 - 4ac}$ is a real number; if $b^2 - 4ac < 0$, then $\sqrt{b^2 - 4ac}$ is not a real number. Thus the sign of the discriminant determines whether the roots of a quadratic equation are real numbers or nonreal complex numbers.

THE DISCRIMINANT AND ROOTS OF A QUADRATIC EQUATION

The quadratic equation $ax^2 + bx + c = 0$, with real coefficients and $a \neq 0$, has discriminant $b^2 - 4ac$.

If $b^2 - 4ac > 0$, then the quadratic equation has *two distinct real roots.*

If $b^2 - 4ac = 0$, then the quadratic equation has *a real root* that is a double root.

If $b^2 - 4ac < 0$, then the quadratic equation has *two distinct complex roots* that are not real. These roots are conjugates of each other.

By examining the discriminant, it is possible to determine whether the roots of a quadratic equation are real numbers without actually finding the roots.

EXAMPLE 6   Use the Discriminant to Classify Roots

Classify the roots of each quadratic equation as real numbers or nonreal complex numbers.

a. $2x^2 - 5x + 1 = 0$ **b.** $3x^2 + 6x + 7 = 0$ **c.** $x^2 + 6x + 9 = 0$

Solution

a. $2x^2 - 5x + 1 = 0$ has coefficients $a = 2$, $b = -5$, and $c = 1$.

$$b^2 - 4ac = (-5)^2 - 4(2)(1) = 25 - 8 = 17$$

Because the discriminant 17 is *positive*, $2x^2 - 5x + 1 = 0$ has *two distinct real roots.*

b. $3x^2 + 6x + 7 = 0$ has coefficients $a = 3$, $b = 6$, and $c = 7$.

$$b^2 - 4ac = 6^2 - 4(3)(7) = 36 - 84 = -48$$

Because the discriminant -48 is *negative*, $3x^2 + 6x + 7 = 0$ has *two distinct nonreal complex roots.*

c. $x^2 + 6x + 9 = 0$ has coefficients $a = 1$, $b = 6$, and $c = 9$.

$$b^2 - 4ac = 6^2 - 4(1)(9) = 36 - 36 = 0$$

Because the discriminant is 0, $x^2 + 6x + 9 = 0$ has *a real root.* The root is a double root.

> **TRY EXERCISE 58, EXERCISE SET 2.3**

> **TAKE NOTE**
>
> Some students state, "If the discriminant is negative, then the roots of the quadratic equation are complex numbers." Although this statement is true, it is not precise, because the real numbers are a subset of the complex numbers.

APPLICATIONS

A **right triangle** contains one 90° angle. The side opposite the 90° angle is called the **hypotenuse.** The two other sides are called **legs.** See **Figure 2.5.**

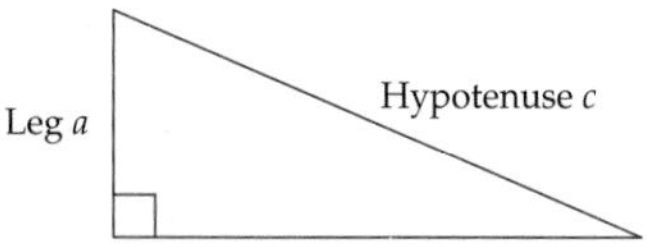

Figure 2.5

THE PYTHAGOREAN THEOREM

If a and b denote the lengths of the legs of a right triangle and c the length of the hypotenuse, then

$$c^2 = a^2 + b^2$$

The Pythagorean Theorem states that the square of the length of the hypotenuse of a right triangle is equal to the sum of the squares of the lengths of the two legs. This theorem is often used to solve applications that involve right triangles.

EXAMPLE 7   *Solve a Construction Application*

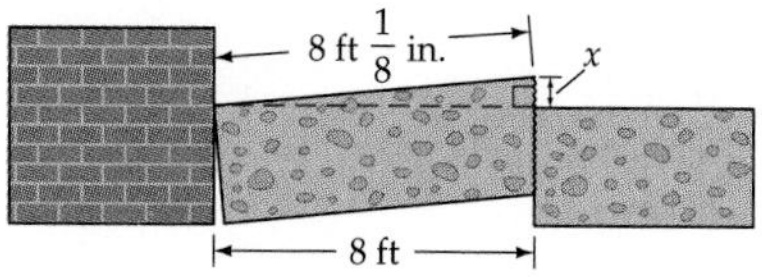

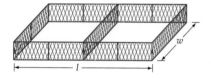

Figure 2.6

Concrete slabs often crack and buckle if proper expansion joints are not installed. Suppose a concrete slab expands as a result of an increase in temperature, as shown in **Figure 2.6**. Determine the height x, to the nearest inch, to which the concrete will rise as a consequence of this expansion.

Solution

Use the Pythagorean Theorem.

$$\left(8 \text{ feet} + \frac{1}{8} \text{ inch}\right)^2 = x^2 + (8 \text{ feet})^2$$

$$(96.125)^2 = x^2 + (96)^2 \qquad \bullet \textbf{ Change units to inches.}$$

$$(96.125)^2 - (96)^2 = x^2$$

$$\sqrt{(96.125)^2 - (96)^2} = x \qquad \bullet \textbf{ Only the positive root is}$$
$$\qquad\qquad\qquad\qquad\qquad\qquad \textbf{taken because } x > 0.$$

$$4.9 \approx x$$

Thus, to the nearest inch, the concrete will rise 5 inches.

TRY EXERCISE 68, EXERCISE SET 2.3

EXAMPLE 8   *Solve a Geometric Application*

Figure 2.7

A veterinarian wishes to use 132 feet of chain-link fencing to enclose a rectangular region and subdivide the region into two smaller rectangles, as shown in **Figure 2.7**. If the total enclosed area is 576 square feet, find the dimensions of the enclosed region.

Solution

Let w be the width of the enclosed region. Then $3w$ represents the amount of fencing used to construct the three widths. The amount of fencing left for the two lengths is $132 - 3w$. Thus each length must be half of the remaining fencing, or $(132 - 3w)/2$.

Now we have variable expressions in w for both the width and the length. Substituting these into the area formula $lw = A$ produces

$$\left(\frac{132 - 3w}{2}\right)w = 576 \qquad \bullet \textbf{ Substitute } \frac{132 - 3w}{2} \textbf{ for } l.$$

$$132w - 3w^2 = 1152 \qquad \bullet \textbf{ Simplify.}$$

$$-3w^2 + 132w - 1152 = 0$$

$$w^2 - 44w + 384 = 0 \qquad \bullet \textbf{ Divide each term by } -3.$$

Continued •➤

Although this quadratic equation can be solved by factoring, the following solution makes use of the quadratic formula.

$$w = \frac{-(-44) \pm \sqrt{(-44)^2 - 4(1)(384)}}{2(1)} \qquad \bullet \text{ Apply the quadratic formula.}$$

$$w = \frac{44 \pm \sqrt{400}}{2} = \frac{44 \pm 20}{2} = 12 \text{ or } 32$$

Thus there are two solutions to the problem:

1. If the width $w = 12$ feet, then the length is $\dfrac{132 - 3(12)}{2} = 48$ feet.

2. If the width $w = 32$ feet, then the length is $\dfrac{132 - 3(32)}{2} = 18$ feet.

TRY EXERCISE 74, EXERCISE SET 2.3

QUESTION   In Example 8, what reason can you give for using the quadratic formula, rather than factoring, to solve $w^2 - 44w + 384 = 0$?

TOPICS FOR DISCUSSION

1. If A and B are algebraic expressions and the product AB equals 0, then $A = 0$ or $B = 0$. Do you agree with this statement? Explain.

2. If A and B are algebraic expressions and the product AB equals 10, then $A = 10$ or $B = 10$. Do you agree with this statement? Explain.

3. Every quadratic equation has two distinct solutions. Do you agree with this statement? Explain.

4. Every quadratic equation $ax^2 + bx + c = 0$ (with real coefficients and $a \neq 0$) can be solved by applying the quadratic formula. Do you agree? Explain.

5. Every quadratic equation of the form $ax^2 + bx = 0$ ($a \neq 0, b \neq 0$) has two distinct real solutions. Do you agree? Explain.

EXERCISE SET 2.3

In Exercises 1 to 16, solve each quadratic equation by factoring and applying the zero product property.

1. $x^2 - 2x - 15 = 0$ **2.** $y^2 + 3y - 10 = 0$

3. $8y^2 + 189y - 72 = 0$ **4.** $12w^2 - 41w + 24 = 0$

5. $3x^2 - 7x = 0$ **6.** $5x^2 = -8x$

7. $8 + 14t - 15t^2 = 0$ **8.** $12 - 26w + 10w^2 = 0$

9. $12 - 21s - 6s^2 = 0$ **10.** $-144 + 320y + 9y^2 = 0$

11. $(x - 5)^2 - 9 = 0$ **12.** $(3x + 4)^2 - 16 = 0$

13. $(2x - 5)^2 - (4x - 11)^2 = 0$

14. $(5x + 3)^2 - (x + 7)^2 = 0$

15. $14x = x^2 + 49$ **16.** $41x = 12x^2 + 35$

ANSWER   Factoring $w^2 - 44w + 384$ may be time-consuming because 384 has several integer factors.

In Exercises 17 to 28, use the square root theorem to solve each quadratic equation.

17. $x^2 = 81$ **18.** $y^2 = 225$

19. $2x^2 = 48$ **20.** $3x^2 = 144$

21. $3x^2 + 12 = 0$ **22.** $4y^2 + 20 = 0$

23. $(x - 5)^2 = 36$ **24.** $(x + 4)^2 = 121$

25. $(x - 8)^2 = (x + 1)^2$ **26.** $(x + 5)^2 = (2x + 1)^2$

27. $x^2 = (x + 1)^2$ **28.** $4x^2 = (2x + 3)^2$

In Exercises 29 to 42, solve by completing the square.

29. $x^2 + 6x + 1 = 0$ **30.** $x^2 + 8x - 10 = 0$

31. $x^2 - 2x - 15 = 0$ **32.** $x^2 + 2x - 8 = 0$

33. $x^2 + 10x = 0$ **34.** $x^2 - 6x = 0$

35. $x^2 + 3x - 1 = 0$ **36.** $x^2 + 7x - 2 = 0$

37. $2x^2 + 4x - 1 = 0$ **38.** $2x^2 + 10x - 3 = 0$

39. $3x^2 - 8x + 1 = 0$ **40.** $4x^2 - 4x + 15 = 0$

41. $5 - 6x - 3x^2 = 0$ **42.** $2 + 10x - 5x^2 = 0$

In Exercises 43 to 56, solve by using the quadratic formula.

43. $x^2 - 2x - 15 = 0$ **44.** $x^2 - 5x - 24 = 0$

45. $x^2 + x - 1 = 0$ **46.** $x^2 + x + 1 = 0$

47. $2x^2 + 4x + 1 = 0$ **48.** $2x^2 + 4x - 1 = 0$

49. $3x^2 - 5x + 3 = 0$ **50.** $3x^2 - 5x + 4 = 0$

51. $\dfrac{1}{2}x^2 + \dfrac{3}{4}x - 1 = 0$ **52.** $\dfrac{2}{3}x^2 - 5x + \dfrac{1}{2} = 0$

53. $\sqrt{2}x^2 + 3x + \sqrt{2} = 0$ **54.** $2x^2 + \sqrt{5}x - 3 = 0$

55. $x^2 = 3x - 5$ **56.** $-x^2 = 7x - 1$

In Exercises 57 to 62, determine the discriminant of the quadratic equation, and then classify the roots of the equation as *a.* two distinct real numbers, *b.* one real number (which is a double root), or *c.* two distinct nonreal complex numbers. Do not solve the equations.

57. $2x^2 - 5x - 7 = 0$ **58.** $x^2 + 3x - 11 = 0$

59. $3x^2 - 2x + 10 = 0$ **60.** $x^2 + 3x + 3 = 0$

61. $x^2 - 20x + 100 = 0$ **62.** $4x^2 + 12x + 9 = 0$

In Exercises 63 to 66, find all values of *k* such that each quadratic equation has exactly one real root. *Hint:* The quadratic equation $ax^2 + bx + c = 0$ has exactly one real root if and only if $b^2 - 4ac = 0$.

63. $16x^2 + kx + 9 = 0$ **64.** $x^2 + kx + 81 = 0$

65. $y^2 - 7y + k = 0$ **66.** $x^2 + 15x + k = 0$

67. **Geometry** The length of each side of a square is 54 inches. Find the length of the diagonal of the square. Round to the nearest tenth of an inch.

68. **Construction** A concrete slab cracks and expands as a result of an increase in temperature, as shown in the following figure. Determine the height *x*, to the nearest inch, to which the concrete will rise as a consequence of this expansion.

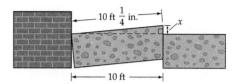

69. **Geometry** The length of each side of an equilateral triangle is 31 centimeters. Find the altitude of the triangle. Round to the nearest tenth of a centimeter.

70. **Baseball** How far, to the nearest foot, is it from home-plate to second base on a baseball diamond? *Hint:* The distance between home plate and first base is 90 feet.

71. **Geometry** The perimeter of a rectangle is 27 centimeters and its area is 35 square centimeters. Find the length and the width of the rectangle.

72. **Geometry** The perimeter of a rectangle is 34 feet and its area is 60 square feet. Find the length and the width of the rectangle.

73. **Rectangular Enclosure** A gardener wishes to use 600 feet of fencing to enclose a rectangular region and subdivide the region into two smaller rectangles. The total enclosed area is 15,000 square feet. Find the dimensions of the enclosed region.

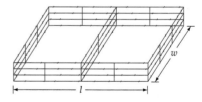

74. **Rectangular Enclosure** A farmer wishes to use 400 yards of fencing to enclose a rectangular region and subdivide the region into three smaller rectangles. The total enclosed area is 4800 square yards. Find the dimensions of the enclosed region.

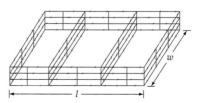

75. CONSECUTIVE EVEN INTEGERS The sum of the squares of two consecutive positive even integers is 244. Find the numbers.

76. CONSECUTIVE INTEGERS The sum of the squares of three consecutive integers is 302. Find the numbers.

77. RECIPROCALS Find a positive real number that is 5 larger than its reciprocal.

78. RECIPROCALS Find a positive real number that is 2 smaller than its reciprocal.

79. AVERAGE RATE A salesperson drove the first 105 miles of a trip in 1 hour more than it took to drive the last 90 miles. The average rate during the last 90 miles was 10 mph faster than the average rate during the first 105 miles. Find the average rate for each portion of the trip.

80. AVERAGE RATE A car and a bus both completed a 240-mile trip. The car averaged 10 mph faster than the bus and completed the trip in 48 minutes less time than the bus. Find the average rate, in miles per hour, of the bus.

81. INDIVIDUAL TIME A mason can build a wall in 6 hours less than an apprentice. Together they can build the wall in 4 hours. How long would it take the apprentice, working alone, to build the wall?

82. INDIVIDUAL TIME Pump A can fill a pool in 2 hours less time than pump B. Together the pumps can fill the pool in 2 hours 24 minutes. Find how long it takes pump A to fill the pool.

83. CLIFF DIVING In feet, the height s above the water of a diver t seconds after diving off a 65-foot cliff is given by

$$s = -16t^2 + 65$$

How long will it take the diver to hit the water?

84. FOOTBALL A football player kicks a football downfield. The height s in feet of the football t seconds after it leaves the kicker's foot is given by

$$s = -16t^2 + 88t + 2$$

Find the "hang time."

SUPPLEMENTAL EXERCISES

The following theorem is known as the *sum and product of the roots theorem*.

Let $ax^2 + bx + c = 0$, $a \neq 0$, be a quadratic equation. Then r_1 and r_2 are roots of the equation if and only if

$$r_1 + r_2 = -\frac{b}{a} \text{ and } r_1 r_2 = \frac{c}{a}.$$

In Exercises 85 to 90, use the sum and product of the roots theorem given in the preceding column to determine whether the given numbers are roots of the quadratic equation.

85. $x^2 - 5x - 24 = 0$; $-3, 8$

86. $x^2 + 4x - 21 = 0$; $-7, 3$

87. $2x^2 - 7x - 30 = 0$; $-5/2, 6$

88. $9x^2 - 12x - 1 = 0$; $(2 + \sqrt{5})/3, (2 - \sqrt{5})/3$

89. $x^2 - 2x + 2 = 0$; $1 + i, 1 - i$

90. $x^2 - 4x + 12 = 0$; $2 + 3i, 2 - 3i$

In Exercises 91 to 98, use the quadratic formula to solve each equation for the indicated variable in terms of the other variables. Assume that none of the denominators is zero.

91. $s = -\dfrac{1}{2}gt^2 + v_0 t + s_0$, for t

92. $S = 2\pi rh + 2\pi r^2$, for r

93. $-xy^2 + 4y + 3 = 0$, for y

94. $D = \dfrac{n}{2}(n - 3)$, for n

95. $3x^2 + xy + 4y^2 = 0$, for x

96. $3x^2 + xy + 4y^2 = 0$, for y

97. $x = y^2 + y - 8$, for y

98. $P = \dfrac{E^2 R}{(r + R)^2}$, for R

99. SUM OF NATURAL NUMBERS The sum S of the first n natural numbers $1, 2, 3, \ldots, n$ is given by the formula

$$S = \frac{n}{2}(n + 1)$$

How many consecutive natural numbers starting with 1 produce a sum of 253?

100. NUMBER OF DIAGONALS The number of diagonals D of a polygon with n sides is given by the formula

$$D = \frac{n}{2}(n - 3)$$

Determine the number of sides of a polygon with 464 diagonals.

PROJECTS

1. **A GOLDEN RECTANGLE** A rectangle is a "golden rectangle" provided its length l and its width w satisfy the equation

$$\frac{l}{w} = \frac{w}{l - w}$$

 a. Solve this formula for w.

 b. If the length l of a golden rectangle measures 101 feet, what is the width of the rectangle?

2. **THE SUM AND PRODUCT OF THE ROOTS THEOREM** Use the quadratic formula to prove the sum and product of the roots theorem stated just before Exercise 85.

3. **ATTRACTORS** Consider the quadratic equation

$$y = 2x(1 - x)$$

 Let x be any number between 0 and 1. Evaluate y for that value of x, substitute that value of y back in for x, and evaluate for the next y value. Continuing this process over and over will produce a sequence of numbers that are *attracted* to 0.5. Verify the above statements for

 a. $x = 0.2$ **b.** $x = 0.713$

 c. Write an essay on *attractors*. Explain how the topic of attractors is related to *chaos*. An excellent source of information on attractors is *The Mathematical Tourist* by Ivars Peterson (New York: Freeman, 1988).

4. **Visual Insight**

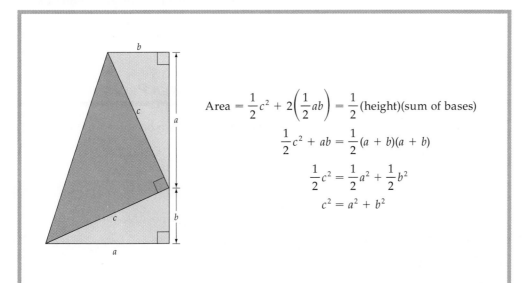

$$\text{Area} = \frac{1}{2}c^2 + 2\left(\frac{1}{2}ab\right) = \frac{1}{2}(\text{height})(\text{sum of bases})$$

$$\frac{1}{2}c^2 + ab = \frac{1}{2}(a + b)(a + b)$$

$$\frac{1}{2}c^2 = \frac{1}{2}a^2 + \frac{1}{2}b^2$$

$$c^2 = a^2 + b^2$$

President James A. Garfield is credited with the above proof of the Pythagorean Theorem. Write the supporting reasons for each of the steps in this proof.

2.4 OTHER TYPES OF EQUATIONS

Some equations that are neither linear nor quadratic can be solved by the various techniques presented in this section. For instance, the **third-degree equation**, or **cubic equation**, in Example 1 can be solved by factoring the polynomial on the left side of the equation and using the zero product property.

TAKE NOTE

If you attempt to solve Example 1 by dividing each side by x, you will produce the equation $x^2 - 16 = 0$, which has roots of only -4 and 4. In this case the division of each side of the equation by the variable x does not produce an equivalent equation. To avoid this common mistake, factor out any variable factors that are common to each term instead of dividing each side of the equation by the factor.

EXAMPLE 1   Solve an Equation by Factoring

Solve: $x^3 - 16x = 0$

Solution

$$x^3 - 16x = 0$$
$$x(x^2 - 16) = 0 \qquad \bullet \text{ Factor out the GCF, } x.$$
$$x(x + 4)(x - 4) = 0 \qquad \bullet \text{ Factor the difference of squares.}$$

Set each factor equal to zero.

$$x = 0 \quad \text{or} \quad x + 4 = 0 \quad \text{or} \quad x - 4 = 0$$
$$x = 0 \quad \text{or} \quad x = -4 \quad \text{or} \quad x = 4$$

A check will show that -4, 0, and 4 are roots of the original equation.

TRY EXERCISE 6, EXERCISE SET 2.4

SOLVING EQUATIONS BY USING THE POWER PRINCIPLE

Some equations that involve radical expressions can be solved by using the following result.

THE POWER PRINCIPLE

If P and Q are algebraic expressions and n is a positive integer, then every solution of $P = Q$ is a solution of $P^n = Q^n$.

EXAMPLE 2   Solve a Radical Equation

Use the power principle to solve $\sqrt{x + 4} = 3$.

Solution

$$\sqrt{x + 4} = 3$$

$$(\sqrt{x + 4})^2 = 3^2 \qquad \bullet \textbf{ Apply the power principle with } n = 2.$$

$$x + 4 = 9$$

$$x = 5$$

Check: $\sqrt{x + 4} = 3$

$$\sqrt{5 + 4} \stackrel{?}{=} 3 \qquad \bullet \textbf{ Substitute 5 for } x.$$

$$\sqrt{9} \stackrel{?}{=} 3$$

$$3 = 3 \qquad \bullet \textbf{ 5 checks.}$$

The only solution is 5.

TRY EXERCISE 14, EXERCISE SET 2.4

Some care must be taken when using the power principle, because the equation $P^n = Q^n$ may have more solutions than the original equation $P = Q$. As an example, consider $x = 3$. The only solution is the real number 3. Square each side of the equation to produce $x^2 = 9$, which has both 3 and -3 as solutions. The -3 is called an *extraneous solution* because it is not a solution of the original equation $x = 3$.

EXTRANEOUS SOLUTIONS

Any solution of $P^n = Q^n$ that is not a solution of $P = Q$ is called an **extraneous solution.** Extraneous solutions *may* be introduced whenever we raise each side of an equation to an *even* power.

EXAMPLE 3   *Solve a Radical Equation*

Solve $x = 2 + \sqrt{2 - x}$. Check all proposed solutions.

Solution

$$x = 2 + \sqrt{2 - x}$$

$$x - 2 = \sqrt{2 - x} \qquad \bullet \textbf{ Isolate the radical.}$$

$$(x - 2)^2 = (\sqrt{2 - x})^2 \qquad \bullet \textbf{ Square each side of the equation.}$$

$$x^2 - 4x + 4 = 2 - x$$

$$x^2 - 3x + 2 = 0 \qquad \bullet \textbf{ Collect and combine like terms.}$$

$$(x - 2)(x - 1) = 0 \qquad \bullet \textbf{ Factor.}$$

$$x - 2 = 0 \quad \text{or} \quad x - 1 = 0$$

$$x = 2 \quad \text{or} \quad x = 1 \qquad \bullet \textbf{ Proposed solutions}$$

Continued •➤

Check for $x = 2$: $x = 2 + \sqrt{2 - x}$

$\qquad 2 \overset{?}{=} 2 + \sqrt{2 - (2)}$ • Substitute 2 for x.

$\qquad 2 \overset{?}{=} 2 + \sqrt{0}$

$\qquad 2 = 2$ • **2 is a solution.**

Check for $x = 1$: $x = 2 + \sqrt{2 - x}$

$\qquad 1 \overset{?}{=} 2 + \sqrt{2 - (1)}$ • Substitute 1 for x.

$\qquad 1 \overset{?}{=} 2 + \sqrt{1}$

$\qquad 1 \neq 3$ • **1 is not a solution.**

The check shows that 1 is not a solution. It is an extraneous solution that we created by squaring each side of the equation. The only solution is 2.

> **TRY EXERCISE 16, EXERCISE SET 2.4**

In Example 4 it will be necessary to square $(1 + \sqrt{2x - 5})$. Recall the special product formula $(x + y)^2 = x^2 + 2xy + y^2$. Using this special product formula to square $(1 + \sqrt{2x - 5})$ produces

$$(1 + \sqrt{2x - 5})^2 = 1 + 2\sqrt{2x - 5} + (2x - 5)$$

> **EXAMPLE 4** *Solve a Radical Equation*

Solve $\sqrt{x + 1} - \sqrt{2x - 5} = 1$. Check all proposed solutions.

Solution

First write an equivalent equation in which one radical is isolated on one side of the equation.

$$\sqrt{x + 1} - \sqrt{2x - 5} = 1$$
$$\sqrt{x + 1} = 1 + \sqrt{2x - 5}$$

The next step is to square each side. Using the result from the discussion preceding this example, we have

$$(\sqrt{x + 1})^2 = (1 + \sqrt{2x - 5})^2$$
$$x + 1 = 1 + 2\sqrt{2x - 5} + (2x - 5)$$
$$-x + 5 = 2\sqrt{2x - 5} \qquad \text{• Isolate the remaining radical.}$$

The right side still contains a radical, so we square each side again.

$$(-x + 5)^2 = (2\sqrt{2x + 5})^2$$
$$x^2 - 10x + 25 = 4(2x - 5)$$
$$x^2 - 10x + 25 = 8x - 20$$
$$x^2 - 18x + 45 = 0$$
$$(x - 3)(x - 15) = 0$$
$$x = 3 \quad \text{or} \quad x = 15 \qquad \text{• Proposed solutions}$$

3 checks as a solution, but 15 does not. Therefore, 3 is the only solution.

> **TRY EXERCISE 20, EXERCISE SET 2.4**

Some equations that involve fractional exponents can be solved by raising each side to a reciprocal power. For example, to solve $x^{1/3} = 4$, raise each side to the third power to find that $x = 64$. Be sure to check all proposed solutions to determine whether they are actual solutions or extraneous solutions.

EXAMPLE 5	*Solve Equations That Involve Fractional Exponents*

Solve: $(x^2 + 4x + 52)^{3/2} = 512$

Solution

Because the equation involves a three-halves power, start by raising each side of the equation to the two-thirds power.

$$[(x^2 + 4x + 52)^{3/2}]^{2/3} = 512^{2/3}$$ • **The reciprocal of 3/2 is 2/3.**

$$x^2 + 4x + 52 = 64$$ • **Think: $512^{2/3} = (\sqrt[3]{512})^2 = 8^2 = 64$**

$$x^2 + 4x - 12 = 0$$ • **Subtract 64 from each side.**

$$(x - 2)(x + 6) = 0$$ • **Factor.**

$$x - 2 = 0 \quad \text{or} \quad x + 6 = 0$$

$$x = 2 \quad \text{or} \quad x = -6$$

A check will verify that 2 and -6 are both solutions of the original equation.

TRY EXERCISE 32, EXERCISE SET 2.4

SOLVING EQUATIONS THAT ARE QUADRATIC IN FORM

The equation $4x^4 - 25x^2 + 36 = 0$ is said to be **quadratic in form,** which means it can be written in the form

$$au^2 + bu + c = 0 \qquad a \neq 0$$

where u is an algebraic expression involving x. For example, if we make the substitution $u = x^2$ (which implies $u^2 = x^4$), then our original equation can be written as

$$4u^2 - 25u + 36 = 0$$

This quadratic equation can be solved for u, and then, using the relationship $u = x^2$, we can find the solutions of the original equation.

EXAMPLE 6	*Solve an Equation That Is Quadratic in Form*

Solve: $4x^4 - 25x^2 + 36 = 0$

Continued •➤

Solution

Make the substitutions $u = x^2$ and $u^2 = x^4$ to produce the quadratic equation $4u^2 - 25u + 36 = 0$. Factor the quadratic polynomial on the left side of the equation.

$$(4u - 9)(u - 4) = 0$$

$$4u - 9 = 0 \qquad \text{or} \qquad u - 4 = 0$$

$$u = \frac{9}{4} \qquad \text{or} \qquad u = 4$$

Substitute x^2 for u to produce

$$x^2 = \frac{9}{4} \qquad \text{or} \qquad x^2 = 4$$

$$x = \pm\sqrt{\frac{9}{4}} \qquad \text{or} \qquad x = \pm\sqrt{4}$$

$$x = \pm\frac{3}{2} \qquad \text{or} \qquad x = \pm 2 \qquad \bullet \textbf{ Check as before.}$$

The solutions are -2, $-\dfrac{3}{2}$, $\dfrac{3}{2}$, and 2.

TRY EXERCISE 42, EXERCISE SET 2.4

Following is a table of equations that are quadratic in form. Each is accompanied by an appropriate substitution that will enable it to be written in the form $au^2 + bu + c = 0$.

Equations That Are Quadratic in Form

Original Equation	Substitution	$au^2 + bu + c = 0$ Form
$x^4 - 8x^2 + 15 = 0$	$u = x^2$	$u^2 - 8u + 15 = 0$
$x^6 + x^3 - 12 = 0$	$u = x^3$	$u^2 + u - 12 = 0$
$x^{1/2} - 9x^{1/4} + 20 = 0$	$u = x^{1/4}$	$u^2 - 9u + 20 = 0$
$2x^{2/3} + 7x^{1/3} - 4 = 0$	$u = x^{1/3}$	$2u^2 + 7u - 4 = 0$
$15x^{-2} + 7x^{-1} - 2 = 0$	$u = x^{-1}$	$15u^2 + 7u - 2 = 0$

EXAMPLE 7   *Solve an Equation That Is Quadratic in Form*

Solve: $3x^{2/3} - 5x^{1/3} - 2 = 0$

Solution

Substituting u for $x^{1/3}$ gives us

$$3u^2 - 5u - 2 = 0$$

$$(3u + 1)(u - 2) = 0 \qquad \bullet \textbf{ Factor.}$$

$$3u + 1 = 0 \qquad \text{or} \qquad u - 2 = 0$$

$$u = -\frac{1}{3} \qquad \text{or} \qquad u = 2$$

$$x^{1/3} = -\frac{1}{3} \qquad \text{or} \qquad x^{1/3} = 2 \qquad \text{• Replace } u \text{ with } x^{1/3}.$$

$$x = -\frac{1}{27} \qquad \text{or} \qquad x = 8 \qquad \text{• Cube each side.}$$

A check will verify that both $-1/27$ and 8 are solutions.

TRY EXERCISE 52, EXERCISE SET 2.4

It is possible to solve equations that are quadratic in form without making a formal substitution. For example, to solve $x^4 + 5x^2 - 36 = 0$, factor the equation and apply the zero product property.

$$x^4 + 5x^2 - 36 = 0$$
$$(x^2 + 9)(x^2 - 4) = 0$$
$$x^2 + 9 = 0 \qquad \text{or} \qquad x^2 - 4 = 0$$
$$x^2 = -9 \qquad \text{or} \qquad x^2 = 4$$
$$x = \pm 3i \qquad \text{or} \qquad x = \pm 2$$

TOPICS FOR DISCUSSION

1. If P and Q are algebraic expressions and n is a positive integer, then the equation $P^n = Q^n$ is equivalent to the equation $P = Q$. Do you agree? Explain.

2. Consider the equation $(x^2 - 1)(x - 2) = 3(x - 2)$. Dividing each side of the equation by $x - 2$ yields $x^2 - 1 = 3$. Is this second equation equivalent to the first equation?

3. A tutor claims that cubing each side of $(4x - 1)^{1/3} = -2$ will not introduce any extraneous solutions. Do you agree?

4. A tutor claims that the equation $x^{-2} - 2/x = 15$ is quadratic in form. Do you agree? If so, what would be an appropriate substitution that would enable you to write the equation as a quadratic?

5. A classmate solves the equation $x^2 + y^2 = 25$ for y and produces the equation $y = \sqrt{25 - x^2}$. Do you agree with this result?

EXERCISE SET 2.4

In Exercises 1 to 10, factor to solve each equation.

1. $x^3 - 25x = 0$

2. $x^3 - x = 0$

3. $x^3 - 2x^2 - x + 2 = 0$

4. $x^3 - 4x^2 - 2x + 8 = 0$

5. $2x^5 - 18x^3 = 0$

6. $x^4 - 36x^2 = 0$

7. $x^4 - 3x^3 - 40x^2 = 0$

8. $x^4 + 3x^3 - 8x - 24 = 0$

9. $x^4 - 16x^2 = 0$

10. $x^4 - 16 = 0$

In Exercises 11 and 12, solve each equation by factoring and by using the quadratic formula.

11. $x^3 - 8 = 0$ **12.** $x^3 + 8 = 0$

In Exercises 13 to 30, use the power principle to solve each radical equation. Check all proposed solutions.

13. $\sqrt{x - 4} - 6 = 0$ **14.** $\sqrt{10 - x} = 4$

15. $x = 3 + \sqrt{3 - x}$ **16.** $x = \sqrt{5 - x} + 5$

17. $\sqrt{3x - 5} - \sqrt{x + 2} = 1$ **18.** $\sqrt{6 - x} + \sqrt{5x + 6} = 6$

19. $\sqrt{2x + 11} - \sqrt{2x - 5} = 2$

20. $\sqrt{x + 7} - 2 = \sqrt{x - 9}$

21. $\sqrt{x + 7} + \sqrt{x - 5} = 6$ **22.** $x = \sqrt{12x - 35}$

23. $2x = \sqrt{4x + 15}$ **24.** $\sqrt[3]{7x - 3} = \sqrt[3]{2x + 7}$

25. $\sqrt[3]{2x^2 + 5x - 3} = \sqrt[3]{x^2 + 3}$

26. $\sqrt[4]{x^2 + 20} = \sqrt[4]{9x}$

27. $\sqrt{3\sqrt{5x + 16}} = \sqrt{5x - 2}$

28. $\sqrt{4\sqrt{2x - 5}} = \sqrt{x + 5}$

29. $\sqrt{3x + 1} + \sqrt{2x - 1} = \sqrt{10x - 1}$

30. $\sqrt{x - 3} + \sqrt{x + 3} = \sqrt{9 - x}$

In Exercises 31 to 40, solve each equation that involves fractional exponents. Check all proposed solutions.

31. $(3x + 5)^{1/3} = (-2x + 15)^{1/3}$

32. $(4z + 7)^{1/3} = 2$

33. $(x + 4)^{2/3} = 9$ **34.** $(x - 5)^{3/2} = 125$

35. $(4x)^{2/3} = (30x + 4)^{1/3}$ **36.** $z^{2/3} = (3z - 2)^{1/3}$

37. $4x^{3/4} = x^{1/2}$ **38.** $x^{3/5} = 2x^{1/5}$

39. $(3x - 5)^{2/3} + 6(3x - 5)^{1/3} = -8$

40. $2(x + 1)^{1/2} - 11(x + 1)^{1/4} + 12 = 0$

In Exercises 41 to 60, find all the real solutions of each equation by first rewriting each equation as a quadratic equation.

41. $x^4 - 9x^2 + 14 = 0$ **42.** $x^4 - 10x^2 + 9 = 0$

43. $2x^4 - 11x^2 + 12 = 0$ **44.** $6x^4 - 7x^2 + 2 = 0$

45. $x^6 + x^3 - 6 = 0$ **46.** $6x^6 + x^3 - 15 = 0$

47. $21x^6 + 22x^3 = 8$ **48.** $-3x^6 + 377x^3 - 250 = 0$

49. $x^{1/2} - 3x^{1/4} + 2 = 0$ **50.** $2x^{1/2} - 5x^{1/4} - 3 = 0$

51. $3x^{2/3} - 11x^{1/3} - 4 = 0$ **52.** $6x^{2/3} - 7x^{1/3} - 20 = 0$

53. $9x^4 = 30x^2 - 25$ **54.** $4x^4 - 28x^2 = -49$

55. $x^{2/5} - 1 = 0$ **56.** $2x^{2/5} - x^{1/5} = 6$

57. $\dfrac{1}{x^2} + \dfrac{3}{x} - 10 = 0$

58. $10\left(\dfrac{x - 2}{x}\right)^2 + 9\left(\dfrac{x - 2}{x}\right) - 9 = 0$

59. $9x - 52\sqrt{x} + 64 = 0$ **60.** $8x - 38\sqrt{x} + 9 = 0$

In Exercises 61 to 64, solve each equation. Round each solution to the nearest hundredth.

61. $x^4 - 3x^2 + 1 = 0$ **62.** $x - 4\sqrt{x} + 1 = 0$

63. $x^2 - \sqrt{9x^2 - 1} = 0$ **64.** $2x^2 = \sqrt{10x^2 - 3}$

SUPPLEMENTAL EXERCISES

In Exercises 65 to 70, solve for x in terms of the other variables.

65. $x^2 + y^2 = 9$ **66.** $\dfrac{x^2}{a^2} + \dfrac{y^2}{b^2} = 1$

67. $\sqrt{x} - \sqrt{y} = \sqrt{z}$ **68.** $x - y = \sqrt{x^2 + y^2 + 5}$

69. $x + y = \sqrt{x^2 - y^2 + 7}$ **70.** $x + \sqrt{x} = -y$

71. Solve $(\sqrt{x} - 2)^2 - 5\sqrt{x} + 14 = 0$ for x. (*Hint:* Use the substitution $u = \sqrt{x} - 2$, and then rewrite so that the equation is quadratic in terms of the variable u.)

72. Solve $(\sqrt[3]{x} + 3)^2 - 8\sqrt[3]{x} = 12$ for x. (*Hint:* Use the substitution $u = \sqrt[3]{x} + 3$, and then rewrite so that the equation is quadratic in terms of the variable u.)

73. **RADIUS OF A CONE** A conical funnel has a height h of 4 inches and a lateral surface area L of 15π square inches. Find the radius r of the cone. (*Hint:* Use the formula $L = \pi r\sqrt{r^2 + h^2}$.)

74. **DIAMETER OF A CONE** As flour is poured onto a table, it forms a right circular cone whose height is one-third the diameter of the base. What is the diameter of the base when the cone has a volume of 192 cubic inches?

75. **SPHERES** A silver sphere has a diameter of 8 millimeters, and a second silver sphere has a diameter of 12 millimeters. The spheres are melted down and recast to form a single cube. What is the length s of each edge of the cube? Round your answer to the nearest tenth of a millimeter.

76. **PENDULUM** The period of a pendulum T is the time it takes the pendulum to complete one swing from left to right and back. For a pendulum near the surface of the earth,

$$T = 2\pi \sqrt{\frac{L}{32}}$$

where T is measured in seconds and L is the length of the pendulum in feet. Find the length of a pendulum that has a period of 4 seconds. Round to the nearest tenth of a foot.

77. **Distance to the Horizon** On a ship, the distance d that you can see to the horizon is given by $d = 1.5\sqrt{h}$, where h is the height of your eye measured in feet above sea level and d is measured in miles. How high is the eye level of a navigator who can see 14 miles to the horizon? Round to the nearest foot.

78. **Radius of a Circle** The radius r of a circle inscribed in a triangle with sides of length a, b, and c is given by

$$r = \sqrt{\frac{(s - a)(s - b)(s - c)}{s}}$$

where $s = \frac{1}{2}(a + b + c)$.

a. Find the length of the radius of a circle inscribed in a triangle with sides of 5 inches, 6 inches, and 7 inches.

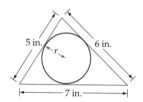

b. The radius of a circle inscribed in an equilateral triangle measures 2 inches. What is the length of each side of the equilateral triangle?

79. **Radius of a Circle** The radius r of a circle that is circumscribed about a triangle with sides a, b, and c is given by

$$r = \frac{abc}{4\sqrt{s(s - a)(s - b)(s - c)}}$$

where $s = \frac{1}{2}(a + b + c)$.

a. Find the radius of a circle that is circumscribed about a triangle with sides of 7 inches, 10 inches, and 15 inches. See figure at top of next column.

b. A circle with radius 5 inches is circumscribed about an equilateral triangle. What is the length of each side of the equilateral triangle?

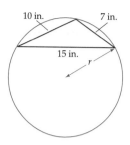

In Exercises 80 and 81, consider that the depth s from the opening of a well to the water can be determined by measuring the total time between the instant you drop a stone and the time you hear it hit the water. The time (in seconds) it takes the stone to hit the water is given by $\sqrt{s}/4$, where s is measured in feet. The time (also in seconds) required for the sound of the impact to travel up to your ears is given by $s/1100$. Thus the total time T (in seconds) between the instant you drop a stone and the moment you hear its impact is

$$T = \frac{\sqrt{s}}{4} + \frac{s}{1100}$$

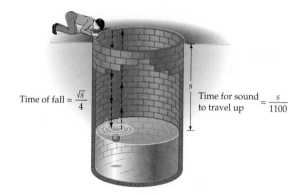

80. **Time of Fall** One of the world's deepest water wells is 7320 feet deep. Find the time between the instant you drop a stone and the time you hear it hit the water if the surface of the water is 7100 feet below the opening of the well. Round your answer to the nearest tenth of a second.

81. Solve $T = \dfrac{\sqrt{s}}{4} + \dfrac{s}{1100}$ for s.

82. **Depth of a Well** Use the result of Exercise 81 to determine the depth from the opening of a well to the water level if the time between the instant you drop a stone and the moment you hear its impact is 3 seconds. Round your answer to the nearest foot.

PROJECTS

1. 🏠 **THE REDUCED CUBIC** The mathematician Francois Vieta knew a method of solving the "reduced cubic" $x^3 + mx + n = 0$ by using the substitution $x = m/(3z) - z$.

    a. Show that this substitution results in the equation $z^6 - nz^3 - m^3/27 = 0$.

    b. Show that the equation in **a.** is quadratic in form.

    c. Solve the equation in **a.** for z.

    d. Use your solution from **c.** to find the real solution of the equation $x^3 + 3x = 14$.

2. ✏️ **FERMAT'S LAST THEOREM** One of the most famous theorems is known as *Fermat's Last Theorem*. Write an essay on Fermat's Last Theorem. Include information about

- the history of Fermat's Last Theorem.

- the relationship between Fermat's Last Theorem and the Pythagorean Theorem.

- Dr. Andrew Wiles's proposed proof of Fermat's Last Theorem.

The following list includes a few of the sources you may wish to consult.

- *The Last Problem* by Eric Temple Bell. The Mathematical Association of America, 1990.

- "Andrew Wiles: A Math Whiz Battles 350-Year-Old Puzzle" by Gina Kolata, *Math Horizons,* Winter 1993, pp. 8–11. The Mathematical Association of America.

- "Introduction to Fermat's Last Theorem," by David A. Cox, *The American Mathematical Monthly,* vol. 101, no. 1 (January 1994), pp. 3–14.

- "The Evidence: Fermat's Last Theorem," by S. Wagon, *The Mathematical Intelligencer,* vol. 8, no. 1, 1986, pp. 59–61.

SECTION

2.5 INEQUALITIES

In Section 1.2 we used the concept of an inequality to describe the order of real numbers on the real number line, and we also used inequalities to represent subsets of real numbers. In this section we consider inequalities that involve a variable. In particular, we consider how to determine which real values of the variable make the inequality a true statement.

The set of all solutions of an inequality is called the **solution set of the inequality.** For example, the solution set of $x + 1 > 4$ is the set of all real numbers greater than 3. **Equivalent inequalities** have the same solution set. We can solve an inequality by producing *simpler* but equivalent inequalities until the solutions are found. To produce these *simpler* but equivalent inequalities, we often apply one or more of the following properties.

PROPERTIES OF INEQUALITIES

Let a, b, and c be real numbers.

1. **Addition Property**   Adding the same real number to each side of an inequality preserves the direction of the inequality symbol.

    $a < b$ and $a + c < b + c$ are equivalent inequalities.

2. **Multiplication Properties**
    a. Multiplying each side of an inequality by the same *positive* real number *preserves* the direction of the inequality symbol.

    If $c > 0$, then $a < b$ and $ac < bc$ are equivalent inequalities.

    b. Multiplying each side of an inequality by the same *negative* real number *changes* the direction of the inequality symbol.

    If $c < 0$, then $a < b$ and $ac > bc$ are equivalent inequalities.

Note the difference between Property 2a and Property 2b. Property 2a states that an equivalent inequality is produced when each side of a given inequality is multiplied by the same *positive* real number and that the direction of the inequality symbol is *not* changed. By contrast, Property 2b states that when each side of a given inequality is multiplied by a *negative* real number, we must *reverse* the direction of the inequality to produce an equivalent inequality.

For instance, $-2b < 6$ and $b > -3$ are equivalent inequalities. (We multiplied each side of the first inequality by $-1/2$, and we changed the "less than" symbol to a "greater than" symbol.)

Because subtraction is defined in terms of addition, subtracting the same real number from each side of an inequality does not change the direction of the inequality symbol.

Because division is defined in terms of multiplication, dividing each side of the inequality by the same *positive* real number does *not* change the direction of the inequality symbol, and dividing each side of an inequality by a *negative* real number *changes* the direction of the inequality symbol.

EXAMPLE 1   *Solve an Inequality*

Solve $2(x + 3) < 4x + 10$. Write the solution set in set notation.

Solution

$2(x + 3) < 4x + 10$
$2x + 6 < 4x + 10$ • **Use the distributive property.**
$-2x < 4$ • **Subtract 4x and 6 from each side of the inequality.**
$x > -2$ • **Divide each side by −2 and reverse the inequality symbol.**

Thus the original inequality is true for all real numbers greater than -2. The solution set is $\{x \mid x > -2\}$.

TRY EXERCISE 8, EXERCISE SET 2.5

TAKE NOTE

Solutions of inequalities are often stated using set notation or interval notation. For instance, the real numbers that are solutions of the inequality in Example 1 can be written in set notation as $\{x \mid x > -2\}$ or in interval notation as $(-2, \infty)$.

EXAMPLE 2	*Solve an Application That Involves an Inequality*

You can rent a car from Company A for $46 per day plus $0.09 a mile. Company B charges $32 per day plus $0.14 a mile. Find the number of miles for which it is cheaper to rent from Company A if you rent a car for 1 day.

Solution

Let m equal the number of miles the car is to be driven. Then the cost of renting the car will be

$$\$46 + \$0.09m \quad \text{from Company A}$$

$$\$32 + \$0.14m \quad \text{from Company B}$$

If renting from Company A is to be cheaper than renting from Company B, then we must have

$$46 + 0.09m < 32 + 0.14m$$

Solving for m produces

$$14 < 0.05m$$

$$\frac{14}{0.05} < m$$

$$280 < m$$

Renting from Company A is cheaper if you drive over 280 miles per day.

TRY EXERCISE 12, EXERCISE SET 2.5

COMPOUND INEQUALITIES

A **compound inequality** is formed by joining two inequalities with the connective word *and* or *or*. The inequalities shown below are compound inequalities.

$$x + 1 > 3 \quad \text{and} \quad 2x - 11 < 7$$

$$x + 3 > 5 \quad \text{or} \quad x - 1 < 9$$

The solution set of a compound inequality with the connective word *or* is the *union* of the solution sets of the two inequalities. The solution set of a compound inequality with the connective word *and* is the *intersection* of the solution sets of the two inequalities.

EXAMPLE 3	*Solve Compound Inequalities*

Solve each compound inequality. Write each solution in set-builder notation.

a. $2x < 10 \quad \text{or} \quad x + 1 > 9$ **b.** $x + 3 > 4 \quad \text{and} \quad 2x + 1 > 15$

Solution

a. $2x < 10$ or $x + 1 > 9$

   $x < 5$ $x > 8$ • Solve each inequality.

   $\{x \mid x < 5\}$ $\{x \mid x > 8\}$ • Write each solution
                                                          as a set.

   $\{x \mid x < 5\} \cup \{x \mid x > 8\} = \{x \mid x < 5 \text{ or } x > 8\}$ • Write the union of
                                                                                  the solution sets.

b. $x + 3 > 4$ and $2x + 1 > 15$

   $x > 1$ $2x > 14$ • Solve each inequality.

                              $x > 7$

   $\{x \mid x > 1\}$ $\{x \mid x > 7\}$ • Write each solution as
                                                          a set.

   $\{x \mid x > 1\} \cap \{x \mid x > 7\} = \{x \mid x > 7\}$ • Write the intersection of
                                                                the solution sets.

TRY EXERCISE 16, EXERCISE SET 2.5

TAKE NOTE

We reserve the notation $a < b < c$ to mean $a < b$ and $b < c$. Thus the solution set of $2 > x > 5$ is the empty set, because there are no numbers less than 2 and greater than 5.

The inequality given by

$$12 < x + 5 < 19$$

is equivalent to the compound inequality $12 < x + 5$ *and* $x + 5 < 19$. You can solve $12 < x + 5 < 19$ by either of the following methods.

Method 1 Find the intersection of the solution sets of the inequalities $12 < x + 5$ and $x + 5 < 19$.

$$12 < x + 5 \qquad \text{and} \qquad x + 5 < 19$$
$$7 < x \qquad \text{and} \qquad x < 14$$

The solution set is $\{x \mid x > 7\} \cap \{x \mid x < 14\} = \{x \mid 7 < x < 14\}$.

Method 2 Subtract 5 from each of the three parts of the inequality.

$$12 < \quad x + 5 \quad < 19$$
$$12 - 5 < x + 5 - 5 < 19 - 5$$
$$7 < \quad x \quad < 14$$

The solution set is $\{x \mid 7 < x < 14\}$.

Caution The compound inequality $a < b$ and $b < c$ can be written in the compact form $a < b < c$. However, the compound inequality $a < b$ or $b > c$ cannot be expressed in a compact form.

EXAMPLE 4   *Solve an Application That Involves a Compound Inequality*

A photographic developer needs to be kept at a temperature between 15°C and 25°C. What is that temperature range in degrees Fahrenheit (°F)?

Continued •▶

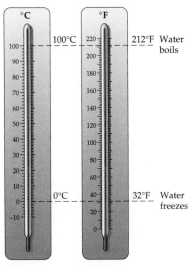

Figure 2.8

Solution

Figure 2.8 shows how some temperatures are related on the Fahrenheit and Celsius scales. The formula that relates the Celsius temperature (C) to the Fahrenheit temperature (F) is

$$C = \frac{5}{9}(F - 32)$$

We are given that

$$15 < C < 25$$

Substituting $\frac{5}{9}(F - 32)$ for C yields

$$15 < \frac{5}{9}(F - 32) < 25$$

$$27 < \quad F - 32 \quad < 45 \qquad \bullet \text{ Multiply each of the three parts of the inequality by 9/5.}$$

$$59 < \qquad F \qquad < 77 \qquad \bullet \text{ Add 32 to each of the three parts of the inequality.}$$

Thus the developer needs to be kept between 59°F and 77°F.

TRY EXERCISE 28, EXERCISE SET 2.5

SOLVING INEQUALITIES BY THE CRITICAL VALUE METHOD

Any value of x that causes a polynomial in x to equal zero is called a **zero of the polynomial.** For example, -4 and 1 are both zeros of the polynomial $x^2 + 3x - 4$, because $(-4)^2 + 3(-4) - 4 = 0$ and $1^2 + 3 \cdot 1 - 4 = 0$.

A SIGN PROPERTY OF POLYNOMIALS

Nonzero polynomials in x have the property that for any value of x between two consecutive real zeros, either all values of the polynomial are positive or all values of the polynomial are negative.

In our work with inequalities that involve polynomials, the real zeros of the polynomial are also referred to as **critical values of the inequality,** because on a number line they separate the real numbers that make an inequality involving a polynomial true from those that make it false. In Example 5 we use critical values and the sign property of polynomials to solve an inequality.

EXAMPLE 5   *Solve a Polynomial Inequality*

Solve: $x^2 + 3x - 4 < 0$

Solution

Factoring the polynomial $x^2 + 3x - 4$ produces the equivalent inequality

$$(x + 4)(x - 1) < 0$$

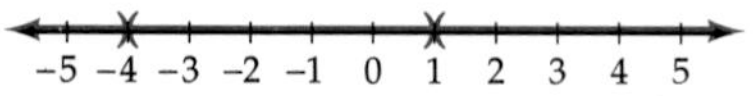

Thus the zeros of the polynomial $x^2 + 3x - 4$ are -4 and 1. They are the critical values of the inequality $x^2 + 3x - 4 < 0$. They separate the real number line into the three intervals shown in **Figure 2.9**.

To determine the intervals on which $x^2 + 3x - 4 < 0$, pick a number called a **test value** from each of the three intervals and then determine whether $x^2 + 3x - 4 < 0$ for each of these test values. For example, in the interval $(-\infty, -4)$, pick a test value of, say, -5. Then

$$x^2 + 3x - 4 = (-5)^2 + 3(-5) - 4 = 6$$

Because 6 is not less than 0, by the sign property of polynomials, no number in the interval $(-\infty, -4)$ makes $x^2 + 3x - 4 < 0$.

Now pick a test value from the interval $(-4, 1)$, say, 0. When $x = 0$,

$$x^2 + 3x - 4 = 0^2 + 3(0) - 4 = -4$$

Because -4 is less than 0, by the sign property of polynomials, all numbers in the interval $(-4, 1)$ make $x^2 + 3x - 4 < 0$.

If we pick a test value of 2 from the interval $(1, \infty)$, then

$$x^2 + 3x - 4 = (2)^2 + 3(2) - 4 = 6$$

Because 6 is not less than 0, by the sign property of polynomials, no number in the interval $(1, \infty)$ makes $x^2 + 3x - 4 < 0$.

The following table is a summary of our work.

Interval	Test Value x	$x^2 + 3x - 4 \overset{?}{<} 0$
$(-\infty, -4)$	-5	$(-5)^2 + 3(-5) - 4 < 0$ $6 < 0$ False
$(-4, 1)$	0	$(0)^2 + 3(0) - 4 < 0$ $-4 < 0$ True
$(1, \infty)$	2	$(2)^2 + 3(2) - 4 < 0$ $6 < 0$ False

In interval notation the solution set of $x^2 + 3x - 4 < 0$ is $(-4, 1)$. The solution set is graphed in **Figure 2.10**. Note that in this case the critical values -4 and 1 are not included in the solution set because they do not make $x^2 + 3x - 4$ less than 0.

Figure 2.9

Figure 2.10

TRY EXERCISE 36, EXERCISE SET 2.5

To avoid the arithmetic in Example 5, we often use a *sign diagram.* For example, note that the factor $(x + 4)$ is negative for all $x < -4$ and positive for all $x > -4$. The factor $(x - 1)$ is negative for all $x < 1$ and positive for all $x > 1$. These results are shown in **Figure 2.11**.

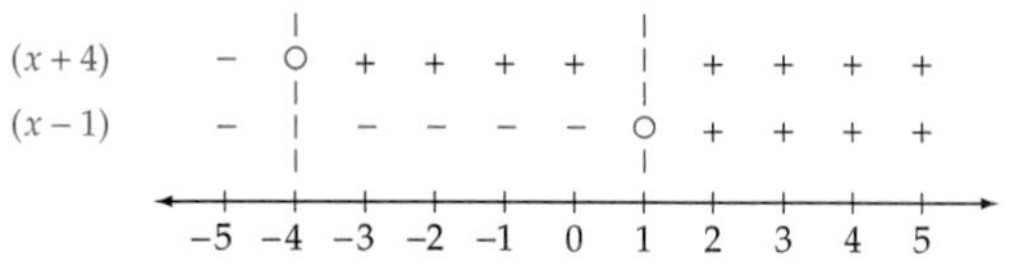

Figure 2.11
Sign diagram for $(x + 4)(x - 1)$

To determine on which intervals the product $(x + 4)(x - 1)$ is negative, we examine the sign diagram to see where the factors have opposite signs. This occurs only on the interval $(-4, 1)$, where $(x + 4)$ is positive and $(x - 1)$ is negative, so the original inequality is true only on the interval $(-4, 1)$.

Following is a summary of the steps used to solve polynomial inequalities by the critical value method.

SOLVING A POLYNOMIAL INEQUALITY BY THE CRITICAL VALUE METHOD

1. Write the inequality so that one side of the inequality is a nonzero polynomial and the other side is 0.

2. Find the real zeros of the polynomial.[3] They are the critical values of the original inequality.

3. Use test values to determine which of the intervals formed by the critical values are to be included in the solution set.

EXAMPLE 6   *Use the Critical Value Method to Solve an Application*

A manufacturer of tennis racquets finds that the yearly revenue R from a particular type of racquet is given by $R = 160x - x^2$, where x is the price in dollars of each racquet. Find the interval, in terms of x, for which the yearly revenue is greater than $6000. That is, solve

$$160x - x^2 > 6000$$

Solution

Write the inequality in such a way that 0 appears on the right side of the inequality.

$160x - x^2 - 6000 > 0$

$x^2 - 160x + 6000 < 0$ • **Arrange the terms in descending powers. Multiply each side of the inequality by −1.**

$(x - 60)(x - 100) < 0$ • **Factor the left side.**

Use the zero product property to find the zeros.

$(x - 60)(x - 100) = 0$ • **Replace the inequality with an equals sign.**

$x = 60$ or $x = 100$ • **Set each factor equal to 0 and solve for x.**

The zeros are 60 and 100. They separate the real number line into the intervals $(-\infty, 60)$, $(60, 100)$, and $(100, \infty)$. The sign diagram in **Figure 2.12**

---

[3] In Chapter 4, additional ways to find the zeros of a polynomial are developed. For the present, however, we will find the zeros by factoring or by using the quadratic formula.

shows that the inequality $(x - 60)(x - 100) < 0$ is true on the interval $(60, 100)$ and that it is false on the other intervals. Thus the revenue is greater than \$6000 per year when the price of each racquet is between \$60 and \$100.

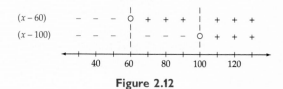

Figure 2.12

TRY EXERCISE 44, EXERCISE SET 2.5

RATIONAL INEQUALITIES

A rational expression is the quotient of two polynomials. **Rational inequalities** involve rational expressions, and they can be solved by an extension of the critical value method.

CRITICAL VALUES OF A RATIONAL EXPRESSION

The **critical values of a rational expression** are the numbers that cause the numerator of the rational expression to equal zero or the denominator of the rational expression to equal zero.

Rational expressions also have the property that they remain either positive for all values of the variable between consecutive critical values or negative for all values of the variable between consecutive critical values.

Following is a summary of the steps used to solve rational inequalities by the critical value method.

SOLVING A RATIONAL INEQUALITY BY THE CRITICAL VALUE METHOD

1. Write the inequality so that one side of the inequality is a rational expression and the other side is 0.

2. Find the real zeros of the numerator of the rational expression and the real zeros of its denominator. They are the critical values of the inequality.

3. Use test values to determine which of the intervals formed by the critical values are to be included in the solution set.

EXAMPLE 7   *Solve a Rational Inequality*

Solve: $\dfrac{(x - 2)(x + 3)}{x - 4} \geq 0$

Solution

The critical values include the zeros of the numerator, which are 2 and -3, and the zero of the denominator, which is 4. These three critical numbers separate the real number line into four intervals.

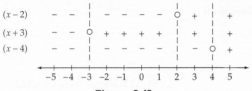

Figure 2.13

The sign diagram in **Figure 2.13** shows the sign of each of the factors $(x - 2)$, $(x + 3)$, and $(x - 4)$ on each of the four intervals. The sign diagram shows that the rational expression is positive on the two intervals $(-3, 2)$ and $(4, \infty)$. The critical values -3 and 2 are solutions because they satisfy the original inequality. However, the critical value 4 is not a solution because the denominator $x - 4$ is zero when $x = 4$. Therefore, in interval notation the inequality's solution set is $[-3, 2] \cup (4, \infty)$. The graph of the solution set is shown in **Figure 2.14.**

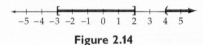

Figure 2.14

TRY EXERCISE 46, EXERCISE SET 2.5

EXAMPLE 8   *Solve a Rational Inequality*

Solve: $\dfrac{3x + 4}{x + 1} \leq 2$

Solution

Write the inequality so that 0 appears on the right side of the inequality.

$$\frac{3x + 4}{x + 1} \leq 2$$

$$\frac{3x + 4}{x + 1} - 2 \leq 0$$

Write the left side as a rational expression.

$$\frac{3x + 4}{x + 1} - \frac{2(x + 1)}{x + 1} \le 0 \qquad \bullet \text{ The LCD is } x + 1.$$

$$\frac{3x + 4 - 2x - 2}{x + 1} \le 0 \qquad \bullet \text{ Simplify.}$$

$$\frac{x + 2}{x + 1} \le 0$$

The critical values of this inequality are -2 and -1 because the numerator $x + 2$ is equal to zero when $x = -2$, and the denominator $x + 1$ is equal to zero when $x = -1$. The critical values -2 and -1 separate the real number line into the three intervals $(-\infty, -2)$, $(-2, -1)$, and $(-1, \infty)$.

All values of x on the interval $(-2, -1)$ make $(x + 2)/(x + 1)$ negative, as desired. On the other intervals, the quotient $(x + 2)/(x + 1)$ is positive. See the sign diagram in **Figure 2.15**.

Figure 2.15

The solution set is $[-2, -1)$. The graph of the solution set is shown in **Figure 2.16**. Note that -2 is included in the solution set because $(x + 2)/(x + 1) = 0$ when $x = -2$. However, -1 is not included in the solution set because the denominator $(x + 1)$ is zero when $x = -1$.

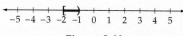

Figure 2.16

TRY EXERCISE 54, EXERCISE SET 2.5

TOPICS FOR DISCUSSION

1. If $x < y$, then $y > x$. Do you agree?

2. A student claims that the solution set of the compound inequality
    $$x < -3 \qquad \text{or} \qquad x > 5$$
    can be expressed as $-3 > x > 5$. Do you agree? Explain.

3. If $-a < b$, then it must be true that $a > -b$. Do you agree? Explain.

4. Do the inequalities $x < 4$ and $x^2 < 4^2$ both have the same solution set? Explain.

5. A tutor claims that if $k < 0$, then $|k| = -k$. Do you agree?

EXERCISE SET 2.5

In Exercises 1 to 10, use the properties of inequalities to solve each inequality. Write the solution set in set notation.

1. $2x + 3 < 11$

2. $3x - 5 > 16$

3. $x + 4 > 3x + 16$

4. $5x + 6 < 2x + 1$

5. $-6x + 1 \geq 19$

6. $-5x + 2 \leq 37$

7. $-3(x + 2) \leq 5x + 7$

8. $-4(x - 5) \geq 2x + 15$

9. $-4(3x - 5) > 2(x - 4)$

10. $3(x + 7) \leq 5(2x - 8)$

11. **PERSONAL FINANCE** A bank offers two checking account plans. The monthly fee and charge per check for each plan are shown below. Under what conditions is it less expensive to use the LowCharge plan?

Account Plan	Monthly Fee	Charge per Check
LowCharge	$5.00	$.01
FeeSaver	$1.00	$.08

12. **PERSONAL FINANCE** You can rent a car for the day from Company A for $29.00 plus $0.12 a mile. Company B charges $22.00 plus $0.21 a mile. Find the number of miles m (to the nearest mile) per day for which it is cheaper to rent from Company A.

13. **PERSONAL FINANCE** A sales clerk has a choice between two payment plans. Plan A pays $100.00 a week plus $8.00 a sale. Plan B pays $250.00 a week plus $3.50 a sale. How many sales per week must be made for plan A to yield the greater paycheck?

14. **PERSONAL FINANCE** A video store offers two rental plans. The yearly membership fee and the daily charge per video for each plan are shown below. How many videos can be rented per year if the No-fee plan is to be the least expensive of the plans?

THE VIDEO STORE

Rental Plan	Yearly Fee	Daily Charge per Video
Low-rate	$15.00	$1.49
No-fee	None	$1.99

In Exercises 15 to 26, solve each compound inequality. Write the solution set using set notation.

15. $4x + 1 > -2$ and $4x + 1 \leq 17$

16. $2x + 5 > -16$ and $2x + 5 < 9$

17. $10 \geq 3x - 1 \geq 0$

18. $0 \leq 2x + 6 \leq 54$

19. $20 > 8x - 2 \geq -5$

20. $4 \leq 10x + 1 \leq 51$

21. $-4x + 5 > 9$ or $4x + 1 < 5$

22. $2x - 7 \leq 15$ or $3x - 1 \leq 5$

23. $3x - 3 > 9$ or $-x + 2 < 3$

24. $4x - 3 \leq -5$ or $4x - 3 \geq 5$

25. $x > -6$ or $-3x + 7 > 19$

26. $2x - 1 > 5$ or $2x - 1 < -5$

27. **AVERAGE TEMPERATURES** The average daily minimum-to-maximum temperature range for the city of Palm Springs during the month of September is 68 to 104 degrees Fahrenheit. What is the corresponding temperature range measured on the Celsius temperature scale? (*Hint:* Let F be the average daily temperature. Then $68 \leq F \leq 104$. Now substitute $\frac{9}{5}C + 32$ for F and solve the resulting inequality for C.)

28. **AVERAGE TEMPERATURES** The average daily minimum-to-maximum temperature range for the city of Palm Springs during the month of January is 41 to 68 degrees Fahrenheit. What is the corresponding temperature range measured on the Celsius temperature scale?

29. **CONSECUTIVE EVEN INTEGERS** The sum of three consecutive even integers is between 36 and 54. Find all possible sets of integers that satisfy these conditions.

30. **CONSECUTIVE ODD INTEGERS** The sum of three consecutive odd integers is between 63 and 81. Find all possible sets of integers that satisfy these conditions.

In Exercises 31 to 42, use the critical value method to solve each inequality. Use interval notation to write each solution set.

31. $x^2 + 7x > 0$

32. $x^2 - 5x \leq 0$

33. $x^2 - 16 \leq 0$

34. $x^2 - 49 > 0$

35. $x^2 + 7x + 10 < 0$

36. $x^2 + 5x + 6 < 0$

37. $x^2 - 3x \geq 28$

38. $x^2 < -x + 30$

39. $6x^2 - 4 \le 5x$

40. $12x^2 + 8x \ge 15$

41. $8x^2 \ge 2x + 15$

42. $12x^2 - 16x < -5$

43. **REVENUE** The monthly revenue R for a product is given by $R = 420x - 2x^2$, where x is the price in dollars of each unit produced. Find the interval, in terms of x, for which the monthly revenue is greater than zero.

44. **REVENUE** A shoe manufacturer finds that the monthly revenue R from a particular style of aerobics shoe is given by $R = 312x - 3x^2$, where x is the price in dollars of each pair of shoes sold. Find the interval, in terms of x, for which the monthly revenue is greater than or equal to $5925.

In Exercises 45 to 60, use the critical value method to solve each inequality. Write each solution set in interval notation.

45. $\dfrac{x + 4}{x - 1} < 0$

46. $\dfrac{x - 2}{x + 3} > 0$

47. $\dfrac{x - 5}{x + 8} \ge 3$

48. $\dfrac{x - 4}{x + 6} \le 1$

49. $\dfrac{x}{2x + 7} \ge 4$

50. $\dfrac{x}{3x - 5} \le -5$

51. $\dfrac{(x + 1)(x - 4)}{x - 2} < 0$

52. $\dfrac{x(x - 4)}{x + 5} > 0$

53. $\dfrac{x + 2}{x - 5} \le 2$

54. $\dfrac{3x + 1}{x - 2} \ge 4$

55. $\dfrac{6x^2 - 11x - 10}{x} > 0$

56. $\dfrac{3x^2 - 2x - 8}{x - 1} \ge 0$

57. $\dfrac{x^2 - 6x + 9}{x - 5} \le 0$

58. $\dfrac{x^2 + 10x + 25}{x + 1} \ge 0$

59. $\dfrac{x^2 - 3x - 4}{x + 1} \ge 0$

60. $\dfrac{x^2 + 6x + 9}{x + 3} \le 0$

In Exercises 61 to 70, determine the set of all real numbers x such that y will be a real number. (Hint: $\sqrt{a}$ is a real number if and only if $a \ge 0$.)

61. $y = \sqrt{x + 9}$

62. $y = \sqrt{x - 3}$

63. $y = \sqrt{9 - x^2}$

64. $y = \sqrt{25 - x^2}$

65. $y = \sqrt{x^2 - 16}$

66. $y = \sqrt{x^2 - 81}$

67. $y = \sqrt{x^2 - 2x - 15}$

68. $y = \sqrt{x^2 + 4x - 12}$

69. $y = \sqrt{x^2 + 1}$

70. $y = \sqrt{x^2 - 1}$

SUPPLEMENTAL EXERCISES

In Exercises 71 to 74, use the critical value method to solve each inequality. Use interval notation to write each solution set.

71. $\dfrac{(x - 3)^2}{(x - 6)^2} > 0$

72. $\dfrac{(x - 1)^2}{(x - 4)^4} \ge 0$

73. $\dfrac{(x - 4)^2}{(x + 3)^3} \ge 0$

74. $\dfrac{(2x - 7)}{(x - 1)^2(x + 2)^2} \ge 0$

In Exercises 75 to 80, determine the set of all real numbers x such that y will be a real number.

75. $y = \sqrt[4]{x^3 - 3x}$

76. $y = \sqrt[4]{x^4 - 4x^3 + 4x^2}$

77. $y = \sqrt[6]{5 + x^2}$

78. $y = \sqrt[6]{(x + 3)^6}$

79. $y = \sqrt{x(x + 2)(x - 5)}$

80. $y = \sqrt{\dfrac{x - 3}{(x + 2)(x - 4)}}$

In Exercises 81 to 84, find the values of k such that the given equation will have at least one real solution.

81. $x^2 + kx + 6 = 0$

82. $x^2 + kx + 11 = 0$

83. $2x^2 + kx + 7 = 0$

84. $-3x^2 + kx - 4 = 0$

85. **HEIGHT OF A PROJECTILE** The equation

$$s = -16t^2 + v_0 t + s_0$$

gives the height s in feet above ground level, at the time t seconds, of an object thrown directly upward from a height s_0 feet above the ground and with an initial velocity of v_0 feet per second. A ball is thrown directly upward from ground level with an initial velocity of 64 feet per second. Find the time interval for which the ball has a height of more than 48 feet.

86. **HEIGHT OF A PROJECTILE** A ball is thrown directly upward from a height of 32 feet above the ground with an initial velocity of 80 feet per second. Find the time interval for which the ball will be more than 96 feet above the ground. (*Hint:* See Exercise 85.)

PROJECTS

1. **TRIANGLES** In any triangle, the sum of the lengths of the two shorter sides must be greater than the length of the longest side. Find all possible values of x if a triangle has sides of length

   a. $x, x + 5,$ and $x + 9$ **b.** $x, x^2 + x,$ and $2x^2 + x$ **c.** $\dfrac{1}{x + 2}, \dfrac{1}{x + 1},$ and $\dfrac{1}{x}$

SECTION

2.6   ABSOLUTE VALUE EQUATIONS AND INEQUALITIES

$|x| = 3$

Figure 2.17

The absolute value of a real number x is the distance between the number x and 0 on the real number line. For example, the solution set of $|x| = 3$ is the set of all real numbers that are 3 units from 0. Therefore, the solution set of $|x| = 3$ is $x = 3$ or $x = -3$. See **Figure 2.17**.

The following property is used to solve absolute value equations.

A PROPERTY OF ABSOLUTE VALUE EQUATIONS

For any variable expression E and any nonnegative real number k,

$$|E| = k \quad \text{if and only if} \quad E = k \quad \text{or} \quad E = -k$$

TAKE NOTE

Some absolute value equations have an empty solution set. For example, $|x + 2| = -5$ is false for all values of x. Because an absolute value is always nonnegative, the equation is never true.

EXAMPLE 1   *Solve an Absolute Value Equation*

Solve: $|2x - 5| = 21$

Solution

$|2x - 5| = 21$ implies $2x - 5 = 21$ or $2x - 5 = -21$. Solving each of these equations produces

$$
\begin{array}{lll}
2x - 5 = 21 & \text{or} & 2x - 5 = -21 \\
2x = 26 & & 2x = -16 \\
x = 13 & & x = -8
\end{array}
$$

Therefore, the solutions of $|2x - 5| = 21$ are -8 and 13.

TRY EXERCISE 10, EXERCISE SET 2.6

ABSOLUTE VALUE INEQUALITIES

The solution set of the absolute value inequality $|x - 1| < 3$ is the set of all real numbers whose distance from 1 is *less than* 3. Therefore, the solution set con-

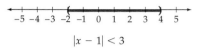

$$|x - 1| < 3$$

Figure 2.18

$$|x - 1| > 3$$

Figure 2.19

sists of all numbers between -2 and 4. See **Figure 2.18**. In interval notation, the solution set is $(-2, 4)$.

The solution set of the absolute value inequality $|x - 1| > 3$ is the set of all real numbers whose distance from 1 is *greater than* 3. Therefore, the solution set consists of all real numbers less than -2 *or* greater than 4. See **Figure 2.19**. In interval notation, the solution set is $(-\infty, -2) \cup (4, \infty)$.

The following properties are used to solve absolute value inequalities.

PROPERTIES OF ABSOLUTE VALUE INEQUALITIES

For any variable expression E and any nonnegative real number k,

$$|E| \leq k \quad \text{if and only if} \quad -k \leq E \leq k$$

$$|E| \geq k \quad \text{if and only if} \quad E \leq -k \quad \text{or} \quad E \geq k$$

EXAMPLE 2   *Solve an Absolute Value Inequality*

Solve: $|2 - 3x| < 7$

Solution

$|2 - 3x| < 7$ implies $-7 < 2 - 3x < 7$. Solve this compound inequality.

$$-7 < 2 - 3x < 7$$
$$-9 < \quad -3x \quad < 5 \qquad \bullet \text{ Subtract 2 from each of the three parts of}$$
$$\qquad\qquad\qquad\qquad\qquad \text{ the inequality.}$$
$$3 > \quad x \quad > -\frac{5}{3} \qquad \bullet \text{ Multiply each part of the inequality by } -1/3$$
$$\qquad\qquad\qquad\qquad\qquad \text{ and reverse the inequality symbols.}$$

In interval notation, the solution set is given by $(-5/3, 3)$. See **Figure 2.20**.

TRY EXERCISE 30, EXERCISE SET 2.6

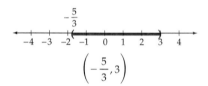

$$\left(-\frac{5}{3}, 3\right)$$

Figure 2.20

TAKE NOTE

Some inequalities have a solution set that consists of all real numbers. For example, $|x + 9| \geq 0$ is true for all values of *x*. Because an absolute value is always nonnegative, the equation is always true.

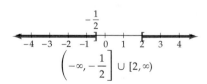

$$\left(-\infty, -\frac{1}{2}\right] \cup [2, \infty)$$

Figure 2.21

EXAMPLE 3   *Solve an Absolute Value Inequality*

Solve: $|4x - 3| \geq 5$

Solution

$|4x - 3| \geq 5$ implies $4x - 3 \leq -5$ or $4x - 3 \geq 5$. Solving each of these inequalities produces

$$4x - 3 \leq -5 \qquad \text{or} \qquad 4x - 3 \geq 5$$
$$4x \leq -2 \qquad\qquad\qquad 4x \geq 8$$
$$x \leq -\frac{1}{2} \qquad\qquad\qquad x \geq 2$$

Therefore, the solution set is $(-\infty, -1/2] \cup [2, \infty)$. See **Figure 2.21**.

TRY EXERCISE 34, EXERCISE SET 2.6

TOLERANCE

The **tolerance** of a dimension is the acceptable amount by which the dimension may differ from a given standard. For example, a machinist has been instructed to make a ball bearing with a diameter of 1.40 centimeters and a tolerance of 0.01 centimeter. The maximum diameter of the bearing is 1.40 centimeters + 0.01 centimeter = 1.41 centimeters. The minimum diameter of the bearing is 1.40 centimeters − 0.01 centimeter = 1.39 centimeters. Thus the machinist must produce bearings with a diameter d such that 1.39 centimeters $\leq d \leq$ 1.41 centimeters. This can be expressed concisely by the following absolute value inequality:

$$|d - 1.40| \leq 0.01$$

In general, if t is a given tolerance, s is a given standard, and m is a measurement that falls within the tolerance, then

$$|m - s| \leq t$$

Figure 2.22

EXAMPLE 4   *Solve a Tolerance Application*

A beaker has an inner radius of 4 centimeters. How high h (to the nearest 0.1 centimeter) should we fill the beaker if we need to measure 1 liter (1000 cubic centimeters) of a solution with an error of 10 cubic centimeters or less? See **Figure 2.22.**

Solution

The volume of the solution in the beaker is given by $V = \pi(4^2)h$. The desired standard is 1000 cubic centimeters. Our tolerance is 10 cubic centimeters. In absolute value notation this is expressed as

$$|V - 1000| \leq 10$$

Substitute $\pi(4^2)h$ for V and solve for h.

$$|\pi(4^2)h - 1000| \leq 10$$

$$|16\pi h - 1000| \leq 10 \qquad \text{• Simplify.}$$

$$-10 \leq 16\pi h - 1000 \leq 10 \qquad \text{• Solve for } h.$$

$$990 \leq \quad 16\pi h \quad \leq 1010$$

$$\frac{990}{16\pi} \leq \quad h \quad \leq \frac{1010}{16\pi}$$

$$19.7 \leq \quad h \quad \leq 20.1 \qquad \text{• Round to the nearest 0.1 centimeter.}$$

To ensure that we have 1 liter of solution within a tolerance of 10 cubic centimeters, we should fill the beaker to a height of at least 19.7 centimeters but not more than 20.1 centimeters.

TRY EXERCISE 48, EXERCISE SET 2.6

We can also solve absolute value inequalities by using the critical value method. To solve an absolute value inequality of the form $|P| > k$, where P is a polynomial such as $x^2 - 5$, we first find all values of x such that the left side of the inequality is *equal* to the right side.

EXAMPLE 5   *Solve an Absolute Value Inequality*

Solve: $|x^2 - 5| > 4$

Solution

Begin by finding the solutions of $|x^2 - 5| = 4$. $|x^2 - 5| = 4$ implies $x^2 - 5 = 4$ or $x^2 - 5 = -4$. Solving each of these equations produces

$$x^2 - 5 = 4 \qquad \text{or} \qquad x^2 - 5 = -4$$
$$x^2 = 9 \qquad\qquad\qquad x^2 = 1$$
$$x = \pm 3 \qquad\qquad\qquad x = \pm 1$$

The four values -3, -1, 1, and 3 separate a real number line into the five intervals shown in **Figure 2.23**. They are the critical values of the inequality because they separate the real numbers that make the inequality true from those that make it false.

We can use a test value from each of the intervals to determine on which intervals the original inequality is true. For example, if we choose -4 from the interval $(-\infty, -3)$, then the inequality $|x^2 - 5| > 4$ is true because

$$|(-4)^2 - 5| = |16 - 5| = |11| = 11 > 4$$

Therefore, the interval $(-\infty, -3)$ is part of the solution set. Continuing in a similar manner produces the results shown in the following table:

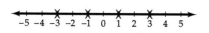

Figure 2.23

Interval	Test Value x	$\lvert x^2 - 5 \rvert \overset{?}{>} 4$
$(-\infty, -3)$	-4	$\lvert (-4)^2 - 5 \rvert > 4$ $11 > 4$ True
$(-3, -1)$	-2	$\lvert (-2)^2 - 5 \rvert > 4$ $1 > 4$ False
$(-1, 1)$	0	$\lvert (0)^2 - 5 \rvert > 4$ $5 > 4$ True
$(1, 3)$	2	$\lvert (2)^2 - 5 \rvert > 4$ $1 > 4$ False
$(3, \infty)$	4	$\lvert (4)^2 - 5 \rvert > 4$ $11 > 4$ True

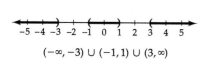

$(-\infty, -3) \cup (-1, 1) \cup (3, \infty)$

Figure 2.24

Thus the solution set is $(-\infty, -3) \cup (-1, 1) \cup (3, \infty)$. See **Figure 2.24**.

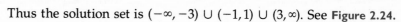

TRY EXERCISE 50, EXERCISE SET 2.6

TOPICS FOR DISCUSSION

1. A student claims that the inequality $|x| < -3$ has no solutions because the absolute value of any real number is never a negative number. Do you agree? Explain.

2. A student claims that the absolute value of any real number is a positive number. A second student claims that the first student is wrong. Which student is correct? Explain.

3. The solution of the inequality $|x - 5| \le 2$ consists of the points in an interval centered at $x = 5$ and extending 2 units to the right and 2 units to the left of 5. Do you agree?

4. The theorem

$$|E| \ge k \quad \text{if and only if} \quad E \le -k \text{ or } E \ge k$$

can also be stated as

$$|E| \ge k \quad \text{if and only if} \quad E \ge k \text{ or } -E \ge k$$

Do you agree? Explain.

5. Consider the equation $|x + y| = |x| + |y|$. Is this equation true for all values of x and y, true for some values of x and y, or never true?

EXERCISE SET 2.6

In Exercises 1 to 22, solve each absolute value equation for x.

1. $|x| = 4$
2. $|x| = 7$
3. $|x - 5| = 2$
4. $|x - 8| = 3$
5. $|x + 6| = 1$
6. $|x + 9| = 5$
7. $|x + 14| = 20$
8. $|x - 3| = 14$
9. $|2x - 5| = 11$
10. $|2x - 3| = 21$
11. $|2x + 6| = 10$
12. $|2x + 14| = 60$
13. $\left| \dfrac{x - 4}{2} \right| = 8$
14. $\left| \dfrac{x + 3}{4} \right| = 6$
15. $|2x + 5| = -8$
16. $|4x - 1| = -17$
17. $2|x + 3| + 4 = 34$
18. $3|x - 5| - 16 = 2$
19. $|2x - a| = b \quad (b > 0)$
20. $3|x - d| = c \quad (c > 0)$
21. $|x - a| = \delta \quad (\delta > 0)$
22. $|x + m| = m \quad (m > 0)$

In Exercises 23 to 46, use interval notation to express the solution set of each inequality.

23. $|x| < 4$
24. $|x| > 2$
25. $|x - 1| < 9$
26. $|x - 3| < 10$
27. $|x + 3| > 30$
28. $|x + 4| < 2$
29. $|2x - 1| > 4$
30. $|2x - 9| < 7$
31. $|x + 3| \ge 5$
32. $|x - 10| \ge 2$
33. $|3x - 10| \le 14$
34. $|2x - 5| \ge 1$
35. $|4 - 5x| \ge 24$
36. $|3 - 2x| \le 5$
37. $|x - 5| \ge 0$
38. $|x - 7| \ge 0$
39. $|x - 4| \le 0$
40. $|2x + 7| \le 0$
41. $|5x - 1| < -4$
42. $|2x - 1| < -9$
43. $|2x + 7| \ge -5$
44. $|3x + 11| \ge -20$
45. $|x - 3| < b \quad (b > 0)$
46. $|x - c| < d \quad (d > 0)$

47. Tolerance A machinist is producing a circular cylinder on a lathe. The circumference of the cylinder must be 28 inches, with a tolerance of 0.15 inch. What maximum and minimum radii (to the nearest 0.001 inch) must the machinist stay between to produce an acceptable cylinder?

48. Tolerance A tall, narrow beaker has an inner radius of 2 centimeters. How high h (to the nearest 0.1 centimeter) should we fill the beaker if we need to measure 3/4 liter (750 cubic centimeters) of a solution with an error of 15 cubic centimeters or less?

2 cm

In Exercises 49 to 56, use the critical value method to solve each inequality. Use interval notation to write the solution sets.

49. $|x^2 - 1| < 1$ **50.** $|x^2 - 2| > 1$

51. $|x^2 - 10| < 6$ **52.** $|x^2 + 4| \geq 10$

53. $|x^2 + 7x + 11| \geq 1$ **54.** $|x^2 - 5x + 6| \leq 1$

55. $|x^2 - 21.5| \geq 4.5$ **56.** $|x^2 - 6.5| \leq 2.5$

In Exercises 57 to 62, determine whether the statement is true or false. If it is false, explain why.

57. $|x + 2| = |x| + |2|$ **58.** $|x - 5| = |x| - |5|$

59. $|x - 7| \geq 0$ **60.** $|x||5| = |5x|$

61. If $t < 0$, then $|t| = -t$.

62. The absolute value of any real number is a positive number.

Supplemental Exercises

In Exercises 63 to 70, find the values of x that make each equation true.

63. $|x + 4| = x + 4$ **64.** $|x - 1| = x - 1$

65. $|x + 7| = -(x + 7)$ **66.** $|x - 3| = -(x - 3)$

67. $|2x + 7| = 2x + 7$ **68.** $|3x - 11| = -3x + 11$

69. $|x - 2| + |x + 4| = 8$ **70.** $|x + 1| - |x + 3| = 4$

In Exercises 71 to 82, use interval notation to express the solution set of each inequality.

71. $1 < |x| < 5$ **72.** $2 < |x| < 3$

73. $3 \leq |x| < 7$ **74.** $0 < |x| \leq 3$

75. $0 < |x - a| < \delta$ $(\delta > 0)$ **76.** $0 < |x - 5| < 2$

77. $2 < |x - 6| < 4$ **78.** $1 \leq |x - 3| < 5$

79. $\left|1 - \dfrac{3x}{4}\right| \geq 6$ **80.** $\left|2 + \dfrac{3x}{5}\right| < 10$

81. $|x| > |x - 1|$ **82.** $|x - 2| \leq |x + 4|$

83. Write an absolute value inequality to represent all real numbers within
  a. 8 units of 3
  b. k units of j (assume $k > 0$)

84. Write an absolute value inequality to represent all real numbers that are more than
  a. 5 units away from 1
  b. k units away from j (assume $k > 0$)

85. The length of the sides of a square has been measured accurately to within 0.01 foot. This measured length is 4.25 feet.
  a. Write an absolute value inequality that describes the relationship between the actual length of each side of the square s and its measured length.
  b. Solve for s the absolute value inequality you found in **a.**

Projects

1. Fair Coins A coin is considered a **fair** coin if it has an equal chance of landing heads up or tails up. To decide whether a coin is a fair coin, a statistician tosses it 1000 times and records the number of tails t. The statistician is prepared to state that the coin is a fair coin if

$$\left| \frac{t - 500}{15.81} \right| \le 2.33$$

a. Determine what values of t will cause the statistician to state that the coin is a fair coin.

b. Pick a coin and test it according to the criteria above to see whether it is a fair coin.

EXPLORING CONCEPTS WITH TECHNOLOGY

Can You Trust Your Calculator?

You may think that your calculator always produces correct results in a *predictable* manner. However, the following experiment may change your opinion.

First note that the algebraic expression

$$p + 3p(1 - p)$$

is equal to the expression

$$4p - 3p^2$$

Use a graphing calculator to evaluate both of these expressions with $p = 0.05$. You should find that both expressions equal 0.1925. So far we do not observe any unexpected results. Now replace p in each expression with the current value of that expression (0.1925 in this case). This is called *feedback* because we are feeding our outputs back into each expression as inputs. Each new evaluation is referred to as an *iteration*. This time each expression takes on the value 0.65883125. Still no surprises. Continue the feedback process. That is, replace p in each expression *with the current value of that expression*. Now each expression takes on the value 1.33314915207, as shown in the following table. The iterations were performed on a *TI-85* calculator.

Iteration	$p + 3p(1 - p)$	$4p - 3p^2$
1	0.1925	0.1925
2	0.65883125	0.65883125
3	1.33314915207	1.33314915207

The following table shows that if we continue this feedback process on a calculator, the expressions $p + 3p(1 - p)$ and $4p - 3p^2$ will start to take on different values starting with the fourth iteration. By the 37th iteration, the values do not even agree to two decimal places.

Iteration	$p + 3p(1 - p)$	$4p - 3p^2$
4	7.366232839E–4	7.366232838E–4
5	0.002944865294	0.002944865294
6	0.011753444481	0.0117534448
7	0.046599347553	0.046599347547
20	1.12135618652	1.12135608405
30	0.947163304835	0.947033128433
37	0.285727963839	0.300943417861

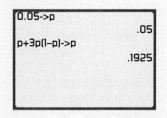

1. Use a calculator to find the first 20 iterations of $p + 3p(1 - p)$ and $4p - 3p^2$, with the initial value of $p = 0.5$.

2. Write a report on chaos and fractals. Include information on the *"butterfly effect."* An excellent source is *Chaos and Fractals, New Frontiers of Science* by Heinz-Otto Peitgen, Hartmut Jurgens, and Dietmar Saupe (New York: Springer-Verlag, 1992).

3. Equations of the form $p_{n+1} = p_n + rp_n(1 - p_n)$ are called Verhulst population models. Write a report on Verhulst population models.

CHAPTER 2 REVIEW

2.1 Linear Equations

• A number is said to satisfy an equation if substituting the number for the variable results in an equation that is a true statement. To solve an equation means to find all values of the variable that satisfy the equation. These values that make the equation true are called solutions or roots of the equation. Equivalent equations have the same solution(s).

• A linear equation in the single variable x is an equation that can be written in the form $ax + b = 0$, where a and b are real numbers, with $a \neq 0$.

2.2 Formulas and Applications

• A formula is an equation that expresses known relationships between two or more variables. Application problems are best solved by using the guidelines developed in this section.

2.3 Quadratic Equations

• A quadratic equation in x is an equation that can be written in the form $ax^2 + bx + c = 0$, where $a \neq 0$. If the quadratic polynomial in a quadratic equation is factorable over the set of integers, then the equation can be solved by factoring and using the zero product property (see page 88). Every quadratic equation can be solved by completing the square or by using the quadratic formula.

• **The Quadratic Formula**

    If $ax^2 + bx + c = 0$, $a \neq 0$, then $x = \dfrac{-b \pm \sqrt{b^2 - 4ac}}{2a}$.

2.4 Other Types of Equations

• **The Power Principle**

    If P and Q are algebraic expressions and n is a positive integer, then every solution of $P = Q$ is a solution of $P^n = Q^n$.

• An equation is said to be quadratic in form if it can be written in the form $au^2 + bu + c = 0$, $a \neq 0$, where u is an algebraic expression involving x.

2.5 Inequalities

• The set of all solutions of an inequality is the solution set of the inequality. Equivalent inequalities have the same solution set. To solve an inequality, use the properties of inequalities or the critical value method.

2.6 Absolute Value Equations and Inequalities

• Absolute value equations and inequalities can be solved by applying the following properties:

    For any variable expression E and any nonnegative real number k,

$\lvert E \rvert = k$	if and only if	$E = k$ or $E = -k$
$\lvert E \rvert \leq k$	if and only if	$-k \leq E \leq k$
$\lvert E \rvert \geq k$	if and only if	$E \leq -k$ or $E \geq k$

CHAPTER 2 TRUE/FALSE EXERCISES

In Exercises 1 to 10, answer true or false. If the statement is false, give an example to show that the statement is false.

1. If $x^2 = 9$, then $x = 3$.

2. The equations

$$x = \sqrt{12 - x} \quad \text{and} \quad x^2 = 12 - x$$

are equivalent equations.

3. Adding the same constant to each side of a given equation produces an equation that is equivalent to the given equation.

4. If $a > b$, then $-a < -b$.

5. If $a \neq 0$, $b \neq 0$, and $a > b$, then $1/a > 1/b$.

6. The discriminant of $ax^2 + bx + c = 0$ is $\sqrt{b^2 - 4ac}$.

7. If $\sqrt{a} + \sqrt{b} = c$, then $a + b = c^2$.

8. The solution set of $|x - a| < b$ with $b > 0$ is given by the interval $(a - b, a + b)$.

9. The only quadratic equation that has roots of 4 and -4 is $x^2 - 16 = 0$.

10. Every quadratic equation $ax^2 + bx + c = 0$ with real coefficients such that $ac < 0$ has two distinct real roots.

CHAPTER 2 REVIEW EXERCISES

In Exercises 1 to 30, solve each equation.

1. $x - 2(5x - 3) = -3(-x + 4)$

2. $3x - 5(2x - 7) = -4(5 - 2x)$

3. $\dfrac{4x}{3} - \dfrac{4x - 1}{6} = \dfrac{1}{2}$

4. $\dfrac{3x}{4} - \dfrac{2x - 1}{8} = \dfrac{3}{2}$

5. $\dfrac{x}{x + 2} + \dfrac{1}{4} = 5$

6. $\dfrac{y - 1}{y + 1} - 1 = \dfrac{2}{y}$

7. $x^2 - 5x + 6 = 0$

8. $6x^2 + x - 12 = 0$

9. $3x^2 - x - 1 = 0$

10. $x^2 - x + 1 = 0$

11. $3x^3 - 5x^2 = 0$

12. $2x^3 - 8x = 0$

13. $6x^4 - 23x^2 + 20 = 0$

14. $3x + 16\sqrt{x} - 12 = 0$

15. $\sqrt{x^2 - 15} = \sqrt{-2x}$

16. $\sqrt{x^2 - 24} = \sqrt{2x}$

17. $\sqrt{3x + 4} + \sqrt{x - 3} = 5$

18. $\sqrt{2x + 2} - \sqrt{x + 2} = \sqrt{x - 6}$

19. $\sqrt{4 - 3x} - \sqrt{5 - x} = \sqrt{5 + x}$

20. $\sqrt{3x + 9} - \sqrt{2x + 4} = \sqrt{x + 1}$

21. $\dfrac{1}{(y + 3)^2} = 1$

22. $\dfrac{1}{(2s - 5)^2} = 4$

23. $|x - 3| = 2$

24. $|x + 5| = 4$

25. $|2x + 1| = 5$

26. $|3x - 7| = 8$

27. $(x + 2)^{1/2} + x(x + 2)^{3/2} = 0$

28. $x^2(3x - 4)^{1/4} + (3x - 4)^{5/4} = 0$

29. $(2x - 1)^{2/3} + (2x - 1)^{1/3} = 12$

30. $6(x + 1)^{1/2} - 7(x + 1)^{1/4} - 3 = 0$

In Exercises 31 to 48, solve each inequality. Express your solution sets by using interval notation.

31. $-3x + 4 \geq -2$

32. $-2x + 7 \leq 5x + 1$

33. $x^2 + 3x - 10 \leq 0$

34. $x^2 - 2x - 3 > 0$

35. $61 \leq \dfrac{9}{5}C + 32 \leq 95$

36. $30 < \dfrac{5}{9}(F - 32) < 65$

37. $x^3 - 7x^2 + 12x \leq 0$

38. $x^3 + 4x^2 - 21x > 0$

39. $\dfrac{x + 3}{x - 4} > 0$

40. $\dfrac{x(x - 5)}{x + 7} \leq 0$

41. $\dfrac{2x}{3 - x} \leq 10$

42. $\dfrac{x}{5 - x} \geq 1$

43. $|3x - 4| < 2$

44. $|2x - 3| \geq 1$

45. $0 < |x| < 2$

46. $0 < |x| \leq 1$

47. $0 < |x - 2| < 1$

48. $0 < |x - a| < b \quad (b > 0)$

In Exercises 49 to 54, solve each equation for the indicated unknown.

49. $V = \pi r^2 h$,   for h

50. $P = \dfrac{A}{1 + rt}$,   for t

51. $A = \dfrac{h}{2}(b_1 + b_2)$,   for b_1

52. $P = 2(l + w)$,   for w

53. $e = mc^2$,   for m

54. $F = G\dfrac{m_1 m_2}{s^2}$,   for m_1

55. UNKNOWN NUMBER One-half of a number minus one-fourth of the number is four more than one-fifth of the number. What is the number?

56. RECTANGULAR REGION The length of a rectangle is 9 feet less than twice the width of the rectangle. The perimeter of the rectangle is 54 feet. Find the width and the length.

57. DISTANCE TO AN ISLAND A motorboat left a harbor and traveled to an island at an average rate of 8 knots. The average speed on the return trip was 6 knots. If the total trip took 7 hours, how far is it from the harbor to the island?

58. PRICE OF SUBSCRIPTION The price of a magazine subscription rose 5% this year. If the subscription now costs $21, how much did the subscription cost last year?

59. INVESTMENT A total of $5500 was deposited into two simple interest accounts. On one account the annual simple interest rate is 4%, and on the second account the annual simple interest rate is 6%. The amount of interest earned for 1 year was $295. How much was invested in each account?

60. INDIVIDUAL PRICE A calculator and a battery together sell for $21. The price of the calculator is $20 more than the price of the battery. Find the price of the calculator and the price of the battery.

61. MAINTENANCE COST Eighteen owners share the maintenance cost of a condominium complex. If six more units are sold, the maintenance cost will be reduced by $12 per month for each of the present owners. What is the total monthly maintenance cost for the condominium complex?

62. RECTANGULAR REGION The perimeter of a rectangle is 40 inches and its area is 96 square inches. Find the length and the width of the rectangle.

63. BUILD A WALL A mason can build a wall in 9 hours less than an apprentice. Together they can build the wall in 6 hours. How long would it take the apprentice, working alone, to build the wall?

64. COMMERCE An art show brought in $33,196 on the sale of 4526 tickets. The adult tickets sold for $8 and the

student tickets sold for $2. How many of each type of ticket were sold?

65. DIAMETER OF A CONE As sand is poured from a chute, it forms a right circular cone whose height is one-fourth the diameter of the base. What is the diameter of the base when the cone has a volume of 144 cubic feet?

66. REVENUE A manufacturer of calculators finds that the monthly revenue R from a particular style of calculator is given by $R = 72x - 2x^2$, where x is the price in dollars of each calculator. Find the interval, in terms of x, for which the monthly revenue is greater than $576.

CHAPTER 2 TEST

1. Solve: $3 - \dfrac{x}{4} = \dfrac{3}{5}$

2. Solve: $\dfrac{3}{x+2} - \dfrac{3}{4} = \dfrac{5}{x+2}$

3. Solve $ax - c = c(x - d)$ for x.

4. Solve $x^2 + 4x - 1 = 0$ by completing the square.

5. Solve: $3x^2 + 2x - 9 = 0$

6. Solve: $x^4 + 4x^3 - x - 4 = 0$

7. Solve: $\sqrt{x - 2} - 1 = \sqrt{3 - x}$

8. Solve: $3x^{2/3} + 10x^{1/3} - 8 = 0$

9. Solve: $(x - 3)^{2/3} = 16$

10. Solve: $-3(x + 2) \le 4 - 7x$

11. Solve: $\dfrac{x^2 + x - 12}{x + 1} \ge 0$

12. Solve: $-3 \le 3x - 9 \le 4$

13. Solve: $|2x + 7| = 5$

14. Solve: $|x + 4| < 3$

15. Solve: $|3x - 2| \ge 7$

16. A boat has a speed of 5 mph in still water. The boat can travel 21 miles with the current in the same time in which it can travel 9 miles against the current. Find the rate of the current.

17. A total of $9000 was deposited into two simple interest accounts. On one account the annual simple interest rate is 8.2%, and on the second account the annual simple interest rate is 6.5%. The amount of interest earned for 1 year was $695.50. How much was invested in each account?

18. A radiator contains 6 liters of a 20% antifreeze solution. How much should be drained and replaced with pure antifreeze to produce a 50% antifreeze solution?

19. A worker can cover a parking lot with asphalt in 10 hours. With the help of an assistant, the work can be done in 6 hours. How long would it take the assistant, working alone, to cover the parking lot with asphalt?

20. You can rent a car for the day from Company A for $28 plus $0.10 a mile. Company B charges $20 plus $0.18 a mile. At what point, in terms of miles driven per day, is it cheaper to rent from Company A?

FUNCTIONS AND GRAPHS

3

Stockbrokers display market trends using Cartesian graphs.

A line graph is a concise way to translate a large amount of raw data.

At a business meeting, different types of graphs prove useful in displaying a range of information.

A Simple but Powerful Concept

This chapter is concerned with the concepts of a function, the graph of a function, and elementary analytic geometry. The following two quotes lend support to the power of these ideas.

> *As long as algebra and geometry proceeded along separate paths, their advance was slow and their applications limited. But when the sciences joined company, they drew from each other's vitality and thence forward marched on at a rapid pace toward perfection.*
>
> —*Joseph Louis Lagrange*

> *[Analytic geometry], far more than any of his metaphysical speculations, immortalized the name of Descartes, and constitutes the greatest single step ever made in the progress of the exact sciences.*
>
> —*John Stuart Mill*

Two types of graphs that are used in this chapter are shown below.

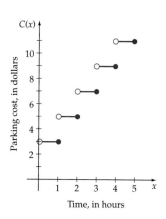

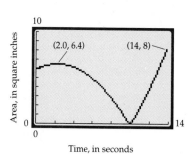

3.1 A TWO-DIMENSIONAL COORDINATE SYSTEM AND GRAPHS

CARTESIAN COORDINATE SYSTEMS

Each point on a coordinate axis is associated with a number called its coordinate. Each point on a flat, two-dimensional surface, called a **coordinate plane** or *xy*-plane, is associated with an **ordered pair** of numbers called **coordinates** of the point. Ordered pairs are denoted by (a, b), where the real number a is the **x-coordinate** or **abscissa** and the real number b is the **y-coordinate** or **ordinate**.

The coordinates of a point are determined by the point's position relative to a horizontal coordinate axis called the **x-axis** and a vertical coordinate axis called the **y-axis**. The axes intersect at the point $(0, 0)$, called the **origin**. In **Figure 3.1**, the axes are labeled such that positive numbers appear to the right of the origin on the *x*-axis and above the origin on the *y*-axis. The four regions formed by the axes are called **quadrants** and are numbered counterclockwise. This two-dimensional coordinate system is referred to as a **Cartesian coordinate system** in honor of René Descartes (1596–1650).

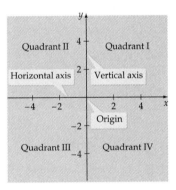

Figure 3.1

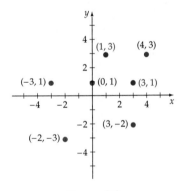

Figure 3.2

To **plot a point** $P(a, b)$ means to draw a dot at its location in the coordinate plane. In **Figure 3.2** we have plotted the points $(4, 3)$, $(-3, 1)$, $(-2, -3)$, $(3, -2)$, $(0, 1)$, $(1, 3)$, and $(3, 1)$. The order in which the coordinates of an ordered pair are listed is important. **Figure 3.2** shows that $(1, 3)$ and $(3, 1)$ do not denote the same point.

Data are often displayed in a visual form as a set of points called a *scattergram* or *scatter plot*. For instance, the scattergram shown in **Figure 3.3** illustrates the number of cars sold per year in the United States from 1980 to 1996. The point with coordinates $(1994, 6500)$ indicates that there were 6,500,000 cars sold in 1994. The zig-zag on the horizontal axis indicates that no data for the years 0 to 1979 have been shown. The line segments that connect the points in **Figure 3.3** help illustrate trends.

QUESTION According to the graph in **Figure 3.3**, between what years was the trend in car sales decreasing?

Cars Sold in the U.S.

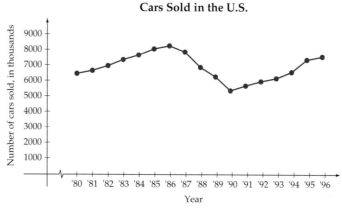

Figure 3.3

EQUALITY OF ORDERED PAIRS

The ordered pairs (a, b) and (c, d) are equal if and only if

$$a = c \quad \text{and} \quad b = d$$

TAKE NOTE

The notation (a, b) was used earlier to denote an interval on a one-dimensional number line. In this section, (a, b) denotes an ordered pair in a two-dimensional plane. This should not cause confusion in future sections, because as each mathematical topic is introduced, it will be clear whether a one-dimensional or a two-dimensional coordinate system is involved.

THE DISTANCE AND MIDPOINT FORMULAS

The Cartesian coordinate system makes it possible to combine the concepts of algebra and geometry into a branch of mathematics called *analytic geometry*.

The distance between two points on a horizontal line is the absolute value of the difference between the x-coordinates of the two points. The distance between two points on a vertical line is the absolute value of the difference between the y-coordinates of the two points. For example, as shown in **Figure 3.4**, the distance d between the points with coordinates $(1, 2)$ and $(1, -3)$ is $d = |2 - (-3)| = 5$.

If two points are not on a horizontal or vertical line, then a *distance formula* for the distance between the two points can be developed as follows.

The distance between the points $P_1(x_1, y_1)$ and $P_2(x_2, y_2)$ in **Figure 3.5** is the length of the hypotenuse of a right triangle whose sides are horizontal and vertical line segments that measure $|x_2 - x_1|$ and $|y_2 - y_1|$, respectively. Applying the Pythagorean Theorem to this triangle produces

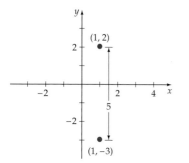

Figure 3.4

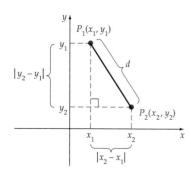

Figure 3.5

$$d^2 = |x_2 - x_1|^2 + |y_2 - y_1|^2$$
$$d = \sqrt{|x_2 - x_1|^2 + |y_2 - y_1|^2}$$

- **The square root theorem. Because d is nonnegative, the negative root is not listed.**

$$= \sqrt{(x_2 - x_1)^2 + (y_2 - y_1)^2}$$

- **Because $|x_2 - x_1|^2 = (x_2 - x_1)^2$ and $|y_2 - y_1|^2 = (y_2 - y_1)^2$**

Thus we have established the following theorem.

ANSWER Car sales decreased between the years 1986 and 1990.

THE DISTANCE FORMULA

The distance d between the points $P_1(x_1, y_1)$ and $P_2(x_2, y_2)$ is

$$d = \sqrt{(x_2 - x_1)^2 + (y_2 - y_1)^2}$$

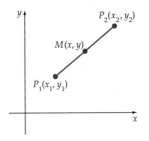

Figure 3.6

The distance d between the points whose coordinates are $P_1(x_1, y_1)$ and $P_2(x_2, y_2)$ is denoted by $d(P_1, P_2)$. To find the distance $d(P_1, P_2)$ between the points $P_1(-3, 4)$ and $P_2(7, 2)$, we apply the distance formula with $x_1 = -3$, $y_1 = 4$, $x_2 = 7$, and $y_2 = 2$.

$$\begin{aligned} d(P_1, P_2) &= \sqrt{(x_2 - x_1)^2 + (y_2 - y_1)^2} \\ &= \sqrt{[7 - (-3)]^2 + (2 - 4)^2} \\ &= \sqrt{104} = 2\sqrt{26} \approx 10.2 \end{aligned}$$

The **midpoint** M of a line segment is the point on the line segment that is equidistant from the endpoints $P_1(x_1, y_1)$ and $P_2(x_2, y_2)$ of the segment. See Figure 3.6.

POINT OF INTEREST

The midpoint formula is a special case of the following theorem. The coordinates (x, y) of the point P dividing the line segment from $P_1(x_1, y_1)$ to $P_2(x_2, y_2)$ into the ratio

$$\frac{d(P_1, P)}{d(P, P_2)} = r$$

are given by the formulas

$$x = \frac{x_1 + rx_2}{1 + r} \quad \text{and}$$

$$y = \frac{y_1 + ry_2}{1 + r}$$

See Exercises 93 to 96, Exercise Set 3.1.

THE MIDPOINT FORMULA

The midpoint M of the line segment from $P_1(x_1, y_1)$ to $P_2(x_2, y_2)$ is given by

$$\left(\frac{x_1 + x_2}{2}, \frac{y_1 + y_2}{2} \right)$$

The midpoint formula states that the x-coordinate of the midpoint of a line segment is the *average* of the x-coordinates of the endpoints of the line segment and that the y-coordinate of the midpoint of a line segment is the average of the y-coordinates of the endpoints of the line segment.

The midpoint M of the line segment connecting $P_1(-2, 6)$ and $P_2(3, 4)$ is

$$M = \left(\frac{x_1 + x_2}{2}, \frac{y_1 + y_2}{2} \right) = \left(\frac{(-2) + 3}{2}, \frac{6 + 4}{2} \right) = \left(\frac{1}{2}, 5 \right)$$

GRAPH OF AN EQUATION

The equations below are equations in two variables.

$$y = 3x^3 - 4x + 2 \qquad x^2 + y^2 = 25 \qquad y = \frac{x}{x + 1}$$

The solution of an equation in two variables is an ordered pair (x, y) whose co-ordinates satisfy the equation. For instance, the ordered pairs $(3, 4)$, $(4, -3)$, and $(0, 5)$ are some of the solutions of $x^2 + y^2 = 25$. Generally, there are an infinite number of solutions of an equation in two variables. These solutions can be displayed in a *graph*.

GRAPH OF AN EQUATION

The **graph of an equation** in the two variables x and y is the set of all points whose coordinates satisfy the equation.

Consider $y = 2x - 1$. Substituting various values of x into the equation and solving for y produces some of the ordered pairs of the equation. It is convenient to record the results in a table similar to the one shown below. The graph of the ordered pairs is shown in **Figure 3.7**.

x	$y = 2x - 1$	y	(x, y)
-2	$2(-2) - 1$	-5	$(-2, -5)$
-1	$2(-1) - 1$	-3	$(-1, -3)$
0	$2(0) - 1$	-1	$(0, -1)$
1	$2(1) - 1$	1	$(1, 1)$
2	$2(2) - 1$	3	$(2, 3)$

Choosing some noninteger values of x produces more ordered pairs to graph, such as $(-3/2, -4)$ and $(5/2, 4)$, as shown in **Figure 3.8**. Using still other values of x would result in more and more ordered pairs being graphed. The result would be so many dots that the graph would appear as the straight line shown in **Figure 3.9**, which is the graph of $y = 2x - 1$.

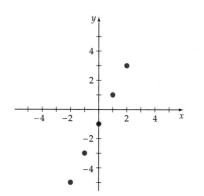

Figure 3.7

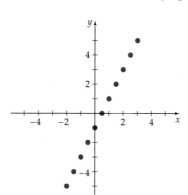

Figure 3.8

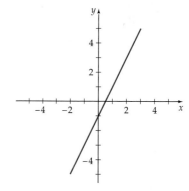

Figure 3.9

EXAMPLE 1   *Draw a Graph by Plotting Points*

Graph: $-x^2 + y = 1$

Solution

Solve the equation for y.

$$y = x^2 + 1$$

Continued • ➤

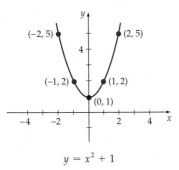

$$y = x^2 + 1$$

Figure 3.10

Select values of x and use the equation to calculate y. Choose enough values of x so that an accurate graph can be drawn. Plot the points and draw a curve through them. See **Figure 3.10**.

x	$y = x^2 + 1$	y	(x, y)
-2	$(-2)^2 + 1$	5	$(-2, 5)$
-1	$(-1)^2 + 1$	2	$(-1, 2)$
0	$(0)^2 + 1$	1	$(0, 1)$
1	$(1)^2 + 1$	2	$(1, 2)$
2	$(2)^2 + 1$	5	$(2, 5)$

TRY EXERCISE 26, EXERCISE SET 3.1

Some graphing calculators, such as the *TI-82*, have a TABLE feature that allows you to create a table similar to the one shown in Example 1. Enter the equation to be graphed, the first value for x, and the increment (the difference between successive values of x). For instance, entering $y_1 = x^2 + 1$, an initial value of x as -2, and an increment of 1 yields a display similar to the one in **Figure 3.11**. Changing the initial value to -6 and the increment to 2 gives the table in **Figure 3.12**.

X	Y₁		
-2	5		
-1	2		
0	1		
1	2		
2	5		
3	10		
4	17		

X=-2

X	Y₁		
-6	37		
-4	17		
-2	5		
0	1		
2	5		
4	17		
6	37		

X=-6

Figure 3.11   **Figure 3.12**

With some calculators, you may scroll through the table by using the up- or down-arrow keys. In this way, you can determine many more ordered pairs of the graph.

EXAMPLE 2   *Graph by Plotting Points*

Graph: $y = |x - 2|$

Solution

This equation is already solved for y, so start by choosing an x value and using the equation to determine the corresponding y value. For

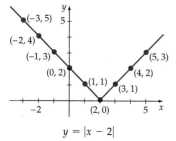

$y = |x - 2|$

Figure 3.13

example, if $x = -3$, then $y = |(-3) - 2| = |-5| = 5$. Continuing in this manner produces the following table:

When x is	−3	−2	−1	0	1	2	3	4	5
y is	5	4	3	2	1	0	1	2	3

Now plot the points listed in the table. Connecting the points forms a V shape, as shown in **Figure 3.13.**

TRY EXERCISE 30, EXERCISE SET 3.1

EXAMPLE 3   *Graph by Plotting Points*

Graph: $y^2 = x$

Solution

Solving this equation for y yields

$$y = \pm\sqrt{x}$$

Choose several x-values and use the equation to determine the corresponding y values.

When x is	0	1	4	9	16
y is	0	±1	±2	±3	±4

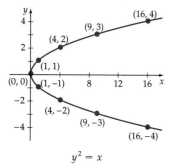

$y^2 = x$

Figure 3.14

Plot the points as shown in **Figure 3.14.** The graph is a *parabola.*

TRY EXERCISE 32, EXERCISE SET 3.1

A graphing calculator or computer graphing software can be used to draw the graphs in Example 2 and Example 3. These graphing utilities graph a curve in much the same way as you would, by selecting values of x and calculating the corresponding values of y. A curve is then drawn through the points.

If you use a graphing utility to graph $y = |x - 2|$, you will need to use the *absolute value* function that is built-in to the utility. The equation you enter will look similar to Y₁=abs(X−2).

To graph the equation in Example 3, you will enter two equations. The equations you enter will be similar to

$$Y_1 = \sqrt{X}$$
$$Y_2 = -\sqrt{X}$$

The first equation will graph the top half of the curve and the second equation will graph the bottom half.

INTERCEPTS

Any point that has an x- or a y-coordinate of zero is called an **intercept** of the graph of an equation, because it is at these points that the graph intersects the x- or the y-axis.

DEFINITION OF x-INTERCEPTS AND y-INTERCEPTS

If $(x_1, 0)$ satisfies an equation, then the point $(x_1, 0)$ is called an **x-intercept** of the graph of the equation.

If $(0, y_1)$ satisfies an equation, then the point $(0, y_1)$ is called a **y-intercept** of the graph of the equation.

To find the x-intercepts of the graph of an equation, let $y = 0$ and solve the equation for x. To find the y-intercepts of the graph of an equation, let $x = 0$ and solve the equation for y.

EXAMPLE 4   *Find Intercepts and Graph an Equation*

Graph: $y = x^2 - 2x - 3$

Solution

To find any y-intercepts, let $x = 0$ and solve for y.

$$y = 0^2 - 2(0) - 3 = -3$$

The y intercept is $(0, -3)$. To find the x-intercepts, let $y = 0$ and solve for x.

$$0 = x^2 - 2x - 3$$
$$0 = (x - 3)(x + 1)$$
$$(x - 3) = 0 \quad \text{or} \quad (x + 1) = 0$$
$$x = 3 \quad \text{or} \qquad x = -1$$

Thus the x-intercepts are $(3, 0)$ and $(-1, 0)$. We use the equation to find a few additional points on the graph.

When x is	1	2	4
y is	−4	−3	5

Drawing a smooth curve through the intercepts and the points $(1, -4)$, $(2, -3)$, and $(4, 5)$ produces the graph of the parabola in **Figure 3.15**.

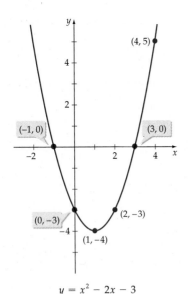

$y = x^2 - 2x - 3$

Figure 3.15

TRY EXERCISE 40, EXERCISE SET 3.1

CIRCLES, THEIR EQUATIONS, AND THEIR GRAPHS

Frequently you will sketch graphs by plotting points. However, some graphs can be sketched by merely recognizing the form of the equation. A *circle* is an example of a curve whose graph you can sketch after you have inspected its equation.

DEFINITION OF A CIRCLE

A **circle** is the set of points in a plane that are a fixed distance from a specified point. The distance is the **radius** of the circle, and the specified point is the **center** of the circle.

STANDARD FORM OF THE EQUATION OF A CIRCLE

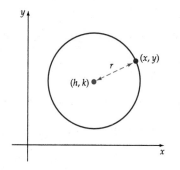

Figure 3.16

The standard form of the equation of a circle is derived by using this definition. To derive the standard form, we use the distance formula. **Figure 3.16** is a circle with center (h, k) and radius r. The point (x, y) is on the circle if and only if it is a distance of r units from the center (h, k). Thus (x, y) is on the circle if and only if

$$\sqrt{(x - h)^2 + (y - k)^2} = r$$
$$(x - h)^2 + (y - k)^2 = r^2 \qquad \bullet \text{ Square each side.}$$

STANDARD FORM OF THE EQUATION OF A CIRCLE

The **standard form of the equation of a circle** with center at (h, k) and radius r is

$$(x - h)^2 + (y - k)^2 = r^2$$

For example, the equation $(x - 3)^2 + (y + 1)^2 = 4$ is the equation of a circle. The standard form of the equation is

$$(x - 3)^2 + (y - (-1))^2 = 2^2$$

from which it can be determined that $h = 3, k = -1$, and $r = 2$. Thus the graph is a circle centered at $(3, -1)$ with a radius of 2.

If a circle is centered at the origin $(0, 0)$ (that is, if $h = 0$ and $k = 0$), then the standard form of the equation of the circle simplifies to

$$x^2 + y^2 = r^2$$

For example, the graph of $x^2 + y^2 = 9$ is a circle with center at the origin and radius of 3.

QUESTION   What are the radius and the coordinates of the center of the circle with equation $x^2 + (y - 2)^2 = 10$?

EXAMPLE 5   Find the Standard Form of the Equation of a Circle

Find the standard form of the equation of a circle that has center $C(-4, -2)$ and contains the point $P(-1, 2)$.

Solution

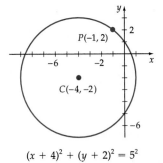

$(x + 4)^2 + (y + 2)^2 = 5^2$

Figure 3.17

See the graph of the circle in **Figure 3.17**. Because the point P is on the circle, the radius r of the circle must equal the distance from C to P. Thus

$$r = \sqrt{(-1 - (-4))^2 + (2 - (-2))^2}$$
$$= \sqrt{9 + 16} = \sqrt{25} = 5$$

Using the standard form with $h = -4$, $k = -2$, and $r = 5$, we obtain

$$(x + 4)^2 + (y + 2)^2 = 5^2$$

TRY EXERCISE 66, EXERCISE SET 3.1

GENERAL FORM

If we rewrite $(x + 4)^2 + (y + 2)^2 = 5^2$ by squaring and combining like terms, we produce

$$x^2 + 8x + 16 + y^2 + 4y + 4 = 25$$
$$x^2 + y^2 + 8x + 4y - 5 = 0$$

This form of the equation is known as the **general form of the equation of a circle**. By completing the square, it is always possible to write the general form $x^2 + y^2 + Ax + By + C = 0$ in the standard form

$$(x - h)^2 + (y - k)^2 = s$$

for some number s. If $s > 0$, the graph is a circle with radius $r = \sqrt{s}$. If $s = 0$, the graph is the point (h, k), and if $s < 0$, the equation has no real solutions and there is no graph.

EXAMPLE 6   Find the Center and Radius of a Circle by Completing the Square

Find the center and the radius of the circle that is given by

$$x^2 + y^2 - 6x + 4y - 3 = 0$$

ANSWER   The radius is $\sqrt{10}$ and the coordinates of the center are $(0, 2)$.

Solution

First rearrange and group the terms as shown.

$$(x^2 - 6x) + (y^2 + 4y) = 3$$

Now complete the square of $(x^2 - 6x)$ and $(y^2 + 4y)$.

$$(x^2 - 6x + 9) + (y^2 + 4y + 4) = 3 + 9 + 4 \qquad \bullet \text{ Add 9 and 4 to each}$$
$$(x - 3)^2 + (y + 2)^2 = 16 \qquad\qquad \text{side of the equation.}$$
$$(x - 3)^2 + (y - (-2))^2 = 4^2$$

The above standard form indicates that the graph of the original equation is a circle with center $(3, -2)$ and radius 4. See **Figure 3.18**.

> **TRY EXERCISE 68, EXERCISE SET 3.1**

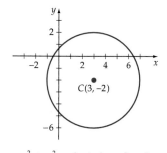

$$x^2 + y^2 - 6x + 4y - 3 = 0$$

Figure 3.18

TOPICS FOR DISCUSSION

1. The distance formula states that the distance d between the points $P_1(x_1, y_1)$ and $P_2(x_2, y_2)$ is $d = \sqrt{(x_2 - x_1)^2 + (y_2 - y_1)^2}$. Can the distance formula also be written as follows? Explain.

$$d = \sqrt{(x_1 - x_2)^2 + (y_1 - y_2)^2}$$

2. A tutor states that the equation $(x - 3)^2 + (y + 4)^2 = -6$ does not graph to be a circle. Do you agree? Explain.

3. Explain why the graph of $|x| + |y| = 1$ does not contain any points that have

    a. a y-coordinate that is greater than 1 or less than -1.

    b. an x-coordinate that is greater than 1 or less than -1.

4. A tutor claims that the graph of $xy = 0$ consists of the x-axis and the y-axis. Do you agree?

5. Explain how to determine the x- and y-intercepts of a graph (without using the graph).

EXERCISE SET 3.1

In Exercises 1 to 2, plot the points whose coordinates are given on a Cartesian coordinate system.

1. $(2, 4)$, $(0, -3)$, $(-2, 1)$, $(-5, -3)$

2. $(-3, -5)$, $(-4, 3)$, $(0, 2)$, $(-2, 0)$

3.   **OLYMPIC SPRINT** The winning times in the women's Olympic 100-meter dash are given on page 142.

    a. Construct a scatter plot for the years from 1928 to 1968.

b. Construct a scatter plot for the years from 1972 to 1992.

c. Explain why the winner in 1968 may have run a faster race than the winner in 1984.

Women's Olympic 100-meter Dash

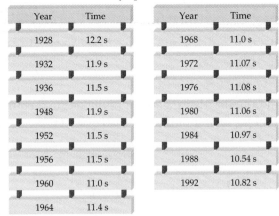

Year	Time		Year	Time
1928	12.2 s		1968	11.0 s
1932	11.9 s		1972	11.07 s
1936	11.5 s		1976	11.08 s
1948	11.9 s		1980	11.06 s
1952	11.5 s		1984	10.97 s
1956	11.5 s		1988	10.54 s
1960	11.0 s		1992	10.82 s
1964	11.4 s			

4. **OLYMPIC FREESTYLE** The winning times in the men's Olympic 200-meter freestyle are listed below.

a. Construct a scatter plot for the data.

b. What rule change took place after the 1968 Olympics and before the 1972 Olympics?

Men's Olympic 200-meter Freestyle

Year	Time
1968	1:55.2
1972	1:52.78
1976	1:50.29
1980	1:49.81
1984	1:47.44
1988	1:47.25
1992	1:46.70

In Exercises 5 to 16, find the distance between the points whose coordinates are given.

5. $(6, 4)$, $(-8, 11)$

6. $(-5, 8)$, $(-10, 14)$

7. $(-4, -20)$, $(-10, 15)$

8. $(40, 32)$, $(36, 20)$

9. $(5, -8)$, $(0, 0)$

10. $(0, 0)$, $(5, 13)$

11. $(\sqrt{3}, \sqrt{8})$, $(\sqrt{12}, \sqrt{27})$

12. $(\sqrt{125}, \sqrt{20})$, $(6, 2\sqrt{5})$

13. (a, b), $(-a, -b)$

14. $(a - b, b)$, $(a, a + b)$

15. $(x, 4x)$, $(-2x, 3x)$ given that $x < 0$

16. $(x, 4x)$, $(-2x, 3x)$ given that $x > 0$

17. Find all points on the x-axis that are 10 units from $(4, 6)$. (*Hint:* First write the distance formula with $(4, 6)$ as one of the points and $(x, 0)$ as the other point.)

18. Find all points on the y-axis that are 12 units from $(5, -3)$.

In Exercises 19 to 24, find the midpoint of the line segment with the following endpoints.

19. $(1, -1)$, $(5, 5)$

20. $(-5, -2)$, $(6, 10)$

21. $(6, -3)$, $(6, 11)$

22. $(4, 7)$, $(-10, 7)$

23. $(1.75, 2.25)$, $(-3.5, 5.57)$

24. $(-8.2, 10.1)$, $(-2.4, -5.7)$

In Exercises 25 to 38, graph each equation by plotting points that satisfy the equation.

25. $x - y = 4$

26. $2x + y = -1$

27. $y = 0.25x^2$

28. $3x^2 + 2y = -4$

29. $y = -2|x - 3|$

30. $y = |x + 3| - 2$

31. $y = x^2 - 3$

32. $y = x^2 + 1$

33. $y = \frac{1}{2}(x - 1)^2$

34. $y = 2(x + 2)^2$

35. $y = x^2 + 2x - 8$

36. $y = x^2 - 2x - 8$

37. $y = -x^2 + 2$

38. $y = -x^2 - 1$

In Exercises 39 to 48, find the x- and the y-intercepts of the graph of each equation. Use the intercepts to draw the graph of the equation.

39. $2x + 5y = 12$

40. $3x - 4y = 15$

41. $x = -y^2 + 5$

42. $x = y^2 - 6$

43. $x = |y| - 4$

44. $x = y^3 - 2$

45. $x^2 + y^2 = 4$

46. $x^2 = y^2$

47. $|x| + |y| = 4$

48. $|x - 4y| = 8$

In Exercises 49 to 58, determine the center and radius of the circle with the given equation.

49. $x^2 + y^2 = 36$

50. $x^2 + y^2 = 49$

51. $x^2 + y^2 = 10^2$

52. $x^2 + y^2 = 4^2$

53. $(x - 1)^2 + (y - 3)^2 = 7^2$

54. $(x - 2)^2 + (y - 4)^2 = 5^2$

55. $(x + 2)^2 + (y + 5)^2 = 25$

56. $(x + 3)^2 + (y + 5)^2 = 121$

57. $(x - 8)^2 + y^2 = \frac{1}{4}$

58. $x^2 + (y - 12)^2 = 1$

In Exercises 59 to 66, find an equation of a circle that satisfies the given conditions. Write your answer in standard form.

59. Center $(4, 1)$, radius $r = 2$

60. Center $(5, -3)$, radius $r = 4$

61. Center $(1/2, 1/4)$, radius $r = \sqrt{5}$

62. Center $(0, 2/3)$, radius $r = \sqrt{11}$

63. Center $(0, 0)$, passing through $(-3, 4)$

64. Center $(0, 0)$, passing through $(5, 12)$

65. Center $(1, 3)$, passing through $(4, -1)$

66. Center $(-2, 5)$, passing through $(1, 7)$

In Exercises 67 to 76, find the center and the radius of each circle. The equations of the circles are written in the general form.

67. $x^2 + y^2 - 6x + 5 = 0$

68. $x^2 + y^2 - 6x - 4y + 12 = 0$

69. $x^2 + y^2 - 4x - 10y + 20 = 0$

70. $x^2 + y^2 + 4x - 2y - 11 = 0$

71. $x^2 + y^2 - 14x + 8y + 56 = 0$

72. $x^2 + y^2 - 10x + 2y + 25 = 0$

73. $4x^2 + 4y^2 + 4x - 63 = 0$

74. $9x^2 + 9y^2 - 6y - 17 = 0$

75. $x^2 + y^2 - x + \dfrac{2}{3}y + \dfrac{1}{3} = 0$

76. $x^2 + y^2 - 2x + 2y + \dfrac{7}{4} = 0$

SUPPLEMENTAL EXERCISES

In Exercises 77 to 88, graph the set of all points whose x- and y-coordinates satisfy the given conditions.

77. $x = 3$ **78.** $y = 2$

79. $x = 1, y \geq 1$ **80.** $y = -3, x \geq -2$

81. $y \leq 3$ **82.** $x \geq 2$

83. $xy \geq 0$ **84.** $|y| \geq 1, \dfrac{x}{y} \leq 0$

85. $|x| = 2, |y| = 3$ **86.** $|x| = 4, |y| = 1$

87. $|x| \leq 2, y \geq 2$ **88.** $x \geq 1, |y| \leq 3$

In Exercises 89 to 92, find the other endpoint of the line segment that has the given endpoint and midpoint.

89. Endpoint $(5, 1)$, midpoint $(9, 3)$

90. Endpoint $(4, -6)$, midpoint $(-2, 11)$

91. Endpoint $(-3, -8)$, midpoint $(2, -7)$

92. Endpoint $(5, -4)$, midpoint $(0, 0)$

The coordinates (x, y) of the point P that divides the line segment from $P_1(x_1, y_1)$ to $P_2(x_2, y_2)$ into the ratio

$$\frac{d(P_1, P)}{d(P, P_2)} = r$$

are given by the formulas

$$x = \frac{x_1 + rx_2}{1 + r} \quad \text{and} \quad y = \frac{y_1 + ry_2}{1 + r}$$

In Exercises 93 to 96 use these formulas to find the indicated point on the line segment with endpoints P_1 and P_2.

93. Find the point three-fourths of the way from $P_1(-8, 11)$ to $P_2(6, 20)$.

94. Find the point two-fifths of the way from $P_1(-9, 6)$ to $P_2(2, 4)$.

95. Find the point seven-eighths of the way from $P_1(6, 10)$ to $P_2(2, 1)$.

96. Find the point nine-tenths of the way from $P_1(-1, 8)$ to $P_2(0, 3)$.

97. Use the distance formula to determine whether the points given by $(1, -4)$, $(3, 2)$, $(-3, 4)$, and $(-5, -2)$ are the vertices of a square.

98. Use the distance formula to determine whether the points given by $(2, -1)$, $(5, 0)$, $(6, 3)$, and $(3, 2)$ are the vertices of a parallelogram, a rhombus, or a square.

99. Find a formula for the set of all points (x, y) for which the distance from (x, y) to $(3, 4)$ is 5.

100. Find a formula for the set of all points (x, y) for which the distance from (x, y) to $(-5, 12)$ is 13.

101. Find a formula for the set of all points (x, y) for which the sum of the distances from (x, y) to $(4, 0)$ and from (x, y) to $(-4, 0)$ is 10.

102. Find a formula for the set of all points for which the absolute value of the difference of the distances from (x, y) to $(0, 4)$ and from (x, y) to $(0, -4)$ is 6.

103. Find an equation of a circle that has a diameter with endpoints $(2, 3)$ and $(-4, 11)$. Write your answer in standard form.

104. Find an equation of a circle that has a diameter with endpoints $(7, -2)$ and $(-3, 5)$. Write your answer in standard form.

105. Find an equation of a circle that has its center at $(7, 11)$ and is tangent to the x-axis. Write your answer in standard form.

106. Find an equation of a circle that has its center at $(-2, 3)$ and is tangent to the y-axis. Write your answer in standard form.

107. Find an equation of a circle that is tangent to both axes, has its center in the second quadrant, and has a radius of 3.

108. Find an equation of a circle that is tangent to both axes, has its center in the third quadrant, and has a diameter of $\sqrt{5}$.

PROJECTS

1. VERIFY A GEOMETRIC THEOREM Use the midpoint formula and the distance formula to prove that the midpoint M of the hypotenuse of a right triangle is equidistant from each of the vertices of the triangle. (*Hint:* Label the vertices of the triangle as shown in the figure at the right.)

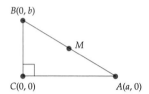

2. SOLVE A QUADRATIC EQUATION GEOMETRICALLY In the 17th century, Descartes (and others) solved equations by using both algebra and geometry. This project outlines the method Descartes used to solve certain quadratic equations.

   a. Consider the equation $x^2 = 2ax + b^2$. Construct a right triangle ABC with $d(A, C) = a$ and $d(C, B) = b$. Now draw a circle with center at A and radius a. Let P be the point at which the circle intersects the hypotenuse of the right triangle and Q the point where an extension of the hypotenuse intersects the circle. Your drawing should be similar to the one at the right.

   b. Show that a solution of the equation $x^2 = 2ax + b^2$ is $d(Q, B)$.

   c. Show that $d(P, B)$ is a solution of the equation $x^2 = -2ax + b^2$.

   d. Construct a line parallel to AC and passing through B. Let S and T be the points at which the line intersects the circle. Show that $d(S, B)$ and $d(T, B)$ are solutions of the equation $x^2 = 2ax - b^2$.

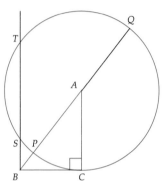

SECTION 3.2 INTRODUCTION TO FUNCTIONS

RELATIONS

In many situations in science, business, and mathematics, a correspondence exists between two sets. The correspondence is often defined by a *table,* an *equation,* or a *graph,* each of which can be viewed from a mathematical perspective as a set of ordered pairs. In mathematics, any set of ordered pairs is called a **relation.**

   Table 3.1 defines a correspondence between a set of percent scores and a set of letter grades. For each score from 0 to 100, there corresponds only one letter grade. The score 94% corresponds to the letter grade of A. Using ordered-pair notation we record this correspondence as (94, A).

   The *equation* $d = 16t^2$ indicates the distance d that a rock falls (neglecting air resistance) corresponds to the time t that it has been falling. For each non-

Table 3.1

Score	Grade
[90, 100]	A
[80, 90)	B
[70, 80)	C
[60, 70)	D
[0, 60)	F

negative value t, the equation assigns only one value for the distance d. According to this equation, in 3 seconds a rock will fall 144 feet, which we record as $(3, 144)$. Some of the other ordered pairs determined by $d = 16t^2$ are $(0, 0)$, $(1, 16)$, $(2, 64)$, and $(2.5, 100)$.

$$\textbf{Equation:} \qquad d = 16t^2$$
$$\text{If } t = 3, \text{ then} \quad d = 16(3)^2 = 144$$

The *graph* in **Figure 3.19** defines a correspondence between the length of a pendulum and the time it takes the pendulum to complete one oscillation. For each nonnegative pendulum length, the graph yields only one time. According to the graph, a pendulum length of 2 feet yields an oscillation time of 1.6 seconds, and a length of 4 feet yields an oscillation time of 2.2 seconds, where the time is measured to the nearest tenth of a second. These results can be recorded as the ordered pairs $(2, 1.6)$ and $(4, 2.2)$.

Graph: A pendulum's oscillation time

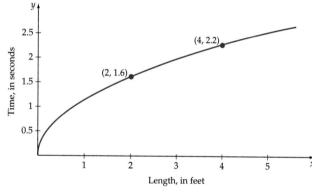

Figure 3.19

FUNCTIONS

The preceding table, equation, and graph each determines a special type of relation called a *function*.

DEFINITION OF A FUNCTION

A **function** is a set of ordered pairs in which no two ordered pairs that have the same first coordinate have different second coordinates.

Although every function is a relation, not every relation is a function. For instance, consider $(94, A)$ from the grading correspondence. The first coordinate, 94, is paired with a second coordinate of A. It would not make sense to have 94 paired with A, $(94, A)$, and 94 paired with B, $(94, B)$. The same first coordinate would be paired with two different second coordinates. This would mean that two students with the same score received different grades, one student an A and the other a B!

Functions may have ordered pairs with the same second coordinate. For instance, (94, A) and (95, A) are both ordered pairs that belong to the function defined by Table 3.1. A function may have different first coordinates and the same second coordinate.

The equation $d = 16t^2$ represents a function because for each value of t there is only one value of d. Not every equation, however, represents a function. For instance, $y^2 = 25 - x^2$ does not represent a function. The ordered pairs $(-3, 4)$ and $(-3, -4)$ are both solutions of the equation. But these ordered pairs do not satisfy the definition of a function; there are two ordered pairs with the same first coordinate but *different* second coordinates.

QUESTION   Does the set {(0, 0), (1, 0), (2, 0), (3, 0), (4, 0)} define a function?

The **domain** of a function is the set of all the first coordinates of the ordered pairs. The **range** of a function is the set of all the second coordinates. In the function determined by the grading correspondence in Table 3.1, the domain is the interval [0, 100]. The range is {A, B, C, D, F}. In a function, each domain element is paired with one and only one range element.

If a function is defined by an equation, the variable that represents elements of the domain is the **independent variable.** The variable that represents elements of the range is the **dependent variable.** In the free-fall experiment, we used the equation $d = 16t^2$. The elements of the domain represented the time the rock fell, and the elements of the range were used for the distance the rock fell. Thus in $d = 16t^2$, the independent variable is t and the dependent variable is d.

The specific letters used for the independent and the dependent variable are not important. For example, $y = 16x^2$ represents the same function as $d = 16t^2$. Traditionally, x is used for the independent variable, and y for the dependent variable. Anytime we use the phrase "y is a function of x" or a similar phrase with different letters, the variable that follows "function of" is the independent variable.

FUNCTIONAL NOTATION

Functions can be named by using a letter or a combination of letters, such as f, g, A, log, or tan. If x is an element of the domain of f, then $f(x)$, which is read "f of x" or "the value of f at x," is the element in the range of f that corresponds with the domain element x. The notation "f" and the notation "$f(x)$" mean different things. "f" is the name of the function, whereas "$f(x)$" is the value of the function at x. Finding the value of $f(x)$ is referred to as *evaluating* f at x. To evaluate $f(x)$ at $x = a$, substitute a for x, and simplify.

ANSWER   Yes. There are no two ordered pairs with the same first coordinate that have different second coordinates.

EXAMPLE 1   *Evaluate Functions*

Let $f(x) = x^2 - 1$, and evaluate.

a. $f(-5)$ **b.** $f(3b)$ **c.** $3f(b)$ **d.** $f(a + 3)$ **e.** $f(a) + f(3)$

Solution

a. $f(-5) = (-5)^2 - 1 = 25 - 1 = 24$ • **Substitute −5 for x, and simplify.**

b. $f(3b) = (3b)^2 - 1 = 9b^2 - 1$ • **Substitute 3b for x, and simplify.**

c. $3f(b) = 3(b^2 - 1) = 3b^2 - 3$ • **Substitute b for x, and simplify.**

d. $f(a + 3) = (a + 3)^2 - 1$ • **Substitute a + 3 for x.**
$= a^2 + 6a + 8$ • **Simplify.**

e. $f(a) + f(3) = (a^2 - 1) + (3^2 - 1)$ • **Substitute a for x; substitute 3 for x.**

$= a^2 + 7$ • **Simplify.**

TRY EXERCISE 2, EXERCISE SET 3.2

> **TAKE NOTE**
>
> In Example 1, observe that
> $$f(3b) \neq 3f(b)$$
> and that
> $$f(a + 3) \neq f(a) + f(3)$$

Piecewise-defined functions are functions represented by more than one expression. The function C in Example 2 is a piecewise-defined function. To evaluate a piecewise-defined function for a domain value a, substitute a in the proper expression and simplify. We determine the appropriate expression to evaluate by examining the domain given to the right of each expression.

EXAMPLE 2   *Evaluate a Piecewise-Defined Function*

The *bar graph* in **Figure 3.20** shows the number of students per computer in public high schools from 1985 to 1996.

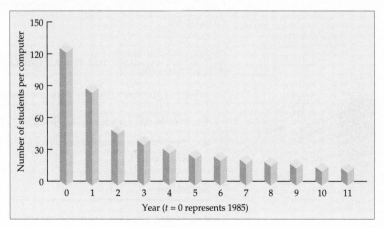

Figure 3.20

Continued •➤

The data in the bar graph can be approximated by

$$C(t) = \begin{cases} 5.1607t^2 - 45.2607t + 124.679 & \text{if } 0 \le t \le 5 \\ -2.0357t + 34.6429 & \text{if } 6 \le t \le 11 \end{cases}$$

where $C(t)$ is the number of students per computer for the year t. The variable t is a whole number and $t = 0$ corresponds to the year 1985. Use this function to estimate (to the nearest student) the number of students per computer in the following years.

a. 1988   **b.** 1992

Solution

a. The year 1988 corresponds to a value of $t = 3$. Because 3 is in the interval $0 \le t \le 5$, evaluate $5.1607t^2 - 45.2607t + 124.679$ at 3.

$$5.1607(3^2) - 45.2607(3) + 124.679 \approx 35$$

The model predicts that there were 35 students per computer in 1988.

b. The year 1992 corresponds to a value of $t = 7$. Because 7 is in the interval $6 \le t \le 11$, evaluate $-2.0357t + 34.6429$ at 7.

$$-2.0357(7) + 34.6429 \approx 20$$

The model predicts that there were 20 students per computer in 1992.

TRY EXERCISE 10, EXERCISE SET 3.2

IDENTIFYING FUNCTIONS

Recall that although every function is a relation, not every relation is a function. In the next example, we examine four relations to determine which are functions.

EXAMPLE 3   *Identify Functions*

Which relations define y as a function of x?

a. $\{(2, 3), (4, 1), (4, 5)\}$ **b.** $3x + y = 1$ **c.** $-4x^2 + y^2 = 9$

d. The correspondence between the x values and the y values in **Figure 3.21.**

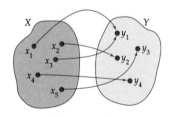

Figure 3.21

Solution

a. There are two ordered pairs, $(4, 1)$ and $(4, 5)$, with the same first coordinate and different second coordinates. This set does not define y as a function of x.

b. Solving $3x + y = 1$ for y yields $y = -3x + 1$. Because $-3x + 1$ is a unique real number for each x, this equation defines y as a function of x.

c. Solving $-4x^2 + y^2 = 9$ for y yields $y = \pm\sqrt{4x^2 + 9}$. The right side $\pm\sqrt{4x^2 + 9}$ produces two values of y for each value of x. For example, when $x = 0$, $y = 3$ or $y = -3$. Thus $-4x^2 + y^2 = 9$ does not define y as a function of x.

d. Each x is paired with one and only one y. The correspondence in **Figure 3.21** defines y as a function of x.

TRY EXERCISE 14, EXERCISE SET 3.2

Sometimes the domain of a function is explicitly stated. For example, each of f, g, and h below is given by an equation, followed by a statement that indicates the domain of the function.

$$f(x) = x^2, x > 0; \qquad g(t) = \frac{1}{t^2 + 4}, 0 \le t \le 5; \qquad h(x) = x^2, x = 1, 2, 3$$

Although f and h have the same equation, they are different functions because they have different domains. If the domain of a function is not explicitly stated, then its domain is determined by the following convention.

DOMAIN OF A FUNCTION

Unless otherwise stated, the domain of a function is the set of all real numbers for which the function makes sense and yields real numbers.

EXAMPLE 4 *Determine the Domain of a Function*

Determine the domain of each function.

a. $G(t) = \dfrac{1}{t - 4}$ **b.** $f(x) = \sqrt{x + 1}$

c. $A(s) = s^2$, where $A(s)$ is the area of a square whose sides are s units.

Solution

a. The number 4 is not an element of the domain because G is undefined when the denominator $t - 4$ equals 0. The domain of G is all real numbers except 4. In interval notation the domain is $(-\infty, 4) \cup (4, \infty)$.

Continued ▶

TAKE NOTE

If a function is defined by an equation, then it may be difficult to determine the *range* of the function by examining its equation. To determine the range of a function generally requires a graph of the function or the application of a theorem such as the theorem on horizontal asymptotes that is stated in Section 4.5. Thus for the present we will concentrate on using an equation only to determine the domain of a function.

TAKE NOTE

You may indicate the domain of a function using set notation, interval notation, a graph, or words. For instance, the domain of $f(x) = \sqrt{x - 3}$ may be given in each of the following ways.

Set notation: $\{x \mid x \ge 3\}$

Interval notation: $[3, \infty)$

A graph:

$$\underset{\substack{\,\\-3 \quad 0 \quad 3 \quad 6}}{\longleftrightarrow}$$

Words: All real numbers x, where x is greater than or equal to 3.

b. The radical $\sqrt{x + 1}$ is a real number only when $x + 1 \geq 0$ or when $x \geq -1$. Thus, in set notation, the domain of f is $\{x \mid x \geq -1\}$.

c. Because s represents the length of the side of a square, s must be positive. In set notation the domain of A is $\{s \mid s > 0\}$.

TRY EXERCISE 28, EXERCISE SET 3.2

GRAPHS OF FUNCTIONS

If a is an element of the domain of a function, then $(a, f(a))$ is an ordered pair that *belongs* to the function.

> ### GRAPH OF A FUNCTION
>
> The **graph of a function** is the graph of all the ordered pairs that belong to the function.

TAKE NOTE

Knowing how many ordered pairs to plot and how to connect the points requires specific knowledge about the function. Much of the rest of this chapter will be concerned with this topic.

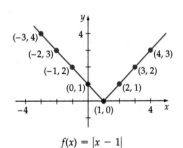

$f(x) = |x - 1|$

Figure 3.22

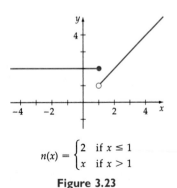

$n(x) = \begin{cases} 2 & \text{if } x \leq 1 \\ x & \text{if } x > 1 \end{cases}$

Figure 3.23

EXAMPLE 5   *Graph a Function by Plotting Points*

Graph each function. State the domain and the range of each function.

a. $f(x) = |x - 1|$ b. $n(x) = \begin{cases} 2 & \text{if } x \leq 1 \\ x & \text{if } x > 1 \end{cases}$

Solution

a. The domain of f is the set of all real numbers. Write the function as $y = |x - 1|$. Evaluate the function for several domain values. We have used $x = -3, -2, -1, 0, 1, 2, 3$, and 4.

x	-3	-2	-1	0	1	2	3	4
$y = |x - 1|$	4	3	2	1	0	1	2	3

Plot the points determined by the ordered pairs. Connect the points to form the graph in **Figure 3.22**.

Because $|x - 1| \geq 0$, we can conclude that the graph of f extends from a height of 0 upward, so the range is $\{y \mid y \geq 0\}$.

b. The domain is the union of the inequalities $x \leq 1$ and $x > 1$. Thus the domain of n is the set of all real numbers. For $x \leq 1$, graph $n(x) = 2$. This results in the horizontal ray in **Figure 3.23**. The solid circle indicates that the point $(1, 2)$ *is* part of the graph. For $x > 1$, graph $n(x) = x$. This produces the second ray in **Figure 3.23**. The open circle indicates that the point $(1, 1)$ *is not* part of the graph.

Examination of the graph shows that it includes only points whose y values are greater than 1. Thus the range of n is $\{y \mid y > 1\}$.

TRY EXERCISE 40, EXERCISE SET 3.2

A graphing utility can also be used to draw the graph of a function. For instance, to graph $f(x) = x^2 - 1$, you will enter an equation similar to $Y_1 = x^2 - 1$. The graph is shown in **Figure 3.24.**

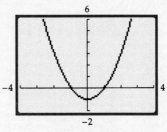

Figure 3.24

The definition that a function is a set of ordered pairs in which no two ordered pairs that have the same first coordinate have different second coordinates implies that any vertical line intersects the graph of a function at no more than one point. This is known as the *vertical line test.*

THE VERTICAL LINE TEST FOR FUNCTIONS

A graph is the graph of a function if and only if no vertical line intersects the graph at more than one point.

EXAMPLE 6 *Apply the Vertical Line Test*

Which of the following graphs are graphs of functions?

a.

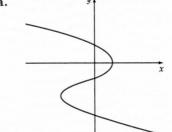

b.

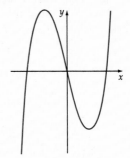

Solution

a. This graph *is not* the graph of a function because some vertical lines intersect the graph in more than one point.

b. This graph *is* the graph of a function because every vertical line intersects the graph in at most one point.

TRY EXERCISE 50, EXERCISE SET 3.2

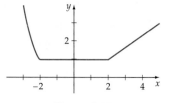

Figure 3.25

Consider the graph in **Figure 3.25.** As a point on the graph moves from left to right, this graph falls for values of $x \leq -2$, it remains the same height from $x = -2$ to $x = 2$, and it rises for $x \geq 2$. The function represented by the graph is said to be *decreasing* on the interval $(-\infty, -2]$, *constant* on the interval $[-2, 2]$, and *increasing* on the interval $[2, \infty)$.

DEFINITION OF INCREASING, DECREASING, AND CONSTANT FUNCTIONS

If a and b are elements of an interval I that is a subset of the domain of a function f, then

- f is **increasing** on I if $f(a) < f(b)$ whenever $a < b$.
- f is **decreasing** on I if $f(a) > f(b)$ whenever $a < b$.
- f is **constant** on I if $f(a) = f(b)$ for all a and b.

Recall that a function is a relation in which no two ordered pairs that have the same first coordinate have different second coordinates. This means that given any x, there is only one y that can be paired with that x. A **one-to-one function** satisfies the additional condition that given any y, there is only one x that can be paired with that given y. In a manner similar to applying the vertical line test, we can apply a *horizontal line test* to identify one-to-one functions.

HORIZONTAL LINE TEST FOR A ONE-TO-ONE FUNCTION

If every horizontal line intersects the graph of a function at most once, then the graph is the graph of a one-to-one function.

For example, some horizontal lines intersect the graph in **Figure 3.26** at more than one point. It is *not* the graph of a one-to-one function. Every horizontal line intersects the graph in **Figure 3.27** at most once. This is the graph of a one-to-one function.

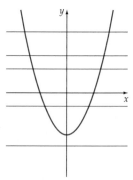

Figure 3.26

Some horizontal lines intersect this graph at more than one point. It is *not* the graph of a one-to-one function.

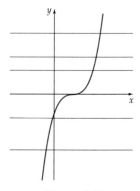

Figure 3.27

Every horizontal line intersects this graph at most once. It is the graph of a one-to-one function.

THE GREATEST INTEGER FUNCTION

The notation $[\![x]\!]$ or int(x) is defined to be the greatest integer less than or equal to x. For example:

$$\text{int}(-3) = -3 \qquad \text{int}(-7.4) = -8 \qquad \text{int}(\pi) = 3$$

The function defined by $y = [\![x]\!]$ or $y = \text{int}(x)$ is known as the **greatest integer function**. The greatest integer function is a **step function**. **Figure 3.28** shows that its graph resembles a series of steps. The graph of $y = \text{int}(x)$ has a break or *discontinuity* whenever x is an integer. The domain of $y = \text{int}(x)$ is the set of all real numbers, and its range is the set of integers.

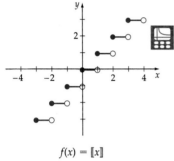

$f(x) = [\![x]\!]$

Figure 3.28

The greatest integer function is programmed into many graphing utilities. A typical display might be as shown in **Figure 3.29**.

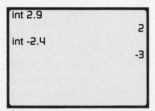

int 2.9

2

int -2.4

-3

Figure 3.29

By entering Y₁=int(X), a graph of the greatest integer function can also be produced.

EXAMPLE 7 | *Use the Greatest Integer Function to Model Expenses*

The cost of parking in a garage is \$3 for the first hour or any part of the hour and \$2 for each additional hour or any part of the hour thereafter. If x is the time in hours that you park your car, then the cost is given by

$$C(x) = 3 - 2 \text{ int}(1 - x), \quad x > 0$$

a. Evaluate $C(2)$ and $C(2.5)$. **b.** Graph $y = C(x)$ for $0 < x \le 5$.

Solution

a.
$$
\begin{aligned}
C(2) &= 3 - 2 \text{ int}(1 - 2) \\
&= 3 - 2 \text{ int}(-1) \\
&= 3 - 2(-1) \\
&= \$5
\end{aligned}
\qquad
\begin{aligned}
C(2.5) &= 3 - 2 \text{ int}(1 - 2.5) \\
&= 3 - 2 \text{ int}(-1.5) \\
&= 3 - 2(-2) \\
&= \$7
\end{aligned}
$$

Continued ▸►

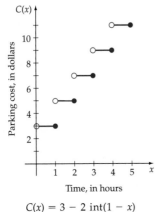

$C(x) = 3 - 2 \text{ int}(1 - x)$

Figure 3.32

b. 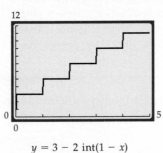 You could construct the graph by plotting several points, but the graph in **Figure 3.30** was constructed by using a graphing utility. Because the graphing utility was in "connected" mode, the graph does not show the discontinuities that occur whenever x is an integer.

The graph in **Figure 3.31** was constructed by graphing $y = C(x)$ in "dot" mode. This graph is a better representation of C because it shows the discontinuities at $x = 1$, $x = 2$, $x = 3$, and $x = 4$.

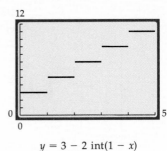

$y = 3 - 2 \text{ int}(1 - x)$

Figure 3.30

$y = 3 - 2 \text{ int}(1 - x)$

Figure 3.31

Because $C(1) = 3$, $C(2) = 5$, $C(3) = 7$, $C(4) = 9$, and $C(5) = 11$, we can use a solid circle at the right endpoint of each "step" and an open circle at the left endpoint of each "step" to better indicate the true nature of the function C as shown in **Figure 3.32**.

TRY EXERCISE 48, EXERCISE SET 3.2

 Example 7 illustrates that a graphing calculator may not produce a graph that is a good representation of a function. You may be required to *make adjustments* in the MODE, SET UP, or WINDOW of the graphing calculator so that it will produce a better representation of the function. Some graphs may also require some *fine tuning*, such as open or solid circles at particular points, to accurately represent the function.

APPLICATIONS

EXAMPLE 8   *Solve an Application*

A car was purchased for $16,500. Assuming the car depreciates at a constant rate of $2200 per year (*straight-line depreciation*) for the first 7 years, write the value v of the car as a function of time, and calculate the value of the car 3 years after purchase.

Solution

Let t represent the number of years that have passed since the car was purchased. Then $2200t$ is the amount that the car has depreciated after t years. The value of the car at time t is given by

$$v(t) = 16{,}500 - 2200t, \quad 0 \le t \le 7$$

When $t = 3$, the value of the car is

$$v(3) = 16{,}500 - 2200(3) = 16{,}500 - 6600 = \$9900$$

TRY EXERCISE 66, EXERCISE SET 3.2

Often in applied mathematics, formulas are used to determine the functional relationship that exists between two variables.

EXAMPLE 9   *Solve an Application*

A lighthouse is 2 miles south of a port. A ship leaves port and sails east at a rate of 7 mph. Express the distance d between the ship and the lighthouse as a function of time, given that the ship has been sailing for t hours.

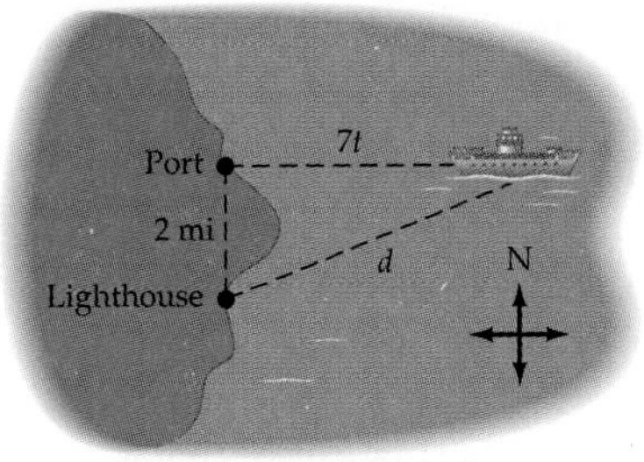

Solution

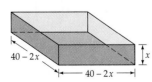

Figure 3.33

Draw a diagram and label it as shown in **Figure 3.33**. Note that because distance = (rate)(time) and the rate is 7, in t hours the ship has sailed a distance of $7t$.

$$[d(t)]^2 = (7t)^2 + 2^2 \qquad \bullet \text{ The Pythagorean Theorem}$$
$$[d(t)]^2 = 49t^2 + 4$$
$$d(t) = \sqrt{49t^2 + 4} \qquad \bullet \text{ The } \pm \text{ sign is not used because } d \text{ must be nonnegative.}$$

TRY EXERCISE 72, EXERCISE SET 3.2

EXAMPLE 10   *Solve an Application*

An open box is to be made from a square piece of cardboard that measures 40 inches on each side. To construct the box, squares that measure x inches on each side are cut from each corner of the cardboard as shown in **Figure 3.34**.

a. Express the volume V of the box as a function of x.

b. Determine the domain of V.

x

$40 - 2x$

x

x | $40 - 2x$ | x
40 in.

$40 - 2x$ $\quad x$
$40 - 2x$

Figure 3.34

Continued • ▶

Solution

a. The length l of the box is $40 - 2x$. The width w is also $40 - 2x$. The height of the box is x. The volume V of a box is a product of its length, its width, and its height. Thus

$$V = (40 - 2x)^2 x$$

b. The squares that are cut from each corner require x to be larger than 0 inches but less than 20 inches. Thus the domain is $\{x \mid 0 < x < 20\}$.

TRY EXERCISE 68, EXERCISE SET 3.2

TOPICS FOR DISCUSSION

1. Discuss the definition of *function.* Give some examples of relationships that are functions and some that are not functions.

2. What is the difference between the domain and range of a function?

3. How many y-intercepts can a function have? How many x-intercepts can a function have?

4. Discuss how the vertical line test is used to determine whether or not a graph is the graph of a function. Explain why the vertical line test works.

5. What is the domain of $f(x) = \dfrac{\sqrt{1 - x}}{x^2 - 9}$? Explain.

6. Is 2 in the range of $g(x) = \dfrac{6x - 5}{3x + 1}$? Explain the process you used to make your decision.

7. Suppose that f is a function and that $f(a) = f(b)$. Does this imply that $a = b$? Explain your answer.

EXERCISE SET 3.2

In Exercises 1 to 8, evaluate each function.

1. Given $f(x) = 3x - 1$, find
   a. $f(2)$ **b.** $f(-1)$ **c.** $f(0)$
   d. $f\left(\dfrac{2}{3}\right)$ **e.** $f(k)$ **f.** $f(k + 2)$

2. Given $g(x) = 2x^2 + 3$, find
   a. $g(3)$ **b.** $g(-1)$ **c.** $g(0)$
   d. $g\left(\dfrac{1}{2}\right)$ **e.** $g(c)$ **f.** $g(c + 5)$

3. Given $A(w) = \sqrt{w^2 + 5}$, find
   a. $A(0)$ **b.** $A(2)$ **c.** $A(-2)$
   d. $A(4)$ **e.** $A(r + 1)$ **f.** $A(-c)$

4. Given $J(t) = 3t^2 - t$, find
   a. $J(-4)$ **b.** $J(0)$ **c.** $J\left(\dfrac{1}{3}\right)$
   d. $J(-c)$ **e.** $J(x + 1)$ **f.** $J(x + h)$

5. Given $f(x) = \dfrac{1}{|x|}$, find
   a. $f(2)$ **b.** $f(-2)$ **c.** $f\left(\dfrac{-3}{5}\right)$
   d. $f(2) + f(-2)$ **e.** $f(c^2 + 4)$ **f.** $f(2 + h)$

6. Given $T(x) = 5$, find
   a. $T(-3)$ **b.** $T(0)$ **c.** $T\left(\dfrac{2}{7}\right)$
   d. $T(3) + T(1)$ **e.** $T(x + h)$ **f.** $T(3k + 5)$

7. Given $s(x) = \dfrac{x}{|x|}$, find

   a. $s(4)$ b. $s(5)$ c. $s(-2)$

   d. $s(-3)$ e. $s(t),\ \ t > 0$ f. $s(t),\ \ t < 0$

8. Given $r(x) = \dfrac{x}{x + 4}$, find

   a. $r(0)$ b. $r(-1)$ c. $r(-3)$

   d. $r\left(\dfrac{1}{2}\right)$ e. $r(0.1)$ f. $r(10,000)$

In Exercises 9 and 10, evaluate each piecewise-defined function for the indicated values.

9. $P(x) = \begin{cases} 3x + 1 & \text{if } x < 2 \\ -x^2 + 11 & \text{if } x \geq 2 \end{cases}$

   a. $P(-4)$ b. $P(\sqrt{5})$

   c. $P(c),\ \ c < 2$ d. $P(k + 1),\ \ k \geq 1$

10. $Q(t) = \begin{cases} 4 & \text{if } 0 \leq t \leq 5 \\ -t + 9 & \text{if } 5 < t \leq 8 \\ \sqrt{t - 7} & \text{if } 8 < t \leq 11 \end{cases}$

   a. $Q(0)$ b. $Q(e),\ \ 6 < e < 7$

   c. $Q(n),\ \ 1 < n < 2$ d. $Q(m^2 + 7),\ \ 1 < m \leq 2$

In Exercises 11 to 20, identify the equations that define y as a function of x.

11. $2x + 3y = 7$ 12. $5x + y = 8$

13. $-x + y^2 = 2$ 14. $x^2 - 2y = 2$

15. $y = 4 \pm \sqrt{x}$ 16. $x^2 + y^2 = 9$

17. $y = \sqrt[3]{x}$ 18. $y = |x| + 5$

19. $y^2 = x^2$ 20. $y^3 = x^3$

In Exercises 21 to 26, identify the sets of the ordered pairs (x, y) that define y as a function of x.

21. $\{(2, 3), (5, 1), (-4, 3), (7, 11)\}$

22. $\{(5, 10), (3, -2), (4, 7), (5, 8)\}$

23. $\{(4, 4), (6, 1), (5, -3)\}$

24. $\{(2, 2), (3, 3), (7, 7)\}$

25. $\{(1, 0), (2, 0), (3, 0)\}$

26. $\left\{\left(-\dfrac{1}{3}, \dfrac{1}{4}\right), \left(-\dfrac{1}{4}, \dfrac{1}{3}\right), \left(\dfrac{1}{4}, \dfrac{2}{3}\right)\right\}$

In Exercises 27 to 38, determine the domain of the function represented by the given equation.

27. $f(x) = 3x - 4$ 28. $f(x) = -2x + 1$

29. $f(x) = x^2 + 2$ 30. $f(x) = 3x^2 + 1$

31. $f(x) = \dfrac{4}{x + 2}$ 32. $f(x) = \dfrac{6}{x - 5}$

33. $f(x) = \sqrt{7 + x}$ 34. $f(x) = \sqrt{4 - x}$

35. $f(x) = \sqrt{4 - x^2}$ 36. $f(x) = \sqrt{12 - x^2}$

37. $f(x) = \dfrac{1}{\sqrt{x + 4}}$ 38. $f(x) = \dfrac{1}{\sqrt{5 - x}}$

In Exercises 39 to 46, graph each function. State the domain of each function. Insert solid circles or hollow circles where necessary to indicate the true nature of the function.

39. $f(x) = \begin{cases} |x| & \text{if } x \leq 1 \\ 2 & \text{if } x > 1 \end{cases}$

40. $g(x) = \begin{cases} -4 & \text{if } x \leq 0 \\ x^2 - 4 & \text{if } 0 < x \leq 1 \\ -x & \text{if } x > 1 \end{cases}$

41. $J(x) = \begin{cases} 4 & \text{if } x = -4, -3, \text{ or } -2 \\ x^2 & \text{if } x = -1, 0, \text{ or } 1 \\ -x + 6 & \text{if } x = 2, 3, \text{ or } 4 \end{cases}$

42. $K(x) = \begin{cases} 1 & \text{if } x = -5, -4, -3, \text{ or } -2 \\ x^2 - 3 & \text{if } x = -1, 0, 1, \text{ or } 2 \\ \dfrac{1}{2}x & \text{if } x = 3, 4, 5, \text{ or } 6 \end{cases}$

43. $L(x) = \left[\!\left[\dfrac{1}{3}x\right]\!\right]$ for $-6 \leq x \leq 6$

44. $M(x) = [\![x]\!] + 2$ for $0 \leq x \leq 4$

45. $N(x) = \text{int}(-x)$ for $-3 \leq x \leq 3$

46. $P(x) = \text{int}(x) + x$ for $0 \leq x \leq 4$

47. **MAILING CHARGES** The cost of mailing a parcel is given by

$$C(w) = 0.29 - 0.29\,\text{int}(1 - w), \quad w > 0$$

where C is in dollars and w is the weight of the parcel in ounces.

   a. Evaluate $C(3.97)$.

   b. Graph $y = C(w)$ for $0 < w \leq 5$.

48. **TELEPHONE CHARGES** The cost of a long-distance telephone call is given by

$$C(t) = 0.85 - 0.50\,\text{int}(1 - t), \quad t > 0$$

where C is in dollars and t is the length of the call in minutes.

   a. Evaluate $C(4.75)$.

   b. Graph $y = C(t)$ for $0 < t \leq 6$.

49. Use the vertical line test to determine which of the following graphs are graphs of functions.

a.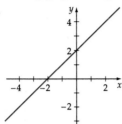

b.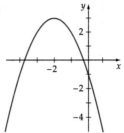

c.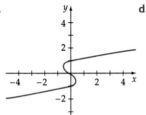

d.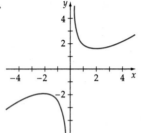

50. Use the vertical line test to determine which of the following graphs are graphs of functions.

a.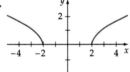

b.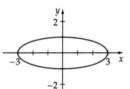

c.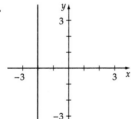

d.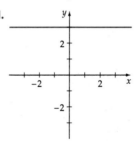

In Exercises 51 to 60, use the indicated graph to identify the intervals over which the function is increasing, constant, or decreasing.

51.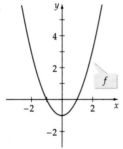

52.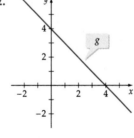

53.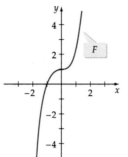

54.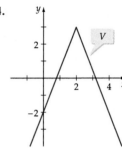

55.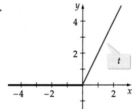

56.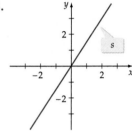

57.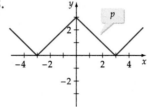

58.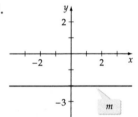

59.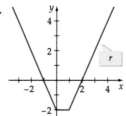

60.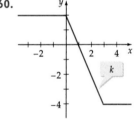

61. Use the horizontal line test to determine which of the following functions are one-to-one.

  f as shown in Exercise 51
  g as shown in Exercise 52
  F as shown in Exercise 53
  V as shown in Exercise 54
  p as shown in Exercise 55

62. Use the horizontal line test to determine which of the following functions are one-to-one.

  s as shown in Exercise 56
  t as shown in Exercise 57
  m as shown in Exercise 58
  r as shown in Exercise 59
  k as shown in Exercise 60

63. A rectangle has a length of l feet and a perimeter of 50 feet.

  a. Write the width w of the rectangle as a function of its length.

  b. Write the area A of the rectangle as a function of its length.

64. The sum of two numbers is 20. Let x represent one of the numbers.

  a. Write the second number y as a function of x.

  b. Write the product P of the two numbers as a function of x.

65. **DEPRECIATION** A bus was purchased for $80,000. Assuming the bus depreciates at a rate of $6500 per year (*straight-line depreciation*) for the first 10 years, write the value v of the bus as a function of the time t (measured in years) for $0 \leq t \leq 10$.

66. **DEPRECIATION** A boat was purchased for $44,000. Assuming the boat depreciates at a rate of $4200 per year (*straight-line depreciation*) for the first 8 years, write the value v of the boat as a function of the time t (measured in years) for $0 \leq t \leq 8$.

67. **COST, REVENUE, AND PROFIT** A manufacturer produces a product at a cost of $22.80 per unit. The manufacturer has a fixed cost of $400.00 per day. Each unit retails for $37.00. Let x represent the number of units produced in a 5-day period.

  a. Write the total cost C as a function of x.

  b. Write the revenue R as a function of x.

  c. Write the profit P as a function of x. (*Hint:* The profit function is given by $P(x) = R(x) - C(x)$.)

68. **VOLUME OF A BOX** An open box is to be made from a square piece of cardboard having dimensions 30 inches by 30 inches by cutting out squares of area x^2 from each corner, as shown in the figure.

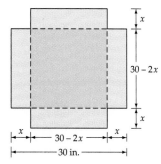
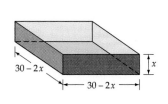

  a. Express the volume V of the box as a function of x.

  b. State the domain of V.

69. **HEIGHT OF AN INSCRIBED CYLINDER** A cone has an altitude of 15 centimeters and a radius of 3 centimeters. A right circular cylinder of radius r and height h is inscribed in the cone as shown in the figure. Use similar triangles to write h as a function of r.

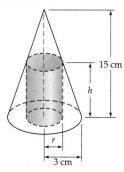

70. **VOLUME OF WATER** Water is running out of a conical funnel that has an altitude of 20 inches and a radius of 10 inches, as shown in the figure.

  a. Write the radius r of the water as a function of its depth h.

  b. Write the volume V of the water as a function of its depth h.

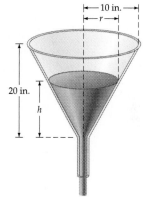

71. **DISTANCE FROM A BALLOON** For the first minute of flight, a hot air balloon rises vertically at a rate of 3 meters per second. If t is the time in seconds that the balloon has been airborne, write the distance d between the balloon and a point on the ground 50 meters from the point of lift-off as a function of t.

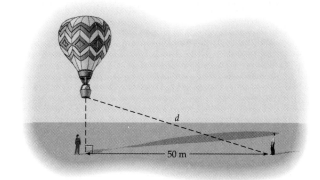

72. **TIME FOR A SWIMMER** An athlete swims from point A to point B at the rate of 2 mph and runs from point B to point C at a rate of 8 mph. Use the dimensions in the figure to write the time t required to reach point C as a function of x.

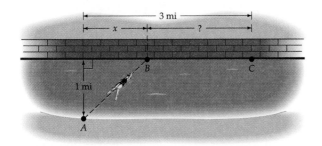

73. **DISTANCE BETWEEN SHIPS** At 12:00 noon Ship A is 45 miles due south of Ship B and is sailing north at a rate of 8 mph. Ship B is sailing east at a rate of 6 mph. Write the distance d between the ships as a function of the time t where $t = 0$ represents 12:00 noon.

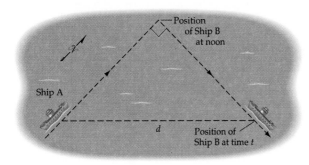

74. **SALES VS. PRICE** A business finds that the number of feet f of pipe it can sell per week is a function of the price p in cents per foot as given by

$$f(p) = \frac{320{,}000}{p + 25}, \quad 40 \le p \le 90$$

Complete the following table by evaluating f (to the nearest 100 feet) for the indicated values of p.

p	40	50	60	75	90
$f(p)$					

75. **MODEL YIELD** The yield Y of apples per tree is related to the amount x of a particular type of fertilizer applied (in pounds per year) by the function

$$Y(x) = 400[1 - 5(x - 1)^{-2}], \quad 5 \le x \le 20$$

Complete the following table by evaluating Y (to the nearest apple) for the indicated applications.

x	5	10	12.5	15	20
$Y(x)$					

76. **MODEL COST** A manufacturer finds that the cost C in dollars of producing x items of a product is given by

$$C(x) = (225 + 1.4\sqrt{x})^2, \quad 100 \le x \le 1000$$

Complete the following table by evaluating C (to the nearest dollar) for the indicated numbers of items.

x	100	200	500	750	1000
$C(x)$					

77. If $f(x) = x^2 - x - 5$ and $f(c) = 1$, find c.

78. If $g(x) = -2x^2 + 4x - 1$ and $g(c) = -4$, find c.

79. Determine whether 1 is in the range of $f(x) = \dfrac{x - 1}{x + 1}$.

80. Determine whether 0 is in the range of $g(x) = \dfrac{1}{x - 3}$.

 In Exercises 81 to 86, use a graphing utility.

81. Graph $f(x) = \dfrac{[\![x]\!]}{|x|}$ for $-4 \le x \le 4$ and $x \ne 0$.

82. Graph $f(x) = \dfrac{[\![2x]\!]}{|x|}$ for $-4 \le x \le 4$ and $x \ne 0$.

83. Graph: $f(x) = x^2 - 2|x| - 3$

84. Graph: $f(x) = x^2 - |2x - 3|$

85. Graph: $f(x) = |x^2 - 1| - |x - 2|$

86. Graph: $f(x) = |x^2 - 2x| - 3$

SUPPLEMENTAL EXERCISES

The notation $f(x)|_a^b$ is used to denote the difference $f(b) - f(a)$. That is,

$$f(x)|_a^b = f(b) - f(a)$$

In Exercises 87 to 90, evaluate $f(x)|_a^b$ for the given function f and the indicated values of a and b.

87. $f(x) = x^2 - x; f(x)|_2^3$

88. $f(x) = -3x + 2; f(x)|_4^7$

89. $f(x) = 2x^3 - 3x^2 - x; f(x)\big|_0^2$

90. $f(x) = \sqrt{8 - x}; f(x)\big|_0^8$

In Exercises 91 to 94, each function has two or more independent variables.

91. Given $f(x, y) = 3x + 5y - 2$, find

  a. $f(1, 7)$ **b.** $f(0, 3)$ **c.** $f(-2, 4)$

  d. $f(4, 4)$ **e.** $f(k, 2k)$ **f.** $f(k + 2, k - 3)$

92. Given $g(x, y) = 2x^2 - |y| + 3$, find

  a. $g(3, -4)$ **b.** $g(-1, 2)$

  c. $g(0, -5)$ **d.** $g\left(\dfrac{1}{2}, -\dfrac{1}{4}\right)$

  e. $g(c, 3c), c > 0$ **f.** $g(c + 5, c - 2), c < 0$

93. AREA OF A TRIANGLE The area of a triangle with sides a, b, and c is given by the function

$$A(a, b, c) = \sqrt{s(s - a)(s - b)(s - c)}$$

where s is the semiperimeter

$$s = \frac{a + b + c}{2}$$

Find $A(5, 8, 11)$.

94. COST OF A PAINTER The cost in dollars to hire a house painter is given by the function

$$C(h, g) = 15h + 14g$$

where h is the number of hours it takes to paint the house and g is the number of gallons of paint required to paint the house. Find $C(18, 11)$.

A *fixed point* of a function is a number a such that $f(a) = a$. **In Exercises 95 and 96, find all fixed points for the given function.**

95. $f(x) = x^2 + 3x - 3$ **96.** $g(x) = \dfrac{x}{x + 5}$

In Exercises 97 and 98, sketch the graph of the piecewise-defined function.

97. $s(x) = \begin{cases} 1 & \text{if } x \text{ is an integer} \\ 2 & \text{if } x \text{ is not an integer} \end{cases}$

98. $v(x) = \begin{cases} 2x - 2 & \text{if } x \neq 3 \\ 1 & \text{if } x = 3 \end{cases}$

PROJECTS

1. DAY OF THE WEEK A formula known as Zeller's Congruence makes use of the greatest integer function $[\![x]\!]$ to determine the day of the week on which a given date fell or will fall. To use Zeller's Congruence, we first compute the integer z given by

$$z = \left[\!\!\left[\frac{13m - 1}{5}\right]\!\!\right] + \left[\!\!\left[\frac{y}{4}\right]\!\!\right] + \left[\!\!\left[\frac{c}{4}\right]\!\!\right] + d + y - 2c$$

The variables c, y, d, and m are defined as follows:

c = the century
y = the year
d = the day of the month
m = the month, using 1 for March, 2 for April,..., 10 for December. January and February are assigned the values 11 and 12 of the previous year.

For example, for the date September 12, 1991, we use $c = 19$, $y = 91$, $d = 12$, and $m = 7$. The remainder of z divided by 7 gives the day of the week. A remainder of 0 represents a Sunday, a remainder of 1 a Monday,..., a remainder of 6 a Saturday.

a. Verify that December 7, 1941 was a Sunday.

b. Verify that January 1, 2000 will fall on a Saturday.

c. Determine on what day of the week Independence Day (July 4, 1776) fell.

d. Determine on what day of the week you were born.

SECTION

3.3 LINEAR FUNCTIONS

The following function has many applications.

> **DEFINITION OF A LINEAR FUNCTION**
>
> A function of the form
>
> $$f(x) = mx + b, \quad m \neq 0$$
>
> where m and b are real numbers, is a **linear function** of x.

SLOPES OF LINES

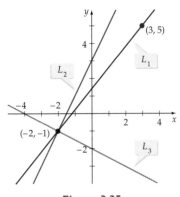

Figure 3.35

The graph of $f(x) = mx + b$, or $y = mx + b$, is a nonvertical straight line.

The graphs shown in **Figure 3.35** are the graphs of $f(x) = mx + b$ for various values of m. The graphs intersect at the point $(-2, -1)$, but they differ in *steepness*. The steepness of a line is called the *slope* of the line and is denoted by the symbol m. The slope of a line is the ratio of the change in the y values of any two points on the line to the change in the x values of the same two points. For example, the graph of the line L_1 in **Figure 3.35** passes through the points $(-2, -1)$ and $(3, 5)$. The change in the y values is determined by subtracting the two y-coordinates.

$$\text{Change in } y = 5 - (-1) = 6$$

The change in the x values is determined by subtracting the two x-coordinates.

$$\text{Change in } x = 3 - (-2) = 5$$

The slope m of L_1 is the ratio of the change in the y values of the two points to the change in the x values of the two points. That is,

$$m = \frac{\text{change in } y}{\text{change in } x} = \frac{6}{5}$$

Because the slope of a nonvertical line can be calculated by using any two arbitrary points on the line, we have the following formula.

> **SLOPE OF A NONVERTICAL LINE**
>
> The **slope** m of the line passing through the points $P_1(x_1, y_1)$ and $P_2(x_2, y_2)$ with $x_1 \neq x_2$ is given by
>
> $$m = \frac{y_2 - y_1}{x_2 - x_1}$$

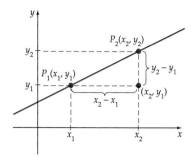

Figure 3.36

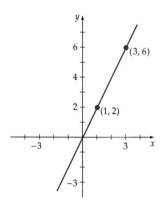

Figure 3.37

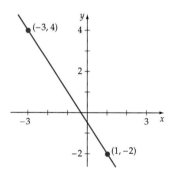

Figure 3.38

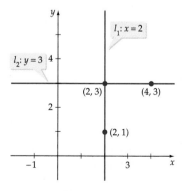

Figure 3.39

Because the numerator $y_2 - y_1$ is the vertical **rise** and the denominator $x_2 - x_1$ is the horizontal **run** from P_1 to P_2, slope is often referred to as the *rise over the run* or the *change in y divided by the change in x*. See **Figure 3.36**. Lines that have a positive slope slant upward from left to right. Lines that have a negative slope slant downward from left to right.

EXAMPLE 1   *Find the Slope of a Line*

Find the slope of the line passing through the points whose coordinates are given.

a. $(1, 2)$ and $(3, 6)$ **b.** $(-3, 4)$ and $(1, -2)$

Solution

a. The slope of the line passing through $(1, 2)$ and $(3, 6)$ is

$$m = \frac{y_2 - y_1}{x_2 - x_1} = \frac{6 - 2}{3 - 1} = \frac{4}{2} = 2$$

Because $m > 0$, the line slants upward from left to right. See **Figure 3.37**.

b. The slope of the line passing through $(-3, 4)$ and $(1, -2)$ is

$$m = \frac{y_2 - y_1}{x_2 - x_1} = \frac{-2 - 4}{1 - (-3)} = \frac{-6}{4} = -\frac{3}{2}$$

Because $m < 0$, the line slants downward from left to right. See **Figure 3.38**.

TRY EXERCISE 2, EXERCISE SET 3.3

The definition of slope does not apply to vertical lines. Consider, for example, the points $(2, 1)$ and $(2, 3)$ on the vertical line l_1 in **Figure 3.39**. Applying the definition of slope to this line produces

$$m = \frac{3 - 1}{2 - 2}$$

which is undefined because it requires division by zero. Because division by zero is undefined, we say that the slope of any vertical line is undefined.

QUESTION   Is the graph of a vertical line the graph of a function?

All horizontal lines have 0 slope. For example, the line l_2 through $(2, 3)$ and $(4, 3)$ in **Figure 3.39** is a horizontal line. Its slope is given by

$$m = \frac{3 - 3}{4 - 2} = \frac{0}{2} = 0$$

ANSWER   No. For example, the vertical line passing through $x = 2$ contains the ordered pairs $(2, 3)$ and $(2, -5)$. Thus, there are two ordered pairs with the same first coordinate but different second coordinates.

When computing the slope of a line, it does not matter which point we label P_1 and which P_2 because

$$\frac{y_2 - y_1}{x_2 - x_1} = \frac{y_1 - y_2}{x_1 - x_2}$$

In functional notation, the points P_1 and P_2 can be represented by

$$(x_1, f(x_1)) \quad \text{and} \quad (x_2, f(x_2))$$

In this notation, the slope formula

$$m = \frac{y_2 - y_1}{x_2 - x_1} \quad \text{is expressed as} \quad m = \frac{f(x_2) - f(x_1)}{x_2 - x_1} \qquad (1)$$

If $m = 0$, then $f(x) = mx + b$ can be written as $f(x) = b$, or $y = b$. The graph of $y = b$ is the horizontal line through $(0, b)$. See **Figure 3.40**. Because every point on the graph of $y = b$ has a y-coordinate of b, the function $y = b$ is called a **constant function**.

Every point on the vertical line through $(a, 0)$ has an x-coordinate of a. The equation of the vertical line through $(a, 0)$ is $x = a$. See **Figure 3.40**.

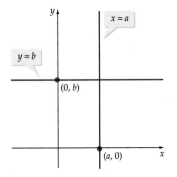

Figure 3.40

HORIZONTAL LINES AND VERTICAL LINES

The graph of $y = b$ is a horizontal line through $(0, b)$.

The graph of $x = a$ is a vertical line through $(a, 0)$.

The equation $f(x) = mx + b$ is called the **slope-intercept form** of the equation of a line because of the following theorem.

SLOPE-INTERCEPT FORM

The graph of $f(x) = mx + b$ is a line with slope m and y-intercept $(0, b)$.

Proof The slope of the graph of $f(x) = mx + b$ is given by Equation (1).

$$\frac{f(x_2) - f(x_1)}{x_2 - x_1} = \frac{(mx_2 + b) - (mx_1 + b)}{x_2 - x_1} = \frac{m(x_2 - x_1)}{x_2 - x_1} = m, \quad x_1 \neq x_2$$

The y-intercept of the graph of $f(x) = mx + b$ is found by letting $x = 0$.

$$f(0) = m(0) + b = b$$

Thus $(0, b)$ is the y-intercept, and m is the slope, of the graph of $f(x) = mx + b$.
◆

If a function is written in the form $f(x) = mx + b$, then its graph can be drawn by first plotting the y-intercept $(0, b)$ and then using its slope m to determine another point on the line.

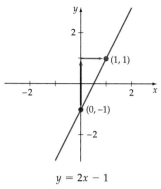

$y = 2x - 1$

Figure 3.41

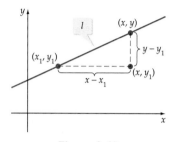

Figure 3.42

The slope of line l is $m = \dfrac{y - y_1}{x - x_1}$.

EXAMPLE 2   *Graph a Linear Function*

Graph: $f(x) = 2x - 1$

Solution

The equation $y = 2x - 1$ is in slope-intercept form, with $b = -1$ and $m = 2$ or $2/1$. To graph the equation, first plot the y-intercept $(0, -1)$ and then use the slope to plot a second point, which is two units up and one unit to the right of the y-intercept. See **Figure 3.41**.

TRY EXERCISE 16, EXERCISE SET 3.3

We can find an equation of a line, provided we know its slope and at least one point on the line. **Figure 3.42** suggests that if (x_1, y_1) is a point on a line l of slope m, and (x, y) is *any other* point on the line, then

$$\frac{y - y_1}{x - x_1} = m$$

Multiplying each side by $x - x_1$ produces $y - y_1 = m(x - x_1)$. This equation is called the **point-slope form** of the equation of line l.

POINT-SLOPE FORM

The graph of

$$y - y_1 = m(x - x_1)$$

is a line that has slope m and passes through (x_1, y_1).

EXAMPLE 3   *Use the Point-Slope Form*

Find an equation of a line with slope -3 that passes through $(-1, 4)$.

Solution

Use the point-slope form with $m = -3$, $x_1 = -1$, and $y_1 = 4$.

$$y - y_1 = m(x - x_1)$$
$$y - 4 = -3[x - (-1)] \qquad \bullet \text{ Substitute.}$$
$$y - 4 = -3x - 3 \qquad \bullet \text{ Solve for } y.$$
$$y = -3x + 1 \qquad \bullet \text{ Slope-intercept form}$$

TRY EXERCISE 28, EXERCISE SET 3.3

An equation of the form $Ax + By + C = 0$, where A, B, and C are real numbers and both A and B are not zero, is called the **general form of the**

TAKE NOTE

To determine an equation of a nonvertical line that passes through two points, first determine the slope of the line and then use the coordinates of either one of the points in the point-slope form.

equation of a line. For example, the equation $y = -3x + 1$ in Example 3 can be written in general form as $3x + y - 1 = 0$.

Although we are mainly concerned with linear functions in this section, the following theorem applies to all functions. It illustrates a powerful relationship between the real solutions of $f(x) = 0$ and the x-intercepts of the graph of $y = f(x)$.

REAL SOLUTIONS AND x-INTERCEPTS THEOREM

For every function f, the real number c is a solution of $f(x) = 0$ if and only if $(c, 0)$ is an x-intercept of the graph of $y = f(x)$.

The real solutions and x-intercepts theorem tells us that we can find real solutions of $f(x) = 0$ by graphing. The following example illustrates the theorem for a linear function of x.

EXAMPLE 4   Verify the Real Solutions and x-Intercepts Theorem

Let $f(x) = -2x + 6$. Find the real solution of $f(x) = 0$ and then graph $y = f(x)$. Compare the solution of $f(x) = 0$ with the x-intercept of the graph of f.

Solution

To find the real solution of $f(x) = 0$, replace $f(x)$ by $-2x + 6$ and solve for x.

$$f(x) = 0$$
$$-2x + 6 = 0$$
$$-2x = -6$$
$$x = 3$$

Now graph $f(x) = -2x + 6$. See **Figure 3.43**. The x-intercept of the graph is $(3, 0)$. The x-coordinate is 3. This is the real solution of $f(x) = 0$.

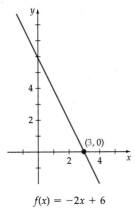

$f(x) = -2x + 6$

Figure 3.43

TRY EXERCISE 40, EXERCISE SET 3.3

Let $f_1(x) = ax + b$ and $f_2(x) = cx + d$, with the restriction that $a \neq c$. The solution of $f_1(x) = f_2(x)$ can be determined by solving $ax + b = cx + d$ for x. Graphically, the solution of this equation is the x-coordinate of the intersection of the graphs of $y = f_1(x)$ and $y = f_2(x)$.

EXAMPLE 5   Solve ax + b = cx + d

Let $f_1(x) = 2x - 1$ and $f_2(x) = -x + 11$. Solve $f_1(x) = f_2(x)$ using

a. an algebraic method   **b.** a graph

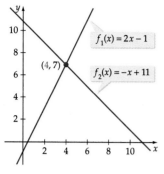

Figure 3.44

$f_1(x) = 2x - 1$

$f_2(x) = -x + 11$

$(4, 7)$

Solution

a. $f_1(x) = f_2(x)$

$2x - 1 = -x + 11$ • **Substitute.**

$3x = 12$ • **Simplify.**

$x = 4$

b. Graph $y = f_1(x)$ and $y = f_2(x)$ on the same coordinate axes. See **Figure 3.44**. The point of intersection is $(4, 7)$. The x-coordinate 4 is the same as the solution obtained in **a.**

TRY EXERCISE 44, EXERCISE SET 3.3

APPLICATIONS

EXAMPLE 6   *Find a Linear Model of Data*

The bar graph in **Figure 3.45** illustrates the number of new car dealerships in the United States for the years 1970 through 1995. Let $t = 70$ represent the year 1970. The ordered pair $(70, 31.5)$ indicates that there were 31,500 new car dealerships in the U.S. in 1970. The ordered pair $(95, 23.4)$ indicates that there were 23,400 new car dealerships in 1995.

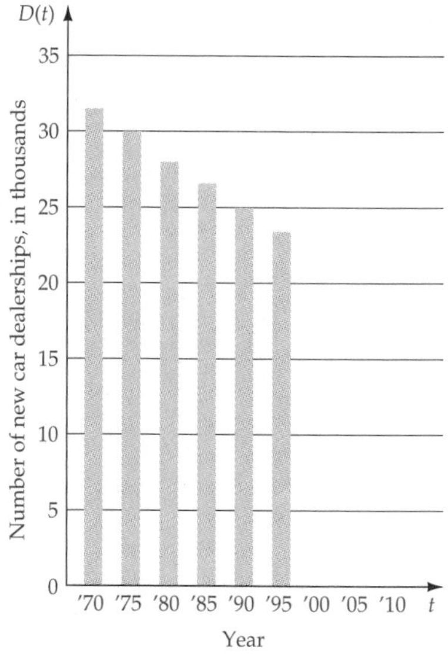

a. Use the ordered pairs $(70, 31.5)$ and $(95, 23.4)$ to find a linear function that models the data in the graph.

b. What number of car dealerships (to the nearest hundred) does the model predict for the year 1992?

c. In what year (to the nearest year) does the model predict that the number of car dealerships will be 20,000?

$D(t)$

Number of new car dealerships, in thousands

35

30

25

20

15

10

5

0

'70 '75 '80 '85 '90 '95 '00 '05 '10 t

Year

Figure 3.45

Solution

a. First calculate the slope of the line and then use the point-slope formula to find the equation of the line.

$$m = \frac{D_2 - D_1}{t_2 - t_1} = \frac{23.4 - 31.5}{95 - 70} = -0.324$$ • **Calculate the slope.**

$$D - D_1 = m(t - t_1)$$ • **Use the point-slope formula.**

$$D - 31.5 = -0.324(t - 70)$$ • **Substitute 31.5 for D_1,**
$$D = -0.324t + 54.18$$ **-0.324 for m, and 70 for t_1.**

In functional notation, the linear model is $D(t) = -0.324t + 54.18$.

Continued • ➤

b. Evaluate $D(t)$ when $t = 92$.

$$D(t) = -0.324t + 54.18$$
$$D(92) = -0.324(92) + 54.18 = 24.372 \approx 24.4$$

Because the units on the vertical axis are in thousands, the model predicts approximately 24,400 dealerships in 1992.

TAKE NOTE

The model in Example 6 is based on using t as the last two digits of a year. You can think of a year as 1900 + t. In part c., t = 105 represents the year 1900 + 105 = 2005.

c. To predict the year in which the number of new car dealerships will be 20,000, let $D(t) = 20$ and solve for t.

$$D(t) = -0.324t + 54.18$$
$$20 = -0.324t + 54.18$$
$$-34.18 = -0.324t$$
$$105 \approx t$$

The model predicts that in 2005 there will be 20,000 new car dealerships.

TRY EXERCISE 48, EXERCISE SET 3.3

The procedure in Example 6b used a linear function to determine a point between two known points. This procedure is referred to as **linear interpolation**. In Example 6c, the linear function was used to estimate a value in the *future*. This is called **linear extrapolation**. See **Figure 3.46**.

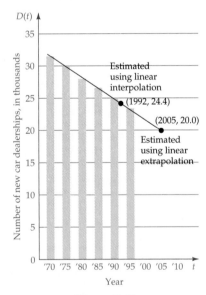

Figure 3.46

If a manufacturer produces x units of a product that sells for p dollars per unit, then the **cost function** C, the **revenue function** R, and the **profit function** P are defined as follows:

$C(x)$ = cost of producing and selling x units

$R(x) = xp$ = revenue from the sale of x units at p dollars each

$P(x)$ = profit from selling x units

Because profit equals the revenue less the cost, we have

$$P(x) = R(x) - C(x)$$

The value of x for which $R(x) = C(x)$ is called the **break-even point.** At the break-even point, $P(x) = 0$.

EXAMPLE 7	*Find the Profit Function and the Break-even Point*

A manufacturer finds that the costs incurred in the manufacture and sale of a particular type of calculator are \$180,000 plus \$27 per calculator.

a. Determine the profit function P, given that x calculators are manufactured and sold at \$59 each.

b. Determine the break-even point.

Solution

a. The cost function is $C(x) = 27x + 180,000$. The revenue function is $R(x) = 59x$. Thus the profit function is

$$P(x) = R(x) - C(x)$$
$$= 59x - (27x + 180,000)$$
$$= 32x - 180,000, \quad x \geq 0 \text{ and } x \text{ is an integer}$$

b. At the break-even point, $R(x) = C(x)$.

$$59x = 27x + 180,000$$
$$32x = 180,000$$
$$x = 5625$$

The manufacturer will break even when 5625 calculators are sold.

TRY EXERCISE 50, EXERCISE SET 3.3

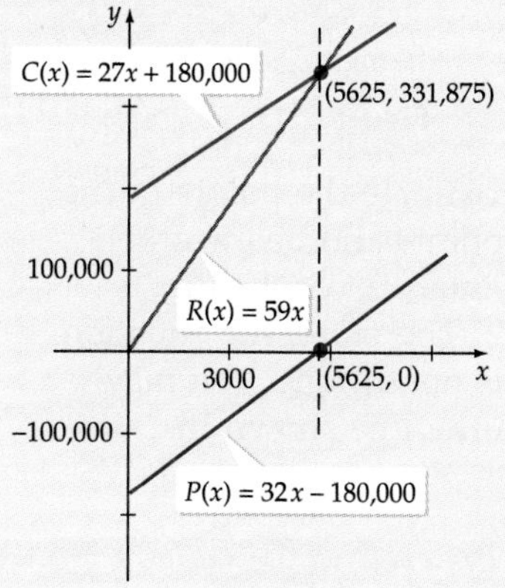

The scatter plot in the next example appears to indicate a linear relationship between the variables t and E. But what linear function best models the relationship? The "Exploring Concepts with Technology" section at the end of this chapter illustrates an analytic method that can be used to find a linear function that models data. However, in the next example, we find a linear model by sketching the line that appears to fit the data better than any other line.

EXAMPLE 8	*Find a Model by Sketching a Line That Fits the Data*

The table in **Figure 3.47** shows the total expenditures (rounded to the nearest billion dollars) for purchased meals and beverages in the United States for each of the years from 1990 to 1996.

a. Construct a scatter plot for the data. Use the variable t to represent the year and the variable E to represent the expenditures. Let $t = 0$ represent the year 1990 and $t = 6$ represent 1996.

TAKE NOTE

The graphs of *C*, *R*, and *P* are shown below. Observe that the graphs of *C* and *R* intersect at the break-even point, where $x = 5625$ and $P(5625) = 0$.

Meals and Beverages Purchased in the U.S.

Year, t	Expenditures, E
1990	191
1991	199
1992	209
1993	220
1994	229
1995	237
1996	244

Figure 3.47

Continued •➤

b. Sketch a line that you think best approximates the points in the scatter plot. Write the equation of the line.

Solution

a. See **Figure 3.48.**

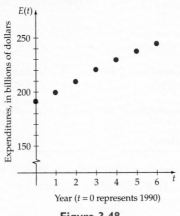

Figure 3.48 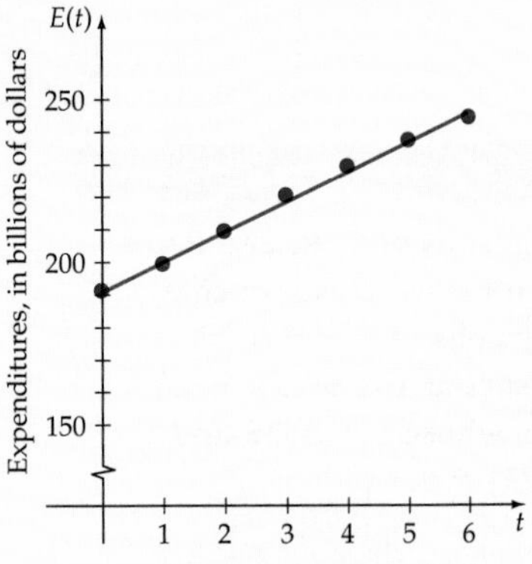   Figure 3.49

b. **Figure 3.49** shows a line that seems to approximate the points in the scatter plot. The line is above some of the points and below others, but it appears to provide a good fit. The y-intercept of the line is about 190. The slope of the line is about 8.2. Therefore, the equation of the line is $y = 8.2t + 190$. This answer is not unique. You may have drawn a line that varies slightly from the line shown in **Figure 3.49.**

> **TRY EXERCISE 60, EXERCISE SET 3.3**

PARALLEL AND PERPENDICULAR LINES

Two nonintersecting lines in a plane are **parallel.** All vertical lines are parallel to each other. All horizontal lines are parallel to each other.

Two lines are **perpendicular** if and only if they intersect and form adjacent angles each of which measures 90°. In a plane, vertical and horizontal lines are perpendicular to one another.

> ### PARALLEL AND PERPENDICULAR LINES
>
> Let l_1 be the graph of $f_1(x) = m_1x + b$ and l_2 be the graph of $f_2(x) = m_2x + b$. Then
>
> - l_1 and l_2 are parallel if and only if $m_1 = m_2$.
>
> - l_1 and l_2 are perpendicular if and only if $m_1 = -\dfrac{1}{m_2}$.

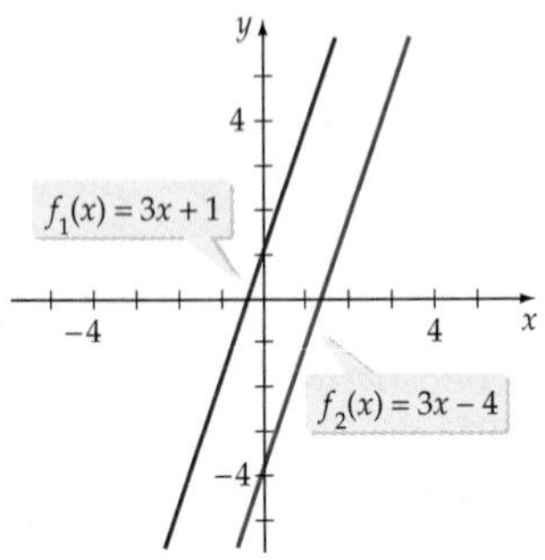

$f_1(x) = 3x + 1$

$f_2(x) = 3x - 4$

Figure 3.50

The graphs of $f_1(x) = 3x + 1$ and $f_2(x) = 3x - 4$ are shown in **Figure 3.50.** Because $m_1 = m_2 = 3$, the lines are parallel.

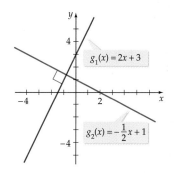

Figure 3.51

If $m_1 = -\dfrac{1}{m_2}$, then m_1 and m_2 are negative reciprocals of each other. The graphs of $g_1(x) = 2x + 3$ and $g_2(x) = -\dfrac{1}{2}x + 1$ are shown in **Figure 3.51**. Because 2 and $-\dfrac{1}{2}$ are negative reciprocals of each other, the lines are perpendicular. The symbol $\sqsupset$ indicates an angle of 90°. In **Figure 3.51** it is used to indicate that the lines are perpendicular.

EXAMPLE 9   *Determine a Point of Impact*

A rock is whirled horizontally in a circular counterclockwise path about the origin. See **Figures 3.52** and **3.53**. When the string breaks, the rock travels on a linear path perpendicular to the radius $\overline{OP}$ and hits a wall located at

$$y = x + 12 \qquad (2)$$

If the string breaks when the rock is at $P(4, 3)$, determine the point where the rock hits the wall.

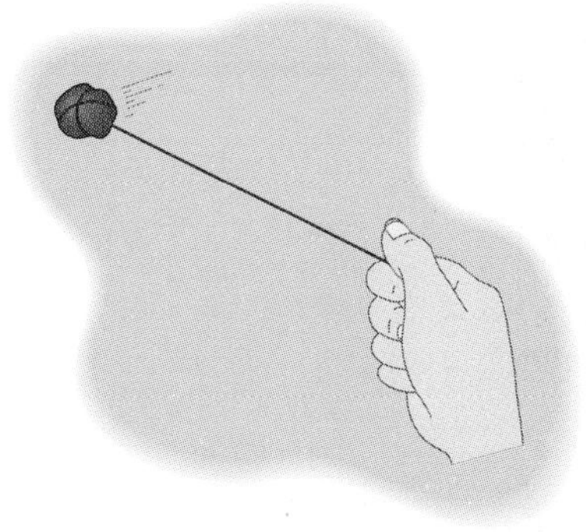

Solution

The slope of the radius from $(0, 0)$ to $(4, 3)$ is $3/4$. The negative reciprocal of $3/4$ is $-4/3$. Therefore, the linear path of the rock is given by

$$y - 3 = -\frac{4}{3}(x - 4)$$

$$y = -\frac{4}{3}x + \frac{25}{3} \qquad (3)$$

To find the point where the rock hits the wall, set the right side of Equation (2) equal to the right side of Equation (3) and solve for x. This is the procedure explained in Example 5.

$$-\frac{4}{3}x + \frac{25}{3} = x + 12$$
$$-4x + 25 = 3x + 36 \qquad \bullet \textbf{ Multiply all terms by 3.}$$
$$-7x = 11$$
$$x = -\frac{11}{7}$$

For every point on the wall, x and y are related by $y = x + 12$. Therefore, substituting $-11/7$ for x in $y = x + 12$ yields $y = -11/7 + 12 = 73/7$, and the rock hits the wall at $(-11/7, 73/7)$.

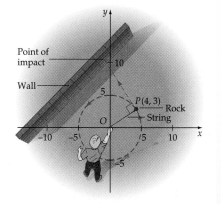

Figure 3.52

Point of impact
Wall

$P(4, 3)$
Rock
String
O

Figure 3.53

TRY EXERCISE 66, EXERCISE SET 3.3

TOPICS FOR DISCUSSION

1. If a linear function has a negative y-intercept and a negative slope, then its graph does not contain any points in Quadrant I. Do you agree? Explain.

2. Is a "break-even point" a point or a number? Explain.

3. A tutor states that some perpendicular lines do not have the property that their slopes are negative reciprocals of each other. Explain why the tutor is correct.

4. The real solutions and x-intercepts theorem applies only to linear functions. Do you agree?

5. Explain why the function $f(x) = x$ is referred to as the identity function.

EXERCISE SET 3.3

In Exercises 1 to 10, find the slope of the line that passes through the given points.

1. $(3, 4)$ and $(1, 7)$

2. $(-2, 4)$ and $(5, 1)$

3. $(4, 0)$ and $(0, 2)$

4. $(-3, 4)$ and $(2, 4)$

5. $(0, 0)$ and $(0, 4)$

6. $(0, 0)$ and $(3, 0)$

7. $(-3, 4)$ and $(-4, -2)$

8. $(-5, -1)$ and $(-3, 4)$

9. $\left(-4, \dfrac{1}{2}\right)$ and $\left(\dfrac{7}{3}, \dfrac{7}{2}\right)$

10. $\left(\dfrac{1}{2}, 4\right)$ and $\left(\dfrac{7}{4}, 2\right)$

In Exercises 11 to 14, find the slope of the line that passes through the given points.

11. $(3, f(3))$ and $(3 + h, f(3 + h))$

12. $(-2, f(-2 + h))$ and $(-2 + h, f(-2 + h))$

13. $(0, f(0))$ and $(h, f(h))$

14. $(a, f(a))$ and $(a + h, f(a + h))$

In Exercises 15 to 26, graph y as a function of x by finding the slope and y-intercept of each.

15. $y = 2x - 4$

16. $y = -x + 1$

17. $y = -\dfrac{1}{3}x + 4$

18. $y = \dfrac{2}{3}x - 2$

19. $y = 3$

20. $y = x$

21. $y = 2x$

22. $y = -3x$

23. $2x + y = 5$

24. $x - y = 4$

25. $4x + 3y - 12 = 0$

26. $2x + 3y + 6 = 0$

In Exercises 27 to 38, find the equation of the indicated line. Write the equation in the form $y = mx + b$.

27. y-intercept $(0, 3)$, slope 1

28. y-intercept $(0, 5)$, slope -2

29. y-intercept $\left(0, \dfrac{1}{2}\right)$, slope $\dfrac{3}{4}$

30. y-intercept $\left(0, \dfrac{3}{4}\right)$, slope $-\dfrac{2}{3}$

31. y-intercept $(0, 4)$, slope 0

32. y-intercept $(0, -1)$, slope $\dfrac{1}{2}$

33. Through $(-3, 2)$, slope -4

34. Through $(-5, -1)$, slope -3

35. Through $(3, 1)$ and $(-1, 4)$

36. Through $(5, -6)$ and $(2, -8)$

37. Through $(7, 11)$ and $(2, -1)$

38. Through $(-5, 6)$ and $(-3, -4)$

In Exercises 39 to 42, verify that the solution of $f(x) = 0$ is the same as the x-coordinate of the x-intercept of the graph of $y = f(x)$.

39. $f(x) = 3x - 12$

40. $f(x) = -2x - 4$

41. $f(x) = \dfrac{1}{4}x + 5$

42. $f(x) = -\dfrac{1}{3}x + 2$

In Exercises 43 to 46, solve $f_1(x) = f_2(x)$ by an algebraic method and by graphing.

43. $f_1(x) = 4x + 5$ \qquad $f_2(x) = x + 6$

44. $f_1(x) = -2x - 11$ \qquad $f_2(x) = 3x + 7$

45. $f_1(x) = 2x - 4$ \qquad $f_2(x) = -x + 12$

46. $f_1(x) = \dfrac{1}{2}x + 5$ \qquad $f_2(x) = \dfrac{2}{3}x - 7$

47. Linear Model in Health Science In 1986 there were about 900 recorded cases of women with AIDS and in 1994 the number had increased to approximately 10,500. The graph below shows that the number of recorded cases of women with AIDS has increased in a near linear pattern for the years from 1986 to 1994.

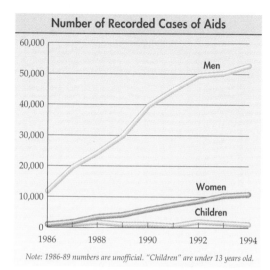

Number of Recorded Cases of Aids

Note: 1986-89 numbers are unofficial. "Children" are under 13 years old.

Source: *U.S. News and World Report,* February 12, 1996

a. Let $t = 0$ represent the year 1986 and $t = 8$ represent the year 1994. Use the data given above to determine a linear function that models the number of recorded cases of women with AIDS for the years from 1986 to 1994.

b. Use the linear function to estimate the number of recorded cases of women with AIDS in 1992.

c. Assuming that the number of recorded cases of women with AIDS continues to increase in a linear manner, what will be the number of recorded cases of women with AIDS in the year 2000?

d. What is the approximate rate at which the number of recorded cases of women with AIDS is increasing per year?

48. Linear Model in Health Science In 1986 there were about 11,000 recorded cases of men with AIDS and in 1994 the number had increased to approximately 52,000. The graph in Exercise 47 shows that the number of recorded cases of men with AIDS has increased in a near linear pattern for the years from 1986 to 1994.

a. Let $t = 0$ represent the year 1986 and $t = 8$ represent the year 1994. Use the data given above to determine a linear function that models the number of recorded cases of men with AIDS for the years from 1986 to 1994.

b. Use the linear function to estimate the number of recorded cases of men with AIDS in 1989.

c. Assuming that the number of recorded cases of men with AIDS continues to increase in a linear manner, what will be the number of recorded cases of men with AIDS in the year 2000?

d. What is the approximate rate at which the number of recorded cases of men with AIDS is increasing per year?

In Exercises 49 to 52, determine the profit function for the given revenue function and cost function. Also determine the break-even point.

49. $R(x) = 92.50x$; $C(x) = 52x + 1782$

50. $R(x) = 124x$; $C(x) = 78.5x + 5005$

51. $R(x) = 259x$; $C(x) = 180x + 10{,}270$

52. $R(x) = 14{,}220x$; $C(x) = 8010x + 1{,}602{,}180$

53. Marginal Cost In business, *marginal cost* is a phrase used to represent the rate of change or slope of a cost function that relates the cost C to the number of units x produced. If a cost function is given by $C(x) = 8x + 275$, find

a. $C(0)$ **b.** $C(1)$ **c.** $C(10)$ **d.** marginal cost

54. Marginal Revenue In business, *marginal revenue* is a phrase used to represent the rate of change or slope of a revenue function that relates the revenue R to the number of units x sold. If a revenue function is given by the function $R(x) = 210x$, find

a. $R(0)$ **b.** $R(1)$ **c.** $R(10)$ **d.** marginal revenue

55. Break-Even Point for a Rental Truck A rental company purchases a truck for $19,500. The truck requires an average of $6.75 per day in maintenance.

a. Find the linear function that expresses the total cost C of owning the truck after t days.

b. The truck rents for $55.00 a day. Find the linear function that expresses the revenue R when the truck has been rented for t days.

c. The profit after t days, $P(t)$, is given by the function $P(t) = R(t) - C(t)$. Find the linear function $P(t)$.

d. Use the function $P(t)$ that you obtained in **c.** to determine how many days it will take the company to break even on the purchase of the truck. Assume that the truck is always in use.

56. Break-Even Point for a Publisher A magazine company had a profit of $98,000 per year when it had 32,000 subscribers. When it obtained 35,000 subscribers, it had a profit of $117,500. Assume that the profit P is a linear function of the number of subscribers s.

a. Find the function P.

b. What will the profit be if the company obtains 50,000 subscribers?

c. What is the number of subscribers needed to break even?

57. OLYMPIC SPRINT The winning times in the men's Olympic 200-meter dash from 1948 to 1992 are shown in the following scatter plot, where $t = 0$ represents the year 1948 and $t = 44$ represents the year 1992.

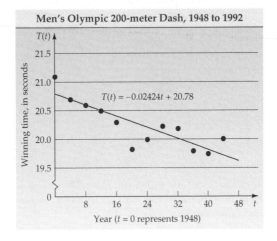

Men's Olympic 200-meter Dash, 1948 to 1992

$T(t) = -0.02424t + 20.78$

Winning time, in seconds

Year ($t = 0$ represents 1948)

The linear function T approximates the data in the scatter plot. According to T, what will be the winning time for this event in the year 1996?

58. OLYMPIC RELAY The winning times in the women's Olympic 400-meter relay from 1948 to 1992 are shown in the following scatter plot, where $t = 0$ represents the year 1948 and $t = 44$ represents the year 1992.

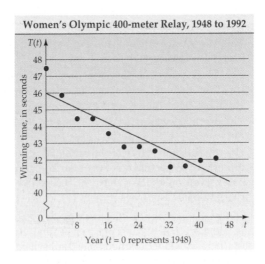

Women's Olympic 400-meter Relay, 1948 to 1992

Winning time, in seconds

Year ($t = 0$ represents 1948)

a. The line approximates the data in the scatter plot. Find an equation of the line. (Because it is not possible to find the exact slope or y-intercept from a graph, answers will vary.)

b. Use the equation (or the line) to predict the winning time for this Olympic event in the year 2000.

59. OLYMPIC 800-METER RUN The winning times in the men's Olympic 800-meter run from 1948 to 1992 are shown in the following scatter plot ($t = 0$ represents the year 1948 and $t = 44$ represents the year 1992).

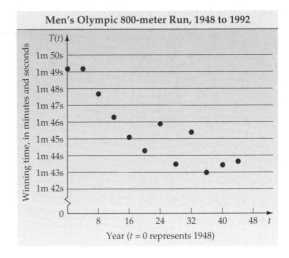

Men's Olympic 800-meter Run, 1948 to 1992

Winning time, in minutes and seconds

Year ($t = 0$ represents 1948)

a. Draw the line that you think best approximates the points in the scatter plot. (Answers will vary.)

b. Find an equation of the line.

c. Use the equation to predict the winning time for this Olympic event in the year 2000.

60. OLYMPIC SPRINT The winning times in the women's Olympic 200-meter dash are given in the following table.

Women's Olympic 200-meter Dash

Year	Time	Year	Time
1948	24.2 s	1972	22.40 s
1952	23.7 s	1976	22.37 s
1956	23.4 s	1980	22.03 s
1960	24.0 s	1984	21.81 s
1964	23.0 s	1988	21.34 s
1968	22.5 s	1992	21.81 s

a. Construct a scatter plot of the data. Let $t = 0$ represent the year 1948 and $t = 44$ represent the year 1992.

b. Sketch the line that you think best approximates the points in the scatter plot. Write the equation of the line. (Answers will vary.)

c. Use the equation (or the line) to predict the winning time for this Olympic event in the year 2000.

In Exercises 61 to 64, find the equation of the indicated line. Write the equation in the form y = mx + b.

61. Through $(1, 3)$ and parallel to $3x + 4y = -24$

62. Through $(2, -1)$ and parallel to $x + y = 10$

63. Through $(1, 2)$ and perpendicular to $x + y = 4$

64. Through $(-3, 4)$ and perpendicular to $2x - y = 7$

65. POINT OF IMPACT A rock is whirled horizontally, in a counterclockwise circular path with radius 5 feet, about the origin. When the string breaks, the rock travels on a linear path perpendicular to the radius $\overline{OP}$ and hits a wall located at $y = 10$ feet.

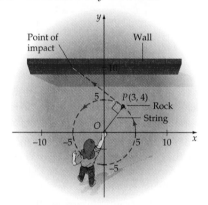

If the string breaks when the rock is at $P(3$ feet, 4 feet), find the x-coordinate of the point where the rock hits the wall.

66. POINT OF IMPACT A rock is whirled horizontally, in a counterclockwise circular path with radius 4 feet, about the origin. When the string breaks, the rock travels on a linear path perpendicular to the radius $\overline{OP}$ and hits a wall located at $y = 14$ feet. If the string breaks when the rock is at $P(\sqrt{15}$ feet, 1 foot), find the x-coordinate of the point where the rock hits the wall.

67. SLOPE OF A SECANT LINE The graph of $y = x^2 + 1$ is shown in the following figure.

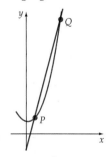

$P(2, 5)$ and $Q(2 + h, [2 + h]^2 + 1)$ are both points on the graph.

a. If $h = 1$, determine the coordinates of Q and the slope of the line PQ.

b. If $h = 0.1$, determine the coordinates of Q and the slope of the line PQ.

c. If $h = 0.01$, determine the coordinates of Q and the slope of the line PQ.

d. As h approaches 0, what value does the slope of the line PQ seem to be approaching?

e. Verify that the slope of the line passing through $(2, 5)$ and $(2 + h, [2 + h]^2 + 1)$ is $4 + h$.

68. SLOPE OF A SECANT LINE The graph of $y = 3x^2$ is shown in the following figure.

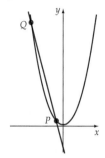

$P(-1, 3)$ and $Q(-1 + h, 3[-1 + h]^2)$ are both points on the graph.

a. If $h = 1$, determine the coordinates of Q and the slope of the line PQ.

b. If $h = 0.1$, determine the coordinates of Q and the slope of the line PQ.

c. If $h = 0.01$, determine the coordinates of Q and the slope of the line PQ.

d. As h approaches 0, what value does the slope of the line PQ seem to be approaching?

e. Verify that the slope of the line passing through $(-1, 3)$ and $(-1 + h, 3[-1 + h]^2)$ is $-6 + 3h$.

69. Verify that the slope of the line passing through (x, x^2) and $(x + h, [x + h]^2)$ is $2x + h$.

70. Verify that the slope of the line passing through $(x, 4x^2)$ and $(x + h, 4[x + h]^2)$ is $8x + 4h$.

SUPPLEMENTAL EXERCISES

71. THE TWO-POINT FORM Use the point-slope form to derive the following equation, which is called the two-point form.

$$y - y_1 = \left(\frac{y_2 - y_1}{x_2 - x_1}\right)(x - x_1)$$

72. THE INTERCEPT FORM Use the two-point form from Exercise 71 to show that the line with intercepts $(a, 0)$ and $(0, b)$, $a \neq 0$ and $b \neq 0$, has the equation

$$\frac{x}{a} + \frac{y}{b} = 1$$

In Exercises 73 and 74, use the two-point form to find an equation of the line that passes through the indicated points. Write your answers in slope-intercept form.

73. $(5, 1)$, $(4, 3)$ **74.** $(2, 7)$, $(-1, 6)$

In Exercises 75 to 82, use the equation from Exercise 72 (called the intercept form) to write an equation of a line with the indicated intercepts.

75. x-intercept $(3, 0)$, y-intercept $(0, 5)$

76. x-intercept $(-2, 0)$, y-intercept $(0, 7)$

77. x-intercept $(a, 0)$, y-intercept $(0, 3a)$, point on the line $(5, 2)$, $a \neq 0$

78. x-intercept $(-b, 0)$, y-intercept $(0, 2b)$, point on the line $(-3, 10)$, $b \neq 0$

79. Verify that the slope of the line passing through $(1, 3)$ and $(1 + h, 3[1 + h]^3)$ is $9 + 9h + 3h^2$.

80. Find the two points on the circle given by $x^2 + y^2 = 25$ such that the slope of the radius from $(0, 0)$ to each point is 0.5.

81. Find a point on the graph of the equation $y = x^2$ such that the slope of the line through the point $(3, 9)$ is 15/2.

82. Determine whether there is a point on the graph of the equation $y = \sqrt{x + 1}$ such that the slope of the line through the point $(3, 2)$ is 3/4.

83. ▮▮ **COSMETIC SURGERY** The following graph shows that men are seeking cosmetic surgery in greater numbers.

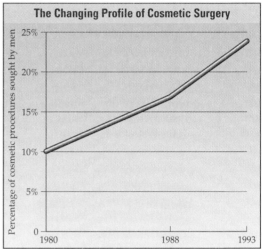

Data: American Academy of Cosmetic Surgery

a. Find a piecewise-defined function that involves two linear functions and closely approximates the information provided by the graph. Let $t = 0$ represent the year 1980.

b. Assuming the trend shown by the graph over the period from 1988 to 1993 continues, to the nearest percent, what percent of cosmetic surgeries will be performed on men in 1996?

PROJECTS

1. Visual Insight

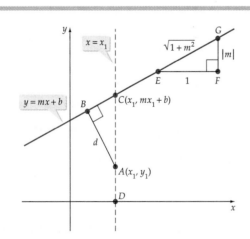

The distance d between the point $A(x_1, y_1)$ and the line $y = mx + b$ is

$$d = \frac{|mx_1 + b - y_1|}{\sqrt{1 + m^2}}$$

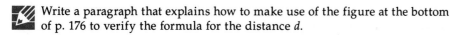

Write a paragraph that explains how to make use of the figure at the bottom of p. 176 to verify the formula for the distance d.

2. **VERIFY GEOMETRIC THEOREMS**

   a. Prove that in any triangle, the line segment that joins the midpoints of two sides of the triangle is parallel to the third side. (*Hint:* Assign coordinates to the vertices of the triangle as shown in the figure at the left below.)

   b. Prove that in any square, the diagonals are perpendicular bisectors of each other. (*Hint:* Assign coordinates to the vertices of the square as shown in the figure at the right below.)

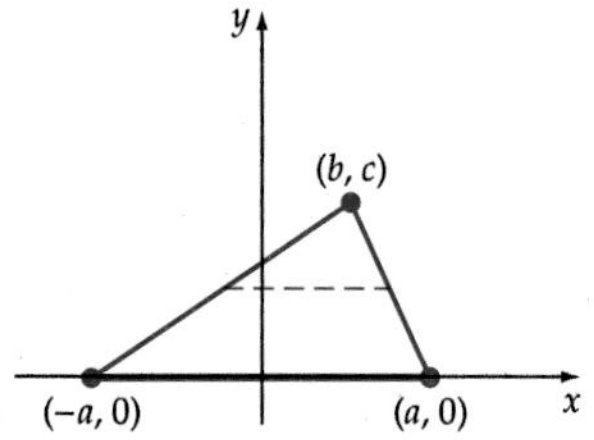

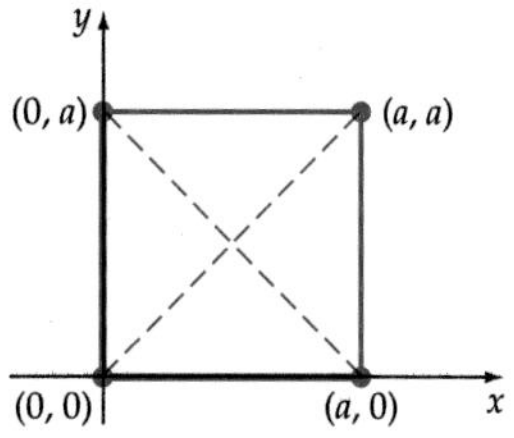

SECTION 3.4 QUADRATIC FUNCTIONS

Some applications can be modeled by a *quadratic function.*

DEFINITION OF A QUADRATIC FUNCTION

A **quadratic function** of x is a function that can be represented by an equation of the form

$$f(x) = ax^2 + bx + c$$

where a, b, and c are real numbers and $a \neq 0$.

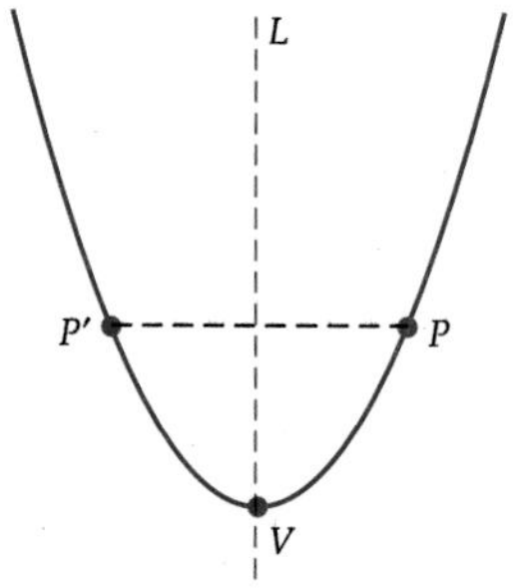

Figure 3.54

The graph of $f(x) = ax^2 + bx + c$ is a *parabola.* The graph opens up if $a > 0$, and it opens down if $a < 0$. The **vertex of a parabola** is the lowest point on a parabola that opens up or the highest point on a parabola that opens down. Point V is the vertex of the parabola in **Figure 3.54.**

The graph of $f(x) = ax^2 + bx + c$ is *symmetric* with respect to a vertical line through its vertex.

DEFINITION OF SYMMETRY WITH RESPECT TO A LINE

A graph is **symmetric with respect to a line** L if for each point P on the graph there is a point P' on the graph such that the line L is the perpendicular bisector of the line segment PP'.

In **Figure 3.54**, the parabola is symmetric with respect to the line L. The line L is called the **axis of symmetry**. The points P and P' are reflections or images of each other with respect to the axis of symmetry.

If $b = 0$ and $c = 0$, then $f(x) = ax^2 + bx + c$ simplifies to $f(x) = ax^2$. The graph of $f(x) = ax^2$ $(a \neq 0)$ is a parabola with its vertex at the origin, and the y-axis is its axis of symmetry. The graph of $f(x) = ax^2$ can be constructed by plotting a few points and drawing a smooth curve that passes through these points with the origin as the vertex and the y-axis as its axis of symmetry. The graphs of $f(x) = x^2$, $g(x) = 2x^2$, and $h(x) = -\dfrac{1}{2}x^2$ are shown in **Figure 3.55**.

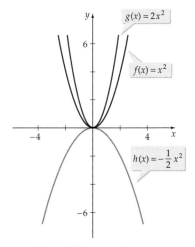

Figure 3.55

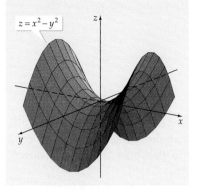
STANDARD FORM OF QUADRATIC FUNCTIONS

Every quadratic function f given by $f(x) = ax^2 + bx + c$ can be written in the **standard form of a quadratic function**:

$$f(x) = a(x - h)^2 + k, \quad a \neq 0$$

The graph of f is a parabola with vertex (h, k). The parabola opens up if $a > 0$, and it opens down if $a < 0$. The vertical line $x = h$ is the axis of symmetry of the parabola.

The standard form is useful because it readily gives information about the vertex of the parabola and its axis of symmetry. For example, the graph of $f(x) = 2(x - 4)^2 - 3$ is a parabola. The coordinates of the vertex are $(4, -3)$, and the line $x = 4$ is its axis of symmetry. Because a is the positive number 2, the parabola opens upward.

EXAMPLE 1 *Find the Standard Form of a Quadratic Function*

Use the technique of completing the square to find the standard form of $g(x) = 2x^2 - 12x + 19$. Sketch the graph.

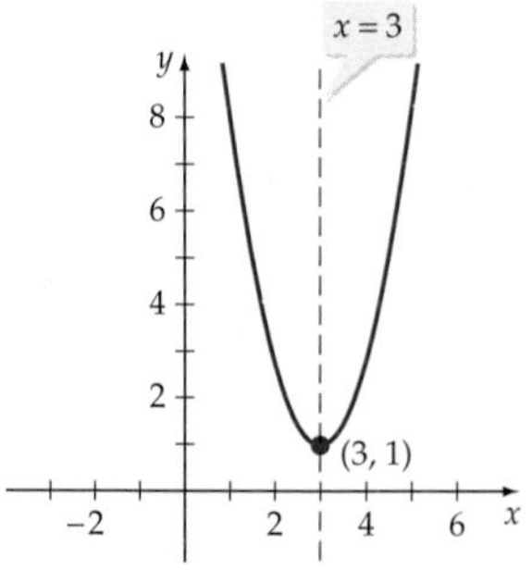

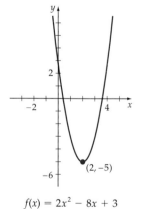

$g(x) = 2x^2 - 12x + 19$

Figure 3.56

Solution

$$g(x) = 2x^2 - 12x + 19$$
$$= 2(x^2 - 6x) + 19 \qquad \bullet \textbf{ Factor 2 from the variable terms.}$$
$$= 2(x^2 - 6x + 9 - 9) + 19 \qquad \bullet \textbf{ Complete the square.}$$
$$= 2(x^2 - 6x + 9) - 2(9) + 19 \qquad \bullet \textbf{ Regroup.}$$
$$= 2(x - 3)^2 - 18 + 19 \qquad \bullet \textbf{ Factor and simplify.}$$
$$= 2(x - 3)^2 + 1 \qquad \bullet \textbf{ Standard form.}$$

The vertex is $(3, 1)$. The axis of symmetry is $x = 3$. Because $a > 0$, the parabola opens up. See **Figure 3.56**.

> **TRY EXERCISE 10, EXERCISE SET 3.4**

VERTEX OF A PARABOLA

By completing the square of $ax^2 + bx + c$, the x-coordinate of the vertex of the graph of $f(x) = ax^2 + bx + c$ can be shown to be $-\dfrac{b}{2a}$. The y-coordinate of the vertex is $f\left(-\dfrac{b}{2a}\right)$. This result is summarized by the following formula.

TAKE NOTE

The vertex formula can be used to write the standard form of a quadratic function. Use the formulas

$$h = -\frac{b}{2a} \quad \text{and} \quad k = f\left(-\frac{b}{2a}\right)$$

VERTEX FORMULA

The vertex of the graph of $f(x) = ax^2 + bx + c$ is $\left(-\dfrac{b}{2a}, f\left(-\dfrac{b}{2a}\right)\right)$.

EXAMPLE 2   **Find the Vertex and Standard Form of a Quadratic Function**

Use the vertex formula to find the vertex and standard form of $f(x) = 2x^2 - 8x + 3$. See **Figure 3.57**.

Solution

$$f(x) = 2x^2 - 8x + 3 \qquad \bullet \, a = 2, b = -8, c = 3$$
$$h = -\frac{b}{2a} = -\frac{-8}{2(2)} = 2 \qquad \bullet \textbf{ x-coordinate of the vertex}$$
$$k = f\left(-\frac{b}{2a}\right) = 2(2)^2 - 8(2) + 3 = -5 \qquad \bullet \textbf{ y-coordinate of the vertex}$$

The vertex is $(2, -5)$. Substituting into the standard form $f(x) = a(x - h)^2 + k$ yields the standard form $f(x) = 2(x - 2)^2 - 5$.

$f(x) = 2x^2 - 8x + 3$

Figure 3.57

> **TRY EXERCISE 20, EXERCISE SET 3.4**

The following theorem can be used to determine the maximum value or the minimum value of a quadratic function.

MAXIMUM OR MINIMUM VALUE OF A QUADRATIC FUNCTION

If $a > 0$, then the vertex (h, k) is the lowest point on the graph of $f(x) = a(x - h)^2 + k$, and the y-coordinate k of the vertex is the **minimum value** of the function f. See **Figure 3.58a**.

If $a < 0$, then the vertex (h, k) is the highest point on the graph of $f(x) = a(x - h)^2 + k$, and the y-coordinate k is the **maximum value** of the function f. See **Figure 3.58b**.

In either case, the maximum or minimum is achieved when $x = h$.

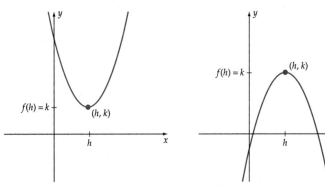

a. k is the minimum value of f.   b. k is the maximum value of f.

Figure 3.58

EXAMPLE 3   *Find the Maximum or Minimum of a Quadratic Function*

Find the maximum or minimum value of each quadratic function. State whether the value is a maximum or a minimum.

a. $F(x) = -2x^2 + 8x - 1$ b. $G(x) = x^2 - 3x + 1$

Solution

The maximum or minimum value of a quadratic function is the y-coordinate of the vertex of the graph of the function.

a. $h = -\dfrac{b}{2a} = -\dfrac{8}{2(-2)} = 2$ • **x-coordinate of the vertex**

$k = F\left(-\dfrac{b}{2a}\right) = -2(2)^2 + 8(2) - 1 = 7$ • **y-coordinate of the vertex**

Because $a < 0$, the function has a maximum value. The maximum value is 7. See **Figure 3.59**.

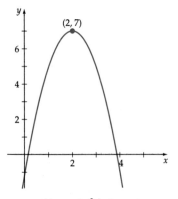

$F(x) = -2x^2 + 8x - 1$

Figure 3.59

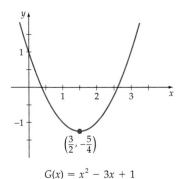

$G(x) = x^2 - 3x + 1$

Figure 3.60

b. $h = -\dfrac{b}{2a} = -\dfrac{-3}{2(1)} = \dfrac{3}{2}$ • **x-coordinate of the vertex**

$k = G\left(-\dfrac{b}{2a}\right) = \left(\dfrac{3}{2}\right)^2 - 3\left(\dfrac{3}{2}\right) + 1$

$= -\dfrac{5}{4}$ • **y-coordinate of the vertex**

Because $a > 0$, the function has a minimum value. The minimum value is $-5/4$. See **Figure 3.60**.

TRY EXERCISE 30, EXERCISE SET 3.4

APPLICATIONS

EXAMPLE 4   *Find the Maximum of a Quadratic Function*

A long sheet of tin 20 inches wide is to be made into a trough by bending up two sides until they are perpendicular to the bottom. How many inches should be turned up so that the trough will achieve its maximum carrying capacity?

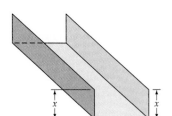

Figure 3.61

Solution

The trough is shown in **Figure 3.61**. If x is the number of inches to be turned up on each side, then the width of the base is $20 - 2x$ inches. The maximum carrying capacity of the trough will occur when the cross-sectional area is a maximum. The cross-sectional area $A(x)$ is given by

$$A(x) = x(20 - 2x) \qquad \text{• } \textbf{Area = (length)(width)}$$
$$= -2x^2 + 20x$$

To find when A obtains its maximum value, find the x-coordinate of the vertex of the graph of A. Using the vertex formula with $a = -2$ and $b = 20$, we have

$$x = -\dfrac{b}{2a} = -\dfrac{20}{2(-2)} = 5$$

Therefore, the maximum carrying capacity will be achieved when $x = 5$ inches are turned up. See **Figure 3.62**.

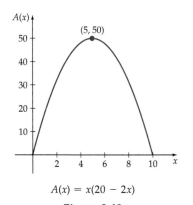

$A(x) = x(20 - 2x)$

Figure 3.62

TRY EXERCISE 40, EXERCISE SET 3.4

EXAMPLE 5   *Solve a Business Application*

The owners of a travel agency have determined that they can sell all 160 tickets for a tour if they charge $8 (their cost) for each ticket. For

Continued • ▶

each \$0.25 increase in the price of a ticket, they estimate they will sell 1 ticket less. A business manager determines that their cost function is $C(x) = 8x$ and that the customer's price per ticket is

$$p(x) = 8 + 0.25(160 - x) = 48 - 0.25x$$

where x represents the number of tickets sold. Determine the maximum profit and the cost per ticket that yields the maximum profit.

Solution

The profit from selling x tickets is $P(x) = R(x) - C(x)$, where P, R, and C are the profit function, the revenue function, and the cost function as defined in Section 3.3. Thus

$$
\begin{aligned}
P(x) &= R(x) - C(x) \\
&= x[p(x)] - C(x) \\
&= x(48 - 0.25x) - 8x \\
&= 40x - 0.25x^2
\end{aligned}
$$

The graph of the profit function is a parabola that opens down. Thus the maximum profit occurs when

$$x = -\frac{b}{2a} = -\frac{40}{2(-0.25)} = 80$$

The maximum profit is determined by evaluating $P(x)$ with $x = 80$.

$$P(80) = 40(80) - 0.25(80)^2 = 1600$$

To find the price per ticket that yields the maximum profit, we evaluate $p(x)$ with $x = 80$.

$$p(80) = 48 - 0.25(80) = 28$$

Thus the travel agency can expect a maximum profit of \$1600 when 80 people take the tour at a ticket price of \$28 per person. The graph of the profit function is shown in **Figure 3.63**.

QUESTION   In **Figure 3.63**, why have we shown only the portion of the graph that lies in Quadrant I?

TRY EXERCISE 54, EXERCISE SET 3.4

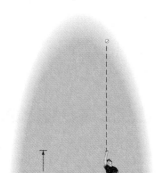

$P(x) = 40x - 0.25x^2$

Figure 3.63

EXAMPLE 6   *Solve a Projectile Application*

In **Figure 3.64**, a ball is thrown vertically upward with an initial velocity of 48 feet per second. If the ball started its flight at a height of 8 feet, then its height s at time t can be determined by $s(t) = -16t^2 + 48t + 8$,

Figure 3.64

8 ft

ANSWER   Since x represents the number of tickets sold, x must be greater than or equal to zero but less than or equal to 160. $P(x)$ is nonnegative for $0 \le x \le 160$.

where $s(t)$ is measured in feet above ground level and t is the number of seconds of flight.

a.   Determine the time it takes the ball to attain its maximum height.

b.   Determine the maximum height the ball attains.

c.   Determine the time it takes the ball to hit the ground.

Solution

a.   The graph of $s(t) = -16t^2 + 48t + 8$ is a parabola that opens downward. See **Figure 3.65**. Therefore, s will attain its maximum value at the vertex of its graph. Using the vertex formula with $a = -16$ and $b = 48$, we get

$$t = -\frac{b}{2a} = -\frac{48}{2(-16)} = \frac{3}{2}$$

Therefore, the ball attains its maximum height $1\frac{1}{2}$ seconds into its flight.

b.   When $t = 3/2$, the height of the ball is

$$s\left(\frac{3}{2}\right) = -16\left(\frac{3}{2}\right)^2 + 48\left(\frac{3}{2}\right) + 8 = 44 \text{ feet}$$

c.   The ball will hit the ground when its height $s(t) = 0$. Therefore, solve $-16t^2 + 48t + 8 = 0$ for t.

$$-16t^2 + 48t + 8 = 0$$

$$-2t^2 + 6t + 1 = 0 \qquad \bullet \text{ **Divide each side by 8.**}$$

$$t = \frac{-(6) \pm \sqrt{6^2 - 4(-2)(1)}}{2(-2)} \qquad \bullet \text{ **Use the quadratic formula.**}$$

$$= \frac{-6 \pm \sqrt{44}}{-4} = \frac{-3 \pm \sqrt{11}}{-2}$$

Using a calculator to approximate the positive root, we find that the ball will hit the ground in $t \approx 3.16$ seconds. This is also the value of the x-coordinate of the x-intercept in **Figure 3.65**.

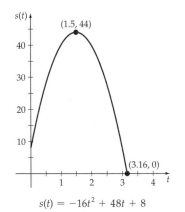

$s(t) = -16t^2 + 48t + 8$

Figure 3.65

TRY EXERCISE 56, EXERCISE SET 3.4

TOPICS FOR DISCUSSION

1.   A classmate states that the graph of every quadratic function of the form

$$f(x) = ax^2 + bx + c$$

has a y-intercept. Do you agree? Explain.

2.   The graph of $f(x) = -x^2 + 6x + 11$ has a vertex of $(3, 20)$. Is this vertex point the highest point or the lowest point on the graph of f?

3. A tutor states that the graph of $f(x) = ax^2 + bx + c \ (a \neq 0)$ is a parabola and that its axis of symmetry is $y = -\dfrac{b}{2a}$. Do you agree?

4. Every quadratic function of the form $f(x) = ax^2 + bx + c$ has a domain of all real numbers. Do you agree?

5. A classmate states that the graph of every quadratic function of the form

$$f(x) = ax^2 + bx + c$$

must contain points from at least two quadrants. Do you agree?

EXERCISE SET 3.4

In Exercises 1 to 8, match each graph in _a._ through _h._ with the proper quadratic function.

1. $f(x) = x^2 - 3$ **2.** $f(x) = x^2 + 2$

3. $f(x) = (x - 4)^2$ **4.** $f(x) = (x + 3)^2$

5. $f(x) = -2x^2 + 2$ **6.** $f(x) = -\dfrac{1}{2}x^2 + 3$

7. $f(x) = (x + 1)^2 + 3$ **8.** $f(x) = -2(x - 2)^2 + 2$

a.

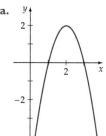

b.

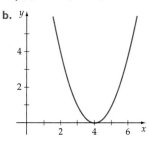

c.

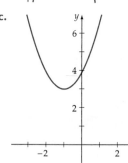

d.

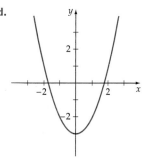

e.

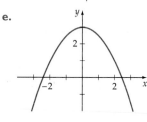

f.

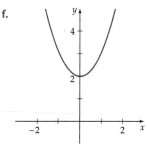

g.

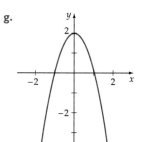

h.

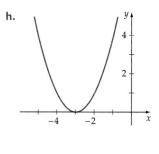

In Exercises 9 to 18, use the method of completing the square to find the standard form of the function, and then sketch its graph. Label its vertex and axis of symmetry.

9. $f(x) = x^2 + 4x + 1$ **10.** $f(x) = x^2 + 6x - 1$

11. $f(x) = x^2 - 8x + 5$ **12.** $f(x) = x^2 - 10x + 3$

13. $f(x) = x^2 + 3x + 1$ **14.** $f(x) = x^2 + 7x + 2$

15. $f(x) = -x^2 + 4x + 2$ **16.** $f(x) = -x^2 - 2x + 5$

17. $f(x) = -3x^2 + 3x + 7$ **18.** $f(x) = -2x^2 - 4x + 5$

In Exercises 19 to 28, use the vertex formula to determine the vertex of the graph of the function and write the function in standard form.

19. $f(x) = x^2 - 10x$ **20.** $f(x) = x^2 - 6x$

21. $f(x) = x^2 - 10$ **22.** $f(x) = x^2 - 4$

23. $f(x) = -x^2 + 6x + 1$ **24.** $f(x) = -x^2 + 4x + 1$

25. $f(x) = 2x^2 - 3x + 7$ **26.** $f(x) = 3x^2 - 10x + 2$

27. $f(x) = -4x^2 + x + 1$ **28.** $f(x) = -5x^2 - 6x + 3$

In Exercises 29 to 38, find the maximum or minimum value of the function. State whether this value is a maximum or a minimum.

29. $f(x) = x^2 + 8x$ **30.** $f(x) = -x^2 - 6x$

31. $f(x) = -x^2 + 6x + 2$ **32.** $f(x) = -x^2 + 10x - 3$

33. $f(x) = 2x^2 + 3x + 1$ **34.** $f(x) = 3x^2 + x - 1$

35. $f(x) = 5x^2 - 11$ **36.** $f(x) = 3x^2 - 41$

37. $f(x) = -\dfrac{1}{2}x^2 + 6x + 17$ **38.** $f(x) = -\dfrac{3}{4}x^2 - \dfrac{2}{5}x + 7$

39. **HEIGHT OF AN ARCH** The height of an arch is given by the equation

$$h(x) = -\frac{3}{64}x^2 + 27, \quad -24 \le x \le 24$$

where $|x|$ is the horizontal distance in feet from the center of the arch.

   a. What is the maximum height of the arch?

   b. What is the height of the arch 10 feet to the right of center?

   c. How far from the center is the arch 8 feet tall?

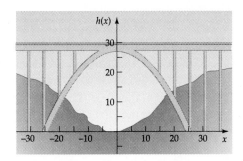

40. The sum of the length l and the width w of a rectangular area is 240 meters.

   a. Write w as a function of l.

   b. Write the area A as a function of l.

   c. Find the dimensions that produce the greatest area.

41. **RECTANGULAR ENCLOSURE** A veterinarian uses 600 feet of chain-link fencing to enclose a rectangular region and also to subdivide the region into two smaller rectangular regions by placing a fence parallel to one of the sides, as shown in the figure.

   a. Write the width w as a function of the length l.

   b. Write the total area A as a function of l.

   c. Find the dimensions that produce the greatest enclosed area.

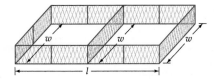

42. **RECTANGULAR ENCLOSURE** A farmer uses 1200 feet of fence to enclose a rectangular region and also to subdivide the region into three smaller rectangular regions by placing the fences parallel to one of the sides. Find the dimensions that produce the greatest enclosed area.

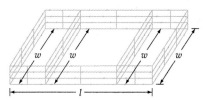

In Exercises 43 to 46, determine the y- and x-intercepts (if any) of the quadratic function.

43. $f(x) = x^2 + 6x$ **44.** $f(x) = -x^2 + 4x$

45. $f(x) = -3x^2 + 5x - 6$ **46.** $f(x) = 2x^2 + 3x + 4$

In Exercises 47 and 48, determine the number of units x that produce a maximum revenue for the given revenue function. Also determine the maximum revenue.

47. $R(x) = 296x - 0.2x^2$ **48.** $R(x) = 810x - 0.6x^2$

In Exercises 49 and 50, determine the number of units x that produce a maximum profit for the given profit function. Also determine the maximum profit.

49. $P(x) = -0.01x^2 + 1.7x - 48$

50. $P(x) = -\dfrac{x^2}{14,000} + 1.68x - 4000$

In Exercises 51 and 52, determine the profit function for the given revenue function and cost function. Also determine the break-even point(s).

51. $R(x) = x(102.50 - 0.1x)$; $C(x) = 52.50x + 1840$

52. $R(x) = x(210 - 0.25x)$; $C(x) = 78x + 6399$

53. **TOUR COST** A charter bus company has determined that its cost of providing x people a tour is

$$C(x) = 180 + 2.50x$$

A full tour consists of 60 people. The ticket price per person is \$15 plus \$0.25 for each unsold ticket. Determine

   a. the revenue function   **b.** the profit function

   c. the company's maximum profit   **d.** the number of ticket sales that yields the maximum profit

54. **DELIVERY COST** An air freight company has determined that its cost of delivering x parcels per flight is

$$C(x) = 2025 + 7x$$

The price it charges to send x parcels is

$$p(x) = 22 - 0.01x$$

Determine

   a. the revenue function   **b.** the profit function

   c. the company's maximum profit   **d.** the price per parcel that yields the maximum profit   **e.** the minimum number of parcels the air freight company must ship to break even

55. PROJECTILE If the initial velocity of a projectile is 128 feet per second, then its height h in feet is a function of time t in seconds, given by the equation $h(t) = -16t^2 + 128t$.

    a. Find the time t when the projectile achieves its maximum height.

    b. Find the maximum height of the projectile.

    c. Find the time t when the projectile hits the ground.

56. PROJECTILE The height in feet of a projectile with an initial velocity of 64 feet per second and an initial height of 80 feet is a function of time t in seconds, given by

$$h(t) = -16t^2 + 64t + 80$$

    a. Find the maximum height of the projectile.

    b. Find the time t when the projectile achieves its maximum height.

    c. Find the time t when the projectile has a height of 0 feet.

57. NORMAN WINDOW A Norman window has the shape of a rectangle surmounted by a semicircle. The perimeter of the window shown in the figure is 48 feet. Find the height h and the radius r that will allow the maximum amount of light to enter the window. (*Hint:* Write the area of the window as a quadratic function of the radius r.)

58. NORMAN WINDOW Assume the semicircle in Exercise 57 permits only 1/3 as much light per square unit to enter as does the rectangular portion of the window. Find the height h and the radius r that will allow the maximum amount of light to enter the window.

SUPPLEMENTAL EXERCISES

59. Find the quadratic function of x whose graph has a minimum at $(2, 1)$ and passes through $(0, 4)$.

60. Find the quadratic function of x whose graph has a maximum at $(-3, 2)$ and passes through $(0, -5)$.

61. AREA OF A RECTANGLE A wire 32 inches long is bent so that it has the shape of a rectangle. The length of the rectangle is x and the width is w.

    a. Write w as a function of x.

    b. Write the area A of the rectangle as a function of x.

62. MAXIMIZE AREA Use the function A from **b.** in Exercise 61 to prove that the area A is greatest if the rectangle is a square.

63. Show that the function $f(x) = x^2 + bx - 1$ has a real zero for any value b.

64. Show that the function $g(x) = -x^2 + bx + 1$ has a real zero for any value b.

65. What effect does increasing the constant c have on the graph of $f(x) = ax^2 + bx + c$?

66. If $a > 0$, what effect does decreasing the coefficient a have on the graph of $f(x) = ax^2 + bx + c$?

67. Find two numbers whose sum is 8 and whose product is a maximum.

68. Find two numbers whose difference is 12 and whose product is a minimum.

69. Verify that the slope of the line passing through (x, x^3) and $(x + h, [x + h]^3)$ is $3x^2 + 3xh + h^2$.

70. Verify that the slope of the line passing through $(x, 4x^3 + x)$ and $(x + h, 4[x + h]^3 + [x + h])$ is given by $12x^2 + 12xh + 4h^2 + 1$.

PROJECTS

1. ![icon] **THE CUBIC FORMULA** Write an essay on the development of the cubic formula. An excellent source of information is the chapter "Cardano and the Solution of the Cubic" in *Journey Through Genius*, by William Dunham (New York: Wiley, 1990).

2. **SIMPSON'S RULE** In calculus a procedure known as *Simpson's Rule* is often used to approximate the area under a curve. The figure at the right shows the graph of a parabola that passes through $P_0(-h, y_0)$, $P_1(0, y_1)$, and $P_2(h, y_2)$. The equation of the parabola is of the form $y = Ax^2 + Bx + C$. Using calculus procedures, we can show that the area bounded by the parabola, the x-axis, and the vertical lines $x = -h$ and $x = h$ is

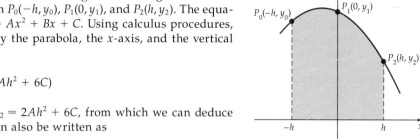

$$\frac{h}{3}(2Ah^2 + 6C)$$

Use algebra to show that $y_0 + 4y_1 + y_2 = 2Ah^2 + 6C$, from which we can deduce that the area of the bounded region can also be written as

$$\frac{h}{3}(y_0 + 4y_1 + y_2)$$

(*Hint:* Evaluate $Ax^2 + Bx + C$ at $x = -h$, $x = 0$, and $x = h$ to determine values of y_0, y_1, and y_2, respectively. Then compute $y_0 + 4y_1 + y_2$.)

SYMMETRY

Figure 3.66

The graph in **Figure 3.66** is symmetric with respect to the line l. Note that the graph has the property that if the paper is folded along the dotted line l, the point A' will coincide with the point A, the point B' will coincide with the point B, and the point C' will coincide with the point C. One part of the graph is a *mirror image* of the rest of the graph across the line l.

A graph is **symmetric with respect to the y-axis** if, whenever the point given by (x, y) is on the graph, then $(-x, y)$ is also on the graph. The graph in **Figure 3.67** is symmetric with respect to the y-axis. A graph is **symmetric with respect to the x-axis** if, whenever the point given by (x, y) is on the graph, then $(x, -y)$ is also on the graph. The graph in **Figure 3.68** is symmetric with respect to the x-axis.

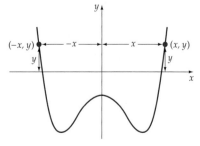

Figure 3.67
Syummetry with respect to the y-axis

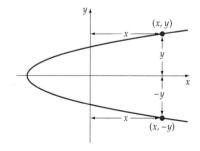

Figure 3.68
Symmetry with respect to the x-axis

The graph of an equation is symmetric with respect to

- the y-axis if the replacement of x with $-x$ leaves the equation unaltered.

- the x-axis if the replacement of y with $-y$ leaves the equation unaltered.

EXAMPLE 1   *Determine Symmetries of a Graph*

Determine whether the graph of the given equations has symmetry with respect to either the x- or the y-axis.

a. $y = x^2 + 2$ b. $x = |y| - 2$

Solution

a. The equation $y = x^2 + 2$ is unaltered by the replacement of x with $-x$. That is, the simplification of $y = (-x)^2 + 2$ yields the original equation $y = x^2 + 2$. Thus the graph of $y = x^2 + 2$ is symmetric with respect to the y-axis. However, the equation $y = x^2 + 2$ *is altered* by the replacement of y with $-y$. That is, the simplification of $-y = x^2 + 2$, which is $y = -x^2 - 2$, *does not* yield the original equation $y = x^2 + 2$. The graph of $y = x^2 + 2$ is not symmetric with respect to the x-axis. See **Figure 3.69**.

b. The equation $x = |y| - 2$ *is altered* by the replacement of x with $-x$. That is, the simplification of $-x = |y| - 2$, which is $x = -|y| + 2$, *does not* yield the original equation $x = |y| - 2$. This implies that the graph of $x = |y| - 2$ is not symmetric with respect to the y-axis. However, the equation $x = |y| - 2$ is unaltered by the replacement of y with $-y$. That is, the simplification of $x = |-y| - 2$ yields the original equation $x = |y| - 2$. The graph of $x = |y| - 2$ is symmetric with respect to the x-axis. See **Figure 3.70**.

TRY EXERCISE 14, EXERCISE SET 3.5

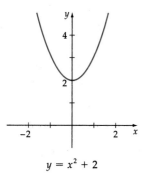

$y = x^2 + 2$

Figure 3.69

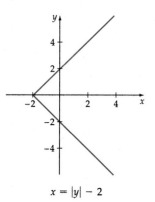

$x = |y| - 2$

Figure 3.70

A graph is **symmetric with respect to a point** Q if for each point P on the graph there is a point P' on the graph such that Q is the midpoint of the line segment PP'.

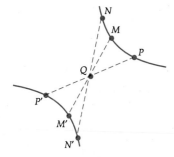

Figure 3.71

The graph in **Figure 3.71** is symmetric with respect to the point Q. For any point P on the graph, there exists a point P' on the graph such that Q is the midpoint of $P'P$.

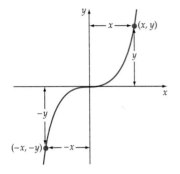

Symmetry with respect to the point Q

Figure 3.72

When we discuss symmetry with respect to a point, we frequently use the origin. A graph is symmetric with respect to the origin if, whenever the point given by (x, y) is on the graph, then $(-x, -y)$ is also on the graph. The graph in **Figure 3.72** is symmetric with respect to the origin.

TEST FOR SYMMETRY WITH RESPECT TO THE ORIGIN

The graph of an equation is symmetric with respect to the origin if the replacement of x with $-x$ and of y with $-y$ leaves the equation unaltered.

EXAMPLE 2   *Determine Symmetry with Respect to the Origin*

Determine whether the graph of each equation has symmetry with respect to the origin.

a. $xy = 4$ **b.** $y = x^3 + 1$

Solution

a. The equation $xy = 4$ is unaltered by the replacement of x with $-x$ and of y with $-y$. That is, the simplification of $(-x)(-y) = 4$ yields the original equation $xy = 4$. Thus the graph of $xy = 4$ is symmetric with respect to the origin. See **Figure 3.73**.

b. The equation $y = x^3 + 1$ *is altered* by the replacement of x with $-x$ and of y with $-y$. That is, the simplification of $-y = (-x)^3 + 1$, which is $y = x^3 - 1$, *does not* yield the original equation $y = x^3 + 1$. Thus the graph $y = x^3 + 1$ is not symmetric with respect to the origin. See **Figure 3.74**.

TRY EXERCISE 24, EXERCISE SET 3.5

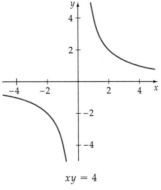

$xy = 4$

Figure 3.73

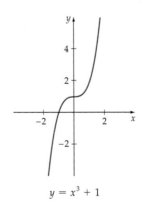

$y = x^3 + 1$

Figure 3.74

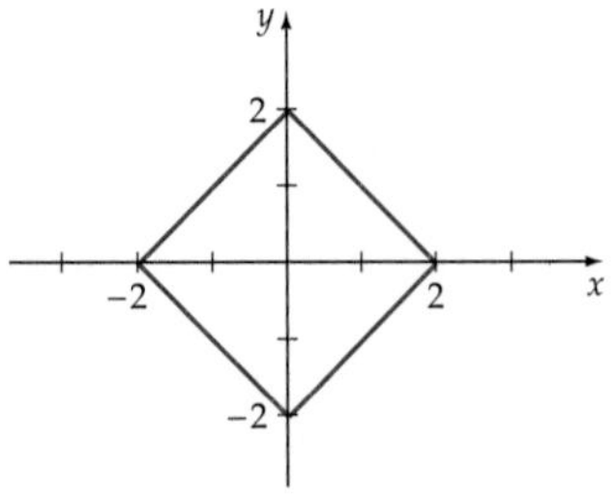

$|x| + |y| = 2$

Figure 3.75

Some graphs have more than one symmetry. For example, the graph of $|x| + |y| = 2$ has symmetry with respect to the x-axis, the y-axis, and the origin. **Figure 3.75** is the graph of $|x| + |y| = 2$.

EVEN AND ODD FUNCTIONS

Some functions are classified as either *even* or *odd*.

DEFINITION OF EVEN AND ODD FUNCTIONS

The function f is an **even function** if

$$f(-x) = f(x) \quad \text{for all } x \text{ in the domain of } f$$

The function f is an **odd function** if

$$f(-x) = -f(x) \quad \text{for all } x \text{ in the domain of } f$$

EXAMPLE 3   *Identify Even or Odd Functions*

Determine whether each function is even, odd, or neither.

a. $f(x) = x^3$ **b.** $F(x) = |x|$ **c.** $h(x) = x^4 + 2x$

Solution

Replace x with $-x$ and simplify.

a. $f(-x) = (-x)^3 = -x^3 = -(x^3) = -f(x)$

This function is an odd function because $f(-x) = -f(x)$.

b. $F(-x) = |-x| = |x| = F(x)$

This function is an even function because $F(-x) = F(x)$.

c. $h(-x) = (-x)^4 + 2(-x) = x^4 - 2x$

This function is neither an even nor an odd function because

$$h(-x) = x^4 - 2x,$$

which is not equal to either $h(x)$ or $-h(x)$.

TRY EXERCISE 44, EXERCISE SET 3.5

The following properties are a result of the tests for symmetry.

- The graph of an even function is symmetric with respect to the y-axis.

- The graph of an odd function is symmetric with respect to the origin.

The graph of f in **Figure 3.76** is symmetric with respect to the y-axis. It is the graph of an even function. The graph of g in **Figure 3.77** is symmetric with respect to the origin. It is the graph of an odd function. The graph of h in **Figure 3.78** is not symmetric to the y-axis and is not symmetric to the origin. It is neither an even nor an odd function.

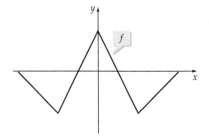

Figure 3.76

The graph of an even function is symmetric with respect to the y-axis.

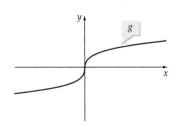

Figure 3.77

The graph of an odd function is symmetric with respect to the origin.

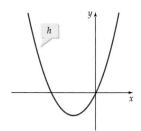

Figure 3.78

If the graph of a function is not symmetric to the y-axis or to the origin, then the function is neither even nor odd.

TRANSLATIONS OF GRAPHS

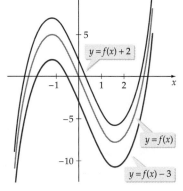

Figure 3.79

The shape of a graph may be exactly the same as the shape of another graph; only their position in the xy-plane may differ. For example, the graph of $y = f(x) + 2$ is the graph of $y = f(x)$ with each point moved up vertically 2 units. The graph of $y = f(x) - 3$ is the graph of $y = f(x)$ with each point moved down vertically 3 units. See **Figure 3.79.**

The graphs of $y = f(x) + 2$ and $y = f(x) - 3$ in **Figure 3.79** are called *vertical translations* of the graph of $y = f(x)$.

VERTICAL TRANSLATIONS

If f is a function and c is a positive constant, then the graph of

- $y = f(x) + c$ is the graph of $y = f(x)$ shifted up *vertically* c units.
- $y = f(x) - c$ is the graph of $y = f(x)$ shifted down *vertically* c units.

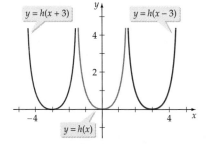

Figure 3.80

In **Figure 3.80,** the graph of $y = h(x + 3)$ is the graph of $y = h(x)$ with each point shifted to the left horizontally 3 units. Similarly, the graph of $y = h(x - 3)$ is the graph of $y = h(x)$ with each point shifted to the right horizontally 3 units.

The graphs of $y = h(x + 3)$ and $y = h(x - 3)$ in **Figure 3.80** are called *horizontal translations* of the graph of $y = h(x)$.

HORIZONTAL TRANSLATIONS

If f is a function and c is a positive constant, then the graph of

- $y = f(x + c)$ is the graph of $y = f(x)$ shifted left *horizontally* c units.
- $y = f(x - c)$ is the graph of $y = f(x)$ shifted right *horizontally* c units.

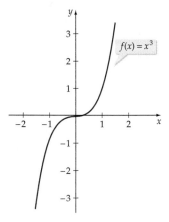

Figure 3.81

EXAMPLE 4   *Graph by Using Translations*

Use vertical and horizontal translations of the graph of $f(x) = x^3$, shown in **Figure 3.81**, to graph

a. $g(x) = x^3 - 2$ **b.** $h(x) = (x + 1)^3$

Solution

a. The graph of $g(x) = x^3 - 2$ is the graph of $f(x) = x^3$ shifted down vertically 2 units. See **Figure 3.82**.

b. The graph of $h(x) = (x + 1)^3$ is the graph of $f(x) = x^3$ shifted to the left horizontally 1 unit. See **Figure 3.83**.

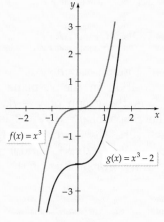

Figure 3.82

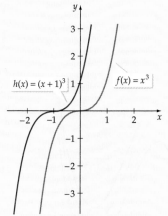

Figure 3.83

TRY EXERCISE 58, EXERCISE SET 3.5

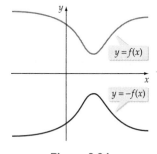

Figure 3.84

REFLECTIONS OF GRAPHS

The graph of $y = -f(x)$ cannot be obtained from the graph of $y = f(x)$ by a combination of vertical and/or horizontal shifts. **Figure 3.84** illustrates that the graph of $y = -f(x)$ is the reflection of the graph of $y = f(x)$ across the x-axis.

The graph of $y = f(-x)$ is the reflection of the graph of $y = f(x)$ across the y-axis as shown in **Figure 3.85**.

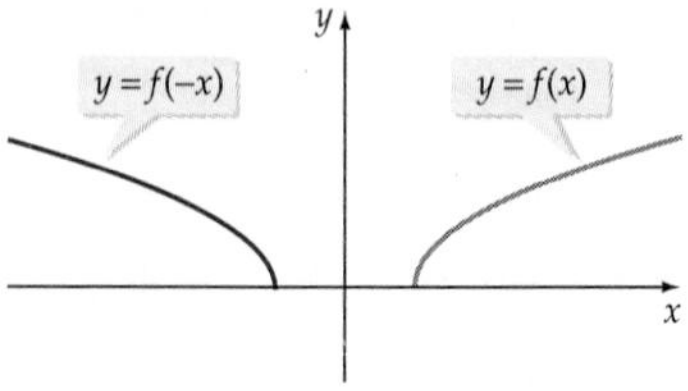

Figure 3.85

REFLECTIONS

The graph of

- $y = -f(x)$ is the graph of $y = f(x)$ reflected across the x-axis.

- $y = f(-x)$ is the graph of $y = f(x)$ reflected across the y-axis.

EXAMPLE 5 *Graph by Using Reflections*

Use reflections of the graph of $f(x) = \sqrt{x-1} + 1$, shown in **Figure 3.86,** to graph

a. $g(x) = -(\sqrt{x-1} + 1)$ **b.** $h(x) = \sqrt{-x-1} + 1$

Solution

a. Because $g(x) = -f(x)$, the graph of g is the graph of f reflected across the x-axis. See **Figure 3.87.**

b. Because $h(x) = f(-x)$, the graph of h is the graph of f reflected across the y-axis. See **Figure 3.88.**

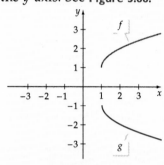

Figure 3.86

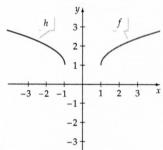

Figure 3.87   Figure 3.88

TRY EXERCISE 60, EXERCISE SET 3.5

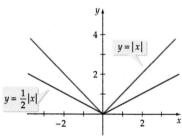

Figure 3.89

Some graphs of functions can be constructed by using a combination of translations and reflections. For instance, the graph of $y = -f(x) + 3$ in **Figure 3.89** was obtained by reflecting the graph of $y = f(x)$ in **Figure 3.89** with respect to the x-axis and then shifting that graph up vertically 3 units.

SHRINKING AND STRETCHING OF GRAPHS

The graph of the equation $y = c \cdot f(x)$ for $c \neq 1$ vertically shrinks or stretches the graph of $y = f(x)$. To determine the points on the graph of $y = c \cdot f(x)$, multiply each y-coordinate of the points on the graph of $y = f(x)$ by c. For example, **Figure 3.90** shows that the graph of $y = \frac{1}{2}|x|$ can be obtained by plotting points that have a y-coordinate that is one-half of the y-coordinate of those found on the graph of $y = |x|$.

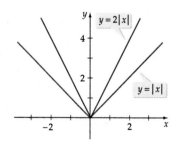

Figure 3.90

If $0 < c < 1$, then the graph of $y = c \cdot f(x)$ is obtained by *shrinking* the graph of $y = f(x)$. **Figure 3.90** illustrates the vertical shrinking of the graph of $y = |x|$ toward the x-axis to form the graph of $y = \frac{1}{2}|x|$.

If $c > 1$, then the graph of $y = c \cdot f(x)$ is obtained by *stretching* the graph of $y = f(x)$. For example, if $f(x) = |x|$, then we obtain the graph of

$$y = 2f(x) = 2|x|$$

by stretching the graph of f away from the x-axis. See **Figure 3.91.**

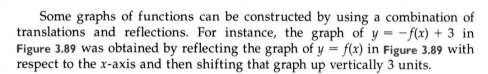

Figure 3.91

EXAMPLE 6   Graph by Using Vertical Shrinking and Shifting

Graph: $H(x) = \dfrac{1}{4}|x| - 3$

Solution

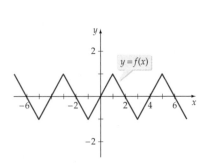

Figure 3.92

The graph of $y = |x|$ has a V shape that has its lowest point at $(0, 0)$ and passes through $(4, 4)$ and $(-4, 4)$. The graph of $y = \frac{1}{4}|x|$ is a shrinking of the graph of $y = |x|$. The y-coordinates $(0, 0)$, $(4, 1)$, and $(-4, 1)$ are obtained by multiplying the y-coordinates of the ordered pairs $(0, 0)$, $(4, 4)$, and $(4, -4)$ by 1/4. To find the points on the graph of H, we still need to subtract 3 from each y-coordinate. Thus the graph of H is a V shape that has its lowest point at $(0, -3)$ and passes through $(4, -2)$ and $(-4, -2)$. See **Figure 3.92**.

TRY EXERCISE 62, EXERCISE SET 3.5

Some functions can be graphed by using a horizontal shrinking or stretching of a given graph. The procedure makes use of the following concept.

HORIZONTAL SHRINKING AND STRETCHING

If $a > 0$ and the graph of $y = f(x)$ contains the point (x, y), then the graph of $y = f(ax)$ contains the point $\left(\dfrac{1}{a}x, y\right)$.

If $a > 1$, then the graph of $y = f(ax)$ is a *horizontal shrinking* of the graph of $y = f(x)$. If $0 < a < 1$, then the graph of $y = f(ax)$ is a *horizontal stretching* of the graph of $y = f(x)$.

EXAMPLE 7   Graph by Using Horizontal Shrinking and Stretching

Use the graph of $y = f(x)$ shown in **Figure 3.93** to graph

a. $y = f(2x)$ **b.** $y = f\left(\dfrac{1}{3}x\right)$

Solution

Figure 3.93

a. Because $2 > 1$, the graph of $y = f(2x)$ is a horizontal contraction (shrinking) of the graph of $y = f(x)$. The graph of $y = f(2x)$ can be constructed by contracting each point on the graph of $y = f(x)$ toward the y-axis by a factor of 1/2. For example, the point $(2, 0)$ on the graph of $y = f(x)$ becomes the point $(1, 0)$ on the graph of $y = f(2x)$. See **Figure 3.94**.

b. Since $0 < \dfrac{1}{3} < 1$, the graph of $y = f\left(\dfrac{1}{3}x\right)$ is a horizontal dilation (stretching) of the graph of $y = f(x)$. The graph of $y = f\left(\dfrac{1}{3}x\right)$ can be constructed by moving each point on the graph of $y = f(x)$ away from the y-axis by a factor of 3. For example, the point $(1, 1)$ on the graph of $y = f(x)$ becomes the point $(3, 1)$ on the graph of $y = f\left(\dfrac{1}{3}x\right)$. See **Figure 3.95**.

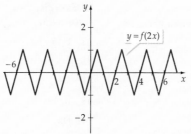

Figure 3.94

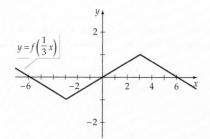

Figure 3.95

TRY EXERCISE 64, EXERCISE SET 3.5

TOPICS FOR DISCUSSION

1. Discuss the meaning of symmetry of a graph with respect to a line. How do you determine whether a graph has symmetry with respect to the x-axis? with respect to the y-axis?

2. Discuss the meaning of symmetry of a graph with respect to a point. How do you determine whether a graph has symmetry with respect to the origin?

3. What does it mean to reflect a graph across the x-axis or across the y-axis?

4. Explain how the graphs of $y_1 = 2x^3 - x^2$ and $y_2 = 2(-x)^3 - (-x)^2$ are related.

5. Given the graph of $y_3 = f(x)$, explain how to obtain the graph of $y_4 = f(x - 3) + 1$.

6. The graph of the *step function* $y_5 = [\![x]\!]$ has steps that are 1 unit wide. Determine how wide the steps are in the graph of $y_6 = \left[\!\!\left[\dfrac{1}{3}x\right]\!\!\right]$.

EXERCISE SET 3.5

In Exercises 1 to 6, plot the image of the given point with respect to

a. the y-axis. Label this point A.

b. the x-axis. Label this point B.

c. the origin. Label this point C.

1. $P(5, -3)$ **2.** $Q(-4, 1)$ **3.** $R(-2, 3)$

4. $S(-5, 3)$ **5.** $T(-4, -5)$ **6.** $U(5, 1)$

In Exercises 7 and 8, sketch a graph that is symmetric to the given graph with respect to the x-axis.

7.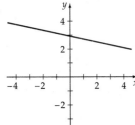

8.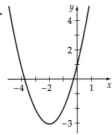

In Exercises 9 and 10, sketch a graph that is symmetric to the given graph with respect to the y-axis.

9.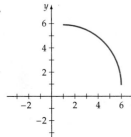

10.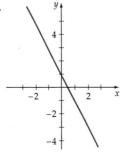

In Exercises 11 and 12, sketch a graph that is symmetric to the given graph with respect to the origin.

11.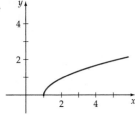

12.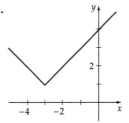

In Exercises 13 to 21, determine whether the graph of each equation is symmetric with respect to the *a.* **x-axis,** *b.* **y-axis.**

13. $y = 2x^2 - 5$ 14. $x = 3y^2 - 7$ 15. $y = x^3 + 2$

16. $y = x^5 - 3x$ 17. $x^2 + y^2 = 9$ 18. $x^2 - y^2 = 10$

19. $x^2 = y^4$ 20. $xy = 8$ 21. $|x| - |y| = 6$

In Exercises 22 to 30, determine whether the graph of each equation is symmetric with respect to the origin.

22. $y = x + 1$ 23. $y = 3x - 2$ 24. $y = x^3 - x$

25. $y = -x^3$ 26. $y = \dfrac{9}{x}$ 27. $x^2 + y^2 = 10$

28. $x^2 - y^2 = 4$ 29. $y = \dfrac{x}{|x|}$ 30. $|y| = |x|$

In Exercises 31 to 42, graph the given equations. Label each intercept. Use the concept of symmetry to confirm that the graph is correct.

31. $y = x^2 - 1$ 32. $x = y^2 - 1$

33. $y = x^3 - x$ 34. $y = -x^3$

35. $xy = 4$ 36. $xy = -8$

37. $y = 2|x - 4|$ 38. $y = |x - 2| - 1$

39. $y = (x - 2)^2 - 4$ 40. $y = (x - 1)^2 - 4$

41. $y = x - |x|$ 42. $|y| = |x|$

In Exercises 43 to 56, identify whether the given function is an even function, an odd function, or neither.

43. $g(x) = x^2 - 7$ 44. $h(x) = x^2 + 1$

45. $F(x) = x^5 + x^3$ 46. $G(x) = 2x^5 - 10$

47. $H(x) = 3|x|$ 48. $T(x) = |x| + 2$

49. $f(x) = 1$ 50. $k(x) = 2 + x + x^2$

51. $r(x) = \sqrt{x^2 + 4}$ 52. $u(x) = \sqrt{3 - x^2}$

53. $s(x) = 16x^2$ 54. $v(x) = 16x^2 + x$

55. $w(x) = 4 + \sqrt[3]{x}$ 56. $z(x) = \dfrac{x^3}{x^2 + 1}$

57. Use the graph of $f(x) = \sqrt{4 - x^2}$ to sketch the graph of
   a. $y = f(x) + 3$ **b.** $y = f(x - 3)$

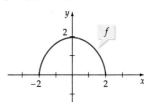

58. Use the graph of $g(x) = |x|$ to sketch the graph of
   a. $y = g(x) - 2$ **b.** $y = g(x - 3)$

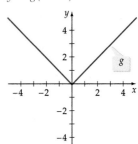

59. Use the graph of $F(x) = (x - 1)^{2/3}$ to sketch the graph of
   a. $y = -F(x)$ **b.** $y = F(-x)$

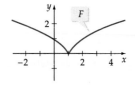

60. Use the graph of $E(x) = |x - 1| + 1$ to sketch the graph of
  a. $y = -E(x)$ **b.** $y = E(-x)$

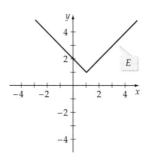

61. Use the graph of $m(x) = x^2 - 2x - 3$ to sketch the graph of $y = -\dfrac{1}{2}m(x) + 3$.

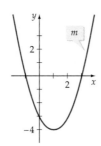

62. Use the graph of $n(x) = -x^2 - 2x + 8$ to sketch the graph of $y = \dfrac{1}{2}n(x) + 1$.

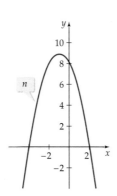

63. Use the graph of $y = f(x)$ to sketch the graph of
  a. $y = f(2x)$ **b.** $y = f\left(\dfrac{1}{3}x\right)$

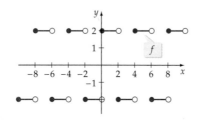

64. Use the graph of $y = g(x)$ to sketch the graph of
  a. $y = g(2x)$ **b.** $y = g\left(\dfrac{1}{2}x\right)$

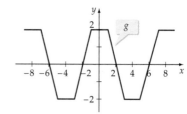

65. Use the graph of $y = h(x)$ to sketch the graph of
  a. $y = h(2x)$ **b.** $y = h\left(\dfrac{1}{2}x\right)$

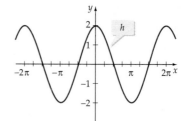

66. Use the graph of $y = j(x)$ to sketch the graph of
  a. $y = j(2x)$ **b.** $y = j\left(\dfrac{1}{3}x\right)$

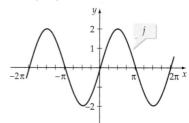

In Exercises 67 to 74, use a graphing utility.

67. On the same coordinate axes, graph
$$G(x) = \sqrt[3]{x} + c$$
  for $c = 0, -1,$ and 3.

68. On the same coordinate axes, graph
$$H(x) = \sqrt[3]{x + c}$$
  for $c = 0, -1,$ and 3.

69. On the same coordinate axes, graph
$$J(x) = |2(x + c) - 3| - |x + c|$$
  for $c = 0, -1,$ and 2.

70. On the same coordinate axes, graph
$$K(x) = |x - 1| - |x| + c$$
  for $c = 0, -1,$ and 2.

71. On the same coordinate axes, graph

$$L(x) = cx^2$$

for $c = 1, 1/2,$ and 2.

72. On the same coordinate axes, graph

$$M(x) = c\sqrt{x^2 - 4}$$

for $c = 1, 1/3,$ and 3.

73. On the same coordinate axes, graph

$$S(x) = c(|x - 1| - |x|)$$

for $c = 1, 1/4,$ and 4.

74. On the same coordinate axes, graph

$$T(x) = c\left(\frac{x}{|x|}\right)$$

for $c = 1, 2/3,$ and 3/2.

75. Graph $V(x) = [\![cx]\!], 0 \le x \le 6,$ for each value of c.

  a. $c = 1$ **b.** $c = 1/2$ **c.** $c = 2$

76. Graph $W(x) = [\![cx]\!] - cx, 0 \le x \le 6,$ for each value of c.

  a. $c = 1$ **b.** $c = 1/3$ **c.** $c = 3$

SUPPLEMENTAL EXERCISES

77. Use the graph of $f(x) = 2/(x^2 + 1)$ to determine an equation for the graphs shown in **a.** and **b.**

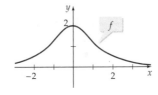

a.
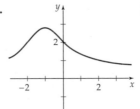

b.
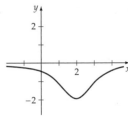

78. Use the graph of $f(x) = x\sqrt{2 + x}$ to determine an equation for the graphs shown in **a.** and **b.**

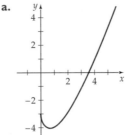

a.
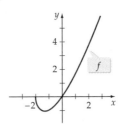

b.
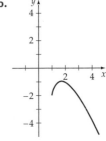

PROJECTS

1. **DIRICHLET FUNCTION** We owe our present-day definition of a function to the German mathematician Peter Gustav Dirichlet (1805–1859). He created the following unusual function, which is now known as the *Dirichlet function*.

$$f(x) = \begin{cases} 0 & \text{if } x \text{ is a rational number} \\ 1 & \text{if } x \text{ is an irrational number} \end{cases}$$

Answer the following questions about the Dirichlet function.

  a. What is its domain?   **b.** What is its range?

  c. What are its x-intercepts?   **d.** What is its y-intercept?

  e. Is it an even or an odd function?

  f. Explain why a graphing calculator cannot be used to produce an accurate graph of the function.

  g. Write a sentence or two that describes its graph.

2.   **ISOLATED POINT** Consider the function given by

$$y = \sqrt{(x-1)^2(x-2)} + 1$$

Verify that the point $(1, 1)$ is a solution of the equation. Now use a graphing utility to graph the function. Does your graph include the isolated point at $(1, 1)$, as shown at the right? If the graphing utility you used failed to include the point $(1, 1)$, explain at least one reason for the omission of this isolated point.

3. **A LINE WITH A HOLE** The function

$$f(x) = \frac{(x-2)(x+1)}{(x-2)}$$

graphs as a line with a y-intercept of 1, a slope of 1, and a hole at $(2, 3)$. Use a graphing utility to graph f. Explain why a graphing utility might not show the hole at $(2, 3)$.

4. **FINDING A COMPLETE GRAPH** Use a graphing utility to graph the function $f(x) = 3x^{5/3} - 6x^{4/3} + 2$ for $-2 \le x \le 10$. Compare your graph with the graph at the right. Does your graph include the part to the left of the y-axis? If not, how might you enter the function in such a way that the graphing utility you used would include this part?

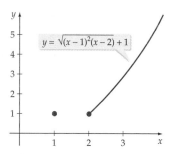

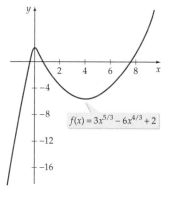

SECTION

3.6 THE ALGEBRA OF FUNCTIONS

Functions can be defined in terms of other functions. For example, the function defined by $h(x) = x^2 + 8x$ is the sum of

$$f(x) = x^2 \quad \text{and} \quad g(x) = 8x$$

Thus if we are given any two functions, f and g, we can define the four new functions $f + g, f - g, fg$, and f/g as follows.

> **OPERATIONS ON FUNCTIONS**
>
> For all values of x for which both $f(x)$ and $g(x)$ are defined, we define the following functions.
>
> Sum $\quad\quad (f + g)(x) = f(x) + g(x)$
>
> Difference $\quad (f - g)(x) = f(x) - g(x)$
>
> Product $\quad (fg)(x) = f(x) \cdot g(x)$
>
> Quotient $\quad \left(\dfrac{f}{g}\right)(x) = \dfrac{f(x)}{g(x)}, \quad g(x) \ne 0$

DOMAIN OF $f + g$, $f - g$, fg, f/g

For the given functions f and g, the domains of $f + g$, $f - g$, and $f \cdot g$ consist of all real numbers formed by the intersection of the domains of f and g. The domain of f/g is the set of all real numbers formed by the intersection of the domains of f and g, except for those real numbers x such that $g(x) = 0$.

EXAMPLE 1　Determine the Domain of a Function

If $f(x) = \sqrt{x - 1}$ and $g(x) = x^2 - 4$, find the domain of $f + g$, of $f - g$, of fg, and of f/g.

Solution

Note that f has the domain $\{x \mid x \geq 1\}$ and g has the domain of all real numbers. Therefore, the domain of $f + g$, $f - g$, and fg is $\{x \mid x \geq 1\}$. Because $g(x) = 0$ when $x = -2$ or $x = 2$, neither -2 nor 2 is in the domain of f/g. The domain of f/g is $\{x \mid x \geq 1 \text{ and } x \neq 2\}$.

TRY EXERCISE 10, EXERCISE SET 3.6

EXAMPLE 2　Evaluate Functions

Let $f(x) = x^2 - 9$ and $g(x) = 2x + 6$. Find

a. $(f + g)(5)$　　**b.** $(fg)(-1)$　　**c.** $\left(\dfrac{f}{g}\right)(4)$

Solution

a. $(f + g)(x) = f(x) + g(x) = (x^2 - 9) + (2x + 6) = x^2 + 2x - 3$

Therefore, $(f + g)(5) = (5)^2 + 2(5) - 3 = 25 + 10 - 3 = 32$.

b. $(fg)(x) = f(x) \cdot g(x) = (x^2 - 9)(2x + 6) = 2x^3 + 6x^2 - 18x - 54$

Therefore, $(fg)(-1) = 2(-1)^3 + 6(-1)^2 - 18(-1) - 54$
$$= -2 + 6 + 18 - 54 = -32.$$

c. $\left(\dfrac{f}{g}\right)(x) = \dfrac{f(x)}{g(x)} = \dfrac{x^2 - 9}{2x + 6} = \dfrac{\cancel{(x + 3)}(x - 3)}{2\cancel{(x + 3)}} = \dfrac{x - 3}{2}, \quad x \neq -3$

Therefore, $\left(\dfrac{f}{g}\right)(4) = \dfrac{4 - 3}{2} = \dfrac{1}{2}$.

TRY EXERCISE 14, EXERCISE SET 3.6

THE DIFFERENCE QUOTIENT

The expression

$$\frac{f(x + h) - f(x)}{h}, \quad h \neq 0$$

is called the **difference quotient** of f. It enables us to study the manner in which a function changes in value as the independent variable changes.

EXAMPLE 3 *Determine a Difference Quotient*

Determine the difference quotient of $f(x) = x^2 + 7$.

Solution

$$\begin{aligned}\frac{f(x + h) - f(x)}{h} &= \frac{[(x + h)^2 + 7] - [x^2 + 7]}{h} \\ &= \frac{[x^2 + 2xh + h^2 + 7] - [x^2 + 7]}{h} \\ &= \frac{x^2 + 2xh + h^2 + 7 - x^2 - 7}{h} \\ &= \frac{2xh + h^2}{h} = \frac{\cancel{h}(2x + h)}{\cancel{h}} = 2x + h\end{aligned}$$

• Apply the difference quotient.

TRY EXERCISE 30, EXERCISE SET 3.6

The difference quotient $2x + h$ of $f(x) = x^2 + 7$ from Example 3 is the slope of the secant line through the points

$$(x, f(x)) \quad \text{and} \quad (x + h, f(x + h))$$

For instance, let $x = 1$ and $h = 1$. Then the difference quotient is

$$2x + h = 2(1) + 1 = 3$$

This is the slope of the secant line l_2 through $(1, 8)$ and $(2, 11)$, as shown in **Figure 3.96.** If we let $x = 1$ and $h = 0.1$ then the difference quotient is

$$2x + h = 2(1) + 0.1 = 2.1$$

This is the slope of the secant line l_1 through $(1, 8)$ and $(1.1, 8.21)$.

The difference quotient

$$\frac{f(x + h) - f(x)}{h}$$

can be used to compute *average velocities.* In such cases it is traditional to replace f with s (for distance), the variable x with the variable a (for the time at the start of an observed interval of time), and the variable h with Δt, where Δt is the difference between the time at the end of an interval and the time at the start of the interval. For example, if an experiment is observed over the time interval

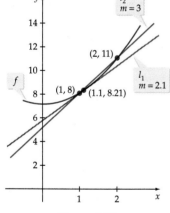

Figure 3.96

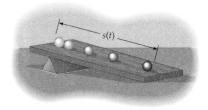

Figure 3.97

from $t = 3$ seconds to $t = 5$ seconds, then the time interval is denoted as $[3, 5]$ with $a = 3$, and $\Delta t = 5 - 3 = 2$. Thus if the distance traveled by a ball that rolls down a ramp is given by $s(t)$, where t is the time in seconds after the ball is released (see **Figure 3.97**), then the **average velocity** of the ball over the interval $t = a$ to $t = a + \Delta t$ is the difference quotient

$$\frac{s(a + \Delta t) - s(a)}{\Delta t}$$

EXAMPLE 4 *Evaluate Average Velocities*

The distance traveled by a ball rolling down a ramp is given by $s(t) = 4t^2$, where t is the time in seconds after the ball is released, and $s(t)$ is measured in feet. Evaluate the average velocity of the ball for each time interval.

a. $[3, 5]$ **b.** $[3, 4]$ **c.** $[3, 3.5]$ **d.** $[3, 3.01]$

Solution

a. In this case, $a = 3$ and $\Delta t = 2$. Thus the average velocity over this interval is

$$\frac{s(a + \Delta t) - s(a)}{\Delta t} = \frac{s(3 + 2) - s(3)}{2} = \frac{s(5) - s(3)}{2} = \frac{100 - 36}{2}$$

$$= 32 \text{ feet per second}$$

b. Let $a = 3$ and $\Delta t = 4 - 3 = 1$.

$$\frac{s(a + \Delta t) - s(a)}{\Delta t} = \frac{s(3 + 1) - s(3)}{1} = \frac{s(4) - s(3)}{1} = \frac{64 - 36}{1}$$

$$= 28 \text{ feet per second}$$

c. Let $a = 3$ and $\Delta t = 3.5 - 3 = 0.5$.

$$\frac{s(a + \Delta t) - s(a)}{\Delta t} = \frac{s(3 + 0.5) - s(3)}{0.5} = \frac{49 - 36}{0.5} = 26 \text{ feet per second}$$

d. Let $a = 3$ and $\Delta t = 3.01 - 3 = 0.01$.

$$\frac{s(a + \Delta t) - s(a)}{\Delta t} = \frac{s(3 + 0.01) - s(3)}{0.01} = \frac{36.2404 - 36}{0.01}$$

$$= 24.04 \text{ feet per second}$$

TRY EXERCISE 72, EXERCISE SET 3.6

COMPOSITION OF FUNCTIONS

Composition of functions is yet another method of constructing a function from two given functions. The process consists of using the range element of one function as the domain element of another function.

Composite functions occur in many situations. For example, suppose the manufacturing cost (in dollars) per compact disc player is given by

$$m(x) = \frac{180x + 2600}{x}$$

where x is the number of compact disc players to be manufactured. An electronics outlet agrees to sell the compact discs by marking up the manufacturing cost per player $m(x)$ by 30%. Note that the selling price s will be a function of $m(x)$. More specifically,

$$s[m(x)] = 1.30[m(x)]$$

Simplifying $s[m(x)]$ produces

$$s[m(x)] = 1.30\left(\frac{180x + 2600}{x}\right) = 1.30(180) + 1.30\frac{2600}{x} = 234 + \frac{3380}{x}$$

The function produced in this manner is referred to as the composition of m by s. The notation $s \circ m$ is used to denote this composition function. That is,

$$(s \circ m)(x) = 234 + \frac{3380}{x}$$

COMPOSITION OF FUNCTIONS

For the functions f and g, the **composite function** or **composition** of f by g is given by

$$(g \circ f)(x) = g[f(x)]$$

for all x in the domain of f such that $f(x)$ is in the domain of g.

If f and g are specified by equations, you can use substitution to find equations that specify $(g \circ f)$ and $(f \circ g)$.

EXAMPLE 5   *Form Composite Functions*

If $f(x) = x^2 - 3x$ and $g(x) = 2x + 1$, find   **a.** $(g \circ f)$ **b.** $(f \circ g)$

Solution

a. $(g \circ f) = g[f(x)] = 2(f(x)) + 1$ • Substitute $f(x)$ for x in g.
          $= 2(x^2 - 3x) + 1$ • $f(x) = x^2 - 3x$
          $= 2x^2 - 6x + 1$

b. $(f \circ g) = f[g(x)] = (g(x))^2 - 3(g(x))$ • Substitute $g(x)$ for x in f.
          $= (2x + 1)^2 - 3(2x + 1)$ • $g(x) = 2x + 1$
          $= 4x^2 - 2x - 2$

Try Exercise 38, Exercise Set 3.6

Note that in this example $(f \circ g) \neq (g \circ f)$. In general, the composition of functions is not a commutative operation.

Caution Some care must be used when forming the composition of functions. For instance, if $f(x) = x + 1$ and $g(x) = \sqrt{x - 4}$, then

$$(g \circ f)(2) = g[f(2)] = g(3) = \sqrt{3 - 4} = \sqrt{-1}$$

which is not a real number. We can avoid this problem by imposing suitable restrictions on the domain of f so that the range of f is part of the domain of g. If the domain of f is restricted to $[3, \infty)$, then the range of f is $[4, \infty)$. But this is precisely the domain of g. Note that $2 \notin [3, \infty)$, and thus we avoid the problem of $(g \circ f)(2)$ not being a real number.

To evaluate $(f \circ g)(c)$ for some constant c, you can use either of the following methods.

Method 1 First evaluate $g(c)$. Then substitute this result for x in $f(x)$.

Method 2 First determine $f[g(x)]$ and then substitute c for x.

EXAMPLE 6 *Evaluate a Composite Function*

Evaluate $(f \circ g)(3)$, where $f(x) = 2x - 7$ and $g(x) = x^2 + 4$.

Solution

Method 1
$$\begin{aligned}
(f \circ g)(3) &= f[g(3)] \\
&= f[(3)^2 + 4] \quad \bullet \text{ Evaluate } g(3). \\
&= f(13) \\
&= 2(13) - 7 = 19 \quad \bullet \text{ Substitute in } f.
\end{aligned}$$

Method 2
$$\begin{aligned}
(f \circ g)(x) &= 2[g(x)] - 7 \quad \bullet \text{ Form } f[g(x)]. \\
&= 2[x^2 + 4] - 7 \\
&= 2x^2 + 1 \\
(f \circ g)(3) &= 2(3)^2 + 1 = 19 \quad \bullet \text{ Substitute 3 for } x.
\end{aligned}$$

TRY EXERCISE 50, EXERCISE SET 3.6

TAKE NOTE

In Example 6, both Method 1 and Method 2 produce the same result. Although Method 2 is longer, it is the better method if you must evaluate $(f \circ g)(x)$ for several values of x.

Figures **3.98** and **3.99** graphically illustrate the difference between Method 1 and Method 2.

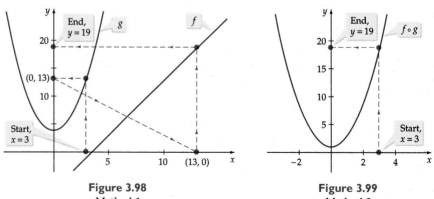

Figure 3.98
Method 1

Figure 3.99
Method 2

EXAMPLE 7   *Use a Composite Function to Solve an Application*

A graphic artist has drawn a 3-inch by 2-inch rectangle on a computer screen. The artist has been scaling the size of the rectangle for t seconds in such a way that the upper right corner of the original rectangle is moving to the right at the rate of 0.5 inch per second and downward at the rate of 0.2 inch per second. See **Figure 3.100**.

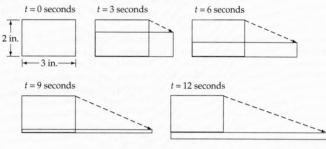

Figure 3.100

a. Write the length l and the width w of the scaled rectangle as functions of t.

b. Write the area A of the scaled rectangle as a function of t.

c. Find the intervals for which A is an increasing function on $0 \le t \le 14$. Also find the intervals where A is a decreasing function.

d. Find the value of t (where $0 \le t \le 14$) that maximizes $A(t)$.

Solution

a. Because *distance = rate · time*, we see that the change in l is given by $0.5t$. Therefore, the length at any time t is $l = 3 + 0.5t$. For $0 \le t \le 10$, the width is given by $w = 2 - 0.2t$. For $10 \le t \le 14$, the width is $w = -2 + 0.2t$. In either case the width can be determined by finding $w = |2 - 0.2t|$. (The absolute value symbol is needed to keep the width positive for $10 < t \le 14$.)

b. $A = lw = (3 + 0.5t)|2 - 0.2t|$

c. Use a graphing utility to determine that A is increasing on $[0, 2]$ and on $[10, 14]$ and that A is decreasing on $[2, 10]$. See **Figure 3.101**.

d. The highest point on the graph of A occurs when $t = 14$ seconds. See **Figure 3.101**.

TRY EXERCISE 66, EXERCISE SET 3.6

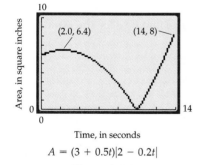

$A = (3 + 0.5t)|2 - 0.2t|$

Figure 3.101

You may be inclined to think that if the area of a rectangle is decreasing, then its perimeter is also decreasing, but this is not always the case. For example, the area of the scaled rectangle in Example 7 was shown to decrease on $[2, 10]$ even though its perimeter is always increasing. See Exercise 68 in Exercise Set 3.6.

TOPICS FOR DISCUSSION

1. The domain of $f + g$ consists of all real numbers formed by the *union* of the domain of f and the domain of g. Do you agree?

2. Given $f(x) = 3x - 2$ and $g(x) = \frac{1}{3}x + \frac{2}{3}$, determine $f \circ g$ and $g \circ f$. Does this show that composition of functions is a commutative operation?

3. A tutor states that the difference quotient of $f(x) = x^2$ and the difference quotient of $g(x) = x^2 + 4$ are the same. Do you agree?

4. A classmate states that the difference quotient of any linear function $f(x) = mx + b$ is always m. Do you agree?

5. When we use a difference quotient to determine an average velocity, we generally replace the variable h with the variable Δt. What does Δt represent?

EXERCISE SET 3.6

In Exercises 1 to 12, use the given functions f and g to find $f + g$, $f - g$, fg, and f/g. State the domain of each.

1. $f(x) = x^2 - 2x - 15$, $g(x) = x + 3$

2. $f(x) = x^2 - 25$, $g(x) = x - 5$

3. $f(x) = 2x + 8$, $g(x) = x + 4$

4. $f(x) = 5x - 15$, $g(x) = x - 3$

5. $f(x) = x^3 + 2x^2 + 7x$, $g(x) = x$

6. $f(x) = x^2 - 5x - 8$, $g(x) = -x$

7. $f(x) = 2x^2 + 4x - 7$, $g(x) = 2x^2 + 3x - 5$

8. $f(x) = 6x^2 + 10$, $g(x) = 3x^2 + x - 10$

9. $f(x) = \sqrt{x - 3}$, $g(x) = x$

10. $f(x) = \sqrt{x - 4}$, $g(x) = -x$

11. $f(x) = \sqrt{4 - x^2}$, $g(x) = 2 + x$

12. $f(x) = \sqrt{x^2 - 9}$, $g(x) = x - 3$

In Exercises 13 to 28, evaluate the indicated function, where $f(x) = x^2 - 3x + 2$ and $g(x) = 2x - 4$.

13. $(f + g)(5)$

14. $(f + g)(-7)$

15. $(f + g)\left(\frac{1}{2}\right)$

16. $(f + g)\left(\frac{2}{3}\right)$

17. $(f - g)(-3)$

18. $(f - g)(24)$

19. $(f - g)(-1)$

20. $(f - g)(0)$

21. $(fg)(7)$

22. $(fg)(-3)$

23. $(fg)\left(\frac{2}{5}\right)$

24. $(fg)(-100)$

25. $\left(\dfrac{f}{g}\right)(-4)$

26. $\left(\dfrac{f}{g}\right)(11)$

27. $\left(\dfrac{f}{g}\right)\left(\dfrac{1}{2}\right)$

28. $\left(\dfrac{f}{g}\right)\left(\dfrac{1}{4}\right)$

In Exercises 29 to 36, find the difference quotient of the given function.

29. $f(x) = 2x + 4$

30. $f(x) = 4x - 5$

31. $f(x) = x^2 - 6$

32. $f(x) = x^2 + 11$

33. $f(x) = 2x^2 + 4x - 3$

34. $f(x) = 2x^2 - 5x + 7$

35. $f(x) = -4x^2 + 6$

36. $f(x) = -5x^2 - 4x$

In Exercises 37 to 48, find $g \circ f$ and $f \circ g$ for the given functions f and g.

37. $f(x) = 3x + 5$, $g(x) = 2x - 7$

38. $f(x) = 2x - 7$, $g(x) = 3x + 2$

39. $f(x) = x^2 + 4x - 1$, $g(x) = x + 2$

40. $f(x) = x^2 - 11x$, $g(x) = 2x + 3$

41. $f(x) = x^3 + 2x$, $g(x) = -5x$

42. $f(x) = -x^3 - 7$, $g(x) = x + 1$

43. $f(x) = \dfrac{2}{x + 1}$, $g(x) = 3x - 5$

44. $f(x) = \sqrt{x + 4}$, $g(x) = \dfrac{1}{x}$

45. $f(x) = \dfrac{1}{x^2}$, $g(x) = \sqrt{x - 1}$

46. $f(x) = \dfrac{6}{x - 2}$, $g(x) = \dfrac{3}{5x}$

47. $f(x) = \dfrac{3}{|5 - x|}$, $g(x) = -\dfrac{2}{x}$

48. $f(x) = |2x + 1|$, $g(x) = 3x^2 - 1$

In Exercises 49 to 64, evaluate each composite function, where $f(x) = 2x + 3$, $g(x) = x^2 - 5x$, and $h(x) = 4 - 3x^2$.

49. $(g \circ f)(4)$

50. $(f \circ g)(4)$

51. $(f \circ g)(-3)$

52. $(g \circ f)(-1)$

53. $(g \circ h)(0)$

54. $(h \circ g)(0)$

55. $(f \circ f)(8)$

56. $(f \circ f)(-8)$

57. $(h \circ g)\left(\dfrac{2}{5}\right)$

58. $(g \circ h)\left(-\dfrac{1}{3}\right)$

59. $(g \circ f)(\sqrt{3})$

60. $(f \circ g)(\sqrt{2})$

61. $(g \circ f)(2c)$

62. $(f \circ g)(3k)$

63. $(g \circ h)(k + 1)$

64. $(h \circ g)(k - 1)$

65. **WATER TANK** A water tank has the shape of a right circular cone, with height 16 feet and radius 8 feet. Water is running into the tank so that the radius r (in feet) of the surface of the water is given by $r = 1.5t$, where t is the time (in minutes) that the water has been running.

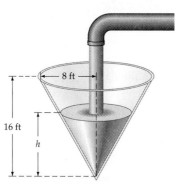

a. The area A of the surface of the water is $A = \pi r^2$. Find $A(t)$ and use it to determine the area of the surface of the water when $t = 2$ minutes.

b. The volume V of the water is given by $V = \dfrac{1}{3}\pi r^2 h$.

Find $V(t)$ and use it to determine the volume of the water when $t = 3$ minutes. (*Hint:* The height of the water in the cone is always twice the radius of the water.)

66. **SCALING A RECTANGLE** Work Example 7 of this section with the scaling as follows. The upper right corner of the original rectangle is pulled to the *left* at 0.5 inch per second and downward at 0.2 inch per second.

67. **TOWING A BOAT** A boat is towed by a rope that runs through a pulley that is 4 feet above the point where the rope is tied to the boat. The length (in feet) of the rope from the boat to the pulley is given by $s = 48 - t$, where t is the time in seconds that the boat has been in tow. The horizontal distance from the pulley to the boat is d.

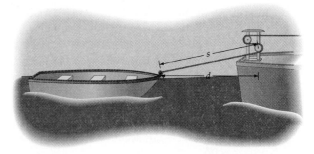

a. Find $d(t)$. b. Evaluate $s(35)$ and $d(35)$.

68. **PERIMETER OF A SCALED RECTANGLE** Show by a graph that the perimeter

$$P = 2(3 + 0.5t) + 2|2 - 0.2t|$$

of the scaled rectangle in Example 7 of this section is an increasing function over $0 \leq t \leq 14$.

69. **CONVERSION FUNCTIONS** The function $F(x) = x/12$ converts x inches to feet. The function $Y(x) = x/3$ converts x feet to yards. Explain the meaning of $(Y \circ F)(x)$.

70. **CONVERSION FUNCTIONS** The function $F(x) = 3x$ converts x yards to feet. The function $I(x) = 12x$ converts x feet to inches. Explain the meaning of $(I \circ F)(x)$.

71. **CONCENTRATION OF A MEDICATION** The concentration $C(t)$ (in milligrams per liter) of a medication in a patient's blood is given by the data in the following table.

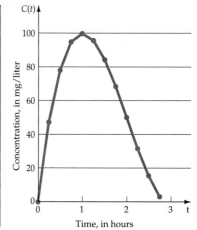

Concentration of Medication in Patient's Blood

t hours	$C(t)$ mg/l
0	0
0.25	47.3
0.50	78.1
0.75	94.9
1.00	99.8
1.25	95.7
1.50	84.4
1.75	68.4
2.00	50.1
2.25	31.6
2.50	15.6
2.75	4.3

The **average rate of change** of the concentration over the time interval from $t = a$ to $t = a + \Delta t$ is

$$\frac{C(a + \Delta t) - C(a)}{\Delta t}$$

Use the data in the table to evaluate the average rate of change for each of the following time intervals.

a. $[0, 1]$ (*Hint:* In this case, $a = 0$ and $\Delta t = 1$.) Compare this result to the slope of the line through $(0, C(0))$ and $(1, C(1))$.

b. $[0, 0.5]$ **c.** $[1, 2]$ **d.** $[1, 1.5]$ **e.** $[1, 1.25]$

f. The data in the table can be modeled by the function $Con(t) = 25t^3 - 150t^2 + 225t$. Use $Con(t)$ to verify that the average rate of change over $[1, 1 + \Delta t]$ is $-75(\Delta t) + 25(\Delta t)^2$. What does the average rate of change over $[1, 1 + \Delta t]$ seem to approach as Δt approaches 0?

72. BALL ROLLING ON A RAMP The distance traveled by a ball rolling down a ramp is given by $s(t) = 6t^2$, where t is the time in seconds after the ball is released, and $s(t)$ is measured in feet. The ball travels 6 feet in 1 second and it travels 24 feet in 2 seconds. Use the difference quotient for average velocity given on page 202 to evaluate the average velocity for each of the following time intervals.

a. $[2, 3]$ (*Hint:* In this case, $a = 2$ and $\Delta t = 1$.) Compare this result to the slope of the line through $(2, f(2))$ and $(3, f(3))$.

b. $[2, 2.5]$ **c.** $[2, 2.1]$

d. $[2, 2.01]$ **e.** $[2, 2.001]$

f. Verify that the average velocity over $[2, 2 + \Delta t]$ is $24 + 6(\Delta t)$. What does the average velocity seem to approach as Δt approaches 0?

SUPPLEMENTAL EXERCISES

In Exercises 73 to 78, show that

$$(g \circ f)(x) = x \quad \text{and} \quad (f \circ g)(x) = x$$

73. $f(x) = 2x + 3, \quad g(x) = \dfrac{x - 3}{2}$

74. $f(x) = 4x - 5, \quad g(x) = \dfrac{x + 5}{4}$

75. $f(x) = \dfrac{4}{x + 1}, \quad g(x) = \dfrac{4 - x}{x}$

76. $f(x) = \dfrac{2}{1 - x}, \quad g(x) = \dfrac{x - 2}{x}$

77. $f(x) = x^3 - 1, \quad g(x) = \sqrt[3]{x + 1}$

78. $f(x) = -x^3 + 2, \quad g(x) = \sqrt[3]{2 - x}$

79. Let x be the number of computer monitors to be manufactured. The manufacturing cost (in dollars) per computer monitor is given by the function

$$m(x) = \frac{60x + 34{,}000}{x}$$

A computer store will sell the monitors by marking up the manufacturing cost per monitor $m(x)$ by 45%. Thus the selling price s is a function of $m(x)$ given by the equation

$$s[m(x)] = 1.45[m(x)]$$

a. Express the selling price as a function of the number of monitors to be manufactured. That is, find $(s \circ m)(x)$.

b. Find $(s \circ m)(24{,}650)$.

PROJECTS

1. **A GRAPHING UTILITY PROJECT** For any two different real numbers x and y, the larger of the two numbers is given by

$$\text{Maximum}(x, y) = \frac{x + y}{2} + \frac{|x - y|}{2} \qquad (1)$$

a. Verify Equation (1) for $x = 5$ and $y = 9$.

b. Verify Equation (1) for $x = 201$ and $y = 80$.

For any two different functional values $f(x)$ and $g(x)$, the larger of the two is given by

$$\text{Maximum}(f(x), g(x)) = \frac{f(x) + g(x)}{2} + \frac{|f(x) - g(x)|}{2} \qquad (2)$$

To illustrate how we might make use of Equation (2), consider the functions $y_1 = x^2$ and $y_2 = \sqrt{x}$ on the interval from Xmin = -1 to Xmax = 6. The graphs of y_1 and y_2 are shown at the left and in the middle below.

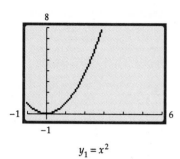

$y_1 = x^2$

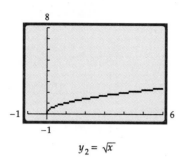

$y_2 = \sqrt{x}$

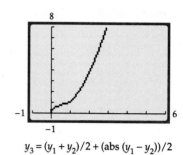

$y_3 = (y_1 + y_2)/2 + (\text{abs}\,(y_1 - y_2))/2$

Now consider the function $y_3 = (y_1 + y_2)/2 + (\text{abs}(y_1 - y_2))/2$, where "abs" represents the absolute value function. The graph of y_3 is shown at the right above.

c. Write a sentence or two that explains why the graph of y_3 is as shown.

d. What is the domain of y_1? of y_2? of y_3? Write a sentence that explains how to determine the domain of y_3, given the domain of y_1 and the domain of y_2.

e. Determine a formula for the function Minimum $(f(x), g(x))$.

2. **THE NEVER-NEGATIVE FUNCTION** The author J. D. Murray describes a function f_+ that is defined in the following manner.[1]

$$f_+ = \begin{cases} f & \text{if } f \geq 0 \\ 0 & \text{if } f < 0 \end{cases}$$

We will refer to this function as a **never-negative** function. Never-negative functions can be graphed by using Equation (2) in Exercise 1. For example, if we let $g(x) = 0$, then Equation (2) simplifies to

$$\text{Maximum}(f(x), 0) = \frac{f(x)}{2} + \frac{|f(x)|}{2} \qquad (3)$$

The graph of $y = \text{Maximum}(f(x), 0)$ is the graph of $y = f(x)$ provided that $f(x) \geq 0$, and it is the graph of $y = 0$ provided that $f(x) < 0$.

An application: The mosquito population per area of a large resort is controlled by spraying on a monthly basis. A biologist has determined that the mosquito population can be approximated by the never-negative function M_+ with

$$M(t) = -35{,}400(t - \text{int}(t))^2 + 35{,}400(t - \text{int}(t)) - 4000$$

Here t represents the month, and $t = 0$ corresponds to June 1, 1996.

a. Use a graphing utility to graph M for $0 \leq t \leq 3$.

b. Use a graphing utility to graph M_+ for $0 \leq t \leq 3$.

c. Write a sentence or two that explains how the graph of M_+ differs from the graph of M.

d. What is the maximum mosquito population per acre for $0 \leq t \leq 3$? When does this maximum mosquito population occur?

e. Explain when would be the best time to visit the resort, provided that you wished to minimize your exposure to mosquitoes.

[1] *Mathematical Biology* (New York: Springer-Verlag, 1989), p. 101.

3.7 INVERSE FUNCTIONS

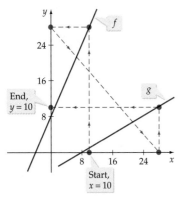

Figure 3.102

Some functions are inverses of each other in the sense that one undoes the other. The function $g(x) = \frac{1}{2}x - 4$ undoes the function $f(x) = 2x + 8$. To illustrate, let $x = 10$. Then

$$f(x) = f(10) = 2(10) + 8 = 28$$

Now evaluate $g[f(10)] = g(28)$.

$$g[f(10)] = g(28) = \frac{1}{2}(28) - 4 = 14 - 4 = 10$$

Thus we started with $x = 10$, we evaluated $f(10)$, we evaluated $g[f(10)]$, and the end result was the number that we started with, 10. This was not a coincidence. See **Figure 3.102**.

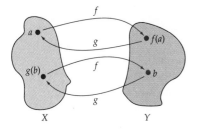

Figure 3.103

DEFINITION OF AN INVERSE FUNCTION

If f is a one-to-one function with domain X and range Y, and g is a function with domain Y and range X, then g is the **inverse function** of f if and only if

$$(f \circ g)(x) = x \quad \text{for all } x \text{ in the domain of } g$$

and

$$(g \circ f)(x) = x \quad \text{for all } x \text{ in the domain of } f$$

Figure 3.103 illustrates a function f and its inverse g. The set X is both the domain of f and the range of g. The set Y is both the domain of g and the range of f.

EXAMPLE 1   *Verify That Functions Are Inverse Functions*

Verify that $g(x) = \frac{1}{2}x - 4$ is the inverse function of $f(x) = 2x + 8$.

Solution

We need to show that $(f \circ g)(x) = x$ and that $(g \circ f)(x) = x$.

$$(f \circ g)(x) = f[g(x)] = f\left[\frac{1}{2}x - 4\right] = 2\left[\frac{1}{2}x - 4\right] + 8 = x - 8 + 8 = x$$

$$(g \circ f)(x) = g[f(x)] = g[2x + 8] = \frac{1}{2}[2x + 8] - 4 = x + 4 - 4 = x$$

Therefore, the function g is the inverse function of f. This work also shows that f is the inverse function of g.

TRY EXERCISE 2, EXERCISE SET 3.7

The definition of an inverse function requires f to be a one-to-one function. The reason for this restriction is now explained.

If a one-to-one function is given as a set of ordered pairs, then its inverse is the set of ordered pairs with their components interchanged. For example, the inverse of

$$\{(4,7), (5,2), (6,11)\} \quad \text{is} \quad \{(7,4), (2,5), (11,6)\}$$

Now consider the function j defined by $j(x) = x^2 - 1$. Some of the ordered pairs of j are

$$(-2,3), \quad (-1,0), \quad (0,-1), \quad (1,0), \quad \text{and} \quad (2,3)$$

The inverse of j contains the ordered pairs

$$(3,-2), \quad (0,-1), \quad (-1,0), \quad (0,1), \quad \text{and} \quad (3,2)$$

This set of ordered pairs does *not* satisfy the definition of a function, because there are ordered pairs with the same first component and *different* second components. For example, the ordered pairs $(3,-2)$ and $(3,2)$ both have 3 as their first component, but they have different second components. This example illustrates that not all functions have inverses that are functions.

Figure 3.104 is the graph of the function j. The horizontal line test indicates that j is *not* the graph of a *one-to-one* function. The horizontal line test can be used to show that the function $h(x) = \dfrac{1}{2}x^3$ is a one-to-one function. See **Figure 3.105**. Some of the ordered pairs of h are

$$(-2,-4), \quad \left(-1, -\frac{1}{2}\right), \quad (0,0), \quad \left(1, \frac{1}{2}\right), \quad \text{and} \quad (2,4)$$

Because h is a one-to-one function, given any y in the range of h, there corresponds exactly one x in the domain of h. Thus interchanging the coordinates of each ordered pair defined by h yields a set of ordered pairs that is a function. This function with the coordinates interchanged is the inverse function of h. The one-to-one property is exactly what is required for a function to have an inverse function.

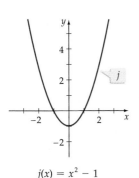

$j(x) = x^2 - 1$

Figure 3.104

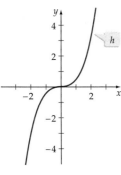

$h(x) = \dfrac{1}{2}x^3$

Figure 3.105

TAKE NOTE

The notation f^{-1} for an inverse function does not mean $1/f$. The function denoted by $1/f$ is called the *reciprocal function* and is an entirely different function from f^{-1}. For instance, in Example 1 we showed that

$$f^{-1}(x) = g(x) = \frac{1}{2}x - 4,$$

whereas $1/f(x) = \dfrac{1}{2x+8}$.

CONDITION FOR A FUNCTION TO HAVE AN INVERSE FUNCTION

A function f has an inverse function if and only if it is a one-to-one function.

FIND AN INVERSE FUNCTION

The inverse of the function f is often denoted by f^{-1}. In Example 1, we verified that g was the inverse of f, so in this case the function g could be written as f^{-1}.

If a one-to-one function f is defined by an equation, then we use the following method to find the equation of the inverse f^{-1}.

FIND THE EQUATION FOR f^{-1}

To find the inverse f^{-1} of the one-to-one function f:

1. Substitute y for $f(x)$.

2. Interchange x and y.

3. Solve, if possible, for y in terms of x.

4. Substitute $f^{-1}(x)$ for y.

5. Verify that the domain of f is the range of f^{-1} and that the range of f is the domain of f^{-1}.

EXAMPLE 2   **Find the Inverse of a One-to-One Function**

Find the inverse of the one-to-one function $f(x) = 2x - 6$.

Solution

TAKE NOTE

We use different procedures to verify that two functions are inverse functions than we use to find an inverse function of a given function.

Begin by substituting y for $f(x)$.

$$y = 2x - 6$$
$$x = 2y - 6 \qquad \text{• Interchange } x \text{ and } y.$$
$$x + 6 = 2y \qquad \text{• Solve for } y.$$
$$\frac{x + 6}{2} = y$$

This equation can be written as

$$y = \frac{1}{2}x + 3$$

In inverse notation,

$$f^{-1}(x) = \frac{1}{2}x + 3$$

In this example, the function f has a domain of all real numbers and a range of all real numbers, so the inverse f^{-1} also has a domain of all real numbers and a range of all real numbers.

TRY EXERCISE 10, EXERCISE SET 3.7

EXAMPLE 3   **Find the Inverse of a One-to-One Function**

Find the inverse of the function defined by $g(x) = \dfrac{2x}{x + 3}$.

Solution

$$y = \frac{2x}{x + 3} \qquad \text{• Replace } g(x) \text{ with } y.$$

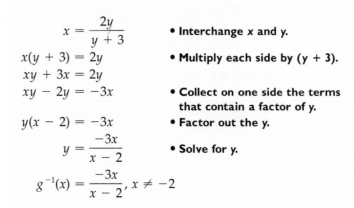

$$x = \frac{2y}{y + 3}$$ • **Interchange x and y.**

$$x(y + 3) = 2y$$ • **Multiply each side by (y + 3).**

$$xy + 3x = 2y$$

$$xy - 2y = -3x$$ • **Collect on one side the terms that contain a factor of y.**

$$y(x - 2) = -3x$$ • **Factor out the y.**

$$y = \frac{-3x}{x - 2}$$ • **Solve for y.**

$$g^{-1}(x) = \frac{-3x}{x - 2}, \; x \neq -2$$

TRY EXERCISE 18, EXERCISE SET 3.7

TAKE NOTE

Because $\dfrac{-3x}{x - 2} = \dfrac{3x}{2 - x}$, we can also write the inverse of g as $g^{-1}(x) = \dfrac{3x}{2 - x}$.

The graph of the function defined by $f(x) = x^2 - 4x$ is a parabola that opens upward. It is not a one-to-one function and therefore does not have an inverse function. However, the function $G(x) = x^2 - 4x$ with domain restricted to $\{x \mid x \geq 2\}$ is a one-to-one function. It has an inverse function denoted by G^{-1}.

EXAMPLE 4 | *Find the Inverse Function and State Its Domain and Range*

Find the inverse G^{-1} of the function $G(x) = x^2 - 4x$, for $x \geq 2$. State the domain and range of both G and G^{-1}.

Solution

First note that the domain of G is given as $\{x \mid x \geq 2\}$. The graph of G in **Figure 3.106** shows that G has the range $\{y \mid y \geq -4\}$. Because the domain of G^{-1} is the range of G and the range of G^{-1} is the domain of G, G^{-1} has the domain $\{x \mid x \geq -4\}$ and the range $\{y \mid y \geq 2\}$.

Now we proceed to find G^{-1}. The method shown uses the technique of completing the square.

$$G(x) = x^2 - 4x \quad \text{for } x \geq 2$$

$$y = x^2 - 4x$$

$$x = y^2 - 4y$$ • **Interchange x and y.**

$$x + 4 = y^2 - 4y + 4$$ • **To complete the square of $y^2 - 4y$, we need to add 4 to each side.**

$$x + 4 = (y - 2)^2$$ • **Factor.**

$$\pm\sqrt{x + 4} = y - 2$$ • **Apply the Square Root Theorem.**

$$2 \pm \sqrt{x + 4} = y$$

The range of G^{-1} is $\{y \mid y \geq 2\}$. Recall that the radical $\sqrt{x + 4}$ is a non-negative number. Therefore, to make $G^{-1}(x) = 2 \pm \sqrt{x + 4}$ a real number greater than or equal to 2 requires that we consider only the nonnegative square root. Thus G^{-1} is given by

$$G^{-1}(x) = 2 + \sqrt{x + 4}$$

TRY EXERCISE 32, EXERCISE SET 3.7

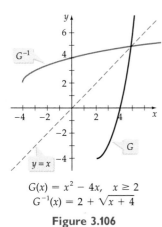

$G(x) = x^2 - 4x, \; x \geq 2$

$G^{-1}(x) = 2 + \sqrt{x + 4}$

Figure 3.106

GRAPHS OF INVERSE FUNCTIONS

The graphs of G and G^{-1} from Example 4 are shown in **Figure 3.106**. The graphs are symmetric with respect to the line $y = x$. This is always the case for the graph of a function and its inverse.

SYMMETRY PROPERTY OF f AND f^{-1}

The graph of a function f and the graph of the inverse function f^{-1} are symmetric with respect to the line given by $y = x$.

The symmetry property of f and f^{-1} can be used to graph the inverse of a one-to-one function.

EXAMPLE 5   *Graph the Inverse of a Function*

Graph f^{-1} if f is the function defined by the graph in **Figure 3.107**.

Solution

Sketch the graph of f^{-1} by drawing the reflection of f with respect to the line given by $y = x$. See **Figure 3.108**.

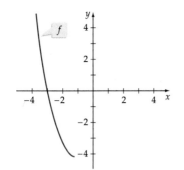

Figure 3.107

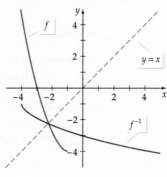

Figure 3.108

TRY EXERCISE 36, EXERCISE SET 3.7

TAKE NOTE

In **Figure 3.108** the diagonal line given by $y = x$ is not a part of the graph of f or its inverse f^{-1}. It is included to illustrate that the graphs of f and f^{-1} are symmetric with respect to this diagonal line.

TOPICS FOR DISCUSSION

1. The notation f^{-1} for an inverse function is also written as $\dfrac{1}{f}$. Do you agree?

2. Every odd function has an inverse function. Do you agree?

3. The domain of f^{-1} is the range of f, and the domain of f is the range of f^{-1}. Do you agree?

4. A student finds it difficult to determine the range of the function $f(x) = 2x/(5x - 3)$. A tutor suggests that the student determine f^{-1} and then use the domain of f^{-1} as the range of f. Is this a feasible approach?

5. The function $y = -x$ is its own inverse. Determine at least three other functions that are their own inverses.

EXERCISE SET 3.7

In Exercises 1 to 8, verify that f and g are inverse functions by showing that $(f \circ g)(x) = x$ and $(g \circ f)(x) = x$.

1. $f(x) = 2x + 1$, $g(x) = \dfrac{x - 1}{2}$

2. $f(x) = \dfrac{1}{2}x - 3$, $g(x) = 2x + 6$

3. $f(x) = 3x - 5$, $g(x) = \dfrac{x + 5}{3}$

4. $f(x) = -2x + 1$, $g(x) = -\dfrac{1}{2}x + \dfrac{1}{2}$

5. $f(x) = \dfrac{1}{x + 1}$, $g(x) = \dfrac{1 - x}{x}$

6. $f(x) = \dfrac{1}{x} + 1$, $g(x) = \dfrac{1}{x - 1}$

7. $f(x) = \sqrt[3]{x - 1}$, $g(x) = x^3 + 1$

8. $f(x) = x^3 - 2$, $g(x) = \sqrt[3]{x + 2}$

In Exercises 9 to 24, find the inverse of the given function.

9. $f(x) = 4x + 1$

10. $g(x) = \dfrac{2}{3}x + 4$

11. $F(x) = -6x + 1$

12. $h(x) = -3x - 2$

13. $j(t) = 2t + 1$

14. $m(s) = -3s + 8$

15. $f(v) = 1 - v^3$

16. $u(t) = 2t^3 + 5$

17. $f(x) = \dfrac{-3x}{x + 4}$

18. $G(x) = \dfrac{3x}{x - 5}$

19. $M(t) = \dfrac{t - 5}{t}$

20. $P(v) = \dfrac{2v}{v + 1}$

21. $r(t) = \dfrac{1}{t^2}$, $t < 0$

22. $F(x) = \dfrac{1}{x}$, $x > 0$

23. $J(x) = x^2 + 4$, $x \geq 0$

24. $N(x) = 2x^2 + 1$, $x \leq 0$

In Exercises 25 to 34, find the inverse of f. State the domain and range of both f and f^{-1}.

25. $f(x) = x^2 + 3$, $x \geq 0$

26. $f(x) = x^2 - 4$, $x \geq 0$

27. $f(x) = \sqrt{x}$, $x \geq 0$

28. $f(x) = \sqrt{16 - x}$, $x \leq 16$

29. $f(x) = \sqrt{9 - x^2}$, $0 \leq x \leq 3$

30. $f(x) = \sqrt{16 - x^2}$, $-4 \leq x \leq 0$

31. $f(x) = x^2 - 4x + 1$, $x \geq 2$

32. $f(x) = x^2 + 6x - 6$, $x \geq -3$

33. $f(x) = x^2 + 8x - 9$, $x \leq -4$

34. $f(x) = x^2 - 2x - 2$, $x \leq 1$

In Exercises 35 to 40, graph f^{-1} if f is the function defined by the graph.

35.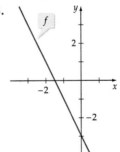

36.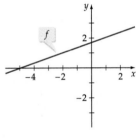

37.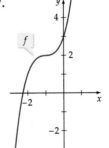

38.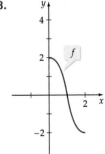

39.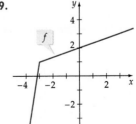

40.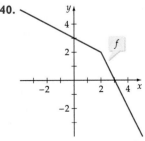

In Exercises 41 to 48, graph each function f and its inverse f^{-1} on the same coordinate plane. Note that the graphs are symmetric with respect to the line $y = x$.

41. $f(x) = 3x + 3, f^{-1}(x) = \dfrac{1}{3}x - 1$

42. $f(x) = x - 4, f^{-1}(x) = x + 4$

43. $f(x) = \dfrac{1}{2}x, f^{-1}(x) = 2x$

44. $f(x) = 2x - 4, f^{-1}(x) = \dfrac{1}{2}x + 2$

45. $f(x) = x^2 + 2, x \geq 0, f^{-1}(x) = \sqrt{x - 2}, x \geq 2$

46. $f(x) = x^2 - 3, x \geq 0, f^{-1}(x) = \sqrt{x + 3}, x \geq -3$

47. $f(x) = (x - 2)^2, x \leq 2, f^{-1}(x) = 2 - \sqrt{x}, x \geq 0$

48. $f(x) = (x + 3)^2, x \geq -3, f^{-1}(x) = \sqrt{x} - 3, x \geq 0$

SUPPLEMENTAL EXERCISES

In Exercises 49 to 52, find the inverse of the given function.

49. $f(x) = ax + b, a \neq 0$

50. $f(x) = ax^2 + bx + c; a \neq 0, x > -\dfrac{b}{2a}$

51. $f(x) = \dfrac{x - 1}{x + 1}, x \neq -1$

52. $f(x) = \dfrac{2 - x}{x + 2}, x \neq -2$

Only one-to-one functions have inverses that are functions. In Exercises 53 to 60, determine whether or not the given function is a one-to-one function.

53. $f(x) = x^2 + 8$ **54.** $v(s) = s^2 - 4$

55. $p(t) = \sqrt{9 - t}$ **56.** $v(t) = \sqrt{16 + t}$

57. $G(x) = -\sqrt{x}$ **58.** $K(x) = 1 - \sqrt{x - 5}$

59. $F(x) = |x| + x$ **60.** $T(x) = |x| - x$

In Exercises 61 to 64, assume that the given function has an inverse function.

61. If $f(5) = 2$, find $f^{-1}(2)$.   **62.** If $v(3) = 11$, find $v^{-1}(11)$.

63. If $s(4) = 60$, find $s^{-1}(60)$.   **64.** If $F(-8) = 5$, find $F^{-1}(5)$.

65. Graph $f(x) = -x + 3$. Use the graph to explain why f is its own inverse.

PROJECTS

1. **SYMMETRY WITH RESPECT TO $y = x$** If the ordered pair (a, b) belongs to the graph of the function f, then (b, a) belongs to the graph of f^{-1}. Prove that the points $P(a, b)$ and $Q(b, a)$ are symmetric with respect to the graph of the line $y = x$ by using the definition of symmetry with respect to a line.

SECTION

3.8 VARIATION AND APPLICATIONS

DIRECT VARIATION

Many real-life situations involve variables that are related by a type of function called a **variation.** For example, a fish jumping or a stone thrown into a pond generates circular ripples whose circumference and diameter are increasing. The equation $C = \pi d$ expresses the relationship between the circumference C of a circle and its diameter d. If d increases, then C increases. In fact, if d doubles in size, then C also doubles in size. The circumference C is said to *vary directly* as the diameter d.

DEFINITION OF DIRECT VARIATION

The variable y **varies directly** as the variable x, or y is **directly proportional** to x, if and only if

$$y = kx$$

where k is a constant called the **constant of proportionality** or the **variation constant**.

Direct variations occur in many daily applications. For example, the cost of a newspaper is 25 cents. The cost C to purchase n newspapers is directly proportional to the number n. That is, $C = 25n$. In this example the variation constant is 25.

To solve a problem that involves a variation, we typically write a general equation that relates the variables and then use given information to solve for the variation constant.

EXAMPLE 1 *Solve a Direct Variation*

The distance sound travels varies directly as the time it travels. If sound travels 1340 meters in 4 seconds, find the distance sound will travel in 5 seconds.

Solution

Write an equation that relates the distance d to the time t. Because d varies directly as t, our equation is $d = kt$. Because $d = 1340$ when $t = 4$, we obtain

$$1340 = k \cdot 4 \quad \text{which implies} \quad k = \frac{1340}{4} = 335$$

Therefore, the specific equation that relates the distance d sound travels in t seconds is $d = 335t$. To find the distance sound travels in 5 seconds, replace t with 5 to produce

$$d = 335(5) = 1675$$

Under the same conditions, sound will travel 1675 meters in 5 seconds. See **Figure 3.109.**

TRY EXERCISE 22, EXERCISE SET 3.8

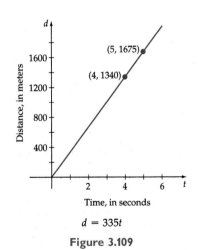

$d = 335t$

Figure 3.109

> **DIRECT VARIATION AS THE *n*TH POWER**
>
> If *y* **varies directly as the *n*th power** of *x*, then
>
> $$y = kx^n$$
>
> where *k* is a constant.

EXAMPLE 2   *Solve a Variation of the Form* $y = kx^2$

The distance *s* that an object falls from rest (neglecting air resistance) varies directly as the square of the time *t* that it has been falling. If an object falls 64 feet in 2 seconds, how far will it fall in 10 seconds?

Solution

Because *s* varies directly as the square of *t*, $s = kt^2$. The variable *s* is 64 when *t* is 2, so

$$64 = k \cdot 2^2 \qquad \text{which implies} \qquad k = \frac{64}{4} = 16$$

The specific equation that relates the distance *s* an object falls in *t* seconds is $s = 16t^2$. Letting $t = 10$ yields

$$s = 16(10^2) = 16(100) = 1600$$

Under the same conditions, the object will fall 1600 feet in 10 seconds. See **Figure 3.110**.

> **TRY EXERCISE 24, EXERCISE SET 3.8**

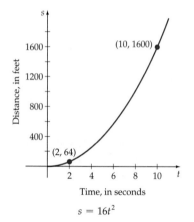

Figure 3.110

INVERSE VARIATION

Two variables can also vary *inversely*.

> **DEFINITION OF INVERSE VARIATION**
>
> The variable *y* **varies inversely** as the variable *x*, or *y* is **inversely proportional** to *x*, if and only if
>
> $$y = \frac{k}{x}$$
>
> where *k* is the variation constant.

In 1661, Robert Boyle made a study of the *compressibility* of gases. **Figure 3.111** shows that he used a J-shaped tube to demonstrate the inverse relationship between the volume of a gas at a given temperature and the applied pressure. The J-shaped tube on the left shows that the volume of a gas at normal atmospheric pressure is 60 milliliters. If the pressure is doubled by adding

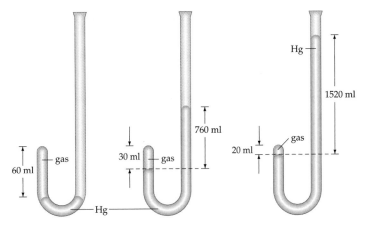

Figure 3.111

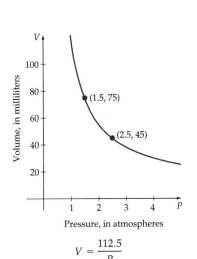

$$V = \frac{112.5}{P}$$

Figure 3.112

mercury (Hg), as shown in the middle tube, the volume of the gas is halved to 30 milliliters. Tripling the pressure decreases the volume of the gas to 20 milliliters, as shown in the tube at the right.

EXAMPLE 3 **Solve an Inverse Variation**

Boyle's Law states that the volume V of a sample of gas (at a constant temperature) varies inversely as the pressure P. The volume of a gas in a J-shaped tube is 75 milliliters when the pressure is 1.5 atmospheres. Find the volume of the gas when the pressure is increased to 2.5 atmospheres.

Solution

The volume V varies inversely as the pressure P, so $V = k/P$. The volume V is 75 milliliters when the pressure is 1.5 atmospheres, so

$$75 = \frac{k}{1.5} \quad \text{and} \quad k = (75)(1.5) = 112.5$$

Thus $V = 112.5/P$. When the pressure is 2.5 atmospheres, we have

$$V = \frac{112.5}{2.5} = 45 \text{ milliliters}$$

See **Figure 3.112**.

TRY EXERCISE 28, EXERCISE SET 3.8

TAKE NOTE

Because the volume *V* varies inversely as the pressure *P*, the function *V* = 112.5/*P* is a decreasing function, as shown in Figure 3.112.

INVERSE VARIATION AS THE nTH POWER

If *y* **varies inversely as the nth power** of *x*, then

$$y = \frac{k}{x^n}$$

where *k* is a constant.

JOINT VARIATION AND COMBINED VARIATION

Some variations involve more than two variables.

> **DEFINITION OF JOINT VARIATION**
>
> The variable z **varies jointly** as the variables x and y if and only if
>
> $$z = kxy$$
>
> where k is a constant.

EXAMPLE 4   Solve a Joint Variation

The cost of insulating the ceiling of a house varies jointly with the thickness of the insulation and the area of the ceiling. It costs $175 to insulate a 2100-square-foot ceiling with insulation 4 inches thick. Find the cost of insulating a 2400-square-foot ceiling with insulation that is 6 inches thick.

Solution

Because the cost C varies jointly as the area A of the ceiling and the thickness T of the insulation, we know $C = kAT$. Using the fact that $C = 175$ when $A = 2100$ and $T = 4$ gives us

$$175 = k(2100)(4) \quad \text{which implies} \quad k = \frac{175}{(2100)(4)} = \frac{1}{48}$$

Consequently, the specific formula for C is $C = \frac{1}{48} AT$. Now when $A = 2400$ and $T = 6$, we have

$$C = \frac{1}{48}(2400)(6) = 300$$

Thus the cost of insulating the 2400-square-foot ceiling with 6-inch insulation is $300.

TRY EXERCISE 30, EXERCISE SET 3.8

Combined variations involve more than one type of variation.

EXAMPLE 5   Solve a Combined Variation

The weight that a horizontal beam with a rectangular cross section can safely support varies jointly as the width and square of the depth of the cross section and inversely as the length of the beam. See **Figure 3.113**. If a 4-inch by 4-inch beam 10 feet long safely supports a load of 256 pounds, what load L can be safely supported by a beam made of the same material and with a width w of 4 inches, a depth d of 6 inches, and a length l of 16 feet?

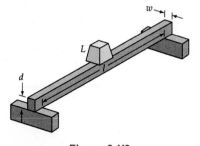

Figure 3.113

Solution

The general variation equation is $L = k\frac{wd^2}{l}$. Using the given data yields

$$256 = k\frac{4(4^2)}{10}$$

Solving for k produces $k = 40$, so the specific formula for L is

$$L = 40\frac{wd^2}{l}$$

Substituting 4 for w, 6 for d, and 16 for l gives

$$L = 40\frac{4(6^2)}{16} = 360 \text{ pounds}$$

Try Exercise 34, Exercise Set 3.8

Topics for Discussion

1. The area A of a trapezoid varies jointly as the product of its height h and the sum of its bases b and B. State an equation that represents this variation. Given that $A = 15$ square inches when $h = 6$ inches, $b = 2$ inches, and $B = 3$ inches, explain how you would determine the value of the variation constant.

2. Given that the variation constant $k > 0$ and that A varies directly as b, then A _____ when b increases, and A _____ when b decreases.

3. Given that the variation constant $k > 0$ and that S varies inversely as d, then S _____ when d increases, and S _____ when d decreases.

4. All direct variations can be written in the form $y =$ _____, and all inverse variations can be written in the form $y =$ _____ .

5. The volume V of a right circular cylinder varies jointly as the square of the radius r and the height h. Tell what happens to V when
   a. h is tripled   **b.** r is tripled
   c. r is doubled and h is decreased to $\frac{1}{2}h$

6. Give some examples of real situations where one quantity varies inversely as a second quantity.

Exercise Set 3.8

In Exercises 1 to 12, write an equation that represents the relationship between the given variables. Use k as the variation constant.

1. d varies directly as t.

2. r varies directly as the square of s.

3. y varies inversely as x.

4. p is inversely proportional to q.

5. m varies jointly as n and p.

6. t varies jointly as r and the cube of s.

7. V varies jointly as l, w, and h.

8. u varies directly as v and inversely as the square of w.

9. A is directly proportional to the square of s.

10. A varies jointly as h and the square of r.

11. F varies jointly as m_1 and m_2 and inversely as the square of d.

12. T varies jointly as t and r and the square of a.

In Exercises 13 to 20, write the equation that expresses the relationship between the variables, and then use the given data to solve for the variation constant.

13. y varies directly as x, and $y = 64$ when $x = 48$.

14. m is directly proportional to n, and $m = 92$ when $n = 23$.

15. r is directly proportional to the square of t, and $r = 144$ when $t = 108$.

16. C varies directly as r, and $C = 94.2$ when $r = 15$.

17. T varies jointly as r and the square of s, and $T = 210$ when $r = 30$ and $s = 5$.

18. u varies directly as v and inversely as the square root of w, and $u = 0.04$ when $v = 8$ and $w = 0.04$.

19. V varies jointly as l, w, and h, and $V = 240$ when $l = 8$ and $w = 6$ and $h = 5$.

20. t varies directly as the cube of r and inversely as the square root of s, and $t = 10$ when $r = 5$ and $s = 0.09$.

21. **CHARLES'S LAW** *Charles's Law* states that the volume V occupied by a gas (at a constant pressure) is directly proportional to its absolute temperature T. An experiment with a balloon shows that the volume of the balloon is 0.85 liter at 270 K (absolute temperature).[2] What will the volume of the balloon be when its temperature is 324 K?

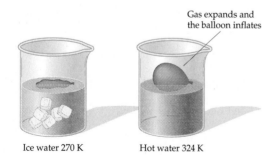

Gas expands and the balloon inflates

Ice water 270 K   Hot water 324 K

[2] Absolute temperature is measured on the Kelvin scale. A unit (called a kelvin) on the Kelvin scale is the same measure as a degree on the Celsius scale; however, 0 on the Kelvin scale corresponds to -273 on the Celsius scale.

22. **HOOKE'S LAW** *Hooke's Law* states that the distance a spring stretches varies directly as the weight on the spring. A weight of 80 pounds stretches a spring 6 inches. How far will a weight of 100 pounds stretch the spring?

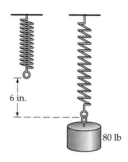

6 in.

80 lb

23. **PRESSURE AND DEPTH** The pressure a liquid exerts at a given point on a submarine is directly proportional to the depth of the point below the surface of the liquid. If the pressure at a depth of 3 feet is 187.5 pounds per square foot, find the pressure at a depth of 7 feet.

24. **MOTORCYCLE JUMP** The range of a projectile is directly proportional to the square of its velocity. If a motorcyclist can make a jump of 140 feet by coming off a ramp at 60 mph, find the distance the motorcyclist could expect to jump if the speed coming off the ramp were increased to 65 mph.

25. **PERIOD OF A PENDULUM** The period T (the time it takes a pendulum to make one complete oscillation) varies directly as the square root of its length L. A pendulum 3 feet long has a period of 1.8 seconds.

a. Find the period of a pendulum 10 feet long.

b. What is the length of a pendulum that *beats seconds* (that is, it has a 2-second period)?

26. **AREA OF A PROJECTED PICTURE** The area of a projected picture on a movie screen varies directly as the square of the distance from the projector to the screen. If a distance of 20 feet produces a picture with an area of 64 square feet, what distance produces an area of 100 square feet?

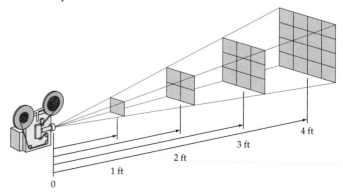

0   1 ft   2 ft   3 ft   4 ft

27. Decibels The loudness, measured in decibels, of a stereo speaker is inversely proportional to the square of the distance of the listener from the speaker. The loudness is 28 decibels at a distance of 8 feet. What is the loudness when the listener is 4 feet from the speaker?

28. Illumination The illumination a source of light provides is inversely proportional to the square of the distance from the source. If the illumination at a distance of 10 feet from the source is 50 footcandles, what is the illumination at a distance of 15 feet from the source?

29. Volume Relationships The volume V of a right circular cone varies jointly as the square of the radius r and the height h. Tell what happens to V when

a. r is tripled

b. h is tripled

c. both r and h are tripled

30. Safe Load The load L that a horizontal beam can safely support varies jointly as the width w and the square of the depth d. If a beam with width 2 inches and depth 6 inches safely supports up to 200 pounds, how many pounds can a beam of the same length that has width 4 inches and depth 4 inches be expected to support?

31. Ideal Gas Law The *Ideal Gas Law* states that the volume V of a gas varies jointly as the number of moles of gas n and the absolute temperature T and inversely as the pressure P. What happens to V when n is tripled and P is reduced by a factor of one-half?

32. Maximum Load The maximum load a cylindrical column of circular cross section can support varies directly as the fourth power of the diameter and inversely as the square of the height. If a column 2 feet in diameter and 10 feet high supports up to 6 tons, how much of a load does a column 3 feet in diameter and 14 feet high support?

33. Astronomy A meteorite approaching the earth has a velocity that varies inversely as the square root of the distance from the center of the earth. The meteorite has a velocity of 3 miles per second at 4900 miles from the center of the earth. Find the velocity of the meteorite when it is 4225 miles from the center of the earth.

34. Safe Load The load L a horizontal beam can safely support varies jointly as the width w and the square of the depth d and inversely as the length l. If a 12-foot beam with width 4 inches and depth 8 inches safely supports 800 pounds, how many pounds can a 16-foot beam that has width 3.5 inches and depth 6 inches be expected to support?

35. Force, Speed, and Radius Relationships The force needed to keep a car from skidding on a curve varies jointly as the weight of the car and the square of the speed and inversely as the radius of the curve. It takes 2800 pounds of force to keep an 1800-pound car from skidding on a curve with radius 425 feet at 45 mph. What force is needed to keep the same car from skidding when it takes a similar curve with radius 450 feet at 55 mph?

36. Stiffness of a Beam A cylindrical log is to be cut so that it will yield a beam that has a rectangular cross section of depth d and width w. The stiffness of a beam of given length is directly proportional to the width and the cube of the depth. The diameter of the log is 18 inches.

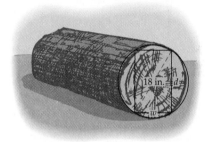

Which depth will yield the "stiffest" beam, $d = 10$ inches, $d = 12$ inches, $d = 14$ inches, or $d = 16$ inches?

SUPPLEMENTAL EXERCISES

37. Kepler's Third Law *Kepler's Third Law* states that the time T needed for a planet to make one complete revolution about the sun is directly proportional to the 3/2 power of the average distance d between the planet and the sun. The earth, which averages 93 million miles from the sun, completes one revolution in 365 days. Find the average distance from the sun to Mars if Mars completes one revolution about the sun in 686 days.

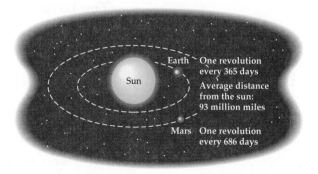

38. Illumination The illumination a light source provides is directly proportional to the strength of the source and inversely proportional to the square of the distance from the source. Two light sources are 10 feet apart. The strength of the light source at point B is 8 times the strength of the light source at point A.

Which position will receive the least amount of illumination, the point on line segment AB where $x = 2.5$ feet, where $x = 3$ feet, where $x = 3.3$ feet, or where $x = 3.5$ feet?

PROJECTS

1. **A DIRECT VARIATION FORMULA** If $f(x)$ varies directly as x, prove that $f(x_2) = f(x_1)\dfrac{x_2}{x_1}$.

   Use this formula to solve the following direct variation *without* solving for the variation constant. The distance a spring stretches varies directly as the force applied. An experiment shows that a force of 17 kilograms stretches the spring 8.5 centimeters. How far will a 22-kilogram force stretch the spring?

2. **AN INVERSE VARIATION FORMULA** Given that $f(x)$ varies inversely as x, prove that $f(x_2) = f(x_1)\dfrac{x_1}{x_2}$. Use this formula to solve the following inverse variation *without* solving for the variation constant. The volume of a gas varies inversely as pressure (assuming the temperature remains constant). An experiment shows that a particular gas has a volume of 2.4 liters under a pressure of 280 grams per square centimeter. What volume will the gas have when a pressure of 330 grams per square centimeter is applied?

EXPLORING CONCEPTS WITH TECHNOLOGY

The Least-Squares Regression Line

If a scatter plot appears to indicate a "linear relationship" between the independent variable and the dependent variable, then it is natural to model the relationship by a linear function. But what linear function should be used? In statistics we use the linear function whose graph is called the *line of best fit* or the *least-squares regression line*.

> **DEFINITION OF THE LEAST-SQUARES REGRESSION LINE**
>
> The **least-squares regression line** is the line that minimizes the sum of the squares of the vertical deviations from each data point to the line.

The least-squares regression line is often referred to as the least-squares line. The least-squares line for the data $\{(1,2), (2,3), (3,3), (4,4), (5,7)\}$ is the line

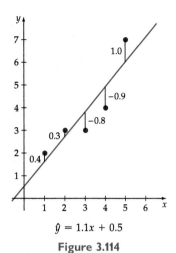

$\hat{y} = 1.1x + 0.5$

Figure 3.114

given by $\hat{y} = 1.1x + 0.5$. **Figure 3.114** shows the graph of $\hat{y} = 1.1x + 0.5$. The vertical deviations from the data points to the line are 0.4, 0.3, −0.8, −0.9, and 1.0. The sum of the squares of the vertical deviations is

$$0.4^2 + 0.3^2 + (-0.8)^2 + (-0.9)^2 + 1.0^2 = 2.7$$

The line given by $\hat{y} = 1.1x + 0.5$ is the least-squares line because for all other lines the sum of the squares of the vertical deviations is greater than 2.7.

The next theorem can be used to determine the equation of the least-squares line for a set of data points.

THE FORMULA FOR THE LEAST-SQUARES LINE

The least-squares line for the set of n data points

$$\{(x_1, y_1), (x_2, y_2), (x_3, y_3), \ldots, (x_n, y_n)\}$$

is $\hat{y} = bx + a$, where

$$b = \frac{n \sum xy - (\sum x)(\sum y)}{n \sum x^2 - (\sum x)^2} \quad \text{and} \quad a = \bar{y} - b\bar{x}$$

The symbol $\sum x$ is the sum of all the x-values, $\sum y$ is the sum of all the y-values, and $\sum xy$ is the sum of the n products $x_1 y_1, x_2 y_2, \ldots, x_n y_n$. The symbol $\bar{x}$ represents the arithmetic mean of the x values, and $\bar{y}$ is the arithmetic mean of the y values. The following example illustrates a procedure that can be used to calculate efficiently the sums needed to find the equation of the least-squares line for a given set of data points.

EXAMPLE 1   *Find the Least-Squares Line*

Determine the least-squares line for

$$\{(1.0, 2.1), (2.0, 3.5), (2.5, 4.2), (3.0, 6.4), (3.7, 7.3)\}$$

Solution

In this example $n = 5$. Organize the data in four columns as shown below. Calculate the sum of each column.

x	y	x^2	xy
1.0	2.1	1.00	2.10
2.0	3.5	4.00	7.00
2.5	4.2	6.25	10.50
3.0	6.4	9.00	19.20
3.7	7.3	13.69	27.01
$\sum x = 12.2$	$\sum y = 23.5$	$\sum x^2 = 33.94$	$\sum xy = 65.81$

Continued ▸

Determine the slope b.

$$b = \frac{n \sum xy - (\sum x)(\sum y)}{n \sum x^2 - (\sum x)^2} = \frac{5(65.81) - (12.2)(23.5)}{5(33.94) - (12.2)^2} \approx 2.03$$

Calculate $\bar{x}$ and $\bar{y}$.

$$\bar{x} = \frac{(12.2)}{5} = 2.44, \quad \bar{y} = \frac{(23.5)}{5} = 4.7$$

Determine the y-intercept a.

$$a = \bar{y} - b\bar{x} \approx -0.25$$

The equation of the least-squares regression line is

$$\hat{y} = bx + a = 2.03x - 0.25$$

The data points and the least-squares line are shown in **Figure 3.115.**

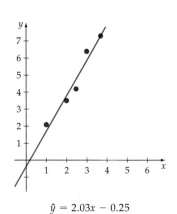

$\hat{y} = 2.03x - 0.25$

Figure 3.115

Figure 3.116 shows the necessary keystrokes for using a *TI-82* graphing calculator to find the least-squares line for the points in the above example. The number $r = 0.9747022599$ shown in the calculator display is called the *linear coefficient of correlation*. It is a measure of the strength of the linear relationship between the x values and the y values of the given data.

TI-82 Keystrokes

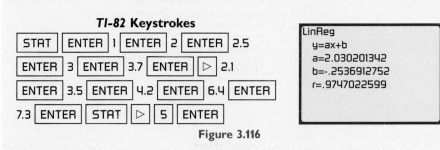

Figure 3.116

1. Determine the least-squares line for $\{(2, 10), (4, 7), (5, 5), (6, 4)\}$.

2. Determine the least-squares line for $\{(3, 5), (4, 5), (4, 3), (5, 6), (7, 8)\}$.

3. Verify that $\hat{y} = 1.1x + 0.5$ is the equation of the least-squares line for the points $(1, 2), (2, 3), (3, 3), (4, 4)$, and $(5, 7)$ as shown in **Figure 3.114.** The sum of the squares of the vertical deviations from the points to the least-squares line was shown to be 2.7. Draw another line that appears to fit the points. Show that for this line, the sum of the squares of the deviations is greater than 2.7.

4. In Exercise 57 of Section 3.3 the data points are

$(0, 21.1)$, $(4, 20.7)$, $(8, 20.6)$, $(12, 20.5)$, $(16, 20.3)$, $(20, 19.83)$,
$(24, 20.00)$, $(28, 20.23)$, $(32, 20.19)$, $(36, 19.80)$, $(40, 19.75)$, $(44, 20.01)$

Use a graphing utility to verify that the line given by

$$T(t) = -0.02424t + 20.78$$

is the least-squares line for the data points. (The decimals in the equation for T are accurate to four significant digits.)

5. In Exercise 58 of Section 3.3, the data points are

$(0, 47.5)$, $(4, 45.9)$, $(8, 44.5)$, $(12, 44.5)$, $(16, 43.6)$, $(20, 42.8)$,
$(24, 42.81)$, $(28, 42.55)$, $(32, 41.60)$, $(36, 41.65)$, $(40, 41.98)$, $(44, 42.11)$

Use a graphing utility to verify that the line given by

$$T(t) = -0.1155t + 46.00$$

is the least-squares line for the data points. (The decimals in the equation for T are accurate to four significant digits.)

CHAPTER 3 REVIEW

3.1 A Two-Dimensional Coordinate System and Graphs

- **The Distance Formula** The distance d between the points represented by (x_1, y_1) and (x_2, y_2) is

$$d = \sqrt{(x_2 - x_1)^2 + (y_2 - y_1)^2}$$

- The midpoint of the line segment from $P_1(x_1, y_1)$ to $P_2(x_2, y_2)$ is

$$\left(\frac{x_1 + x_2}{2}, \frac{y_1 + y_2}{2} \right)$$

- The standard form of the equation of a circle with center at (h, k) and radius r is $(x - h)^2 + (y - k)^2 = r^2$.

3.2 Introduction to Functions

- **Definition of a Function**
  A function is a set of ordered pairs in which no two ordered pairs that have the same first coordinate have different second coordinates.

- A graph is the graph of a function if and only if no vertical line intersects the graph at more than one point. If any horizontal line intersects the graph of a function at most once, then the graph is the graph of a one-to-one function.

3.3 Linear Functions

- A function is a linear function of x if it can be written in the form $f(x) = mx + b$, where m and b are real numbers and $m \neq 0$.

- The slope m of the line passing through the points $P_1(x_1, y_1)$ and $P_2(x_2, y_2)$ with $x_1 \neq x_2$ is given by

$$m = \frac{y_2 - y_1}{x_2 - x_1}.$$

- The graph of the equation $f(x) = mx + b$ has slope m and y intercept $(0, b)$.

- Two nonvertical lines are parallel if and only if their slopes are equal. Two lines with slopes m_1 and m_2 are perpendicular if and only if $m_1 = -\dfrac{1}{m_2}$.

3.4 Quadratic Functions

- A quadratic function of x is a function that can be represented by an equation of the form $f(x) = ax^2 + bx + c$, where a, b, and c are real numbers and $a \neq 0$.

- The vertex of the graph of $f(x) = ax^2 + bx + c$ is

$$\left(-\frac{b}{2a}, f\left(-\frac{b}{2a} \right) \right)$$

- Every quadratic function $f(x) = ax^2 + bx + c$ can be written in the standard form $f(x) = a(x - h)^2 + k$, $a \neq 0$. The graph of f is a parabola with vertex (h, k). The parabola is symmetric with respect to the vertical line $x = h$, which is called the axis of symmetry of the parabola. The parabola opens up if $a > 0$; it opens down if $a < 0$.

3.5 Properties of Graphs

- The graph of an equation is symmetric with respect to

  the y-axis if the replacement of x with $-x$ leaves the equation unaltered.

  the x-axis if the replacement of y with $-y$ leaves the equation unaltered.

  the origin if the replacement of x with $-x$ and of y with $-y$ leaves the equation unaltered.

- If f is a function and c is a positive constant, then

  $y = f(x) + c$ is the graph of $y = f(x)$ shifted up *vertically* c units.

  $y = f(x) - c$ is the graph of $y = f(x)$ shifted down *vertically* c units.

  $y = f(x + c)$ is the graph of $y = f(x)$ shifted left *horizontally* c units.

  $y = f(x - c)$ is the graph of $y = f(x)$ shifted right *horizontally* c units.

- The graph of

  $y = -f(x)$ is the graph of $y = f(x)$ reflected across the x-axis.

  $y = f(-x)$ is the graph of $y = f(x)$ reflected across the y-axis.

3.6 The Algebra of Functions

- For all values of x for which both $f(x)$ and $g(x)$ are defined, we define the following functions.

Sum	$(f + g)(x) = f(x) + g(x)$
Difference	$(f - g)(x) = f(x) - g(x)$

Product $(fg)(x) = f(x) \cdot g(x)$

Quotient $\left(\dfrac{f}{g}\right)(x) = \dfrac{f(x)}{g(x)}$, $g(x) \neq 0$

- The expression

$$\frac{f(x + h) - f(x)}{h}, \quad h \neq 0$$

is called the difference quotient of f. The difference quotient is an important function because it can be used to compute the *average rate of change* of f over the time interval $[x, x + h]$.

- For the functions f and g, the composite function, or composition, of f by g is given by $(g \circ f)(x) = g[f(x)]$ for all x in the domain of f such that $f(x)$ is in the domain of g.

3.7 Inverse Functions

- If f is a one-to-one function with domain X and range Y, and g is a function with domain Y and range X, then g is the inverse function of f if and only if $(f \circ g)(x) = x$ for all x in the domain of g and $(g \circ f)(x) = x$ for all x in the domain of f.

- A function f has an inverse function if and only if it is a one-to-one function. The graph of a function f and the graph of the inverse function f^{-1} are symmetric with respect to the line given by $y = x$.

3.8 Variation and Applications

- The variable y varies directly as the variable x if and only if $y = kx$, where k is a constant called the variation constant.

- The variable y varies inversely as the variable x if and only if $y = k/x$, where k is the variation constant.

- The variable z varies jointly as the variables x and y if and only if $z = kxy$, where k is the variation constant.

CHAPTER 3 TRUE/FALSE EXERCISES

In Exercises 1 to 12, answer true or false. If the statement is false, give an example to show that the statement is false.

1. Let f be any function. Then $f(a) = f(b)$ implies that $a = b$.

2. Every function has an inverse function.

3. If $(f \circ g)(a) = a$ and $(g \circ f)(a) = a$ for some constant a, then f and g are inverse functions.

4. Let f be a function such that $f(x) = f(x + 4)$ for all real numbers x. If $f(2) = 3$, then $f(18) = 3$.

5. For all functions f, $[f(x)]^2 = f[f(x)]$.

6. Let f be any function. Then for all a and b in the domain of f such that $f(b) \neq 0$ and $b \neq 0$,

$$\frac{f(a)}{f(b)} = \frac{a}{b}$$

7. The **identity function** $f(x) = x$ is its own inverse.

8. If f is defined by $f(x) = |x|$, then $f(a + b) = f(a) + f(b)$ for all real numbers a and b.

9. If f is defined by $f(x) = |x|$, then $f(ab) = f(a)f(b)$ for all real numbers a and b.

10. If f is a one-to-one function and a and b are real numbers in the domain of f with $a < b$, then $f(a) \neq f(b)$.

11. The coordinates of a point on the graph of $y = f(x)$ are (a, b). If k is a positive constant, then (a, kb) are the coordinates of a point on the graph of $y = kf(x)$.

12. For every function f, the real number c is a solution of $f(x) = 0$ if and only if $(c, 0)$ is an x-intercept of the graph of $y = f(x)$.

CHAPTER 3 REVIEW EXERCISES

In Exercises 1 and 2, find the distance between the points whose coordinates are given.

1. $(-3, 2)$ $(7, 11)$

2. $(5, -4)$ $(-3, -8)$

In Exercises 3 and 4, find the midpoint of the line segment with the given endpoints.

3. $(2, 8)$ $(-3, 12)$

4. $(-4, 7)$ $(8, -11)$

In Exercises 5 and 6, determine the center and radius of the circle with the given equation.

5. $(x - 3)^2 + (y + 4)^2 = 81$

6. $x^2 + y^2 + 10x + 4y + 20 = 0$

In Exercises 7 and 8, find the equation in standard form of a circle that satisfies the given conditions.

7. Center $C = (2, -3)$, radius $r = 5$

8. Center $C = (-5, 1)$, passing through $(3, 1)$

9. If $f(x) = 3x^2 + 4x - 5$, find

   a. $f(1)$ **b.** $f(-3)$ **c.** $f(t)$

   d. $f(x + h)$ **e.** $3f(t)$ **f.** $f(3t)$

10. If $g(x) = \sqrt{64 - x^2}$, find

   a. $g(3)$ **b.** $g(-5)$ **c.** $g(8)$

   d. $g(-x)$ **e.** $2g(t)$ **f.** $g(2t)$

11. If $f(x) = x^2 + 4x$ and $g(x) = x - 8$, find

   a. $(f \circ g)(3)$ **b.** $(g \circ f)(-3)$

   c. $(f \circ g)(x)$ **d.** $(g \circ f)(x)$

12. If $f(x) = 2x^2 + 7$ and $g(x) = |x - 1|$, find

   a. $(f \circ g)(-5)$ **b.** $(g \circ f)(-5)$

   c. $(f \circ g)(x)$ **d.** $(g \circ f)(x)$

13. If $f(x) = 4x^2 - 3x - 1$, find the difference quotient

$$\frac{f(x + h) - f(x)}{h}$$

14. If $g(x) = x^3 - x$, find the difference quotient

$$\frac{g(x + h) - g(x)}{h}$$

In Exercises 15 to 20, sketch the graph of f. Find the interval(s) in which f is a. increasing, b. constant, c. decreasing.

15. $f(x) = |x - 3| - 2$

16. $f(x) = x^2 - 5$

17. $f(x) = |x + 2| - |x - 2|$

18. $f(x) = [\![x + 3]\!]$

19. $f(x) = \dfrac{1}{2}x - 3$

20. $f(x) = \sqrt[3]{x}$

In Exercises 21 to 24, determine the domain of the function represented by the given equation.

21. $f(x) = -2x^2 + 3$

22. $f(x) = \sqrt{6 - x}$

23. $f(x) = \sqrt{25 - x^2}$

24. $f(x) = \dfrac{3}{x^2 - 2x - 15}$

In Exercises 25 and 26, find the slope-intercept form of the equation of the line through the two points.

25. $(-1, 3)$ $(4, -7)$

26. $(0, 0)$ $(7, 11)$

27. Find the slope-intercept form of the equation of the line that is parallel to the graph of $3x - 4y = 8$ and passes through $(2, 11)$.

28. Find the slope-intercept form of the equation of the line that is perpendicular to the graph of $2x = -5y + 10$ and passes through $(-3, -7)$.

In Exercises 29 to 34, use the method of completing the square to write each quadratic equation in its standard form.

29. $f(x) = x^2 + 6x + 10$ **30.** $f(x) = 2x^2 + 4x + 5$

31. $f(x) = -x^2 - 8x + 3$ **32.** $f(x) = 4x^2 - 6x + 1$

33. $f(x) = -3x^2 + 4x - 5$ **34.** $f(x) = x^2 - 6x + 9$

In Exercises 35 to 38, find the vertex of the graph of the quadratic function.

35. $f(x) = 3x^2 - 6x + 11$ **36.** $h(x) = 4x^2 - 10$

37. $k(x) = -6x^2 + 60x + 11$ **38.** $m(x) = 14 - 8x - x^2$

39. Use the formula

$$d = \frac{|mx_1 + b - y_1|}{\sqrt{1 + m^2}}$$

to find the distance from the point $(1, 3)$ to the line given by $y = 2x - 3$.

40. A freight company has determined that its cost of delivering x parcels per delivery is

$$C(x) = 1050 + 0.5x$$

The price it charges to send a parcel is $13.00 per parcel. Determine

a. the revenue function

b. the profit function

c. the minimum number of parcels the company must ship to break even

In Exercises 41 and 42, sketch a graph that is symmetric to the given graph with respect to the *a.* **x-axis,** *b.* **y-axis,** *c.* **origin.**

41.

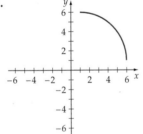

42.

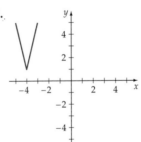

In Exercises 43 to 50, determine whether the graph of each equation is symmetric with respect to the *a.* **x-axis,** *b.* **y-axis,** *c.* **origin.**

43. $y = x^2 - 7$ **44.** $x = y^2 + 3$

45. $y = x^3 - 4x$ **46.** $y^2 = x^2 + 4$

47. $\dfrac{x^2}{3^2} + \dfrac{y^2}{4^2} = 1$ **48.** $xy = 8$

49. $|y| = |x|$ **50.** $|x + y| = 4$

In Exercises 51 to 56, sketch the graph of g. *a.* **Find the domain and the range of g.** *b.* **State whether g is even, odd, or neither even nor odd.**

51. $g(x) = -x^2 + 4$ **52.** $g(x) = -2x - 4$

53. $g(x) = |x - 2| + |x + 2|$ **54.** $g(x) = \sqrt{16 - x^2}$

55. $g(x) = x^3 - x$ **56.** $g(x) = 2[\![x]\!]$

In Exercises 57 to 62, first write the function in standard form, and then make use of translations to graph the function.

57. $F(x) = x^2 + 4x - 7$ **58.** $A(x) = x^2 - 6x - 5$

59. $P(x) = 3x^2 - 4$ **60.** $G(x) = 2x^2 - 8x + 3$

61. $W(x) = -4x^2 - 6x + 6$ **62.** $T(x) = -2x^2 - 10x$

63. On the same set of coordinate axes, sketch the graph of $p(x) = \sqrt{x} + c$ for $c = 0, -1,$ and 2.

64. On the same set of coordinate axes, sketch the graph of $q(x) = \sqrt{x + c}$ for $c = 0, -1,$ and 2.

65. On the same set of coordinate axes, sketch the graph of $r(x) = c\sqrt{9 - x^2}$ for $c = 1, 1/2,$ and -2.

66. On the same set of coordinate axes, sketch the graph of $s(t) = [\![cx]\!]$ for $c = 1, 1/4,$ and 4.

In Exercises 67 and 68, graph each piecewise-defined function.

67. $f(x) = \begin{cases} x & \text{if } x \le 0 \\ \dfrac{1}{2}x & \text{if } x > 0 \end{cases}$

68. $g(x) = \begin{cases} -2 & \text{if } x < -3 \\ \dfrac{2}{3}x & \text{if } -3 \le x \le 3 \\ 2 & \text{if } x > 3 \end{cases}$

In Exercises 69 and 70, use the given functions f and g to find f + g, f − g, fg, and f/g. State the domain of each.

69. $f(x) = x^2 - 9,\quad g(x) = x + 3$

70. $f(x) = x^3 + 8,\quad g(x) = x^2 - 2x + 4$

In Exercises 71 to 74, determine whether the given functions are inverses.

71. $F(x) = 2x - 5\qquad G(x) = \dfrac{x + 5}{2}$

72. $h(x) = \sqrt{x}$ $k(x) = x^2$, $x \geq 0$

73. $l(x) = \dfrac{x + 3}{x}$ $m(x) = \dfrac{3}{x - 1}$

74. $p(x) = \dfrac{x - 5}{2x}$ $q(x) = \dfrac{2x}{x - 5}$

In Exercises 75 to 78, find the inverse of the function. Sketch the graph of the function and its inverse on the same set of coordinates axes.

75. $f(x) = 3x - 4$ **76.** $g(x) = -2x + 3$

77. $h(x) = -\dfrac{1}{2}x - 2$ **78.** $k(x) = \dfrac{1}{x}$

79. Find two numbers whose sum is 50 and whose product is a maximum.

80. Find two numbers whose difference is 10 and the sum of whose squares is a minimum.

81. The distance traveled by a ball rolling down a ramp is given by $s(t) = 3t^2$, where t is the time in seconds after the ball is released and $s(t)$ is measured in feet. Evaluate the average velocity of the ball for each of the following time intervals.

a. [2, 4]   **b.** [2, 3]   **c.** [2, 2.5]   **d.** [2, 2.01]

e. What appears to be the average velocity of the ball for the time interval $[2, 2 + \Delta t]$ if Δt approaches 0?

82. The distance traveled by a ball that is pushed down a ramp is given by $s(t) = 2t^2 + t$, where t is the time in seconds after the ball is released and $s(t)$ is measured in feet. Evaluate the average velocity of the ball for each of the following time intervals.

a. [3, 5]   **b.** [3, 4]   **c.** [3, 3.5]   **d.** [3, 3.01]

e. What appears to be the average velocity of the ball for the time interval $[3, 3 + \Delta t]$ if Δt approaches 0?

CHAPTER 3 TEST

1. Find the midpoint and the length of the line segment with endpoints $(-2, 3)$ and $(4, -1)$.

2. Determine the x- and y-intercepts, and then graph the equation $x = 2y^2 - 4$.

3. Graph the equation $y = |x + 2| + 1$.

4. Find the center and radius of the circle that has the general form $x^2 - 4x + y^2 + 2y - 4 = 0$.

5. Determine the domain of the function

$$f(x) = -\sqrt{x^2 - 16}$$

6. Use the formula

$$d = \frac{|mx_1 + b - y_1|}{\sqrt{1 + m^2}}$$

to find the distance from the point $(3, 4)$ to the line given by $y = 3x + 1$.

7. Graph $f(x) = -2|x - 2| + 1$. Identify the intervals over which the function is

a. increasing

b. constant

c. decreasing

8. Graph the function $f(x) = x^2 + 2$. From the graph, find the domain and range of the function.

9. An air freight company has determined that its cost of delivering x parcels per flight is

$$C(x) = 875 + 0.75x$$

The price it charges to send a parcel is $12.00 per parcel. Determine

a. the revenue function

b. the profit function

c. the minimum number of parcels the company must ship to break even

10. Use the graph of $f(x) = |x|$ to graph $y = -f(x + 2) - 1$.

11. Classify each of the following as either an even function, an odd function, or neither an even nor an odd function.

a. $f(x) = x^4 - x^2$ **b.** $f(x) = x^3 - x$ **c.** $f(x) = x - 1$

12. Find the slope-intercept form of the equation of the line that passes through $(4, -2)$ and is perpendicular to the graph of $3x - 2y = 4$.

13. Find the maximum or minimum value of the function $f(x) = x^2 - 4x - 8$. State whether this value is a maximum or a minimum value.

14. Let $f(x) = x^2 - 1$ and $g(x) = x - 2$. Find $(f + g)$ and (f/g).

15. Find the difference quotient of the function

$$f(x) = x^2 + 1$$

16. Evaluate $(f \circ g)$, where

$$f(x) = x^2 - 2x + 1 \quad \text{and} \quad g(x) = \sqrt{x - 2}$$

17. Find the inverse of $f(x) = x^2 - 9$, $x \geq 0$. State the domain and range of both f and f^{-1}.

18. Find the inverse of the function given by the equation $f(x) = 2x - 3$. Graph f and f^{-1} on the same coordinate axes.

19. The distance traveled by a ball rolling down a ramp is given by $s(t) = 5t^2$, where t is the time in seconds after the ball is released and $s(t)$ is measured in feet. Evaluate the average velocity of the ball for each of the following time intervals.

a. $[2, 3]$ **b.** $[2, 2.5]$ **c.** $[2, 2.01]$

20. The illumination that a source of light provides is inversely proportional to the square of the distance from the source. If the illumination at a distance of 8 feet is 20 lumens, what is the illumination (to the nearest 0.01) at a distance of 15 feet from the source?

POLYNOMIAL AND RATIONAL FUNCTIONS

1929 Ford Model A
gas mileage: 20 mpg.

1960 Chevrolet Impala
gas mileage: 18 mpg.

1996 Jeep Grand
Cherokee gas mileage:
17 mpg.

Consumption of Gasoline by Automobiles

Suppose your car currently gets 20 miles per gallon (mpg) and you purchase a new car that gets 25 mpg. If you drive 12,000 miles per year, then the number of gallons *fewer* that you will use per year is

$$\frac{12,000}{20} - \frac{12,000}{25} = 120 \text{ gallons}$$

If the car you purchased gets m mpg more than your previous car, then the number, $g(m)$, of gallons fewer per year is given by

$$g(m) = \frac{12,000}{20} - \frac{12,000}{20 + m}$$

Simplifying this expression gives

$$g(m) = \frac{12,000m}{400 + 20m}$$

This equation represents a *rational function*. Rational functions are one of the topics of this chapter. For each value of m, $g(m)$ represents the number of gallons of gasoline saved per year.

As another example of a situation that leads to a rational function, suppose that car manufacturers can increase the average gasoline mileage of cars in the United States by 10 mpg. Let x represent the current average gasoline mileage of U.S. cars. Then, assuming (conservatively) that 6×10^{12} miles are driven annually in the United States and that the average price of gasoline is $1.15 per gallon, the cost savings, $C(x)$, to consumers is given by

$$C(x) = \frac{1.15(6 \times 10^{12})}{x(x + 10)}$$

For instance, if the average for all U.S. cars is currently 22 mpg, then increasing the average to 32 mpg (22 + 10), would save consumers

$$C(22) = \frac{1.15(6 \times 10^{12})}{22(22 + 10)} \approx 9.8 \times 10^9 = \$9.8 \text{ billion}$$

4.1 POLYNOMIAL DIVISION AND SYNTHETIC DIVISION

If $P(x)$ is a polynomial, then the values of x for which $P(x)$ is equal to 0 are called the **zeros** of $P(x)$ or the **roots** of $P(x) = 0$. Much of the work in this chapter concerns finding the zeros of a polynomial. Sometimes the zeros of a polynomial can be determined by dividing the polynomial by another polynomial. Dividing a polynomial by another polynomial is similar to the long-division process used for dividing positive integers. For example, to divide $(x^2 + 9x - 16)$ by $(x - 3)$, we use the following procedure.

POINT OF INTEREST

**Apart from René Descartes'
work with analytic geometry,
he was also (in his book
Discours de la Méthode) the
first to write powers of x
($x, xx, x^3, x^4, \ldots$) as we do
today. Note that for some
reason, he wrote xx instead
of x^2.**

$$
\begin{array}{r}
x + 12 \\
x - 3 \overline{)\, x^2 + 9x - 16} \\
\underline{x^2 - 3x} \\
12x - 16 \\
\underline{12x - 36} \\
20
\end{array}
$$

Thus $(x^2 + 9x - 16) \div (x - 3) = x + 12$, with a remainder of 20.

In this example, $x^2 + 9x - 16$ is called the **dividend,** $x - 3$ is the **divisor,** $x + 12$ is the **quotient,** and 20 is the **remainder.** The dividend is equal to the product of the divisor and the quotient, plus the remainder. That is,

$$
\underbrace{x^2 + 9x - 16}_{\text{Dividend}} = \underbrace{(x - 3)}_{\text{Divisor}} \cdot \underbrace{(x + 12)}_{\text{Quotient}} + \underbrace{20}_{\text{Remainder}}
$$

The above result is a special case of a theorem known as the *Division Algorithm for Polynomials.*

THE DIVISION ALGORITHM FOR POLYNOMIALS

If $P(x)$ and $D(x)$ are polynomials such that $D(x) \neq 0$, then there exist unique polynomials $Q(x)$ and $R(x)$ such that $P(x) = D(x)Q(x) + R(x)$, where either $R(x) = 0$ or the degree of $R(x)$ is less than the degree of $D(x)$.

The polynomial $P(x)$ is the dividend, $D(x)$ is the divisor, $Q(x)$ is the quotient, and the polynomial $R(x)$ is the remainder.

$$
\underbrace{P(x)}_{\text{Dividend}} = \underbrace{D(x)}_{\text{Divisor}} \cdot \underbrace{Q(x)}_{\text{Quotient}} + \underbrace{R(x)}_{\text{Remainder}}
$$

Multiplying both sides of $P(x) = D(x)Q(x) + R(x)$ by $1/D(x)$ produces the fractional form

$$\frac{P(x)}{D(x)} = Q(x) + \frac{R(x)}{D(x)}$$

EXAMPLE 1 *Divide Polynomials*

Perform the indicated division.

$$\frac{x^4 + 3x^2 - 6x - 10}{x^2 + 3x - 5}$$

Solution

$$
\begin{array}{r}
x^2 - 3x\ + 17 \\
x^2 + 3x - 5 \overline{)\, x^4 + 0x^3 +\ \ 3x^2 -\ \ 6x - 10} \\
\underline{x^4 + 3x^3 -\ \ 5x^2} \\
-3x^3 +\ \ 8x^2 -\ 6x \\
\underline{-3x^3 -\ \ 9x^2 + 15x} \\
17x^2 - 21x - 10 \\
\underline{17x^2 + 51x - 85} \\
-72x + 75
\end{array}
$$

• **Writing $0x^3$ for the missing term helps us align like terms in the same column.**

Thus $\dfrac{x^4 + 3x^2 - 6x - 10}{x^2 + 3x - 5} = x^2 - 3x + 17 + \dfrac{-72x + 75}{x^2 + 3x - 5}.$

TRY EXERCISE 6, EXERCISE SET 4.1

SYNTHETIC DIVISION

The procedure for dividing a polynomial by a binomial of the form $x - c$ can be condensed by a method called **synthetic division.** To understand the synthetic division method, consider the following division.

$$
\begin{array}{r}
3x^2 - 2x\ + 3 \\
x - 2 \overline{)\, 3x^3 - 8x^2 + 7x + 2} \\
\underline{3x^3 - 6x^2} \\
-2x^2 + 7x \\
\underline{-2x^2 + 4x} \\
3x + 2 \\
\underline{3x - 6} \\
8
\end{array}
$$

No essential data are lost by omitting the variables, because the position of a term indicates the power of the term.

$$
\begin{array}{r}
3 \quad -2 \quad 3 \qquad\quad \\
-2\overline{)3 \quad -8 \quad 7 \quad\ \ 2} \\
\underline{3 \quad -6 \qquad\qquad} \\
-2 \quad 7 \qquad \\
\underline{-2 \quad 4 \qquad} \\
3 \quad\ \ 2 \\
\underline{3 \quad -6} \\
8
\end{array}
$$

The coefficients shown in red are duplicates of those directly above them. Omitting these repeated coefficients (in red) enables us to condense the vertical spacing.

$$
\begin{array}{r}
3 \quad -2 \quad\ \ 3 \qquad\quad \\
-2\overline{)3 \quad -8 \quad\ \ 7 \qquad 2} \\
\underline{-6 \quad\ \ 4 \quad -6} \\
-2 \quad\ \ 3 \qquad 8
\end{array}
$$

The coefficients in blue in the top row can be omitted because they are duplicates of those in the bottom row. The leading coefficient of the quotient (top row) can be written in the bottom row with the coefficients of the other terms in order to condense the vertical spacing even more.

$$
\begin{array}{r|rrrr}
-2 & 3 & -8 & 7 & 2 \\
& & -6 & 4 & -6 \\
\hline
& 3 & -2 & 3 & 8
\end{array}
$$

So that we may add the numbers in each column instead of subtracting them, we change the sign of the divisor. This changes the sign of each number in the second row.

$$
\begin{array}{r|rrrr}
2 & 3 & -8 & 7 & 2 \\
& & 6 & -4 & 6 \\
\hline
& 3 & -2 & 3 & 8
\end{array}
$$

Coefficients of the quotient ———↗   └— Remainder

The following example illustrates step by step the synthetic division procedure, used here to find the quotient

$$
\frac{2x^3 - 9x^2 + 5}{x - 3}
$$

TAKE NOTE

The synthetic division method shown at the right is used only to divide by a polynomial of the form $x - c$, where the coefficient of x is 1. To divide a polynomial by a polynomial that is not a binomial, use the long division method.

Coefficients of the dividend

$$3 \,|\, \begin{array}{cccc} 2 & -9 & 0 & 5 \end{array}$$

- Synthetic-division form with 0 inserted for the missing x term.

$$3 \,|\, \begin{array}{cccc} 2 & -9 & 0 & 5 \\ \hline 2 \end{array}$$

- Bring down the leading coefficient 2.

$$3 \,|\, \begin{array}{cccc} 2 & -9 & 0 & 5 \\ & 6 & & \\ \hline 2 \end{array}$$

- Multiply $3 \cdot 2$ and place the product (6) in the middle row and in the next column to the right.

$$3 \,|\, \begin{array}{cccc} 2 & -9 & 0 & 5 \\ & 6 & & \\ \hline 2 & -3 \end{array}$$

- Add -9 and 6 and place the sum (-3) in the bottom row.

$$3 \,|\, \begin{array}{cccc} 2 & -9 & 0 & 5 \\ & 6 & -9 & -27 \\ \hline 2 & -3 & -9 & -22 \end{array}$$

- Repeat the previous steps for columns 3 and 4.

Remainder

Coefficients of the quotient

$$\frac{2x^3 - 9x^2 + 5}{x - 3} = 2x^2 - 3x - 9 + \frac{-22}{x - 3}$$

EXAMPLE 2 *Use Synthetic Division to Divide Polynomials*

Use synthetic division to perform the indicated division.

$$\frac{x^4 - 4x^2 + 7x + 15}{x + 4}$$

Solution

Because the divisor is $x + 4$, we perform the synthetic division with $c = -4$.

$$-4 \,|\, \begin{array}{ccccc} 1 & 0 & -4 & 7 & 15 \\ & -4 & 16 & -48 & 164 \\ \hline 1 & -4 & 12 & -41 & 179 \end{array}$$

The quotient is $x^3 - 4x^2 + 12x - 41$ and the remainder is 179.

$$\frac{x^4 - 4x^2 + 7x + 15}{x + 4} = x^3 - 4x^2 + 12x - 41 + \frac{179}{x + 4}$$

TRY EXERCISE 12, EXERCISE SET 4.1

The following theorem shows that synthetic division can be used to find the value $P(c)$ for any polynomial function P and constant c.

THE REMAINDER THEOREM

If a polynomial $P(x)$ is divided by $x - c$, then the remainder is $P(c)$.

Proof   The Division Algorithm states that

$$P(x) = (x - c)Q(x) + R(x)$$

where $R(x)$ is zero or the degree of $R(x)$ is less than the degree of $x - c$. Because the degree of $x - c$ is 1, the remainder $R(x)$ must be some constant—say, r. Therefore,

$$P(x) = (x - c)Q(x) + r$$

This equality evaluated at $x = c$ produces

$$P(c) = (c - c)Q(c) + r = (0)Q(c) + r = r \qquad \blacklozenge$$

EXAMPLE 3   *Use the Remainder Theorem to Evaluate a Polynomial*

Use the Remainder Theorem to evaluate $P(x) = 2x^3 + 3x^2 + 2x - 2$ when $x = -2$ and $x = 1/2$.

Solution

Perform the synthetic division and examine the remainders.

$$
\begin{array}{r|rrrr}
-2 & 2 & 3 & 2 & -2 \\
   &   & -4 & 2 & -8 \\
\hline
   & 2 & -1 & 4 & -10
\end{array}
$$

The remainder is -10. By the Remainder Theorem, $P(-2) = -10$. From **Figure 4.1**, the point with coordinates $(-2, -10)$ is on the graph of P.

$$
\begin{array}{r|rrrr}
\frac{1}{2} & 2 & 3 & 2 & -2 \\
            &   & 1 & 2 & 2 \\
\hline
            & 2 & 4 & 4 & 0
\end{array}
$$

The remainder is 0. By the Remainder Theorem, $P(1/2) = 0$. **Figure 4.1** shows that the point whose coordinates are $(1/2, 0)$ is on the graph of P.

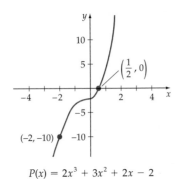

$P(x) = 2x^3 + 3x^2 + 2x - 2$

Figure 4.1

TRY EXERCISE 32, EXERCISE SET 4.1

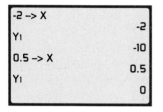

Figure 4.2

Recall that a graphing calculator can be used to evaluate a polynomial for many different values of x. Use (for instance) y_1 as the name of the variable that contains the polynomial. For the polynomial in Example 3 we would enter

$$Y_1=2X^3+3X^2+2X-2$$

Once you have entered the polynomial, you can evaluate it for various values of x by first storing the value of x and then recalling y_1. The display in **Figure 4.2** shows the results of evaluating the polynomial for $x = -2$ and $x = 0.5$.

Note from Example 3 that $P(1/2) = 0$. The number $1/2$ is called a *zero* of the polynomial because the value of the polynomial is 0 when $x = 1/2$.

ZERO OF A POLYNOMIAL

If $P(x)$ is a polynomial and a is a number for which $P(a) = 0$, then a is called a **zero** of $P(x)$.

The following theorem is a result of the Remainder Theorem.

THE FACTOR THEOREM

A polynomial $P(x)$ has a factor $(x - c)$ if and only if $P(c) = 0$. That is, $(x - c)$ is a factor of $P(x)$ if and only if c is a zero of P.

Proof   **Part 1:** Given that $P(x)$ has a factor of $(x - c)$, show that $P(c) = 0$. If $(x - c)$ is a factor of $P(x)$, then $P(x) = (x - c) \cdot Q(x)$ for some $Q(x)$. Thus the division of $P(x)$ by $(x - c)$ has a remainder of zero, and the Remainder Theorem implies that $P(c) = 0$.

   Part 2: Given $P(c) = 0$, show that $(x - c)$ is a factor of $P(x)$. The division algorithm applied to the polynomial $P(x)$ with divisor $(x - c)$ produces

$$P(x) = (x - c)Q(x) + R(x)$$

Because $P(c) = 0$, the Remainder Theorem implies that $R(x) = 0$. Thus

$$P(x) = (x - c)Q(x)$$

which shows that $(x - c)$ is a factor of $P(x)$.   ◆

EXAMPLE 4   *Find a Factor of a Polynomial*

Determine whether $(x + 5)$ is a factor of

$$P(x) = x^4 + x^3 - 21x^2 - x + 20$$

Solution

$$
\begin{array}{r|rrrrr}
-5 & 1 & 1 & -21 & -1 & 20 \\
   &   & -5 & 20 & 5 & -20 \\
\hline
   & 1 & -4 & -1 & 4 & 0
\end{array}
$$

The remainder 0 implies that $(x + 5)$ is a factor of $P(x)$.

TRY EXERCISE 42, EXERCISE SET 4.1

QUESTION   From the result of Example 4, is -5 a zero of $P(x)$?

From Example 4, $(x + 5)$ is a factor of $P(x) = x^4 + x^3 - 21x^2 - x + 20$, and the quotient is $Q(x) = x^3 - 4x^2 - x + 4$ (from the last line of the synthetic division). Thus

$$P(x) = (x + 5)(x^3 - 4x^2 - x + 4)$$

The polynomial $Q(x) = x^3 - 4x^2 - x + 4$ is called a **reduced polynomial** because it is 1 degree less than the degree of $P(x)$. Reduced polynomials play an important role in our work in Section 4.3.

EXAMPLE 5   *Solve an Application Concerning Security Codes*

A security panel with n buttons is disarmed when three different buttons are pushed in the correct order. The total number of three-button sequences in which no button is pushed twice is given by

$$P(n) = n^3 - 3n^2 + 2n$$

where n is an integer greater than or equal to 3. For security reasons, the manufacturer of the panels requires that $P(n) \geq 10,000$. Find the least number of buttons that will meet the manufacturer's requirement.

ANSWER   Yes. Because $(x + 5)$ is a factor of $P(x)$, the Factor Theorem states that $P(-5) = 0$, and thus -5 is a zero of P.

Solution

The polynomial $P(n)$ can be written as $n^3 - 3n^2 + 2n + 0$. Experiment with synthetic division and various values of n. Eventually you will try synthetic division with $n = 22$ and $n = 23$.

$$
\begin{array}{r|rrrr}
22 & 1 & -3 & 2 & 0 \\
   &   & 22 & 418 & 9240 \\
\hline
   & 1 & 19 & 420 & 9240
\end{array}
\qquad
\begin{array}{r|rrrr}
23 & 1 & -3 & 2 & 0 \\
   &   & 23 & 460 & 10{,}626 \\
\hline
   & 1 & 20 & 462 & 10{,}626
\end{array}
$$

By the Remainder Theorem, $P(22) = 9240$ and $P(23) = 10{,}626$. Synthetic division with $n = 3, 4, \ldots, 21$, cannot yield a remainder greater than or equal to 10,000 because deleting a button decreases the number of possible three-button sequences. Thus the least number of buttons required to meet the manufacturer's requirement is 23.

> **TRY EXERCISE 64, EXERCISE SET 4.1**

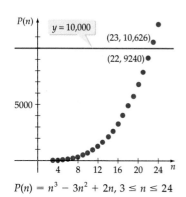

$P(n) = n^3 - 3n^2 + 2n,\ 3 \le n \le 24$

Figure 4.3

The graph of $P(n)$ in **Figure 4.3** also shows that $n = 23$ is the smallest positive integer for which $P(n) \ge 10{,}000$.

TOPICS FOR DISCUSSION

1. Explain the meaning of the phrase *zero of a polynomial*.

2. If $P(x)$ is a polynomial of degree $n \ge 2$, what is the degree of the quotient $\dfrac{P(x)}{x - a}$?

3. Discuss how the Factor Theorem can be used to determine whether a number is a zero of a polynomial.

4. A zero of $P(x) = x^3 - x^2 - 14x + 24$ is -4. Discuss how this information and the Factor Theorem can be used to solve $x^3 - x^2 - 14x + 24 = 0$.

5. Discuss the advantages and disadvantages of using synthetic division rather than a calculator to evaluate a polynomial.

EXERCISE SET 4.1

In Exercises 1 to 10, use long division to divide the first polynomial by the second.

1. $5x^3 + 6x^2 - 17x + 20,\ \ x + 3$

2. $6x^3 + 15x^2 - 8x + 2,\ \ x + 4$

3. $2x^4 + 15x^3 + 7x^2 - 135x - 225,\ \ 2x + 5$

4. $6x^4 + 3x^3 - 11x^2 - 3x + 9,\ \ 2x - 3$

5. $3x^4 + x^3 - 99x^2 - 30,\ \ 3x^2 + x + 1$

6. $2x^4 - x^3 - 23x^2 + 9x + 45,\ \ 2x^2 - x - 5$

7. $20x^4 - 3x^2 + 9,\ \ 5x^2 - 2$

8. $24x^5 + 20x^3 - 16x^2 - 15,\ \ 6x^2 + 5$

9. $x^3 + 5x^2 + 6x - 19,\ \ x^2 + x - 4$

10. $2x^4 + 3x^3 - 7x - 10,\ \ x^2 - 2x - 5$

In Exercises 11 to 30, use synthetic division to divide the first polynomial by the second.

11. $4x^3 - 5x^2 + 6x - 7, \quad x - 2$

12. $5x^3 + 6x^2 - 8x + 1, \quad x - 5$

13. $4x^3 - 2x + 3, \quad x + 1$

14. $6x^3 - 4x^2 + 17, \quad x + 3$

15. $x^5 - 10x^3 + 5x - 1, \quad x - 4$

16. $6x^4 - 2x^3 - 3x^2 - x, \quad x - 5$

17. $x^5 - 1, \quad x - 1$

18. $x^4 + 1, \quad x + 1$

19. $8x^3 - 4x^2 + 6x - 3, \quad x - \dfrac{1}{2}$

20. $12x^3 + 5x^2 + 5x + 6, \quad x + \dfrac{3}{4}$

21. $x^8 + x^6 + x^4 + x^2 + 4, \quad x - 2$

22. $-x^7 - x^5 - x^3 - x - 5, \quad x + 1$

23. $x^6 + x - 10, \quad x + 3$

24. $2x^5 - 3x^4 - 5x^2 - 10, \quad x - 4$

25. $3x^2 - 4x + 5, \quad x - 0.3$

26. $2x^2 - 12x + 1, \quad x + 0.4$

27. $2x^3 - 11x^2 - 17x + 3, \quad x$

28. $5x^4 - 2x^2 + 6x - 1, \quad x$

29. $x + 8, \quad x + 2$

30. $3x - 17, \quad x - 3$

In Exercises 31 to 40, use the Remainder Theorem to find $P(c)$.

31. $P(x) = 3x^3 + x^2 + x - 5, c = 2$

32. $P(x) = 2x^3 - x^2 + 3x - 1, c = 3$

33. $P(x) = 4x^4 - 6x^2 + 5, c = -2$

34. $P(x) = 6x^3 - x^2 + 4x, c = -3$

35. $P(x) = -2x^3 - 2x^2 - x - 20, c = 10$

36. $P(x) = -x^3 + 3x^2 + 5x + 30, c = 8$

37. $P(x) = -x^4 + 1, c = 3$

38. $P(x) = x^5 - 1, c = 1$

39. $P(x) = x^4 - 10x^3 + 2, c = 3$

40. $P(x) = x^5 + 20x^2 - 1, c = -5$

In Exercises 41 to 52, use synthetic division and the Factor Theorem to determine whether the given binomial is a factor of $P(x)$.

41. $P(x) = x^3 + 2x^2 - 5x - 6, x - 2$

42. $P(x) = x^3 + 4x^2 - 27x - 90, x + 6$

43. $P(x) = 2x^3 + x^2 - 2x - 1, x + 1$

44. $P(x) = 3x^3 + 4x^2 - 27x - 36, x - 4$

45. $P(x) = x^4 - 25x^2 + 144, x + 3$

46. $P(x) = x^4 - 25x^2 + 144, x - 3$

47. $P(x) = x^5 + 2x^4 - 22x^3 - 50x^2 - 75x, x - 5$

48. $P(x) = 9x^4 - 6x^3 - 23x^2 - 4x + 4, x + 1$

49. $P(x) = 16x^4 - 8x^3 + 9x^2 + 14x - 4, x - \dfrac{1}{4}$

50. $P(x) = 10x^4 + 9x^3 - 4x^2 + 9x + 6, x + \dfrac{1}{2}$

51. $P(x) = x^2 - 4x - 1, x - (2 + \sqrt{5})$

52. $P(x) = x^2 - 4x - 1, x - (2 - \sqrt{5})$

In Exercises 53 to 62, use synthetic division to show that c is a zero of $P(x)$.

53. $P(x) = 3x^3 - 8x^2 - 10x + 28, c = 2$

54. $P(x) = 4x^3 - 10x^2 - 8x + 6, c = 3$

55. $P(x) = x^4 - 1, c = 1$

56. $P(x) = x^3 + 8, c = -2$

57. $P(x) = 3x^4 + 8x^3 + 10x^2 + 2x - 20, c = -2$

58. $P(x) = x^4 - 2x^2 - 100x - 75, c = 5$

59. $P(x) = 2x^3 - 18x^2 - 50x + 66, c = 11$

60. $P(x) = 2x^4 - 34x^3 + 70x^2 - 153x + 45, c = 15$

61. $P(x) = 3x^2 - 8x + 4, c = \dfrac{2}{3}$

62. $P(x) = 5x^2 + 12x + 4, c = -\dfrac{2}{5}$

63. SECURITY CODES A security panel with n buttons is disarmed when four different buttons are pushed in the correct order. The total number of four-button sequences in which no button is pushed more than one time is given by

$$P(n) = n^4 - 6n^3 + 11n^2 - 6n$$

where n is an integer greater than or equal to 4. For security reasons, the manufacturer of the panels requires that $P(n) \geq 14{,}000$. Find the least number of buttons that will meet the manufacturer's requirement.

64. SECURITY CODES A security panel with n buttons is disarmed when five different buttons are pushed in the correct order. The total number of five-button sequences in which no button is pushed more than one time is given by

$$P(n) = n^5 - 10n^4 + 35n^3 - 50n^2 + 24n$$

where n is an integer greater than or equal to 5. For security reasons the manufacturer of the panels requires that $P(n) \geq 15{,}000$. Find the least number of buttons that will meet the manufacturer's requirement.

You can use a graph to factor some polynomials. For example, the graph of $y = x^2 - x - 12$ intersects the x-axis at $x = -3$ and $x = 4$. Thus $x^2 - x - 12$ has -3 and 4 as zeros. Hence $x^2 - x - 12$ has factors of $(x + 3)$ and $(x - 4)$. Use a graphing utility to factor each polynomial in Exercises 65 to 68.

65. $x^3 - 7x + 6$ **66.** $x^3 + 6x^2 + 3x - 10$

67. $x^4 + 2x^3 - 13x^2 - 38x - 24$

68. $x^4 + 2x^3 - 7x^2 - 8x + 12$

SUPPLEMENTAL EXERCISES

69. Use the Factor Theorem to prove that for any positive odd integer n, $x^n + 1$ has $x + 1$ as a factor.

70. Use the Factor Theorem to prove that for any positive integer n, $x^n - 1$ has $x - 1$ as a factor.

71. Find the remainder of $5x^{48} + 6x^{10} - 5x + 7$ divided by $x - 1$.

72. Find the remainder of $18x^{80} - 6x^{50} + 4x^{20} - 2$ divided by $x + 1$.

73. Prove that $P(x) = 4x^4 + 7x^2 + 12$ has no factor of the form $x - c$, where c is a real number.

74. Prove that $P(x) = -5x^6 - 4x^2 - 10$ has no factor of the form $x - c$, where c is a real number.

75. Use synthetic division to show that $(x - i)$ is a factor of $x^3 - 3x^2 + x - 3$.

76. Use synthetic division to show that $(x + 2i)$ is a factor of $x^4 - 2x^3 + x^2 - 8x - 12$.

PROJECTS

1. HORNER'S METHOD AND SYNTHETIC DIVISION A method of factoring a polynomial, called Horner's Method, was described in the Topics for Discussion in Section 1.6. Show how this method is essentially synthetic division.

SECTION

4.2 INTRODUCTION TO POLYNOMIAL FUNCTIONS

Table 4.1 summarizes information developed in Chapter 3 about graphs of polynomial functions of degree 0, 1, or 2. Polynomial functions of degree 3 or higher can be graphed by the technique of plotting points. However, some additional knowledge about polynomial functions will make graphing easier.

Table 4.1

Polynomial Function $P(x)$	Graph
$P(x) = a$ (degree 0), $a \neq 0$	Horizontal line through $(0, a)$
$P(x) = ax + b$ (degree 1)	Line with y-intercept $(0, b)$ and slope a
$P(x) = ax^2 + bx + c$ (degree 2)	Parabola with vertex $\left(-\dfrac{b}{2a}, P\left(-\dfrac{b}{2a}\right)\right)$

All polynomial functions have graphs that are **smooth continuous curves.** The terms *smooth* and *continuous* are defined rigorously in calculus, but for the present, a smooth curve is a curve that does not have sharp corners, as shown in **Figure 4.4a.** A continuous curve does not have a break or hole, as shown in Figure 4.4b.

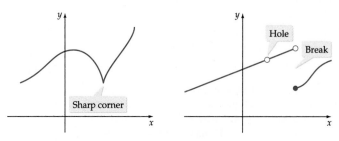

a. Continuous, but not smooth   b. Not continuous

Figure 4.4

THE LEADING TERM TEST

The graph of a polynomial function may have several up and down fluctuations; however, the graph of every polynomial function will eventually increase or decrease without bound as the graph moves far to the left or far to the right. The **leading term** $a_n x^n$ is said to **dominate** the polynomial function $P(x) = a_n x^n + a_{n-1} x^{n-1} + \cdots + a_1 x + a_0$ as $|x|$ becomes large, because the absolute value of $a_n x^n$ will be much larger than the absolute value of any of the other terms. Because of this condition, you can determine the far-left and far-right behavior of the polynomial by examining the leading coefficient a_n and the degree n of the polynomial.

Table 4.2 indicates the far-left and the far-right behavior of a polynomial function $P(x)$ with leading term $a_n x^n$.

Table 4.2	Far-Right and Far-Left Behavior of a Polynomial with Leading Term $a_n x^n$	
	n is even	**n is odd**
$a_n > 0$	Up to left and up to right	Down to left and up to right
$a_n < 0$	Down to left and down to right	Up to left and down to right

EXAMPLE 1 *Determine the Far-Left and Far-Right Behavior of a Polynomial Function*

Examine the leading term to determine the far-left and the far-right behavior of the graph of each polynomial function.

a. $P(x) = x^3 - x$ b. $S(x) = \dfrac{1}{2}x^4 - \dfrac{5}{2}x^2 + 2$

c. $T(x) = -2x^3 + x^2 + 7x - 6$ d. $U(x) = -x^4 + 8x^2 + 9$

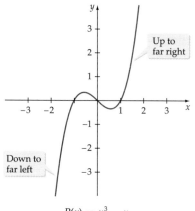

Up to
far right

Down to
far left

$$P(x) = x^3 - x$$

Figure 4.5

Solution

a. Because $a_n = 1$ is *positive* and $n = 3$ is *odd*, the graph of P goes down to its far left and up to its far right. See **Figure 4.5**.

b. Because $a_n = \dfrac{1}{2}$ is *positive* and $n = 4$ is *even*, the graph of S goes up to its far left and up to its far right. See **Figure 4.6**.

c. Because $a_n = -2$ is *negative* and $n = 3$ is *odd*, the graph of T goes up to its far left and down to its far right. See **Figure 4.7**.

d. Because $a_n = -1$ is *negative* and $n = 4$ is *even*, the graph of U goes down to its far left and down to its far right. See **Figure 4.8**.

TRY EXERCISE 2, EXERCISE SET 4.2

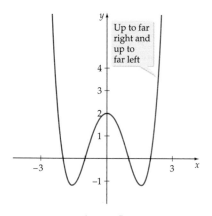

Up to far
right and
up to
far left

$$S(x) = \frac{1}{2}x^4 - \frac{5}{2}x^2 + 2$$

Figure 4.6

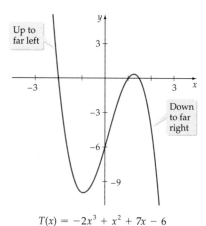

Up to
far left

Down
to far
right

$$T(x) = -2x^3 + x^2 + 7x - 6$$

Figure 4.7

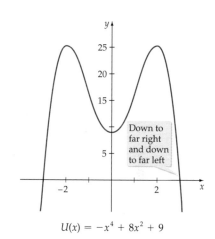

Down to
far right
and down
to far left

$$U(x) = -x^4 + 8x^2 + 9$$

Figure 4.8

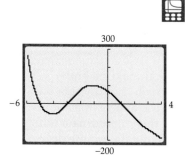

Figure 4.9

The far-left and far-right behavior of the graph of a polynomial can be used as an estimation check for the graph of a polynomial. For instance, suppose you use a graphing utility to produce the graph of a fourth-degree polynomial and the result is the graph in **Figure 4.9**. Then either the viewing window has not been set correctly or the equation for the graph has not been entered correctly, because the graph does not have the correct far-right and far-left behavior for a fourth-degree polynomial.

Figure 4.10 on page 246 illustrates the graph of a polynomial function of degree 3 with two **turning points,** points where the function changes from an increasing function to a decreasing function or vice versa. In general, the graph of a polynomial function of degree n has at most $n - 1$ turning points. Determining the exact location of turning points generally requires concepts and techniques from calculus.

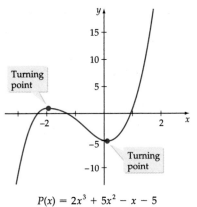

$P(x) = 2x^3 + 5x^2 - x - 5$

Figure 4.10

Turning points can be related to the concepts of maximum and minimum value of a function. These concepts were introduced in the discussion of graphs of second-degree equations in two variables earlier in the text. Recall that the minimum value of a function f is the smallest range value of f. It is often called the **absolute minimum**. For the function whose graph is shown in **Figure 4.11**, the y value of point E is the absolute minimum. There are no y values less than y_5.

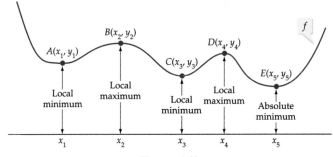

Figure 4.11

Now consider y_1, the y value of turning point A in **Figure 4.11**. It is not the smallest y value of every point on the graph of f; however, it is the smallest y value if we *localize* our field of view to a small neighborhood or open interval containing x_1. It is for this reason that we refer to y_1 as a *local minimum*, or *relative minimum*, of f. The y value of point C is also a local minimum of f.

The function does not have an absolute maximum because it goes up both to its far left and to its far right.

The y value of the point B is a local maximum, as is the y value of point D. The formal definitions of *local maximum* and *local minimum* are presented below.

LOCAL MAXIMUM AND LOCAL MINIMUM

Let f be a function defined on the open interval I, and let c be an element of I. Then

- $f(c)$ is a **local minimum** of f on I if $f(c) \leq f(x)$ for all x in I.

- $f(c)$ is a **local maximum** of f on I if $f(c) \geq f(x)$ for all x in I.

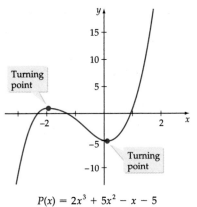

$P(x) = 2x^3 - 3x^2 - 12x + 6$

Figure 4.12

A graphing utility can be used to estimate the local maxima and minima of a function. One way to do this is to graph the function and then use the TRACE feature. Move the cursor to the approximate location of the maximum or minimum. The value of the y-coordinate will be the maximum or minimum of the function. To obtain a more accurate value, use the ZOOM feature. Depending on the type of graphing utility you have, there may be alternative methods for determining a maximum or minimum.

The graph in **Figure 4.12** is of $P(x) = 2x^3 - 3x^2 - 12x + 6$. By tracing along the curve, we were able to determine the local maximum and minimum of the function. Because of the far-left and far-right behavior of a cubic function, this function does not have an absolute maximum or an absolute minimum.

The following example illustrates the role a local maximum may play in an application.

EXAMPLE 2   *Solve a Maximization Application Problem*

A rectangular piece of cardboard measures 12 inches by 16 inches. An open box is formed by cutting congruent squares that measure x by x from each of the corners of the cardboard and folding as shown in **Figure 4.13**.

a. Express the volume V of the box as a function of x.

b. Determine (to the nearest 0.1 inch) the x value that maximizes the volume.

Solution

a. The lengths of the sides of the open box are x, $12 - 2x$, and $16 - 2x$. The volume is given by

$$V(x) = x(12 - 2x)(16 - 2x)$$
$$= 4x^3 - 56x^2 + 192x$$

b. Use a graphing utility to graph $y = V(x)$. The graph is shown in **Figure 4.14**. Note that we are interested only in the part of the graph for which $0 \leq x \leq 6$. This conclusion is a result of the following. The length of each side of the box must be nonnegative. Hence,

$$x \geq 0, \quad 12 - 2x \geq 0, \quad \text{and} \quad 16 - 2x \geq 0$$

The domain of V is the intersection of the solution sets of the three inequalities. Thus the domain is $\{x \mid 0 \leq x \leq 6\}$.

Now use the TRACE and ZOOM features to find that V attains its maximum of 194.028 when $x = 2.3$. See **Figures 4.15** and **4.16**.

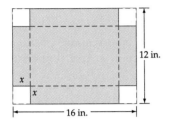

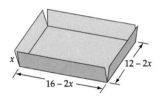

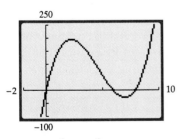

Figure 4.13

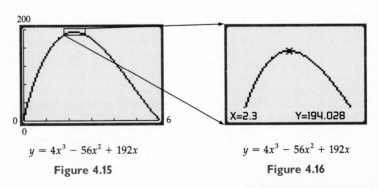

$y = 4x^3 - 56x^2 + 192x$

Figure 4.14

$y = 4x^3 - 56x^2 + 192x$

Figure 4.15

X=2.3   Y=194.028

$y = 4x^3 - 56x^2 + 192x$

Figure 4.16

TRY EXERCISE 26, EXERCISE SET 4.2

REAL ZEROS OF A POLYNOMIAL

The graph of every polynomial P is a smooth continuous curve, and if the value of P changes sign on an interval, then $P(c)$ must equal zero for at least one c in the interval. This result is known as the *Zero Location Theorem*.

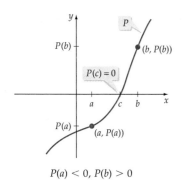

$P(a) < 0, P(b) > 0$

Figure 4.17

THE ZERO LOCATION THEOREM

Let $P(x)$ be a polynomial. If $a < b$, and if $P(a)$ and $P(b)$ have opposite signs, then there is at least one value c between a and b such that $P(c) = 0$.

If, for instance, the value of a polynomial P is negative at $x = a$ and positive at $x = b$, then there is at least one real number c between a and b such that $P(c) = 0$. See **Figure 4.17**.

EXAMPLE 3   Apply the Zero Location Theorem

Verify that $S(x) = x^3 - x - 2$ has a real zero between 1 and 2.

Solution

Use synthetic division to evaluate S for $x = 1$ and $x = 2$.

$$
\begin{array}{r|rrrr}
1 & 1 & 0 & -1 & -2 \\
  &   & 1 & 1 & 0 \\
\hline
  & 1 & 1 & 0 & -2
\end{array}
$$
• $S(1)$ is negative.

$$
\begin{array}{r|rrrr}
2 & 1 & 0 & -1 & -2 \\
  &   & 2 & 4 & 6 \\
\hline
  & 1 & 2 & 3 & 4
\end{array}
$$
• $S(2)$ is positive.

Because $S(1)$ is negative and $S(2)$ is positive, by the Zero Location Theorem there is a real zero between 1 and 2. Note in **Figure 4.18** that the graph of S has an x-intercept between 1 and 2.

TRY EXERCISE 34, EXERCISE SET 4.2

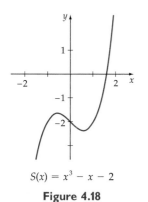

$S(x) = x^3 - x - 2$

Figure 4.18

The following theorem expresses important relationships among the real zeros of a polynomial, the x-intercepts of its graph, and its factors that can be written in the form $(x - c)$, where c is a real number.

POLYNOMIALS, REAL ZEROS, GRAPHS, AND FACTORS $(x - c)$

If P is a polynomial and c is a real number, then all of the following statements are equivalent in the sense that if any one statement is true, then they are all true, and if any one statement is false, then they are all false.

• $(x - c)$ is a factor of P.

• $x = c$ is a real solution of $P(x) = 0$.

• $x = c$ is a real zero of P.

• $(c, 0)$ is an x-intercept of the graph of $y = P(x)$.

Sometimes it is possible to make use of the preceding theorem and a graph of a polynomial to factor the polynomial. For example, the graph of

$$S(x) = x^3 - 2x^2 - 5x + 6$$

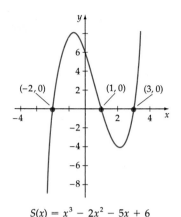

$$S(x) = x^3 - 2x^2 - 5x + 6$$

Figure 4.19

is shown in **Figure 4.19**. The x-intercepts are $(-2, 0)$, $(1, 0)$, and $(3, 0)$. Hence -2, 1, and 3 are zeros of S, and $[x - (-2)]$, $(x - 1)$, and $(x - 3)$ are all factors of S. Thus, in factored form,

$$S(x) = (x + 2)(x - 1)(x - 3)$$

TOPICS FOR DISCUSSION

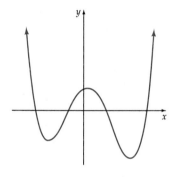

Figure 4.20

1. Discuss the meaning of the phrase *polynomial function*. Give examples of polynomial functions and of functions that are not polynomials.

2. Is it possible for the graph of the polynomial shown in **Figure 4.20** to have a degree that is an odd number? If so, explain how. If not, explain why not.

3. Explain the difference between a local minimum and an absolute minimum and the difference between a local maximum and an absolute maximum.

4. Discuss how the Zero Location Theorem can be used to find a real zero of a polynomial.

5. A complex number may be a zero of a polynomial. For instance, i is a zero of $P(x) = x^2 + 1$ because $P(i) = 0$. Explain why the Zero Location Theorem cannot be used to find the complex number zeros of a polynomial.

6. Let $P(x)$ be a polynomial with real coefficients. Explain the relationship among a real zero of a polynomial, the x-coordinate of the x-intercept of the graph of the polynomial, and the solution of the equation $P(x) = 0$.

EXERCISE SET 4.2

In Exercises 1 to 10, examine the leading term and determine the far-left and far-right behavior of the graph of the polynomial function.

1. $P(x) = 3x^4 - 2x^2 - 7x + 1$

2. $P(x) = -2x^3 - 6x^2 + 5x - 1$

3. $P(x) = 5x^5 - 4x^3 - 17x^2 + 2$

4. $P(x) = -6x^4 - 3x^3 + 5x^2 - 2x + 5$

5. $P(x) = 2 - 3x - 4x^2$

6. $P(x) = -16 + x^4$

7. $P(x) = \dfrac{1}{2}(x^3 + 5x^2 - 2)$

8. $P(x) = -\dfrac{1}{4}(x^4 + 3x^3 - 2x + 6)$

9. $P(x) = -\dfrac{2}{3}(x + 1)^3$

10. $P(x) = \dfrac{1}{5}(x - 1)^4$

In Exercises 11 to 14, use your knowledge of the vertex of a parabola to find the maximum or minimum of each quadratic function.

11. $P(x) = x^2 + 4x - 1$

12. $P(x) = x^2 + 6x + 1$

13. $P(x) = -x^2 - 8x + 1$

14. $P(x) = -2x^2 + 8x - 1$

15. Maximize an Area Function A farmer has $1000 to spend to fence a rectangular corral. Because extra reinforcing is needed for one side, the corral cost $6 per foot along that side. It costs $2 per foot to fence the remaining three sides. What dimensions of the corral will maximize the area of the corral?

16. Maximize a Profit A manufacturer has determined that the revenue received from selling x items of a product is given by $R(x) = -\dfrac{1}{10}x^2 + 90x$ and the cost to produce x items of the product is $C(x) = 40x + 2000$. Assuming that all of the products produced can be sold, how many should be produced to maximize profit? *Suggestion:* profit = revenue − cost

17. Minimize a Sum Let $S = (3 - x)^2 + (7 - x)^2 + (8 - x)^2$. Find the number x for which the value of S is a minimum. Now show that x is the average of 3, 7, and 8.

18. Minimize a Distance Find the coordinates (x, y) of the point on the graph of $y = -x + 4$ that is closest to the origin. *Suggestion:* The *square* of the distance from the point is $D^2 = x^2 + y^2$. Replace y by $-x + 4$ and find the point that minimizes the square of the distance. Then use the fact that the point that minimizes the *square* of the distance also minimizes the distance.

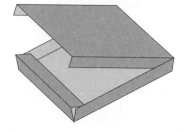 In Exercises 19 to 24, use a graphing utility to graph each polynomial. Now use the **TRACE** feature to estimate, to the nearest tenth, the coordinates of the local maximum and local minimum. The number in parentheses to the right of the polynomial is the total number of local maxima and minima.

19. $P(x) = x^3 + x^2 - 9x - 9$ (2)

20. $P(x) = x^3 + 4x^2 - 4x - 16$ (2)

21. $P(x) = x^3 - 3x^2 - 24x + 3$ (2)

22. $P(x) = -2x^3 - 3x^2 + 12x + 1$ (2)

23. $P(x) = x^4 - 4x^3 - 2x^2 + 12x - 5$ (3)

24. $P(x) = x^4 - 10x^2 + 9$ (3)

25. **Maximizing Volume** An open box is to be constructed from a rectangular sheet of cardboard that measures 16 inches by 22 inches. It is made by cutting along the long side of the rectangles that measure x inches by $2x$ inches at each corner of the rectangle, as shown at the top of the next column. What value of x (to the nearest 0.001 inch) will produce a box with maximum volume? What is the maximum volume (to the nearest 0.1 cubic inch)?

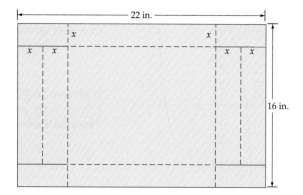

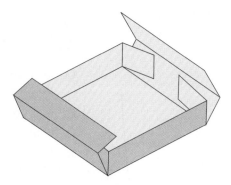

26. **Maximizing Volume** A closed box is to be constructed from a rectangular sheet of cardboard that measures 18 inches by 42 inches. The box is made by cutting rectangles that measure x inches by $2x$ inches from two of the corners and by cutting two squares that measure x inches by x inches from the top and from the bottom of the rectangle, as shown in the following figure. What value of x (to the nearest 0.001 inch) will produce a box with maximum volume? What is the maximum volume (to the nearest 0.1 cubic inch)?

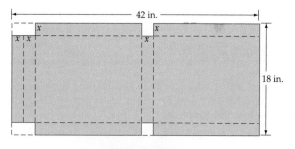

27. ▥ **MAXIMIZING VELOCITY** A car traveled at a velocity (kilometers per hour) of

$$v(t) = 0.038(t + 2)(t - 18)(t - 21) + 48$$

where t is the time in minutes after 1 P.M. What is the maximum velocity (to the nearest 0.1 kilometer per hour) that the car attained during the period $0 \leq t \leq 24$? At what time (to the nearest 0.01 minute) did it reach this maximum velocity?

28. ▥ **MINIMIZING VELOCITY** A plane traveled at a velocity (mph) of

$$v(t) = 0.00182(t + 2)(t - 48)(t - 31)(t + 9) + 378$$

where t is the time in minutes after 8 A.M. What is the minimum velocity (to the nearest mph) that the plane attained during the period $0 \leq t \leq 50$? At what time (to the nearest 0.1 minute) did it reach this minimum velocity?

Exercises 29 and 30 refer to the following mathematical model. The Fahrenheit temperature T in a city during a particular 12-hour period can be modeled by

$$T(t) = 0.051(t - 1)(t - 11)(t - 14) + 44$$

where t is the time in hours after 9 A.M.

29. ▥ **MAXIMIZING TEMPERATURE** What was the maximum temperature (to the nearest degree), and at what time (to the nearest minute) was it achieved?

30. ▥ **MINIMIZING TEMPERATURE** What was the minimum temperature (to the nearest degree), and at what time (to the nearest minute) was it achieved?

31. ▥ **PHYSIOLOGY** The velocity of the air that is expelled during a cough can be modeled by $v = 0.6r^2 - r^3$, where v is measured in centimeters per second and r is the radius of the trachea in centimeters. Find the radius of the trachea, to the nearest 0.1 centimeter, that maximizes the velocity of the air. Physical considerations of the structure of the trachea indicate that $0.3 \leq r \leq 0.6$ is an appropriate domain for r.

32. ▥ **PHYSIOLOGY** The rate at which a patient's blood pressure changes as a medication is absorbed into the blood can be modeled by $R = (8 - A)A$, where A is the number of milligrams of medication absorbed and R is the rate, in millimeters of mercury per minute, at which a patient's blood pressure is changing. At what amount of absorption will the patient's blood pressure change most rapidly?

In Exercises 33 to 38, use the Zero Location Theorem to verify that P has a zero between a and b.

33. $P(x) = 2x^3 + 3x^2 - 23x - 42; \quad a = 3, b = 4$

34. $P(x) = 4x^3 - x^2 - 6x + 1; \quad a = 0, b = 1$

35. $P(x) = 3x^3 + 7x^2 + 3x + 7; \quad a = -3, b = -2$

36. $P(x) = 2x^3 - 21x^2 - 2x + 21; \quad a = 10, b = 11$

37. $P(x) = 4x^4 + 7x^3 - 11x^2 + 7x - 15; \quad a = 1, b = 1\frac{1}{2}$

38. $P(x) = 5x^3 - 16x^2 - 20x + 64; \quad a = 3, b = 3\frac{1}{2}$

In Exercises 39 to 48, find the real zeros of the polynomial function.

39. $P(x) = (x + 2)(x - 3)(2x + 7)$

40. $P(x) = (x - 5)(x - 1)(4x + 1)$

41. $P(x) = x(x - 1)(5x - 2)$

42. $P(x) = x(x - 4)(x - 1)(x + 7)$

43. $P(x) = (3x + 7)(2x - 11)(x + 5)^2$

44. $P(x) = (x - 3)^2(2x - 1)$

45. $P(x) = x^3 + x^2 - 6x$ **46.** $P(x) = 2x^3 - 7x^2 - 15x$

47. $P(x) = x^3 - 1$ **48.** $P(x) = x^3 + 8$

▥ **In Exercises 49 to 52, use a graphing utility to estimate the real zeros of the polynomial by finding the x-coordinate of the x-intercept to the nearest tenth. Then verify the result by using the quadratic formula. *Suggestion:* When finding intercepts with a graphing utility, you can set Ymin = −1 and Ymax = 1, because you are interested only in the behavior of the graph as it passes through the x-axis.**

49. $P(x) = x^2 + 2x - 1$ **50.** $P(x) = x^2 - 2x - 2$

51. $P(x) = x^2 + 6x - 1$ **52.** $P(x) = x^2 - x - 4$

SUPPLEMENTAL EXERCISES

53. On the same set of coordinate axes, sketch the graph of $f(x) = x^n$ for $-1 \leq x \leq 1$ for each value of n.

     a. $n = 2$ **b.** $n = 4$ **c.** $n = 6$

54. ▨ Use the result of Exercise 53 to make a conjecture about the graph of $y = x^n$ over the interval $-1 \leq x \leq 1$, where n is a large positive even integer.

55. ▨ Graph $P(x) = (x - 2)^n$ for $n = 1, 2, 3, 4, 5$, and 6. Explain the relationship between the exponent and the way the graph intersects or touches the x-axis.

56. Based on Exercise 55, determine where the graph of each function intersects the x-axis or is tangent to the x-axis.

     a. $f(x) = (x + 2)^3(x - 1)^2$

     b. $f(x) = (x - 1)(x + 2)^2(x - 3)^3$

57. Graph $P(x) = x^3 - x - 25$ and determine between which two consecutive integers P has a real zero.

58. Graph $P(x) = 4x^4 - 12x^3 + 13x^2 - 12x + 9$ and determine between which two consecutive integers P has a real zero.

59. Consider the following conjecture. Let $P(x)$ be a polynomial. If $a < b$, $P(a) > 0$, and $P(b) > 0$, then $P(x)$ does not have a real zero between a and b. Is this conjecture true or false? Support your answer.

60. Let $f(x) = x^3 + c$. On the same coordinate axes, sketch the graph of f for each value of c.

  a. $c = 0$ **b.** $c = 2$ **c.** $c = -3$

61. Let $f(x) = ax^3$. On the same coordinate axes, sketch the graph of f for each value of a.

  a. $a = 2$ **b.** $a = 1/2$ **c.** $a = -1$

62. Let $f(x) = (x - h)^3$. On the same coordinate axes, sketch the graph of f for each value of h.

  a. $h = 2$ **b.** $h = -1$ **c.** $h = -5$

63. Explain how the graph of the equation $f(x) = a(x - h)^3 + c$ compares with the graph of $g(x) = x^3$.

PROJECTS

1. **REAL ZEROS AND THE DEGREE OF A POLYNOMIAL** This project examines the connection between the number of real zeros of a polynomial and its degree.

  a. Graph the polynomials

$$R(x) = x^4 + 1, Q(x) = x^4 - x^3 - 11x^2 - x - 12$$

  and

$$P(x) = x^4 - 2x^3 - 13x^2 + 14x + 24$$

  How many real zeros does each of the polynomials $P(x)$, $Q(x)$, and $R(x)$ have? Graph some other polynomials of degree 4. On the basis of your graphs, make a conjecture about the number of real zeros a polynomial of degree 4 can have.

  b. Graph the polynomials

$$R(x) = x^5, P(x) = x^5 - 3x^4 - 11x^3 + 27x^2 + 10x - 24$$

  and

$$Q(x) = x^5 - 2x^4 - 10x^3 + 10x^2 - 11x + 12$$

  How many real zeros do the polynomials $P(x)$, $Q(x)$, and $R(x)$ have? Graph some other polynomials of degree 5. On the basis of your graphs, make a conjecture about the number of real zeros a polynomial of degree 5 can have.

  c. Graph some second-degree and third-degree polynomials, and note the number of real zeros of the polynomials. On the basis of all your graphs and the graphs in **a.** and **b.**, make a conjecture about the number of real zeros a polynomial of degree n can have.

SECTION

4.3 ZEROS OF POLYNOMIAL FUNCTIONS

Recall that if $P(x)$ is a polynomial function, then the values of x for which $P(x)$ is equal to 0 are called the *zeros* of $P(x)$ or the *roots* of the equation $P(x) = 0$. A zero of a polynomial may be a **multiple zero**. For example, the polynomial $x^2 + 6x + 9$ can be expressed in factored form as $(x + 3)(x + 3)$. Setting each factor equal to zero yields $x = -3$ in both cases. Thus $x^2 + 6x + 9$ has a zero of -3 that occurs twice. The following definition will be most useful when we are discussing multiple zeros.

DEFINITION OF MULTIPLE ZEROS OF A POLYNOMIAL

If a polynomial $P(x)$ has $(x - r)$ as a factor exactly k times, then r is a **zero of multiplicity k** of the polynomial $P(x)$.

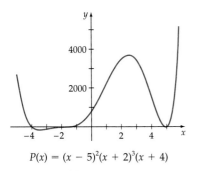

$$P(x) = (x - 5)^2(x + 2)^3(x + 4)$$

Figure 4.21

The graph of the polynomial

$$P(x) = (x - 5)^2(x + 2)^3(x + 4)$$

is shown in **Figure 4.21.** This polynomial has

- 5 as a zero of multiplicity 2
- -2 as a zero of multiplicity 3
- -4 as a zero of multiplicity 1

A zero of multiplicity 1 is generally referred to as a **simple zero.**

When searching for the zeros of a polynomial function, it is important that we know how many zeros to expect. This question is answered completely in Section 4.4. For the work in this section, the following result is valuable.

NUMBER OF ZEROS OF A POLYNOMIAL FUNCTION

A polynomial function P of degree n has at most n zeros, where each zero of multiplicity k is counted k times.

The rational zeros of polynomials with integer coefficients can be found with the aid of the following theorem.

THE RATIONAL ZERO THEOREM

If $P(x) = a_nx^n + a_{n-1}x^{n-1} + \cdots + a_1x + a_0$ has integer coefficients, and p/q (where p and q have no common prime factors) is a rational zero of $P(x)$, then p is a factor of a_0 and q is a factor of a_n.

The Rational Zero Theorem often is used to make a list of all possible rational zeros of a polynomial. The list consists of all rational numbers of the form p/q, where p is an integer factor of the constant term a_0, and q is an integer factor of the leading coefficient a_n.

> **EXAMPLE 1**   *Apply the Rational Zero Theorem*

Use the Rational Zero Theorem to list all possible rational zeros of

$$P(x) = 4x^4 + x^3 - 40x^2 + 38x + 12$$

Solution

List all integers p that are factors of 12 and all integers q that are factors of 4.

$$p: \quad \pm 1, \pm 2, \pm 3, \pm 4, \pm 6, \pm 12$$
$$q: \quad \pm 1, \pm 2, \pm 4$$

Form all possible rational numbers using ± 1, ± 2, ± 3, ± 4, ± 6, and ± 12 for the numerator and ± 1, ± 2, and ± 4 for the denominator. By the Rational Zero Theorem, the possible rational zeros are

$$\pm 1, \pm \frac{1}{2}, \pm \frac{1}{4}, \pm 2, \pm 3, \pm \frac{3}{2}, \pm \frac{3}{4}, \pm 4, \pm 6, \pm 12$$

> **TRY EXERCISE 12, EXERCISE SET 4.3**

It is not necessary to list a factor that is already listed in reduced form. For example, $\pm 6/4$ is not listed because it is equal to $\pm 3/2$.

TAKE NOTE

The Rational Zero Theorem gives the *possible* rational zeros of a polynomial. That is, if P has a rational zero, then it must be one indicated by the theorem. However, P may not have any rational zeros. In the case of the polynomial in Example 1, the only rational zeros are $-1/4$ and 2. The remaining rational numbers in the list are not zeros of P.

UPPER AND LOWER BOUNDS FOR REAL ZEROS

A real number b is called an **upper bound** of the zeros of the polynomial function P if no zero is greater than b. A real number a is called a **lower bound** of the zeros of P if no zero is less than a. The following theorem is often used to find positive upper bounds and negative lower bounds for the real zeros of a polynomial function.

> **UPPER- AND LOWER-BOUND THEOREM**
>
> **Upper bound**   If $b > 0$ and all the numbers in the bottom row of the synthetic division of P by $x - b$ are either positive or zero, then b is an upper bound for the real zeros of P.
>
> **Lower bound**   If $a < 0$ and the numbers in the bottom row of the synthetic division of P by $x - a$ alternate in sign (the number zero can be considered positive or negative), then a is a lower bound for the real zeros of P.

Upper and lower bounds are not unique. For example, if b is an upper bound for the real zeros of P, then any number greater than b is also an upper bound. Also, if a is a lower bound for the real zeros of P, then any number less than a is also a lower bound.

EXAMPLE 2 *Find Upper and Lower Bounds*

According to the Upper- and Lower-Bound Theorem, what is the smallest positive integer that is an upper bound and the largest negative integer that is a lower bound of the zeros of $P(x) = 2x^3 + 7x^2 - 4x - 14$?

Solution

To find the smallest positive-integer upper bound, use synthetic division with $1, 2, \ldots,$ as test values.

$$
\begin{array}{r|rrrr}
1 & 2 & 7 & -4 & -14 \\
  &   & 2 & 9 & 5 \\
\hline
  & 2 & 9 & 5 & -9
\end{array}
\qquad
\begin{array}{r|rrrr}
2 & 2 & 7 & -4 & -14 \\
  &   & 4 & 22 & 36 \\
\hline
  & 2 & 11 & 18 & 22
\end{array}
$$
• **All positive signs**

Thus 2 is the smallest positive-integer upper bound.
 Now find the largest negative-integer lower bound.

$$
\begin{array}{r|rrrr}
-1 & 2 & 7 & -4 & -14 \\
   &   & -2 & -5 & 9 \\
\hline
   & 2 & 5 & -9 & -5
\end{array}
\qquad
\begin{array}{r|rrrr}
-2 & 2 & 7 & -4 & -14 \\
   &   & -4 & -6 & 20 \\
\hline
   & 2 & 3 & -10 & 6
\end{array}
$$

$$
\begin{array}{r|rrrr}
-3 & 2 & 7 & -4 & -14 \\
   &   & -6 & -3 & 21 \\
\hline
   & 2 & 1 & -7 & 7
\end{array}
\qquad
\begin{array}{r|rrrr}
-4 & 2 & 7 & -4 & -14 \\
   &   & -8 & 4 & 0 \\
\hline
   & 2 & -1 & 0 & -14
\end{array}
$$
• **Alternating signs**

Thus -4 is the largest negative-integer lower bound.

TRY EXERCISE 24, EXERCISE SET 4.3

You can use the Upper- and Lower-Bound Theorem in conjunction with a graphing utility to determine Xmin (the lower bound) and Xmax (the upper bound) for the viewing window. This will ensure that all the real zeros, which are the x-coordinates of the x-intercepts of the polynomial, will be shown. Note in **Figure 4.22** that the x-intercepts of the graph of $P(x) = 2x^3 + 7x^2 - 4x - 14$ (the zeros of $P(x)$) are between -4 (the lower bound) and 2 (the upper bound).

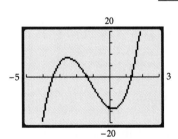

$P(x) = 2x^3 + 7x^2 - 4x - 14$

Figure 4.22

DESCARTES' RULE OF SIGNS

Descartes' Rule of Signs is another theorem often used to obtain information about the zeros of a polynomial. In Descartes' Rule of Signs, the number of *variations in sign* of the coefficients of a polynomial $P(x)$ or $P(-x)$ refers to sign changes of the coefficients from positive to negative or from negative to positive that we find when we examine successive terms of the polynomial. The terms of the polynomial are assumed to appear in the order of descending powers of x. For example, the polynomial

$$P(x) = +3x^4 - 5x^3 - 7x^2 + x - 7$$

has three variations of sign. The polynomial

$$P(-x) = +3(-x)^4 - 5(-x)^3 - 7(-x)^2 + (-x) - 7$$
$$= +\ 3x^4 + 5x^3 - 7x^2 - x - 7$$

has one variation in sign.

Terms that have a coefficient of 0 are not counted as a variation of sign and may be ignored. For example,

$$P(x) = -x^5 + 4x^2 + 1$$

has one variation in sign.

DESCARTES' RULE OF SIGNS

Let $P(x)$ be a polynomial with real coefficients and with the terms arranged in the order of decreasing powers of x.

1. The number of positive real zeros of $P(x)$ is equal to the number of variations in sign of $P(x)$ or is equal to that number decreased by an even integer.

2. The number of negative real zeros of $P(x)$ is equal to the number of variations in sign of $P(-x)$ or is equal to that number decreased by an even integer.

EXAMPLE 3   *Apply Descartes' Rule of Signs*

Determine both the number of possible positive and the number of possible negative real zeros of each polynomial.

a. $x^4 - 5x^3 + 5x^2 + 5x - 6$ **b.** $2x^5 + 3x^3 + 5x^2 + 8x + 7$

Solution

a.
$$P(x) = +x^4 - 5x^3 + 5x^2 + 5x - 6$$

There are three variations of sign. By Descartes' Rule of Signs, there are either three or one positive real zeros. Now examine the variations of sign of $P(-x)$.

$$P(-x) = x^4 + 5x^3 + 5x^2 - 5x - 6$$

$$1$$

There is one variation of sign of $P(-x)$. By Descartes' Rule of Signs, there is one negative real zero.

b. $P(x) = 2x^5 + 3x^3 + 5x^2 + 8x + 7$ has no variation of sign, so there are no positive real zeros.

$$P(-x) = -2x^5 - 3x^3 + 5x^2 - 8x + 7$$

$$1 \qquad 2 \qquad 3$$

$P(-x)$ has three variations of sign, so there are either three or one negative real zeros.

TRY EXERCISE 36, EXERCISE SET 4.3

In applying Descartes' Rule of Signs, we count each zero of multiplicity k as k zeros. For instance, the polynomial

$$P(x) = x^2 - 10x + 25$$

has two variations in sign. Thus by Descartes' Rule of Signs it must have either two or zero positive real zeros. Factoring the polynomial produces $(x - 5)^2$, from which it can be observed that 5 is a positive zero of multiplicity 2.

EXAMPLE 4   *Find the Zeros of a Polynomial*

Find the zeros of $P(x) = 3x^4 + 23x^3 + 56x^2 + 52x + 16$.

Solution

By Descartes' Rule of Signs there are no positive real zeros, and there are either four, two, or no negative real zeros. By the Rational Zero Theorem, the possible negative rational zeros are

$$\frac{p}{q}: \quad -1, -2, -4, -8, -16, -\frac{1}{3}, -\frac{2}{3}, -\frac{4}{3}, -\frac{8}{3}, -\frac{16}{3}$$

We use synthetic division to test possible zeros. We show only the work for values that are zeros.

$$x = -4: \quad -4 \begin{array}{|rrrrr} 3 & 23 & 56 & 52 & 16 \\ & -12 & -44 & -48 & -16 \\ \hline 3 & 11 & 12 & 4 & 0 \end{array}$$

• -4 is a zero.

Continued • ▶

TAKE NOTE

If you have a graphing utility, you can produce a graph similar to the one below. By looking at the x-intercepts of the graph, you can reject as possible zeros some of the values suggested by the Rational Zero Theorem. This will reduce the amount of work that is necessary to find the zeros of the polynomial.

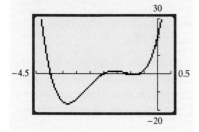

Because -4 is a zero, the factors of $P(x)$ are $(x + 4)$ and the reduced polynomial $(3x^3 + 11x^2 + 12x + 4)$. Thus

$$P(x) = (x + 4)(3x^3 + 11x^2 + 12x + 4)$$

All remaining zeros must be zeros of $3x^3 + 11x^2 + 12x + 4$. The Rational Zero Theorem indicates that the only possible negative rational zeros are

$$\frac{p}{q}: \quad -1, -2, -4, -\frac{1}{3}, -\frac{2}{3}, -\frac{4}{3}$$

Synthetic division is again used to test possible zeros.

$$
\begin{array}{r|rrrr}
-2 & 3 & 11 & 12 & 4 \\
   &   & -6 & -10 & -4 \\
\hline
   & 3 & 5 & 2 & 0
\end{array}
$$
 • **-2 is a zero.**

Because -2 is a zero, $(x + 2)$ is also a factor of P. Thus we have

$$P(x) = (x + 4)(x + 2)(3x^2 + 5x + 2)$$

The remaining zeros are zeros of $3x^2 + 5x + 2$.

$$3x^2 + 5x + 2 = 0$$
$$(3x + 2)(x + 1) = 0$$
$$x = -\frac{2}{3} \quad \text{and} \quad x = -1$$

The zeros of $3x^4 + 23x^3 + 56x^2 + 52x + 16$ are $-4, -2, -1, -2/3$.

TRY EXERCISE 48, EXERCISE SET 4.3

TOPICS FOR DISCUSSION

1. What is a multiple zero of a polynomial? Give an example of a polynomial that has -2 as a multiple zero.

2. Discuss how the Rational Zero Theorem is used.

3. In Topics 4 and 5, we talk about polynomials with integer coefficients and those with real coefficients. Discuss the similarities and differences between these two types of polynomials.

4. Let $P(x)$ be a polynomial with real coefficients. Explain why $(a, 0)$ is an x-intercept of the graph of $P(x)$ if a is a real zero of $P(x)$.

5. Let $P(x)$ be a polynomial with integer coefficients. Suppose that the Rational Zero Theorem is applied to $P(x)$ and that after testing each possible rational zero it is determined that the polynomial has no rational zero. Does this mean that all of the zeros of the polynomial are irrational numbers?

EXERCISE SET 4.3

In Exercises 1 to 10, find the zeros of the polynomial and state the multiplicity of each zero.

1. $P(x) = (x - 3)^2(x + 5)$

2. $P(x) = (x + 4)^3(x - 1)^2$

3. $P(x) = x^2(3x + 5)^2$

4. $P(x) = x^3(2x + 1)(3x - 12)^2$

5. $P(x) = (x^2 - 4)(x + 3)^2$

6. $P(x) = (x + 4)^3(x^2 - 9)^2$

7. $P(x) = (x^2 - 3x - 10)^2$

8. $P(x) = (x^3 - 4x)(2x - 7)^2$

9. $P(x) = x^4 - 10x^2 + 9$

10. $P(x) = x^4 - 12x^2 + 32$

In Exercises 11 to 22, use the Rational Zero Theorem to list possible rational zeros for each polynomial.

11. $x^3 + 3x^2 - 6x - 8$ 12. $x^3 - 19x - 30$

13. $2x^3 + x^2 - 25x + 12$ 14. $3x^3 + 11x^2 - 6x - 8$

15. $6x^4 + 23x^3 + 19x^2 - 8x - 4$

16. $6x^4 + 23x^3 + 15x^2 - 23x - 21$

17. $2x^3 + 9x^2 - 2x - 9$

18. $2x^4 + 11x^3 + 21x^2 + 17x + 5$

19. $4x^4 - 12x^3 - 3x^2 + 12x - 7$

20. $x^5 - x^4 - 7x^3 + 7x^2 - 12x - 12$

21. $x^5 - 32$ 22. $x^4 - 1$

In Exercises 23 to 34, find the smallest positive integer and the largest negative integer that, by the Upper- and Lower-Bound Theorem, are upper and lower bounds for the real zeros of each polynomial.

23. $x^3 + 3x^2 - 6x - 6$ 24. $x^3 - 19x - 28$

25. $2x^3 + x^2 - 25x + 10$ 26. $3x^3 + 11x^2 - 6x - 9$

27. $6x^4 + 23x^3 + 19x^2 - 8x - 4$

28. $6x^4 + 23x^3 + 15x^2 - 23x - 21$

29. $2x^3 + 9x^2 - 2x - 9$

30. $2x^4 + 11x^3 + 21x^2 + 17x + 5$

31. $4x^4 - 12x^3 - 3x^2 + 12x - 7$

32. $x^5 - x^4 - 7x^3 + 7x^2 - 12x - 12$

33. $x^5 - 32$ 34. $x^4 - 1$

In Exercises 35 to 46, use Descartes' Rule of Signs to state the number of possible positive and negative real zeros of each polynomial.

35. $x^3 + 3x^2 - 6x - 8$ 36. $x^3 - 19x - 30$

37. $2x^3 + x^2 - 25x + 12$ 38. $3x^3 + 11x^2 - 6x - 8$

39. $6x^4 + 23x^3 + 19x^2 - 8x - 4$

40. $6x^4 + 23x^3 + 15x^2 - 23x - 21$

41. $2x^3 + 9x^2 - 2x - 9$

42. $2x^4 + 11x^3 + 21x^2 + 17x + 5$

43. $4x^4 - 12x^3 - 3x^2 + 12x - 7$

44. $x^5 - x^4 - 7x^3 + 7x^2 - 12x - 12$

45. $x^5 - 32$ 46. $x^4 - 1$

In Exercises 47 to 64, find the zeros of each polynomial. If a zero is a multiple zero, state its multiplicity.

47. $x^3 + 3x^2 - 6x - 8$ 48. $x^3 - 19x - 30$

49. $2x^3 + x^2 - 25x + 12$ 50. $3x^3 + 11x^2 - 6x - 8$

51. $6x^4 + 23x^3 + 19x^2 - 8x - 4$

52. $6x^4 + 23x^3 + 15x^2 - 23x - 21$

53. $2x^3 + 9x^2 - 2x - 9$

54. $2x^4 + 11x^3 + 21x^2 + 17x + 5$

55. $2x^4 - 9x^3 - 2x^2 + 27x - 12$

56. $3x^3 - x^2 - 6x + 2$

57. $x^3 - 3x - 2$

58. $3x^4 - 4x^3 - 11x^2 + 16x - 4$

59. $x^4 - 5x^2 - 2x$

60. $x^3 - 2x + 1$

61. $x^4 + x^3 - 3x^2 - 5x - 2$

62. $6x^4 - 17x^3 - 11x^2 + 42x$

63. $2x^4 - 17x^3 + 4x^2 + 35x - 24$

64. $x^5 + 5x^4 + 10x^3 + 10x^2 + 5x + 1$

🖳 **Recall that a real zero of a polynomial is the x-coordinate of the x-intercept of the graph of the polynomial. Use this fact and a graphing utility to find, to the nearest tenth, the real zeros of each polynomial. The number in parentheses is the number of real zeros.**

65. $P(x) = x^4 + x^3 - 21x^2 - x + 20$; (4)

66. $P(x) = x^4 - x^3 - 16x^2 + 4x + 48$; (4)

67. $P(x) = 4x^4 - 8x^3 - 39x^2 + 43x + 70$; (4)

68. $P(x) = 4x^4 + 4x^3 - 43x^2 - 22x + 21$; (4)

69. $P(x) = 4x^5 - 28x^4 - 3x^3 + 280x^2 - 283x - 210$; (5)

70. $P(x) = 8x^5 + 18x^4 - 99x^3 - 148x^2 + 141x + 180$; (5)

71. $P(x) = 2x^4 - 3x^3 + 6x^2 - 12x - 8$; (2)

72. $P(x) = 2x^4 + 3x^3 + 2x^2 - x - 6$; (2)

SUPPLEMENTAL EXERCISES

In Exercises 73 to 78, verify that each polynomial has no rational zeros.

73. $x^4 - 2x^3 + 11x^2 - 2x + 10$

74. $x^4 - 2x^3 + 21x^2 - 2x + 20$

75. $2x^4 + x^2 + 5$

76. $4x^4 + 14x^2 + 5$

77. $x^4 - 4x^3 + 14x^2 - 4x + 13$

78. $x^6 + 3x^4 + 3x^2 + 1$

In Exercises 79 to 82, determine whether the given polynomial satisfies the following theorem.

Theorem   **Let $P(x) = a_n x^n + a_{n-1} x^{n-1} + \cdots + a_1 x + a_0$ be a polynomial with integer coefficients and $n \geq 2$. If a_n, a_0, and $P(1)$ are all odd, then $P(x)$ has no rational zeros.**

79. $x^5 + 2x^4 + x^3 - x^2 + x + 945$

80. $5x^3 - 2x^2 - x + 1815$

81. $3x^4 - 5x^3 + 6x^2 - 2x + 9009$

82. $15x^7 - 4x^3 + x^2 - 6075$

PROJECTS

1. **WRITE A PROOF** Use the Rational Zero Theorem to prove that $\sqrt{p}$ is an irrational number, where p is a prime number.

SECTION

4.4   THE FUNDAMENTAL THEOREM OF ALGEBRA

The German mathematician Carl Friedrich Gauss (1777–1855) was the first to prove that every polynomial has at least one complex zero. This concept is so basic to the study of algebra that it is called the **Fundamental Theorem of Algebra.** The proof of the Fundamental Theorem is beyond the scope of this text; however, it is important to understand the theorem and its consequences. As you consider each of the following theorems, keep in mind that the terms *complex coefficients* and *complex zeros* include real coefficients and real zeros, because the set of real numbers is a subset of the set of complex numbers.

THE FUNDAMENTAL THEOREM OF ALGEBRA

If $P(x)$ is a polynomial of degree $n \geq 1$ with complex coefficients, then $P(x)$ has at least one complex zero.

Let $P(x)$ be a polynomial of degree $n \geq 1$ with complex coefficients. The Fundamental Theorem implies that $P(x)$ has a complex zero—say, c_1. The Factor Theorem implies that

$$P(x) = (x - c_1)Q(x)$$

where $Q(x)$ is a polynomial of degree one less than the degree of $P(x)$. Recall that the polynomial $Q(x)$ is called a reduced polynomial. Assuming that the degree of $Q(x)$ is 1 or more, the Fundamental Theorem implies that it must also have a zero. A continuation of this reasoning process leads to the following theorem, which is a corollary of the Fundamental Theorem.

THE NUMBER OF ZEROS OF A POLYNOMIAL

If $P(x)$ is a polynomial of degree $n \geq 1$ with complex coefficients, then $P(x)$ has exactly n complex zeros, provided that each zero is counted according to its multiplicity.

Even though every polynomial of nth degree has exactly n zeros, the zeros may not be distinct. For example, the third-degree polynomial

$$x^3 - 5x^2 + 3x + 9$$

factors into

$$(x + 1)(x - 3)(x - 3)$$

which has zeros -1, 3, and 3. The zero 3 is a zero of multiplicity 2.

Although the Fundamental Theorem and its corollary give information about the existence and the number of zeros of a polynomial, they do not provide a method of actually finding the zeros. If a polynomial has real coefficients, then the following theorem can help us determine the zeros of the polynomial.

THE CONJUGATE PAIR THEOREM

If $a + bi$ ($b \neq 0$) is a complex zero of the polynomial $P(z)$, *with real coefficients,* then the conjugate $a - bi$ is also a complex zero of the polynomial.

EXAMPLE 1   *Use the Conjugate Pair Theorem to Find Zeros*

Find all the zeros of $x^4 - 4x^3 + 14x^2 - 36x + 45$ given that $2 + i$ is a zero.

Solution

Because the coefficients are real numbers and $2 + i$ is a zero, the Conjugate Pair Theorem implies that $2 - i$ must also be a zero. Using

Continued •➤

synthetic division with $2 + i$ and then $2 - i$, we have

$$
\begin{array}{r|rrrr}
2+i & 1 & -4 & 14 & -36 & 45 \\
 & & 2+i & -5 & 18+9i & -45 \\
\hline
 & 1 & -2+i & 9 & -18+9i & 0 \\
\end{array}
$$

$$
\begin{array}{r|rrrr}
2-i & 1 & -2+i & 9 & -18+9i \\
 & & 2-i & 0 & 18-9i \\
\hline
 & 1 & 0 & 9 & 0 \\
\end{array}
$$

- The coefficients of the reduced polynomial

- The coefficients of the next reduced polynomial

The resulting reduced polynomial is $x^2 + 9$, which has $3i$ and $-3i$ as zeros. Therefore, the four zeros of $x^4 - 4x^3 + 14x^2 - 36x + 45$ are $2 + i$, $2 - i$, $3i$, and $-3i$.

TRY EXERCISE 2, EXERCISE SET 4.4

Recall that the real zeros of a polynomial P are the x-coordinates of the x-intercepts of the graph of P. This important connection between real zeros of a polynomial and x-intercepts of the graph of the polynomial is the basis for using a graphing utility to solve equations. Careful analysis of the graph of a polynomial and your knowledge of the properties of polynomials can be used to solve polynomial equations.

 EXAMPLE 2   *Solve a Polynomial Equation*

Solve: $x^4 - 5x^3 + 4x^2 + 3x + 9 = 0$

Solution

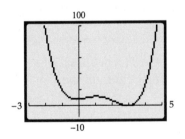

Figure 4.23

Let $P(x) = x^4 - 5x^3 + 4x^2 + 3x + 9$. The x-intercepts of the graph of P are the real solutions of the equation. Use a graphing utility to graph P. See **Figure 4.23**.

From the graph, it appears that $(3, 0)$ is an x-intercept and the only x-intercept. Using synthetic division and the Remainder Theorem, we can verify that 3 is a real zero of P and, therefore, a solution of the equation.

$$
\begin{array}{r|rrrrr}
3 & 1 & -5 & 4 & 3 & 9 \\
 & & 3 & -6 & -6 & -9 \\
\hline
 & 1 & -2 & -2 & -3 & 0 \\
\end{array}
$$

- Coefficients of P

- Remainder is zero. Thus 3 is a zero.

By the Number of Zeros Theorem, there are three more zeros of P. Because the only x-intercept is $(3, 0)$, this suggests that 3 may be a repeated solution of the equation. From the synthetic division, the reduced polynomial is $x^3 - 2x^2 - 2x - 3$. Try synthetic division again, using 3 as a possible zero of the reduced polynomial.

$$3 \; | \; \begin{array}{cccc} 1 & -2 & -2 & -3 \\ & 3 & 3 & 3 \end{array}$$ • **Coefficients of reduced polynomial**

$$\begin{array}{cccc} 1 & 1 & 1 & 0 \end{array}$$ • **Remainder is zero. Thus 3 is a zero.**

We now have 3 as a double root of the equation and, from the last line of the preceding synthetic division, $x^2 + x + 1 = 0$. Use the quadratic formula to solve this equation.

$$x = \frac{-1 \pm \sqrt{1^2 - 4(1)(1)}}{2(1)} = \frac{-1 \pm \sqrt{-3}}{2} = \frac{-1 \pm i\sqrt{3}}{2}$$

The four solutions of the equation are 3, 3, $\dfrac{-1 + i\sqrt{3}}{2}$, and $\dfrac{-1 - i\sqrt{3}}{2}$.

TRY EXERCISE 20, EXERCISE SET 4.4

FACTORS OF A POLYNOMIAL

The following theorem is a result of the Conjugate Pair Theorem.

LINEAR AND QUADRATIC FACTORS OF A POLYNOMIAL

Every polynomial with real coefficients and positive degree n can be written as the product of linear and quadratic factors with real coefficients, where the quadratic factors have no real zeros.

A quadratic factor with no real zeros is said to be **irreducible over the reals.**

EXAMPLE 3 | *Factor a Polynomial into Linear and Quadratic Factors*

Write each polynomial as a product of linear factors and quadratic factors that are irreducible over the reals.

a. $P(x) = x^3 - 3x^2 + x - 3$ **b.** $P(x) = x^3 - 6x^2 + 13x - 10$

Solution

a. Factoring by grouping produces

$$P(x) = x^3 - 3x^2 + x - 3 = (x^3 - 3x^2) + (x - 3)$$
$$= x^2(x - 3) + 1(x - 3) = (x - 3)(x^2 + 1)$$

Because each binomial factor is irreducible over the reals, the factorization is complete.

Continued • ➤

b. Because $x^3 - 6x^2 + 13x - 10$ cannot be factored by grouping, synthetic division is used to determine zeros that also determine factors. By the Rational Zero Theorem, we know that ± 1, ± 2, ± 5, and ± 10 are possible rational zeros. Testing each of these, we find

$$\begin{array}{r|rrr} 2 & 1 & -6 & 13 & -10 \\ & & 2 & -8 & 10 \\ \hline & 1 & -4 & 5 & 0 \end{array} \qquad \bullet\ \textbf{2 is a zero}$$

Using the quadratic formula, we find that the reduced polynomial $x^2 - 4x + 5$ has zeros of $2 \pm i$, so it cannot be factored using real numbers. Thus $x^3 - 6x^2 + 13x - 10$ factors into

$$(x - 2)(x^2 - 4x + 5)$$

which is a product of a linear factor and a quadratic factor that is irreducible over the reals.

Try Exercise 28, Exercise Set 4.4

Many of the problems in this section and in Section 4.3 dealt with the process of finding the zeros of a given polynomial. Example 4 considers the reverse process, finding a polynomial when the zeros are given.

> **EXAMPLE 4**   *Determine a Polynomial Given Its Zeros*

Find each polynomial.

a. A polynomial of degree 3 that has 1, 2, and -3 as zeros

b. A polynomial of degree 4 that has real coefficients and zeros $2i$ and $3 - 7i$

Solution

a. Because 1, 2, and -3 are zeros, $(x - 1)$, $(x - 2)$, and $(x + 3)$ are factors. Multiplying these factors produces a polynomial that has the indicated zeros.

$$(x - 1)(x - 2)(x + 3) = (x^2 - 3x + 2)(x + 3) = x^3 - 7x + 6$$

b. By the Conjugate Pair Theorem, the polynomial also must have $-2i$ and $3 + 7i$ as zeros. The product of the factors $x - 2i$, $x - (-2i)$, $x - (3 - 7i)$, and $x - (3 + 7i)$ produces the desired polynomial.

$$\begin{aligned} (x - 2i)(x + 2i)&[x - (3 - 7i)][x - (3 + 7i)] \\ &= (x^2 + 4)(x^2 - 6x + 58) \\ &= x^4 - 6x^3 + 62x^2 - 24x + 232 \end{aligned}$$

Try Exercise 48, Exercise Set 4.4

A polynomial that has a given set of zeros is not unique. For example, $x^3 - 7x + 6$ has zeros 1, 2, and -3, but so does any nonzero multiple of that polynomial, such as $2x^3 - 14x + 12$. This concept is illustrated in **Figure 4.24.** The graphs of the two polynomials are different, but they have the same x-intercepts.

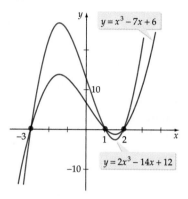

Figure 4.24

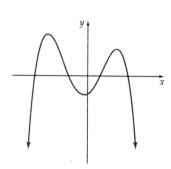

Figure 4.25

TOPICS FOR DISCUSSION

1. What is the Fundamental Theorem of Algebra and why is this theorem so important?

2. Let $P(x)$ be a polynomial of degree n with real coefficients. Discuss the number of *possible* real zeros of this polynomial. Include in your discussion the cases when n is even and when n is odd.

3. Consider the graph of a polynomial in **Figure 4.25.** Is it possible that the degree of the polynomial is 3? Explain.

4. If two polynomials have exactly the same zeros, do the graphs of the polynomials look exactly the same?

5. Does the graph of every polynomial have at least one x-intercept?

EXERCISE SET 4.4

In Exercises 1 to 12, use the given zero to find the remaining zeros of each polynomial.

1. $2x^3 - 5x^2 + 6x - 2$; $1 + i$

2. $3x^3 - 29x^2 + 92x + 34$; $5 + 3i$

3. $x^3 + 3x^2 + x + 3$; $-i$

4. $x^4 - 6x^3 + 71x^2 - 146x + 530$; $2 + 7i$

5. $x^5 - x^4 - 3x^3 + 3x^2 - 10x + 10$; $i\sqrt{2}$

6. $x^4 - 4x^3 + 14x^2 - 4x + 13$; $2 - 3i$

7. $12x^3 - 28x^2 + 23x - 5$; $\frac{1}{3}$

8. $8x^4 - 2x^3 + 199x^2 - 50x - 25$; $-5i$

9. $x^4 - 4x^3 + 19x^2 - 30x + 50$; $1 + 3i$

10. $12x^4 - 52x^3 + 19x^2 - 13x + 4$; $\frac{1}{2}i$

11. $x^5 - x^4 - 4x^3 - 4x^2 - 5x - 3$; i

12. $x^5 - 3x^4 + 7x^3 - 13x^2 + 12x - 4$; $-2i$

In Exercises 13 to 18, find all the zeros of the polynomial. (*Hint:* First determine the rational zeros.)

13. $x^4 + x^3 - 2x^2 + 4x - 24$

14. $x^4 - 3x^3 + 5x^2 - 27x - 36$

15. $2x^4 + x^3 + 39x^2 + 136x - 78$

16. $x^3 - 13x^2 + 65x - 125$

17. $x^5 - 9x^4 + 34x^3 - 58x^2 + 45x - 13$

18. $x^4 - 4x^3 + 53x^2 - 196x + 196$

In Exercises 19 to 26, use a graph and your knowledge of the zeros of polynomial functions to determine the *exact* value of all the solutions of each equation.

19. $2x^3 - x^2 + x - 6 = 0$

20. $4x^3 + 3x^2 + 16x + 12 = 0$

21. $24x^3 - 62x^2 - 7x + 30 = 0$

22. $12x^3 - 52x^2 + 27x + 28 = 0$

23. $x^4 - 4x^3 + 5x^2 - 4x + 4 = 0$

24. $x^4 + 4x^3 + 8x^2 + 16x + 16 = 0$

25. $x^4 + 4x^3 - 2x^2 - 12x + 9 = 0$

26. $x^4 + 3x^3 - 6x^2 - 28x - 24 = 0$

In Exercises 27 to 36, factor each polynomial into linear factors and/or quadratic factors that are irreducible over the reals.

27. $x^3 - x^2 - 2x$

28. $6x^3 - 23x^2 - 4x$

29. $x^3 + 9x$

30. $x^3 + 10x$

31. $x^4 + 2x^2 - 24$

32. $x^4 - 8x^2 - 20$

33. $x^4 + 3x^2 + 2$

34. $x^5 + 11x^3 + 18x$

35. $x^4 - 2x^3 + x^2 - 8x - 12$

36. $x^4 + 2x^3 + 6x^2 + 32x + 40$

In Exercises 37 to 46, find a polynomial of lowest degree that has the given zeros.

37. $4, -3, 2$

38. $-1, 1, -5$

39. $3, 2i, -2i$

40. $0, i, -i$

41. $3 + i, 3 - i, 2 + 5i, 2 - 5i$

42. $2 + 3i, 2 - 3i, -5, 2$

43. $6 + 5i, 6 - 5i, 2, 3, 5$

44. $\frac{1}{2}, 4 - i, 4 + i$

45. $\frac{3}{4}, 2 + 7i, 2 - 7i$

46. $\frac{1}{4}, -\frac{1}{5}, i, -i$

In Exercises 47 to 54, find a polynomial $P(x)$ with real coefficients that has the indicated zeros and satisfies the given conditions.

47. Zeros: $2 - 5i, -4$, degree 3

48. Zeros: $3 + 2i, 7$, degree 3

49. Zeros: $4 + 3i, 5 - i$, degree 4

50. Zeros: $i, 3 - 5i$, degree 4

51. Zeros: $-1, 2, 3$, degree 3, $P(1) = 12$

52. Zeros: $3i, 2$, degree 3, $P(3) = 27$

53. Zeros: $3, -5, 2 + i$, degree 4, $P(1) = 48$

54. Zeros: $\frac{1}{2}, 1 - i$, degree 3, $P(4) = 140$

SUPPLEMENTAL EXERCISES

55. Verify that $x^3 - x^2 - ix^2 - 9x + 9 + 9i$ has $1 + i$ as a zero and that its conjugate $1 - i$ is not a zero. Explain why this does not contradict the Conjugate Pair Theorem.

56. Verify that $x^3 - x^2 - ix^2 - 20x + ix + 20i$ has a zero of i and that its conjugate $-i$ is not a zero. Explain why this does not contradict the Conjugate Pair Theorem.

57. Show that 2 is a zero of multiplicity 3 of

$$P(x) = x^5 - 6x^4 + 21x^3 - 62x^2 + 108x - 72$$

and express $P(x)$ as a product of linear factors and/or quadratic factors that are irreducible over the reals.

58. Show that -1 is a zero of multiplicity 4 of

$$P(x) = x^6 + 5x^5 + 11x^4 + 14x^3 + 11x^2 + 5x + 1$$

and express $P(x)$ as a product of linear factors and/or quadratic factors that are irreducible over the reals.

59. Find a polynomial $P(x)$ of degree 5 such that 1 is a zero of multiplicity 2, 2 is a zero of multiplicity 3, and $P(-1) = -54$.

60. Find a polynomial $P(x)$ of degree 5 such that -4 is a zero of multiplicity 4, 1/2 is a zero of multiplicity 1, and $P(1) = 125$.

PROJECTS

1. **INVESTIGATE THE ROOTS OF A CUBIC EQUATION** Hieronimo Cardano, using a trick he stole from Nicolo Tartaglia, was able to solve some cubic equations.

   a. Show that the cubic equation $x^3 + bx^2 + cx + d = 0$ can be transformed into the "reduced" cubic $y^3 + my = n$, where m and n are constants depending on b, c, and d, by using the substitution $x = y - b/3$.

   b. Cardano then showed that a solution of the reduced cubic is given by

   $$\sqrt[3]{\frac{n}{2} + \sqrt{\frac{n^2}{4} + \frac{m^3}{27}}} - \sqrt[3]{-\frac{n}{2} + \sqrt{\frac{n^2}{4} + \frac{m^3}{27}}}$$

   Use Cardano's procedure to solve the equation $x^3 - 6x^2 + 20x - 33 = 0$.

SECTION 4.5 RATIONAL FUNCTIONS AND THEIR GRAPHS

If $P(x)$ and $Q(x)$ are polynomials, then the function F given by

$$F(x) = \frac{P(x)}{Q(x)}$$

is called a **rational function**. The domain of F is the set of all real numbers except those for which $Q(x) = 0$. For example, the domain of

$$F(x) = \frac{x^2 - x - 5}{x(2x - 5)(x + 3)}$$

is the set of all real numbers except 0, 5/2, and −3.

The graph of $G(x) = \dfrac{x + 1}{x - 2}$ is given in **Figure 4.26.** The graph shows that G has the following properties:

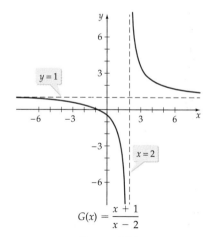

$$G(x) = \frac{x + 1}{x - 2}$$

Figure 4.26

- The graph has an x-intercept at $(-1, 0)$ and a y-intercept at $(0, -1/2)$.

- The graph does not exist when $x = 2$.

Note the behavior of the graph as x takes on values that are close to 2 but *less* than 2. Mathematically, we say that "x approaches 2 from the left."

x	1.9	1.95	1.99	1.995	1.999
G(x)	−29	−59	−299	−599	−2999

From this table and the graph, it appears that as x approaches 2 from the left, the functional values $G(x)$ decrease without bound.

- We say that "$G(x)$ approaches negative infinity."

Now observe the behavior of the graph as x takes on values that are close to 2 but *greater* than 2. Mathematically, we say that "x approaches 2 from the right."

x	2.1	2.05	2.01	2.005	2.001
G(x)	31	61	301	601	3001

From this table and the graph, it appears that as x approaches 2 from the right, the functional values $G(x)$ increase without bound.

● In this case, we say that "$G(x)$ approaches positive infinity."

Now consider the values of $G(x)$ as x *increases* without bound. The table below indicates this for selected values of x.

x	1000	5000	10,000	50,000	100,000
G(x)	1.00301	1.00060	1.00030	1.00006	1.00003

● As x increases without bound, the values of $G(x)$ are becoming closer to 1.

Now let the values of x *decrease* without bound. The table below gives the value of $G(x)$ for selected values of x.

x	−1000	−5000	−10,000	−50,000	−100,000
G(x)	0.997006	0.999400	0.997001	0.999940	0.999970

● As x decreases without bound, the values of $G(x)$ are becoming closer to 1.

When we are discussing graphs that increase or decrease without bound, it is convenient to use mathematical notation. The notation

$$f(x) \to \infty \quad \text{as} \quad x \to a^{+}$$

means that the functional values $f(x)$ increase without bound as x approaches a from the right. Recall that the symbol ∞ does not represent a real number but is used merely to describe the concept of a variable taking on larger and larger values without bound. See **Figure 4.27a.**

The notation

$$f(x) \to \infty \text{ as } x \to a^{-}$$

means that the functional values $f(x)$ increase without bound as x approaches a from the left. See **Figure 4.27b.**

The notation

$$f(x) \to -\infty \text{ as } x \to a^{+}$$

means that the functional values $f(x)$ decrease without bound as x approaches a from the right. See **Figure 4.27c.**

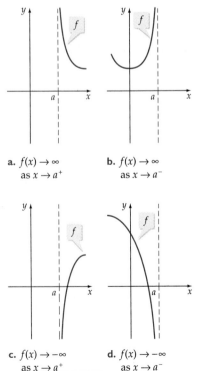

a. $f(x) \to \infty$
as $x \to a^{+}$

b. $f(x) \to \infty$
as $x \to a^{-}$

c. $f(x) \to -\infty$
as $x \to a^{+}$

d. $f(x) \to -\infty$
as $x \to a^{-}$

Figure 4.27

The notation

$$f(x) \to -\infty \text{ as } x \to a^-$$

means that the functional values $f(x)$ decrease without bound as x approaches a from the left. See **Figure 4.27d.**

ASYMPTOTES

Each graph in **Figure 4.27** approaches a vertical line through $(a, 0)$ as $x \to a^+$ or a^-. The line is said to be a *vertical asymptote* to the graph.

DEFINITION OF A VERTICAL ASYMPTOTE

The line $x = a$ is a **vertical asymptote** of the graph of a function F provided that

$$F(x) \to \infty \quad \text{or} \quad F(x) \to -\infty$$

as x approaches a from either left or right.

In **Figure 4.26,** the line $x = 2$ is a vertical asymptote of the graph of G. Note that the graph of G in **Figure 4.26** also approaches the horizontal line $y = 1$ as $x \to \infty$ and as $x \to -\infty$. The line $y = 1$ is a *horizontal asymptote* of the graph of G.

DEFINITION OF A HORIZONTAL ASYMPTOTE

The line $y = b$ is a **horizontal asymptote** of the graph of a function F provided that

$$F(x) \to b \quad \text{as} \quad x \to \infty \quad \text{or} \quad x \to -\infty$$

Figure 4.28 illustrates some of the ways in which the graph of a rational function may approach its horizontal asymptote. It is common practice to

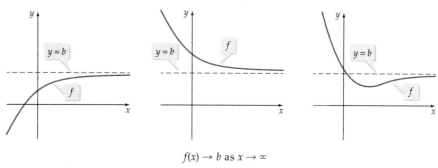

$f(x) \to b$ as $x \to \infty$

Figure 4.28

display the asymptotes of the graph of a rational function by using dashed lines. Although a rational function may have several vertical asymptotes, it can have at most one horizontal asymptote. The graph may intersect its horizontal asymptote.

QUESTION   Can a graph of a rational function cross its vertical asymptote? Why or why not?

Geometrically, a line is an asymptote to a curve if the distance between the line and a point $P(x, y)$ on the curve approaches zero as the distance between the origin and the point P increases without bound.

Vertical asymptotes of the graph of a rational function can be found by using the following theorem.

THEOREM ON VERTICAL ASYMPTOTES

If the real number a is a zero of the denominator $Q(x)$, then the graph of $F(x) = P(x)/Q(x)$, where $P(x)$ and $Q(x)$ have no common factors, has the vertical asymptote $x = a$.

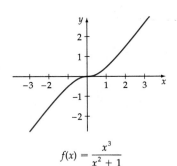

$$f(x) = \frac{x^3}{x^2 + 1}$$

Figure 4.29

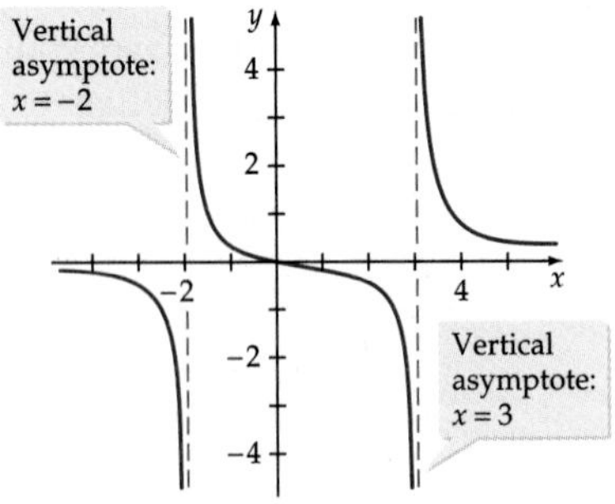

EXAMPLE 1   **Find the Vertical Asymptotes of a Rational Function**

Find the vertical asymptotes of each rational function.

a. $f(x) = \dfrac{x^3}{x^2 + 1}$ **b.** $g(x) = \dfrac{x}{x^2 - x - 6}$

Solution

a. To find the vertical asymptotes, set the denominator equal to zero. The denominator $x^2 + 1$ has no real zeros, so the graph of f has no vertical asymptotes. See **Figure 4.29**.

b. The denominator $x^2 - x - 6 = (x - 3)(x + 2)$ has zeros of 3 and -2. The numerator has no common factors with the denominator, so $x = 3$ and $x = -2$ are both vertical asymptotes of the graph of g, as shown in **Figure 4.30**.

TRY EXERCISE 2, EXERCISE SET 4.5

Vertical asymptote: $x = -2$

Vertical asymptote: $x = 3$

$$g(x) = \frac{x}{x^2 - x - 6}$$

Figure 4.30

The following theorem implies that a horizontal asymptote can be determined by examining the leading terms of the numerator and the denominator of a rational function.

ANSWER   No. If $x = a$ is a vertical asymptote of a rational function R, then $R(a)$ is undefined.

THEOREM ON HORIZONTAL ASYMPTOTES

Let
$$F(x) = \frac{a_n x^n + a_{n-1} x^{n-1} + \cdots + a_1 x + a_0}{b_m x^m + b_{m-1} x^{m-1} + \cdots + b_1 x + b_0}$$

be a rational function with numerator of degree n and denominator of degree m.

1. If $n < m$, then the x-axis is the horizontal asymptote of the graph of F.

2. If $n = m$, then the line $y = a_n/b_m$ is the horizontal asymptote of the graph of F.

3. If $n > m$, the graph of F has no horizontal asymptote.

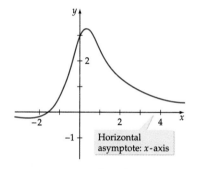

$$f(x) = \frac{2x + 3}{x^2 + 1}$$

Figure 4.31

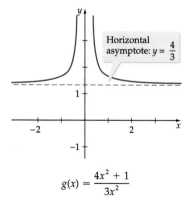

$$g(x) = \frac{4x^2 + 1}{3x^2}$$

Figure 4.32

EXAMPLE 2   *Find the Horizontal Asymptote of a Rational Function*

Find the horizontal asymptote of each rational function.

a. $f(x) = \dfrac{2x + 3}{x^2 + 1}$ b. $g(x) = \dfrac{4x^2 + 1}{3x^2}$ c. $h(x) = \dfrac{x^3 + 1}{x - 2}$

Solution

a. The degree of the numerator $2x + 3$ is less than the degree of the denominator $x^2 + 1$. By the Theorem on Horizontal Asymptotes, the x-axis is the horizontal asymptote of f. See the graph of f in **Figure 4.31.**

b. The numerator $4x^2 + 1$ and the denominator $3x^2$ of g are both of degree 2. By the Theorem on Horizontal Asymptotes, the line $y = 4/3$ is the horizontal asymptote of g. See the graph of g in **Figure 4.32.**

c. The degree of the numerator $x^3 + 1$ is larger than the degree of the denominator $x - 2$, so by the Theorem on Horizontal Asymptotes, the graph of h has no horizontal asymptote.

TRY EXERCISE 6, EXERCISE SET 4.5

The proof of the Theorem on Horizontal Asymptotes makes use of the technique employed in the following verification. To verify that

$$y = \frac{5x^2 + 4}{3x^2 + 8x + 7}$$

has a horizontal asymptote of $y = 5/3$, divide the numerator and the denominator by the largest power of the variable x (x^2 in this case).

$$y = \frac{\dfrac{5x^2 + 4}{x^2}}{\dfrac{3x^2 + 8x + 7}{x^2}} = \frac{5 + \dfrac{4}{x^2}}{3 + \dfrac{8}{x} + \dfrac{7}{x^2}}, \quad x \neq 0$$

As x increases without bound or decreases without bound, the fractions $4/x^2$, $8/x$, and $7/x^2$ approach zero. Thus

$$y \rightarrow \frac{5 + 0}{3 + 0 + 0} = \frac{5}{3} \quad \text{as} \quad x \rightarrow \pm\infty$$

and hence the line $y = 5/3$ is a horizontal asymptote of the graph.

A SIGN PROPERTY OF RATIONAL FUNCTIONS

The zeros and vertical asymptotes of a rational function F divide the x-axis into intervals. In each interval,

- $F(x)$ is positive for all x in the interval, or
- $F(x)$ is negative for all x in the interval.

For example, consider the rational function

$$g(x) = \frac{x + 1}{x^2 + 2x - 3}$$

which has vertical asymptotes of $x = -3$ and $x = 1$ and a zero of -1. These three numbers divide the x-axis into the four intervals $(-\infty, -3)$, $(-3, -1)$, $(-1, 1)$, and $(1, \infty)$. Note in **Figure 4.33** that the graph of g is

- negative for all x such that $x < -3$.
- positive for all x such that $-3 < x < -1$.
- negative for all x such that $-1 < x < 1$.
- positive for all x such that $x > 1$.

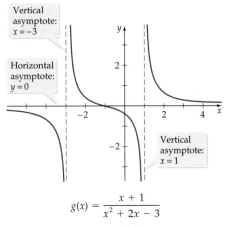

$$g(x) = \frac{x + 1}{x^2 + 2x - 3}$$

Figure 4.33

GENERAL GRAPHING PROCEDURE

If $F(x) = P(x)/Q(x)$, where $P(x)$ and $Q(x)$ are polynomials that have no common factor, then the following general procedure offers useful guidelines for graphing F.

GENERAL PROCEDURE FOR GRAPHING RATIONAL FUNCTIONS THAT HAVE NO COMMON FACTORS

1. **Asymptotes**   Find the real zeros of the denominator $Q(x)$. For each zero a, draw the dashed line $x = a$. Each line is a vertical asymptote of the graph of F. Graph any horizontal asymptotes. These can be found by using the Theorem on Horizontal Asymptotes. If the degree of the numerator $P(x)$ is larger than the degree of the denominator $Q(x)$, then the graph of F does not have a horizontal asymptote.

2. **Intercepts**   Find the real zeros of the numerator $P(x)$. For each zero a, plot the point $(a, 0)$. Each such point is an x-intercept of the graph of F. Evaluate $F(0)$. Plot $(0, F(0))$, the y-intercept of the graph of F.

3. **Symmetry**   Use the tests for symmetry to determine whether the graph of the function has symmetry with respect to the y-axis or symmetry with respect to the origin.

4. **Additional points**   Plot at least two points that lie in the intervals between and beyond the vertical asymptotes and the x-intercepts.

5. **Behavior near asymptotes**   If $x = a$ is a vertical asymptote, determine whether $F(x) \to \infty$ or $F(x) \to -\infty$ as $x \to a^-$ and also as $x \to a^+$.

6. **Complete the sketch**   Use all the information obtained above to sketch the graph of F. Plot additional points if necessary to gain additional knowledge about the function.

EXAMPLE 3   *Graph a Rational Function*

Sketch a graph of $f(x) = \dfrac{2x^2 - 18}{x^2 + 3}$.

Solution

Asymptotes   The denominator $x^2 + 3$ has no real zeros, so the graph of f has no vertical asymptotes. The numerator and denominator both have degree 2. The leading coefficients are 2 and 1, respectively. By the Theorem on Horizontal Asymptotes, the graph of f has a horizontal asymptote $y = 2/1 = 2$.

Continued ●➤

Intercepts The zeros of the numerator occur when $2x^2 - 18 = 0$ or, solving for x, when $x = -3$ and $x = 3$. Therefore, the x-intercepts are $(-3, 0)$ and $(3, 0)$. To find the y-intercept, evaluate f when $x = 0$. This gives $y = -6$. Therefore, the y-intercept is $(0, -6)$.

Symmetry Below we show that $f(-x) = f(x)$, which means that f is an even function and therefore its graph is symmetric with respect to the y-axis.

$$f(-x) = \frac{2(-x)^2 - 18}{(-x)^2 + 3} = \frac{2x^2 - 18}{x^2 + 3} = f(x)$$

Because $f(x) = f(-x)$, f is an even function.

Additional points The intervals determined by the x-intercepts are $x < -3$, $-3 < x < 3$, and $x > 3$. Generally, it is necessary to determine points in all intervals. However, because f is an even function, its graph is symmetric with respect to the y-axis. The following table lists a few points for $x > 0$. Symmetry can be used to locate corresponding points for $x < 0$.

x	1	2	6
$f(x)$	-4	$-\dfrac{10}{7} \approx -1.43$	$\dfrac{18}{13} \approx 1.38$

Behavior near asymptotes As x increases or decreases without bound, $f(x)$ approaches the horizontal asymptote $y = 2$.

To determine whether the graph of f intersects the horizontal asymptote at any point, solve the equation $f(x) = 2$.

There are no solutions of $f(x) = 2$ because

$$\frac{2x^2 - 18}{x^2 + 3} = 2 \quad \text{implies} \quad 2x^2 - 18 = 2x^2 + 6$$

This is not possible. Thus the graph of f does not intersect the horizontal asymptote but approaches it from below as x increases or decreases without bound.

Complete the sketch Use the summary in Table 4.3 to finish the sketch. The completed graph is shown in **Figure 4.34**.

Table 4.3

Vertical Asymptote	None
Horizontal Asymptote	$y = 2$
x-Intercepts	$(-3, 0)$, $(3, 0)$
y-Intercept	$(0, -6)$
Additional Points	$(1, -4)$, $(2, -1.43)$, $(6, 1.38)$

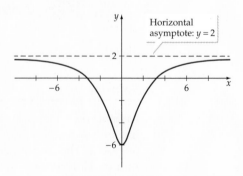

Figure 4.34

TRY EXERCISE 10, EXERCISE SET 4.5

EXAMPLE 4 *Graph a Rational Function*

Sketch the graph of $h(x) = \dfrac{x^2 + 1}{x^2 + x - 2}$.

Solution

Asymptotes The denominator $x^2 + x - 2 = (x + 2)(x - 1)$ has zeros -2 and 1; because there are no common factors of the numerator and the denominator, the lines $x = -2$ and $x = 1$ are vertical asymptotes.

The numerator and denominator both have degree 2. The leading coefficients of the numerator and denominator are both 1. Thus h has the horizontal asymptote $y = 1/1 = 1$.

Intercept(s) The numerator $x^2 + 1$ has no real zeros, so the graph of h has no x-intercepts. Because $h(0) = -0.5$, h has the y-intercept $(0, -0.5)$.

Symmetry By applying the tests for symmetry, we can determine that the graph of h is not symmetric with respect to the origin or to the y-axis.

Additional points The intervals determined by the vertical asymptotes are $(-\infty, -2)$, $(-2, 1)$, and $(1, \infty)$. Plot a few points from each interval:

x	-5	-3	-1	0.5	2	3	4
$h(x)$	$\dfrac{13}{9}$	2.5	-1	-1	$\dfrac{5}{4}$	1	$\dfrac{17}{18}$

The graph of h will intersect the horizontal asymptote $y = 1$ exactly once. This can be determined by solving the equation $h(x) = 1$.

$$\frac{x^2 + 1}{x^2 + x - 2} = 1$$
$$x^2 + 1 = x^2 + x - 2 \qquad \text{• Multiply both sides by } x^2 + x - 2.$$
$$1 = x - 2$$
$$3 = x$$

The only solution is $x = 3$. Therefore, the graph of h intersects the horizontal asymptote at $(3, 1)$.

Behavior near asymptotes As x approaches -2 from the left, the denominator $(x + 2)(x - 1)$ approaches 0 but remains positive. The numerator $x^2 + 1$ approaches 5, which is positive, so the quotient $h(x)$ increases without bound. Stated in mathematical notation,

$$h(x) \to \infty \quad \text{as} \quad x \to -2^-$$

Similarly, it can be determined that

$$h(x) \to -\infty \quad \text{as} \quad x \to -2^+$$
$$h(x) \to -\infty \quad \text{as} \quad x \to 1^-$$
$$h(x) \to \infty \quad \text{as} \quad x \to 1^+$$

Continued •➤

Table 4.4

Vertical Asymptotes	$x = -2, x = 1$
Horizontal Asymptote	$y = 1$
x-Intercept	None
y-Intercept	$(0, -0.5)$
Additional Points	$(-5, 1.\overline{4}), (-3, 2.5),$ $(-1, -1), (0.5, -1),$ $(2, 1.25), (3, 1),$ $(4, 0.9\overline{4})$

Complete the sketch   Use the summary in Table 4.4 to obtain the graph sketched in **Figure 4.35**.

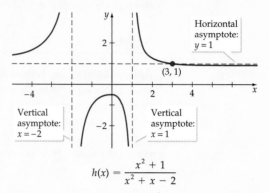

$$h(x) = \frac{x^2 + 1}{x^2 + x - 2}$$

Figure 4.35

TRY EXERCISE 26, EXERCISE SET 4.5

SLANT ASYMPTOTES

Some rational functions have an asymptote that is neither vertical nor horizontal but slanted.

THEOREM ON SLANT ASYMPTOTES

The rational function given by $F(x) = P(x)/Q(x)$, where $P(x)$ and $Q(x)$ have no common factors, has a **slant asymptote** if the degree of the polynomial $P(x)$ in the numerator is one greater than the degree of the polynomial $Q(x)$ in the denominator.

To find the slant asymptote, use division to express $F(x)$ in the form

$$F(x) = \frac{P(x)}{Q(x)} = (mx + b) + \frac{r(x)}{Q(x)}$$

where the degree of $r(x)$ is less than the degree of $Q(x)$. Because

$$\frac{r(x)}{Q(x)} \to 0 \quad \text{as} \quad x \to \pm\infty$$

we know that $F(x) \to mx + b$ as $x \to \pm\infty$.

The line represented by $y = mx + b$ is called the slant asymptote of the graph of F.

EXAMPLE 5 *Find the Slant Asymptote of a Rational Function*

Find the slant asymptote of $f(x) = \dfrac{2x^3 + 5x^2 + 1}{x^2 + x + 3}$.

Solution

Because the degree of the numerator $2x^3 + 5x^2 + 1$ is exactly one larger than the degree of the denominator $x^2 + x + 3$ and f is in simplest form, f has a slant asymptote. To find the asymptote, divide $2x^3 + 5x^2 + 1$ by $x^2 + x + 3$.

$$
\begin{array}{r}
2x + 3 \\
x^2 + x + 3 \overline{)\,2x^3 + 5x^2 + 0x + 1} \\
\underline{2x^3 + 2x^2 + 6x} \\
3x^2 - 6x + 1 \\
\underline{3x^2 + 3x + 9} \\
-9x - 8
\end{array}
$$

Therefore,

$$f(x) = \frac{2x^3 + 5x^2 + 1}{x^2 + x + 3} = (2x + 3) + \frac{-9x - 8}{x^2 + x + 3}$$

and the line $y = 2x + 3$ is the slant asymptote for the graph of f. **Figure 4.36** shows the graph of f and its slant asymptote.

TRY EXERCISE 34, EXERCISE SET 4.5

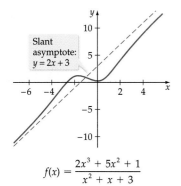

Slant asymptote: $y = 2x + 3$

$f(x) = \dfrac{2x^3 + 5x^2 + 1}{x^2 + x + 3}$

Figure 4.36

The function f in Example 5 does not have a vertical asymptote because the denominator $x^2 + x + 3$ does not have any real zeros. However, the function

$$g(x) = \frac{2x^2 - 4x + 5}{3 - x}$$

has both a slant asymptote and a vertical asymptote. The vertical asymptote is $x = 3$, and the slant asymptote is $y = -2x - 2$. **Figure 4.37** shows the graph of g and its asymptotes.

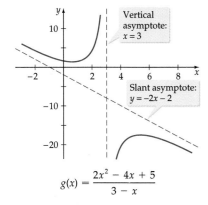

Vertical asymptote: $x = 3$

Slant asymptote: $y = -2x - 2$

$g(x) = \dfrac{2x^2 - 4x + 5}{3 - x}$

Figure 4.37

EXAMPLE 6 *Graph a Rational Function That Has a Slant Asymptote*

Sketch the graph of $j(x) = \dfrac{x^2 - 1}{x}$.

Continued ▶

Solution

Asymptotes The denominator x has 0 as its only zero. Because there are no common factors of the numerator and the denominator, the y-axis is the vertical asymptote of the graph of j.

The degree of the numerator $x^2 - 1$ is exactly one more than the degree of the denominator x, so j has a slant asymptote. Dividing $x^2 - 1$ by x shows that j can be expressed as

$$j(x) = \frac{x^2 - 1}{x} = \frac{x^2}{x} - \frac{1}{x} = x - \frac{1}{x}$$

From this we can conclude that $j(x) \to x$ as $x \to \pm\infty$. Therefore, the graph of j has a slant asymptote of $y = x$.

Intercepts By setting the numerator of j to zero, we can determine that the zeros of j are $x = -1$ and $x = 1$.

Symmetry As shown below, $j(-x) = -j(x)$, and therefore j is an odd function. The graph of an odd function is symmetric with respect to the origin.

$$j(-x) = \frac{(-x)^2 - 1}{-x} = \frac{x^2 - 1}{-x} = -\left(\frac{x^2 - 1}{x}\right) = -j(x)$$

Additional points The intervals determined by the x-intercepts and the vertical asymptote are $x < -1$, $-1 < x < 0$, $0 < x < 1$, and $x > 1$. The following table lists a few points from each interval.

x	-5	-2	-0.5	0.5	2	5
$j(x)$	-4.8	-1.5	1.5	-1.5	1.5	4.8

Complete the sketch Use all the previous information to complete the sketch of j as shown in **Figure 4.38.**

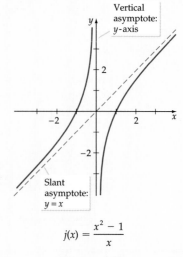

$$j(x) = \frac{x^2 - 1}{x}$$

Figure 4.38

TRY EXERCISE 40, EXERCISE SET 4.5

EXAMPLE 7   *Solve an Application*

A cylindrical soft drink can is to be constructed so that it will have a volume of 21.6 cubic inches. See **Figure 4.39**.

a. Write the total surface area A of the can as a function of r, where r is the radius of the can in inches.

b. Use a graphing utility to estimate the value of r (to the nearest 0.1 inch) that produces the minimum surface area.

Figure 4.39

Solution

a. The formula for the volume of a cylinder is $V = \pi r^2 h$, where r is the radius and h is the height. Because we are given that the volume is 21.6 cubic inches, we have

$$21.6 = \pi r^2 h$$

$$\frac{21.6}{\pi r^2} = h \qquad \text{• Solve for } h.$$

The surface area of the cylinder is given by

$$A = 2\pi r^2 + 2\pi rh$$

$$A = 2\pi r^2 + 2\pi r\left(\frac{21.6}{\pi r^2}\right) \qquad \text{• Substitute for } h.$$

$$A = 2\pi r^2 + \frac{2(21.6)}{r} \qquad \text{• Simplify.}$$

$$A = \frac{2\pi r^3 + 43.2}{r} \qquad (1)$$

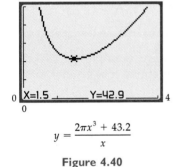

$$y = \frac{2\pi x^3 + 43.2}{x}$$

Figure 4.40

b. Use Equation (1) with $y = A$ and $x = r$ and a graphing utility to determine that A is a minimum when $r = 1.5$ inches. See **Figure 4.40**.

> **TRY EXERCISE 68, EXERCISE SET 4.5**

If a rational function has a numerator and denominator that have a common factor, then you should reduce the rational function to lowest terms before you apply the general procedure for sketching the graph of a rational function.

EXAMPLE 8   *Graph a Rational Function That Has a Common Factor*

Sketch the graph of $f(x) = \dfrac{x^2 - 3x - 4}{x^2 - 6x + 8}$.

Continued •➤

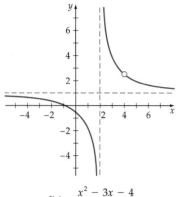

$$f(x) = \frac{x^2 - 3x - 4}{x^2 - 6x + 8}$$

Figure 4.41

Solution

Factor the numerator and denominator to obtain

$$f(x) = \frac{x^2 - 3x - 4}{x^2 - 6x + 8} = \frac{(x + 1)(x - 4)}{(x - 2)(x - 4)}, \quad x \neq 2, x \neq 4$$

Thus for all x values other than $x = 4$, the graph of f is the same as the graph of

$$G(x) = \frac{x + 1}{x - 2}$$

Figure 4.26 shows a graph of G. The graph of f will be the same as this graph, except that it will have an open circle at $(4, 2.5)$ to indicate that it is undefined for $x = 4$. See the graph of f in **Figure 4.41**.

> **TRY EXERCISE 50, EXERCISE SET 4.5**

QUESTION   Does $F(x) = \dfrac{x^2 - x - 6}{x^2 - 9}$ have a vertical asymptote when $x = 3$?

TOPICS FOR DISCUSSION

1. Discuss the meaning of a rational function. Give examples of functions that are rational functions and examples of functions that are not rational functions.

2. Does the graph of every rational function have at least one vertical asymptote? If so, explain why. If not, give an example of a rational function without a vertical asymptote.

3. Does the graph of every rational function have a horizontal asymptote? If so, explain why. If not, give an example of a rational function without a horizontal asymptote.

4. What conditions must exist to ensure that a rational function has a slant asymptote? Give an example of a rational function that has a slant asymptote.

5. Can the graph of a polynomial function have a vertical asymptote? a horizontal asymptote?

EXERCISE SET 4.5

In Exercises 1 to 4, find all vertical asymptotes of each rational function.

1. $F(x) = \dfrac{2x - 1}{x^2 + 3x}$

2. $F(x) = \dfrac{3x^2 + 5}{x^2 - 4}$

3. $F(x) = \dfrac{x^2 + 11}{6x^2 - 5x - 4}$

4. $F(x) = \dfrac{3x - 5}{x^3 - 8}$

ANSWER   No. $F(x) = \dfrac{x^2 - x - 6}{x^2 - 9} = \dfrac{(x - 3)(x + 2)}{(x - 3)(x + 3)} = \dfrac{x + 2}{x + 3}$. As x approaches 3, $F(x)$ approaches $\dfrac{5}{6}$.

In Exercises 5 to 8, find the horizontal asymptote of each rational function.

5. $F(x) = \dfrac{4x^2 + 1}{x^2 + x + 1}$

6. $F(x) = \dfrac{3x^3 - 27x^2 + 5x - 11}{x^5 - 2x^3 + 7}$

7. $F(x) = \dfrac{15,000x^3 + 500x - 2000}{700 + 500x^3}$

8. $F(x) = 6000\left(1 - \dfrac{25}{(x + 5)^2}\right)$

In Exercises 9 to 32, determine the vertical and horizontal asymptotes and sketch the graph of the rational function F. Label all intercepts and asymptotes.

9. $F(x) = \dfrac{1}{x + 4}$

10. $F(x) = \dfrac{1}{x - 2}$

11. $F(x) = \dfrac{-4}{x - 3}$

12. $F(x) = \dfrac{-3}{x + 2}$

13. $F(x) = \dfrac{4}{x}$

14. $F(x) = \dfrac{-4}{x}$

15. $F(x) = \dfrac{x}{x + 4}$

16. $F(x) = \dfrac{x}{x - 2}$

17. $F(x) = \dfrac{x + 4}{2 - x}$

18. $F(x) = \dfrac{x + 3}{1 - x}$

19. $F(x) = \dfrac{1}{x^2 - 9}$

20. $F(x) = \dfrac{-2}{x^2 - 4}$

21. $F(x) = \dfrac{1}{x^2 + 2x - 3}$

22. $F(x) = \dfrac{1}{x^2 - 2x - 8}$

23. $F(x) = \dfrac{x}{9 - x^2}$

24. $F(x) = \dfrac{x}{x^2 - 16}$

25. $F(x) = \dfrac{x^2}{x^2 + 4x + 4}$

26. $F(x) = \dfrac{x^2}{x^2 - 6x + 9}$

27. $F(x) = \dfrac{10}{x^2 + 2}$

28. $F(x) = \dfrac{-20}{x^2 + 4}$

29. $F(x) = \dfrac{2x^2 - 2}{x^2 - 9}$

30. $F(x) = \dfrac{6x^2 - 5}{2x^2 + 6}$

31. $F(x) = \dfrac{x^2 + x + 4}{x^2 + 2x - 1}$

32. $F(x) = \dfrac{2x^2 - 14}{x^2 - 6x + 5}$

In Exercises 33 to 36, find the slant asymptote of each rational function.

33. $F(x) = \dfrac{3x^2 + 5x - 1}{x + 4}$

34. $F(x) = \dfrac{x^3 - 2x^2 + 3x + 4}{x^2 - 3x + 5}$

35. $F(x) = \dfrac{x^3 - 1}{x^2}$

36. $F(x) = \dfrac{4000 + 20x + 0.0001x^2}{x}$

In Exercises 37 to 46, determine the vertical and slant asymptotes and sketch the graph of the rational function F.

37. $F(x) = \dfrac{x^2 - 4}{x}$

38. $F(x) = \dfrac{x^2 + 10}{2x}$

39. $F(x) = \dfrac{x^2 - 3x - 4}{x + 3}$

40. $F(x) = \dfrac{x^2 - 4x - 5}{2x + 5}$

41. $F(x) = \dfrac{2x^2 + 5x + 3}{x - 4}$

42. $F(x) = \dfrac{4x^2 - 9}{x + 3}$

43. $F(x) = \dfrac{x^2 - x}{x + 2}$

44. $F(x) = \dfrac{x^2 + x}{x - 1}$

45. $F(x) = \dfrac{x^3 + 1}{x^2 - 4}$

46. $F(x) = \dfrac{x^3 - 1}{3x^2}$

In Exercises 47 to 56, sketch the graph of the rational function F. (*Hint:* First examine the numerator and denominator to determine whether there are any common factors.)

47. $F(x) = \dfrac{x^2 + x}{x + 1}$

48. $F(x) = \dfrac{x^2 - 3x}{x - 3}$

49. $F(x) = \dfrac{2x^3 + 4x^2}{2x + 4}$

50. $F(x) = \dfrac{x^2 - x - 12}{x^2 - 2x - 8}$

51. $F(x) = \dfrac{-2x^3 + 6x}{2x^2 - 6x}$

52. $F(x) = \dfrac{x^3 + 3x^2}{x(x + 3)(x - 1)}$

53. $F(x) = \dfrac{x^2 - 3x - 10}{x^2 + 4x + 4}$

54. $F(x) = \dfrac{2x^2 + x - 3}{x^2 - 2x + 1}$

55. $F(x) = \dfrac{x^3 + x^2 - 14x - 24}{x + 2}$

56. $F(x) = \dfrac{2x^3 + 5x^2 - 4x - 3}{x - 1}$

In Exercises 57 to 64, use a graphing utility to graph each function and determine, to the nearest 0.1, the equations of the vertical asymptotes.

57. $R(x) = \dfrac{x^2 + 4}{x^2 - x - 3}$

58. $R(x) = \dfrac{2x^2 - x}{x^2 + x - 4}$

59. $G(x) = \dfrac{x^2 - 4}{x^2 + 2x + 5}$

60. $Y(x) = \dfrac{3x^2 + 4x - 1}{2x^2 + 3x + 5}$

61. $P(x) = \dfrac{x^3 - x - 3}{x^3 - 2x^2 - 5x + 6}$

62. $V(x) = \dfrac{2x^3 + x + 1}{x^3 - 2x^2 - 11x + 12}$

63. $R(x) = \dfrac{x^3 - x^2 - x - 2}{x^3 - 3x^2 - 10x + 24}$

64. $Z(x) = \dfrac{x^3 - 2x^2 - 2x - 3}{x^3 + 3x^2 - 10x - 24}$

65. DESALINIZATION The cost C in dollars to remove $p\%$ of the salt in a tank of sea water is given by

$$C(p) = \frac{2000p}{100 - p}, \quad 0 \le p < 100$$

a. Find the cost of removing 40% of the salt.

b. Find the cost of removing 80% of the salt.

c. Sketch the graph of C.

66. FOOD SCIENCE The temperature F (measured in degrees Fahrenheit) of a dessert placed in a freezer for t hours is given by the rational function

$$F(t) = \frac{60}{t^2 + 2t + 1}, \quad t \ge 0$$

a. Find the temperature of the dessert after it has been in the freezer for 1 hour.

b. Find the temperature of the dessert after 4 hours.

c. Sketch the graph of F.

67. MANUFACTURING A large electronics firm finds that the number of computers it can produce per week after t weeks of production is approximated by

$$C(t) = \frac{2000t^2 + 20{,}000t}{t^2 + 10t + 25}, \quad 0 \le t \le 50$$

a. Find the number of computers it produced during the first week.

b. Find the number of computers it produced during the tenth week.

c. What is the equation of the horizontal asymptote of the graph of C?

d. Sketch the graph of C, and then use the graph to estimate how many weeks pass until the firm can produce 1900 computers in a single week.

68. PRODUCTION COSTS The cost of publishing x books is given by

$$C(x) = 40{,}000 + 20x + 0.0001x^2$$

The average cost per book is given by

$$A(x) = \frac{C(x)}{x} = \frac{40{,}000 + 20x + 0.0001x^2}{x}$$

where $1000 \le x \le 100{,}000$.

a. What is the average cost per book if 5000 books are published?

b. What is the average cost per book if 10,000 books are published?

c. What is the equation of the slant asymptote of the graph of the average cost function?

d. Graph A, and estimate the number of books that should be published to minimize the average cost per book.

69. PHYSIOLOGY One of Poiseuille's Laws states that the resistance R encountered by blood flowing through a blood vessel is given by the rational function

$$R(r) = C\frac{L}{r^4}$$

where C is a positive constant determined by the viscosity of the blood, L is the length of the blood vessel, and r is the radius.

a.  Explain the meaning of $R(r) \to \infty$ as $r \to 0$.

b. Explain the meaning of $R(r) \to 0$ as $r \to \infty$.

c. Graph R for $0 < r \le 4$ millimeters, given that $C = 1$ and $L = 100$ millimeters.

70. MINIMIZING A CYLINDRICAL CONTAINER A cylindrical soft drink can is to be made so that it will have a volume of 354 milliliters. If r is the radius of the can in centimeters, then the total surface area A of the can is given by the rational function

$$A(r) = \frac{2\pi r^3 + 708}{r}$$

a. Use the graph of A to estimate the value of r that produces the minimum value of A.

b. Does the graph of A have a slant asymptote?

c. Explain the meaning of the following statement as it applies to the graph of A.

$$\text{As } r \to \infty, \ A \to 2\pi r^2.$$

SUPPLEMENTAL EXERCISES

71. Determine the point where the graph of

$$F(x) = \frac{2x^2 + 3x + 4}{x^2 + 4x + 7}$$

intersects its horizontal asymptote.

72. Determine the point where the graph of

$$F(x) = \frac{3x^3 + 2x^2 - 8x - 12}{x^2 + 4}$$

intersects its slant asymptote.

73. Determine the two points where the graph of

$$F(x) = \frac{x^3 + x^2 + 4x + 1}{x^3 + 1}$$

intersects its horizontal asymptote.

74. Give an example of a rational function that intersects its slant asymptote at two points.

PROJECTS

1. **INVESTIGATE VERTICAL ASYMPTOTES** The Theorem on Vertical Asymptotes given in this section requires that the numerator and denominator of a rational function have no common factors. Note from the graphs of rational functions in this section that in the case of all vertical asymptotes, the values of the function approach plus or minus infinity as x approaches the vertical asymptote.

a. Because $Q(x) = \dfrac{x + 1}{x - 2}$ has no common factors, by the Theorem on Vertical Asymptotes the graph of Q has a vertical asymptote when $x = 2$. Show that $Q(x)$ approaches plus or minus infinity as x approaches 2 by completing the following tables.

x	2.1	2.01	2.001	2.0001	2.00001
Q(x)					

On the basis of the table results, complete the following sentence. "As x approaches 2 from the right, $Q(x)$ approaches _____ infinity." Now complete the table below.

x	1.9	1.99	1.999	1.9999	1.99999
Q(x)					

On the basis of the table results, complete the following sentence. "As x approaches 2 from the left, $Q(x)$ approaches _____ infinity."

b. Now consider the rational function $R(x) = \dfrac{x^2 + 2x - 8}{x^2 + x - 6}$. In this case, $x - 2$ is a common factor. Verify this! Complete the table below.

x	2.1	2.01	2.001	2.0001	2.00001
R(x)					

On the basis of the table results, complete the following sentence. "As x approaches 2 from the right, $R(x)$ approaches _____." Now complete the table below.

x	1.9	1.99	1.999	1.9999	1.99999
$R(x)$					

On the basis of the table results, complete the following sentence. "As x approaches 2 from the left, $R(x)$ approaches _____." From your work on the rational function R, does the graph of R have a vertical asymptote at $x = 2$?

c. Explain the conditions under which a rational function will be undefined at $x = a$ but whose graph will not have an asymptote at $x = a$.

d. Explain the conditions under which a rational function will be undefined at $x = a$ and whose graph will have an asymptote at $x = a$.

EXPLORING CONCEPTS WITH TECHNOLOGY

Finding Zeros of a Polynomial using *Mathematica*

Computer algebra systems (CAS) are computer programs that are used to solve equations, graph functions, simplify algebraic expressions, and help us perform many other mathematical tasks. In this exploration, we will demonstrate how to use one of these programs, *Mathematica*, to find zeros of a polynomial.

Recall that a zero of a function P is a number, x, for which $P(x) = 0$. The idea behind finding a zero of a polynomial by using a CAS is to solve the polynomial equation $P(x) = 0$ for x.

Two commands in *Mathematica* that can be used to solve an equation are Solve and NSolve. (*Mathematica* is very sensitive about syntax (the way in which an expression is typed.) You *must* use upper-case and lower-case letters as we indicate.) Solve will attempt to find an *exact* solution of the equation; NSolve attempts to find *approximate* solutions. Here are some examples.

To find the exact values of the zeros of $P(x) = x^3 + 5x^2 + 11x + 15$, input the following. *Note:* The two equals signs are necessary.

$$\text{Solve[x^3+5x^2+11x+15==0]}$$

Press ⌈Enter⌉. The result should be

$$\{\{x->-3\}, \{x->-1-2\ \text{I}\}, \{x->-1+2\ \text{I}\}\}$$

Thus the three zeros of P are -3, $-1 - 2i$, and $-1 + 2i$.

To find the approximate values of the zeros of $P(x) = x^4 - 3x^3 + 4x^2 + x - 4$, input the following.

$$\text{NSolve[x^4-3x^3+4x^2+x-4==0]}$$

Press ⌈Enter⌉. The result should be

$$\{\{x->-0.821746\}, \{x->1.2326\}, \{x->1.29457-1.50771\ \text{I}\},$$
$$\{x->1.29457+1.50771\ \text{I}\}\}$$

The four zeros are (approximately) -0.821746, 1.2326, $1.29457 - 1.50771i$, and $1.29457 + 1.50771i$.

Not all polynomial equations can be solved exactly. This means that Solve will not always give solutions with *Mathematica*. Consider the two examples below.

Input   NSolve[x^5−3x^3+2x^2−5==0]

Output   {{x−>−1.80492}, {x−>−1.12491}, {x−>0.620319−1.03589 ɪ}, {x−>0.620319+1.03589 ɪ}, {x−>1.68919}}

These are the approximate zeros of the polynomial.

Input   Solve[x^5−3x^3+2x^2−5==0]

Output   {ToRules[Roots[2x²−3x³+x⁵==5]]}

In this case, no exact solution could be found. In general, there are no formulas like the quadratic formula, for instance, that yield exact solutions for fifth- or higher-degree polynomial equations.

Use *Mathematica* (or another CAS) to find the zeros of each polynomial.

1. $P(x) = x^4 - 3x^3 + x - 5$ **2.** $P(x) = 3x^3 - 4x^2 + x - 3$

3. $P(x) = 4x^5 - 3x^3 + 2x^2 - x + 2$ **4.** $P(x) = -3x^4 - 6x^3 + 2x - 8$

CHAPTER 4 REVIEW

4.1   Polynomial Division and Synthetic Division

- *The Remainder Theorem*   If a polynomial $P(x)$ is divided by $(x - c)$, then the remainder is $P(c)$.

- *The Factor Theorem*   A polynomial $P(x)$ has a factor $(x - c)$ if and only if $P(c) = 0$.

4.2   Introduction to Polynomial Functions

- Characteristics and properties used in graphing polynomial functions:

  1. Continuity—Polynomial functions are smooth continuous curves.

  2. Leading term test—Determines the behavior of the graph of a polynomial function at the far right or the far left.

  3. Zeros of the function determine the x-intercepts.

- *The Zero Location Theorem*   Let $P(x)$ be a polynomial with real coefficients. If $a < b$, and if $P(a)$ and $P(b)$ have opposite signs, then there is at least one value c between a and b such that $P(c) = 0$.

4.3   Zeros of Polynomial Functions

- Values of x that satisfy $P(x) = 0$ are called zeros of P.

- Definition of Multiple Zeros of a Polynomial   If a polynomial $P(x)$ has $(x - r)$ as a factor exactly k times, then r is said to be a zero of multiplicity k of the polynomial $P(x)$.

- Number of Zeros of a Polynomial Function   A polynomial function P of degree n has at most n zeros, where each zero of multiplicity k is counted k times.

- *The Rational Zero Theorem*   If
  $$P(x) = a_n x^n + a_{n-1} x^{n-1} + \cdots + a_1 x + a_0$$
  has integer coefficients, and p/q (where p and q have no common prime factors) is a rational zero of $P(x)$, then p is a factor of a_0 and q is a factor of a_n.

4.4   The Fundamental Theorem of Algebra

- *The Fundamental Theorem of Algebra*   If $P(x)$ is a polynomial of degree $n \geq 1$ with complex coefficients, then $P(x)$ has at least one complex zero.

- *The Number of Zeros of a Polynomial*   If $P(x)$ is a polynomial of degree $n \geq 1$ with complex coefficients, then $P(x)$ has exactly n complex zeros, provided that each zero is counted according to its multiplicity.

- *The Conjugate Pair Theorem*   If $a + bi$ ($b \neq 0$) is a complex zero of the polynomial $P(x)$, with real coefficients, then the conjugate $a - bi$ is also a complex zero of the polynomial.

4.5   Rational Functions and Their Graphs

- If $P(x)$ and $Q(x)$ are polynomials, then the function F given by
  $$F(x) = \frac{P(x)}{Q(x)}$$
  is called a rational function.

- **General Procedure for Graphing Rational Functions That Have No Common Factors**

  1. Find the real zeros of the denominator. For each zero a the vertical line $x = a$ will be a vertical asymptote. The vertical asymptotes will occur at these points. Use the Theorem on Horizontal Asymptotes to determine the equation of any horizontal asymptote. Graph any horizontal asymptotes.

  2. Find the real zeros of the numerator. For each zero a, plot $(a, 0)$. These points will give the x-intercepts.

  3. Use the tests for symmetry to determine whether the graph has symmetry with respect to the x-axis or to the origin.

  4. Find additional points that lie in the intervals between the x-intercepts and the vertical asymptotes.

  5. Determine the behavior near the asymptotes.

CHAPTER 4 TRUE/FALSE EXERCISES

In Exercises 1 to 14, answer true or false. If the statement is false, give an example.

1. The complex zeros of a polynomial with complex coefficients always occur in conjugate pairs.

2. Descartes' Rule of Signs indicates that $x^3 - x^2 + x - 1$ must have three positive zeros.

3. The polynomial $2x^5 + x^4 - 7x^3 - 5x^2 + 4x + 10$ has two variations in sign.

4. If 4 is an upper bound of the zeros of the polynomial P, then 5 is also an upper bound of the zeros of P.

5. The graph of every rational function has a vertical asymptote.

6. The graph of the rational function
$$F(x) = \frac{x^2 - 4x + 4}{x^2 - 5x + 6}$$
has a vertical asymptote of $x = 2$.

7. If 7 is a zero of the polynomial P, then $x - 7$ is a factor of P.

8. According to the Zero Location Theorem, the polynomial function $P(x) = x^3 + 6x - 2$ has a real zero between 0 and 1.

9. Synthetic division can be used to show that $3i$ is a zero of $x^3 - 2x^2 + 9x - 18$.

10. Every fourth-degree polynomial with complex coefficients has exactly four complex zeros, provided that each zero is counted according to its multiplicity.

11. The graph of a rational function never intersects any of its vertical asymptotes.

12. The graph of a rational function can have at most one horizontal asymptote.

13. Descartes' Rule of Signs indicates that the polynomial function $P(x) = x^3 + 2x^2 + 4x - 7$ does have a positive zero.

14. Every polynomial has at least one real zero.

CHAPTER 4 REVIEW EXERCISES

In Exercises 1 to 6, use long division to divide the first polynomial by the second.

1. $x^3 + 5x^2 + 2x - 17, \ x^2 + x + 3$

2. $2x^3 - 5x + 1, \ x^2 + 4$

3. $-x^4 + 2x^2 - 12x - 3, \ x^3 + x$

4. $x^3 - 5x^2 - 6x - 11, \ x^2 - 6x - 1$

5. $6x^4 + 8x^3 - 47x^2 + 19x + 5, \ 2x^2 + 6x - 5$

6. $x^4 + 3x^3 - 6x^2 - 13x + 15, \ x^2 + 2x - 3$

In Exercises 7 to 12, use synthetic division to divide the first polynomial by the second.

7. $4x^3 - 11x^2 + 5x - 2, \ x - 3$

8. $5x^3 - 18x + 2, \ x - 1$

9. $3x^3 - 5x + 1, \ x + 2$

10. $2x^3 + 7x^2 + 16x - 10, \ x - \dfrac{1}{2}$

11. $3x^3 - 10x^2 - 36x + 55, \ x - 5$

12. $x^4 + 9x^3 + 6x^2 - 65x - 63, \ x + 7$

In Exercises 13 to 16, use the Remainder Theorem to find $P(c)$.

13. $P(x) = x^3 + 2x^2 - 5x + 1, \ c = 4$

14. $P(x) = -4x^3 - 10x + 8, \ c = -1$

15. $P(x) = 6x^4 - 12x^2 + 8x + 1, \ c = -2$

16. $P(x) = 5x^5 - 8x^4 + 2x^3 - 6x^2 - 9, \ c = 3$

In Exercises 17 to 20, use synthetic division to show that c is a zero of the given polynomial.

17. $x^3 + 2x^2 - 26x + 33, c = 3$

18. $2x^4 + 8x^3 - 8x^2 - 31x + 4, c = -4$

19. $x^5 - x^4 - 2x^2 + x + 1, c = 1$

20. $2x^3 + 3x^2 - 8x + 3, c = \dfrac{1}{2}$

In Exercises 21 to 26, graph the polynomial function.

21. $P(x) = x^3 - x$

22. $P(x) = -x^3 - x^2 + 8x + 12$

23. $P(x) = x^4 - 6$ 24. $P(x) = x^5 - x$

25. $P(x) = x^4 - 10x^2 + 9$ 26. $P(x) = x^5 - 5x^3$

In Exercises 27 to 32, use the Rational Zero Theorem to list all possible rational zeros for each polynomial.

27. $x^3 - 7x - 6$ 28. $2x^3 + 3x^2 - 29x - 30$

29. $15x^3 - 91x^2 + 4x + 12$

30. $x^4 - 12x^3 + 52x^2 - 96x + 64$

31. $x^3 + x^2 - x - 1$ 32. $6x^5 + 3x - 2$

In Exercises 33 to 36, use Descartes' Rule of Signs to state the number of possible positive and negative real zeros of each polynomial.

33. $x^3 + 3x^2 + x + 3$

34. $x^4 - 6x^3 - 5x^2 + 74x - 120$

35. $x^4 - x - 1$

36. $x^5 - 4x^4 + 2x^3 - x^2 + x - 8$

In Exercises 37 to 42, find the zeros of the polynomial.

37. $x^3 + 6x^2 + 3x - 10$ 38. $x^3 - 10x^2 + 31x - 30$

39. $6x^4 + 35x^3 + 72x^2 + 60x + 16$

40. $2x^4 + 7x^3 + 5x^2 + 7x + 3$

41. $x^4 - 4x^3 + 6x^2 - 4x + 1$ 42. $2x^3 - 7x^2 + 22x + 13$

43. Find a third-degree polynomial with zeros of 4, -3, and 1/2.

44. Find a fourth-degree polynomial with zeros of 2, -3, i, and $-i$.

45. Find a fourth-degree polynomial with real coefficients that has zeros of 1, 2, and $5i$.

46. Find a fourth-degree polynomial with real coefficients that has -2 as a zero of multiplicity 2 and also has $1 + 3i$ as a zero.

In Exercises 47 to 50, find the vertical, horizontal, and slant asymptotes for each rational function.

47. $f(x) = \dfrac{3x + 5}{x + 2}$ 48. $f(x) = \dfrac{2x^2 + 12x + 2}{x^2 + 2x - 3}$

49. $f(x) = \dfrac{2x^2 + 5x + 11}{x + 1}$ 50. $f(x) = \dfrac{6x^2 - 1}{2x^2 + x + 7}$

In Exercises 51 to 58, graph each rational function.

51. $f(x) = \dfrac{3x - 2}{x}$ 52. $f(x) = \dfrac{x + 4}{x - 2}$

53. $f(x) = \dfrac{6}{x^2 + 2}$ 54. $f(x) = \dfrac{4x^2}{x^2 + 1}$

55. $f(x) = \dfrac{2x^3 - 4x + 6}{x^2 - 4}$ 56. $f(x) = \dfrac{x}{x^3 - 1}$

57. $f(x) = \dfrac{3x^2 - 6}{x^2 - 9}$ 58. $f(x) = \dfrac{-x^3 + 6}{x^2}$

In Exercises 59 to 62, the given polynomials have one zero that satisfies the given condition. Use a graphing utility and the Zero Location Theorem to approximate the zero to the nearest 0.001.

59. $x^3 - x - 1 = 0, \quad x > 0$

60. $x^3 - 3x - 6 = 0, \quad x > 0$

61. $x^4 + x^2 - 1 = 0, \quad x > 0$

62. $x^4 - 2x^2 - 2 = 0, \quad x > 0$

CHAPTER 4 TEST

1. Use synthetic division to divide:

$$(3x^3 + 5x^2 + 4x - 1) \div (x + 2)$$

2. Use the Remainder Theorem to find $P(-2)$ if

$$P(x) = -3x^3 + 7x^2 + 2x - 5$$

3. Show that $x - 1$ is a factor of

$$x^4 - 4x^3 + 7x^2 - 6x + 2$$

4. Examine the leading term of the function given by the equation $P(x) = -3x^3 + 2x^2 - 5x + 2$ and determine the far-left and far-right behavior of the graph of the polynomial function.

5. Find the real solutions of $3x^3 + 7x^2 - 6x = 0$.

6. Use the Zero Location Theorem to verify that

$$P(x) = 2x^3 - 3x^2 - x + 1$$

has a zero between 1 and 2.

7. Find the zeros of the polynomial

$$P(x) = (x^2 - 4)^2(2x - 3)(x + 1)^3$$

and state the multiplicity of each.

8. Use the Rational Zero Theorem to list the possible rational zeros for the polynomial

$$P(x) = 6x^3 - 3x^2 + 2x - 3$$

9. Find, by using the Upper- and Lower-Bound Theorem, the smallest positive integer and the largest negative integer that are upper and lower bounds for the polynomial

$$P(x) = 2x^4 + 5x^3 - 23x^2 - 38x + 24$$

10. Use Descartes' Rule of Signs to state the number of possible positive and negative real zeros of

$$P(x) = x^4 - 3x^3 + 2x^2 - 5x + 1$$

11. Find the zeros of the polynomial

$$P(x) = 2x^3 - 3x^2 - 11x + 6$$

12. Given that $-i$ is a zero of $P(x) = 2x^4 - 3x^3 - 3x - 2$, find the remaining zeros.

13. Find all the zeros of the polynomial

$$P(x) = x^5 - 6x^4 + 14x^3 - 14x^2 + 5x$$

14. Find a polynomial of lowest degree that has real coefficients and zeros $1 + i$, 3, and 0.

15. Find all vertical asymptotes of the graph of

$$f(x) = \frac{3x^2 - 2x + 1}{x^2 - 5x + 6}$$

16. Find all horizontal asymptotes of the graph of

$$f(x) = \frac{3x^2 - 2x + 1}{2x^2 - 1}$$

17. Graph: $f(x) = \dfrac{x^2 - 1}{x^2 - 2x - 3}$

18. Graph: $f(x) = \dfrac{2x^2 + 2x + 1}{x + 1}$

19. Graph: $f(x) = \dfrac{x}{x^2 + 1}$

20. Use a graphing utility to approximate the zero in the interval $1 < x < 2$ of the polynomial $P(x) = x^3 - 5x + 3$ to within one-tenth of a unit.

EXPONENTIAL AND LOGARITHMIC FUNCTIONS

The shape of the Eiffel Tower can be modeled by an equation involving the number e.

The number e is instrumental in determining the length of the reeds of a panpipe.

Logarithms can be used to model and recreate the spiral in a nautilus shell.

An Optical Illusion

The St. Louis Gateway Arch is one of the largest optical illusions ever created. As you look at the arch, it seems to be much taller than it is wide. However, the two distances are exactly the same, 630 feet. You can measure the arch on this page and see for yourself that its width and height are equal.

The equation of the curve described by the Gateway Arch is approximated by

$$y = 693.8597 - 68.7672\left(\frac{e^{0.0100333x} + e^{-0.0100333x}}{2}\right)$$

This equation involves *exponential functions*, one of the topics of this chapter.

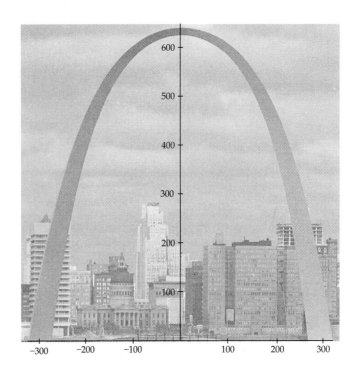

5.1 EXPONENTIAL FUNCTIONS AND THEIR GRAPHS

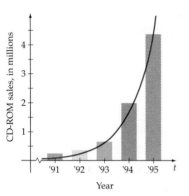

Figure 5.1

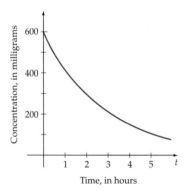

Figure 5.2

The bar graph in **Figure 5.1** depicts the growth of CD-ROM sales. (CD-ROM stands for compact disk, read-only memory.) The curve is a mathematical model of the sales growth. This model is based on an *exponential function.*

The effectiveness of sodium pentabaritol, which is used for surgical anesthesia, during an operation depends on the concentration of the drug in the patient. Through natural body chemistry, the amount of this drug in the body decreases over time. The graph in **Figure 5.2** shows the decrease. This graph is another example of an *exponential model.*

DEFINITION OF AN EXPONENTIAL FUNCTION

The **exponential function** f with base b is defined by

$$f(x) = b^x$$

where $b > 0$, $b \neq 1$, and x is any real number.

In the definition of an exponential function, b, the base, is required to be positive. If the base of an exponential function were a negative number, the value of the function would be a complex number for some values of x. For instance, the value of $f(x) = (-4)^x$ when $x = 1/2$ is

$$f\left(\frac{1}{2}\right) = (-4)^{1/2} = \sqrt{-4} = 2i$$

For this reason, the base of an exponential function is a defined to be positive number. If $b = 1$, then $1^x = 1$ for all values of x. If $b = 0$, then $0^x = 0$ for $x > 0$ and is undefined for $x \leq 0$.

Examples of exponential functions are

$$f(x) = 2^x, \qquad g(x) = \left(\frac{2}{3}\right)^x, \qquad \text{and} \qquad h(x) = \pi^x$$

The value of $f(x) = 2^x$ when $x = 3$ is $f(3) = 2^3 = 8$.

The value of $g(x) = \left(\frac{2}{3}\right)^x$ when $x = -2$ is $g(-2) = \left(\frac{2}{3}\right)^{-2} = \left(\frac{3}{2}\right)^2 = \frac{9}{4}$.

To evaluate an exponential function for an irrational number such as $\sqrt{3}$ or π, an approximation to the value of the function can be obtained by approximating the irrational number. For instance, the value of $f(x) = 4^x$ when $x = \sqrt{5}$ can be approximated by using an approximation of $\sqrt{5}$.

$$f(\sqrt{5}) = 4^{\sqrt{5}} \approx 4^{2.236068} \approx 22.194587$$

Because $f(x) = b^x$ is a real number for both rational and irrational numbers x, the domain of f is all real numbers. Because $b > 0$, $b^x > 0$ for all values of x;

therefore, the range of f is the set of positive real numbers. Other properties of the exponential function can be determined from its graph.

The graph of $f(x) = (3/2)^x$ is shown in **Figure 5.3.** The coordinates of some of the points on the curve are given in Table 5.1.

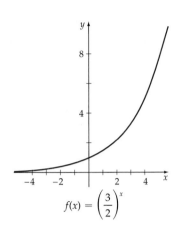

$$f(x) = \left(\frac{3}{2}\right)^x$$

Figure 5.3

Table 5.1

x	−2	−1	0	1	2
y	$\dfrac{4}{9}$	$\dfrac{2}{3}$	1	$\dfrac{3}{2}$	$\dfrac{9}{4}$

Observe the following properties of the graph of an exponential function for which b is greater than 1. (In **Figure 5.3,** $b = 3/2 > 1$.)

● The y-intercept is $(0, 1)$.

● $f(x) \to 0$ as $x \to -\infty$. Therefore, the x-axis is a horizontal asymptote.

● The graph of f is the graph of an increasing function. By the horizontal line test, f is a one-to-one function.

When using a graphing utility to draw the graph of $f(x) = \left(\dfrac{3}{2}\right)^x$, remember the Order of Operations Agreement. Entering **3/2^X** will result in $\dfrac{3}{2^x}$, which is not the same as $\left(\dfrac{3}{2}\right)^x$. You must use parentheses to ensure that division is performed before exponentiation.

Now consider the graph of an exponential function for which $0 < b < 1$. The graph of $g(x) = (1/2)^x$ is shown in **Figure 5.4.** The coordinates of some of the points on the curve are given in Table 5.2.

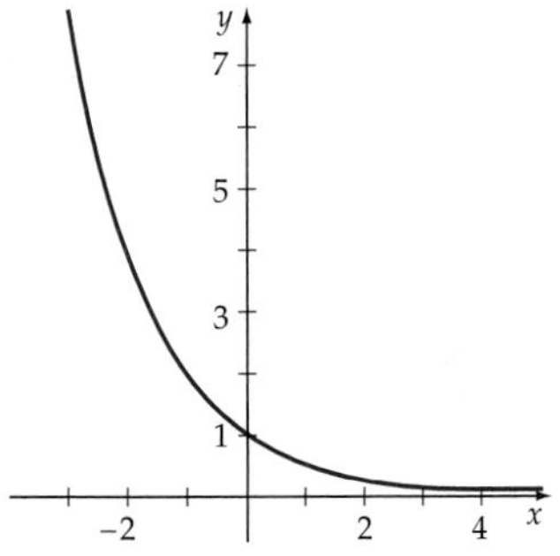

$$g(x) = \left(\frac{1}{2}\right)^x$$

Figure 5.4

Table 5.2

x	−2	−1	0	1	2	3
y	4	2	1	$\dfrac{1}{2}$	$\dfrac{1}{4}$	$\dfrac{1}{8}$

Observe the following properties of the graph of an exponential function for which b is between 0 and 1. (In **Figure 5.4,** $b = 1/2$ and $0 < 1/2 < 1$.)

● The y-intercept is $(0, 1)$.

● $g(x) \to 0$ as $x \to \infty$. Therefore, the x-axis is a horizontal asymptote.

● The graph of g is the graph of a decreasing function. By the horizontal line test, g is a one-to-one function.

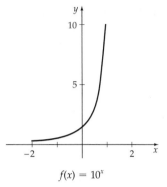

$f(x) = 10^x$

Figure 5.5

To graph an exponential function over a certain portion of its domain, you may need to use different scales on the x- and y-axes. For example, a graph of $f(x) = 10^x$, for $-2 \le x \le 1$, is shown in **Figure 5.5**. Observe that each space between tick marks on the y-axis represents a distance of 5 units and that each space between tick marks on the x-axis represents 1 unit.

PROPERTIES OF $f(x) = b^x$

For positive real numbers b, $b \ne 1$, the exponential function defined by $f(x) = b^x$ has the following properties:

1. f has the set of real numbers as its domain.
2. f has the set of positive real numbers as its range.
3. f has a graph with a y-intercept of $(0, 1)$.
4. f has a graph asymptotic to the x-axis.
5. f is a one-to-one function.
6. f is an increasing function if $b > 1$. See **Figure 5.6a**.
7. f is a decreasing function if $0 < b < 1$. See **Figure 5.6b**.

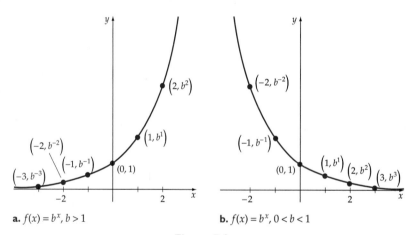

a. $f(x) = b^x, b > 1$ **b.** $f(x) = b^x, 0 < b < 1$

Figure 5.6

EXAMPLE 1   *Graph an Exponential Function Using Translations*

Sketch the graph of each function.

a. $F(x) = 2^x - 3$ **b.** $G(x) = 2^{x-3}$

Solution

a. The graph of $F(x) = 2^x - 3$ is the graph of $f(x) = 2^x$ shifted down 3 units as shown in **Figure 5.7**.

b. The graph of $G(x) = 2^{x-3}$ is the graph of $g(x) = 2^x$ shifted 3 units to the right, as shown in **Figure 5.8**.

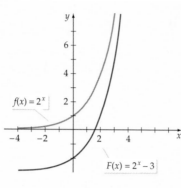

Figure 5.7

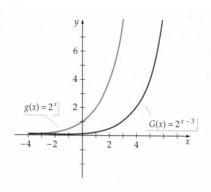

Figure 5.8

TRY EXERCISE 36, EXERCISE SET 5.1

When a graphing utility is used to draw a graph, it is important to look at the graph and ask yourself, "Does this graph have the characteristics of the function I intended to graph?" What you must guard against is incorrectly entering the function. The graphing utility will graph only what you enter—not what you intended to enter.

For instance, suppose we want to draw the graph of $f(x) = 3(2^{-x^2})$ and produce the graph shown in **Figure 5.9**. First note that

$$f(-x) = 3(2^{-(-x)^2}) = 3(2^{-x^2}) = f(x)$$

Thus f is an even function, and its graph should be symmetric with respect to the y-axis. This appears to be the case from our graph. Next, if we write f as

$$f(x) = 3/2^{x^2}$$

we observe that as $|x|$ increases without bound, the denominator increases without bound. Therefore, $f(x)$ is approaching zero. This is also consistent with our graph. For another observation, recall that the value of a fraction with a constant numerator is as large as possible when the denominator is as small as possible. For our function, the smallest denominator occurs when $x = 0$. In that case $f(0) = 3$, which is again consistent with our graph. It appears that we have entered the function correctly and produced the desired graph. As a final check, we can evaluate the function for various values of x and compare those values to values found by using the TRACE feature of the graphing utility.

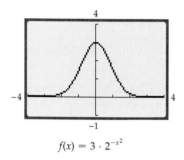

$f(x) = 3 \cdot 2^{-x^2}$

Figure 5.9

EXAMPLE 2 *Graph a Function of the Form* $f(x) = ab^{p(x)}$

Use a graphing utility to graph $f(x) = 2^{|x|}$. Then use your knowledge of functions to verify the accuracy of the graph.

Continued ▸▶

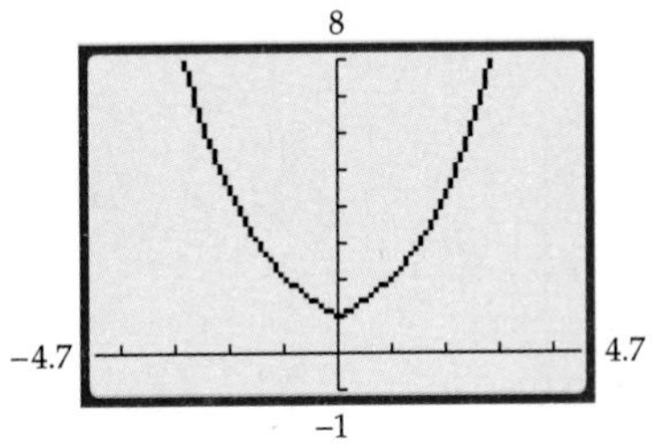

$f(x) = 2^{|x|}$

Figure 5.10

Solution

The graph as drawn with a graphing utility is shown in **Figure 5.10**. The minimum value of f occurs when $|x| = 0$, which means $x = 0$. At $x = 0$, $f(x) = 2^{|0|} = 1$. This is consistent with the graph. As shown below, f is an even function.

$$f(x) = 2^{|x|}$$
$$f(-x) = 2^{|-x|} = 2^{|x|} = f(x)$$

Because f is an even function, the graph of f should be symmetric with respect to the y-axis. This is consistent with the graph of f in **Figure 5.10**. As a final check, evaluate f at a few values of x and compare them to the corresponding values you find by tracing along the curve.

TRY EXERCISE 40, EXERCISE SET 5.1

THE NATURAL EXPONENTIAL FUNCTION

Table 5.3

Value of n	Value of $\left(1 + \dfrac{1}{n}\right)^n$
1	2
10	2.59374246
100	2.704813829
1000	2.716923932
10,000	2.718145927
100,000	2.718268237
1,000,000	2.718280469
10,000,000	2.718281693

The irrational number π is often used in applications that involve circles. Another irrational number called e is useful for applications that involve the growth of a population or radioactive decay. Using techniques developed in calculus, it can be shown that as $n \to \infty$,

$$\left(1 + \frac{1}{n}\right)^n \to e$$

The letter e was chosen in honor of Leonhard Euler (1707–1783). He was able to compute e to several places by using large values of n to evaluate $(1 + 1/n)^n$. The entries in Table 5.3 illustrate the process. The value of e accurate to eight places is 2.71828183.

THE NATURAL EXPONENTIAL FUNCTION

For all real numbers x, the function defined by

$$f(x) = e^x$$

is called the **natural exponential function.**

To evaluate e^x for specific values of x, you use a calculator with an $\boxed{e^x}$ key. For example,

$$e^2 \approx 7.389056099$$
$$e^{4.21} \approx 67.35653981$$
$$e^{-1.8} \approx 0.165298888$$

To graph the natural exponential function, use a calculator to approximate e^x for the desired domain values. Then plot the resulting points and connect them with a smooth curve.

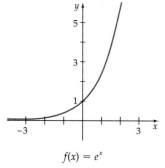

$f(x) = e^x$

Figure 5.11

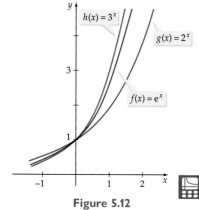

Figure 5.12

EXAMPLE 3   *Graph the Natural Exponential Function*

Graph: $f(x) = e^x$

Solution

The values in the table below have been rounded to the nearest hundredth. Plot the points and then connect the points with a smooth curve. Because $e > 1$, we know by the properties of exponential functions that the graph of $f(x) = e^x$ is an increasing function. To the far left the graph is asymptotic to the x-axis. The y-intercept is $(0, 1)$. See **Figure 5.11**.

x	-3	-2	-1	1	2
$f(x) = e^x$	0.05	0.14	0.37	2.72	7.39

TRY EXERCISE 44, EXERCISE SET 5.1

Note in **Figure 5.12** how the graph of $f(x) = e^x$ compares with the graphs of $g(x) = 2^x$ and $h(x) = 3^x$. You may have anticipated that the graph of f would be between the graph of g and that of h because e is between 2 and 3.

EXAMPLE 4   *Graph a Combination of Exponential Functions*

Use a graphing utility graph $S(x) = \dfrac{e^x - e^{-x}}{2}$.

Solution

The graph as drawn with a graphing utility is shown in **Figure 5.13**. The graph is symmetric about the origin because S, as shown below, is an odd function.

$$S(x) = \frac{e^x - e^{-x}}{2}$$

$$S(-x) = \frac{e^{-x} - e^{-(-x)}}{2} = \frac{e^{-x} - e^x}{2} = -\left(\frac{e^x - e^{-x}}{2}\right) = -S(x)$$

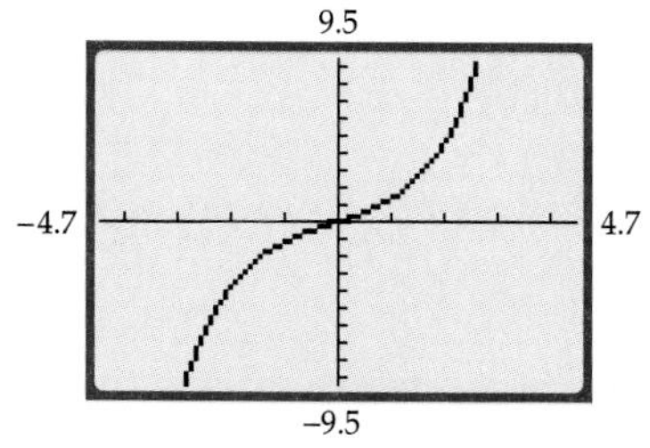

As another check, evaluate S for a few values of x and compare them to the corresponding values you find by tracing along the curve. For instance, $S(0) = \dfrac{e^0 - e^{-0}}{2} = \dfrac{1 - 1}{2} = 0$. Thus $(0, 0)$ is on the graph. This result also is consistent with the graph.

TRY EXERCISE 48, EXERCISE SET 5.1

$S(x) = \dfrac{e^x - e^{-x}}{2}$

Figure 5.13

Recall that a zero of a function is a value x for which $f(x) = 0$, and when $f(x) = 0$, the graph of f intersects the x-axis. Thus the real number zeros of a function f can be approximated by determining the x-intercepts of its graph.

EXAMPLE 5 *Determine the Zero of a Function*

Use a graphing utility to determine the zero of $f(x) = -\dfrac{1}{3}e^x + 2$ to the nearest hundredth.

Solution

Graph f and use the features of your graphing utility to find the x-coordinate of the x-intercept to the nearest hundredth. From the graph in **Figure 5.14** the x-coordinate of the x-intercept is approximately 1.79. Therefore, the zero of f to the nearest hundredth is 1.79.

You can verify that 1.79 approximates the zero by evaluating f at 1.79.

$$f(1.79) = -\frac{1}{3}e^{1.79} + 2 \approx 0.00352 \approx 0$$

TRY EXERCISE 50, EXERCISE SET 5.1

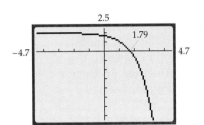

$f(x) = -\dfrac{1}{3}e^x + 2$

Figure 5.14

TOPICS FOR DISCUSSION

1. The definition of an exponential function as $f(x) = b^x$ requires that $b > 0$ and $b \neq 1$. Discuss why these conditions are imposed.

2. Discuss the properties of the graph of $f(x) = b^x$ when $b > 1$.

3. Discuss the properties of the graph of $f(x) = b^x$ when $0 < b < 1$.

4. What is the base of the natural exponential function? How is it calculated? What is its approximate value?

EXERCISE SET 5.1

In Exercises 1 to 12, evaluate each power accurate to six significant digits.

1. $3^{\sqrt{2}}$ **2.** $5^{\sqrt{3}}$ **3.** $10^{\sqrt{7}}$ **4.** $10^{\sqrt{11}}$

5. $\sqrt{3}^{\sqrt{2}}$ **6.** $\sqrt{5}^{\sqrt{7}}$ **7.** $e^{5.1}$ **8.** $e^{-3.2}$

9. $e^{\sqrt{3}}$ **10.** $e^{\sqrt{5}}$ **11.** $e^{-0.031}$ **12.** $e^{-0.42}$

In Exercises 13 to 24, evaluate each functional value accurate to six significant digits, given that $f(x) = 3^x$ and $g(x) = e^x$.

13. $f(\sqrt{15})$ **14.** $f(\pi)$ **15.** $f(e)$

16. $f(-\sqrt{15})$ **17.** $g(\sqrt{7})$ **18.** $g(\pi)$

19. $g(e)$ **20.** $g(-3.4)$ **21.** $f[g(2)]$

22. $f[g(-1)]$ **23.** $g[f(2)]$ **24.** $g[f(-1)]$

In Exercises 25 to 48, sketch the graph of each function.

25. $f(x) = 3^x$ **26.** $f(x) = 4^x$ **27.** $f(x) = \left(\dfrac{3}{2}\right)^x$

28. $f(x) = \left(\dfrac{4}{3}\right)^x$ **29.** $f(x) = \left(\dfrac{1}{3}\right)^x$ **30.** $f(x) = \left(\dfrac{2}{3}\right)^x$

31. $f(x) = \left(\dfrac{1}{2}\right)^{-x}$ **32.** $f(x) = \left(\dfrac{1}{3}\right)^{-x}$ **33.** $f(x) = \dfrac{5^x}{2}$

34. $f(x) = \dfrac{10^x}{10}$ **35.** $f(x) = 2^{x+2}$ **36.** $f(x) = 2^{x+3}$

37. $f(x) = 3^x - 1$ **38.** $f(x) = 3^x + 1$ **39.** $f(x) = 3^{x^2}$

40. $f(x) = 2^{-|x|}$ **41.** $f(x) = \dfrac{3^x + 3^{-x}}{2}$ **42.** $f(x) = 4 \cdot 3^{-x^2}$

43. $f(x) = e^{-x}$ **44.** $f(x) = 2e^x$

45. $f(x) = -e^x$ **46.** $f(x) = 0.5e^{-x}$

47. $f(x) = e^{-x^2}$ **48.** $f(x) = \dfrac{e^x + e^{-x}}{2}$

In Exercises 49 to 56, use a graphing utility to determine the zero of f to the nearest hundredth.

49. $f(x) = 2^x - 3$ **50.** $f(x) = 3^{-x} - 4$

51. $f(x) = 1 - 2e^{-x}$ **52.** $f(x) = 3 - 4e^x$

53. $f(x) = e^x + x - 3$ **54.** $f(x) = 2x - 2^{-\frac{1}{2}x}$

55. $f(x) = 3^x + 2x - 4$ **56.** $f(x) = e^x + x - 5$

57. Graph $f(x) = x^x$ on $(0, 3]$. Estimate

   a. the minimum value of f to the nearest ten thousandth on this interval

   b. the behavior of f as x approaches 0 from the right

58. Graph the **normal distribution curve** defined by

$$f(x) = \frac{1}{\sqrt{2\pi}}e^{-x^2/2}$$

Estimate the maximum value of f (to the nearest tenth).

59. The number of bacteria present in a culture is given by $N(t) = 10,000(2^t)$, where $N(t)$ is the number of bacteria present after t hours. Find the number of bacteria present for each value of t.

   a. $t = 1$ hour   **b.** $t = 2$ hours   **c.** $t = 5$ hours

60. The production function for an oil well is given by the function $B(t) = 100,000(e^{-0.2t})$, where $B(t)$ is the number of barrels of oil the well can produce per month after t years. Find the number of barrels of oil the well can produce per month for each value of t.

   a. $t = 1$ year   **b.** $t = 2$ years   **c.** $t = 5$ years

61. Calculating Monthly Car Payments A formula to determine the monthly payment (PMT) for a car loan, home mortgage, or other installment loan is given by

$$\text{PMT} = P\left(\frac{i/12}{1 - (1 + i/12)^{-n}}\right)$$

where P (called the present value) is the amount borrowed, i is the annual interest rate, and n is the total number of payments.

   a. An accountant purchases a car and secures a loan for $9000 at an annual interest rate of 10% for a term of 4 years. Find the monthly car payment.

   b. If the accountant makes all 48 payments, how much money will be repaid?

   c. How much of the total repaid is interest?

62. Calculating Monthly Car Payments Using the formula in Exercise 61, answer the following questions.

   a. A nurse purchases a car and secures a loan for $15,000 at an annual interest rate of 9% for a term of 5 years. Find the monthly car payment.

   b. If the nurse makes all 60 payments, how much money will be repaid?

   c. How much of the total repaid is interest?

63. Calculating the Payoff of a Car Loan The formula in Exercise 61 can be solved for P with the following result:

$$P = \text{PMT}\left(\frac{1 - (1 + i/12)^{-n}}{i/12}\right)$$

This formula gives the present value (amount owed on a loan) for the *remaining n* payments of an installment loan.

   a. Suppose a person has a loan that has a monthly payment of $258, has an annual interest rate of 9%, and runs for 5 years. What is the loan amount?

   b. How much is owed after the borrower has made 12 payments?

   c. After how many months will the present value first be less than one-half the original present value?

64. Musical Scales Starting on the left of a standard 88-key piano, the frequency of the nth note is given by $f(n) = (27.5)2^{(n-1)/12}$.

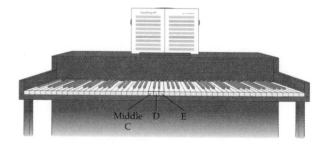

   a. Using this formula, determine the frequency of middle C, key number 40 on an 88-key piano.

   b. Is the difference in frequency between middle C and the next note, D, the same as the difference in frequency between D and the next note, E? Explain why this is so.

c. Some animals make a sound by blowing air through a nasal passage. The frequency f of the sound is approximately given by $f = 170/x$, where x is the length of the nasal passage in meters. Estimate the length of the nasal passage of an animal that emits a sound that has the frequency of the 70th key (from the left) on a piano.

SUPPLEMENTAL EXERCISES

In Exercises 65 to 72, graph f. State the domain and range of f using interval notation. When necessary, estimate values to the nearest tenth. Also state whether f is an even function, an odd function, or neither an even nor an odd function.

65. $f(x) = \dfrac{e^x - e^{-x}}{e^x + e^{-x}}$

66. $f(x) = x^2 e^x$

67. $f(x) = \dfrac{4x^2}{e^{|x|}}$

68. $f(x) = \sqrt{\dfrac{|x|}{1 + e^x}}$

69. $f(x) = \dfrac{e^{|x|}}{1 + e^x}$

70. $f(x) = 1 + e^{(x^3 - x^2 - 2x)}$

71. $f(x) = \sqrt{1 - e^x}$

72. $f(x) = \sqrt{e^x - e^{-x}}$

In Exercises 73 and 74, use $f(x) = e^x$ and $g(x) = e^{-x}$ to graph the given equations. State the domain and range of y using interval notation.

73. a. $y = (f + g)(x)$ **b.** $y = (f - g)(x)$

74. a. $y = (f \cdot g)(x)$ **b.** $y = (f/g)(x)$

75. Evaluate $h(x) = (-2)^x$ for each value of x.

    a. $x = 1$ **b.** $x = 2$ **c.** $x = 1.5$

    d. Explain what is meant by the statement that h is not a *real-valued* exponential function.

76. Graph $j(x) = 1^x$. Explain why j is not an exponential function.

77. Graph $f(x) = e^x$, and then sketch the graph of f reflected about the graph of the line given by $y = x$.

78. Graph $g(x) = 10^x$, and then sketch the graph of g reflected about the graph of the line given by $y = x$.

79. Prove that the hyperbolic sine function

$$\sinh(x) = \frac{e^x - e^{-x}}{2}$$

is an odd function.

80. Prove that $G(x) = e^x$ is neither an odd function nor an even function.

81. Which of the numbers e^π and π^e is larger?

82. Let $f(x) = x^{(x^x)}$ and $g(x) = (x^x)^x$. Which is larger, $f(3)$ or $g(3)$?

In Exercises 83 to 88, determine the a. vertical and horizontal asymptotes, b. x-intercepts, and c. graph of f.

83. $f(x) = \dfrac{2}{1 + e^x}$

84. $f(x) = \dfrac{4}{1 + 2^x}$

85. $f(x) = \dfrac{x + 1}{1 - 2^x}$

86. $f(x) = \dfrac{x - 1}{2 - e^x}$

87. $f(x) = \dfrac{1 - 2x}{1 - e^x}$

88. $f(x) = \dfrac{1 - x}{1 - 2^x}$

PROJECTS

1. PROPERTIES OF AN EXPONENTIAL FUNCTION Graph $f(x) = b^x$ for $b = 2, 3, 4$, and 10. On the basis of the graphs, complete the following statements.

    a. As $x \to \infty$, $f(x) \to$ _____?_____.   **b.** As $x \to -\infty$, $f(x) \to$ _____?_____.

    c. Let $x_1 = 1$ and $x_2 = 3$. For each value of b, calculate the slope of the line through $P_1(x_1, f(x_1))$ and $P_2(x_2, f(x_2))$. On the basis of your answers, complete the following statement: As b increases, the slope of the line through P_1 and P_2 _____.

    d. Calculate the slope of the line through $(x_1, f(x_1))$ and $(x_2, f(x_2))$, where x_1 and x_2 are consecutive integers. Is the slope of each line the same? If not, what relationship exists between the successive slopes?

SECTION

5.2 LOGARITHMS AND LOGARITHMIC PROPERTIES

Every exponential function of the form $f(x) = b^x$ is a one-to-one function and therefore has an inverse function. Sometimes we can determine the inverse of a function represented by an equation by interchanging the variables and then solving for the dependent variable. If we attempt to use this procedure for the function defined by $f(x) = b^x$, we get

$$f(x) = b^x$$
$$y = b^x$$
$$x = b^y \qquad \bullet \textbf{ Interchange the variables.}$$

None of our previous methods can be used to solve the equation $x = b^y$ for the exponent y. Thus we must develop a new procedure. One method would be merely to write

$$y = \text{the power of } b \text{ that produces } x$$

This procedure would work, but it is not concise. We need compact notation to represent y as the exponent of b that produces x. For historical reasons, we use the notation in the following definition.

DEFINITION OF A LOGARITHM

If $x > 0$ and b is a positive constant ($b \neq 1$), then

$$y = \log_b x \qquad \text{if and only if} \qquad b^y = x$$

In the equation $y = \log_b x$, y is referred to as the **logarithm,** b is the **base,** and x is the **argument.**

The notation $\log_b x$ is read "the logarithm (or log) base b of x." The definition of a logarithm indicates that *a logarithm is an exponent.*

The equations

$$y = \log_b x \qquad \text{and} \qquad b^y = x$$

are different ways of expressing the same thing.

$$y = \log_b x \text{ is the logarithmic form of } b^y = x$$
$$b^y = x \text{ is the exponential form of } y = \log_b x$$

EXAMPLE 1 *Change from Logarithmic to Exponential Form*

Write each equation in its exponential form.

a. $2 = \log_7 x$ **b.** $3 = \log_{10}(x + 8)$ **c.** $\log_5 125 = x$

Solution

Use the definition $y = \log_b x$ if and only if $b^y = x$.

a.

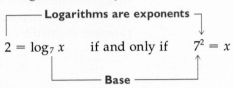

b. $3 = \log_{10}(x + 8)$ if and only if $10^3 = (x + 8)$.

c. $\log_5 125 = x$ if and only if $5^x = 125$.

TRY EXERCISE 2, EXERCISE SET 5.2

EXAMPLE 2 *Change from Exponential to Logarithmic Form*

Write $x = 25^{1/2}$ in its logarithmic form.

Solution

Use $x = b^y$ if and only if $y = \log_b x$.

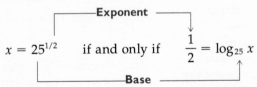

TRY EXERCISE 12, EXERCISE SET 5.2

Some logarithms can be evaluated by using the definition of a logarithm and the following theorem.

EQUALITY OF EXPONENTS THEOREM

If b is a positive real number ($b \neq 1$) such that $b^x = b^y$, then $x = y$.

EXAMPLE 3   *Evaluate Logarithms*

Evaluate: $\log_2 32 = x$

Solution

$\log_2 32 = x$ if and only if $2^x = 32$ • **Change to exponential form.**

$\qquad\qquad\qquad\qquad\qquad\qquad\quad 2^x = 2^5$ • **Factor.**

$\qquad\qquad\qquad\qquad\qquad\qquad\quad\ \ x = 5$ • **Equality of Exponents Theorem**

TRY EXERCISE 22, EXERCISE SET 5.2

PROPERTIES OF LOGARITHMS

Because logarithms are exponents, they have many properties that can be established by using the properties of exponents.

PROPERTIES OF LOGARITHMS

In the following properties, b, M, and N are positive real numbers ($b \neq 1$), and p is any real number.

$$\log_b b = 1$$

$$\log_b 1 = 0$$

An inverse property $\log_b (b^p) = p$

Product property $\log_b (MN) = \log_b M + \log_b N$

Quotient property $\log_b \dfrac{M}{N} = \log_b M - \log_b N$

Power property $\log_b (M^p) = p \log_b M$

One-to-one property $\log_b M = \log_b N$ implies $M = N$

Logarithm-of-each-side property $M = N$ implies $\log_b M = \log_b N$

An inverse property $b^{\log_b p} = p$ (for $p > 0$)

The properties of logarithms are often used to rewrite logarithms and expressions that involve logarithms.

EXAMPLE 4   *Rewrite Logarithmic Expressions*

Use the properties of logarithms to express the following logarithms in terms of logarithms of x, y, and z.

a. $\log_b (xy^2)$ **b.** $\log_b \dfrac{x^2 \sqrt{y}}{z^5}$

Solution

a. $\log_b (xy^2) = \log_b x + \log_b y^2$ • **Product property**

  $= \log_b x + 2 \log_b y$ • **Power property**

b. $\log_b \dfrac{x^2 \sqrt{y}}{z^5} = \log_b (x^2 \sqrt{y}) - \log_b z^5$ • **Quotient property**

  $= \log_b x^2 + \log_b \sqrt{y} - \log_b z^5$ • **Product property**

  $= 2 \log_b x + \dfrac{1}{2} \log_b y - 5 \log_b z$ • **Power property**

TRY EXERCISE 32, EXERCISE SET 5.2

The properties of logarithms are also used to rewrite expressions that involve logarithms as a single logarithm.

EXAMPLE 5   *Rewrite Logarithmic Expressions*

Use the properties of logarithms to rewrite each expression as a single logarithm.

a. $2 \log_b x + \dfrac{1}{2} \log_b (x + 4)$ **b.** $4 \log_b (x + 2) - 3 \log_b (x - 5)$

Solution

a. $2 \log_b x + \dfrac{1}{2} \log_b (x + 4)$

  $= \log_b x^2 + \log_b (x + 4)^{1/2}$ • **Power property**

  $= \log_b [x^2(x + 4)^{1/2}]$ • **Product property**

b. $4 \log_b (x + 2) - 3 \log_b (x - 5)$

  $= \log_b (x + 2)^4 - \log_b (x - 5)^3$ • **Power property**

  $= \log_b \dfrac{(x + 2)^4}{(x - 5)^3}$ • **Quotient property**

TRY EXERCISE 42, EXERCISE SET 5.2

> **DEFINITION OF COMMON LOGARITHM**
>
> Logarithms with a base of 10 are called **common logarithms.** It is customary to write $\log_{10} x$ as $\log x$.

> **DEFINITION OF NATURAL LOGARITHM**
>
> Logarithms with a base of e are called **natural logarithms.** They are often used in calculus. It is customary to write $\log_e x$ as $\ln x$.

Most scientific calculators have a key marked $\boxed{\log}$ for evaluating common logarithms and a key marked $\boxed{\ln}$ for evaluating natural logarithms. For example,

$$\log 24 \approx 1.380211242$$

$$\ln 81 \approx 4.394449155$$

$$\log 0.58 \approx -0.236572006$$

If you use a scientific calculator to try to evaluate the logarithm of a negative number, it may give you an error indication. Recall that the definition of $y = \log_b x$ required x to be a positive real number.

Logarithms that are not common logarithms or natural logarithms can be evaluated by using the following theorem.

> **CHANGE-OF-BASE FORMULA**
>
> If x, a, and b are positive real numbers with $a \neq 1$ and $b \neq 1$, then
>
> $$\log_b x = \frac{\log_a x}{\log_a b}$$

EXAMPLE 6   *Use the Change-of-Base Formula*

Evaluate each logarithm.   **a.** $\log_3 18$ **b.** $\log_{12} 400$

Solution

In each case we use the change-of-base formula with $a = 10$. That is, we will evaluate these logarithms by using the $\boxed{\log}$ key on a scientific calculator.

a. $\log_3 18 = \dfrac{\log 18}{\log 3} \approx 2.63093$ **b.** $\log_{12} 400 = \dfrac{\log 400}{\log 12} \approx 2.41114$

TRY EXERCISE 52, EXERCISE SET 5.2

> **TAKE NOTE**
>
> In Example 6, we could also have evaluated the logarithms by using the $\boxed{\ln}$ key. For example, for a.,
>
> $$\log_3 18 = \frac{\ln 18}{\ln 3} \approx 2.63093$$

ANTILOGARITHMS

Given $M = \log N$, it is often necessary to determine the value of N. In this case the number N is called the **antilogarithm** of M.

DEFINITION OF ANTILOGARITHMS

If M and N are real numbers with $N > 0$, such that

$$\log_b N = M$$

then N is the **antilogarithm** of M for the base b.

Rewriting $\log_b N = M$ as $N = b^M$, we have a formula to evaluate N, the antilogarithm of M.

$$N = b^M$$

For instance,

$$\text{if } \log_4 N = 1.2251 \qquad \text{then} \qquad N = 4^{1.2251} \approx 5.4649$$

$$\text{if } \log_7 N = -1.3041 \qquad \text{then} \qquad N = 7^{-1.3041} \approx 0.0791$$

For the special cases of common logarithms and natural logarithms, we have

$$\text{if } \log N = 2.3571 \qquad \text{then} \qquad N = 10^{2.3571} \approx 227.5621$$

$$\text{if } \ln N = 1.0892 \qquad \text{then} \qquad N = e^{1.0892} \approx 2.9719$$

TOPICS FOR DISCUSSION

1. The definition of a logarithm as $f(x) = \log_b x$ requires that $x > 0$ and that $b \neq 1$. Discuss why these conditions are imposed.

2. Let a and b be bases for a logarithm. Discuss the relationship between $\log_b x$ and $\log_a x$ for any positive real number x.

3. Discuss the relationship between $f(x) = \log_b x$ and $g(x) = b^x$.

4. What are the product, quotient, and power rules for logarithms?

5. If $f(x) = \log_b x$ and $f(a) = f(b)$, can we conclude that $a = b$? Is there an analogous statement for any function? That is, if f is some function and $f(a) = f(b)$, does it follow that $a = b$?

EXERCISE SET 5.2

In Exercises 1 to 10, change each equation to its exponential form.

1. $\log_{10} 100 = 2$

2. $\log_{10} 1000 = 3$

3. $\log_5 125 = 3$

4. $\log_5 \dfrac{1}{25} = -2$

5. $\log_3 81 = 4$

6. $\log_3 1 = 0$

7. $\log_b r = t$

8. $\log_b (s + t) = r$

9. $-3 = \log_3 \dfrac{1}{27}$

10. $-1 = \log_7 \dfrac{1}{7}$

In Exercises 11 to 20, change each equation to its logarithmic form.

11. $2^4 = 16$

12. $3^5 = 243$

13. $7^3 = 343$

14. $7^{-4} = \dfrac{1}{2401}$

15. $10{,}000 = 10^4$

16. $\dfrac{1}{1000} = 10^{-3}$

17. $b^k = j$

18. $p = m^n$

19. $b^1 = b$

20. $b^0 = 1$

In Exercises 21 to 30, evaluate each logarithm. Do not use a calculator.

21. $\log_{10} 1{,}000{,}000$

22. $\log_{10} \dfrac{1}{1000}$

23. $\log_2 32$

24. $\log_3 243$

25. $\log_{3/2} \dfrac{27}{8}$

26. $\log_{0.5} 16$

27. $\log_5 \dfrac{1}{25}$

28. $\log_{0.3} \dfrac{100}{9}$

29. $\log_b 1$

30. $\log_b b$

In Exercises 31 to 40, write the given logarithm in terms of logarithms of x, y, and z.

31. $\log_b (xyz)$

32. $\log_b (x^2 y^3)$

33. $\log_3 \dfrac{x}{z^4}$

34. $\log_5 \dfrac{x^2}{yz^3}$

35. $\log_b \dfrac{\sqrt{x}}{y^3}$

36. $\log_b \dfrac{\sqrt{x}}{\sqrt[3]{z}}$

37. $\log_b x \sqrt[3]{\dfrac{y^2}{z}}$

38. $\log_b \sqrt[3]{x^2 z \sqrt{y}}$

39. $\log_7 \dfrac{\sqrt{xz}}{y^2}$

40. $\log_5 x^2 \left(\dfrac{y}{z^3}\right)^2$

In Exercises 41 to 50, write each logarithmic expression as a single logarithm.

41. $\log_{10} (x + 5) + 2 \log_{10} x$

42. $5 \log_3 x - 4 \log_3 y + 2 \log_3 z$

43. $\dfrac{1}{2} [3 \log_b (x - y) + \log_b (x + y) - \log_b z]$

44. $\log_b (y^3 z^2) - 3 \log_b (x\sqrt{y}) + 2 \log_b \dfrac{x}{z}$

45. $\log_8 (x^2 - y^2) - \log_8 (x - y)$

46. $\log_4 (x^3 - y^3) - \log_4 (x - y)$

47. $4 \ln (x - 3) + 2 \ln x$

48. $3 \ln z - 2 \ln (z + 1)$

49. $\ln x - \ln y + \ln z$

50. $\dfrac{1}{2} \log x + 2 \log y$

In Exercises 51 to 60, use the change-of-base formula to approximate the logarithm accurate to five significant digits.

51. $\log_7 20$

52. $\log_5 37$

53. $\log_{11} 8$

54. $\log_{50} 22$

55. $\log_6 0.045$

56. $\log_4 \sqrt{7}$

57. $\log_{0.5} 5$

58. $\log_{0.2} 17$

59. $\log_\pi e$

60. $\log_\pi \sqrt{15}$

In Exercises 61 to 68, approximate the antilogarithm N accurate to three significant digits.

61. $\log N = 0.4857$

62. $\log N = 7.8476$

63. $\log N = -2.4760$

64. $\log N = -4.3536$

65. $\ln N = 2.001$

66. $\ln N = 0.531$

67. $\ln N = -1.204$

68. $\ln N = -0.511$

SUPPLEMENTAL EXERCISES

In Exercises 69 to 74, find all the real numbers that are solutions of the given inequality. Use interval notation to write your answers.

69. $0 \le \log x \le 1000$

70. $-3 \le \log x \le -2$

71. $e \le \ln x \le e^3$

72. $-2 \le \ln x \le 3$

73. $-\log x > 0$

74. $100 - 10 \log (x + 1) > 0$

75. Verify the quotient property of logarithms.

76. Verify the power property of logarithms.

77. Give the reason for each step in the proof of one of the inverse properties of logarithms.

$$\log_b x = \log_b x \qquad \underline{\qquad ? \qquad}$$
$$b^{\log_b x} = x \qquad \underline{\qquad ? \qquad}$$

78. Give the reason for each step in the proof of the one-to-one property of logarithms.

$$\log_b M = \log_b N \qquad \text{Given}$$
$$b^{\log_b N} = M \qquad \underline{\qquad ? \qquad}$$
$$N = M \qquad \underline{\qquad ? \qquad}$$

PROJECTS

1. **BIOLOGICAL DIVERSITY** To discuss the variety of species that live in a certain environment, a biologist needs a precise definition of *diversity*. Let $p_1, p_2, \ldots, p_n$ be the proportion of n species that live in an environment. The biological diversity, D, of this system is

$$D = -(p_1 \log_2 p_1 + p_2 \log_2 p_2 + \cdots + p_n \log_2 p_n)$$

Suppose an ecosystem has exactly five different varieties of grass: rye (R), bermuda (B), blue (L), fescue (F), and St. Augustine (A). The various proportions of these grasses are as shown in the tables at the right.

a. Calculate the diversity of this ecosystem if the proportions of these grasses are as shown in Table 1.

b. Because bermuda and St. Augustine are virulent grasses, after a time the proportions will be as shown in Table 2. Does this system have more or less diversity than the system given in Table 1?

c. After an even longer time period, the bermuda and St. Augustine completely overrun the environment and the proportions are as shown in Table 3. Calculate the diversity of this system. (*Note:* Although the equation is not technically correct, for purposes of the diversity definition, we may say that $0 \log_2 0 = 0$. By using very small values of p_i, we can demonstrate that this definition makes sense.) Does this system have more or less diversity than the system given in Table 2?

d. Finally, the St. Augustine overruns the bermuda and the proportions are as shown in Table 4. Calculate the diversity of this system. Write a sentence that explains the meaning of the value you obtained.

Table 1

R	B	L	F	A
$\dfrac{1}{5}$	$\dfrac{1}{5}$	$\dfrac{1}{5}$	$\dfrac{1}{5}$	$\dfrac{1}{5}$

Table 2

R	B	L	F	A
$\dfrac{1}{8}$	$\dfrac{3}{8}$	$\dfrac{1}{16}$	$\dfrac{1}{8}$	$\dfrac{5}{16}$

Table 3

R	B	L	F	A
0	$\dfrac{1}{4}$	0	0	$\dfrac{3}{4}$

Table 4

R	B	L	F	A
0	0	0	0	1

SECTION 5.3 LOGARITHMIC FUNCTIONS AND THEIR GRAPHS

Using the concept of a logarithm and the properties of logarithms, we can now introduce the concept of a *logarithmic function*.

DEFINITION OF A LOGARITHMIC FUNCTION

The **logarithmic function f with base b** is defined by

$$f(x) = \log_b x$$

where b is a positive constant, $b \neq 1$, and x is any *positive* real number.

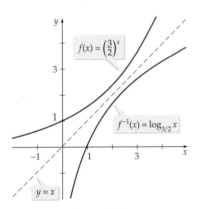

Figure 5.15

Recall that logarithms were defined so that we could write the inverse of $g(x) = b^x$ in a convenient manner. Thus the logarithmic function given by $f(x) = \log_b x$ is the inverse of the exponential function $g(x) = b^x$. The graph of $y = \log_b x$ can be obtained by reflecting the graph of $y = b^x$ across the graph of the line $y = x$. This is illustrated in **Figure 5.15** for the exponential function $f(x) = (3/2)^x$ and its inverse $f^{-1}(x) = \log_{3/2} x$.

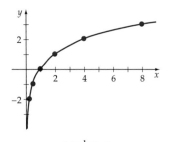

$y = \log_2 x$

Figure 5.16

The graph of a logarithmic function can be drawn by considering its inverse function. For instance, to graph $y = \log_2 x$, consider the equivalent exponential equation $2^y = x$. Choose values of y and calculate the corresponding values of x, as shown in Table 5.4. Now plot the ordered pairs and connect the points with a smooth curve, as shown in **Figure 5.16**.

Table 5.4

$x = 2^y$	$2^{-2} = \dfrac{1}{4}$	$2^{-1} = \dfrac{1}{2}$	$2^0 = 1$	$2^1 = 2$	$2^2 = 4$	$2^3 = 8$
y	-2	-1	0	1	2	3

In a similar manner, we can draw the graph of a logarithmic function with a fractional base. For instance, consider $y = \log_{2/3} x$. Rewriting this in exponential form yields $(2/3)^y = x$. Choose values of y and calculate the corresponding values of x (see Table 5.5), plot the points corresponding to the ordered pairs, and then draw a smooth graph through the points as shown in **Figure 5.17**.

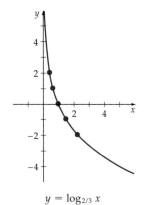

$y = \log_{2/3} x$

Figure 5.17

Table 5.5

$x = \left(\dfrac{2}{3}\right)^y$	$\left(\dfrac{2}{3}\right)^{-2} = \dfrac{9}{4}$	$\left(\dfrac{2}{3}\right)^{-1} = \dfrac{3}{2}$	$\left(\dfrac{2}{3}\right)^0 = 1$	$\left(\dfrac{2}{3}\right)^1 = \dfrac{2}{3}$	$\left(\dfrac{2}{3}\right)^2 = \dfrac{4}{9}$
y	-2	-1	0	1	2

PROPERTIES OF $f(x) = \log_b x$

For all positive real numbers b, $b \neq 1$, the function defined by $f(x) = \log_b x$ has the following properties:

1. f has the set of positive real numbers as its domain.

2. f has the set of real numbers as its range.

3. f has a graph with an x-intercept of $(1, 0)$.

4. f has a graph asymptotic to the y-axis.

5. f is a one-to-one function.

6. f is an increasing function if $b > 1$. See **Figure 5.18a**.

7. f is a decreasing function if $0 < b < 1$. See **Figure 5.18b**.

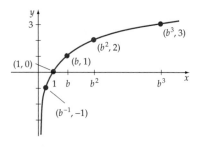

a. $f(x) = \log_b x, b > 1$

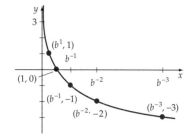

b. $f(x) = \log_b x, 0 < b < 1$

Figure 5.18

The graphs of $y = \log x$ and $y = \ln x$ can be drawn by using the same technique we used earlier to draw the graph of $y = \log_2 x$ and $y = \log_{2/3} x$. However, $\log x$ and $\ln x$ are available in graphing utilities, and the graphs can be produced by using them. The graphs are shown in **Figure 5.19**.

Observe that each graph passes through $(1, 0)$. Also note that as $x \to 0^+$, $y \to -\infty$; thus the y-axis is a vertical asymptote. From the definition of logarithms, x must be a positive number. Thus the domain of $y = \log x$ and $y = \ln x$ is the set of positive real numbers; the range is the set of real numbers.

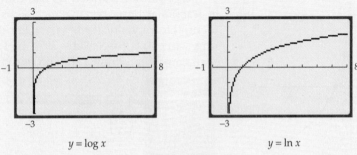

$y = \log x$ $y = \ln x$

Figure 5.19

With the help of the $\boxed{\log}$ or $\boxed{\ln}$ keys and the change-of-base formula, we can draw the graphs of other logarithmic functions. For example, the graph of $y = \log_2 x$ (**Figure 5.20**) can be drawn by using the change-of-base formula. Write $y = \log_2 x = \log x / \log 2$. Now graph $y = \log x / \log 2$.

QUESTION   Could we have written $y = \log_2 x = \ln x / \ln 2$? That is, could we have used the natural log function to produce the graph of $y = \log_2 x$?

You can verify the accuracy of the graph in **Figure 5.20** by using a graphing utility. Use the change-of-base formula and graph $y = \log x / \log 2$. We emphasize again that the natural logarithm function can also be used.

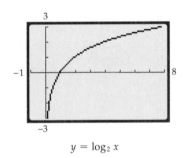

$y = \log_2 x$

Figure 5.20

EXAMPLE 1   *Graph a Function That Is Logarithmic in Form*

Graph: $y = \log_4 (-2x)$

ANSWER   Yes. The change-of-base formula is $\log_b x = \log_a x / \log_a b$. Letting $a = e$ (the base of the natural logarithm) and $b = 2$, we have $y = \log_2 x = \ln x / \ln 2$.

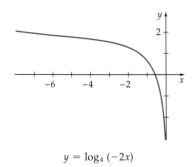

$y = \log_4 (-2x)$

Figure 5.21

Solution

Writing $y = \log_4 (-2x)$ in its exponential form produces $4^y = -2x$ or $(-1/2)4^y = x$. Choosing convenient values for y yields the table below. Connecting the points with a smooth curve produces the graph shown in **Figure 5.21.**

$x = \left(-\dfrac{1}{2}\right)4^y$	$-\dfrac{1}{32}$	$-\dfrac{1}{8}$	$-\dfrac{1}{2}$	-2	-8
y	-2	-1	0	1	2

TRY EXERCISE 12, EXERCISE SET 5.3

EXAMPLE 2   *Graph a Function That Is Logarithmic in Form*

Graph: $f(x) = \log \sqrt{x}$

Solution

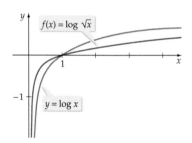

Figure 5.22

Because $f(x) = \log \sqrt{x}$ is equivalent to $f(x) = \dfrac{1}{2} \log x$, the graph of $f(x) = \log \sqrt{x}$ can be obtained by *shrinking* the graph of $y = \log x$. That is, we determine points (a, b) on the graph of $y = \log x$. Then we plot the points $\left(a, \dfrac{1}{2}b\right)$ to sketch the graph of $f(x) = \log \sqrt{x}$.

    Figure 5.22 shows the graph of $f(x) = \log \sqrt{x}$. Note that the point $\left(a, \dfrac{1}{2}b\right)$ is on the graph of $f(x) = \log \sqrt{x}$ if and only if the point (a, b) is on the graph of $f(x) = \log x$.

TRY EXERCISE 14, EXERCISE SET 5.3

    Horizontal and/or vertical translations of the graph of the logarithmic function $f(x) = \log_b x$ sometimes can be used to obtain the graph of functions that involve logarithms.

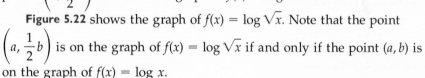

EXAMPLE 3   *Use Translations to Graph*

Graph.   **a.** $f(x) = \log_4 (x + 3)$ **b.** $f(x) = \log_4 x + 3$

Solution

Figure 5.23

a.   The graph of $f(x) = \log_4 (x + 3)$ can be obtained by shifting the graph of $g(x) = \log_4 x$ three units to the left. **Figure 5.23** shows the graph of $g(x) = \log_4 x$ and the graph of $f(x) = \log_4 (x + 3)$.

Continued •➤

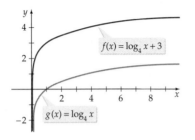

Figure 5.24

b. The graph of $f(x) = \log_4 x + 3$ can be obtained by shifting the graph of $g(x) = \log_4 x$ three units upward. **Figure 5.24** shows the graph of $g(x) = \log_4 x$ and the graph of $f(x) = \log_4 x + 3$.

> **TRY EXERCISE 20, EXERCISE SET 5.3**

APPLICATIONS OF LOGARITHMIC FUNCTIONS

Logarithmic functions and their graphs are useful in classifying data by associating very large differences or very small differences with small positive numbers. An example of mapping very large numbers to small positive numbers is the *Richter scale* for classifying the magnitude of an earthquake.

> **TAKE NOTE**
>
> Observe that in **Figure 5.25**, the scale on the horizontal axis is such that each scale number is 10 times the previous value, so the numbers are not equally spaced. These amounts of energy are mapped to consecutive integers on the vertical axis. Thus an increase of 1 unit on the Richter scale (the vertical axis) corresponds to a tenfold increase in the energy in an earthquake.

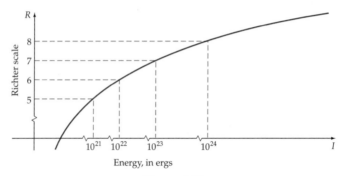

Figure 5.25

The amount of energy released in a moderate earthquake is on the order of 10^{20} ergs. The intervals on the horizontal axis in **Figure 5.25** have been reduced, and the distance between consecutive numbers on the vertical axis has been expanded to illustrate the mapping of the measure of the energy of an earthquake to the Richter number that represents the magnitude of that earthquake.

On the Richter scale, seismologists measure the magnitude R of an earthquake by the formula

$$R = \log \frac{I}{I_0}$$

where I_0 is the measure of a *zero-level earthquake,* and I is the intensity of the earthquake being measured.

Because the Richter scale is a logarithmic scale, a one-unit change in magnitude corresponds to a ten-fold change in intensity. For instance, suppose the intensity I_1 of one earthquake has a magnitude of 6 on the Richter scale and the intensity I_2 of a second earthquake has a magnitude of 3 on the Richter scale. Then

$$6 = \log \frac{I_1}{I_0} = \log I_1 - \log I_0 \quad (1)$$

$$3 = \log \frac{I_2}{I_0} = \log I_2 - \log I_0 \quad (2)$$

$$3 = \log I_1 - \log I_2 = \log \frac{I_1}{I_2} \qquad \bullet \textbf{ Subtract Equation (2) from Equation (1).}$$

Now write the logarithmic equation $3 = \log (I_1/I_2)$ in exponential form and solve for I_1.

$$10^3 = \frac{I_1}{I_2}$$

$$1000I_2 = I_1$$

From this last equation, a magnitude 6 earthquake has 1000 times the intensity of a magnitude 3 earthquake.

Astronomers also use logarithms to measure the distance to a star.

EXAMPLE 4   *An Application to Astronomy*

Astronomers use the *distance modulus* of a star as a method of determining the star's distance from Earth. The formula is $M = 5 \log r - 5$, where M is the distance modulus and r is the star's distance from Earth in parsecs. (One parsec is approximately 3.3 light-years, or 2.1×10^{13} miles.) Graph $M = 5 \log r - 5$ with r on the horizontal axis and M on the vertical axis. Use the graph to estimate, to the nearest integer, the number of parsecs from Earth of a star that has a distance modulus of 3.

Solution

Use a graphing utility with $y = M$ and $x = r$ to graph $M = 5 \log r - 5$ (see **Figure 5.26**). Note that the domain and range for this graph were chosen so that we could determine a distance for a star that has a distance modulus of 3.

One method that can be used to determine the distance r from Earth for a star whose distance modulus is 3 is to trace along the graph until the value of M (the y-coordinate) is 3. The number of parsecs that the star is from Earth is r (the x-coordinate). From the graph, the star whose distance modulus is 3 is approximately 40 parsecs from Earth.

TRY EXERCISE 44, EXERCISE SET 5.3

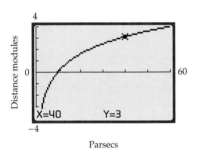

Parsecs

$$y = 5 \log x - 5$$

Figure 5.26

Note: The x and y values shown are rounded to the nearest integer.

TOPICS FOR DISCUSSION

1. If $b > 0$, discuss the characteristics of the graph of $f(x) = \log_b x$.

2. Let b be a positive real number and let a be its reciprocal. Discuss the relationship between the graphs of $f(x) = \log_b x$ and $g(x) = \log_a x$.

3. Explain why the y-axis is a vertical asymptote of the graph of $f(x) = \log_b x$. Does the graph of f have a horizontal asymptote?

4. If $f(x) = \log_b x$, then the graph of $y = f(ax)$ is a stretching or shrinking of $y = f(x)$ (see Chapter 3). However, by the properties of logarithms, $f(ax) = \log_b (ax) = \log_b a + \log_b x$. This relationship indicates that the graph of $f(x) = \log_b x$ is shifted vertically $\log_b a$ units. Discuss why stretching/shrinking and vertical translation are the same for a logarithmic function.

EXERCISE SET 5.3

In Exercises 1 to 20, graph each function.

1. $f(x) = \log_3 x$

2. $f(x) = \log_5 x$

3. $f(x) = \log_{1/2} x$

4. $f(x) = \log_{1/4} x$

5. $f(x) = -2 \ln x$

6. $f(x) = -\log x$

7. $f(x) = \log_4 x^2$

8. $f(x) = 2 \log_5 x$

9. $f(x) = |\ln x|$

10. $f(x) = \ln |x|$

11. $f(x) = -\ln |x|$

12. $f(x) = -|\log_2 x^3|$

13. $f(x) = \log \sqrt[3]{x}$

14. $f(x) = \ln \sqrt{x}$

15. $f(x) = 3 + \log_2 x$

16. $f(x) = -2 + \log_4 x$

17. $f(x) = \log (x + 10)$

18. $f(x) = \ln (x + 3)$

19. $f(x) = \ln (x - 5)$

20. $f(x) = \log_2 (x - 2)$

In Exercises 21 to 28, graph each function. A graphing utility will be helpful.

21. $f(x) = 4x \ln x$

22. $f(x) = -5x \log_3 x$

23. $f(x) = \dfrac{6 \log x}{x}$

24. $f(x) = \dfrac{7 \ln x}{x}$

25. $f(x) = 4x - 2 \ln x$

26. $f(x) = x + \ln x$

27. $f(x) = 5x + 2 \log_2 x$

28. $f(x) = 2x - \ln (x + 1)$

In Exercises 29 to 32, use a graphing utility.

29. Graph $f(x) = \dfrac{\log_2 x}{x}$. What does $f(x)$ approach as $x \to \infty$?

30. Graph $f(x) = x \log x$. Is $x = 0$ a vertical asymptote for the graph of f? Explain your answer.

31. Graph $f(x) = \ln (e^x)$. Is the graph the same as the graph of $y = x$? Explain your answer.

32. Graph $f(x) = e^{\ln x}$. Is the graph the same as the graph of $y = x$? Explain your answer.

33. A question on a true/false exam makes the statement that $1/\log_2 x = \log_2 x^{-1}$ for all positive values of x. What is the correct answer?

34. A question on a true/false exam makes the statement that $\log_3 (1/x) = -\log_3 x$ for all positive values of x. What is the correct answer?

35. **MAINTAINING A MANUAL SKILL** If a skill is not practiced, the proficiency of the person performing the skill diminishes over time. In one study, the average typing speed for students after a 6-week class was 58 words per minute. Every month after the class, the people who did not practice were tested. The results can be modeled by the equation $S = 58 - 6.8 \ln (t + 1)$, where S is the typing speed

and t is the number of months after the class. Use a graphing utility to answer the following questions.

a. What is the average typing speed, to the nearest integer, after 3 months?

b. How many months, to the nearest 0.1, will elapse before the average typing speed falls below 50 words per minute?

36. **INTEREST RATES** General interest rate theory suggests that short-term interest rates (less than 2 years) are lower than long-term rates (more than 10 years) because short-term securities are less risky than long-term ones. In periods of high inflation, however, the situation is reversed and economists discuss *inverted-yield* curves. During the early 1980s inflation was very high in the United States. The rates for short- and long-term U.S. Treasury securities during 1980 are shown in the following table.

US Treasury Securities, 1980	
Term (in years)	Interest rate
0.5	15.0%
1	14.0%
5	13.5%
10	12.8%
20	12.5%

A model of these data is $y = 14.33759 - 0.62561 \ln x$, where x is the term of the security and y is the interest rate.

a. Use a graphing utility to graph this equation.

b. The point whose coordinates are $(15, 12.6)$ is on the graph of this equation. Write a sentence that explains the meaning of this point.

37. **ADVERTISING EXPENDITURES** An advertising agency estimates that the number of sales, N (in thousands), of a certain product is related to the amount A spent on advertising (in thousands of dollars) by the equation $N = 1.6 + 2.3 \ln A$. Graph this equation and use the graph to estimate the amount spent on advertising when 6000 units of the product were sold.

38. **SALARY GROWTH** The inflation-adjusted salary S (in thousands of dollars) of a computer programmer for a corporation can be approximated by the equation $S = 10 \ln (y + 1) + 26$, where y is the number of years the programmer has worked for the corporation.

Graph this equation and use the graph to estimate the number of years, to the nearest tenth of a year, an employee must work to reach an inflation-adjusted salary of $40,000.

39. **CONSUMPTION OF NATURAL RESOURCES** Assuming a constant growth rate of r% in consumption of oil, an equation that models the time before oil resources will be depleted is given by $T = \dfrac{1}{r}\ln(99.47r + 1)$, where T is the number of years before the resource is depleted. Graph this equation and use the graph to determine the rate of consumption, to the nearest tenth of a percent, that would deplete the resource in 40 years.

40. **EARTHQUAKES** What will an earthquake measure on the Richter scale if it has an intensity of $I = 100{,}000I_0$?

41. **EARTHQUAKES** The Colombia earthquake of 1906 had an intensity of $I = 398{,}107{,}000I_0$. What did it measure on the Richter scale?

42. **EARTHQUAKES** If an earthquake has an intensity 1000 times the intensity of a second earthquake, then how much larger is the Richter scale measure of the larger earthquake than that of the smaller?

In Exercises 43 to 46, use a graphing utility to graph the distance modulus equation $M = 5\log r - 5$ and use the graph to approximate, to the nearest tenth, the number of parsecs each star is from Earth.

43. **CALCULATING DISTANCES TO STARS** The approximate distance modulus for the star *Antares* is 5.4. Determine, to the nearest tenth, the number of parsecs *Antares* is from Earth.

44. **CALCULATING DISTANCES TO STARS** The approximate distance modulus for the star *Achernar* is 1.5. Determine, to the nearest tenth, the number of parsecs *Achernar* is from Earth.

45. **CALCULATING DISTANCES TO STARS** The approximate distance modulus for the star *Altair* is −1.5. Determine, to the nearest tenth, the number of parsecs *Altair* is from Earth.

46. **CALCULATING DISTANCES TO STARS** The approximate distance modulus for the star *Vega* is −0.48. Determine, to the nearest tenth, the number of parsecs *Vega* is from Earth.

SUPPLEMENTAL EXERCISES

In Exercises 47 to 50, use a graphing utility.

47. Graph $f(x) = \dfrac{e^x - e^{-x}}{2}$ and $g(x) = \ln(x + \sqrt{x^2 + 1})$ on the same coordinate axes. Use the same scale on both the x- and the y-axis. What appears to be the relationship between f and g?

48. On the same coordinate axes, graph $f(x) = \dfrac{e^x + e^{-x}}{2}$ for $x \geq 0$ and $g(x) = \ln(x + \sqrt{x^2 - 1})$ for $x \geq 1$. Use the same scale on both the x- and the y-axis. What appears to be the relationship between f and g?

49. Graph $f(x) = e^{-x}(\ln x)$ for $1 \leq x \leq e^2$.

50. Graph $g(x) = \log [\![x]\!]$ for $1 \leq x \leq 10$. Recall that $[\![x]\!]$ represents the greatest integer function.

In Exercises 51 to 56, use a graphing utility to graph each function. Determine the domain and range of the function.

51. $f(x) = \sqrt{\log x}$

52. $f(x) = \sqrt{\ln x^3}$

53. $f(x) = 100 - \ln\sqrt{1 - x^2}$

54. $f(x) = 10 + |\ln(x - e)|$

55. $f(x) = \log(\log x)$

56. $f(x) = |\ln(-\ln x)|$

57. Given $f(x) = \ln x$, evaluate
    a. $f(e^3)$ b. $f(e^{\ln 4})$ c. $f(e^{3\ln 3})$

58. Given $f(x) = \log_5 x$, evaluate
    a. $f(5^2)$ b. $f(5^{\log_5 4})$ c. $f(5^{3\log_5 3})$

59. Explain why the graph of $F(x) = \log_b x^2$ and the graph of $G(x) = 2\log_b x$ are not identical.

60. Explain why the graph of $F(x) = |\log_b x|$ and the graph of $G(x) = \log_b |x|$ are not identical.

61. **EARTHQUAKES** The Coalinga, California, earthquake of May 2, 1983, had a Richter scale measure of 6.5. Find the Richter scale measure (to the nearest tenth) of an earthquake that has an intensity 200 times the intensity of the Coalinga quake.

62. **EARTHQUAKES** The earthquake that occurred just south of Concepcion, Chile, on May 22, 1960, had a Richter scale measure of 9.5. Find the Richter scale measure of an earthquake that has an intensity one-half the intensity of this quake.

PROJECTS

1. **NAPIERIAN LOGARITHMS** The original definition that John Napier gave for logarithms is quite different from the modern one. (Napier, a few years before he died, did reconsider his original definition and suggested the one that is used today. He

died before he could complete that work.) In modern form, Napier's original definition can be stated as

$$\text{NLog } x = 10{,}000{,}000 \ln \frac{10{,}000{,}000}{x}$$

where NLog (x) is the Napierian logarithm of a number x.

a. Show that NLog $(10{,}000{,}000) = 0$. Observe that this is different from the natural logarithm function in that $\ln (1) = 0$.

b. Determine which of the following three properties of natural logarithms are also satisfied by Napierian logarithms.

$$\text{NLog } (ab) \stackrel{?}{=} \text{NLog } a + \text{NLog } b$$

$$\text{NLog } \frac{a}{b} \stackrel{?}{=} \text{NLog } a - \text{NLog } b$$

$$\text{NLog } (a^r) \stackrel{?}{=} r \text{ NLog } a$$

c. Graph $y = \text{NLog } x$. Explain how the graph differs from the graph of $y = \ln x$.

5.4 EXPONENTIAL AND LOGARITHMIC EQUATIONS

If a variable appears as an exponent in a term of an equation, then the equation is called an **exponential equation**. Example 1 uses the Equality of Exponents Theorem to solve exponential equations.

EXAMPLE 1 *Solve Exponential Equations*

Solve: $49^{2x} = \dfrac{1}{7}$

Solution

Write each side of the equation as a power of the same base, and then equate the exponents.

$$49^{2x} = \frac{1}{7}$$

$$(7^2)^{2x} = 7^{-1} \qquad \bullet \text{ Write each side as a power of 7.}$$

$$7^{4x} = 7^{-1}$$

$$4x = -1 \qquad \bullet \text{ Equate the exponents.}$$

$$x = -\frac{1}{4}$$

TRY EXERCISE 2, EXERCISE SET 5.4

In Example 1, we were able to write each side of the equation as a power of the same base. If this is difficult to do, then consider taking the logarithm of both sides of the equation.

<table>
<tr><td>

TAKE NOTE

We can find solutions of an equation in the form $f(x) = g(x)$ by finding the x-intercepts of the graph of $y = f(x) - g(x)$. A solution of $5^x = 40$ in Example 2 is a zero of $y = 5^x - 40$. Similarly, a solution of $3^{2x-1} = 5^{x+2}$ is a zero of $y = 3^{2x-1} - 5^{x+2}$. See the graphs below. Also note the scale on the y-axis for the second graph.

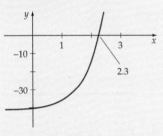

a. $y = 5^x - 40$

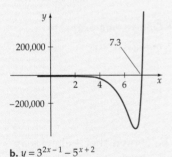

b. $y = 3^{2x-1} - 5^{x+2}$

</td></tr>
</table>

EXAMPLE 2   *Solve Exponential Equations*

Solve each exponential equation.   **a.** $5^x = 40$ **b.** $3^{2x-1} = 5^{x+2}$

Solution

a.
$$5^x = 40$$
$$\log(5^x) = \log 40 \qquad \bullet \text{ Take the logarithm of each side.}$$
$$x \log 5 = \log 40$$
$$x = \frac{\log 40}{\log 5} \qquad \bullet \text{ Exact solution}$$
$$x \approx 2.3 \qquad \bullet \text{ Decimal approximation}$$

b.
$$3^{2x-1} = 5^{x+2}$$
$$\ln 3^{2x-1} = \ln 5^{x+2} \qquad \bullet \text{ Take the natural logarithm of each side.}$$
$$(2x - 1)\ln 3 = (x + 2)\ln 5 \qquad \bullet \text{ Power property}$$
$$2x \ln 3 - \ln 3 = x \ln 5 + 2\ln 5 \qquad \bullet \text{ Distributive property}$$

Collecting terms that involve the variable x on the left side yields

$$2x \ln 3 - x \ln 5 = 2\ln 5 + \ln 3 \qquad \bullet \text{ Solve for x.}$$
$$x(2\ln 3 - \ln 5) = 2\ln 5 + \ln 3$$
$$x = \frac{2\ln 5 + \ln 3}{2\ln 3 - \ln 5} \qquad \bullet \text{ Exact solution}$$
$$x \approx 7.3 \qquad \bullet \text{ Decimal approximation}$$

TRY EXERCISE 10, EXERCISE SET 5.4

LOGARITHMIC EQUATIONS

Equations that involve logarithms are called **logarithmic equations.** The properties of logarithms, along with the definition of a logarithm, are valuable aids to solving a logarithmic equation.

EXAMPLE 3   *Solve a Logarithmic Equation*

Solve: $\log 2x - \log(x - 3) = 1$

Continued • ➤

Solution

$$\log 2x - \log (x - 3) = 1$$

$$\log \frac{2x}{x - 3} = 1 \qquad \bullet \textbf{ Quotient property}$$

$$\frac{2x}{x - 3} = 10^1 \qquad \bullet \textbf{ Definition of logarithm}$$

$$2x = 10x - 30$$

$$-8x = -30$$

$$x = \frac{15}{4}$$

Check the solution by substituting 15/4 into the original equation.

TRY EXERCISE 22, EXERCISE SET 5.4

TAKE NOTE

Example 4 shows that solving a logarithmic equation by using logarithms may introduce extraneous solutions. You must check each proposed solution to ensure that it actually satisfies the equation. Note from the graph that there is only one x-intercept and therefore only one real solution of the equation.

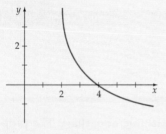

$y = \ln (3x + 8) - \ln (2x + 2) - \ln (x - 2)$

EXAMPLE 4 *Solve a Logarithmic Equation*

Solve: $\ln (3x + 8) = \ln (2x + 2) + \ln (x - 2)$

Solution

$$\ln (3x + 8) = \ln (2x + 2) + \ln (x - 2)$$

$$\ln (3x + 8) = \ln [(2x + 2)(x - 2)] \qquad \bullet \textbf{ Product property}$$

$$\ln (3x + 8) = \ln (2x^2 - 2x - 4)$$

$$3x + 8 = 2x^2 - 2x - 4 \qquad \bullet \textbf{ One-to-one property}$$

$$0 = 2x^2 - 5x - 12 \qquad \quad \textbf{of logarithms}$$

$$0 = (2x + 3)(x - 4)$$

$$x = -\frac{3}{2} \quad \text{or} \quad x = 4$$

Thus $-3/2$ and 4 are possible solutions. It can be shown that 4 checks as a solution of the equation but that $-3/2$ does not check. The only solution is 4.

TRY EXERCISE 30, EXERCISE SET 5.4

QUESTION   Why does $x = -3/2$ not check as a solution of the equation in Example 4?

ANSWER   If $x = -3/2$, the original equation becomes $\ln (7/2) = \ln (-1) + \ln (-7/2)$. This cannot be true, because the function $f(x) = \ln x$ is not defined for negative values of x.

TAKE NOTE

Setting the left side and the right side of the equation in Example 5 equal to y, we have $y = (2^x + 2^{-x})/2$ and $y = 3$. Graphing these two equations with a graphing utility produces the graph shown below. The x-coordinates of the points of intersection are the solutions of the equation.

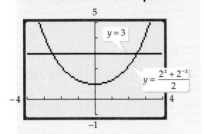

EXAMPLE 5   *Solve an Equation Involving $b^x + b^{-x}$*

Solve: $\dfrac{2^x + 2^{-x}}{2} = 3$

Solution

Multiplying each side by 2 produces

$$2^x + 2^{-x} = 6$$
$$2^{2x} + 2^0 = 6(2^x)$$ • Multiply each side by 2^x to clear negative exponents.

$$(2^x)^2 - 6(2^x) + 1 = 0$$ • Write in quadratic form.
$$(u)^2 - 6(u) + 1 = 0$$ • Substitute u for 2^x.

By the quadratic formula,

$$u = \frac{6 \pm \sqrt{36 - 4}}{2} = \frac{6 \pm 4\sqrt{2}}{2} = 3 \pm 2\sqrt{2}$$

$$2^x = 3 \pm 2\sqrt{2}$$ • Replace u with 2^x.
$$\log 2^x = \log(3 \pm 2\sqrt{2})$$ • Take the common logarithm of each side.
$$x \log 2 = \log(3 \pm 2\sqrt{2})$$ • Power property of logarithms
$$x = \frac{\log(3 \pm 2\sqrt{2})}{\log 2} \approx \pm 2.54$$

The approximate solutions are -2.54 and 2.54.

TRY EXERCISE 34, EXERCISE SET 5.4

THE pH OF A SOLUTION

Whether an aqueous solution is acidic or basic depends on its hydronium-ion concentration. Thus acidity is a function of hydronium-ion concentration. Because these hydronium-ion concentrations may be very small, it is convenient to measure acidity in terms of **pH**, which is defined as the negative of the common logarithm of the molar hydronium-ion concentration, H_3O^+. As a mathematical formula, this is stated as

$$\text{pH} = -\log[H_3O^+]$$

EXAMPLE 6   *Find the pH of a Solution*

Find the pH of each liquid.

a.   Orange juice with $[H_3O^+] = 2.80 \times 10^{-4}$ M

b.   Milk with $[H_3O^+] = 3.97 \times 10^{-7}$ M

Continued ▸

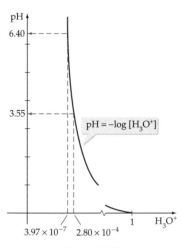

Figure 5.27

Solution

a. $\text{pH} = -\log\,[\text{H}_3\text{O}^+] = -\log\,(2.80 \times 10^{-4}) \approx -(-3.55) = 3.55$

The orange juice has a pH of 3.55 (to the nearest hundredth).

b. $\text{pH} = -\log\,[\text{H}_3\text{O}^+] = -\log\,(3.97 \times 10^{-7}) \approx -(-6.40) = 6.40$

The milk has a pH of 6.40 (to the nearest hundredth).

> **TRY EXERCISE 52, EXERCISE SET 5.4**

The difference in hydronium-ion concentration between the two numbers shown in **Figure 5.27** is

$$0.00028 - 0.000000397 = 0.000279603$$

Figure 5.27 shows how the pH function *maps* small positive numbers that are relatively close together on the hydronium-ion concentration axis into numbers (3.55 and 6.40) that are farther apart on the pH axis. The pH of pure water is 7.0.

EXAMPLE 7   *Find the Hydronium-Ion Concentration*

Determine the hydronium-ion concentration of a sample of blood with $\text{pH} = 7.41$.

Solution

Substitute 7.41 for the pH and solve for H_3O^+.

$$\text{pH} = -\log\,[\text{H}_3\text{O}^+]$$
$$7.41 = -\log\,[\text{H}_3\text{O}^+] \qquad \bullet \textbf{ Substitute 7.41 for pH.}$$
$$-7.41 = \log\,[\text{H}_3\text{O}^+] \qquad \bullet \textbf{ Multiply both sides by } -1.$$
$$10^{-7.41} = [\text{H}_3\text{O}^+] \qquad \bullet \textbf{ Definition of } y = \log_b x$$
$$3.9 \times 10^{-8} \approx [\text{H}_3\text{O}^+]$$

The hydronium-ion concentration of the blood sample is 3.9×10^{-8} M.

> **TRY EXERCISE 54, EXERCISE SET 5.4**

EXAMPLE 8   *Velocity of an Object Experiencing Air Resistance*

The time t in seconds required for an object that is dropped to reach a velocity v (in feet per second) is given by $t = v/32$ when air resistance is not considered. If air resistance is considered, then one possible model is given by $t = 2.43 \ln \dfrac{150 + v}{150 - v}$ for $0 \le v < 150$. Use this equation to determine, to the nearest tenth, the velocity of an object that has been falling for 5 seconds. The graph of the equation has a vertical asymptote at $v = 150$. Explain the meaning of that asymptote in the context of this problem.

Solution

Substitute 5 for t and solve for v.

$$t = 2.43 \ln \frac{150 + v}{150 - v}$$

$$5 = 2.43 \ln \frac{150 + v}{150 - v} \qquad \bullet \text{ **Replace t by 5.**}$$

$$2.0576132 = \ln \frac{150 + v}{150 - v} \qquad \bullet \text{ **Divide by 2.43.**}$$

$$e^{2.0576132} = \frac{150 + v}{150 - v} \qquad \bullet \text{ **If a = ln b, then e^a = b.**}$$

$$e^{2.0576132}(150 - v) = 150 + v \qquad \bullet \text{ **Solve for v.**}$$

$$e^{2.0576132}(150) - e^{2.0576132}v = 150 + v$$

$$-v - e^{2.0576132}v = 150 - e^{2.0576132}(150)$$

$$v(-e^{2.0576132} - 1) = 150 - e^{2.0576132}(150)$$

$$v \approx 116.014$$

After 5 seconds, the velocity of the object will be approximately 116 feet per second. See **Figure 5.28**.

The vertical asymptote $v = 150$ indicates that the object cannot attain a speed greater than 150 feet per second. In **Figure 5.28**, note that as $v \to 150$, $t \to \infty$.

TRY EXERCISE 68, EXERCISE SET 5.4

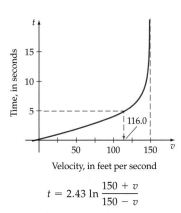

$$t = 2.43 \ln \frac{150 + v}{150 - v}$$

Figure 5.28

TOPICS FOR DISCUSSION

1. Discuss how to solve the equation $a = \log_b x$ for x.

2. What is the domain of $y = \log_4 (2x - 5)$? Explain why this means that the equation $\log_4 (x - 3) = \log_4 (2x - 5)$ has no real number solution.

3. -8 is not a solution of the equation $\log_2 x + \log_2 (x + 6) = 4$. Discuss at which step in the following solution the extraneous solution -8 was introduced.

$$\log_2 x + \log_2 (x + 6) = 4$$

$$\log_2 x(x + 6) = 4$$

$$x(x + 6) = 2^4$$

$$x^2 + 6x = 16$$

$$x^2 + 6x - 16 = 0$$

$$(x + 8)(x - 2) = 0$$

$$x = -8 \quad \text{or} \quad x = 2$$

Exercise Set 5.4

In Exercises 1 to 40, solve for x algebraically.

1. $2^x = 64$

2. $3^x = 243$

3. $49^x = \dfrac{1}{343}$

4. $9^x = \dfrac{1}{243}$

5. $2^{5x+3} = \dfrac{1}{8}$

6. $3^{4x-7} = \dfrac{1}{9}$

7. $\left(\dfrac{2}{5}\right)^x = \dfrac{8}{125}$

8. $\left(\dfrac{2}{5}\right)^x = \dfrac{25}{4}$

9. $5^x = 70$

10. $6^x = 50$

11. $3^{-x} = 120$

12. $7^{-x} = 63$

13. $10^{2x+3} = 315$

14. $10^{6-x} = 550$

15. $e^x = 10$

16. $e^{x+1} = 20$

17. $2^{1-x} = 3^{x+1}$

18. $3^{x-2} = 4^{2x+1}$

19. $2^{2x-3} = 5^{-x-1}$

20. $5^{3x} = 3^{x+4}$

21. $\log_2 x + \log_2 (x - 4) = 2$

22. $\log_3 x + \log_3 (x + 6) = 3$

23. $\log (5x - 1) = 2 + \log (x - 2)$

24. $1 + \log (3x - 1) = \log (2x + 1)$

25. $\log \sqrt{x^3 - 17} = \dfrac{1}{2}$

26. $\log (x^3) = (\log x)^2$

27. $\log (\log x) = 1$

28. $\ln (\ln x) = 2$

29. $\ln (e^{3x}) = 6$

30. $\ln x = \dfrac{1}{2} \ln\left(2x + \dfrac{5}{2}\right) + \dfrac{1}{2} \ln 2$

31. $e^{\ln (x-1)} = 4$

32. $10^{\log (2x+7)} = 8$

33. $\dfrac{10^x - 10^{-x}}{2} = 20$

34. $\dfrac{10^x + 10^{-x}}{2} = 8$

35. $\dfrac{10^x + 10^{-x}}{10^x - 10^{-x}} = 5$

36. $\dfrac{10^x - 10^{-x}}{10^x + 10^{-x}} = \dfrac{1}{2}$

37. $\dfrac{e^x + e^{-x}}{2} = 15$

38. $\dfrac{e^x - e^{-x}}{2} = 15$

39. $\dfrac{1}{e^x - e^{-x}} = 4$

40. $\dfrac{e^x + e^{-x}}{e^x - e^{-x}} = 3$

In Exercises 41 to 50, use a graphing utility to approximate the solutions of the equation to the nearest hundredth.

41. $2^{-x+3} = x + 1$

42. $3^{x-2} = -2x - 1$

43. $e^{3-2x} - 2x = 1$

44. $2e^{x+2} + 3x = 2$

45. $3 \log_2 (x - 1) = -x + 3$

46. $2 \log_3 (2 - 3x) = 2x - 1$

47. $\ln (2x + 4) + \dfrac{1}{2}x = -3$

48. $2 \ln (3 - x) + 3x = 4$

49. $2^{x+1} = x^2 - 1$

50. $\ln x = -x^2 + 4$

51. **Calculating pH** Find the pH of a sample of lemon juice that has a hydronium-ion concentration of 6.3×10^{-3}.

52. **Calculating pH** An *acidic solution* has a pH of less than 7, whereas a *basic solution* has a pH of greater than 7. Household ammonia has a hydronium-ion concentration of 1.26×10^{-12}. Determine the pH of the ammonia, and state whether it is an acid or a base.

53. **Calculating Hydronium-Ion Concentration** Find the hydronium-ion concentration of beer, which has a pH of 4.5.

54. **Calculating Hydronium-Ion Concentration** Normal rain has a pH of 5.6. A recent acid rain had a pH of 3.1. Find the hydronium-ion concentration of this rain.

55. **Population Growth** The population P of a city grows exponentially according to the function

$$P(t) = 8500(1.1)^t, \quad 0 \le t \le 8$$

where t is measured in years.

a. Find the population at time $t = 0$ and also at time $t = 2$.

b. When, to the nearest year, will the population reach 15,000?

56. **Physical Fitness** After a race, a runner's pulse rate R in beats per minute decreases according to the function

$$R(t) = 145e^{-0.092t}, \quad 0 \le t \le 15$$

where t is measured in minutes.

a. Find the runner's pulse rate at the end of the race and also 1 minute after the end of the race.

b. How long, to the nearest minute, after the end of the race will the runner's pulse rate be 80 beats per minute?

57. **Rate of Cooling** A can of soda at 79°F is placed in a refrigerator that maintains a constant temperature of 36°F. The temperature T of the soda t minutes after it is placed in the refrigerator is given by

$$T(t) = 36 + 43e^{-0.058t}$$

a. Find the temperature of the soda 10 minutes after it is placed in the refrigerator.

b. When, to the nearest minute, will the temperature of the soda be 45°F?

58. Medicine During surgery, a patient's circulatory system requires at least 50 milligrams of an anesthetic. The amount of anesthetic present t hours after 80 milligrams of anesthetic is administered is given by

$$T(t) = 80(0.727)^t$$

a. How much of the anesthetic is present in the patient's circulatory system 30 minutes after the anesthetic is administered?

b. How long, to the nearest minute, can the operation last if the patient does not receive additional anesthetic?

59. Psychology Industrial psychologists study employee training programs to assess the effectiveness of the instruction. In one study, the percent score P on a test for a person who has completed t hours of training was given by

$$P = \frac{100}{1 + 30e^{-0.088t}}$$

a. Use a graphing utility to graph the equation for $t \geq 0$.

b. Use the graph to estimate (to the nearest hour) the number of hours of training necessary to achieve a 70% score on the test.

c. From the graph, determine the horizontal asymptote.

d. Write a sentence that explains the meaning of the horizontal asymptote.

60. Psychology An industrial psychologist has determined that the average percent score for an employee on a test of the employee's knowledge of the company's product is given by

$$P = \frac{100}{1 + 40e^{-0.1t}}$$

where t is the number of weeks on the job and P is the percent score.

a. Use a graphing utility to graph the equation for $t \geq 0$.

b. Use the graph to estimate (to the nearest week) the number of weeks of employment that are necessary for the average employee to earn a 70% score on the test.

c. Determine the horizontal asymptote of the graph.

d. Write a sentence that explains the meaning of the horizontal asymptote.

61. Ecology In 1993, a herd of bison was placed in a wildlife preserve that can ultimately support 1000 bison. Suppose a population model for the bison is given by

$$B = \frac{1000}{1 + 30e^{-0.127t}}$$

where B is the number of bison in the preserve and t is in years.

a. Use a graphing utility to graph the equation for $t \geq 0$.

b. Use the graph to estimate (to the nearest year) the number of years before the bison population reaches 500.

c. Determine the horizontal asymptote of the graph.

d. Write a sentence that explains the meaning of the horizontal asymptote.

62. Population Growth A yeast culture grows according to the equation

$$Y = \frac{50,000}{1 + 250e^{-0.305t}}$$

where Y is the number of yeast and t is in hours.

a. Use a graphing utility to graph the equation for $t \geq 0$.

b. Use the graph to estimate (to the nearest hour) the number of hours before the yeast population reaches 35,000.

c. From the graph, determine the horizontal asymptote.

d. Write a sentence that explains the meaning of the horizontal asymptote.

63. Consumption of Natural Resources A model for how long our coal resources will last is approximated by

$$T = \frac{1}{r} \ln (300r + 1)$$

where r is the percent increase in consumption from current levels of use and T is the time (in years) before the resource is depleted.

a. Use a graphing utility to graph this equation. (*Hint:* r is a percent, so consider a domain from 0 to 1.)

b. If our consumption of coal increases by 3%, in how many years (to the nearest year) will we deplete our coal resources?

c. What percent increase in consumption of coal will deplete the resource in 100 years? Round to the nearest tenth of a percent.

64. Consumption of Natural Resources A model for how long our aluminum resources will last is approximated by

$$T = \frac{1}{r} \ln (20,500r + 1)$$

where r is the percent increase of consumption from current levels of use and T is the time (in years) before the resource is depleted.

a. Use a graphing utility to graph this equation. (*Hint: r* is a percent, so consider a domain from 0 to 1.)

b. If our consumption of aluminum increases by 5% per year, in how many years (to the nearest year) will we deplete our aluminum resources?

c. What percent increase in consumption of aluminum will deplete the resource in 100 years? Round to the nearest tenth of a percent.

65. **Consumption of Natural Resources** A more accurate model for how long our coal resources will last (see Exercise 63) is given by

$$T = \frac{\ln (300r + 1)}{\ln (r + 1)}$$

where *r* is the percent increase in consumption from current levels of use and *T* is the time (in years) before the resource is depleted.

a. Use a graphing utility to graph this equation.

b. If our consumption of coal increases by 3%, in how many years will we deplete our coal resources?

c. What percent increase in consumption of coal will deplete the resource in 100 years? Round to the nearest tenth of a percent.

d. If our consumption of coal increases by *r*% per year, does this model predict more or less time than the model in Exercise 63 before the resource is depleted? Explain.

66. **Consumption of Natural Resources** A more accurate model for how long our aluminum resources will last (see Exercise 64) is given by

$$T = \frac{\ln (20{,}500r + 1)}{\ln (r + 1)}$$

where *r* is the percent increase in consumption from current levels of use and *T* is the time (in years) before the resource is depleted.

a. Use a graphing utility to graph this equation.

b. If our consumption of aluminum increases by 5% per year, in how many years (to the nearest year) will we deplete our aluminum resources?

c. What percent increase in consumption of aluminum will deplete the resource in 100 years? Round to the nearest tenth of a percent.

d. If our consumption of aluminum resources increases *r*% per year, does this model predict more or less time than the model in Exercise 64 before the resource is depleted? Explain.

67. **Velocity of a Sky Diver** The time *t* in seconds required for a sky diver to reach a velocity *v* in feet per second is given by

$$t = -\frac{175}{32} \ln \left(1 - \frac{v}{175} \right)$$

a. Determine the velocity of the sky diver after 10 seconds.

b. Determine the vertical asymptote for the graph of this function.

c. Write a sentence that describes the meaning of the vertical asymptote in the context of this problem.

68. **Affects of Air Resistance on Velocity** If we assume that air resistance is proportional to the square of the velocity, then the time *t* in seconds required for an object to reach a velocity *v* in feet per second is given by

$$t = \frac{9}{24} \ln \frac{24 + v}{24 - v}$$

a. Determine the velocity of the object after 1.5 seconds.

b. Determine the vertical asymptote for the graph of this function.

c. Write a sentence that describes the meaning of the vertical asymptote in the context of this problem.

69. **Terminal Velocity with Air Resistance** The velocity *v* of an object *t* seconds after it's been dropped from a height above the surface of the earth is given by the equation *v* = 32*t* feet/second, assuming no air resistance. If we assume that air resistance is proportional to the square of the velocity, then the velocity after *t* seconds is given by

$$v = 100 \left(\frac{e^{0.64t} - 1}{e^{0.64t} + 1} \right)$$

a. In how many seconds will the velocity be 50 feet per second?

b. Determine the horizontal asymptote for the graph of this function.

c. Write a sentence that describes the meaning of the horizontal asymptote in the context of this problem.

70. **Terminal Velocity with Air Resistance** If we assume that air resistance is proportional to the square of the velocity, then the velocity *v* in feet per second of an object *t* seconds after it has been dropped is given by

$$v = 50 \left(\frac{e^{1.6t} - 1}{e^{1.6t} + 1} \right)$$

(See Exercise 69. The reason for the difference in the equations is that the proportionality constants are different.)

a. In how many seconds will the velocity be 20 feet per second?

b. Determine the horizontal asymptote for the graph of this function.

c. Write a sentence that describes the meaning of the horizontal asymptote in the context of this problem.

71. **AFFECTS OF AIR RESISTANCE ON DISTANCE** The distance s in feet that the object in Exercise 69 will fall in t seconds is given by

$$s = \frac{100^2}{32} \ln\left(\frac{e^{0.32t} + e^{-0.32t}}{2}\right)$$

a. Use a graphing utility to graph this equation for $t \geq 0$.

b. How long does it take for the object to fall 100 ft? Round to the nearest tenth of a second.

72. **AFFECTS OF AIR RESISTANCE ON DISTANCE** The distance s in feet that the object in Exercise 70 will fall in t seconds is given by

$$s = \frac{50^2}{32} \ln\left(\frac{e^{0.8t} + e^{-0.8t}}{2}\right)$$

a. Use a graphing utility to graph this equation for $t \geq 0$.

b. How long does it take for the object to fall 100 ft? Round to the nearest tenth of a second.

73. **RETIREMENT PLANNING** The retirement account for a graphic designer contains $250,000 on January 1, 1994, and earns interest at a rate of 0.5% per month. On February 1, 1994, the designer withdraws $2000 and plans to continue these withdrawals as retirement income each month. The value V in the account after x months is

$$V = 400,000 - 150,000(1.005)^x$$

a. What is the domain for this equation?

b. If the designer wishes to leave $100,000 to a scholarship foundation, what is the maximum number of withdrawals (to the nearest month) the designer can make from this account and still have $100,000 to donate?

74. **RETIREMENT PLANNING** The retirement account for an assembly line shift manager contains $300,000 on January 1, 1995, and earns interest at a rate of 0.75% per

month. On February 1, 1995, the manager withdraws $3000 from the account, and these withdrawals continue each month thereafter. The value V in the account after x months is given by the equation

$$V = 700,000 - 400,000(1.0075)^x$$

a. What is the domain for this equation?

b. What is the number of monthly withdrawals (to the nearest month) possible at $3000 per month before there is less than $3000 in the account?

SUPPLEMENTAL EXERCISES

75. The following argument shows that $0.125 > 0.25$. Find the first incorrect statement.

$$3 > 2$$
$$3(\log 0.5) > 2(\log 0.5)$$
$$\log 0.5^3 > \log 0.5^2$$
$$0.5^3 > 0.5^2$$
$$0.125 > 0.25$$

76. The following argument shows that $4 = 6$. Find the first incorrect statement.

$$4 = \log_2 16$$
$$4 = \log_2 (8 + 8)$$
$$4 = \log_2 8 + \log_2 8$$
$$4 = 3 + 3$$
$$4 = 6$$

77. A common mistake that students make is to write $\log (x + y)$ as $\log x + \log y$. For what values of x and y does $\log (x + y) = \log x + \log y$? (*Hint:* Solve for x in terms of y.)

78. Which is larger, 500^{501} or 506^{500}? (*Hint:* Let $x = 500^{501}$ and $y = 506^{500}$ and then compare $\ln x$ with $\ln y$.)

79. Explain why the functions $F(x) = 1.4^x$ and $G(x) = e^{0.336x}$ represent essentially the same function.

80. Find the constant k that will make $f(t) = 2.2^t$ and $g(t) = e^{-kt}$ represent essentially the same function.

81. Solve $e^{1/x} > 2$. Write the solution in interval notation.

82. Solve $\log (x^2) > (\log x)^2$. Write the solution in interval notation.

PROJECTS

1. NAVIGATING The pilot of a boat is trying to cross a river to a point O two miles due west of the boat's starting position by always pointing the nose of the boat toward O. Suppose the speed of the current is w miles per hour and the speed of the boat is v miles per hour. If point O is the origin and the boat's starting position is $(2, 0)$ (see the diagram at the right), then the equation of the boat's path is given by

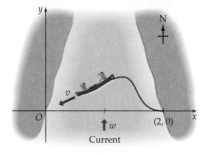

$$y = \left(\frac{x}{2}\right)^{1-(w/v)} - \left(\frac{x}{2}\right)^{1+(w/v)}$$

a. If the speed of the current and the speed of the boat are the same, can the pilot reach point O by always having the nose of the boat pointed toward O? If not, at what point will the pilot arrive? Explain your answer.

b. If the speed of the current is greater than the speed of the boat, can the pilot reach point O by always pointing the nose of the boat toward point O? If not, where will the pilot arrive? Explain.

c. If the speed of the current is less than the speed of the boat, can the pilot reach point O by always pointing the nose of the boat toward point O? If not, where will the pilot arrive? Explain.

SECTION **5.5** APPLICATIONS OF EXPONENTIAL AND LOGARITHMIC FUNCTIONS

In many applications, a quantity N grows or decays according to the function $N(t) = N_0 e^{kt}$. In this function, N is a function of time t, and N_0 is the value of N at time $t = 0$. If k is a *positive* constant, then $N(t) = N_0 e^{kt}$ is called an **exponential growth function**. If k is a *negative* constant, then $N(t) = N_0 e^{kt}$ is called an **exponential decay function**. The following examples illustrate how growth and decay functions arise naturally in the investigation of certain phenomena.

Interest is money paid for the use of money. The interest I is called **simple interest** if it is a fixed percent r, per time period t, of the amount of money invested. The amount of money invested is called the **principal** P. Simple interest is computed using the formula $I = Prt$. For example, if $1000 is invested at 12% for 3 years, the simple interest is

$$I = Prt = \$1000(0.12)(3) = \$360$$

The balance after t years is $B = P + I = P + Prt$. In the previous example, the $1000 invested for 3 years produced $360 interest. Thus the balance after 3 years is $1360.

COMPOUND INTEREST

In many financial transactions, interest is added to the principal at regular intervals so that interest is paid on interest as well as on the principal. Interest earned in this manner is called **compound interest.** For example, if $1000 is invested at 12% annual interest compounded annually for 3 years, then the total interest after 3 years is

First-year interest	$1000(0.12) = $120.00
Second-year interest	$1120(0.12) = $134.40
Third-year interest	$1254.40(0.12) \approx \underline{$150.53}$
	$404.93 • **Total interest**

This method of computing the balance can be tedious and time-consuming. A *compound interest formula* that can be used to determine the balance due after t years of compounding can be developed as follows.

Note that if P dollars is invested at an interest rate of r per year, then the balance after one year is $B_1 = P + Pr = P(1 + r)$, where Pr represents the interest earned for the year. Observe that B_1 is the product of the original principal P and $(1 + r)$. If the amount B_1 is reinvested for another year, then the balance after the second year is

$$B_2 = (B_1)(1 + r) = P(1 + r)(1 + r) = P(1 + r)^2$$

Successive reinvestments lead to the results shown in Table 5.6. The equation $B_t = P(1 + r)^t$ is valid if r is the interest rate paid during each of the t years.

If r is an annual interest rate and n is the number of compounding periods per year, then the interest rate each period is r/n and the number of compounding periods after t years is nt. Thus the compound interest formula is expressed as follows:

Table 5.6

Number of Years	Balance
3	$B_3 = P(1 + r)^3$
4	$B_4 = P(1 + r)^4$
⋮	⋮
n	$B_n = P(1 + r)^n$

THE COMPOUND INTEREST FORMULA

A principal P invested at an annual interest rate r, expressed as a decimal and compounded n times per year for t years, produces the balance

$$B = P\left(1 + \frac{r}{n}\right)^{nt}$$

EXAMPLE 1   *Solve a Compound Interest Application*

Find the balance if $1000 is invested at an annual interest rate of 10% for 2 years compounded   **a.** annually   **b.** daily   **c.** hourly

Continued • ▶

Solution

a. Use the compound interest formula, with $P = 1000$, $r = 0.1$, $t = 2$, and $n = 1$.

$$B = \$1000\left(1 + \frac{0.1}{1}\right)^{1 \cdot 2} = \$1000(1.1)^2 = \$1210.00$$

b. Because there are 365 days in a year, use $n = 365$.

$$B = \$1000\left(1 + \frac{0.1}{365}\right)^{365 \cdot 2} \approx \$1000(1.000273973)^{730} \approx \$1221.37$$

c. Because there are 8760 hours in a year, use $n = 8760$.

$$B = \$1000\left(1 + \frac{0.1}{8760}\right)^{8760 \cdot 2} \approx \$1000(1.000011416)^{17520} \approx \$1221.40$$

TRY EXERCISE 4, EXERCISE SET 5.5

To **compound continuously** means to increase the number of compounding periods without bound.

To derive a continuous compounding interest formula, substitute $1/m$ for r/n in the compound interest formula

$$B = P\left(1 + \frac{r}{n}\right)^{nt} \tag{1}$$

to produce

$$B = P\left(1 + \frac{1}{m}\right)^{nt} \tag{2}$$

This substitution is motivated by the desire to express $(1 + r/n)^n$ as $[(1 + 1/m)^m]^r$, which approaches e^r as m gets large without bound.

Solving the equation $1/m = r/n$ for n yields $n = mr$, so the exponent nt can be written as mrt. Therefore Equation (2) can be expressed as

$$B = P\left(1 + \frac{1}{m}\right)^{mrt} = P\left[\left(1 + \frac{1}{m}\right)^m\right]^{rt} \tag{3}$$

By the definition of e, we know that as m gets larger without bound,

$$\left(1 + \frac{1}{m}\right)^m \qquad \text{approaches} \qquad e$$

Thus, using continuous compounding, Equation (3) simplifies to $B = Pe^{rt}$.

CONTINUOUS COMPOUNDING INTEREST FORMULA

If an account with principal P and annual interest rate r is compounded continuously for t years, then the balance is $B = Pe^{rt}$.

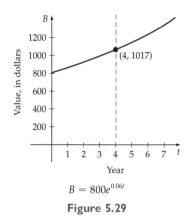

$$B = 800e^{0.06t}$$

Figure 5.29

EXAMPLE 2   *Compound Continuously*

Find the balance after 4 years if $800 is invested at an annual rate of 6% compounded continuously.

Solution

Use the continuous compounding formula.

$$B = Pe^{rt} = 800e^{0.06(4)} = 800e^{0.24}$$

$$\approx \$800(1.27124915) = \$1017.00 \qquad \bullet \text{ To the nearest cent}$$

See **Figure 5.29**.

TRY EXERCISE 6, EXERCISE SET 5.5

EXAMPLE 3   *Doubling a Sum of Money*

Find the time it takes for money invested at an annual rate of r to double.

Solution

Use $B = Pe^{rt}$ with $B = 2P$, twice the principal P.

$$2P = Pe^{rt}$$

$$2 = e^{rt}$$

$$\ln 2 = rt \qquad \bullet \text{ Take the natural logarithm of each side.}$$

$$\frac{\ln 2}{r} = t \qquad \bullet \text{ Solve for t.}$$

The time it takes for money to double when interest is compounded continuously at an annual rate of r is $t = (\ln 2)/r$.

TRY EXERCISE 10, EXERCISE SET 5.5

EXPONENTIAL GROWTH

Given any two points on the graph of $N(t) = N_0 e^{kt}$, you can use the given data to solve for the constants N_0 and k.

EXAMPLE 4   *Find the Exponential Growth Equation That Models Given Data*

Find the exponential growth function for a town whose population was 16,400 in 1980 and 20,200 in 1990.

Continued • ➤

Solution

We need to determine N_0 and k in $N(t) = N_0 e^{kt}$. If we represent the year 1980 by $t = 0$, then our given data are $N(0) = 16{,}400$ and $N(10) = 20{,}200$. Because N_0 is defined to be $N(0)$, we know $N_0 = 16{,}400$. To determine k, substitute $t = 10$ and $N_0 = 16{,}400$ into $N(t) = N_0 e^{kt}$ to produce

$$N(10) = 16{,}400e^{k \cdot 10}$$

$$20{,}200 = 16{,}400e^{10k} \qquad \bullet \textbf{ Substitute 20,200 for } N\textbf{(10).}$$

$$\frac{20{,}200}{16{,}400} = e^{10k}$$

To solve this equation for k, take the natural logarithm of each side.

$$\ln \frac{20{,}200}{16{,}400} = \ln e^{10k}$$

$$\ln \frac{20{,}200}{16{,}400} = 10k \qquad \bullet \textbf{ Use } \log_b (b^p) = p.$$

$$\frac{1}{10} \ln \frac{20{,}200}{16{,}400} = k$$

$$0.0208 \approx k$$

The exponential growth equation is $N(t) = 16{,}400e^{0.0208t}$.

TRY EXERCISE 18, EXERCISE SET 5.5

EXPONENTIAL DECAY

Many radioactive materials *decrease* exponentially over time. This decrease, called radioactive decay, is measured in terms of **half-life,** which is defined as the time required for the disintegration of half the atoms in a sample of a radioactive substance. Table 5.7 shows the half-lives of selected radioactive isotopes.

Table 5.7

Isotope	Half-Life
Carbon (^{14}C)	5730 years
Radium (^{226}Ra)	1660 years
Polonium (^{210}Po)	138 days
Phosphorus (^{32}P)	14 days
Polonium (^{214}Po)	1/10,000th of a second

EXAMPLE 5

Find the Exponential Decay Equation That Models Given Data

Find the exponential decay function for the amount of phosphorus (^{32}P) that remains in a sample after t days.

Solution

When $t = 0$, $N(0) = N_0 e^{k(0)} = N_0$. Thus $N(0) = N_0$. Also, because the phosphorus has a half-life of 14 days (from Table 5.7) $N(14) = 0.5N_0$. To find k, substitute $t = 14$ into $N(t) = N_0 e^{kt}$ and solve for k.

TAKE NOTE

Because $e^{-0.0495} \approx (0.5)^{1/14}$, the decay function $N(t) = N_0 e^{-0.0495t}$ can also be written as $N(t) = N_0 (0.5)^{t/14}$. In this form it is easy to see that if t is increased by **14**, then N will decrease by a factor of **0.5**.

$$N(14) = N_0 \cdot e^{k \cdot 14}$$
$$0.5 N_0 = N_0 e^{14k} \qquad \bullet \text{ Substitute } 0.5N_0 \text{ for } N(14).$$
$$0.5 = e^{14k} \qquad \bullet \text{ Divide each side by } N_0.$$
$$\ln 0.5 = 14k \qquad \bullet \text{ Take the natural logarithm of each side.}$$
$$\frac{1}{14} \ln 0.5 = k \qquad \bullet \text{ Solve for } k.$$
$$-0.0495 \approx k$$

The exponential decay function is $N(t) = N_0 e^{-0.0495t}$.

TRY EXERCISE 20, EXERCISE SET 5.5

EXAMPLE 6   *Application to Air Resistance*

Assuming that air resistance is proportional to the velocity of a falling object, the velocity (in feet per second) of the object t seconds after it has been dropped is given by $v = 82(1 - e^{-0.39t})$.

a. Determine when the velocity will be 70 feet per second.

b. Write a sentence that explains the meaning of the horizontal asymptote, which is $v = 82$, in the context of this problem.

Solution

a.
$$v = 82(1 - e^{-0.39t})$$
$$70 = 82(1 - e^{-0.39t}) \qquad \bullet \text{ Replace } v \text{ by 70.}$$
$$\frac{70}{82} = 1 - e^{-0.39t} \qquad \bullet \text{ Divide each side by 82.}$$
$$e^{-0.39t} = 1 - \frac{70}{82} \qquad \bullet \text{ Solve for } e^{-0.39t}.$$
$$\ln e^{-0.39t} = \ln \frac{6}{41} \qquad \bullet \text{ Take the natural logarithm of each side.}$$
$$-0.39t = \ln \frac{6}{41} \qquad \bullet \text{ Solve for } t.$$
$$t = \frac{\ln (6/41)}{-0.39} \approx 4.9277246$$

The time is approximately 4.9 seconds.

b. A horizontal asymptote of $v = 82$ means that as time increases, the velocity of the object will never exceed 82 feet per second.

TRY EXERCISE 36, EXERCISE SET 5.5

TAKE NOTE

The graph below suggests that our work in Example 6 is correct. With $x = t$ and $y = v$, you can use the TRACE feature of a graphing utility to find that when t (the x-coordinate) is approximately 4.9, v (the y-coordinate) is 70 feet per second.

In Example 6, it was possible to algebraically approximate the solution numerically.

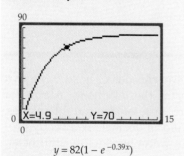

$$y = 82(1 - e^{-0.39x})$$

Note: The x and y values shown are rounded to the nearest tenth.

CARBON DATING

The bone tissue in all living animals contains both carbon-12, which is non-radioactive, and carbon-14, which is radioactive with a half-life of approximately 5730 years. See **Figure 5.30.** As long as the animal is alive, the ratio of carbon-14 to carbon-12 remains constant. When the animal dies ($t = 0$), the carbon-14 begins to decay. Thus a bone that has a smaller ratio of carbon-14 to carbon-12 is older than a bone that has a larger ratio. The percent of carbon-14 present at time t is

$$N(t) = 100(0.5)^{t/5730}$$

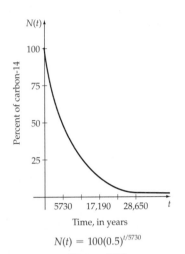

$$N(t) = 100(0.5)^{t/5730}$$

Figure 5.30

EXAMPLE 7   *Application to Archeology*

Determine the age of a bone if it now contains 85% of the carbon-14 it had when $t = 0$.

Solution

Let t be the time at which $N(t) = 0.85N_0$.

$$0.85N_0 = N_0(0.5)^{t/5730}$$
$$0.85 = (0.5)^{t/5730} \quad \bullet \text{ Divide each side by } N_0.$$
$$\ln 0.85 = \ln (0.5)^{t/5730} \quad \bullet \text{ Take the natural logarithm of each side.}$$
$$\ln 0.85 = \frac{t}{5730} \ln 0.5 \quad \bullet \text{ Power property}$$
$$5730\left(\frac{\ln 0.85}{\ln 0.5}\right) = t \quad \bullet \text{ Solve for t.}$$
$$1340 \approx t \quad \bullet \text{ To the nearest 10 years}$$

The bone is about 1340 years old.

TRY EXERCISE 24, EXERCISE SET 5.5

THE DECIBEL SCALE

The range of sound intensities that the human ear can detect is so large that a special *decibel scale* (named after the inventor of the telephone, Alexander Graham Bell) is used to measure and compare different sound intensities. Specifically, the *intensity level N* of sound measured in decibels is directly proportional to the *power I* of the sound measured in watts per square centimeter. That is,

$$N(I) = 10 \log \frac{I}{I_0}$$

where I_0 is the power of sound that is barely audible to the human ear. By international agreement, I_0 is the constant 10^{-16} watts per square centimeter.

EXAMPLE 8   *Application to Noise Levels*

The power of normal conversation is 10^{-10} watts per square centimeter. What is the intensity level N, in decibels, of normal conversation?

Solution

Evaluate $N(10^{-10})$.

$$N(I) = 10 \log \frac{I}{10^{-16}}$$

$$N(10^{-10}) = 10 \log \frac{10^{-10}}{10^{-16}} \qquad \bullet \text{ Substitute } 10^{-10} \text{ for } I.$$

$$= 10 \log 10^{6} \qquad \bullet \text{ Because } 10^{-10}/10^{-16} = 10^{-10-(-16)} = 10^{6}$$

$$= 10(6) = 60$$

The intensity level of normal conversation is 60 decibels.

TRY EXERCISE 28, EXERCISE SET 5.5

THE LOGISTIC MODEL

The population growth model $P(t) = P_0 e^{kt}$ is called the Malthusian model after Robert Malthus (1766–1834), who wrote about population growth in *An Essay on the Principle of Population Growth,* which was published in 1798. This model does not consider that there are only limited resources (for instance, food) and that this limitation will naturally curb population growth.

A model that does consider limited natural resources is called the *logistic model* and is given by

$$P(t) = \frac{mP_0}{P_0 + (m - P_0)e^{-kt}}$$

where m is the maximum population that can be supported and k is a positive constant.

EXAMPLE 9   *Application of the Logistic Model to Population Growth*

In 1976 the world's population was 4 billion, and in 1986 it was 5 billion. Assume that the maximum possible world population is 66 billion and that population growth satisfies the logistic model.

a.   Algebraically determine the value of k in the logistic equation.

b.   Determine algebraically, to the nearest year, when the world's population will reach 20 billion.

Continued •➤

c. Determine the horizontal asymptote for $t > 0$. Write a sentence that describes the meaning of the asymptote in the context of this application.

Solution

a. We must choose a starting value of t. We will use $t = 0$ for the year 1976. For this choice, 1986 corresponds to $t = 10$. Substituting into the logistic model $[P_0 = 4, m = 70, t = 10, P(10) = 5]$ yields

$$P(t) = \frac{mP_0}{P_0 + (m - P_0)e^{-kt}}$$

$$5 = \frac{280}{4 + 66e^{-10k}}$$ • $P_0 = 4, m = 70, t = 10, P(10) = 5$

$$5(4 + 66e^{-10k}) = 280$$ • **Multiply each side by $(4 + 66e^{-10k})$.**

$$20 + 330e^{-10k} = 280$$ • **Simplify.**

$$330e^{-10k} = 260$$ • **Subtract 20 from each side.**

$$e^{-10k} = \frac{260}{330}$$ • **Divide each side by 330.**

$$-10k = \ln\frac{26}{33}$$ • **Take the natural logarithm of each side of the equation.**

$$k \approx 0.0238$$ • **Solve for k.**

b. Using $k = 0.0238$ from **a.**, the logistic equation is

$$P(t) = \frac{70(4)}{4 + (70 - 4)e^{-0.0238t}} = \frac{280}{4 + 66e^{-0.0238t}}$$

To determine the year in which the world's population will first reach 20 billion, replace $P(t)$ with 20 and solve for t.

$$20 = \frac{280}{4 + 66e^{-0.0238t}}$$

$$(4 + 66e^{-0.0238t})20 = 280$$ • **Multiply each side by $(4 + 66e^{-0.0238t})$.**

$$4 + 66e^{-0.0238t} = 14$$ • **Divide each side by 20.**

$$66e^{-0.0238t} = 10$$ • **Subtract 4 from each side.**

$$e^{-0.0238t} = \frac{10}{66}$$ • **Divide each side by 66.**

$$-0.0238t = \ln\frac{5}{33}$$ • **Take the natural logarithm of each side.**

$$t = \frac{\ln(5/33)}{-0.0238} \approx 91$$ • **Solve for t and round to the nearest year.**

When $P(t) = 20$, $t \approx 79$. Thus, according to the model, the world's population will first reach 20 billion in 2055 (1976 + 79).

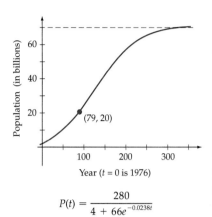

Figure 5.31

$$P(t) = \frac{280}{4 + 66e^{-0.0238t}}$$

c. In the equation $P(t) = 280/(4 + 66e^{-0.0238t})$, the expression $66e^{-0.0238t} \to 0$ as $t \to \infty$. Therefore, $P(t) \to 280/4 = 70$ and the horizontal asymptote is $P(t) = 70$. This means that no matter how many years pass, this model predicts that the world's population will never exceed 70 billion people.

TRY EXERCISE 40, EXERCISE SET 5.5

The graph of $P(t) = 280/(4 + 66e^{-0.0238t})$ is shown in **Figure 5.31**. Note that when $t = 79$, $P(t) = 20$ billion. Also note that as $t \to \infty$, the graph has a horizontal asymptote at $P(t) = 70$.

TOPICS FOR DISCUSSION

1. Explain the difference between compound interest and simple interest.

2. What is an exponential growth model? Give an example of an application for which the exponential growth model might be appropriate.

3. What is an exponential decay model? Give an example of an application for which the exponential decay model might be appropriate.

4. Consider the linear model $y = mx + c$, $m > 0$, and the exponential model $y = ce^{kt}$, $k > 0$. Explain the similarities and differences between the two models.

EXERCISE SET 5.5

1. **COMPOUND INTEREST** If $8000 is invested at an annual interest rate of 5% and compounded annually, find the balance after
   a. 4 years   **b.** 7 years

2. **COMPOUND INTEREST** If $22,000 is invested at an annual interest rate of 4.5% and compounded annually, find the balance after
   a. 2 years   **b.** 10 years

3. **COMPOUND INTEREST** If $38,000 is invested at an annual interest rate of 6.5% for 4 years, find the balance if the interest is compounded
   a. annually   **b.** daily   **c.** hourly

4. **COMPOUND INTEREST** If $12,500 is invested at an annual interest rate of 8% for 10 years, find the balance if the interest is compounded
   a. annually   **b.** daily   **c.** hourly

5. **COMPOUND INTEREST** Find the balance if $15,000 is invested at an annual rate of 10% for 5 years, compounded continuously.

6. **COMPOUND INTEREST** Find the balance if $32,000 is invested at an annual rate of 8% for 3 years, compounded continuously.

7. **COMPOUND INTEREST** How long will it take $4000 to double if it is invested in a certificate of deposit that pays 7.84% annual interest compounded continuously? Round to the nearest tenth of a year.

8. **COMPOUND INTEREST** How long will it take $25,000 to double if it is invested in a savings account that pays 5.88% annual interest compounded continuously? Round to the nearest tenth of a year.

9. **CONTINUOUS COMPOUNDING INTEREST** Use the Continuous Compounding Interest Formula to derive an expression for the time it will take money to triple when

invested at an annual interest rate of r compounded continuously.

10. CONTINUOUS COMPOUNDING INTEREST How long will it take $1000 to triple if it is invested at an annual interest rate of 5.5% compounded continuously? Round to the nearest year.

11. CONTINUOUS COMPOUNDING INTEREST How long will it take $6000 to triple if it is invested in a savings account that pays 7.6% annual interest compounded continuously? Round to the nearest year.

12. CONTINUOUS COMPOUNDING INTEREST How long will it take $10,000 to triple if it is invested in a savings account that pays 5.5% annual interest compounded continuously? Round to the nearest year.

13. POPULATION GROWTH The number of bacteria $N(t)$ present in a culture at time t hours is given by $N(t) = 2200(2)^t$. Find the number of bacteria present when

a. $t = 0$ hours   **b.** $t = 3$ hours

14. POPULATION GROWTH The population of a town grows exponentially according to the function

$$f(t) = 12,400(1.14)^t$$

for $0 \le t \le 5$ years. Find the population of the town when t is

a. 3 years   **b.** 4.25 years

15. POPULATION GROWTH Find the growth function for a town whose population was 22,600 in 1980 and 24,200 in 1985. Use $t = 0$ to represent the year 1980.

16. POPULATION GROWTH Find the growth function for a town whose population was 53,700 in 1992 and 58,100 in 1996. Use $t = 0$ to represent the year 1992.

17. POPULATION GROWTH The growth of the population of Los Angeles, California, for the years 1992 through 1996 can be approximated by the equation

$$P = 10,130e^{0.005t}$$

where $t = 0$ corresponds to January 1, 1992 and P is in thousands.

a. Assuming this growth rate continues, what will be the population of Los Angeles on January 1 in the year 2000?

b. In what year will the population of Los Angeles first exceed 13,000,000?

18. POPULATION GROWTH The growth of the population of Mexico City, Mexico, for the years 1991 through 1995 can be approximated by the equation

$$P = 20,899(1.027)^x$$

where $x = 0$ corresponds to 1991 and P is in thousands.

a. Assuming this growth rate continues, what will be the population of Mexico City in the year 2000?

b. Assuming this growth rate continues, in what year will the population of Mexico City first exceed 25,000,000?

19. MEDICINE Sodium-24 is a radioactive isotope of sodium that is used to study circulatory dysfunction. Assuming that 4 micrograms of sodium-24 is injected into a person, the amount A in micrograms remaining in that person after t hours is given by the equation $A = 4e^{-0.046t}$.

a. Graph this equation.

b. What amount of sodium-24 remains after 5 hours?

c. What is the half-life of sodium-24?

d. In how many hours will the amount of sodium-24 be 1 microgram?

20. RADIOACTIVE DECAY Polonium (^{210}Po) has a half-life of 138 days. Find the decay function for the percent of polonium (^{210}Po) that remains in a sample after t days.

21. GEOLOGY Geologists have determined that Crater Lake in Oregon was formed by a volcanic eruption. Chemical analysis of a wood chip that is assumed to be from a tree that died during the eruption has shown that it contains approximately 45% of its original carbon-14. Determine how long ago the volcanic eruption occurred. Use 5730 years as the half-life of carbon-14.

22. RADIOACTIVE DECAY Use $N(t) = N_0(0.5)^{t/138}$, where t is measured in days, to estimate the percentage of polonium (^{210}Po) that remains in a sample after 2 years.

23. ARCHEOLOGY The Rhind papyrus, named after A. Henry Rhind, contains most of what we know today of ancient Egyptian mathematics. A chemical analysis of a sample from the papyrus has shown that it contains approximately 75% of its original carbon-14. What is the age of the Rhind papyrus? Use 5730 years as the half-life of carbon-14.

24. ARCHEOLOGY Determine the age of a bone if it now contains 65% of its original amount of carbon-14.

25. PHYSICS Newton's Law of Cooling states that if an object at temperature T_0 is placed into an environment at constant temperature A, then the temperature of the object, $T(t)$, after t minutes is given by $T(t) = A + (T_0 - A)e^{-kt}$, where k is a constant that depends on the object.

a. Determine the constant k (to the nearest thousandth) for a canned soda drink that takes 5 minutes to cool from 75°F to 65°F after being placed in a refrigerator that maintains a constant temperature of 34°F.

b. What will be the temperature (to the nearest degree) of the soda drink after 30 minutes?

c. When (to the nearest minute) will the temperature of the soda drink be 36°F?

d. When will the temperature of the soda drink be exactly 34°F?

26. Solve the sound intensity equation $N = 10 \log \dfrac{I}{I_0}$ for I.

27. **SOUND INTENSITY** How much more powerful is a sound that measures 120 decibels than a sound (at the same frequency) that measures 110 decibels?

28. **SOUND INTENSITY** The power of a band is 3.4×10^{-5} watts per square centimeter. What is the band's intensity level N in decibels?

29. **SOUND INTENSITY** If the power of a sound is doubled, what is the increase in the intensity level? (*Hint:* Find $N(2I) - N(I)$.)

30. **PSYCHOLOGY** According to a software company, the users of its typing tutorial can expect to type $N(t)$ words per minute after t hours of practice with the product, according to the function $N(t) = 100(1.04 - 0.99^t)$.

a. How many words per minute can a student expect to type after 2 hours of practice?

b. How many words per minute can a student expect to type after 40 hours of practice?

c. According to the function N, how many hours (to the nearest hour) of practice will be required before a student can expect to type 60 words per minute?

31. **PSYCHOLOGY** In the city of Whispering Palms, the number of people $P(t)$ exposed to a rumor in t hours is given by the function $P(t) = 80{,}000(1 - e^{-0.0005t})$.

a. Find the number of hours until 10% of the population have heard the rumor.

b. Find the number of hours until 50% of the population have heard the rumor.

32. **LAW** A lawyer has determined that the number of people $P(t)$ who have been exposed to a news item after t days is given by the function

$$P(t) = 1{,}200{,}000(1 - e^{-0.03t})$$

a. How many days after a major crime has been reported have 40% of the population heard of the crime?

b. A defense lawyer knows it will be very difficult to pick an unbiased jury after 80% of the population have heard of the crime. After how many days will 80% of the population have heard of the crime?

33. **DEPRECIATION** An automobile depreciates according to the function $V(t) = V_0(1 - r)^t$, where $V(t)$ is the value in dollars after t years, V_0 is the original value, and r is the yearly depreciation rate. A car has a yearly depreciation rate of 20%. Determine in how many years (to the nearest 0.1 year) the car will depreciate to half its original value.

34. **PHYSICS** The current $I(t)$ (measured in amperes) of a circuit is given by the function $I(t) = 6(1 - e^{-2.5t})$, where t is the number of seconds after the switch is closed.

a. Find the current when $t = 0$.

b. Find the current when $t = 0.5$.

c. Solve the equation for t.

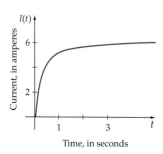

35. **AIR RESISTANCE** Assuming that air resistance is proportional to velocity, the velocity v, in feet per second, of a certain object after t seconds is given by $v = 32(1 - e^{-t})$.

a. Graph this equation for $t \geq 0$.

b. Determine algebraically, to the nearest 0.01 second, when the velocity is 20 feet per second.

c. Determine the horizontal asymptote of the graph of v.

d. ✏️ Write a sentence that explains the meaning of the horizontal asymptote in the context of this application.

36. **AIR RESISTANCE** Assuming that air resistance is proportional to velocity, the velocity v, in feet per second, of a certain object after t seconds is given by $v = 64(1 - e^{-t/2})$.

a. Graph this equation for $t \geq 0$.

b. Determine algebraically, to the nearest 0.1 second, when the velocity is 50 feet per second.

c. Determine the horizontal asymptote of the graph of v.

d. ✏️ Write a sentence that explains the meaning of the horizontal asymptote in the context of this application.

37. ⌨ The distance s (in feet) that the object in Exercise 35 will fall in t seconds is given by $s = 32t + 32(e^{-t} - 1)$.

a. Use a graphing utility to graph this equation for $t \geq 0$.

b. Determine, to the nearest 0.1 second, the time it takes the object to fall 50 feet.

c. Calculate the slope of the secant line through $(1, s(1))$ and $(2, s(2))$.

d. Write a sentence that explains the meaning of the slope of the secant line you calculated in **c.**

38. The distance s (in feet) that the object in Exercise 36 will fall in t seconds is given by $s = 64t + 128(e^{-t/2} - 1)$.

a. Use a graphing utility to graph this equation for $t \geq 0$.

b. Determine, to the nearest 0.1 second, the time it takes the object to fall 50 feet.

c. Calculate the slope of the secant line through $(1, s(1))$ and $(2, s(2))$.

d. Write a sentence that explains the meaning of the slope of the secant line you calculated in **c.**

39. **POPULATION GROWTH** The population of squirrels in a nature preserve satisfies the logistic model with $P_0 = 1500$ in 1985, $m = 16,500$, and $P(2) = 2500$ (the population in 1987).

a. Determine k to the nearest hundredth.

b. Determine the year in which the population first exceeds 10,000.

40. **POPULATION GROWTH** The population of walruses in a colony satisfies the logistic model with $P_0 = 800$ in 1990, $m = 5500$, and $P(1) = 900$ (the population in 1991).

a. Determine k to the nearest hundredth.

b. Determine the year in which the population first exceeds 2000.

41. **RATE OF GROWTH** The *rate of growth* of the squirrel population in Exercise 39 is given by

$$R = \frac{47,850e^{0.29t}}{(10 + e^{0.29t})^2}$$

where R is the rate of growth; that is, the units of R are squirrels per year.

a. Use a graphing utility to graph this equation.

b. What is the maximum value, to the nearest whole number, of this function?

c. What is the t-coordinate, to the nearest hundredth, associated with the maximum value of the function?

d. Write a sentence that explains the meaning of the maximum value of this function in the context of the application.

42. **RATE OF GROWTH** The *rate of growth* of the walrus population in Exercise 40 is given by

$$R = \frac{289,520e^{0.14t}}{(47 + 8e^{0.14t})^2}$$

where R is the rate of growth; that is, the units of R are walruses per year.

a. Use a graphing utility to graph this equation.

b. What is the maximum value, to the nearest whole number, of this function?

c. What is the t-coordinate associated with the maximum value of the function?

d. Write a sentence that explains the meaning of the maximum value of this function in the context of the application.

43. **RATE OF GROWTH** The equation for the *rate of growth* of the bison population in Exercise 61 of Section 5.4 is given by

$$R = \frac{3810e^{0.127t}}{(30 + e^{0.127t})^2}$$

where R is the rate of growth; that is, the units of R are bison per year.

a. Use a graphing utility to graph this equation using the domain you used in Exercise 61 of Section 5.4.

b. What is the maximum value of this function?

c. What is the t-coordinate, to the nearest hundredth, associated with the maximum value of the function?

d. Write a sentence that explains the meaning of the maximum value of this function in the context of the application.

44. **RATE OF GROWTH** The equation for the *rate of growth* of the yeast population in Exercise 62 of Section 5.4 is given by

$$R = \frac{3.8125 \times 10^6 e^{0.305t}}{(250 + e^{0.305t})^2}$$

where R is the rate of growth; that is, the units of R are number of yeast per hour.

a. Use a graphing utility to graph this equation using the domain you used in Exercise 62 of Section 5.4.

b. What is the maximum value of this function?

c. What is the t-coordinate associated with the maximum value of the function?

d. Write a sentence that explains the meaning of the maximum value of this function in the context of the application.

45. **LEARNING THEORY** The logistic model is also used in learning theory. Suppose that historical records from employee training at a company show that the percent score on a product information test is given by

$$P = \frac{100}{1 + 25e^{-0.095t}}$$

where t is the number of hours of training. What is the number of hours (to the nearest hour) of training needed before a new employee will answer at least 75% of the questions correctly?

46. LEARNING THEORY A company provides training in the assembly of a computer circuit to new employees. Past experience has shown that the number of correctly assembled circuits per week can be modeled by

$$N = \frac{250}{1 + 249e^{-0.503t}}$$

where t is the number of weeks of training. What is the number of weeks (to the nearest week) of training needed before a new employee will correctly make 140 circuits?

47. PREDICTING ADEQUACY OF RESOURCES The adequacy of a city's resources can sometimes be modeled by a *gamma density function*. The distribution created from this function enables city planners to determine the probability that certain city services can be maintained. Suppose a city has determined that the probability of its being able to provide more than x million liters of water per day is given by

$$P = \left(\frac{1}{3}x + 1\right)e^{-x/3}$$

a. Use a graphing utility to graph this equation for $x \geq 0$.

b. Determine the probability that a city can supply more than 5 million liters of water per day.

c. The city manager wants to determine the minimum water supply in a reservoir that the city can maintain so that there is less than a 0.25 chance that the city will not be able to meet demand. What must the capacity of the water supply be to meet the goal of the manager?

d. As $x \to \infty$, $P \to 0$. Explain why this makes sense in the context of this application.

48. PREDICTING ADEQUACY OF RESOURCES The probability (see Exercise 47) that a utility company can supply more than x million kilowatt-hours of electricity per day is given by

$$P = \left(\frac{1}{4}x + 1\right)e^{-x/4}$$

a. Use a graphing utility to graph this equation for $x \geq 0$.

b. Determine the probability that this utility company can supply more than 8 million kilowatt-hours of electricity per day.

c. The utility company wants to determine what capacity it must have so that there is less than a 0.50 chance that the utility will not be able to meet demand. What must the capacity of the utility company supply be to meet the goal of the company?

d. As $x \to \infty$, $P \to 0$. Explain why this makes sense in the context of this application.

49. PHYSICS If air resistance is proportional to velocity, then the time t in seconds for a particular object to reach a velocity of v feet per second is given by

$$t = 3.125 \ln \frac{100}{100 - v}$$

a. How long, to the nearest 0.1 second, is required before the velocity is 50 feet per second?

b. There is a vertical asymptote when $v = 100$. Describe the meaning of this asymptote in the context of the application.

50. PHYSICS If air resistance is proportional to velocity, then the time t in seconds for a particular object to reach a velocity of v feet per second is given by

$$t = 2.34 \ln \frac{75}{75 - v}$$

a. How long, to the nearest 0.1 second, is required before the velocity is 30 feet per second?

b. There is a vertical asymptote when $v = 75$. Describe the meaning of this asymptote in the context of the application.

SUPPLEMENTAL EXERCISES

51. PRIME NUMBER THEOREM The Prime Number Theorem states that the number of prime numbers $P(n)$ less than a number n can be approximated by the function

$$P(n) = \frac{n}{\ln n}$$

a. The actual number of prime numbers less than 100 is 25. Compute $P(100)$ and $P(100)/25$.

b. The actual number of prime numbers less than 10,000 is 1229. Compute $P(10,000)$ and $P(10,000)/1229$.

c. The actual number of prime numbers less than 1,000,000 is 78,498. Compute $P(1,000,000)$, and then compute the ratio $P(1,000,000)/78,498$.

52. STIRLING'S FORMULA The number $n!$ (which is read "n factorial") is defined as

$$n! = n(n - 1)(n - 2) \cdots 1$$

for all positive integers n. Thus $4! = 4 \cdot 3 \cdot 2 \cdot 1 = 24$. *Stirling's Formula* (after James Stirling, 1692–1770),

$$n! \approx \left(\frac{n}{e}\right)^n \sqrt{2\pi n}$$

is often used to approximate very large factorials. Use Stirling's Formula to approximate $10!$, and then compute the ratio of Stirling's approximation of $10!$ divided by the actual value of $10!$, which is 3,628,800.

53. AGRICULTURE A farmer knows that planting the same crop in the same field year after year reduces the yield. If the yield on each succeeding year's crop is 90% of the preceding year's yield, then the yield $Y(t)$ at any time t is given by the function $Y(t) = Y_0(0.90)^t$, where Y_0 is the yield when $t = 0$. In how many years (to the nearest year) will the yield be 60% of Y_0?

54. OIL SPILLS Crude oil leaks from a tank at a rate that depends on the amount of oil that remains in the tank. Because 1/8 of the oil in the tank leaks out every 2 hours, the volume of oil $V(t)$ in the tank at t hours is given by $V(t) = V_0(0.875)^{t/2}$, where $V_0 = 350,000$ gallons is the number of gallons in the tank at the time the tank started to leak ($t = 0$).

a. How many gallons does the tank hold after 3 hours?

b. How many gallons does the tank hold after 5 hours?

c. How long will it take until 90% of the oil has leaked from the tank?

55. EARTHQUAKES How many times stronger is an earthquake that measures 5 on the Richter scale than one that measures 3 on the Richter scale?

56. EARTHQUAKES How many times stronger was the Chile earthquake of 1960, which measured 9.5 on the Richter scale, than the San Francisco earthquake of 1906, which measured 8.3 on the Richter scale?

57. NUMBER OF DIGITS IN b^c Logarithms and a function called the integer function (denoted by int) can be used to determine the number of digits in a number written in exponential notation. The number of digits $N(x)$ in the number b^x, with $0 < b < 10$ and x a positive integer, is given by the function $N(x) = \text{int}\,(x \log b) + 1$.

a. Find the number of digits in 3^{200}.

b. Find the number of digits in 7^{4005}.

c. The largest known prime number as of 1980 was the number $2^{44,497} - 1$. Find the number of digits in this prime number.

d. The largest known prime number as of 1983 was $2^{132,049} - 1$. Find the number of digits in this prime number.

e. The largest known Mersenne prime as of January 1994 was $2^{859,433} - 1$. Verify that this number has 258,716 digits.

f. The largest known perfect number as of January 1994 was $2^{859,432}(2^{859,433} - 1)$. Verify that this number has 517,430 digits. (*Hint:* Use the theorem $b^p \cdot b^q = b^{p+q}$.)

58. INFLATION How many years will it take the price of goods to double if the annual rate of inflation is 5% per year? Use continuous compounding.

59. INFLATION Determine what rate of inflation will cause the price of goods to double in the next 10 years. Use continuous compounding.

60. HANGING CABLE The height h in feet of any point P on the cable shown is a function of the horizontal distance in feet from point P to the origin:

$$h(x) = \frac{20}{2}(e^{x/20} + e^{-x/20}), \quad -40 \le x \le 40$$

a. What is the height of the cable at point P if P is directly above the origin?

b. What is the height of the cable at point P if P is 25 feet to the right of the origin?

c. How far to the right or left of the origin is the cable 30 feet in height?

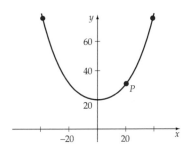

PROJECTS

When you are trying to determine an equation that models certain data, it is sometimes convenient to use a logarithmic scale. The following project uses this technique.

1. AVIATION A pressure altimeter is used to determine the height of a plane above sea level. The table at the top of page 339 gives values for the pressure p and altitude h for an altimeter.

a. Make a scatter plot of these data with pressure along the horizontal axis and height on the vertical axis. *Suggestion:* Choose a scale that will enable you to plot these points fairly accurately.

b. Draw a smooth curve that goes through the points. *Note:* This curve does not have to pass through each point. The idea is to draw a *smooth* curve that approximates the data. Your task is to determine the equation of the curve you have drawn by completing the remaining parts of this project.

c. Make a new graph in which you graph the natural logarithm of each pressure and the corresponding height. That is, plot $(\ln p, h)$ for each value of p.

d. Draw a smooth curve through the new data points. This curve should be very close to a straight line.

e. Assuming that the points in **d.** are approximately on a straight line, find the equation of the line by using the points $(\ln 14.1, 1100)$ and $(\ln 11.0, 7700)$. *Suggestion:* Because the values along the horizontal axis are natural logarithms, the point-slope formula becomes $h - h_1 = m[\ln p - \ln p_1]$.

f. Evaluate the function you found in part **e.** for the pressures given in the table above and verify that the values are approximately the heights of the plane for the given pressures.

g. Explain why the function you found in **e.** is a model for the pressure altimeter.

h. Use your model to predict the height of the plane when the pressure is 11.5 pounds per square inch.

p, in pounds per square inch	h, in feet
14.1	1100
13.8	1700
13.5	2300
12.8	3700
12.1	5200
11.0	7700

EXPLORING CONCEPTS WITH TECHNOLOGY

Using a Semi-Log Graph to Model Exponential Decay

Consider the data in Table 5.8, which shows the viscosity V of SAE 40 motor oil at various temperatures T. The graph of this data is shown below, along with a curve that passes through the points. The graph in **Figure 5.32** appears to have an exponential decay shape.

One way to determine whether the graph in **Figure 5.32** is the graph of an exponential function is to plot the data on *semi-log* graph paper. On this graph paper, the horizontal axis remains the same, but the vertical axis uses a logarithmic scale.

The data in Table 5.8 are graphed again in **Figure 5.33,** but this time the vertical axis is a natural logarithm axis. This graph is approximately a straight line. The slope of the line, to the nearest ten-thousandth, is

$$m = \frac{\ln 500 - \ln 120}{100 - 150} \approx -0.0285$$

Table 5.8

T	V
90	700
100	500
110	350
120	250
130	190
140	150
150	120

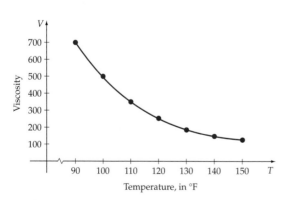

Figure 5.32

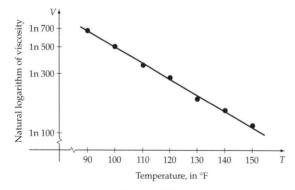

Figure 5.33

Using this slope and the point-slope formula with V replaced by $\ln V$, we have

$$\ln V - \ln 120 = -0.0285(T - 150)$$
$$\ln V \approx -0.0285T + 9.062 \tag{1}$$

Equation (1) is the equation of the line on a semi-log coordinate grid.

Now solve Equation (1) for V.

$$e^{\ln V} = e^{-0.0285T + 9.062}$$
$$V = e^{-0.0285T}e^{9.062}$$
$$V \approx 8621e^{-0.0285T} \tag{2}$$

Equation (2) is a model of the data in the rectangular coordinate system shown in **Figure 5.32.**

Table 5.9

t	A
1	91.77
4	70.92
8	50.30
15	27.57
20	17.95
30	7.60

1. A chemist wishes to determine the decay characteristics of iodine-131. A 100-mg sample of iodine-131 is observed over a 30-day period. Table 5.9 shows the amount A (in milligrams) of iodine-131 remaining after t days.

   a. Graph the ordered pairs (t, A) on semi-log paper. (*Note:* Semi-log paper comes in different varieties. Our calculations are based on semi-log paper that has a natural logarithm scale on the vertical axis.)

   b. Use the points $(4, 4.3)$ and $(15, 3.3)$ to approximate the slope of the line that passes through the points.

   c. Using the slope calculated in **b.** and the point $(4, 4.3)$, determine the equation of the line.

   d. Solve the equation you derived in **c.** for A.

   e. Graph the equation you derived in **d.** in a rectangular coordinate system.

   f. What is the half-life of iodine-131?

Table 5.10

t	B
0	15.5
1	15.7
2	15.9
3	16.2
4	16.7

2. The live birth rates B per thousand births in the United States are given in Table 5.10 for the years 1986 through 1990 ($t = 0$ corresponds to 1986).

   a. Graph the ordered pairs $(t, \ln B)$. (You will need to adjust the scale so that you can discriminate between plotted points. A suggestion is given in **Figure 5.34.**)

   b. Use the points $(1, 2.754)$ and $(3, 2.785)$ to approximate the slope of the line that passes through the points.

   c. Using the slope calculated in **b.** and the point $(1, 2.754)$, determine the equation of the line.

   d. Solve the equation you derived in **c.** for B.

   e. Graph the equation you derived in **d.** in a rectangular coordinate system.

   f. If the birth rate continues as predicted by your model, in what year will the birth rate be 17.5 per thousand?

The difference in graphing strategies between Exercise 1 and Exercise 2 is that in Exercise 1, semi-log paper was used. When a point is graphed on this coordinate paper, the y-coordinate is $\ln y$. In Exercise 2, graphing a point at $(x, \ln y)$ in a rectangular coordinate system has the same effect as graphing (x, y) in a semi-log coordinate system.

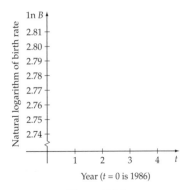

Figure 5.34

CHAPTER 5 REVIEW

5.1 Exponential Functions and Their Graphs

- For all positive real numbers b, $b \neq 1$, the exponential function defined by $f(x) = b^x$ has the following properties:

  1. f has the set of real numbers as its domain.

  2. f has the set of positive real numbers as its range.

  3. f has a graph with a y-intercept of $(0, 1)$.

  4. f has a graph asymptotic to the x-axis.

  5. f is a one-to-one function.

  6. f is an increasing function if $b > 1$.

  7. f is a decreasing function if $0 < b < 1$.

- As n increases without bound, $(1 + 1/n)^n$ approaches an irrational number denoted by e. The value of e accurate to eight decimal places is 2.71828183.

- The function defined by $f(x) = e^x$ is called the natural exponential function.

5.2 Logarithms and Logarithmic Properties

- *Definition of a Logarithm* If $x > 0$ and b is a positive constant ($b \neq 1$), then

$$y = \log_b x \quad \text{if and only if} \quad b^y = x$$

  In the equation $y = \log_b x$, y is referred to as the logarithm, b is the base, and x is the argument.

- *Equality of Exponents Theorem* If b is a positive real number ($b \neq 1$) such that $b^x = b^y$, then $x = y$.

- If b, M, and N are positive real numbers ($b \neq 1$), and p is any real number, then

  $\log_b b = 1$

  $\log_b 1 = 0$

  $\log_b (b^p) = p$

  $\log_b (MN) = \log_b M + \log_b N$

  $\log_b \dfrac{M}{N} = \log_b M - \log_b N$

  $\log_b (M^p) = p \log_b M$

  $\log_b M = \log_b N \quad \text{implies} \quad M = N$

  $M = N \quad \text{implies} \quad \log_b M = \log_b N$

  $b^{\log_b p} = p \quad (\text{for } p > 0)$

- *Change-of-Base Formula* If x, a, and b are positive real numbers with $a \neq 1$ and $b \neq 1$, then

$$\log_b x = \frac{\log_a x}{\log_a b}$$

5.3 Logarithmic Functions and Their Graphs

- For all positive real numbers b, $b \neq 1$, the function defined by $f(x) = \log_b x$ has the following properties:

  1. f has the set of positive real numbers as its domain.

  2. f has the set of real numbers as its range.

  3. f has a graph with an x-intercept of $(1, 0)$.

  4. f has a graph asymptotic to the y-axis.

  5. f is a one-to-one function.

  6. f is an increasing function if $b > 1$.

  7. f is a decreasing function if $0 < b < 1$.

5.4 Exponential and Logarithmic Equations

- Exponential equations of the form $b^x = c$ can be solved by taking either the common logarithm or the natural logarithm of each side of the equation.

- Exponential equations of the form $b^x = b^y$ can be solved by using the Equality of Exponents Theorem.

- Logarithmic equations can often be solved by using the properties of logarithms and the definition of a logarithm.

- The pH of a liquid is given by the formula

$$\text{pH} = -\log [\text{H}_3\text{O}^+]$$

5.5 Applications of Exponential and Logarithmic Functions

- The function defined by $N(t) = N_0 e^{kt}$ is called an exponential growth function if k is a positive constant, and it is called an exponential decay function if k is a negative constant.

- *The Compound Interest Formula* A principal P invested at an annual interest rate r, expressed as a decimal and compounded n times per year for t years, produces the balance

$$B = P\left(1 + \frac{r}{n}\right)^{nt}$$

- *Continuous Compounding Interest Formula* If an account with principal P and annual interest rate r is compounded continuously for t years, then the balance is $B = Pe^{rt}$.

CHAPTER 5 TRUE/FALSE EXERCISES

In Exercises 1 to 14, answer true or false. If the statement is false, given an example to show that the statement is false.

1. If $7^x = 40$, then $\log_7 40 = x$.

2. If $\log_4 x = 3.1$, then $4^{3.1} = x$.

3. If $f(x) = \log x$ and $g(x) = 10^x$, then $f[g(x)] = x$ for all real numbers x.

4. If $f(x) = \log x$ and $g(x) = 10^x$, then $g[f(x)] = x$ for all real numbers x.

5. The exponential function $h(x) = b^x$ is an increasing function.

6. The logarithmic function $j(x) = \log_b x$ is an increasing function.

7. The exponential function $h(x) = b^x$ is a one-to-one function.

8. The logarithmic function $j(x) = \log_b x$ is a one-to-one function.

9. The graph of $f(x) = \dfrac{2^x + 2^{-x}}{2}$ is symmetric with respect to the y-axis.

10. The graph of $f(x) = \dfrac{2^x - 2^{-x}}{2}$ is symmetric with respect to the origin.

11. If $x > 0$ and $y > 0$, then $\log (x + y) = \log x + \log y$.

12. If $x > 0$, then $\log x^2 = 2 \log x$.

13. If M and N are positive real numbers, then

$$\ln \frac{M}{N} = \ln M - \ln N$$

14. For all $p > 0$, $e^{\ln p} = p$.

CHAPTER 5 REVIEW EXERCISES

In Exercises 1 to 12, solve each equation. Do not use a calculator.

1. $\log_5 25 = x$ 2. $\log_3 81 = x$ 3. $\ln e^3 = x$

4. $\ln e^{\pi} = x$ 5. $3^{2x+7} = 27$ 6. $5^{x-4} = 625$

7. $2^x = \dfrac{1}{8}$ 8. $27(3^x) = 3^{-1}$ 9. $\log x^2 = 6$

10. $\dfrac{1}{2} \log |x| = 5$ 11. $10^{\log 2x} = 14$ 12. $e^{\ln x^2} = 64$

In Exercises 13 to 18, use a calculator to evaluate each power. Give your answers accurate to six significant digits.

13. $7^{\sqrt{2}}$ 14. $3^{\sqrt{5}}$ 15. $e^{1.7}$

16. $e^{-2.2}$ 17. $10^{1.135}$ 18. $10^{-\sqrt{10}}$

In Exercises 19 to 32, sketch the graph of each function.

19. $f(x) = (2.5)^x$ 20. $f(x) = \left(\dfrac{1}{4}\right)^x$

21. $f(x) = 3^{|x|}$ 22. $f(x) = 4^{-|x|}$

23. $f(x) = 2^x - 3$ 24. $f(x) = 2^{(x-3)}$

25. $f(x) = \dfrac{4^x + 4^{-x}}{2}$ 26. $f(x) = \dfrac{3^x - 3^{-x}}{2}$

27. $f(x) = \dfrac{1}{3} \log x$ 28. $f(x) = 3 \log x^{1/3}$

29. $f(x) = -x + \log x$ 30. $f(x) = 2^{-x} \log x$

31. $f(x) = -\dfrac{1}{2} \ln x$ 32. $f(x) = -\ln |x|$

In Exercises 33 to 36, change each logarithmic equation to its exponential form.

33. $\log_4 64 = 3$ 34. $\log_{1/2} 8 = -3$

35. $\log_{\sqrt{2}} 4 = 4$ 36. $\ln 1 = 0$

In Exercises 37 to 40, change each exponential equation to its logarithmic form.

37. $5^3 = 125$ 38. $2^{10} = 1024$

39. $10^0 = 1$ 40. $8^{1/2} = 2\sqrt{2}$

In Exercises 41 to 44, write the given logarithm in terms of logarithms of x, y, and z.

41. $\log_b \dfrac{x^2 y^3}{z}$ 42. $\log_b \dfrac{\sqrt{x}}{y^2 z}$

43. $\ln xy^3$ 44. $\ln \dfrac{\sqrt{xy}}{z^4}$

In Exercises 45 to 48, write each logarithmic expression as a single logarithm.

45. $2 \log x + \dfrac{1}{3} \log (x + 1)$ 46. $5 \log x - 2 \log (x + 5)$

47. $\dfrac{1}{2} \ln 2xy - 3 \ln z$ 48. $\ln x - (\ln y - \ln z)$

In Exercises 49 to 52, use the change-of-base formula and a calculator to approximate each logarithm accurate to six significant digits.

49. $\log_5 101$

50. $\log_3 40$

51. $\log_4 0.85$

52. $\log_8 0.3$

In Exercises 53 to 56, use a calculator to approximate N to three significant digits.

53. $\log N = 2.47$

54. $\log N = -0.48$

55. $\ln N = 51$

56. $\ln N = -0.09$

In Exercises 57 to 72, solve each equation for x. Give exact answers. Do not use a calculator.

57. $4^x = 30$

58. $5^{x+1} = 41$

59. $\ln 3x - \ln (x - 1) = \ln 4$

60. $\ln 3x + \ln 2 = 1$

61. $e^{\ln (x+2)} = 6$

62. $10^{\log (2x+1)} = 31$

63. $\dfrac{4^x + 4^{-x}}{4^x - 4^{-x}} = 2$

64. $\dfrac{5^x + 5^{-x}}{2} = 8$

65. $\log (\log x) = 3$

66. $\ln (\ln x) = 2$

67. $\log \sqrt{x - 5} = 3$

68. $\log x + \log (x - 15) = 1$

69. $\log_4 (\log_3 x) = 1$

70. $\log_7 (\log_5 x^2) = 0$

71. $\log_5 x^3 = \log_5 16x$

72. $25 = 16^{\log_4 x}$

73. **CHEMISTRY** Find the pH of tomatoes that have a hydronium-ion concentration of 6.28×10^{-5}.

74. **CHEMISTRY** Find the hydronium-ion concentration of rainwater that has a pH of 5.4.

75. **COMPOUND INTEREST** Find the balance when $16,000 is invested at an annual rate of 8% for 3 years if the interest is compounded
   a. monthly   **b.** continuously

76. **COMPOUND INTEREST** Find the balance when $19,000 is invested at an annual rate of 6% for 5 years if the interest is compounded
   a. daily   **b.** continuously

77. **DEPRECIATION** The scrap value S of a product with an expected life span of n years is given by $S(n) = P(1 - r)^n$, where P is the original purchase price of the product and r is the annual rate of depreciation. A taxicab is purchased for $12,400 and is expected to last 3 years. What is its scrap value if it depreciates at a rate of 29% per year?

78. **MEDICINE** A skin wound heals according to the function given by $N(t) = N_0 e^{-0.12t}$, where N is the number of square centimeters of unhealed skin t days after the injury, and N_0 is the number of square centimeters covered by the original wound.
   a. What percentage of the wound will be healed after 10 days?
   b. How many days will it take for 50% of the wound to heal?
   c. How long will it take for 90% of the wound to heal?

In Exercises 79 to 82, find the exponential growth/decay function $N(t) = N_0 e^{kt}$ that satisfies the given conditions.

79. $N(0) = 1,\ N(2) = 5$

80. $N(0) = 2,\ N(3) = 11$

81. $N(1) = 4,\ N(5) = 5$

82. $N(-1) = 2,\ N(0) = 1$

CHAPTER 5 TEST

1. Given that $f(x) = 2.5^x$, use a calculator to evaluate $f(\sqrt{3})$ accurate to six significant digits.

2. Given that $f(x) = e^{x/2}$, use a calculator to evaluate $f(2.7)$ accurate to six significant digits.

3. Graph: $f(x) = 3^{-x/2}$

4. Graph: $f(x) = \left(\dfrac{3}{2}\right)^x$

5. Graph: $f(x) = e^{x/2}$

6. Write $\log_b (5x - 3) = c$ in exponential form.

7. Write $3^{x/2} = y$ in logarithmic form.

8. Write $\log_b \dfrac{x^2 y^4}{z^3}$ in terms of logarithms of x, y, and z.

9. Write $\log_b \dfrac{z^2}{y^3 \sqrt{x}}$ in terms of logarithms of x, y, and z.

10. Write $\log_{10} (2x + 3) - 3 \log_{10} (x - 2)$ as a single logarithm.

11. Determine the age of a bone (to the nearest 10 years) if it now contains 92% of its original amount of carbon-14. The half-life of carbon-14 is 5730 years.

12. Use the change-of-base formula and a calculator to approximate $\log_4 12$ to five significant digits.

13. Graph: $f(x) = \log (x - 1)$

14. Graph: $f(x) = \log x + 2$

15. Graph: $f(x) = -\ln (x + 1)$

16. Solve: $3^{2x-5} = \dfrac{1}{27}$

17. Solve: $5^x = 22$

18. Solve $\log(x + 99) - \log(3x - 2) = 2$

19. Find the balance if $20,000 is invested at an annual interest rate of 7.8% for 5 years, compounded continuously.

20. The scrap value S of a product with an expected life of n years is given by $S(n) = P(1 - r)^n$, where P is the original purchase price of the product and r is the annual rate of depreciation. A computer system is purchased for $8400 and is expected to last 4 years. What is its scrap value if the computer system depreciates at a rate of 22% per year?

TOPICS IN ANALYTIC GEOMETRY

The planets travel in elliptical orbits around the sun.

The lenses in any telescope, big or small, must have the shape of a conic section.

A technician inspects the parabolic mirrors used in an optical laser system.

Kidney Stones and Ellipses

The conic sections were studied by the ancient Greeks. One of the conic sections is an ellipse. Besides applications of the ellipse to astronomy and optics, the ellipse has an application to medicine. The ellipse is used in a nonsurgical treatment of kidney stones.

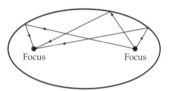

Inside the ellipse, an oval-shaped curve, there are two points called foci of the ellipse. Sound or light waves emitted from one focus are reflected off the surface of the ellipse to the other focus. In an analogous way, think of a billiard table shaped like an ellipse. A ball struck from one focus would pass through the other focus no matter what the direction of the initial shot.

The treatment of kidney stones is based on this reflective property of an ellipse. An electrode is placed at one focus of an elliptic reflector, and the patient is placed such that the kidney stone is at the other focus. When the electrode is discharged, ultrasound waves are produced. These waves hit the walls of the ellipse and are reflected to the kidney stone. Because of the reflective property of the ellipse, there is very little energy loss, and it is as though the electrode actually discharged at the kidney stone. The energy of the discharge pulverizes the kidney stone into fragments.

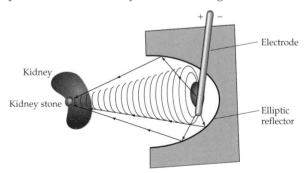

This figure shows a lithotripter, which is used to pulverize kidney stones.

6.1 PARABOLAS

The graph of a parabola, circle, ellipse, or hyperbola can be formed by the intersection of a plane and a cone. Hence these figures are referred to as conic sections. See **Figure 6.1.**

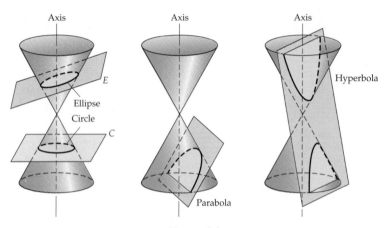

Figure 6.1
Cones intersected by planes

A plane perpendicular to the axis of the cone intersects the cone in a circle (plane *C*). The plane *E*, tilted so that it is not perpendicular to the axis, intersects the cone in an ellipse. When the plane is parallel to a line on the surface of the cone, the plane intersects the cone in a parabola. When the plane intersects both cones, a hyperbola is formed.

PARABOLAS WITH VERTEX AT (0,0)

Besides the geometric description of a conic section just given, a conic can be defined as a set of points. This method uses some specified conditions about the curve to determine which points in a coordinate system are points of the graph. For example, a parabola can be defined by the following set of points.

> **DEFINITION OF A PARABOLA**
>
> A **parabola** is the set of points in the plane that are equidistant from a fixed line (the **directrix**) and a fixed point (the **focus**) not on the directrix.

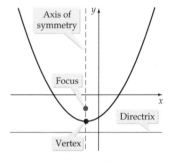

Figure 6.2

The line that passes through the focus and is perpendicular to the directrix is called the **axis of symmetry** of the parabola. The midpoint of the line segment between the focus and directrix on the axis of symmetry is the **vertex** of the parabola, as shown in **Figure 6.2.**

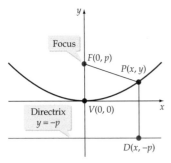

Figure 6.3

Using this definition of a parabola, we can determine an equation of a parabola. Suppose that the coordinates of the vertex of a parabola are $V(0,0)$ and the axis of symmetry is the y-axis. The equation of the directrix is $y = -p$, $p > 0$. The focus lies on the axis of symmetry and is the same distance from the vertex as the vertex is from the directrix. Thus the coordinates of the focus are $F(0, p)$, as shown in **Figure 6.3**.

Let $P(x, y)$ be any point P on the parabola. Then, using the distance formula and the fact that the distance between any point P on the parabola and the focus is equal to the distance from the point P to the directrix, we can write the equation

$$d(P, F) = d(P, D)$$

By the distance formula,

$$\sqrt{(x - 0)^2 + (y - p)^2} = y + p$$

Now, squaring each side and simplifying, we get

$$(\sqrt{(x - 0)^2 + (y - p)^2})^2 = (y + p)^2$$
$$x^2 + y^2 - 2py + p^2 = y^2 + 2py + p^2$$
$$x^2 = 4py$$

This is an equation of a parabola with vertex at the origin and a vertical axis of symmetry. The equation of a parabola with a horizontal axis of symmetry is derived in a similar manner.

STANDARD FORMS OF THE EQUATION OF A PARABOLA WITH VERTEX AT THE ORIGIN

Vertical Axis of Symmetry
The standard form of the equation of a parabola with vertex $(0,0)$ and a vertical axis of symmetry is $x^2 = 4py$. The focus is $(0, p)$, and the equation of the directrix is $y = -p$.

Horizontal Axis of Symmetry
The standard form of the equation of a parabola with vertex $(0,0)$ and a horizontal axis of symmetry is $y^2 = 4px$. The focus is $(p, 0)$, and the equation of the directrix is $x = -p$.

> **TAKE NOTE**
>
> **The tests for y-axis and x-axis symmetry can be used to verify these statements and provide connections to earlier topics on symmetry.**

In the equation $x^2 = 4py$, $x^2 \geq 0$. Therefore, $4py \geq 0$. Thus if $p > 0$, then $y \geq 0$ and the parabola opens up. If $p < 0$, then $y \leq 0$ and the parabola opens down. A similar analysis shows that for $y^2 = 4px$, the parabola opens to the right when $p > 0$ and opens to the left when $p < 0$.

EXAMPLE 1   *Find the Focus and Directrix of a Parabola*

Find the focus and directrix of the parabola given by the equation

$$y = -\frac{1}{2}x^2.$$

Continued ▸

Solution

Because the x term is squared, the standard form of the equation is $x^2 = 4py$.

$$y = -\frac{1}{2}x^2$$

$$x^2 = -2y \qquad \bullet \text{ Write the given equation in standard form.}$$

Comparing this equation with $x^2 = 4py$ gives

$$4p = -2$$

$$p = -\frac{1}{2}$$

Because p is negative, the parabola opens down and the focus is below the vertex $(0, 0)$, as shown in **Figure 6.4.** The coordinates of the focus are $(0, -1/2)$. The equation of the directrix is $y = 1/2$.

TRY EXERCISE 4, EXERCISE SET 6.1

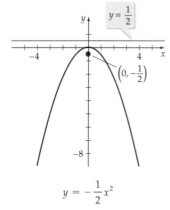

$$y = -\frac{1}{2}x^2$$

Figure 6.4

EXAMPLE 2   *Find the Equation of a Parabola in Standard Form*

Find the equation of the parabola in standard form with vertex at the origin and focus at $(-2, 0)$.

Solution

Because the vertex is at $(0, 0)$ and the focus is at $(-2, 0)$, $p = -2$. The graph of the parabola opens toward the focus, so in this case, the parabola opens to the left. The equation of the parabola in standard form that opens to the left is $y^2 = 4px$. Substitute -2 for p in this equation and simplify.

$$y^2 = 4(-2)x = -8x$$

The equation of the parabola is $y^2 = -8x$.

TRY EXERCISE 28, EXERCISE SET 6.1

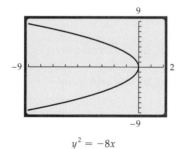

$$y^2 = -8x$$

Figure 6.5

The graph of $y^2 = -8x$ is shown in **Figure 6.5.** Note that the graph is not the graph of a function. To graph $y^2 = -8x$ with a graphing utility, we first solve for y to produce $y = \pm\sqrt{-8x}$. From this equation we can see that for any $x < 0$, there are two values of y. For example, when $x = -2$,

$$y = \pm\sqrt{(-8)(-2)} = \pm\sqrt{16} = \pm 4$$

The graph of $y^2 = -8x$ in **Figure 6.5** was drawn by graphing both $y_1 = \sqrt{-8x}$ and $y_2 = -\sqrt{-8x}$ on the same coordinate system.

PARABOLAS WITH THE VERTEX AT (h, k)

The equation of a parabola with a vertical or horizontal axis and with the vertex at a point (h, k) can be found by using the translations discussed previously. Consider a coordinate system with coordinate axes labeled x' and y' placed so that its origin is at (h, k) of the xy-coordinate system.

The relationship between an ordered pair in the $x'y'$-coordinate system and in the xy-coordinate system is given by the transformation equations

$$\begin{aligned} x' &= x - h \\ y' &= y - k \end{aligned} \tag{1}$$

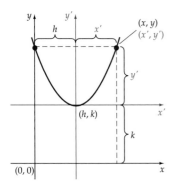

Figure 6.6

Now consider a parabola with vertex at (h, k) as shown in **Figure 6.6**. Place a new coordinate system labeled x' and y' with its origin at (h, k). The equation of a parabola in the $x'y'$-coordinate system is

$$(x')^2 = 4py' \tag{2}$$

Using the transformation Equations (1), we can substitute the expressions for x' and y' into Equation (2). The standard form of the equation of the parabola with vertex (h, k) and a vertical axis of symmetry is

$$(x - h)^2 = 4p(y - k)$$

Similarly, we can derive the standard form of the equation of the parabola with vertex (h, k) and a horizontal axis of symmetry.

STANDARD FORMS OF THE EQUATION OF A PARABOLA WITH VERTEX AT (h, k)

Vertical Axis of Symmetry
The standard form of the equation of the parabola with vertex $V(h, k)$ and a vertical axis of symmetry is

$$(x - h)^2 = 4p(y - k)$$

The focus is $(h, k + p)$, and the equation of the directrix is $y = k - p$. See **Figure 6.7**.

Horizontal Axis of Symmetry
The standard form of the equation of the parabola with vertex (h, k) and a horizontal axis of symmetry is

$$(y - k)^2 = 4p(x - h)$$

The focus is $(h + p, k)$, and the equation of the directrix is $x = h - p$.

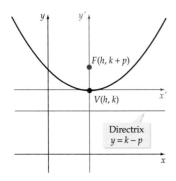

Figure 6.7

EXAMPLE 3   *Find the Focus and Directrix of a Parabola*

Find the equation of the directrix and the coordinates of the vertex and focus of the parabola given by the equation $3x + 2y^2 + 8y - 4 = 0$.

Continued •➤

Solution

Rewrite the equation so that the y terms are on one side of the equation, and then complete the square on y.

$$3x + 2y^2 + 8y - 4 = 0$$
$$2y^2 + 8y = -3x + 4$$
$$2(y^2 + 4y) = -3x + 4$$
$$2(y^2 + 4y + 4) = -3x + 4 + 8 \qquad \bullet \text{ Complete the square. Note that } 2 \cdot 4 = 8 \text{ is added to each side.}$$

$$2(y + 2)^2 = -3(x - 4) \qquad \bullet \text{ Simplify and then factor.}$$

$$(y + 2)^2 = -\frac{3}{2}(x - 4) \qquad \bullet \text{ Write the equation in standard form.}$$

Comparing this equation to $(y - k)^2 = 4p(x - h)$, we have a parabola that opens to the left with vertex $(4, -2)$ and $4p = -3/2$. Thus $p = -3/8$.
The coordinates of the focus are

$$\left(4 + \left(-\frac{3}{8}\right), -2\right) = \left(\frac{29}{8}, -2\right)$$

The equation of the directrix is

$$x = 4 - \left(-\frac{3}{8}\right) = \frac{35}{8}$$

Choosing some values for y and finding the corresponding values for x, we plot a few points. Because the line $y = -2$ is the axis of symmetry, for each point on one side of the axis of symmetry, there is a corresponding point on the other side. Two points are $(-2, 1)$ and $(-2, -5)$. See **Figure 6.8**.

TRY EXERCISE 20, EXERCISE SET 6.1

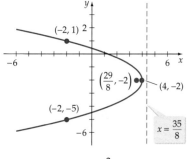

$$(y + 2)^2 = -\frac{3}{2}(x - 4)$$

Figure 6.8

EXAMPLE 4   **Find the Equation in Standard Form of a Parabola**

Find the equation in standard form of the parabola with directrix $x = -1$ and focus $(3, 2)$.

Solution

The vertex is the midpoint of the line segment joining $(3, 2)$ and the point $(-1, 2)$ on the directrix.

$$(h, k) = \left(\frac{-1 + 3}{2}, \frac{2 + 2}{2}\right) = (1, 2)$$

The standard form of the equation is $(y - k)^2 = 4p(x - h)$. The distance from the vertex to the focus is 2. Thus $4p = 4(2) = 8$, and the equation of the parabola in standard form is $(y - 2)^2 = 8(x - 1)$. See **Figure 6.9**.

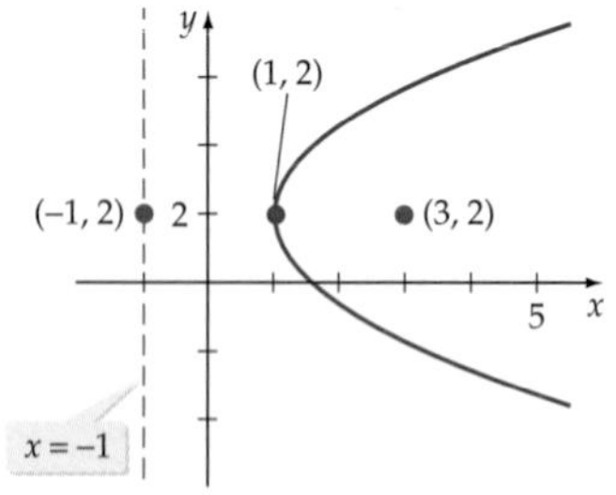

$(y - 2)^2 = 8(x - 1)$

Figure 6.9

TRY EXERCISE 30, EXERCISE SET 6.1

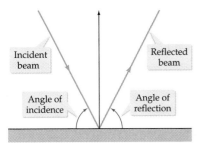

Figure 6.10

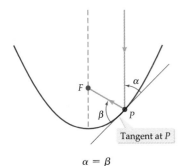

$\alpha = \beta$

Figure 6.11

Applications

A principle of physics states that when light is reflected from a point P on a surface, the angle of incidence (that of the incoming ray) equals the angle of reflection (that of the outgoing ray). See **Figure 6.10**. This principle applied to parabolas has some useful consequences.

Optical Property of a Parabola

The line tangent to a parabola at a point P makes equal angles with the line through P and parallel to the axis of symmetry and the line through P and the focus of the parabola (see **Figure 6.11**).

A cross section of the reflecting mirror of a telescope has the shape of a parabola. The incoming parallel rays of light are reflected from the surface of the mirror and to the focus. See **Figure 6.12**.

Flashlights and car headlights also make use of this property. The light bulb is positioned at the focus of the parabolic reflector, which causes the reflected light to be reflected outward in parallel rays. See **Figure 6.13**.

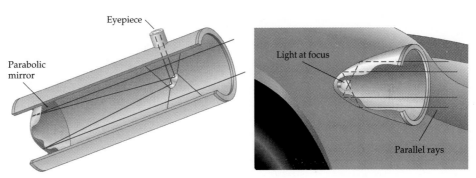

Figure 6.12   Figure 6.13

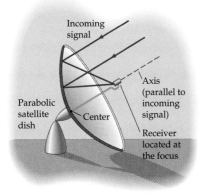

Figure 6.14

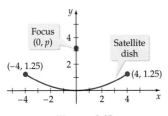

Figure 6.15

EXAMPLE 5   *Locate the Focus of a Satellite Dish*

A satellite dish has the shape of a paraboloid. The signals that it receives are reflected to a receiver that is located at the focus of the paraboloid. If the dish is 8 feet across at its opening and $1\frac{1}{4}$ feet deep at its center, determine the location of its focus.

Solution

Figure 6.14 shows that a cross section of the paraboloid along its axis of symmetry is a parabola. **Figure 6.15** shows this cross section placed in a rectangular coordinate system with the vertex of the parabola at $(0,0)$ and the axis of symmetry of the parabola on the y-axis. The parabola has an equation of the form

$$4py = x^2$$

Continued ▶

Because the parabola contains the point $(4, 1\frac{1}{4})$, this equation is satisfied by the substitutions $x = 4$ and $y = 1\frac{1}{4}$. Thus we have

$$4p\left(1\frac{1}{4}\right) = 4^2$$

$$5p = 16$$

$$p = \frac{16}{5}$$

The focus of the satellite dish is on the axis of symmetry of the dish, and it is $3\frac{1}{5}$ feet above the vertex of the dish.

TRY EXERCISE 36, EXERCISE SET 6.1

TOPICS FOR DISCUSSION

1. Do the graphs of the parabola given by $y = x^2$ and the vertical line given by $x = 10{,}000$ intersect? Explain.

2. A student claims that the focus of the parabola given by $y = 8x^2$ is at $(0, 2)$ because $4p = 8$ implies that $p = 2$. Explain the error in the student's reasoning.

3. The vertex of a parabola is always halfway between its focus and its directrix. Do you agree? Explain.

4. A tutor claims that the graph of $(x - h)^2 = 4p(y - k)$ has a y-intercept of $(0, h^2/(4p) + k)$. Explain why the tutor is correct.

EXERCISE SET 6.1

In Exercises 1 to 26, find the vertex, focus, and directrix of the parabola given by each equation. Sketch the graph.

1. $x^2 = -4y$

2. $2y^2 = x$

3. $y^2 = \frac{1}{3}x$

4. $x^2 = -\frac{1}{4}y$

5. $(x - 2)^2 = 8(y + 3)$

6. $(y + 1)^2 = 6(x - 1)$

7. $(y + 4)^2 = -4(x - 2)$

8. $(x - 3)^2 = -(y + 2)$

9. $(y - 1)^2 = 2x + 8$

10. $(x + 2)^2 = 3y - 6$

11. $(2x - 4)^2 = 8y - 16$

12. $(3x + 6)^2 = 18y - 36$

13. $x^2 + 8x - y + 6 = 0$

14. $x^2 - 6x + y + 10 = 0$

15. $x + y^2 - 3y + 4 = 0$

16. $x - y^2 - 4y + 9 = 0$

17. $2x - y^2 - 6y + 1 = 0$

18. $3x + y^2 + 8y + 4 = 0$

19. $x^2 + 3x + 3y - 1 = 0$

20. $x^2 + 5x - 4y - 1 = 0$

21. $2x^2 - 8x - 4y + 3 = 0$

22. $6x - 3y^2 - 12y + 4 = 0$

23. $2x + 4y^2 + 8y - 5 = 0$

24. $4x^2 - 12x + 12y + 7 = 0$

25. $3x^2 - 6x - 9y + 4 = 0$ 26. $2x - 3y^2 + 9y + 5 = 0$

27. Find the equation in standard form of the parabola with vertex at the origin and focus $(0, -4)$.

28. Find the equation in standard form of the parabola with vertex at the origin and focus $(5, 0)$.

29. Find the equation in standard form of the parabola with vertex at $(-1, 2)$ and focus $(-1, 3)$.

30. Find the equation in standard form of the parabola with vertex at $(2, -3)$ and focus $(0, -3)$.

31. Find the equation in standard form of the parabola with focus $(3, -3)$ and directrix $y = -5$.

32. Find the equation in standard form of the parabola with focus $(-2, 4)$ and directrix $x = 4$.

33. Find the equation in standard form of the parabola that has vertex $(-4, 1)$, has its axis of symmetry parallel to the y-axis, and passes through the point $(-2, 2)$.

34. Find the equation in standard form of the parabola that has vertex $(3, -5)$, has its axis of symmetry parallel to the x-axis, and passes through the point $(4, 3)$.

35. SATELLITE DISH A satellite dish has the shape of a paraboloid. The signals that it receives are reflected to a receiver that is located at the focus of the paraboloid. If the dish is 8 feet across at its opening and 1 foot deep at its vertex, determine the location (distance above the vertex of the dish) of its focus.

36. THE HALE TELESCOPE The parabolic mirror in the Hale telescope at the Palomar Observatory in southern California has a diameter of 200 inches, and it has a concave depth of 3.75375 inches. Determine the location of its focus (to the nearest inch).

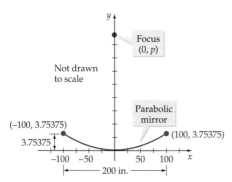

Mirror in the Hale Telescope

37. THE LICK TELESCOPE The parabolic mirror in the Lick telescope at the Lick Observatory on Mount Hamilton has a diameter of 120 inches, and it has a focal length of 600 inches. In the construction of the mirror, workers ground the mirror as shown in the following diagram. Determine the dimension a, which is the concave depth of the mirror.

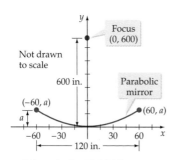

Mirror in the Lick Telescope

38. HEADLIGHT DESIGN A light source is to be placed on the axis of symmetry of the parabolic reflector shown in the figure at the top of the next column. How far to the right of the vertex point should the light source be located if the designer wishes the reflected light rays to form a beam of parallel rays?

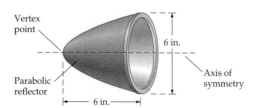

In Exercises 39 to 42, graph each equation, and find the coordinates of the points of intersection of the two graphs to the nearest ten-thousandth.

39. $y = 2x^2 - x - 1$
$y = x$

40. $y = x^2 + 2x - 4$
$y = x - 1$

41. $y = 2x^2 - 1$
$y = x^2 + x + 3$

42. $y = 2x^2 - x - 1$
$y = x^2 - 4$

SUPPLEMENTAL EXERCISES

In Exercises 43 to 45, use the following definition of latus rectum: The line segment that has endpoints on a parabola, passes through the focus of the parabola, and is perpendicular to the axis of symmetry is called the *latus rectum* of the parabola.

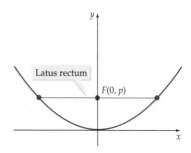

43. Find the length of the latus rectum for the parabola $x^2 = 4y$.

44. Find the length of the latus rectum for the parabola $y^2 = -8x$.

45. Find the length of the latus rectum for any parabola in terms of $|p|$, the distance from the vertex of the parabola to the focus.

The result of Exercise 45 can be stated as the following theorem: Two points on a parabola will be $2|p|$ units on each side of the axis of symmetry on the line through the focus and perpendicular to that axis.

46. Use the theorem to sketch a graph of the parabola given by the equation $(x - 3)^2 = 2(y + 1)$.

47. Use the theorem to sketch a graph of the parabola given by the equation $(y + 4)^2 = -(x - 1)$.

48. By using the definition of a parabola, find the equation in standard form of the parabola with $V(0, 0)$, $F(-c, 0)$, and directrix $x = c$.

49. Sketch a graph of $4(y - 2) = x|x| - 1$.

50. Find the equation of the directrix of the parabola with vertex at the origin and focus at the point $(1, 1)$.

51. Find the equation of the parabola with vertex at the origin and focus at the point $(1, 1)$. (*Hint:* You will need the answer to Exercise 50 and the definition of a parabola.)

PROJECTS

1. PARABOLAS AND TANGENTS Calculus procedures can be used to show that the equation of a tangent line to the parabola $4py = x^2$ at the point (x_0, y_0) is given by

$$y - y_0 = \left(\frac{1}{2p}x_0\right)(x - x_0)$$

Use this equation to verify each of the following statements.

a. If two tangent lines to a parabola intersect at right angles, then the point of intersection of the tangent lines is on the directrix of the parabola.

b. If two tangent lines to a parabola intersect at right angles, then the focus of the parabola is located on the line segment that connects the two points of tangency.

c. The tangent line to the parabola $4py = x^2$ at the point (x_0, y_0) intersects the y-axis at the point $(0, -y_0)$.

6.2 ELLIPSES

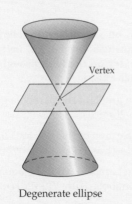

Degenerate ellipse

An ellipse is another of the conic sections formed when a plane intersects a right circular cone. If β is the angle at which the plane intersects the axis of the cone and α is the angle shown in **Figure 6.16,** an ellipse is formed when $\alpha < \beta < 90°$. If $\beta = 90°$, then a circle is formed.

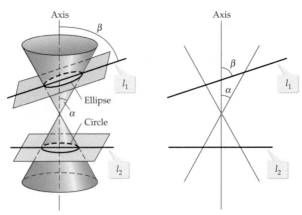

Figure 6.16

As was the case for a parabola, there is a definition for an ellipse in terms of a certain set of points in the plane.

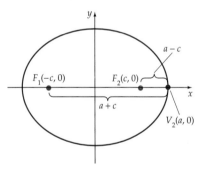

Figure 6.17

DEFINITION OF AN ELLIPSE

An **ellipse** is the set of all points in the plane, the sum of whose distances from two fixed points (**foci**) is a positive constant.

We can use this definition to draw an ellipse, equipped only with a piece of string and two tacks (see **Figure 6.17**). Tack the ends of the string to the foci, and trace a curve with a pencil held tight against the string. The resulting curve is an ellipse. The positive constant mentioned in the definition of an ellipse is the length of the string.

ELLIPSES WITH CENTER AT $(0, 0)$

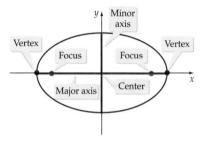

Figure 6.18

The graph of an ellipse has two axes of symmetry (see **Figure 6.18**). The longer axis is called the **major axis.** The foci of the ellipse are on the major axis. The shorter axis is called the **minor axis.** It is customary to denote the length of the major axis as $2a$ and the length of the minor axis as $2b$. The **semiaxes** are one-half the axes in length. Thus the length of the semimajor axis is denoted by a and the length of the semiminor axis by b. The **center** of the ellipse is the midpoint of the major axis. The endpoints of the major axis are the **vertices** (plural of *vertex*) of the ellipse.

Consider the point $V_2(a, 0)$, which is one vertex on an ellipse, and the points $F_2(c, 0)$ and $F_1(-c, 0)$, which are the foci of the ellipse shown in **Figure 6.19**. The distance from V_2 to F_1 is $a + c$. Similarly, the distance from V_2 to F_2 is $a - c$. From the definition of an ellipse, the sum of the distances from any point on the ellipse to the foci is a positive constant. By adding the expressions $a + c$ and $a - c$, we have

$$(a + c) + (a - c) = 2a$$

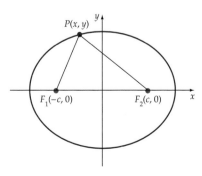

Figure 6.19

Thus the positive constant referred to in the definition of an ellipse is $2a$, the length of the major axis.

Now let $P(x, y)$ be any point on the ellipse (see **Figure 6.20**). By using the definition of an ellipse, we have

$$d(P, F_1) + d(P, F_2) = 2a$$
$$\sqrt{(x + c)^2 + y^2} + \sqrt{(x - c)^2 + y^2} = 2a$$

Subtract the second radical from each side of the equation, and then square each side.

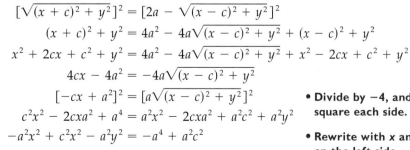

Figure 6.20

$$-(a^2 - c^2)x^2 - a^2y^2 = -a^2(a^2 - c^2)$$ • Factor and let $b^2 = a^2 - c^2$.

$$-b^2x^2 - a^2y^2 = -a^2b^2$$ • Divide each side by $-a^2b^2$.

$$\frac{x^2}{a^2} + \frac{y^2}{b^2} = 1$$ • An equation of an ellipse with center at (0, 0)

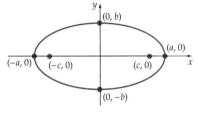

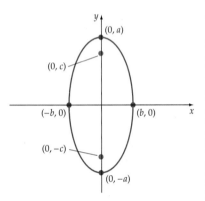

a. Major axis on x-axis

b. Major axis on y-axis

Figure 6.21

STANDARD FORMS OF THE EQUATION OF AN ELLIPSE WITH CENTER AT THE ORIGIN

Major Axis on the x-axis
The standard form of the equation of an ellipse with the center at the origin and major axis on the x-axis (see **Figure 6.21a**) is given by

$$\frac{x^2}{a^2} + \frac{y^2}{b^2} = 1, \quad a > b$$

The length of the major axis is $2a$. The length of the minor axis is $2b$. The coordinates of the vertices are $(a, 0)$ and $(-a, 0)$, and the coordinates of the foci are $(c, 0)$ and $(-c, 0)$, where $c^2 = a^2 - b^2$.

Major Axis on the y-axis
The standard form of the equation of an ellipse with the center at the origin and major axis on the y-axis (see **Figure 6.21b**) is given by

$$\frac{x^2}{b^2} + \frac{y^2}{a^2} = 1, \quad a > b$$

The length of the major axis is $2a$. The length of the minor axis is $2b$. The coordinates of the vertices are $(0, a)$ and $(0, -a)$, and the coordinates of the foci are $(0, c)$ and $(0, -c)$, where $c^2 = a^2 - b^2$.

TAKE NOTE

By looking at the standard forms of the equation of an ellipse and noting that $a > b$, observe that the orientation of the major axis is determined by the larger denominator. When the x^2 term has the larger denominator, the major axis is on the x-axis. When the y^2 term has the larger denominator, the major axis is on the y-axis.

EXAMPLE 1 Find the Vertices and Foci of an Ellipse

Find the vertices and foci of the ellipse given by the equation $\frac{x^2}{25} + \frac{y^2}{49} = 1$. Sketch the graph.

Solution

Because the y^2 term has the larger denominator, the major axis is on the y-axis.

$$a^2 = 49 \qquad b^2 = 25 \qquad c^2 = a^2 - b^2$$
$$a = 7 \qquad\quad b = 5 \qquad\quad = 49 - 25 = 24$$
$$c = \sqrt{24} = 2\sqrt{6}$$

The vertices are $(0, 7)$ and $(0, -7)$. The foci are $(0, 2\sqrt{6})$ and $(0, -2\sqrt{6})$. See **Figure 6.22**.

TRY EXERCISE 20, EXERCISE SET 6.2

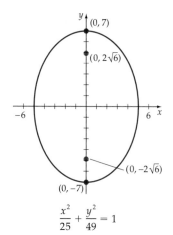

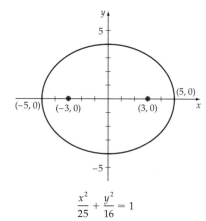

$$\frac{x^2}{25} + \frac{y^2}{49} = 1$$

Figure 6.22

$$\frac{x^2}{25} + \frac{y^2}{16} = 1$$

Figure 6.23

An ellipse with foci $(3,0)$ and $(-3,0)$ and major axis of length 10 is shown in **Figure 6.23.** To find the equation of the ellipse in standard form, we must find a^2 and b^2. Because the foci are on the major axis, the major axis is on the x-axis. The length of the major axis is $2a$. Thus $2a = 10$. Solving for a, we have $a = 5$ and $a^2 = 25$.

Because the foci are $(3,0)$ and $(-3,0)$ and the center of the ellipse is the midpoint between the two foci, the distance from the center of the ellipse to a focus is 3. Therefore, $c = 3$. To find b^2, use the equation

$$c^2 = a^2 - b^2$$
$$9 = 25 - b^2$$
$$b^2 = 16$$

The equation of the ellipse in standard form is $\dfrac{x^2}{25} + \dfrac{y^2}{16} = 1$.

ELLIPSES WITH THE CENTER AT (h, k)

The equation of an ellipse with center (h, k) and with horizontal or vertical major axes can be found by using a translation of coordinates. On a coordinate system with axes labeled x' and y', the standard form of the equation of an ellipse with center at the origin of the $x'y'$-coordinate system is

$$\frac{(x')^2}{a^2} + \frac{(y')^2}{b^2} = 1$$

Now place the origin of the $x'y'$-coordinate system at (h, k) in an xy-coordinate system. See **Figure 6.24.**

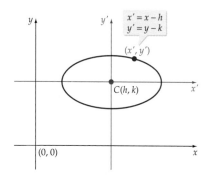

Figure 6.24

The relationship between an ordered pair in the $x'y'$-coordinate system and the xy-coordinate system is given by the transformation equations

$$x' = x - h$$
$$y' = y - k$$

Substitute the expressions for x' and y' into the equation of an ellipse. The equation of the ellipse with center at (h, k) is

$$\frac{(x - h)^2}{a^2} + \frac{(y - k)^2}{b^2} = 1$$

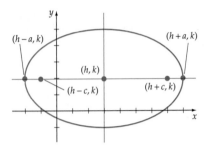

a. Major axis parallel to x-axis

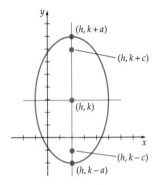

b. Major axis parallel to y-axis

Figure 6.25

STANDARD FORMS OF THE EQUATION OF AN ELLIPSE WITH CENTER AT (h, k)

Major Axis Parallel to the x-axis

The standard form of the equation of an ellipse with the center at (h, k) and major axis parallel to the x-axis (see **Figure 6.25a**) is given by

$$\frac{(x - h)^2}{a^2} + \frac{(y - k)^2}{b^2} = 1 \quad a > b$$

The length of the major axis is $2a$. The length of the minor axis is $2b$. The coordinates of the vertices are $(h + a, k)$ and $(h - a, k)$, and the coordinates of the foci are $(h + c, k)$ and $(h - c, k)$, where $c^2 = a^2 - b^2$.

Major Axis Parallel to the y-axis

The standard form of the equation of an ellipse with the center at (h, k) and major axis parallel to the y-axis (see **Figure 6.25b**) is given by

$$\frac{(x - h)^2}{b^2} + \frac{(y - k)^2}{a^2} = 1 \quad a > b$$

The length of the major axis is $2a$. The length of the minor axis is $2b$. The coordinates of the vertices are $(h, k + a)$ and $(h, k - a)$, and the coordinates of the foci are $(h, k + c)$ and $(h, k - c)$, where $c^2 = a^2 - b^2$.

EXAMPLE 2   Find the Vertices and Foci of an Ellipse

Find the vertices and foci of the ellipse $4x^2 + 9y^2 - 8x + 36y + 4 = 0$. Sketch the graph.

Solution

Write the equation of the ellipse in standard form by completing the square.

$$4x^2 + 9y^2 - 8x + 36y + 4 = 0$$
$$4x^2 - 8x + 9y^2 + 36y = -4 \qquad \bullet \text{ Rearrange terms.}$$
$$4(x^2 - 2x) + 9(y^2 + 4y) = -4 \qquad \bullet \text{ Factor.}$$
$$4(x^2 - 2x + 1) + 9(y^2 + 4y + 4) = -4 + 4 + 36 \qquad \bullet \text{ Complete the square.}$$
$$4(x - 1)^2 + 9(y + 2)^2 = 36 \qquad \bullet \text{ Factor.}$$
$$\frac{(x - 1)^2}{9} + \frac{(y + 2)^2}{4} = 1 \qquad \bullet \text{ Divide by 36.}$$

From the equation of the ellipse in standard form, the coordinates of the center of the ellipse are $(1, -2)$. Because the larger denominator is 9, the major axis is parallel to the x-axis and $a^2 = 9$. Thus $a = 3$. The vertices are $(4, -2)$ and $(-2, -2)$.

To find the coordinates of the foci, we find c.

$$c^2 = a^2 - b^2 = 9 - 4 = 5$$
$$c = \sqrt{5}$$

The foci are $(1 + \sqrt{5}, -2)$ and $(1 - \sqrt{5}, -2)$. See **Figure 6.26**.

$V_2(-2, -2)$ $C(1, -2)$ $V_1(4, -2)$

$F_2(1 - \sqrt{5}, -2)$ $F_1(1 + \sqrt{5}, -2)$

$$\frac{(x - 1)^2}{9} + \frac{(y + 2)^2}{4} = 1$$

Figure 6.26

TRY EXERCISE 26, EXERCISE SET 6.2

EXAMPLE 3 **Find the Equation of an Ellipse**

Find the standard form of the equation of the ellipse with center at $(4, -2)$, foci $F_2(4, 1)$ and $F_1(4, -5)$, and minor axis of length 10, as shown in **Figure 6.27**.

Solution

Because the foci are on the major axis, the major axis is parallel to the y-axis. The distance from the center of the ellipse to a focus is c. The distance between $(4, -2)$ and $(4, 1)$ is 3. Therefore, $c = 3$.

The length of the minor axis is $2b$. Thus $2b = 10$ and $b = 5$.

To find a^2, use the equation $c^2 = a^2 - b^2$.

$$9 = a^2 - 25$$
$$a^2 = 34$$

Thus the equation in standard form is

$$\frac{(x - 4)^2}{25} + \frac{(y + 2)^2}{34} = 1$$

TRY EXERCISE 42, EXERCISE SET 6.2

Figure 6.27

$V_2(4, -2 + \sqrt{34})$

$F_2(4, 1)$

$C(4, -2)$

$F_1(4, -5)$

$V_1(4, -2 - \sqrt{34})$

ECCENTRICITY OF AN ELLIPSE

The graph of an ellipse can be very long and thin, or it can be much like a circle. The **eccentricity** of an ellipse is a measure of its "roundness."

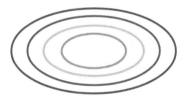

Eccentricity = 0.87

Figure 6.28

ECCENTRICITY (e) OF AN ELLIPSE

The eccentricity e of an ellipse is the ratio of c to a, where c is the distance from the center to the focus and a is one-half the length of the major axis. (See **Figure 6.28.**) That is,

$$e = \frac{c}{a}$$

Because $c < a$, for an ellipse, $0 < e < 1$. When $e \approx 0$, the graph is almost a circle. When $e \approx 1$, the graph is long and thin. See **Figure 6.29.**

$e = 0.60$
$e = 0.80$
$e = 0.92$
$e = 0.98$

Figure 6.29

EXAMPLE 4 **Find the Eccentricity of an Ellipse**

Find the eccentricity of the ellipse given by $8x^2 + 9y^2 = 18$.

Continued ▶

Solution

First, write the equation of the ellipse in standard form. Divide each side of the equation by 18.

$$\frac{8x^2}{18} + \frac{9y^2}{18} = 1$$

$$\frac{4x^2}{9} + \frac{y^2}{2} = 1$$

$$\frac{x^2}{9/4} + \frac{y^2}{2} = 1 \qquad \bullet \frac{4}{9} = \frac{1}{9/4}$$

The last step is necessary because the standard form of the equation has coefficients of 1 in the numerator. Thus

$$a^2 = \frac{9}{4} \quad \text{and} \quad a = \frac{3}{2}$$

Use the equation $c^2 = a^2 - b^2$ to find c.

$$c^2 = \frac{9}{4} - 2 = \frac{1}{4} \quad \text{and} \quad c = \sqrt{\frac{1}{4}} = \frac{1}{2}$$

Now find the eccentricity.

$$e = \frac{c}{a} = \frac{1/2}{3/2} = \frac{1}{3}$$

The eccentricity of the ellipse is 1/3.

TRY EXERCISE 48, EXERCISE SET 6.2

APPLICATIONS

The planets travel around the sun in elliptical orbits. The sun is located at a focus of the orbit. The eccentricities of the orbits for the planets in our solar system are given in Table 6.1.

QUESTION   Which planet has the most nearly circular orbit?

The terms *perihelion* and *aphelion* are used to denote the position of a planet in its orbit around the sun. The perihelion is the point nearest the sun; the aphelion is the point farthest from the sun. See **Figure 6.30**. The length of the semi-major axis of a planet's elliptical orbit is called the *mean distance* of the planet from the sun.

Table 6.1

Planet	Eccentricity
Mercury	0.206
Venus	0.007
Earth	0.017
Mars	0.093
Jupiter	0.049
Saturn	0.051
Uranus	0.046
Neptune	0.005
Pluto	0.250

ANSWER   Neptune has the smallest eccentricity, so it is the planet with the most nearly circular orbit.

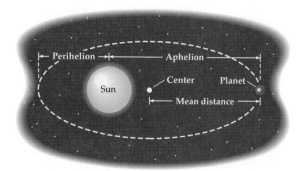

Figure 6.30

| EXAMPLE 5 | *Determine an Equation for the Orbit of Earth* |

Earth has a mean distance of 93 million miles and a perihelion of 91.5 million miles. Find an equation for Earth's orbit.

Solution

A mean distance of 93 million miles implies that the length of the semi-major axis of the orbit is $a = 93$ million miles. Earth's aphelion is the length of the major axis less the length of the perihelion. Thus

$$\text{Aphelion} = 2(93) - 91.5 = 94.5 \text{ million miles}$$

The distance c from the sun to the center of Earth's orbit is

$$c = \text{aphelion} - 93 = 94.5 - 93 = 1.5 \text{ million miles}$$

The length b of the semiminor axis of the orbit is

$$b = \sqrt{a^2 - c^2} = \sqrt{93^2 - 1.5^2} = \sqrt{8646.75}$$

An equation of Earth's orbit is

$$\frac{x^2}{93^2} + \frac{y^2}{8646.75} = 1$$

TRY EXERCISE 54, EXERCISE SET 6.2

ACOUSTIC PROPERTY OF AN ELLIPSE

Sound waves, although different from light waves, have a similar reflective property. When sound is reflected from a point P on a surface, the angle of incidence equals the angle of reflection. Applying this principle to a room with an elliptical ceiling results in what are called whispering galleries. These galleries are based on the following theorem.

THE REFLECTIVE PROPERTY OF AN ELLIPSE

The lines from the foci to a point on an ellipse make equal angles with the tangent line at that point. See **Figure 6.31**.

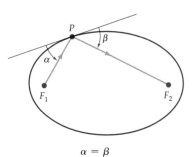

$\alpha = \beta$

Figure 6.31

The Statuary Hall in the Capitol Building in Washington, D.C., is a whispering gallery. Two people standing at the foci of the elliptical ceiling can whisper and yet hear each other even though they are a considerable distance apart. The whisper from one person is reflected to the person standing at the other focus.

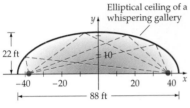

Elliptical ceiling of a whispering gallery

22 ft   10

−40   −20   20   40   x

88 ft

Figure 6.32

EXAMPLE 6   *Locate the Foci of a Whispering Gallery*

A room 88 feet long is constructed to be a whispering gallery. The room has an elliptical ceiling, as shown in **Figure 6.32**. If the height of the ceiling is 22 feet, determine where the foci are located.

Solution

The length a of the semimajor axis of the elliptical ceiling is 44 feet. The height b of the semiminor axis is 22 feet. Thus

$$c^2 = a^2 - b^2$$
$$c^2 = 44^2 - 22^2$$
$$c = \sqrt{44^2 - 22^2} \approx 38.1 \text{ feet}$$

The foci are located 38.1 feet from the center of the elliptical ceiling along its major axis.

TRY EXERCISE 56, EXERCISE SET 6.2

TOPICS FOR DISCUSSION

1. In every ellipse, the length of the semimajor axis a is greater than the length of the semiminor axis b and greater than the distance c from a focus to the center of the ellipse. Do you agree? Explain.

2. How many vertices does an ellipse have?

3. Every ellipse has two y-intercepts. Do you agree? Explain.

4. Explain why the eccentricity of every ellipse is a number between 0 and 1.

EXERCISE SET 6.2

In Exercises 1 to 32, find the vertices and foci of the ellipse given by each equation. Sketch the graph.

1. $\dfrac{x^2}{16} + \dfrac{y^2}{25} = 1$

2. $\dfrac{x^2}{49} + \dfrac{y^2}{36} = 1$

3. $\dfrac{x^2}{9} + \dfrac{y^2}{4} = 1$

4. $\dfrac{x^2}{64} + \dfrac{y^2}{25} = 1$

5. $\dfrac{x^2}{7} + \dfrac{y^2}{9} = 1$

6. $\dfrac{x^2}{5} + \dfrac{y^2}{4} = 1$

7. $\dfrac{4x^2}{9} + \dfrac{y^2}{16} = 1$

8. $\dfrac{x^2}{9} + \dfrac{9y^2}{16} = 1$

9. $\dfrac{(x-3)^2}{25} + \dfrac{(y+2)^2}{16} = 1$

10. $\dfrac{(x+3)^2}{9} + \dfrac{(y+1)^2}{16} = 1$

11. $\dfrac{(x+2)^2}{9} + \dfrac{y^2}{25} = 1$

12. $\dfrac{x^2}{25} + \dfrac{(y-2)^2}{81} = 1$

13. $\dfrac{(x-1)^2}{21} + \dfrac{(y-3)^2}{4} = 1$

14. $\dfrac{(x+5)^2}{9} + \dfrac{(y-3)^2}{7} = 1$

15. $\dfrac{9(x-1)^2}{16} + \dfrac{(y+1)^2}{9} = 1$ **16.** $\dfrac{(x+6)^2}{25} + \dfrac{25y^2}{144} = 1$

17. $3x^2 + 4y^2 = 12$ **18.** $5x^2 + 4y^2 = 20$

19. $25x^2 + 16y^2 = 400$ **20.** $25x^2 + 12y^2 = 300$

21. $64x^2 + 25y^2 = 400$ **22.** $9x^2 + 64y^2 = 144$

23. $4x^2 + y^2 - 24x - 8y + 48 = 0$

24. $x^2 + 9y^2 + 6x - 36y + 36 = 0$

25. $5x^2 + 9y^2 - 20x + 54y + 56 = 0$

26. $9x^2 + 16y^2 + 36x - 16y - 104 = 0$

27. $16x^2 + 9y^2 - 64x - 80 = 0$

28. $16x^2 + 9y^2 + 36y - 108 = 0$

29. $25x^2 + 16y^2 + 50x - 32y - 359 = 0$

30. $16x^2 + 9y^2 - 64x - 54y + 1 = 0$

31. $8x^2 + 25y^2 - 48x + 50y + 47 = 0$

32. $4x^2 + 9y^2 + 24x + 18y + 44 = 0$

In Exercises 33 to 44, find the equation in standard form of each ellipse, given the information provided.

33. Center $(0,0)$, major axis of length 10, foci at $(4,0)$ and $(-4,0)$

34. Center $(0,0)$, minor axis of length 6, foci at $(0,4)$ and $(0,-4)$

35. Vertices $(6,0)$, $(-6,0)$; ellipse passes through $(0,-4)$ and $(0,4)$

36. Vertices $(7,0)$, $(-7,0)$; ellipse passes through $(0,5)$ and $(0,-5)$

37. Major axis of length 12 on the x-axis, center at $(0,0)$; ellipse passes through $(2,-3)$

38. Minor axis of length 8, center at $(0,0)$; ellipse passes through $(-2,2)$

39. Center $(-2,4)$, vertices $(-6,4)$ and $(2,4)$, foci $(-5,4)$ and $(1,4)$

40. Center $(0,3)$, minor axis of length 4, foci $(0,0)$ and $(0,6)$

41. Center $(2,4)$, major axis parallel to the y-axis and of length 10; ellipse passes through the point $(3,3)$

42. Center $(-4,1)$, minor axis parallel to the y-axis and of length 8; ellipse passes through $(0,4)$

43. Vertices $(5,6)$ and $(5,-4)$, foci $(5,4)$ and $(5,-2)$

44. Vertices $(-7,-1)$ and $(5,-1)$, foci $(-5,-1)$ and $(3,-1)$

In Exercises 45 to 52, use the eccentricity of each ellipse to find its equation in standard form.

45. Eccentricity 2/5, major axis on the x-axis and of length 10, center at $(0,0)$

46. Eccentricity 3/4, foci at $(9,0)$ and $(-9,0)$

47. Foci at $(0,-4)$ and $(0,4)$, eccentricity 2/3

48. Foci at $(0,-3)$ and $(0,3)$, eccentricity 1/4

49. Eccentricity 2/5, foci $(-1,3)$ and $(3,3)$

50. Eccentricity 1/4, foci $(-2,4)$ and $(-2,-2)$

51. Eccentricity 2/3, major axis of length 24 on the y-axis, center at $(0,0)$

52. Eccentricity 3/5, major axis of length 15 on the x-axis, center at $(0,0)$

53. THE ORBIT OF SATURN The distance from Saturn to the sun at Saturn's aphelion is 934.34 million miles, and the distance from Saturn to the sun at its perihelion is 835.14 million miles. Find an equation for the orbit of Saturn.

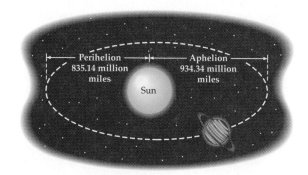

54. THE ORBIT OF VENUS Venus has a mean distance from the sun of 67.08 million miles, and the distance from Venus to the sun at its aphelion is 67.58 million miles. Find an equation for the orbit of Venus.

55. WHISPERING GALLERY An architect wishes to design a large room that will be a whispering gallery. See Example 6. The ceiling of the room has a cross section that is an ellipse, as shown in the following figure.

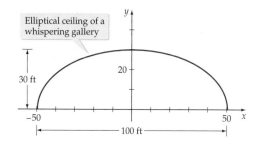

How far to the right and to the left of center are the foci located?

56. WHISPERING GALLERY An architect wishes to design a large room 100 feet long that will be a whispering gallery. The ceiling of the room has a cross section that is an ellipse, as shown in the following figure.

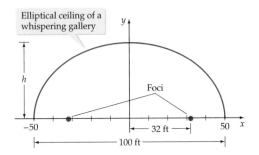

If the foci are to be located 32 feet to the right and to the left of center, find the height h of the elliptical ceiling (to the nearest 0.1 foot).

57. HALLEY'S COMET Find the equation of the path of Halley's comet in astronomical units by letting the sun (one focus) be at the origin and letting the other focus be on the positive x-axis. The length of the major axis of the orbit of Halley's comet is approximately 36 astronomical units (36 AU), and the length of the minor axis is 9 AU (1 AU = 92,600,000 miles).

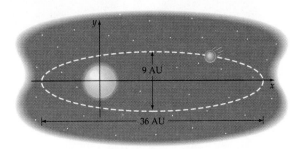

SUPPLEMENTAL EXERCISES

58. Explain why the graph of the equation $4x^2 + 9y - 16x - 2 = 0$ is or is not an ellipse. Sketch the graph of this equation.

In Exercises 59 to 62, find the equation in standard form of each ellipse by using the definition of an ellipse.

59. Find the equation of the ellipse with foci at $(-3, 0)$ and $(3, 0)$ that passes through the point $(3, 9/2)$.

60. Find the equation of the ellipse with foci at $(0, 4)$ and $(0, -4)$ that passes through the point $(9/5, 4)$.

61. Find the equation of the ellipse with foci at $(-1, 2)$ and $(3, 2)$ that passes through the point $(3, 5)$.

62. Find the equation of the ellipse with foci at $(-1, 1)$ and $(-1, 7)$ that passes through the point $(3/4, 1)$.

In Exercises 63 and 64, find the latus rectum of the given ellipse. The line segment with endpoints on the ellipse that is perpendicular to the major axis and passes through a focus is the *latus rectum* of the ellipse.

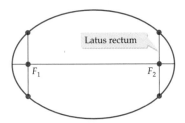

63. Find the length of the latus rectum of the ellipse given by

$$\frac{(x-1)^2}{9} + \frac{(y+1)^2}{16} = 1$$

64. Find the length of the latus rectum of the ellipse given by

$$9x^2 + 16y^2 - 36x + 96y + 36 = 0$$

65. Show that for any ellipse, the length of the latus rectum is $2b^2/a$.

66. Use the definition of an ellipse to find the equation of an ellipse with center at $(0, 0)$ and foci $(0, c)$ and $(0, -c)$.

Recall that a parabola has a directrix that is a line perpendicular to the axis of symmetry. An ellipse has two directrices, both of which are perpendicular to the major axis and outside the ellipse. For an ellipse with center at the origin and whose major axis is the x-axis, the equations of the directrices are $x = a^2/c$ and $x = -a^2/c$.

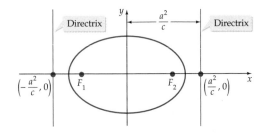

67. Find the directrix of the ellipse in Exercise 3.

68. Find the directrix of the ellipse in Exercise 4.

69. Let $P(x, y)$ be a point on the ellipse $\frac{x^2}{12} + \frac{y^2}{8} = 1$. Show that the distance from the point P to the focus $(2, 0)$ divided by the distance from the point P to the directrix $x = 6$ equals the eccentricity. (*Hint:* Solve the equation of the ellipse for y^2. Substitute this value for y^2 after applying the distance formula.)

70. Let $P(x, y)$ be a point on the ellipse $\dfrac{x^2}{25} + \dfrac{y^2}{16} = 1$. Show that the distance from the point P to the focus $(3, 0)$ divided by the distance from the point P to the directrix $x = 25/3$ equals the eccentricity. (*Hint:* Solve the equation of the ellipse for y^2. Substitute this value for y^2 after applying the distance formula.)

71. Generalize the results of Exercises 69 and 70. That is, show that if $P(x, y)$ is a point on the ellipse $\dfrac{x^2}{a^2} + \dfrac{y^2}{b^2} = 1$, where $F(c, 0)$ is a focus and $x = a^2/c$ is a directrix, then the following equation is true: $e = d(P, F)/d(P, D)$. (*Hint:* Solve the equation of the ellipse for y^2. Substitute this value for y^2 after applying the distance formula.)

PROJECTS

1. AN INSCRIBED RECTANGLE The diagram at the right shows a rectangle of length $2a$ inscribed in the ellipse given by $\dfrac{x^2}{25} + \dfrac{y^2}{16} = 1$.

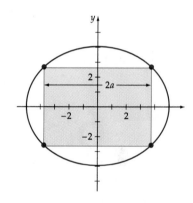

a. Show that the coordinates of the vertices of the rectangle are

$$\left(a, \frac{4}{5}\sqrt{25 - a^2}\right), \quad \left(a, -\frac{4}{5}\sqrt{25 - a^2}\right),$$

$$\left(-a, \frac{4}{5}\sqrt{25 - a^2}\right), \quad \text{and} \quad \left(-a, -\frac{4}{5}\sqrt{25 - a^2}\right)$$

b. Use a graphing utility to determine the dimensions (to the nearest 0.1) of the inscribed rectangle that has the largest area.

2. **I.M. PEI'S OVAL** The poet and architect I. M. Pei suggested that the oval with the most appeal to the eye is given by the equation

$$\left(\frac{x}{a}\right)^{3/2} + \left(\frac{y}{b}\right)^{3/2} = 1$$

Use a graphing utility to graph this equation with $a = 5$ and $b = 3$. Then compare your graph with the graph of

$$\left(\frac{x}{5}\right)^{2} + \left(\frac{y}{3}\right)^{2} = 1$$

3. **KEPLER'S LAWS** The German astronomer Johannes Kepler (1571–1630) derived three laws that describe how the planets orbit the sun. Write an essay that includes biographical information about Kepler and a statement of Kepler's Laws. In addition, use Kepler's Laws to answer the following questions.

a. Where is a planet located in its orbit around the sun when it achieves its greatest velocity?

b. What is the period of Mars if it has a mean distance from the sun of 1.52 astronomical units? (*Hint:* Use Earth as a reference with a period of 1 year and a mean distance from the sun of 1 astronomical unit.)

4. **NEPTUNE** The position of the planet Neptune was discovered by using celestial mechanics and mathematics. Write an essay that tells how, when, and by whom Neptune was discovered.

5. ONE OF RAMANUJAN'S FORMULAS The circumference of a circle with radius r is given by $2\pi r$. There is no simple algebraic expression for the perimeter of an ellipse. In 1914, however, the famous mathematician Srinivasa Ramanujan (1887–1920) discovered the following formula, which closely approximates the perimeter P of an ellipse with semimajor axis of length a and semiminor axis of length b.

$$P \approx \pi(a + b)\left(3 - \frac{\sqrt{(a + 3b)(3a + b)}}{a + b}\right)$$

Ramanujan also determined that the error E in this formula is given by

$$E \approx \pi a \frac{e^{12}}{2^{20}}$$

where e is the eccentricity of the ellipse.

a. Use Ramanujan's formula to estimate the perimeter of an ellipse with semimajor axis of length 4 and semiminor axis of length 3.

b. Use Ramanujan's error formula to estimate the error in the perimeter you calculated in **a.**

c. The area A of an ellipse with semimajor axis of length a and semiminor axis of length b is given by $A = \pi ab$. Use the formulas presented above to show that an ellipse with semimajor axis $a = 10$ and semiminor axis $b = 1$ has a larger perimeter, but a smaller area, than an ellipse with semimajor axis $a = 5$ and semiminor axis $b = 3$.

SECTION

6.3 HYPERBOLAS

The hyperbola is a conic section formed when a plane intersects a right circular cone at a certain angle. If β is the angle at which the plane intersects the axis of the cone and α is the angle shown in **Figure 6.33**, a hyperbola is formed when $0° < \beta < \alpha$ or when the plane is parallel to the axis of the cone.

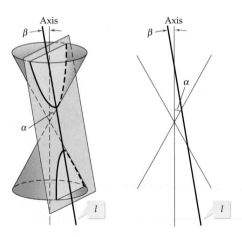

Figure 6.33

TAKE NOTE

If the plane intersects the cone along the axis of the cone, the resulting curve is two intersecting straight lines. This is the *degenerate* form of a hyperbola. See the accompanying figure.

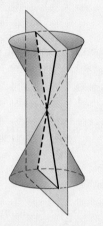

Degenerate hyperbola

As with the other conic sections, there is a definition of a hyperbola in terms of a certain set of points in the plane.

DEFINITION OF A HYPERBOLA

A **hyperbola** is the set of all points in the plane, the difference between whose distances from two fixed points (foci) is a positive constant.

This definition differs from that of an ellipse in that the ellipse was defined in terms of the *sum* of two distances, whereas the hyperbola is defined in terms of the *difference* of two distances.

HYPERBOLAS WITH CENTER AT $(0, 0)$

The **transverse axis** is the line segment joining the intercepts (see **Figure 6.34**). The midpoint of the transverse axis is called the **center** of the hyperbola. The **conjugate axis** passes through the center of the hyperbola and is perpendicular to the transverse axis.

The length of the transverse axis is customarily represented as $2a$, and the distance between the two foci is represented as $2c$. The length of the conjugate axis is represented as $2b$.

The **vertices** of a hyperbola are the points where the hyperbola intersects the transverse axis.

To determine the positive constant stated in the definition of a hyperbola, consider the point $V_1(a, 0)$, which is one vertex of a hyperbola, and the points $F_1(c, 0)$ and $F_2(-c, 0)$, which are the foci of the hyperbola (see **Figure 6.35**). The difference between the distance from $V_1(a, 0)$ to $F_1(c, 0)$, $c - a$, and the distance from $V_1(a, 0)$ to $F_2(-c, 0)$, $c + a$, must be a constant. By subtracting these distances, we find

$$|(c - a) - (c + a)| = |-2a| = 2a$$

Thus the constant is $2a$ and is the length of the transverse axis. The absolute value is used to ensure that the distance is a positive number.

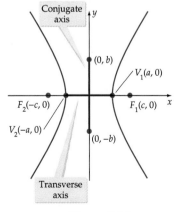

Figure 6.34

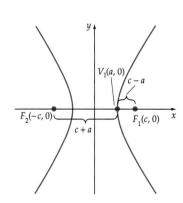

Figure 6.35

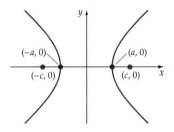

a. Transverse axis on the x-axis

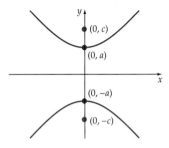

b. Transverse axis on the y-axis

Figure 6.36

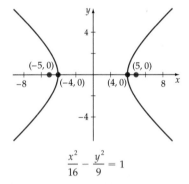

Figure 6.37

Transverse Axis on the x-axis
The standard form of the equation of a hyperbola with the center at the origin and transverse axis on the x-axis (see **Figure 6.36a**) is given by

$$\frac{x^2}{a^2} - \frac{y^2}{b^2} = 1$$

The coordinates of the vertices are $(a, 0)$ and $(-a, 0)$, and the coordinates of the foci are $(c, 0)$ and $(-c, 0)$, where $c^2 = a^2 + b^2$.

Transverse Axis on the y-axis
The standard form of the equation of a hyperbola with the center at the origin and transverse axis on the y-axis (see **Figure 6.36b**) is given by

$$\frac{y^2}{a^2} - \frac{x^2}{b^2} = 1$$

The coordinates of the vertices are $(0, a)$ and $(0, -a)$, and the coordinates of the foci are $(0, c)$ and $(0, -c)$, where $c^2 = a^2 + b^2$.

By looking at the equations, it is possible to determine the location of the transverse axis by finding which term in the equation is positive. When the x^2 term is positive, the transverse axis is on the x-axis. When the y^2 term is positive, the transverse axis is on the y-axis.

Consider the hyperbola given by the equation $\dfrac{x^2}{16} - \dfrac{y^2}{9} = 1$. Because the x^2 term is positive, the transverse axis is on the x-axis, $a^2 = 16$, and thus $a = 4$. The vertices are $(4, 0)$ and $(-4, 0)$. To find the foci, we determine c.

$$c^2 = a^2 + b^2 = 16 + 9 = 25$$
$$c = \sqrt{25} = 5$$

The foci are $(5, 0)$ and $(-5, 0)$. The graph is shown in **Figure 6.37**.

Each hyperbola has two asymptotes that pass through the center of the hyperbola. The asymptotes of the hyperbola are a useful guide to sketching the graph of the hyperbola.

The **asymptotes** of the hyperbola defined by $\dfrac{x^2}{a^2} - \dfrac{y^2}{b^2} = 1$ are given by the equations $y = \dfrac{b}{a}x$ and $y = -\dfrac{b}{a}x$ (see **Figure 6.38a**).

The asymptotes of the hyperbola defined by $\dfrac{y^2}{a^2} - \dfrac{x^2}{b^2} = 1$ are given by the equations $y = \dfrac{a}{b}x$ and $y = -\dfrac{a}{b}x$ (see **Figure 6.38b**).

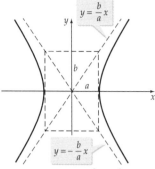

a. Asympyotes of $\dfrac{x^2}{a^2} - \dfrac{y^2}{b^2} = 1$

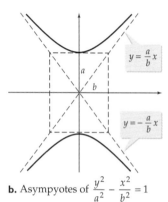

b. Asympyotes of $\dfrac{y^2}{a^2} - \dfrac{x^2}{b^2} = 1$

Figure 6.38

One method for remembering the equations of the asymptotes is to write the equation of a hyperbola in standard form but to replace 1 by 0 and then solve for y.

$$\frac{x^2}{a^2} - \frac{y^2}{b^2} = 0 \quad \text{so} \quad y^2 = \frac{b^2}{a^2}x^2, \text{ or } y = \pm\frac{b}{a}x$$

$$\frac{y^2}{a^2} - \frac{x^2}{b^2} = 0 \quad \text{so} \quad y^2 = \frac{a^2}{b^2}x^2, \text{ or } y = \pm\frac{a}{b}x$$

EXAMPLE 1   *Find the Vertices, Foci, and Asymptotes of a Hyperbola*

Find the vertices, foci, and asymptotes of the hyperbola given by the equation $\dfrac{y^2}{9} - \dfrac{x^2}{4} = 1$. Sketch the graph.

Solution

Because the y^2 term is positive, the transverse axis is on the y-axis. We know $a^2 = 9$; thus $a = 3$. The vertices are $V_1(0, 3)$ and $V_2(0, -3)$.

$$c^2 = a^2 + b^2 = 9 + 4$$
$$c = \sqrt{13}$$

The foci are $F_1(0, \sqrt{13})$ and $F_2(0, -\sqrt{13})$.

Because $a = 3$ and $b = 2$ ($b^2 = 4$), the equations of the asymptotes are $y = \dfrac{3}{2}x$ and $y = -\dfrac{3}{2}x$.

To sketch the graph, we draw a rectangle that has its center at the origin and has dimensions equal to the lengths of the transverse and conjugate axes. The asymptotes are extensions of the diagonals of the rectangle. See **Figure 6.39.**

TRY EXERCISE 4, EXERCISE SET 6.3

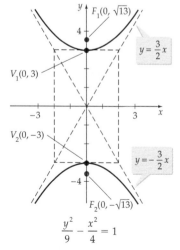

Figure 6.39

HYPERBOLAS WITH THE CENTER AT THE POINT (h, k)

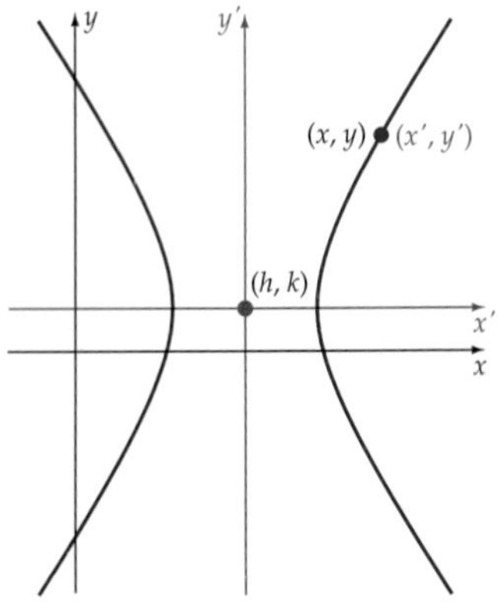

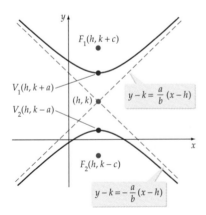

Figure 6.40

Using a translation of coordinates similar to that used for ellipses, we can write the equation of a hyperbola with its center at the point (h, k). Given coordinate axes labeled x' and y', an equation of a hyperbola with center at the origin is

$$\frac{(x')^2}{a^2} - \frac{(y')^2}{b^2} = 1 \qquad (1)$$

Now place the origin of this coordinate system at the point (h, k) of the xy-coordinate system, as shown in **Figure 6.40**. The relationship between an ordered pair in the $x'y'$-coordinate system and the xy-coordinate system is given by the transformation equations

$$x' = x - h$$
$$y' = y - k$$

Substitute the expressions for x' and y' into Equation (1). The equation of the hyperbola with center at (h, k) is

$$\frac{(x - h)^2}{a^2} - \frac{(y - k)^2}{b^2} = 1$$

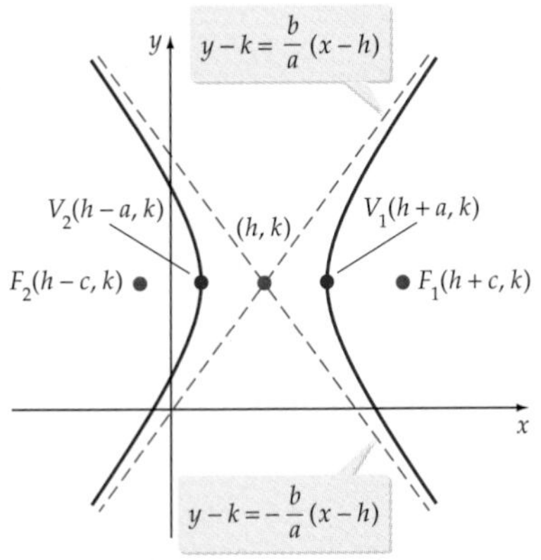

STANDARD FORMS OF THE EQUATION OF A HYPERBOLA WITH CENTER AT (h, k)

Transverse Axis Parallel to the x-axis

The standard form of the equation of a hyperbola with center (h, k) and transverse axis parallel to the x-axis (see **Figure 6.41a**) is given by

$$\frac{(x - h)^2}{a^2} - \frac{(y - k)^2}{b^2} = 1$$

The coordinates of the vertices are $V_1(h + a, k)$ and $V_2(h - a, k)$. The coordinates of the foci are $F_1(h + c, k)$ and $F_2(h - c, k)$ where $c^2 = a^2 + b^2$.

The equations of the asymptotes are $y - k = \pm\dfrac{b}{a}(x - h)$.

Transverse Axis Parallel to the y-axis

The standard form of the equation of a hyperbola with center (h, k) and transverse axis parallel to the y-axis (see **Figure 6.41b**) is given by

$$\frac{(y - k)^2}{a^2} - \frac{(x - h)^2}{b^2} = 1$$

The coordinates of the vertices are $V_1(h, k + a)$ and $V_2(h, k - a)$. The coordinates of the foci are $F_1(h, k + c)$ and $F_2(h, k - c)$, where $c^2 = a^2 + b^2$.

The equations of the asymptotes are $y - k = \pm\dfrac{a}{b}(x - h)$.

a. Transverse axis parallel to the x-axis

b. Transverse axis parallel to the y-axis

Figure 6.41

EXAMPLE 2	*Find the Vertices, Foci, and Asymptotes of a Hyperbola*

Find the vertices, foci, and asymptotes of the hyperbola given by the equation $4x^2 - 9y^2 - 16x + 54y - 29 = 0$. Sketch the graph.

Solution

Write the equation of the hyperbola in standard form by completing the square.

$$4x^2 - 9y^2 - 16x + 54y - 29 = 0$$

$$4x^2 - 16x - 9y^2 + 54y = 29 \qquad \bullet \text{ **Rearrange terms.**}$$

$$4(x^2 - 4x) - 9(y^2 - 6y) = 29 \qquad \bullet \text{ **Factor.**}$$

$$4(x^2 - 4x + 4) - 9(y^2 - 6y + 9) = 29 + 16 - 81 \qquad \bullet \text{ **Complete the square.**}$$

$$4(x - 2)^2 - 9(y - 3)^2 = -36 \qquad \bullet \text{ **Factor.**}$$

$$\frac{(y - 3)^2}{4} - \frac{(x - 2)^2}{9} = 1 \qquad \bullet \text{ **Divide by −36.**}$$

The coordinates of the center are $(2, 3)$. Because the term containing $(y - 3)^2$ is positive, the transverse axis is parallel to the y-axis. We know $a^2 = 4$; thus $a = 2$. The vertices are $(2, 5)$ and $(2, 1)$. See **Figure 6.42**.

$$c^2 = a^2 + b^2 = 4 + 9$$

$$c = \sqrt{13}$$

The foci are $(2, 3 + \sqrt{13})$ and $(2, 3 - \sqrt{13})$. We know $b^2 = 9$; thus $b = 3$. The equations of the asymptotes are $y - 3 = \pm 2/3(x - 2)$ which simplifies to

$$y = \frac{2}{3}x + \frac{5}{3} \qquad \text{and} \qquad y = -\frac{2}{3}x + \frac{13}{3}$$

> **TRY EXERCISE 26, EXERCISE SET 6.3**

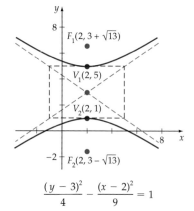

$$\frac{(y - 3)^2}{4} - \frac{(x - 2)^2}{9} = 1$$

Figure 6.42

ECCENTRICITY OF A HYPERBOLA

The graph of a hyperbola can be very wide or very narrow. The **eccentricity** of a hyperbola is a measure of its "wideness."

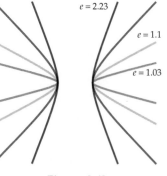

Figure 6.43

ECCENTRICITY (e) OF A HYPERBOLA

The eccentricity e of a hyperbola is the ratio of c to a, where c is the distance from the center to a focus and a is the length of the semitransverse axis.

$$e = \frac{c}{a}$$

For a hyperbola, $c > a$ and therefore $e > 1$. As the eccentricity of the hyperbola increases, the graph becomes wider and wider, as shown in **Figure 6.43** on page 371.

EXAMPLE 3 Find the Equation of a Hyperbola Given Its Eccentricity

Find the standard form of the equation of the hyperbola that has eccentricity 3/2, center at the origin, and a focus (6, 0).

Solution

Because the focus is located at (6, 0) and the center is at the origin, $c = 6$. An extension of the transverse axis contains the foci, so the transverse axis is on the x-axis.

$$e = \frac{3}{2} = \frac{c}{a}$$

$$\frac{3}{2} = \frac{6}{a} \qquad \bullet \text{ Substitute 6 for } c.$$

$$a = 4 \qquad \bullet \text{ Solve for } a.$$

To find b^2, use the equation $c^2 = a^2 + b^2$ and the values for c and a.

$$c^2 = a^2 + b^2$$

$$36 = 16 + b^2$$

$$b^2 = 20$$

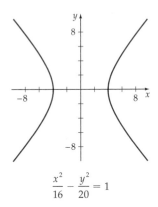

$$\frac{x^2}{16} - \frac{y^2}{20} = 1$$

Figure 6.44

The equation of the hyperbola is $\dfrac{x^2}{16} - \dfrac{y^2}{20} = 1$. See **Figure 6.44**.

TRY EXERCISE 48, EXERCISE SET 6.3

APPLICATIONS

Orbits of Comets In Section 6.2 we noted that the orbits of the planets are elliptical. Some comets have elliptical orbits also, the most notable being Halley's comet, whose eccentricity is 0.97.

Other comets have hyperbolic orbits with the sun at a focus. These comets pass by the sun only once. The velocity of a comet determines whether its orbit is elliptical or hyperbolic. See **Figure 6.45**.

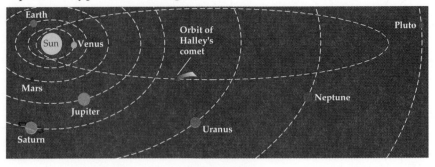

Figure 6.45

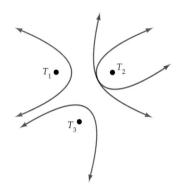

Figure 6.46

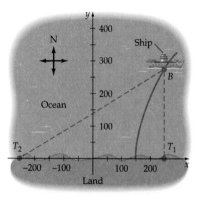

Figure 6.47

Hyperbolas as an Aid to Navigation Consider two radio transmitters, T_1 and T_2, placed some distance apart. A ship with electronic equipment measures the difference between the times it takes signals from the transmitters to reach the ship. Because the difference between the times is proportional to the difference between distances of the ship from the transmitters, the ship must be located on the hyperbola with foci at the two transmitters.

Using a third transmitter, T_3, we can find a second hyperbola with foci T_2 and T_3. The ship lies on the intersection of the two hyperbolas, as shown in **Figure 6.46**.

EXAMPLE 4 *Determine the Position of a Ship*

Using a LORAN (LOng RAnge Navigation) system, a ship determines that a radio signal from transmitter T_1 reaches the ship 1600 microseconds before it receives a simultaneous signal from transmitter T_2.

a. Find an equation of a hyperbola (with foci located at T_1 and T_2) on which the ship lies. See **Figure 6.47**. (Assume the radio signals travel at 0.186 mile per microsecond.)

b. If the ship is directly north of transmitter T_1, determine how far (to the nearest mile) the ship is from the transmitter.

Solution

a. The ship lies on a hyperbola at point B, with foci at T_1 and T_2. The difference of the distances $d(T_2, B)$ and $d(T_1, B)$ is given by

$$\text{Distance} = \text{rate} \times \text{time}$$
$$= 0.186 \text{ mile/microsecond} \times 1600 \text{ microseconds}$$
$$= 297.6 \text{ miles}$$

Thus the ship is located on a hyperbola with transverse axis of 297.6 miles and semitransverse axis $a = 148.8$ miles. **Figure 6.47** shows that the foci are located at $(250, 0)$ and $(-250, 0)$. Thus $c = 250$ miles, and

$$b = \sqrt{c^2 - a^2} = \sqrt{250^2 - 148.8^2} \approx 200.9 \text{ miles}$$

The ship is located on the hyperbola given by

$$\frac{x^2}{148.8^2} - \frac{y^2}{200.9^2} = 1$$

b. If the ship is directly north of T_1, then $x = 250$, and the distance from the ship to the transmitter T_1 is y, where

$$-\frac{y^2}{200.9^2} = 1 - \frac{250^2}{148.8^2}$$

$$y = \frac{200.9}{148.8} \sqrt{250^2 - 148.8^2} \approx 271 \text{ miles}$$

The ship is about 271 miles north of transmitter T_1.

TRY EXERCISE 54, EXERCISE SET 6.3

Hyperbolas also have a reflective property that makes them useful in many applications.

REFLECTIVE PROPERTY OF A HYPERBOLA

A ray of light directed toward one focus of a hyperbolic mirror is reflected toward the other focus. See **Figures 6.48** and **6.49**.

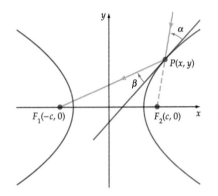

Figure 6.48

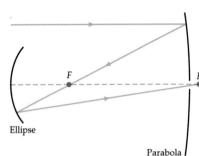

Ellipse

Parabola

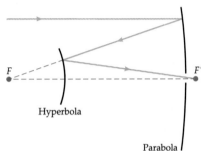

Hyperbola

Parabola

Figure 6.49

TOPICS FOR DISCUSSION

1. In every hyperbola, the distance c from a focus to the center of the hyperbola is greater than the length of the semi-transverse axis a. Do you agree? Explain.

2. How many vertices does a hyperbola have?

3. Explain why the eccentricity of every hyperbola is a number greater than 1.

4. Is the conjugate axis of a hyperbola perpendicular to the transverse axis of the hyperbola?

EXERCISE SET 6.3

In Exercises 1 to 26, find the center, vertices, foci, and asymptotes for the hyperbola given by each equation. Graph each equation.

1. $\dfrac{x^2}{16} - \dfrac{y^2}{25} = 1$

2. $\dfrac{x^2}{16} - \dfrac{y^2}{9} = 1$

3. $\dfrac{y^2}{4} - \dfrac{x^2}{25} = 1$

4. $\dfrac{y^2}{25} - \dfrac{x^2}{36} = 1$

5. $\dfrac{x^2}{7} - \dfrac{y^2}{9} = 1$

6. $\dfrac{x^2}{5} - \dfrac{y^2}{4} = 1$

7. $\dfrac{4x^2}{9} - \dfrac{y^2}{16} = 1$

8. $\dfrac{x^2}{9} - \dfrac{9y^2}{16} = 1$

9. $\dfrac{(x-3)^2}{16} - \dfrac{(y+4)^2}{9} = 1$

10. $\dfrac{(x+3)^2}{25} - \dfrac{y^2}{4} = 1$

11. $\dfrac{(y+2)^2}{4} - \dfrac{(x-1)^2}{16} = 1$

12. $\dfrac{(y-2)^2}{36} - \dfrac{(x+1)^2}{49} = 1$

13. $\dfrac{(x+2)^2}{9} - \dfrac{y^2}{25} = 1$

14. $\dfrac{x^2}{25} - \dfrac{(y-2)^2}{81} = 1$

15. $\dfrac{9(x-1)^2}{16} - \dfrac{(y+1)^2}{9} = 1$

16. $\dfrac{(x+6)^2}{25} - \dfrac{25y^2}{144} = 1$

17. $x^2 - y^2 = 9$

18. $4x^2 - y^2 = 16$

19. $16y^2 - 9x^2 = 144$

20. $9y^2 - 25x^2 = 225$

21. $9y^2 - 36x^2 = 4$ 22. $16x^2 - 25y^2 = 9$

23. $x^2 - y^2 - 6x + 8y - 3 = 0$

24. $4x^2 - 25y^2 + 16x + 50y - 109 = 0$

25. $9x^2 - 4y^2 + 36x - 8y + 68 = 0$

26. $16x^2 - 9y^2 - 32x - 54y + 79 = 0$

In Exercises 27 to 32, use the quadratic formula to solve for y in terms of x. Then use a graphing utility to graph each equation.

27. $4x^2 - y^2 + 32x + 6y + 39 = 0$

28. $x^2 - 16y^2 + 8x - 64y + 16 = 0$

29. $9x^2 - 16y^2 - 36x - 64y + 116 = 0$

30. $2x^2 - 9y^2 + 12x - 18y + 18 = 0$

31. $4x^2 - 9y^2 + 8x - 18y - 6 = 0$

32. $2x^2 - 9y^2 - 8x + 36y - 46 = 0$

In Exercises 33 to 46, find the equation in standard form of the hyperbola that satisfies the stated conditions.

33. Vertices $(3, 0)$ and $(-3, 0)$, foci $(4, 0)$ and $(-4, 0)$

34. Vertices $(0, 2)$ and $(0, -2)$, foci $(0, 3)$ and $(0, -3)$

35. Foci $(0, 5)$ and $(0, -5)$, asymptotes $y = 2x$ and $y = -2x$

36. Foci $(4, 0)$ and $(-4, 0)$, asymptotes $y = x$ and $y = -x$

37. Vertices $(0, 3)$ and $(0, -3)$, passing through $(2, 4)$

38. Vertices $(5, 0)$ and $(-5, 0)$, passing through $(-1, 3)$

39. Asymptotes $y = \dfrac{1}{2}x$ and $y = -\dfrac{1}{2}x$, vertices $(0, 4)$ and $(0, -4)$

40. Asymptotes $y = \dfrac{2}{3}x$ and $y = -\dfrac{2}{3}x$, vertices $(6, 0)$ and $(-6, 0)$

41. Vertices $(6, 3)$ and $(2, 3)$, foci $(7, 3)$ and $(1, 3)$

42. Vertices $(-1, 5)$ and $(-1, -1)$, foci $(-1, 7)$ and $(-1, -3)$

43. Foci $(1, -2)$ and $(7, -2)$, slope of an asymptote $5/4$

44. Foci $(-3, -6)$ and $(-3, -2)$, slope of an asymptote 1

45. Passing through $(9, 4)$, slope of an asymptote $1/2$, center $(7, 2)$, transverse axis parallel to the y-axis

46. Passing through $(6, 1)$, slope of an asymptote 2, center $(3, 3)$, transverse axis parallel to the x-axis

In Exercises 47 to 52, use the eccentricity to find the equation in standard form of each hyperbola.

47. Vertices $(1, 6)$ and $(1, 8)$, eccentricity 2

48. Vertices $(2, 3)$ and $(-2, 3)$, eccentricity $5/2$

49. Eccentricity 2, foci $(4, 0)$ and $(-4, 0)$

50. Eccentricity $4/3$, foci $(0, 6)$ and $(0, -6)$

51. Center $(4, 1)$, conjugate axis of length 4, eccentricity $4/3$ (*Hint:* There are two answers.)

52. Center $(-3, -3)$, conjugate axis of length 6, eccentricity 2. (*Hint:* There are two answers.)

53. **LORAN** Two radio transmitters are positioned along the coast, 250 miles apart. A signal is sent simultaneously from each transmitter. The signal from transmitter T_2 is received by a ship's LORAN 500 microseconds after it receives the signal from T_1. The radio signal travels 0.186 mile per microsecond.

   a. Find an equation of a hyperbola, with foci at T_1 and T_2, on which the ship is located.

   b. If the ship is 100 miles east of the y-axis, determine its distance from the coastline (to the nearest mile).

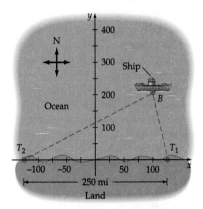

54. **LORAN** Two radio transmitters are positioned along the coast, 300 miles apart. A signal is sent simultaneously from each transmitter. The signal from transmitter T_1 is received by a ship's LORAN 800 microseconds after it receives the signal from T_2. The radio signal travels 0.186 mile per microsecond.

   a. Find an equation of a hyperbola, with foci at T_1 and T_2, on which the ship is located.

   b. If the ship continues to travel so that the difference of 800 microseconds is maintained, determine the point at which the ship will reach the coastline.

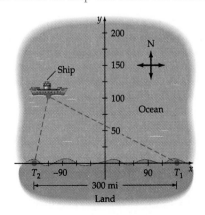

In Exercises 55 to 62, identify the graph of each equation as a parabola, ellipse, or hyperbola. Graph each equation.

55. $4x^2 + 9y^2 - 16x - 36y + 16 = 0$

56. $2x^2 + 3y - 8x + 2 = 0$

57. $5x - 4y^2 + 24y - 11 = 0$

58. $9x^2 - 25y^2 - 18x + 50y = 0$

59. $x^2 + 2y - 8x = 0$

60. $9x^2 + 16y^2 + 36x - 64y - 44 = 0$

61. $25x^2 + 9y^2 - 50x - 72y - 56 = 0$

62. $(x - 3)^2 + (y - 4)^2 = (x + 1)^2$

SUPPLEMENTAL EXERCISES

In Exercises 63 to 66, use the definition of a hyperbola to find the equation of the hyperbola in standard form.

63. Foci $(2, 0)$ and $(-2, 0)$; passes through the point $(2, 3)$

64. Foci $(0, 3)$ and $(0, -3)$; passes through the point $(5/2, 3)$

65. Foci $(0, 4)$ and $(0, -4)$; passes through the point $(7/3, 4)$

66. Foci $(5, 0)$ and $(-5, 0)$; passes through the point $(5, 9/4)$

Recall that an ellipse has two directrices that are lines perpendicular to the line containing the foci. A hyperbola also has two directrices; they are perpendicular to the transverse axis and outside the hyperbola. For a hyperbola with center at the origin and transverse axis on the x-axis, the equations of the directrices are $x = a^2/c$ and $x = -a^2/c$. In Exercises 67 to 71, use this information to solve each exercise.

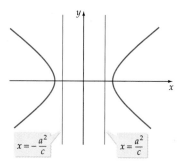

$x = -\dfrac{a^2}{c}$ $x = \dfrac{a^2}{c}$

67. Find the directrices for the hyperbola in Exercise 1.

68. Find the directrices for the hyperbola in Exercise 2.

69. Let $P(x, y)$ be a point on the hyperbola $\dfrac{x^2}{9} - \dfrac{y^2}{16} = 1$.

Show that the distance from the point P to the focus $(5, 0)$ divided by the distance from the point P to the directrix $x = 9/5$ equals the eccentricity.

70. Let $P(x, y)$ be a point on the hyperbola $\dfrac{x^2}{7} - \dfrac{y^2}{9} = 1$.

Show that the distance from the point P to the focus $(4, 0)$ divided by the distance from the point P to the directrix $x = 7/4$ equals the eccentricity.

71. Generalize the results of Exercises 69 and 70. That is, show that if $P(x, y)$ is a point on the hyperbola $\dfrac{x^2}{a^2} - \dfrac{y^2}{b^2} = 1$, $F(c, 0)$ is a focus, and $x = a^2/c$ is a directrix, then the following equation is true:

$$e = d(P, F)/d(P, D)$$

72. Derive the equation of a hyperbola with center at the origin, foci at $(0, c)$ and $(0, -c)$, and vertices $(0, a)$ and $(0, -a)$.

73. Sketch a graph of $\dfrac{x|x|}{16} - \dfrac{y|y|}{9} = 1$.

74. Sketch a graph of $\dfrac{x|x|}{16} + \dfrac{y|y|}{9} = 1$.

PROJECTS

1. **A HYPERBOLIC PARABOLOID** A *hyperbolic paraboloid* is a three-dimensional figure. Some of its cross sections are parabolas and some hyperbolas. Make a drawing of a hyperbolic paraboloid. Explain the relationship that exists between the equations of the parabolic cross sections and the relationship that exists between the equations of the hyperbolic cross sections.

2. **A HYPERBOLOID OF ONE SHEET** Make a sketch of a *hyperboloid of one sheet*. Explain the different cross sections of the hyperboloid of one sheet. Do some research on nuclear power plants, and explain why nuclear cooling towers are designed in the shape of hyperboloids of one sheet.

3.   **A GREATEST INTEGER FUNCTION AND HYPERBOLAS**   Graph

$$y = \frac{[\![x]\!]}{|x|}, \quad -5 \le x \le 5$$

How is this graph related to the graph of a hyperbola? Explain.

EXPLORING CONCEPTS WITH TECHNOLOGY

Using a Graphing Utility to Graph Conic Sections

Example 1   The graph of

$$4x^2 + 9y^2 - 8x + 36y + 4 = 0 \tag{1}$$

is an ellipse. To graph the ellipse with a graphing utility requires that we solve Equation (1) for y. First write Equation (1) as a quadratic equation in y.

$$9y^2 + 36y + (4x^2 - 8x + 4) = 0 \qquad \bullet \textbf{ A quadratic equation in y with A = 9,}$$
$$\textbf{B = 36, and C = 4x}^2 \textbf{ − 8x + 4}$$

Now use the quadratic formula to solve for y.

$$y = \frac{-36 \pm \sqrt{36^2 - 4(9)(4x^2 - 8x + 4)}}{2(9)}$$

The graph of

$$y_1 = \frac{-36 + \sqrt{36^2 - 4(9)(4x^2 - 8x + 4)}}{2(9)}$$

is the part of the ellipse on or above the line $y = -2$. The graph of

$$y_2 = \frac{-36 - \sqrt{36^2 - 4(9)(4x^2 - 8x + 4)}}{2(9)}$$

is the part of the ellipse on or below the line $y = -2$. See **Figure 6.50.**

TAKE NOTE

The procedure illustrated in Example 1 does not require that the given equation be written in standard form. Thus we can use a graphing utility to construct the graph quickly, but the procedure does not indicate where the foci are located.

TAKE NOTE

It is not necessary to simplify the equations for y_1 and y_2 since the graphing utility performs all the necessary calculations.

Example 2   The graph of

$$4x^2 - 9y^2 - 16x + 54y - 29 = 0 \tag{2}$$

is a hyperbola. To graph the hyperbola we first write Equation (2) as a quadratic equation in y.

$$-9y^2 + 54y + (4x^2 - 16x - 29) = 0 \qquad \bullet \textbf{ A quadratic equation with A = −9,}$$
$$\textbf{B = 54, and C = 4x}^2 \textbf{ − 16x − 29}$$

Apply the quadratic formula to solve for y.

$$y = \frac{-54 \pm \sqrt{2916 + 36(4x^2 - 16x - 29)}}{-18}$$

The graph of

$$y_1 = \frac{-54 + \sqrt{2916 + 36(4x^2 - 16x - 29)}}{-18}$$

is the upper branch of the hyperbola in **Figure 6.51**. The graph of

$$y_2 = \frac{-54 - \sqrt{2916 + 36(4x^2 - 16x - 29)}}{-18}$$

is the lower branch of the hyperbola.

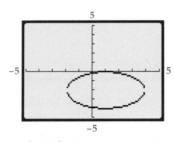

$4x^2 + 9y^2 - 8x + 36y + 4 = 0$

Figure 6.50

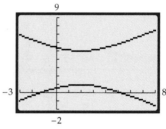

$4x^2 + 9y^2 - 16x + 54y - 29 = 0$

Figure 6.51

Graph some of the conic sections in this chapter by the method demonstrated in Example 1 and Example 2.

CHAPTER 6 REVIEW

6.1 Parabolas

- A parabola is the set of points in the plane that are equidistant from a fixed line (the directrix) and a fixed point (the focus) not on the directrix.

- The equations of a parabola with vertex at (h, k) and axis of symmetry parallel to a coordinate axis are given by

$(x - h)^2 = 4p(y - k)$; focus $(h, k + p)$;
directrix $y = k - p$
$(y - k)^2 = 4p(x - h)$; focus $(h + p, k)$;
directrix $x = h - p$

6.2 Ellipses

- An ellipse is the set of all points in the plane, the sum of whose distances from two fixed points (foci) is a positive constant.

- The equations of an ellipse with center at (h, k) and major axis parallel to a coordinate axis are given by

$\dfrac{(x - h)^2}{a^2} + \dfrac{(y - k)^2}{b^2} = 1$; foci $(h \pm c, k)$; vertices $(h \pm a, k)$

$\dfrac{(x - h)^2}{b^2} + \dfrac{(y - k)^2}{a^2} = 1$; foci $(h, k \pm c)$; vertices $(h, k \pm a)$

For each equation, $a > b$ and $c^2 = a^2 - b^2$.

- The eccentricity e of an ellipse is given by $e = c/a$.

6.3 Hyperbolas

- A hyperbola is the set of all points in the plane, the difference of whose distances from two fixed points (foci) is a positive constant.

- The equations of a hyperbola with center at (h, k) and transverse axis parallel to a coordinate axis are given by

$\dfrac{(x - h)^2}{a^2} - \dfrac{(y - k)^2}{b^2} = 1$; foci $(h \pm c, k)$; vertices $(h \pm a, k)$

$\dfrac{(y - k)^2}{a^2} - \dfrac{(x - h)^2}{b^2} = 1$; foci $(h, k \pm c)$; vertices $(h, k \pm a)$

For each equation, $c^2 = a^2 + b^2$.

- The eccentricity e of a hyperbola is given by $e = c/a$.

Chapter 6 True/False Exercises

In Exercises 1 to 9, answer true or false. If the statement is false, give an example to show that the statement is false.

1. The graph of a parabola is the same shape as that of one branch of a hyperbola.

2. For the two axes of an ellipse, the major axis and the minor axis, the major axis is always the longer axis.

3. For the two axes of a hyperbola, the transverse axis and the conjugate axis, the transverse axis is always the longer axis.

4. If two ellipses have the same foci, they have the same graph.

5. A hyperbola is similar to a parabola in that both curves have asymptotes.

6. If a hyperbola with center at the origin and a parabola with vertex at the origin have the same focus, $(0, c)$, then the two graphs always intersect.

7. The graphs of all the conic sections are not the graphs of functions.

8. If F_1 and F_2 are the two foci of an ellipse and P is a point on the ellipse, then $d(P, F_1) + d(P, F_2) = 2a$, where a is the length of the semimajor axis of the ellipse.

9. The eccentricity of a hyperbola is always greater than 1.

Chapter 6 Review Exercises

In Exercises 1 to 12, find the foci and vertices of each conic. If the conic is a hyperbola, find the asymptotes. Graph each equation.

1. $x^2 - y^2 = 4$

2. $y^2 = 16x$

3. $x^2 + 4y^2 - 6x + 8y - 3 = 0$

4. $3x^2 - 4y^2 + 12x - 24y - 36 = 0$

5. $3x - 4y^2 + 8y + 2 = 0$

6. $3x + 2y^2 - 4y - 7 = 0$

7. $9x^2 + 4y^2 + 36x - 8y + 4 = 0$

8. $11x^2 - 25y^2 - 44x - 50y - 256 = 0$

9. $4x^2 - 9y^2 - 8x + 12y - 144 = 0$

10. $9x^2 + 16y^2 + 36x - 16y - 104 = 0$

11. $4x^2 + 28x + 32y + 81 = 0$

12. $x^2 - 6x - 9y + 27 = 0$

In Exercises 13 to 20, find the equation of the conic that satisfies the given conditions.

13. Ellipse with vertices at $(7, 3)$ and $(-3, 3)$; length of minor axis is 8.

14. Hyperbola with vertices at $(4, 1)$ and $(-2, 1)$; eccentricity 4/3.

15. Hyperbola with foci $(-5, 2)$ and $(1, 2)$; length of transverse axis is 4.

16. Parabola with focus $(2, -3)$ and directrix $x = 6$.

17. Parabola with vertex $(0, -2)$ and passing through the point $(3, 4)$.

18. Ellipse with eccentricity 2/3 and foci $(-4, -1)$ and $(0, -1)$.

19. Hyperbola with vertices $(\pm 6, 0)$ and asymptotes whose equations are $y = \pm\dfrac{1}{9}x$.

20. Parabola passing through the points $(1, 0)$, $(2, 1)$, and $(0, 1)$ with axis of symmetry parallel to the y-axis.

21. Find the equation of the parabola traced by a point $P(x, y)$ that moves in such a way that the distance between $P(x, y)$ and the line $x = 2$ equals the distance between $P(x, y)$ and the point $(-2, 3)$.

22. Find the equation of the parabola traced by a point $P(x, y)$ that moves in such a way that the distance between $P(x, y)$ and the line $y = 1$ equals the distance between $P(x, y)$ and the point $(-1, 2)$.

23. Find the equation of the ellipse traced by a point $P(x, y)$ that moves in such a way that the sum of its distances to $(-3, 1)$ and $(5, 1)$ is 10.

24. Find the equation of the ellipse traced by a pint $P(x, y)$ that moves in such a way that the sum of its distances to $(3, 5)$ and $(3, -1)$ is 8.

CHAPTER 6 TEST

1. Find the vertex, focus, and directrix of the parabola given by the equation $y = \frac{1}{8}x^2$.

2. Find the vertex, focus, and directrix of the parabola given by the equation $x^2 + 4x - 12y + 16 = 0$.

3. Find the equation in standard form of the parabola with directrix $x = 3$ and focus $(-1, -2)$.

4. Graph the parabola with focus $(0, -1)$ and directrix $y = -5$.

5. Find the vertices and foci of the ellipse given by the equation $\dfrac{x^2}{9} + \dfrac{y^2}{64} = 1$.

6. Graph: $\dfrac{x^2}{16} + \dfrac{y^2}{1} = 1$

7. Find the vertices and foci of the ellipse given by the equation $25x^2 - 150x + 9y^2 + 18y + 9 = 0$.

8. Find the equation in standard form of the ellipse with center $(0, -3)$, foci $(-6, -3)$ and $(6, -3)$, and minor axis of length 6.

9. Find the eccentricity of the ellipse given by the equation $9x^2 + 25y^2 = 81$.

10. Graph: $\dfrac{y^2}{25} - \dfrac{x^2}{16} = 1$

11. Find the vertices, foci, and asymptotes of the hyperbola given by the equation $\dfrac{x^2}{36} - \dfrac{y^2}{64} = 1$.

12. Graph: $16y^2 + 32y - 4x^2 - 24x = 84$

13. Find the vertices and foci of the hyperbola given by the equation $\dfrac{(y - 4)^2}{36} - \dfrac{(x + 5)^2}{9} = 1$.

14. Find the equation in standard form of the hyperbola with vertices at $(-2, -3)$ and $(-6, -3)$ and foci $(-4 + \sqrt{34}, -3)$ and $(-4 - \sqrt{34}, -3)$.

15. Find the equation in standard form of the parabola with focus $(-2, 4)$ and directrix $x = 6$.

7

Systems of Equations

Boardwalk? Park Place? How about New York?: A Winning Strategy for Monopoly

A modern Monopoly board.

PENNSYLVANIA
477947

Pennsylvania Railroad— a familiar site on a Monopoly board.

The Boardwalk in Atlantic City-namesake of one of the most coveted Monopoly properties.

Monopoly, invented during the Depression, is still a very popular board game. In fact, Parker Brothers, the maker of Monopoly, has sponsored world championship Monopoly games in Atlantic City, home of Baltic and Mediterranean Avenues.

In the early 1980s, Stephen Heppe, at the time a student, became interested in winning Monopoly strategies. He wanted to know which properties on the Monopoly board paid greater rates of return for each dollar invested. Answering Heppe's question required solving a system of linear equations. Heppe's system of equations contained 123 equations with 123 variables. Here are some of the results obtained from solving this system of equations.

> A player is less likely to land on Mediterranean Avenue during the course of the game than on any other property.

> The chances of a player landing on Illinois Avenue are greater than those of landing on any other property.

Besides knowing which properties have the greatest chance of being occupied, Heppe also wanted to know which properties pay the greatest return for each dollar invested in houses or hotels. Some of his conclusions:

> New York with a hotel has the highest rate of return.

> The lowest rate of return for a property with a hotel is offered by Mediterranean.

> Assuming all the railroads are owned, the B&O railroad has the greatest rate of return of all the railroads. This is because it is more likely that a player will land on this railroad than on the other railroads.

For more information on the mathematics of Monopoly, see "Matrix Mathematics: How to Win at Monopoly" by Dr. Crypton in the September 1985 issue of *Science Digest*.

7.1 SYSTEMS OF LINEAR EQUATIONS IN
TWO VARIABLES

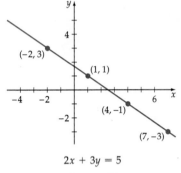

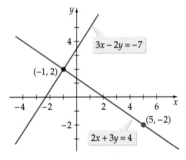

$$2x + 3y = 5$$

Figure 7.1

Recall that an equation of the form $Ax + By = C$ is a linear equation in two variables. A solution of a linear equation in two variables is an ordered pair (x, y) that makes the equation a true statement. For example, $(-2, 3)$ is a solution of the equation

$$2x + 3y = 5 \quad \text{since} \quad 2(-2) + 3(3) = 5$$

The graph of a linear equation, a straight line, is the set of points whose ordered pairs satisfy the equation. **Figure 7.1** is the graph of $2x + 3y = 5$.

A **system of equations** is two or more equations considered together. The following system of equations is a **linear system of equations** in two variables.

$$\begin{cases} 2x + 3y = 4 \\ 3x - 2y = -7 \end{cases}$$

A **solution** of a system of equations in two variables is an ordered pair that is a solution of both equations.

In **Figure 7.2**, the graphs of the two equations in the system of equations above intersect at the point $(-1, 2)$. Because that point lies on both lines, $(-1, 2)$ is a solution of both equations and thus is a solution of the system of equations. The point $(5, -2)$ is a solution of the first equation but not the second equation. Therefore $(5, -2)$ is not a solution of the system of equations.

The graphs of two linear equations in two variables can intersect, be the same line, or be parallel. When the graphs intersect at a single point or are the same line, the system is called a **consistent** system of equations. The system is called an **independent** system of equations when the lines intersect at exactly one point. The system is called a **dependent** system of equations when the equations represent the same line. In this case, the system has an infinite number of solutions. When the graphs of the two lines are parallel, the system is called **inconsistent** and has no solution. See **Figure 7.3.**

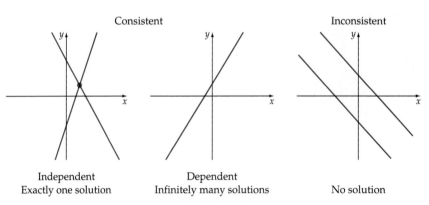

Figure 7.3

SUBSTITUTION METHOD FOR SOLVING A SYSTEM OF LINEAR EQUATIONS

The **substitution method** is one procedure for solving a system of equations. This method is illustrated in Example 1.

EXAMPLE 1	*Solve a System of Equations by the Substitution Method*

Solve: $\begin{cases} 3x - 5y = 7 & \text{(1)} \\ \quad\quad y = 2x & \text{(2)} \end{cases}$

Solution

The solutions of the equation $y = 2x$ are the ordered pairs $(x, 2x)$. For the system of equations to have a solution, ordered pairs of the form $(x, 2x)$ must also be a solution of the equation $3x - 5y = 7$. To determine whether ordered pairs of this form are solutions of Equation (1), substitute $(x, 2x)$ into Equation (1) and solve for x. Think of this as *substituting* $2x$ for y.

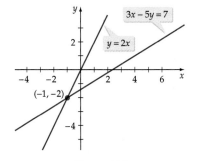

$3x - 5y = 7$ at top, $y = 2x$ labeled, point $(-1, -2)$

Figure 7.4

An independent system of equations

$$3x - 5y = 7$$
$$3x - 5(2x) = 7 \qquad \text{• Substitute } 2x \text{ for } y.$$
$$3x - 10x = 7$$
$$-7x = 7$$
$$x = -1$$
$$y = 2(-1) = -2 \qquad \text{• Substitute } -1 \text{ for } x \text{ in Equation (2).}$$

The only ordered-pair solution of the system of equations is $(-1, -2)$. When a system of equations has a unique solution, the system of equations is independent. See **Figure 7.4**.

TRY EXERCISE 6, EXERCISE SET 7.1

EXAMPLE 2	*Identify an Inconsistent System of Equations*

Solve: $\begin{cases} 3x - y = 6 & \text{(1)} \\ 6x - 2y = 5 & \text{(2)} \end{cases}$

Solution

Solve Equation (1) for y.

$$3x - y = 6$$
$$y = 3x - 6$$

Continued • ➤

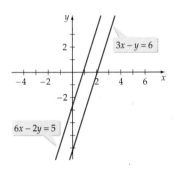

Figure 7.5

An inconsistent system of equations

The solutions of the equation $y = 3x - 6$ are the ordered pairs $(x, 3x - 6)$. For the system of equations to have a solution, ordered pairs of the form $(x, 3x - 6)$ must also be a solution of the equation $6x - 2y = 5$.

To determine whether ordered pairs of this form are solutions of Equation (2), substitute $(x, 3x - 6)$ into Equation (2) and solve for x.

$$6x - 2(3x - 6) = 5 \qquad \bullet \text{ Substitute } 3x - 6 \text{ for } y \text{ in Equation (2).}$$
$$12 = 5 \qquad \bullet \text{ A false equation}$$

Arrival at the false statement $12 = 5$ means that no ordered pair that is a solution of Equation (1) is also a solution of Equation (2). The system of equations has no ordered pairs in common and thus has no solution. The system of equations is inconsistent. See **Figure 7.5.**

TRY EXERCISE 18, EXERCISE SET 7.1

EXAMPLE 3   *Solve a Dependent System of Equations*

Solve: $\begin{cases} 4x - 8y = 16 & \quad (1) \\ x - 2y = 4 & \quad (2) \end{cases}$

Solution

Solve Equation (2) for y.

$$x - 2y = 4$$
$$y = \frac{1}{2}x - 2$$

The solution of $y = \frac{1}{2}x - 2$ is the set of ordered pairs $\left(x, \frac{1}{2}x - 2 \right)$. For the system of equations to have a solution, the ordered pairs $\left(x, \frac{1}{2}x - 2 \right)$ must also be a solution of the equation $4x - 8y = 16$.

To determine whether any ordered pair of this form is a solution of Equation (1), substitute $\left(x, \frac{1}{2}x - 2 \right)$ into Equation (1) and solve for x.

$$4x - 8\left(\frac{1}{2}x - 2 \right) = 16 \qquad \bullet \text{ Substitute } \frac{1}{2}x - 2 \text{ for } y \text{ in Equation (1).}$$
$$16 = 16 \qquad \bullet \text{ A true equation}$$

Arrival at the true statement $16 = 16$ means that the ordered pairs $\left(x, \frac{1}{2}x - 2 \right)$, which are solutions of Equation (2), are also solutions of Equation (1). Because x can be replaced by any real number c, there are an infinite number of ordered pairs $\left(c, \frac{1}{2}c - 2 \right)$ that are solutions of the system of equations. The system of equations is dependent. See **Figure 7.6.**

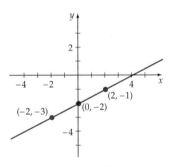

Figure 7.6

A dependent system of equations

TRY EXERCISE 20, EXERCISE SET 7.1

Some of the specific ordered-pair solutions in Example 3 can be found by choosing various values for c. The table below shows the ordered pairs that result from choosing c as -2, 0, and 2. The ordered pairs $(-2, -3)$, $(0, -2)$, and $(2, -1)$ are specific solutions of the system of equations. These points are on the graphs of Equation (1) and Equation (2) as shown in **Figure 7.6**.

c	$\left(c, \dfrac{1}{2}c - 2\right)$	(x, y)
-2	$\left(-2, \dfrac{1}{2}(-2) - 2\right)$	$(-2, -3)$
0	$\left(0, \dfrac{1}{2}(0) - 2\right)$	$(0, -2)$
2	$\left(2, \dfrac{1}{2}(2) - 2\right)$	$(2, -1)$

Before leaving Example 3, note that there is more than one way to represent the ordered-pair solutions. To illustrate this point, solve Equation (2) for x.

$$x - 2y = 4 \qquad \bullet \text{ Equation (2)}$$
$$x = 2y + 4 \qquad \bullet \text{ Solve for } x.$$

Because y can be replaced by any real number b, there are an infinite number of ordered pairs $(2b + 4, b)$ that are solutions of the system of equations. Choosing b as -3, -2, and -1 gives the same ordered pairs: $(-2, -3)$, $(0, -2)$, and $(2, -1)$. There is always more than one way to describe the ordered pairs when writing the solution of a dependent system of equations. For Example 3, either the ordered pairs $\left(c, \dfrac{1}{2}c - 2\right)$ or the ordered pairs $(2b + 4, b)$ would generate all the solutions of the system of equations.

ELIMINATION METHOD FOR SOLVING A SYSTEM OF EQUATIONS

Two systems of equations are **equivalent** if each system has exactly the same solutions. The systems

$$\begin{cases} 3x + 5y = 9 \\ 2x - 3y = -13 \end{cases} \text{ and } \begin{cases} x = -2 \\ y = 3 \end{cases}$$

are equivalent systems of equations. Each system has the solution $(-2, 3)$, as shown in **Figure 7.7**.

A second technique for solving a system of equations is similar to the strategy for solving first-degree equations in one variable. The system of equations is replaced by a series of equivalent systems until the solution is obvious.

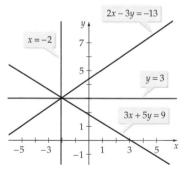

Figure 7.7

OPERATIONS THAT PRODUCE EQUIVALENT SYSTEMS OF EQUATIONS

1. Interchange any two equations.

2. Replace an equation with a nonzero multiple of that equation.

3. Replace an equation with the sum of an equation and a nonzero constant multiple of another equation in the system.

Because the order in which the equations are written does not affect the system of equations, interchanging the equations does not affect its solution. The second operation restates the property that says that multiplying each side of an equation by the same nonzero constant does not change the solutions of the equation.

The third operation can be illustrated as follows. Consider the system of equations

$$\begin{cases} 3x + 2y = 10 & (1) \\ 2x - 3y = -2 & (2) \end{cases}$$

Multiply each side of Equation (2) by 2. (Any nonzero number would work.) Add the resulting equation to Equation (1).

$$
\begin{array}{ll}
3x + 2y = 10 & \text{• Equation (1)} \\
\underline{4x - 6y = -4} & \text{• 2 times Equation (2)} \\
7x - 4y = 6 \quad (3) & \text{• Add the equations.}
\end{array}
$$

Replace Equation (1) with the new Equation (3) to produce the following equivalent system of equations.

$$\begin{cases} 7x - 4y = 6 & (3) \\ 2x - 3y = -2 & (2) \end{cases}$$

The third property states that the resulting system of equations has the same solutions as the original system and is therefore equivalent to the original system of equations. **Figure 7.8** shows the graph of $7x - 4y = 6$. Note that the line passes through the same point at which the lines of the original system of equations intersect, the point $(2, 2)$.

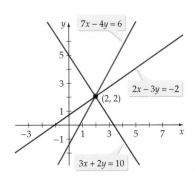

Figure 7.8

| EXAMPLE 4 | Solve a System of Equations by the Elimination Method |

Solve: $\begin{cases} 3x - 4y = 10 & (1) \\ 2x + 5y = -1 & (2) \end{cases}$

Solution

Use the operations that produce equivalent equations to eliminate a variable from one of the equations. We will eliminate x from Equation (2) by multiplying each equation by a different constant so as to have a new system of equations in which the coefficients of x are additive inverses.

$$
\begin{array}{ll}
6x - 8y = 20 & \text{• 2 times Equation (1)} \\
\underline{-6x - 15y = 3} & \text{• -3 times Equation (2)} \\
-23y = 23 & \text{• Add the equations.} \\
y = -1 & \text{• Solve for y.}
\end{array}
$$

Solve Equation (1) for x by substituting -1 for y.

$$3x - 4(-1) = 10$$
$$3x = 6$$
$$x = 2$$

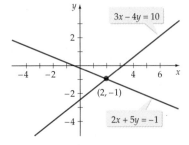

Figure 7.9

The solution of the system of equations is $(2, -1)$. See **Figure 7.9**.

TRY EXERCISE 24, EXERCISE SET 7.1

The method just described is called the **elimination method** for solving a system of equations, because it involves *eliminating* a variable from one of the equations.

You can use a graphing utility to solve a system of equations in two variables. Solve each equation of the system for y and then graph each equation. Adjust the viewing window so that the point of intersection can be seen. Then use the TRACE feature to determine the coordinates of the point of intersection. Because you obtain a graphical solution, your solution may be a numerical approximation to the exact solution.

For the system of equations in Example 4, first solve each equation for y.

$$\textbf{Solve for } \textit{y.}$$

$$3x - 4y = 10 \quad \longrightarrow \quad y = 0.75x - 2.5$$
$$2x + 5y = -1 \quad \longrightarrow \quad y = -0.4x - 0.2$$

The graph of the equations is shown in **Figure 7.10**. The point of intersection is the solution of the system of equations. By tracing along one of the lines, you can approximate the solution of the system of equations. Some graphing utilities, however, have a feature that calculates the point of intersection and jumps the cursor to that point.

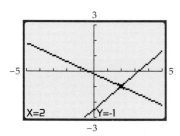

Figure 7.10

EXAMPLE 5   *Solve a Dependent System of Equations*

Solve: $\begin{cases} x - 2y = 2 & (1) \\ 3x - 6y = 6 & (2) \end{cases}$

Solution

Eliminate x by multiplying Equation (2) by $-1/3$ and then adding the result to Equation (1).

$$
\begin{array}{ll}
x - 2y = 2 & \text{• Equation (1)} \\
\underline{-x + 2y = -2} & \text{• } -1/3 \text{ times Equation (2)} \\
0 = 0 & \text{• Add the two equations.}
\end{array}
$$

Replace Equation (2) by $0 = 0$.

$$\begin{cases} x - 2y = 2 \\ 0 = 0 \end{cases} \quad \text{• This is an equivalent system of equations.}$$

Because the equation $0 = 0$ is an identity, an ordered pair that is a solution of Equation (1) is also a solution of $0 = 0$. Thus the solutions are the solutions of $x - 2y = 2$. Solving for y, we find that $y = \dfrac{1}{2}x - 1$.

Because x can be replaced by any real number c, the solutions of the system of equations are the ordered pairs $\left(c, \dfrac{1}{2}c - 1 \right)$. See **Figure 7.11**.

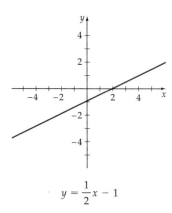

$$y = \frac{1}{2}x - 1$$

Figure 7.11

TAKE NOTE

Referring again to Example 5 and solving Equation (1) for x, we have $x = 2y + 2$. Because y can be any real number b, the ordered-pair solutions of the system of equations can be written also as $(2b + 2, b)$.

TRY EXERCISE 28, EXERCISE SET 7.1

If one equation of the system of equations is replaced by a false equation, the system of equations has no solution. For example, the system of equations

$$\begin{cases} x + y = 4 \\ \quad\quad 0 = 5 \end{cases}$$

has no solution because the second equation is false for any choice of x and y.

APPLICATIONS OF SYSTEMS OF EQUATIONS

As application problems become more difficult, it becomes impossible to represent all unknowns in terms of a single variable. In such cases, a system of equations can be used.

EXAMPLE 6   Solve an Application

A rowing team rowing with the current traveled 18 miles in 2 hours. Against the current, the team rowed 10 miles in 2 hours. Find the rate of the rowing team in calm water and the rate of the current.

Solution

Let r_1 represent the rate of the boat in calm water, and let r_2 represent the rate of the current.

> The rate of the boat *with the current* is $r_1 + r_2$.
> The rate of the boat *against the current* is $r_1 - r_2$.

Because the rowing team traveled 18 miles in 2 hours with the current, we use the equation $d = rt$.

$$d = r \cdot t$$
$$18 = (r_1 + r_2) \cdot 2 \qquad \bullet\, d = 18, t = 2$$
$$9 = r_1 + r_2 \qquad \bullet\, \text{Divide each side by 2.}$$

Because the team rowed 10 miles in 2 hours against the current, we write

$$10 = (r_1 - r_2) \cdot 2 \qquad \bullet\, d = 10, t = 2$$
$$5 = r_1 - r_2 \qquad \bullet\, \text{Divide each side by 2.}$$

Thus we have a system of two linear equations in the variables r_1 and r_2.

$$\begin{cases} 9 = r_1 + r_2 \\ 5 = r_1 - r_2 \end{cases}$$

Solving the system by using the elimination method, we find that r_1 is 7 mph and r_2 is 2 mph. Thus the rate of the boat in calm water is 7 mph and the rate of the current is 2 mph. You should verify these solutions.

TRY EXERCISE 44, EXERCISE SET 7.1

TOPICS FOR DISCUSSION

1. Explain how to use the substitution method to solve a system of equations.

2. Explain how to use the elimination method to solve a system of equations.

3. Give an example of a system of equations in two variables that is

   a. independent   **b.** dependent   **c.** inconsistent

4. If a linear system of equations in two variables has no solution, what does that mean about the graphs of the equations of the system?

5. If $A = \{(x, y) \mid x + y = 5\}$ and $B = \{(x, y) \mid x - y = 3\}$, explain the meaning of $A \cap B$.

EXERCISE SET 7.1

In Exercises 1 to 20, solve each system of equations by the substitution method.

1. $\begin{cases} 2x - 3y = 16 \\ \quad\quad x = 2 \end{cases}$

2. $\begin{cases} 3x - 2y = -11 \\ \quad\quad y = 1 \end{cases}$

3. $\begin{cases} 3x + 4y = 18 \\ \quad y = -2x + 3 \end{cases}$

4. $\begin{cases} 5x - 4y = -22 \\ \quad y = 5x - 2 \end{cases}$

5. $\begin{cases} -2x + 3y = 6 \\ \quad x = 2y - 5 \end{cases}$

6. $\begin{cases} 8x + 3y = -7 \\ \quad x = 3y + 15 \end{cases}$

7. $\begin{cases} 6x + 5y = 1 \\ x - 3y = 4 \end{cases}$

8. $\begin{cases} -3x + 7y = 14 \\ 2x - y = -13 \end{cases}$

9. $\begin{cases} 7x + 6y = -3 \\ \quad y = \dfrac{2}{3}x - 6 \end{cases}$

10. $\begin{cases} 9x - 4y = 3 \\ \quad x = \dfrac{4}{3}y + 3 \end{cases}$

11. $\begin{cases} y = 4x - 3 \\ y = 3x - 1 \end{cases}$

12. $\begin{cases} y = 5x + 1 \\ y = 4x - 2 \end{cases}$

13. $\begin{cases} y = 5x + 4 \\ x = -3y - 4 \end{cases}$

14. $\begin{cases} y = -2x - 6 \\ x = -2y - 2 \end{cases}$

15. $\begin{cases} 3x - 4y = 2 \\ 4x + 3y = 14 \end{cases}$

16. $\begin{cases} 6x + 7y = -4 \\ 2x + 5y = 4 \end{cases}$

17. $\begin{cases} 3x - 3y = 5 \\ 4x - 4y = 9 \end{cases}$

18. $\begin{cases} 3x - 4y = 8 \\ 6x - 8y = 9 \end{cases}$

19. $\begin{cases} 4x + 3y = 6 \\ \quad y = -\dfrac{4}{3}x + 2 \end{cases}$

20. $\begin{cases} 5x + 2y = 2 \\ \quad y = -\dfrac{5}{2}x + 1 \end{cases}$

In Exercises 21 to 40, solve each system of equations by the elimination method.

21. $\begin{cases} 3x - y = 10 \\ 4x + 3y = -4 \end{cases}$

22. $\begin{cases} 3x + 4y = -5 \\ x - 5y = -8 \end{cases}$

23. $\begin{cases} 4x + 7y = 21 \\ 5x - 4y = -12 \end{cases}$

24. $\begin{cases} 3x - 8y = -6 \\ -5x + 4y = 10 \end{cases}$

25. $\begin{cases} 5x - 3y = 0 \\ 10x - 6y = 0 \end{cases}$

26. $\begin{cases} 3x + 2y = 0 \\ 2x + 3y = 0 \end{cases}$

27. $\begin{cases} 6x + 6y = 1 \\ 4x + 9y = 4 \end{cases}$

28. $\begin{cases} 4x + 5y = 2 \\ 8x - 15y = 9 \end{cases}$

29. $\begin{cases} 3x + 6y = 11 \\ 2x + 4y = 9 \end{cases}$

30. $\begin{cases} 4x - 2y = 9 \\ 2x - y = 3 \end{cases}$

31. $\begin{cases} \dfrac{5}{6}x - \dfrac{1}{3}y = -6 \\ \dfrac{1}{6}x + \dfrac{2}{3}y = 1 \end{cases}$

32. $\begin{cases} \dfrac{3}{4}x + \dfrac{2}{5}y = 1 \\ \dfrac{1}{2}x - \dfrac{3}{5}y = -1 \end{cases}$

33. $\begin{cases} \dfrac{3}{4}x + \dfrac{1}{3}y = 1 \\ \dfrac{1}{2}x + \dfrac{2}{3}y = 0 \end{cases}$

34. $\begin{cases} \dfrac{3}{5}x - \dfrac{2}{3}y = 7 \\ \dfrac{2}{5}x - \dfrac{5}{6}y = 7 \end{cases}$

35. $\begin{cases} 2\sqrt{3}x - 3y = 3 \\ 3\sqrt{3}x + 2y = 24 \end{cases}$

36. $\begin{cases} 4x - 3\sqrt{5}y = -19 \\ 3x + 4\sqrt{5}y = 17 \end{cases}$

37. $\begin{cases} 3\pi x - 4y = 6 \\ 2\pi x + 3y = 5 \end{cases}$

38. $\begin{cases} 2x - 5\pi y = 3 \\ 3x + 4\pi y = 2 \end{cases}$

39. $\begin{cases} 3\sqrt{2}x - 4\sqrt{3}y = -6 \\ 2\sqrt{2}x + 3\sqrt{3}y = 13 \end{cases}$

40. $\begin{cases} 2\sqrt{2}x + 3\sqrt{5}y = 7 \\ 3\sqrt{2}x - \sqrt{5}y = -17 \end{cases}$

In Exercises 41 to 55, solve by using a system of equations.

41. **RATE OF WIND** Flying with the wind, a plane traveled 450 miles in 3 hours. Flying against the wind, the plane traveled the same distance in 5 hours. Find the rate of the plane in calm air and the rate of the wind.

42. RATE OF WIND A plane flew 800 miles in 4 hours while flying with the wind. Against the wind, it took the plane 5 hours to travel the 800 miles. Find the rate of the plane in calm air and the rate of the wind.

43. RATE OF CURRENT A motorboat traveled a distance of 120 miles in 4 hours while traveling with the current. Against the current, the same trip took 6 hours. Find the rate of the boat in calm water and the rate of the current.

44. RATE OF CURRENT A canoeist can row 12 miles with the current in 2 hours. Rowing against the current, it takes the canoeist 4 hours to travel the same distance. Find the rate of the canoeist in calm water and the rate of the current.

45. METALLURGY A metallurgist made two purchases. The first purchase, which cost $1080, included 30 kilograms of an iron alloy and 45 kilograms of a lead alloy. The second purchase, at the same prices, cost $372 and included 15 kilograms of the iron alloy and 12 kilograms of the lead alloy. Find the cost per kilogram of the iron and lead alloys.

46. CHEMISTRY For $14.10, a chemist purchased 10 liters of hydrochloric acid and 15 liters of silver nitrate. A second purchase, at the same prices, cost $18.16 and included 12 liters of hydrochloric acid and 20 liters of silver nitrate. Find the cost per liter of each of the two chemicals.

47. COIN PROBLEM A coin bank contains only nickels and dimes. The value of the coins is $1.30. If the nickels were dimes and the dimes were nickels, the value of the coins would be $1.55. Find the original number of nickels and dimes in the bank.

48. COIN PROBLEM The coin drawer of a cash register contains dimes and quarters. The value of the coins is $4.35. If the dimes were quarters and the quarters were dimes, the value of the coins would be $3.00. Find the original number of dimes and quarters in the cash register.

49. NUMBER THEORY The sum of the digits of a two-digit number is 14. If the digits are reversed, the new number is 18 less than the original number. Find the original number.

50. NUMBER THEORY The sum of the digits of a two-digit number is 11. If the digits are reversed, the new number is 63 more than the original number. Find the original number.

51. INVESTMENT A broker invests $25,000 of a client's money in two different municipal bonds. The annual rate of return on one bond is 6%, and the annual rate of return on the second bond is 6.5%. The investor receives a total annual interest payment from the two bonds of $1555. Find the amount invested in each bond.

52. INVESTMENT An investment of $3000 is placed in stocks and bonds. The annual rate of return for the stocks is 4.5%, and the rate of return on the bonds is 8%. The annual interest payment from the stocks and bonds is $177. Find the amount invested in bonds.

53. CHEMISTRY A goldsmith has two gold alloys. The first alloy is 40% gold; the second alloy is 60% gold. How many grams of each should be mixed to produce 20 grams of an alloy that is 52% gold?

54. CHEMISTRY One acetic acid solution is 70% water and another is 30% water. How many liters of each solution should be mixed to produce 20 liters of a solution that is 40% water?

55. CHEMISTRY A chemist wants to make 50 milliliters of a 16% acid solution. How many milliliters each of a 13% acid solution and an 18% acid solution should be mixed to produce the desired solution?

SUPPLEMENTAL EXERCISES

In Exercises 56 to 65, solve for x and y. Use the fact that if $z_1 = a_1 + b_1 i$ and $z_2 = a_2 + b_2 i$ are two complex numbers, then $z_1 = z_2$ if and only if $a_1 = a_2$ and $b_1 = b_2$.

56. $(2 + i)x + (3 - i)y = 7$

57. $(3 + 2i)x + (4 - 3i)y = 2 - 16i$

58. $(4 - 3i)x + (5 + 2i)y = 11 + 9i$

59. $(2 + 6i)x + (4 - 5i)y = -8 - 7i$

60. $(-3 - i)x - (4 + 2i)y = 1 - i$

61. $(5 - 2i)x + (-3 - 4i)y = 12 - 35i$

62. $\begin{cases} 2x + 5y = 11 + 3i \\ 3x + y = 10 - 2i \end{cases}$ **63.** $\begin{cases} 4x + 3y = 11 + 6i \\ 3x - 5y = 1 + 19i \end{cases}$

64. $\begin{cases} 2x + 3y = 11 + 5i \\ 3x - 3y = 9 - 15i \end{cases}$ **65.** $\begin{cases} 5x - 4y = 15 - 41i \\ 3x + 5y = 9 + 5i \end{cases}$

PROJECTS

1. INDEPENDENT AND DEPENDENT CONDITIONS Consider the following problem: "Maria and Michael drove from Los Angeles to New York in 60 hours. How long did Maria drive?" It is difficult to answer this question. She may have driven all 60 hours while Michael relaxed, or she may have relaxed while Michael drove all 60 hours. The difficulty is that there are two unknowns (how long each drove) and only one condition (the total driving time) relating the unknowns. If we added another condition, such as Michael drove 25 hours, then we could determine how long Maria drove, 35 hours. In most cases, an application problem will have a single answer only when there are as many *independent* conditions as there are variables. Conditions are independent if knowing one does *not* allow you to know the other.

Here is an example of conditions that are not independent. "The perimeter of a rectangle is 50 meters. The sum of the width and length is 25 meters." To see that these conditions are dependent, we write the perimeter equation and then divide each side by 2.

$$2w + 2l = 50$$

$$w + l = 25 \qquad \text{• Divide each side by 2.}$$

Note that the resulting equation is the second condition: the sum of the width and length is 25. Thus knowing the first condition allows us to determine the second condition. The conditions are not independent, so there is no one solution to this problem.

For each of the problems below, determine whether the conditions are independent or dependent. For those problems that have independent conditions, find the solution (if possible). For those problems for which the conditions are dependent, find two solutions.

a. The sum of two numbers is 30. The difference between the two numbers is 10. Find the numbers.

b. The area of a square is 25 square meters. Find the length of each side.

c. The area of a rectangle is 25 square meters. Find the length of each side.

d. Emily spent $1000 for carpeting and tile. Carpeting cost $20 per square yard and tile cost $30 per square yard. How many square yards of each did she purchase?

e. The sum of two numbers is 20. Twice the smaller number is 10 minus twice the larger number. Find the two numbers.

f. Make up one word problem for which there are two independent conditions. Solve the problem.

g. Make up one word problem for which there are two dependent conditions. Find at least two solutions.

SECTION

7.2 Systems of Linear Equations in More Than Two Variables

An equation of the form $Ax + By + Cz = D$, with A, B, and C not all zero, is a linear equation in three variables. A solution of an equation in three variables is an **ordered triple** (x, y, z).

The ordered triple $(2, -1, -3)$ is one of the solutions of the equation $2x - 3y + z = 4$. The ordered triple $(3, 1, 1)$ is another solution. In fact, an infinite number of ordered triples are solutions of the equation.

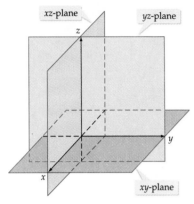

Figure 7.12

Graphing an equation in three variables requires a third coordinate axis perpendicular to the *xy*-plane. This third axis is commonly called the **z-axis.** The result is a three-dimensional coordinate system called the *xyz*-coordinate system (**Figure 7.12**). To help visualize a three-dimensional coordinate system, think of a corner of a room: the floor is the *xy*-plane, one wall is the *yz*-plane, and the other wall is the *xz*-plane.

Graphing an ordered triple requires three moves, the first along the *x*-axis, the second along the *y*-axis, and the third along the *z*-axis. **Figure 7.13** is the graph of the points $(-5, -4, 3)$ and $(4, 5, -2)$.

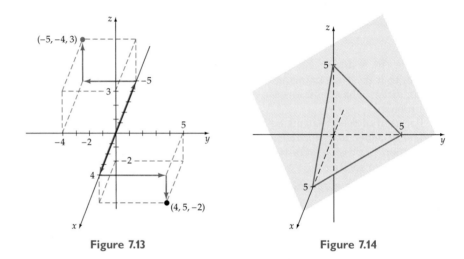

Figure 7.13   Figure 7.14

The graph of a linear equation in three variables is a plane. That is, if all the solutions of a linear equation in three variables were plotted in an *xyz*-coordinate system, the graph would look like a large piece of paper with infinite extent. **Figure 7.14** is the graph of $x + y + z = 5$.

There are different ways in which three planes can be oriented in an *xyz*-coordinate system. **Figure 7.15** illustrates several ways.

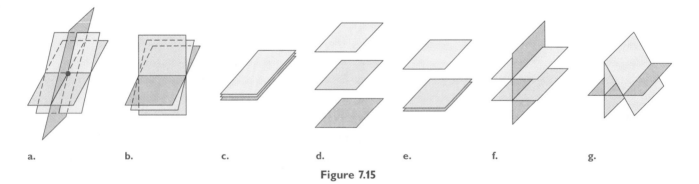

a.   b.   c.   d.   e.   f.   g.

Figure 7.15

For a linear system of equations in three variables to have a solution, the graphs of the planes must intersect at a single point, they must intersect along a common line, or all equations must have a graph that is the same plane. In **Figure 7.15**, the graphs in (a), (b), and (c) represent systems of equations that have a solution. The system of equations represented in **Figure 7.15a** is a consistent system of equations. **Figures 7.15b** and **7.15c** are graphs of a dependent system of equations. The remaining graphs are examples of inconsistent systems of equations.

A system of equations in more than two variables can be solved by using the substitution method or the elimination method. To illustrate the substitution method, consider the system of equations

$$\begin{cases} x - 2y + z = 7 & (1) \\ 2x + y - z = 0 & (2) \\ 3x + 2y - 2z = -2 & (3) \end{cases}$$

Solve Equation (1) for x and substitute the result into Equations (2) and (3).

$$x = 2y - z + 7 \quad (4)$$

$2(2y - z + 7) + y - z = 0$ • Substitute $2y - z + 7$ for x in Equation (2).

$4y - 2z + 14 + y - z = 0$ • Simplify.

$$5y - 3z = -14 \quad (5)$$

$3(2y - z + 7) + 2y - 2z = -2$ • Substitute $2y - z + 7$ for x in Equation (3).

$6y - 3z + 21 + 2y - 2z = -2$ • Simplify.

$$8y - 5z = -23 \quad (6)$$

Now solve the system of equations formed from Equations (5) and (6).

$$\begin{cases} 5y - 3z = -14 & \text{multiply by 8} \longrightarrow & 40y - 24z = -112 \\ 8y - 5z = -23 & \text{multiply by } -5 \longrightarrow & -40y + 25z = 115 \end{cases}$$
$$z = 3$$

Substitute 3 for z into Equation (5) and solve for y.

$$5y - 3z = -14 \qquad \text{• Equation (5)}$$
$$5y - 3(3) = -14$$
$$5y - 9 = -14$$
$$5y = -5$$
$$y = -1$$

Substitute -1 for y and 3 for z into Equation (4) and solve for x.

$$x = 2y - z + 7 = 2(-1) - (3) + 7 = 2$$

The ordered-triple solution is $(2, -1, 3)$. The graphs of the three planes intersect at a single point.

TRIANGULAR FORM

There are many approaches one can take to determine the solution of a system of equations by the elimination method. For consistency, we will always follow a plan that produces an equivalent system of equations in **triangular form**. Three examples of systems of equations in triangular form are

$$\begin{cases} 2x - 3y + z = -4 \\ 2y + 3z = 9 \\ -2z = -2 \end{cases} \quad \begin{cases} w + 3x - 2y + 3z = 0 \\ 2x - y + 4z = 8 \\ -3y - 2z = -1 \\ 3z = 9 \end{cases} \quad \begin{cases} 3x - 4y + z = 1 \\ 3y + 2z = 3 \end{cases}$$

Once a system of equations is written in triangular form, the solution can be found by *back substitution*—that is, by solving the last equation of the system and substituting *back* into the previous equation. This process is continued until the value of each variable has been found.

As an example of solving a system of equations by back substitution, consider the following system of equations in triangular form.

$$\begin{cases} 2x - 4y + z = -3 & (1) \\ 3y - 2z = 9 & (2) \\ 3z = -9 & (3) \end{cases}$$

Solve Equation (3) for z. Substitute the value of z into Equation (2) and solve for y.

$$3z = -9 \qquad \bullet \text{ Equation (3)} \qquad\qquad 3y - 2z = 9 \qquad \bullet \text{ Equation (2)}$$
$$z = -3 \qquad\qquad\qquad\qquad\qquad 3y - 2(-3) = 9 \qquad \bullet \, z = -3$$
$$3y = 3$$
$$y = 1$$

Replace z by -3 and y by 1 in Equation (1) and then solve for x.

$$2x - 4y + z = -3 \qquad \bullet \text{ Equation (1)}$$
$$2x - 4(1) + (-3) = -3$$
$$2x - 7 = -3$$
$$x = 2$$

The solution is the ordered triple $(2, 1, -3)$.

EXAMPLE 1   *Solve an Independent System of Equations*

Solve: $\begin{cases} x + 2y - z = 1 & (1) \\ 2x - y + z = 6 & (2) \\ 2x - y - z = 0 & (3) \end{cases}$

Solution

Eliminate x from Equation (2) by multiplying Equation (1) by -2 and then adding it to Equation (2). Replace Equation (2) by the new equation.

$$\begin{array}{ll} -2x - 4y + 2z = -2 & \bullet \, -2 \text{ times Equation (1)} \\ \underline{2x - y + z = 6} & \bullet \text{ Equation (2)} \\ -5y + 3z = 4 & \bullet \text{ Add the equations.} \end{array}$$

$$\begin{cases} x + 2y - z = 1 & (1) \\ -5y + 3z = -4 & (4) \qquad \bullet \text{ Replace Equation (2).} \\ 2x - y - z = 0 & (3) \end{cases}$$

Eliminate x from Equation (3). Multiply Equation (1) by -2 and add to Equation (3). Replace Equation (3) by the new equation.

$$
\begin{array}{rl}
-2x - 4y + 2z = -2 & \quad \bullet \text{ } -2 \text{ times Equation (1)} \\
\underline{2x - y - z = 0} & \quad \bullet \text{ Equation (3)} \\
-5y + z = -2 & \quad \bullet \text{ Add the equations.}
\end{array}
$$

$$
\left\{
\begin{array}{rll}
x + 2y - z = 1 & (1) & \\
-5y + 3z = 4 & (4) & \\
-5y + z = -2 & (5) & \bullet \text{ Replace Equation (3).}
\end{array}
\right.
$$

Eliminate y from Equation (5) by multiplying Equation (4) by -1 and then adding to Equation (5). Replace Equation (5) by the new equation.

$$
\begin{array}{rl}
5y - 3z = -4 & \quad \bullet \text{ } -1 \text{ times Equation (4)} \\
\underline{-5y + z = -2} & \quad \bullet \text{ Equation (5)} \\
-2z = -6 & \quad \bullet \text{ Add the equations.}
\end{array}
$$

$$
\left\{
\begin{array}{rll}
x + 2y - z = 1 & (1) & \\
-5y + 3z = 4 & (4) & \\
-2z = -6 & (6) & \bullet \text{ Replace Equation (5).}
\end{array}
\right.
$$

The system of equations is now in triangular form. Solve the system of equations by back substitution.

Solve Equation (6) for z. Substitute the value into Equation (4) and then solve for y.

$$
\begin{array}{rl}
-2z = -6 & \quad \bullet \text{ Equation (6)} \\
z = 3 &
\end{array}
\qquad
\begin{array}{rl}
-5y + 3z = 4 & \quad \bullet \text{ Equation (4)} \\
-5y + 3(3) = 4 & \quad \bullet \text{ Replace z by 3.} \\
-5y = -5 & \quad \bullet \text{ Solve for y.} \\
y = 1 &
\end{array}
$$

Replace z by 3 and y by 1 in Equation (1) and solve for x.

$$
\begin{array}{rl}
x + 2y - z = 1 & \quad \bullet \text{ Equation (1)} \\
x + 2(1) - 3 = 1 & \quad \bullet \text{ Replace y by 1; replace z by 3.} \\
x = 2 &
\end{array}
$$

The system of equations is consistent. The solution is the ordered triple $(2, 1, 3)$. See **Figure 7.16**.

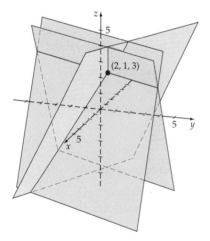

Figure 7.16

| **Try Exercise 12, Exercise Set 7.2** |

EXAMPLE 2   *Solve a Dependent System of Equations*

Solve:
$$
\left\{
\begin{array}{rl}
2x - y - z = -1 & \quad (1) \\
-x + 3y - z = -3 & \quad (2) \\
-5x + 5y + z = -1 & \quad (3)
\end{array}
\right.
$$

Continued • ▶

Solution

Eliminate x from Equation (2) by multiplying Equation (2) by 2 and then adding it to Equation (1). Replace Equation (2) by the new equation.

$$
\begin{array}{ll}
2x - y - z = -1 & \bullet \text{ Equation (1)} \\
\underline{-2x + 6y - 2z = -6} & \bullet \text{ 2 times Equation (2)} \\
5y - 3z = -7 & \bullet \text{ Add the equations.}
\end{array}
$$

$$
\left\{
\begin{array}{ll}
2x - y - z = -1 & (1) \\
5y - 3z = -7 & (4) \\
-5x + 5y + z = -1 & (3)
\end{array}
\right. \qquad \bullet \text{ Replace Equation (2).}
$$

Eliminate x from Equation (3). Multiply Equation (1) by 5 and multiply Equation (3) by 2. Then add. Replace Equation (3) by the new equation.

$$
\begin{array}{ll}
10x - 5y - 5z = -5 & \bullet \text{ 5 times Equation (1)} \\
\underline{-10x + 10y + 2z = -2} & \bullet \text{ 2 times Equation (3)} \\
5y - 3z = -7 & \bullet \text{ Add the equations.}
\end{array}
$$

$$
\left\{
\begin{array}{ll}
2x - y - z = -1 & (1) \\
5y - 3z = -7 & (4) \\
5y - 3z = -7 & (5)
\end{array}
\right. \qquad \bullet \text{ Replace Equation (3).}
$$

Eliminate y from Equation (5) by multiplying Equation (4) by -1 and then adding to Equation (5). Replace Equation (5) by the new equation.

$$
\begin{array}{ll}
-5y + 3z = 7 & \bullet -1 \text{ times Equation (4)} \\
\underline{5y - 3z = -7} & \bullet \text{ Equation (5)} \\
0 = 0 & \bullet \text{ Add the equations.}
\end{array}
$$

$$
\left\{
\begin{array}{ll}
2x - y - z = -1 & (1) \\
5y - 3z = -7 & (4) \\
0 = 0 & (6)
\end{array}
\right. \qquad \bullet \text{ Replace Equation (5).}
$$

Because any ordered triple (x, y, z) is a solution of Equation (6), the solutions of the system of equations will be the ordered triples that are solutions of Equations (1) and (4).

Solve Equation (4) for y.

$$
\begin{aligned}
5y - 3z &= -7 \\
5y &= 3z - 7 \\
y &= \frac{3}{5}z - \frac{7}{5}
\end{aligned}
$$

Substitute $\frac{3}{5}z - \frac{7}{5}$ for y in Equation (1) and solve for x.

$$
\begin{array}{ll}
2x - y - z = -1 & \bullet \text{ Equation (1)} \\
2x - \left(\dfrac{3}{5}z - \dfrac{7}{5}\right) - z = -1 & \bullet \text{ Replace } y \text{ by } \dfrac{3}{5}z - \dfrac{7}{5}.
\end{array}
$$

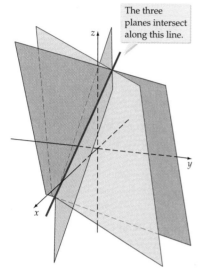

The three planes intersect along this line.

Figure 7.17

$$2x - \frac{8}{5}z + \frac{7}{5} = -1 \qquad \bullet \text{ Simplify and solve for } x.$$

$$2x = \frac{8}{5}z - \frac{12}{5}$$

$$x = \frac{4}{5}z - \frac{6}{5}$$

By choosing any real number c for z, we have $y = \frac{3}{5}c - \frac{7}{5}$ and $x = \frac{4}{5}c - \frac{6}{5}$. The ordered-triple solutions of the equation are $\left(\frac{4}{5}c - \frac{6}{5}, \frac{3}{5}c - \frac{7}{5}, c\right)$ which are the coordinates of the points on the solid blue line shown in **Figure 7.17**.

Try Exercise 16, Exercise Set 7.2

TAKE NOTE

Although the ordered triples

$$\left(\frac{4}{5}c - \frac{6}{5}, \frac{3}{5}c - \frac{7}{5}, c\right)$$

and

$$\left(a, \frac{3}{4}a - \frac{1}{2}, \frac{5}{4}a + \frac{3}{2}\right)$$

appear to be different, they represent exactly the same set of ordered triples. For instance, choosing $c = -1$, we have $(-2, -2, -1)$. Choosing $a = -2$ results in the same ordered triple, $(-2, -2, -1)$.

As in the case of a dependent system of equations in two variables, there is more than one way to represent the solutions of a dependent system of equations in three variables. For instance, from Example 2, let $a = \frac{4}{5}c - \frac{6}{5}$, the x-coordinate of the ordered triple $\left(\frac{4}{5}c - \frac{6}{5}, \frac{3}{5}c - \frac{7}{5}, c\right)$, and solve for c.

$$a = \frac{4}{5}c - \frac{6}{5} \quad \longrightarrow \quad c = \frac{5}{4}a + \frac{3}{2}$$

Substitute this value of c into each component of the ordered triple.

$$\left(\frac{4}{5}\left(\frac{5}{4}a + \frac{3}{2}\right) - \frac{6}{5}, \frac{3}{5}\left(\frac{5}{4}a + \frac{3}{2}\right) - \frac{7}{5}, \frac{5}{4}a + \frac{3}{2}\right) = \left(a, \frac{3}{4}a - \frac{1}{2}, \frac{5}{4}a + \frac{3}{2}\right)$$

Thus the solutions of the system of equation can also be written as

$$\left(a, \frac{3}{4}a - \frac{1}{2}, \frac{5}{4}a + \frac{3}{2}\right)$$

EXAMPLE 3   *Identify an Inconsistent System of Equations*

Solve: $\begin{cases} x + 2y + 3z = 4 & (1) \\ 2x - y - z = 3 & (2) \\ 3x + y + 2z = 5 & (3) \end{cases}$

Solution

Eliminate x from Equation (2) by multiplying Equation (1) by -2 and adding it to Equation (2). Replace Equation 2. Eliminate x from Equa-

Continued •▶

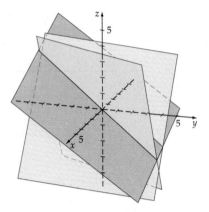

Figure 7.18

tion (3) by multiplying Equation (1) by -3 and adding it to Equation (3). Replace Equation (3). The equivalent system is

$$\begin{cases} x + 2y + 3z = 4 & (1) \\ -5y - 7z = -5 & (4) \\ -5y - 7z = -7 & (5) \end{cases}$$

Eliminate y from Equation (5) by multiplying Equation (4) by -1 and adding it to Equation (5). Replace Equation (5). The equivalent system is

$$\begin{cases} x + 2y + 3z = 4 & (1) \\ -5y - 7z = -5 & (4) \\ 0 = -2 & (6) \end{cases}$$

This system of equations contains a false equation. The system is inconsistent and has no solutions. There is no point on all three planes as shown in **Figure 7.18**.

TRY EXERCISE 18, EXERCISE SET 7.2

NONSQUARE SYSTEMS OF EQUATIONS

The linear systems of equations that we have solved so far contain the same number of variables as equations. These are *square systems of equations*. If there are fewer equations than variables—a *nonsquare system of equations*—the system has either no solution or an infinite number of solutions.

EXAMPLE 4   *Solve a Nonsquare System of Equations*

Solve: $\begin{cases} x - 2y + 2z = 3 & (1) \\ 2x - y - 2z = 15 & (2) \end{cases}$

Solution

Eliminate x from Equation (2) by multiplying Equation (1) by -2 and adding it to Equation (2). Replace Equation (2).

$$\begin{cases} x - 2y + 2z = 3 & (1) \\ 3y - 6z = 9 & (3) \end{cases}$$

Solve Equation (3) for y.

$$3y - 6z = 9$$
$$y = 2z + 3$$

Substitute $2z + 3$ for y into Equation (1) and solve for x.

$$x - 2y + 2z = 3$$
$$x - 2(2z + 3) + 2z = 3 \qquad \bullet\, y = 2z + 3$$
$$x = 2z + 9$$

For each value of z selected, there correspond values for x and y. If z is any real number c, then the solutions of the system are the ordered triples $(2c + 9, 2c + 3, c)$.

TRY EXERCISE 20, EXERCISE SET 7.2

HOMOGENEOUS SYSTEMS OF EQUATIONS

A linear system of equations for which the constant term is zero for all equations is called a **homogeneous system of equations.** Two examples of homogeneous systems of equations are

$$\begin{cases} 3x + 4y = 0 \\ 2x + 3y = 0 \end{cases} \qquad \begin{cases} 2x - 3y + 5z = 0 \\ 3x + 2y + z = 0 \\ x - 4y + 5z = 0 \end{cases}$$

The solution $(0, 0)$ is always a solution of a homogeneous system of equations in two variables, and $(0, 0, 0)$ is always a solution of a homogeneous system of equations in three variables. This solution is called the **trivial solution.**

Sometimes a homogeneous system of equations may have solutions other than the trivial solution. For example, $(1, -1, -1)$ is a solution to the homogeneous system of three equations in three variables above.

If a homogeneous system of equations has a unique solution, the graphs intersect only at the origin. If the homogeneous system of equations has infinitely many solutions, the graphs intersect along a line or plane that passes through the origin.

Solutions to a homogeneous system of equations can be found by using the substitution method or the elimination method.

EXAMPLE 5   *Solve a Homogeneous System of Equations*

Solve: $\begin{cases} x + 2y - 3z = 0 & (1) \\ 2x - y + z = 0 & (2) \\ 3x + y - 2z = 0 & (3) \end{cases}$

Solution

Eliminate x from Equations (2) and (3) and replace these equations by the new equations.

$$\begin{cases} x + 2y - 3z = 0 & (1) \\ -5y + 7z = 0 & (4) \\ -5y + 7z = 0 & (5) \end{cases}$$

Continued ▶

Eliminate y from Equation (5). Replace Equation (5).

$$\begin{cases} x + 2y - 3z = 0 & (1) \\ -5y + 7z = 0 & (4) \\ 0 = 0 & (6) \end{cases}$$

Because Equation (6) is an identity, the solutions of the system are the solutions of Equations (1) and (4).

Solve Equation (4) for y.

$$y = \frac{7}{5}z$$

Substitute the expression for y into Equation (1) and solve for x.

$$x + 2y - 3z = 0 \qquad \bullet \text{ Equation (1)}$$

$$x + 2\left(\frac{7}{5}z\right) - 3z = 0 \qquad \bullet \, y = \frac{7}{5}z$$

$$x = \frac{1}{5}z$$

Letting z be any real number c, we find the solutions of the system are

$$\left(\frac{1}{5}c, \frac{7}{5}c, c\right)$$

TRY EXERCISE 32, EXERCISE SET 7.2

CURVE FITTING

One application of a system of equations is "curve fitting." Given a set of points in the plane, try to find an equation whose graph passes through those points, or "fits" those points.

EXAMPLE 6	*Solve an Application of a System of Equations to Curve Fitting*

Find an equation of the form $y = ax^2 + bx + c$ whose graph passes through the points whose coordinates are $(1, 4)$, $(-1, 6)$, and $(2, 9)$.

Solution

Substitute each of the given ordered pairs into the equation $y = ax^2 + bx + c$. Write the resulting system of equations.

$$\begin{cases} 4 = a(1)^2 + b(1) + c & \text{or} \\ 6 = a(-1)^2 + b(-1) + c & \text{or} \\ 9 = a(2)^2 + b(2) + c & \text{or} \end{cases} \quad \begin{cases} a + b + c = 4 & (1) \\ a - b + c = 6 & (2) \\ 4a + 2b + c = 9 & (3) \end{cases}$$

Solve the resulting system of equations for a, b, and c.

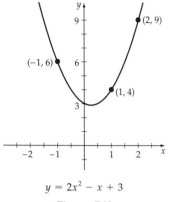

$$y = 2x^2 - x + 3$$

Figure 7.19

Eliminate a from Equation (2) by multiplying Equation (1) by -1 and adding it to Equation (2). Now eliminate a from Equation (3) by multiplying Equation (1) by -4 and adding it to Equation (3). The result is

$$\begin{cases} a + b + c = 4 \\ \quad -2b \quad = 2 \\ \quad -2b - 3c = -7 \end{cases}$$

Although this system of equations is not in triangular form, we can solve the second equation for b and use this value to find a and c.

Solving by substitution, we obtain $a = 2$, $b = -1$, $c = 3$. The equation of the form $y = ax^2 + bx + c$ whose graph passes through the three points is $y = 2x^2 - x + 3$. See **Figure 7.19**.

> **TRY EXERCISE 36, EXERCISE SET 7.2**

TOPICS FOR DISCUSSION

1. Can a system of equations contain more equations than variables? If not, explain why not. If so, give an example.

2. If a linear system of three equations in three variables is dependent, what does that mean about the graphs of the equations of the system?

3. If a linear system of three equations in three variables is inconsistent, what does that mean about the graphs of the equations of the system?

4. The equation of a circle centered at the origin with radius 5 is given by $x^2 + y^2 = 25$. Discuss the shape of $x^2 + y^2 + z^2 = 25$ in an xyz-coordinate system.

5. Consider the plane P given by $2x + 4y - 3z = 12$. The *trace* of the graph of P is obtained by letting one of the variables equal zero. For instance, the trace in the xy-plane is the graph of $2x + 4y = 12$ that is obtained by letting $z = 0$. Determine the traces of P in the xz- and yz-planes, and discuss how the traces can be used to visualize the graph of P.

EXERCISE SET 7.2

In Exercises 1 to 24, solve each system of equations.

1. $\begin{cases} 2x - y + z = 8 \\ \quad 2y - 3z = -11 \\ \quad 3y + 2z = 3 \end{cases}$

2. $\begin{cases} 3x + y + 2z = -4 \\ \quad -3y - 2z = -5 \\ \quad 2y + 5z = -4 \end{cases}$

3. $\begin{cases} x + 3y - 2z = 8 \\ 2x - y + z = 1 \\ 3x + 2y - 3z = 15 \end{cases}$

4. $\begin{cases} x - 2y + 3z = 5 \\ 3x - 3y + z = 9 \\ 5x + y - 3z = 3 \end{cases}$

5. $\begin{cases} 3x + 4y - z = -7 \\ x - 5y + 2z = 19 \\ 5x + y - 2z = 5 \end{cases}$

6. $\begin{cases} 2x - 3y - 2z = 12 \\ x + 4y + z = -9 \\ 4x + 2y - 3z = 6 \end{cases}$

7. $\begin{cases} 2x - 5y + 3z = -18 \\ 3x + 2y - z = -12 \\ x - 3y - 4z = -4 \end{cases}$

8. $\begin{cases} 4x - y + 2z = -1 \\ 2x + 3y - 3z = -13 \\ x + 5y + z = 7 \end{cases}$

9. $\begin{cases} x + 2y - 3z = -7 \\ 2x - y + 4z = 11 \\ 4x + 3y - 4z = -3 \end{cases}$

10. $\begin{cases} x - 3y + 2z = -11 \\ 3x + y + 4z = 4 \\ 5x - 5y + 8z = -18 \end{cases}$

11. $\begin{cases} 2x - 5y + 2z = -4 \\ 3x + 2y + 3z = 13 \\ 5x - 3y - 4z = -18 \end{cases}$

12. $\begin{cases} 3x + 2y - 5z = 6 \\ 5x - 4y + 3z = -12 \\ 4x + 5y - 2z = 15 \end{cases}$

13. $\begin{cases} 2x + y - z = -2 \\ 3x + 2y + 3z = 21 \\ 7x + 4y + z = 17 \end{cases}$

14. $\begin{cases} 3x + y + 2z = 2 \\ 4x - 2y + z = -4 \\ 11x - 3y + 4z = -6 \end{cases}$

15. $\begin{cases} 3x - 2y + 3z = 11 \\ 2x + 3y + z = 3 \\ 5x + 14y - z = 1 \end{cases}$

16. $\begin{cases} 2x + 3y + 2z = 14 \\ x - 3y + 4z = 4 \\ -x + 12y - 6z = 2 \end{cases}$

17. $\begin{cases} 2x - 3y + 6z = 3 \\ x + 2y - 4z = 5 \\ 3x + 4y - 8z = 7 \end{cases}$

18. $\begin{cases} 2x + 3y - 6z = 4 \\ 3x - 2y - 9z = -7 \\ 2x + 5y - 6z = 8 \end{cases}$

19. $\begin{cases} 2x - 3y + 5z = 14 \\ x + 4y - 3z = -2 \end{cases}$

20. $\begin{cases} x - 3y + 4z = 9 \\ 3x - 8y - 2z = 4 \end{cases}$

21. $\begin{cases} 6x - 9y + 6z = 7 \\ 4x - 6y + 4z = 9 \end{cases}$

22. $\begin{cases} 4x - 2y + 6z = 5 \\ 2x - y + 3z = 2 \end{cases}$

23. $\begin{cases} 5x + 3y + 2z = 10 \\ 3x - 4y - 4z = -5 \end{cases}$

24. $\begin{cases} 3x - 4y - 7z = -5 \\ 2x + 3y - 5z = 2 \end{cases}$

In Exercises 25 to 32, solve each homogeneous system of equations.

25. $\begin{cases} x + 3y - 4z = 0 \\ 2x + 7y + z = 0 \\ 3x - 5y - 2z = 0 \end{cases}$

26. $\begin{cases} x - 2y + 3z = 0 \\ 3x - 7y - 4z = 0 \\ 4x - 4y + z = 0 \end{cases}$

27. $\begin{cases} 2x - 3y + z = 0 \\ 2x + 4y - 3z = 0 \\ 6x - 2y - z = 0 \end{cases}$

28. $\begin{cases} 5x - 4y - 3z = 0 \\ 2x + y + 2z = 0 \\ x - 6y - 7z = 0 \end{cases}$

29. $\begin{cases} 3x - 5y + 3z = 0 \\ 2x - 3y + 4z = 0 \\ 7x - 11y + 11z = 0 \end{cases}$

30. $\begin{cases} 5x - 2y - 3z = 0 \\ 3x - y - 4z = 0 \\ 4x - y - 9z = 0 \end{cases}$

31. $\begin{cases} 4x - 7y - 2z = 0 \\ 2x + 4y + 3z = 0 \\ 3x - 2y - 5z = 0 \end{cases}$

32. $\begin{cases} 5x + 2y + 3z = 0 \\ 3x + y - 2z = 0 \\ 4x - 7y + 5z = 0 \end{cases}$

In Exercises 33 to 42, solve a system of equations.

33. CURVE FITTING Find an equation of the form $y = ax^2 + bx + c$ whose graph passes through the points $(2, 3)$, $(-2, 7)$, and $(1, -2)$.

34. CURVE FITTING Find an equation of the form $y = ax^2 + bx + c$ whose graph passes through the points $(1, -2)$, $(3, -4)$, and $(2, -2)$.

35. CURVE FITTING Find the equation of the circle whose graph passes through the points $(5, 3)$, $(-1, -5)$, and $(-2, 2)$. (*Hint:* Use the equation $x^2 + y^2 + ax + by + c = 0$.)

36. CURVE FITTING Find the equation of the circle whose graph passes through the points $(0, 6)$, $(1, 5)$, and $(-7, -1)$. (*Hint:* See Exercise 35.)

37. CURVE FITTING Find the center and radius of the circle whose graph passes through the points $(-2, 10)$, $(-12, -14)$, and $(5, 3)$. (*Hint:* See Exercise 35.)

38. CURVE FITTING Find the center and radius of the circle whose graph passes through the points $(2, 5)$, $(-4, -3)$, and $(3, 4)$. (*Hint:* See Exercise 35.)

39. COIN PROBLEM A coin bank contains only nickels, dimes, and quarters. The value of the coins is $2. There are twice as many nickels as dimes and one more dime than quarters. Find the number of each coin in the bank.

40. COIN PROBLEM A coin bank contains only nickels, dimes, and quarters. The value of the coins is $5.50. The number of nickels is six more than twice the number of quarters. The number of dimes is one-third the number of nickels. Find the number of each coin in the bank.

41. NUMBER THEORY The sum of the digits of a positive three-digit number is 19. The tens digit is four less than twice the hundreds digit. The number is decreased by 99 when the digits are reversed. Find the number.

42. NUMBER THEORY The sum of the digits of a positive three-digit number is 10. The hundreds digit is one less than twice the ones digit. The number is decreased by 198 when the digits are reversed. Find the number.

SUPPLEMENTAL EXERCISES

In Exercises 43 to 48, solve each system of equations.

43. $\begin{cases} 2x + y - 3z + 2w = -1 \\ 2y - 5z - 3w = 9 \\ 3y - 8z + w = -4 \\ 2y - 2z + 3w = -3 \end{cases}$

44. $\begin{cases} 3x - y + 2z - 3w = 5 \\ 2y - 5z + 2w = -7 \\ 4y - 9z + w = -19 \\ 3y + z - 2w = -12 \end{cases}$

45. $\begin{cases} x - 3y + 2z - w = 2 \\ 2x - 5y - 3z + 2w = 21 \\ 3x - 8y - 2z - 3w = 12 \\ -2x + 8y + z + 2w = -13 \end{cases}$

46. $\begin{cases} x - 2y + 3z + 2w = 8 \\ 3x - 7y - 2z + 3w = 18 \\ 2x - 5y + 2z - w = 19 \\ 4x - 8y + 3z + 2w = 29 \end{cases}$

47. $\begin{cases} x + 2y - 2z + 3w = 2 \\ 2x + 5y + 2z + 4w = 9 \\ 4x + 9y - 2z + 10w = 13 \\ -x - y + 8z - 5w = 3 \end{cases}$

48. $\begin{cases} x - 2y + 3z - 2w = -1 \\ 3x - 7y - 2z - 3w = -19 \\ 2x - 5y + 2z - w = -11 \\ -x + 3y - 2z - w = 3 \end{cases}$

In Exercises 49 and 50, use the system of equations

$$\begin{cases} x - 3y - 2z = A^2 \\ 2x - 5y + Az = 9 \\ 2x - 8y + z = 18 \end{cases}$$

49. Find all values of A for which the system has no solutions.

50. Find all values of A for which the system has a unique solution.

In Exercises 51 to 53, use the system of equations

$$\begin{cases} x + 2y + z = A^2 \\ -2x - 3y + Az = 1 \\ 7x + 12y + A^2z = 4A^2 - 3 \end{cases}$$

51. Find all values of A for which the system has a unique solution.

52. Find all values of A for which the system has an infinite number of solutions.

53. Find all values of A for which the system has no solution.

54. Find an equation of a plane that contains the points $(2, 1, 1)$, $(-1, 2, 12)$, and $(3, 2, 0)$. (*Hint:* The equation of a plane can be written as $z = ax + by + c$.)

55. Find an equation of a plane that contains the points $(1, -1, 5)$, $(2, -2, 9)$, and $(-3, -1, -1)$. (*Hint:* The equation of a plane can be written as $z = ax + by + c$.)

Projects

1. **Concept of Dimension** In this chapter we graphed first-degree equations in three variables. If we were to attempt to graph an equation in four variables, we would need a fourth axis perpendicular to the three axes of an *xyz*-coordinate system. It seems impossible to imagine a fourth dimension, but incorporating it is really a quite practical matter in mathematics. In fact, there are some systems that require an infinite-dimensional coordinate system. To gain some insight into the concept of dimension, read the book *Flatland* by Edwin A. Abbott, and then write an essay explaining what this book has to do with dimension.

2. **Abilities of a Four-Dimensional Human** There have been a number of attempts to describe the abilities of a four-dimensional human. Read some of these accounts, and then write an essay on some of the actions a four-dimensional person could perform. Answer the following question in your essay. Can a four-dimensional person remove money from a locked safe without first opening the safe?

Section 7.3 Nonlinear Systems of Equations

A **nonlinear system of equations** is one in which one or more equations of the system are not linear equations. **Figure 7.20** on page 404 shows examples of nonlinear systems of equations and the corresponding graphs of the equations. Each point of intersection of the graphs is a solution of the system of equations. In the third example, the graphs do not intersect; therefore, the system of equations has no real number solution.

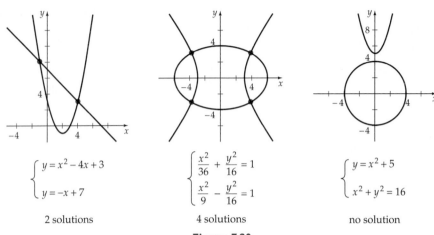

$$\begin{cases} y = x^2 - 4x + 3 \\ y = -x + 7 \end{cases}$$

2 solutions

$$\begin{cases} \dfrac{x^2}{36} + \dfrac{y^2}{16} = 1 \\ \dfrac{x^2}{9} - \dfrac{y^2}{16} = 1 \end{cases}$$

4 solutions

$$\begin{cases} y = x^2 + 5 \\ x^2 + y^2 = 16 \end{cases}$$

no solution

Figure 7.20

To solve a nonlinear system of equations, use the substitution method or the elimination method. The substitution method is usually easier for solving a nonlinear system that contains a linear equation.

EXAMPLE 1	*Solve a Nonlinear System by the Substitution Method*

Solve: $\begin{cases} y = x^2 - x - 1 & (1) \\ 3x - y = 4 & (2) \end{cases}$

Solution

We will use the substitution method. Using the equation $y = x^2 - x - 1$, substitute the expression for y into $3x - y = 4$.

$$3x - y = 4$$
$$3x - (x^2 - x - 1) = 4 \qquad \bullet\ y = x^2 - x - 1$$
$$-x^2 + 4x + 1 = 4 \qquad \bullet\ \textbf{Simplify.}$$
$$x^2 - 4x + 3 = 0 \qquad \bullet\ \textbf{Write the quadratic equation in standard form.}$$
$$(x - 3)(x - 1) = 0 \qquad \bullet\ \textbf{Solve for } x.$$
$$x - 3 = 0 \quad \text{or} \quad x - 1 = 0$$
$$x = 3 \quad \text{or} \qquad x = 1$$

Substitute these values into Equation (1) and solve for y.

$$y = 3^2 - 3 - 1 = 5, \quad \text{or} \quad y = 1^2 - 1 - 1 = -1$$

The solutions are $(3, 5)$ and $(1, -1)$. See **Figure 7.21**.

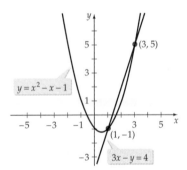

Figure 7.21

TRY EXERCISE 8, EXERCISE SET 7.3

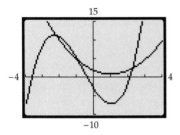

Figure 7.22

A graphing utility can be used to solve some nonlinear systems of equations. For instance, to solve

$$\begin{cases} y = x^2 - 2x + 2 \\ y = x^3 + 2x^2 - 7x - 3 \end{cases}$$

graph each equation and then determine the points of intersection. The graph of the system of equations is shown in **Figure 7.22**. Note that there are three points of intersection. The approximate solutions (to the nearest hundredth) of the system of equations are $(-1, 5)$, $(-2.24, 11.47)$, and $(2.24, 2.53)$. This system of equations also illustrates that you must choose an appropriate viewing window so that all solutions can be determined.

EXAMPLE 2	*Solve a Nonlinear System by the Elimination Method*

Solve: $\begin{cases} 4x^2 + 3y^2 = 48 & (1) \\ 3x^2 + 2y^2 = 35 & (2) \end{cases}$

Solution

We will eliminate the x^2 term. Multiply Equation (1) by -3 and Equation (2) by 4. Then add the two equations.

$$\begin{array}{r} -12x^2 - 9y^2 = -144 \\ \underline{12x^2 + 8y^2 = 140} \\ -y^2 = -4 \\ y^2 = 4 \\ y = \pm 2 \end{array}$$

Substitute 2 for y into Equation (1) and solve for x.

$$4x^2 + 3(2)^2 = 48$$
$$4x^2 = 36$$
$$x^2 = 9$$
$$x = \pm 3$$

Because $(-2)^2 = 2^2$, replacing y by -2 yields the same values of x: $x = 3$ or $x = -3$. The solutions are $(3, 2)$, $(3, -2)$, $(-3, 2)$, and $(-3, -2)$. See **Figure 7.23**.

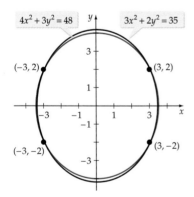

$4x^2 + 3y^2 = 48$ $3x^2 + 2y^2 = 35$

$(-3, 2)$ $(3, 2)$

$(-3, -2)$ $(3, -2)$

Figure 7.23

TRY EXERCISE 16, EXERCISE SET 7.3

EXAMPLE 3 *Identify an Inconsistent System of Equations*

Solve: $\begin{cases} 4x^2 + 9y^2 = 36 & (1) \\ x^2 - y^2 = 25 & (2) \end{cases}$

Solution

Using the elimination method, we will eliminate the x^2 term from each equation. Multiplying Equation (2) by -4 and then adding, we have

$$\begin{array}{rcr} 4x^2 + 9y^2 = & 36 \\ -4x^2 + 4y^2 = & -100 \\ \hline 13y^2 = & -64 \end{array}$$

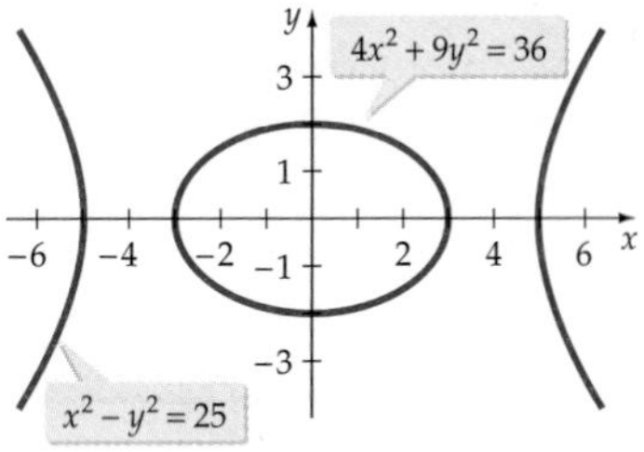

Because the equation $13y^2 = -64$ has no real number solutions, the system of equations has no real solutions. The graphs of the equations do not intersect. See **Figure 7.24.**

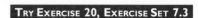

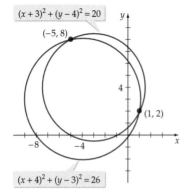

Figure 7.24

TRY EXERCISE 20, EXERCISE SET 7.3

EXAMPLE 4 *Solve a Nonlinear System of Equations*

Solve: $\begin{cases} (x + 3)^2 + (y - 4)^2 = 20 \\ (x + 4)^2 + (y - 3)^2 = 26 \end{cases}$

Solution

Expand the binomials in each equation. Then subtract the two equations and simplify.

$$\begin{array}{rcl} x^2 + 6x + 9 + y^2 - 8y + 16 = & 20 & (1) \\ x^2 + 8x + 16 + y^2 - 6y + 9 = & 26 & (2) \\ \hline -2x - 7 \quad - 2y + 7 = & -6 \\ x + y = & 3 \end{array}$$

Now solve the resulting equation for y.

$$y = -x + 3$$

Substitute $-x + 3$ for y into Equation (1) and solve for x.

$$x^2 + 6x + 9 + (-x + 3)^2 - 8(-x + 3) + 16 = 20$$
$$2(x^2 + 4x - 5) = 0$$
$$2(x + 5)(x - 1) = 0$$
$$x = -5 \quad \text{or} \quad x = 1$$

Substitute -5 and 1 for x into the equation $y = -x + 3$ and solve for y. This yields $y = 8$ or $y = 2$. The solutions of the system of equations are $(-5, 8)$ and $(1, 2)$. See **Figure 7.25.**

Figure 7.25

TRY EXERCISE 28, EXERCISE SET 7.3

TOPICS FOR DISCUSSION

1. What distinguishes a system of linear equations from a system of nonlinear equations? Give an example of both types of systems of equations.

2. Is the system of equations

$$\begin{cases} xy = 1 \\ x + y = 1 \end{cases}$$

a nonlinear system of equations? Why or why not?

3. Can a nonlinear system of equations have no solution? If so, give an example. If not, explain why not.

4. Make up a nonlinear system of equations in two variables that has at least $(2, -3)$ as a solution, contains one nonlinear equation, and contains one linear equation.

EXERCISE SET 7.3

In Exercises 1 to 32, solve the system of equations.

1. $\begin{cases} y = x^2 - x \\ y = 2x - 2 \end{cases}$

2. $\begin{cases} y = x^2 + 2x - 3 \\ y = x + 1 \end{cases}$

3. $\begin{cases} y = 2x^2 - 3x - 3 \\ y = x - 4 \end{cases}$

4. $\begin{cases} y = -x^2 + 2x - 4 \\ y = \dfrac{1}{2}x + 1 \end{cases}$

5. $\begin{cases} y = x^2 - 2x + 3 \\ y = x^2 - x - 2 \end{cases}$

6. $\begin{cases} y = 2x^2 - x + 1 \\ y = x^2 + 2x + 5 \end{cases}$

7. $\begin{cases} x + y = 10 \\ xy = 24 \end{cases}$

8. $\begin{cases} x - 2y = 3 \\ xy = -1 \end{cases}$

9. $\begin{cases} 2x - y = 1 \\ xy = 6 \end{cases}$

10. $\begin{cases} x - 3y = 7 \\ xy = -4 \end{cases}$

11. $\begin{cases} 3x^2 - 2y^2 = 1 \\ y = 4x - 3 \end{cases}$

12. $\begin{cases} x^2 + 3y^2 = 7 \\ x + 4y = 6 \end{cases}$

13. $\begin{cases} y = x^3 + 4x^2 - 3x - 5 \\ y = 2x^2 - 2x - 3 \end{cases}$

14. $\begin{cases} y = x^3 - 2x^2 + 5x + 1 \\ y = x^2 + 7x - 5 \end{cases}$

15. $\begin{cases} 2x^2 + y^2 = 9 \\ x^2 - y^2 = 3 \end{cases}$

16. $\begin{cases} 3x^2 - 2y^2 = 19 \\ x^2 - y^2 = 5 \end{cases}$

17. $\begin{cases} x^2 - 2y^2 = 8 \\ x^2 + 3y^2 = 28 \end{cases}$

18. $\begin{cases} 2x^2 + 3y^2 = 5 \\ x^2 - 3y^2 = 4 \end{cases}$

19. $\begin{cases} 2x^2 + 4y^2 = 5 \\ 3x^2 + 8y^2 = 14 \end{cases}$

20. $\begin{cases} 2x^2 + 3y^2 = 11 \\ 3x^2 + 2y^2 = 19 \end{cases}$

21. $\begin{cases} x^2 - 2x + y^2 = 1 \\ 2x + y = 5 \end{cases}$

22. $\begin{cases} x^2 + y^2 + 3y = 22 \\ 2x + y = -1 \end{cases}$

23. $\begin{cases} (x - 3)^2 + (y + 1)^2 = 5 \\ x - 3y = 7 \end{cases}$

24. $\begin{cases} (x + 2)^2 + (y - 2)^2 = 13 \\ 2x + y = 6 \end{cases}$

25. $\begin{cases} x^2 - 3x + y^2 = 4 \\ 3x + y = 11 \end{cases}$

26. $\begin{cases} x^2 + y^2 - 4y = 4 \\ 5x - 2y = 2 \end{cases}$

27. $\begin{cases} (x - 1)^2 + (y + 2)^2 = 14 \\ (x + 2)^2 + (y - 1)^2 = 2 \end{cases}$

28. $\begin{cases} (x + 2)^2 + (y - 3)^2 = 10 \\ (x - 3)^2 + (y + 1)^2 = 13 \end{cases}$

29. $\begin{cases} (x + 3)^2 + (y - 2)^2 = 20 \\ (x - 2)^2 + (y - 3)^2 = 2 \end{cases}$

30. $\begin{cases} (x - 4)^2 + (y - 5)^2 = 8 \\ (x + 1)^2 + (y + 2)^2 = 34 \end{cases}$

31. $\begin{cases} (x-1)^2 + (y+1)^2 = 2 \\ (x+2)^2 + (y-3)^2 = 3 \end{cases}$

32. $\begin{cases} (x+1)^2 + (y-3)^2 = 4 \\ (x-3)^2 + (y+2)^2 = 2 \end{cases}$

In Exercises 33 to 40, approximate the real number solutions of each system of equations to the nearest ten-thousandth.

33. $\begin{cases} y = 2^x \\ y = x+1 \end{cases}$

34. $\begin{cases} y = \log_2 x \\ y = x-3 \end{cases}$

35. $\begin{cases} y = e^{-x} \\ y = x^2 \end{cases}$

36. $\begin{cases} y = \ln x \\ y = -x+4 \end{cases}$

37. $\begin{cases} y = \sqrt{x} \\ y = \dfrac{1}{x-1} \end{cases}$

38. $\begin{cases} y = \dfrac{6}{x+1} \\ y = \dfrac{x}{x-1} \end{cases}$

39. $\begin{cases} y = |x| \\ y = 2^{-x^2} \end{cases}$

40. $\begin{cases} y = \dfrac{2^x + 2^{-x}}{2} \\ y = \dfrac{2^x - 2^{-x}}{2} \end{cases}$

SUPPLEMENTAL EXERCISES

In Exercises 41 to 46, solve the system of equations for *rational number* ordered pairs.

41. $\begin{cases} y = x^2 + 4 \\ x = y^2 - 24 \end{cases}$

42. $\begin{cases} y = x^2 - 5 \\ x = y^2 - 13 \end{cases}$

43. $\begin{cases} x^2 - 3xy + y^2 = 5 \\ x^2 - xy - 2y^2 = 0 \end{cases}$

(*Hint:* Factor the second equation. Now use the principle of zero products and the substitution principle.)

44. $\begin{cases} x^2 + 2xy - y^2 = 1 \\ x^2 + 3xy + 2y^2 = 0 \end{cases}$

(*Hint:* See Exercise 43.)

45. $\begin{cases} 2x^2 - 4xy - y^2 = 6 \\ 4x^2 - 3xy - y^2 = 6 \end{cases}$

(*Hint:* Subtract the two equations.)

46. $\begin{cases} 3x^2 + 2xy - 5y^2 = 11 \\ x^2 + 3xy + y^2 = 11 \end{cases}$

(*Hint:* Subtract the two equations.)

47. Show that the line $y = mx$ intersects the hyperbola $\dfrac{x^2}{a^2} - \dfrac{y^2}{b^2} = 1$ if and only if $|m| < \left| \dfrac{b}{a} \right|$.

PROJECTS

1. FINDING ZEROS OF A POLYNOMIAL Consider the polynomial $P(x) = x^3 + 2x^2 + Cx - 6$. One zero of this polynomial is the sum of the other two zeros of the polynomial. Find C and the three zeros of $P(x)$.

2. PROVING A GEOMETRY THEOREM Consider the triangle that is shown at the right inscribed in a circle of radius a with one side along the diameter of the circle. Prove that the triangle is a right triangle by completing the following steps.

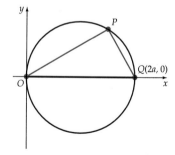

a. Let $y = mx$, $m \geq 0$. Show that the graph of $y = mx$, $m \geq 0$, intersects the circle whose equation is $(x - a)^2 + y^2 = a^2$, $a > 0$, at $P\left(\dfrac{2a}{1 + m^2}, \dfrac{2ma}{1 + m^2}\right)$.

b. Show that the slope of the line through P and $Q(2a, 0)$ is $-1/m$.

c. What is the slope of the line between O and P?

d. Prove that line segment OP is perpendicular to line segment PQ.

e. How can you conclude from the foregoing that triangle OPQ is a right triangle?

7.4 PARTIAL FRACTIONS

An algebraic application of systems of equations is a technique known as *partial fractions.* In Chapter 1, we reviewed the problem of adding two rational expressions. For example,

$$\frac{5}{x - 1} + \frac{1}{x + 2} = \frac{6x + 9}{(x - 1)(x + 2)}$$

Now we will take an opposite approach. That is, given a rational expression, we will find simpler rational expressions whose sum is the given expression. The method by which a more complicated rational expression is written as a sum of rational expressions is called **partial fraction decomposition.** This technique is based on the following theorem.

PARTIAL FRACTION DECOMPOSITION THEOREM

If
$$f(x) = \frac{p(x)}{q(x)}$$

is a rational expression in which the degree of the numerator is less than the degree of the denominator, and $p(x)$ and $q(x)$ have no common factors, then $f(x)$ can be written as a partial fraction decomposition in the form

$$f(x) = f_1(x) + f_2(x) + \cdots + f_n(x)$$

where each $f_i(x)$ has one of the following forms:

$$\frac{A}{(px + q)^m} \quad \text{or} \quad \frac{Bx + C}{(ax^2 + bx + c)^m}$$

The procedure for finding a partial fraction decomposition of a rational expression depends on factorization of the denominator of the rational expression. There are four cases.

Case 1   Nonrepeated Linear Factors

The partial fraction decomposition will contain an expression of the form $A/(x + a)$ for each nonrepeated linear factor of the denominator. Example:

$$\frac{3x - 1}{x(3x + 4)(x - 2)}$$

• **Each linear factor of the denominator occurs only once.**

Partial fraction decomposition:

$$\frac{3x - 1}{x(3x + 4)(x - 2)} = \frac{A}{x} + \frac{B}{3x + 4} + \frac{C}{x - 2}$$

Case 2 Repeated Linear Factors

The partial fraction decomposition will contain an expression of the form

$$\frac{A_1}{(x + a)} + \frac{A_2}{(x + a)^2} + \cdots + \frac{A_m}{(x + a)^m}$$

for each repeated linear factor. Example:

$$\frac{4x + 5}{(x - 2)^2(2x + 1)}$$

- $(x - 2)^2 = (x - 2)(x - 2)$, a repeated linear factor.

Partial fraction decomposition:

$$\frac{4x + 5}{(x - 2)^2(2x + 1)} = \frac{A_1}{x - 2} + \frac{A_2}{(x - 2)^2} + \frac{B}{2x + 1}$$

Case 3 Nonrepeated Quadratic Factors

The partial fraction decomposition will contain an expression of the form

$$\frac{Ax + B}{ax^2 + bx + c}$$

for each quadratic factor irreducible over the real numbers. Example:

$$\frac{x - 4}{(x^2 + x + 1)(x - 4)}$$

- $x^2 + x + 1$ is irreducible over the real numbers.

Partial fraction decomposition:

$$\frac{x - 4}{(x^2 + x + 1)(x - 4)} = \frac{Ax + B}{x^2 + x + 1} + \frac{C}{x - 4}$$

Case 4 Repeated Quadratic Factors

The partial fraction decomposition will contain an expression of the form

$$\frac{A_1x + B_1}{ax^2 + bx + c} + \frac{A_2x + B_2}{(ax^2 + bx + c)^2} + \cdots + \frac{A_mx + B_m}{(ax^2 + bx + c)^m}$$

for each quadratic factor irreducible over the real numbers. Example:

$$\frac{2x}{(x - 2)(x^2 + 4)^2}$$

- $(x^2 + 4)^2$ is a repeated quadratic factor.

Partial fraction decomposition:

$$\frac{2x}{(x - 2)(x^2 + 4)^2} = \frac{A_1x + B_1}{x^2 + 4} + \frac{A_2x + B_2}{(x^2 + 4)^2} + \frac{C}{x - 2}$$

There are various methods for finding the constants of a partial fraction decomposition. One such method is based on a property of polynomials.

EQUALITY OF POLYNOMIALS

If the two polynomials $p(x) = a_nx^n + a_{n-1}x^{n-1} + \cdots + a_1x + a_0$ and $r(x) = b_nx^n + b_{n-1}x^{n-1} + \cdots + b_1x + b_0$ are of degree n, then $p(x) = r(x)$ if and only if $a_0 = b_0, a_1 = b_1, a_2 = b_2, \ldots, a_n = b_n$.

EXAMPLE 1

Find a Partial Fraction Decomposition
Case 1: Nonrepeated Linear Factors

Find a partial fraction decomposition of $\dfrac{x + 11}{x^2 - 2x - 15}$.

Solution

First factor the denominator.

$$x^2 - 2x - 15 = (x + 3)(x - 5)$$

The factors are nonrepeated linear factors. Therefore, the partial fraction decomposition will have the form

$$\frac{x + 11}{(x + 3)(x - 5)} = \frac{A}{x + 3} + \frac{B}{x - 5} \qquad (1)$$

To solve for A and B, multiply each side of the equation by the least common multiple of the denominators, $(x + 3)(x - 5)$.

$$x + 11 = A(x - 5) + B(x + 3)$$
$$x + 11 = (A + B)x + (-5A + 3B) \qquad \bullet \textbf{ Combine like terms.}$$

Using the Equality of Polynomials Theorem, equate coefficients of like powers. The result will be the system of equations

$$\begin{cases} 1 = A + B & \bullet \textbf{ Recall that } x = 1 \cdot x. \\ 11 = -5A + 3B \end{cases}$$

Solving the system of equations for A and B, we have $A = -1$ and $B = 2$. Substituting -1 for A and 2 for B into the form of the partial fraction decomposition (1), we obtain

$$\frac{x + 11}{(x + 3)(x - 5)} = \frac{-1}{x + 3} + \frac{2}{x - 5}$$

You should add the two expressions to verify the equality.

TRY EXERCISE 14, EXERCISE SET 7.4

EXAMPLE 2

Find a Partial Fraction Decomposition
Case 2: Repeated Linear Factors

Find the partial fraction decomposition of $\dfrac{x^2 + 2x + 7}{x(x - 1)^2}$.

Solution

The denominator has one nonrepeated factor and one repeated factor. The partial fraction decomposition will have the form

$$\frac{x^2 + 2x + 7}{x(x - 1)^2} = \frac{A}{x} + \frac{B}{x - 1} + \frac{C}{(x - 1)^2}$$

Continued • ➤

Multiplying each side by the LCD $x(x - 1)^2$, we have

$$x^2 + 2x + 7 = A(x - 1)^2 + B(x - 1)x + Cx$$

Expanding the right side and combining like terms give

$$x^2 + 2x + 7 = (A + B)x^2 + (-2A - B + C)x + A$$

Using the Equality of Polynomials Theorem, equate coefficients of like powers. This will result in the system of equations

$$\begin{cases} 1 = A + B \\ 2 = -2A - B + C \\ 7 = A \end{cases}$$

The solution is $A = 7$, $B = -6$, and $C = 10$. Thus the partial fraction decomposition is

$$\frac{x^2 + 2x + 7}{x(x - 1)^2} = \frac{7}{x} + \frac{-6}{x - 1} + \frac{10}{(x - 1)^2}$$

TRY EXERCISE 22, EXERCISE SET 7.4

EXAMPLE 3   *Find a Partial Fraction Decomposition*
Case 3: Nonrepeated Quadratic Factor

Find the partial fraction decomposition of $\dfrac{3x + 16}{(x - 2)(x^2 + 7)}$.

Solution

Because $(x - 2)$ is a nonrepeated linear factor and $x^2 + 7$ is an irreducible quadratic over the real numbers, the partial fraction decomposition will have the form

$$\frac{3x + 16}{(x - 2)(x^2 + 7)} = \frac{A}{x - 2} + \frac{Bx + C}{x^2 + 7}$$

Multiplying each side by the LCD $(x - 2)(x^2 + 7)$ yields

$$3x + 16 = A(x^2 + 7) + (Bx + C)(x - 2)$$

Expanding the right side and combining like terms, we have

$$3x + 16 = (A + B)x^2 + (-2B + C)x + (7A - 2C)$$

Using the Equality of Polynomials Theorem, equate coefficients of like powers. This will result in the system of equations

$$\begin{cases} 0 = A + B \\ 3 = -2B + C \\ 16 = 7A - 2C \end{cases}$$

• **Think of $3x + 16$ as $0x^2 + 3x + 16$.**

The solution is $A = 2$, $B = -2$, and $C = -1$. Thus the partial fraction decomposition is

$$\frac{3x + 16}{(x - 2)(x^2 + 7)} = \frac{2}{x - 2} + \frac{-2x - 1}{x^2 + 7}$$

TRY EXERCISE 24, EXERCISE SET 7.4

EXAMPLE 4 **Find a Partial Fraction Decomposition**
Case 4: Repeated Quadratic Factors

Find the partial fraction decomposition of $\dfrac{4x^3 + 5x^2 + 7x - 1}{(x^2 + x + 1)^2}$.

Solution

The quadratic factor $(x^2 + x + 1)$ is irreducible over the real numbers and is a repeated factor. The partial fraction decomposition will be of the form

$$\frac{4x^3 + 5x^2 + 7x - 1}{(x^2 + x + 1)^2} = \frac{Ax + B}{x^2 + x + 1} + \frac{Cx + D}{(x^2 + x + 1)^2}$$

Multiplying each side by the LCD $(x^2 + x + 1)^2$ and collecting like terms, we obtain

$$\begin{aligned}
4x^3 + 5x^2 + 7x - 1 &= (Ax + B)(x^2 + x + 1) + Cx + D \\
&= Ax^3 + Ax^2 + Ax + Bx^2 + Bx + B + Cx + D \\
&= Ax^3 + (A + B)x^2 + (A + B + C)x + (B + D)
\end{aligned}$$

Equating coefficients of like powers gives the system of equations

$$\begin{cases}
4 = A \\
5 = A + B \\
7 = A + B + C \\
-1 = \quad B \quad + D
\end{cases}$$

Solving this system, we have $A = 4$, $B = 1$, $C = 2$, and $D = -2$. Thus the partial fraction decomposition is

$$\frac{4x^3 + 5x^2 + 7x - 1}{(x^2 + x + 1)^2} = \frac{4x + 1}{x^2 + x + 1} + \frac{2x - 2}{(x^2 + x + 1)^2}$$

TRY EXERCISE 30, EXERCISE SET 7.4

The Partial Fraction Decomposition Theorem requires that the degree of the numerator be less than the degree of the denominator. If this is *not* the case, use long division to first write the rational expression as a polynomial plus a remainder.

EXAMPLE 5 *Find a Partial Fraction Decomposition When the Degree of the Numerator Exceeds the Degree of the Denominator*

Find the partial fraction decomposition of $F(x) = \dfrac{x^3 - 4x^2 - 19x - 35}{x^2 - 7x}$.

Solution

Because the degree of the denominator is less than the degree of the numerator, use long division first to obtain

$$F(x) = x + 3 + \frac{2x - 35}{x^2 - 7x}$$

The partial fraction decomposition of $\dfrac{2x - 35}{x^2 - 7x}$ will have the form

$$\frac{2x - 35}{x^2 - 7x} = \frac{2x - 35}{x(x - 7)} = \frac{A}{x} + \frac{B}{x - 7}$$

Multiplying each side by $x(x - 7)$ and combining like terms, we have

$$2x - 35 = (A + B)x + (-7A)$$

Equating coefficients of like powers yields

$$\begin{cases} 2 = A + B \\ -35 = -7A \end{cases}$$

The solution of this system is $A = 5$ and $B = -3$. The partial fraction decomposition is

$$\frac{x^3 - 4x^2 - 19x - 35}{x^2 - 7x} = x + 3 + \frac{5}{x} + \frac{-3}{x - 7}$$

TRY EXERCISE 34, EXERCISE SET 7.4

TOPICS FOR DISCUSSION

1. What is the purpose of a partial fraction decomposition?

2. Discuss how the factors of the denominator of a rational expression dictate how a partial fraction decomposition is determined.

3. Discuss the Equality of Polynomials Theorem and how it is used in a partial fraction decomposition.

4. For the rational expression $\dfrac{3x - 1}{x^3 - 2x^2 - x + 2}$, what is the first step you perform to find a partial fraction decomposition? What equation or equations do you solve to find the partial fraction decomposition?

EXERCISE SET 7.4

In Exercises 1 to 10, evaluate the constants A, B, C, and D.

1. $\dfrac{x + 15}{x(x - 5)} = \dfrac{A}{x} + \dfrac{B}{x - 5}$

2. $\dfrac{5x - 6}{x(x + 3)} = \dfrac{A}{x} + \dfrac{B}{x + 3}$

3. $\dfrac{1}{(2x + 3)(x - 1)} = \dfrac{A}{2x + 3} + \dfrac{B}{x - 1}$

4. $\dfrac{6x - 5}{(x + 4)(3x + 2)} = \dfrac{A}{x + 4} + \dfrac{B}{3x + 2}$

5. $\dfrac{x + 9}{x(x - 3)^2} = \dfrac{A}{x} + \dfrac{B}{(x - 3)} + \dfrac{C}{(x - 3)^2}$

6. $\dfrac{2x - 7}{(x + 1)(x - 2)^2} = \dfrac{A}{x + 1} + \dfrac{B}{x - 2} + \dfrac{C}{(x - 2)^2}$

7. $\dfrac{4x^2 + 3}{(x - 1)(x^2 + x + 5)} = \dfrac{A}{x - 1} + \dfrac{Bx + C}{x^2 + x + 5}$

8. $\dfrac{x^2 + x + 3}{(x^2 + 7)(x - 3)} = \dfrac{Ax + B}{x^2 + 7} + \dfrac{C}{x - 3}$

9. $\dfrac{x^3 + 2x}{(x^2 + 1)^2} = \dfrac{Ax + B}{x^2 + 1} + \dfrac{Cx + D}{(x^2 + 1)^2}$

10. $\dfrac{3x^3 + x^2 - x - 5}{(x^2 + 2x + 5)^2} = \dfrac{Ax + B}{x^2 + 2x + 5} + \dfrac{Cx + D}{(x^2 + 2x + 5)^2}$

In Exercises 11 to 36, find the partial fraction decomposition of the given rational expression.

11. $\dfrac{8x + 12}{x(x + 4)}$

12. $\dfrac{x - 14}{x(x - 7)}$

13. $\dfrac{3x + 50}{x^2 - 7x - 18}$

14. $\dfrac{7x + 44}{x^2 + 10x + 24}$

15. $\dfrac{16x + 34}{4x^2 + 16x + 15}$

16. $\dfrac{-15x + 37}{9x^2 - 12x - 5}$

17. $\dfrac{x - 5}{(3x + 5)(x - 2)}$

18. $\dfrac{1}{(x + 7)(2x - 5)}$

19. $\dfrac{x^3 + 3x^2 - 4x - 8}{x^2 - 4}$

20. $\dfrac{x^3 - 13x - 9}{x^2 - x - 12}$

21. $\dfrac{3x^2 + 49}{x(x + 7)^2}$

22. $\dfrac{x - 18}{x(x - 3)^2}$

23. $\dfrac{5x^2 - 7x + 2}{x^3 - 3x^2 + x}$

24. $\dfrac{9x^2 - 3x + 49}{x^3 - x^2 + 10x - 10}$

25. $\dfrac{2x^3 + 9x^2 + 26x + 41}{(x + 3)^2(x^2 + 1)}$

26. $\dfrac{12x^3 - 37x^2 + 48x - 36}{(x - 2)^2(x^2 + 4)}$

27. $\dfrac{3x - 7}{(x - 4)^2}$

28. $\dfrac{5x - 53}{(x - 11)^2}$

29. $\dfrac{3x^3 - x^2 + 34x - 10}{(x^2 + 10)^2}$

30. $\dfrac{2x^3 + 9x + 1}{x^4 + 14x^2 + 49}$

31. $\dfrac{1}{k^2 - x^2}$, where k is a constant

32. $\dfrac{1}{x(k + lx)}$, where k and l are constants

33. $\dfrac{x^3 - x^2 - x - 1}{x^2 - x}$

34. $\dfrac{2x^3 + 5x^2 + 3x - 8}{2x^2 + 3x - 2}$

35. $\dfrac{2x^3 - 4x^2 + 5}{x^2 - x - 1}$

36. $\dfrac{x^4 - 2x^3 - 2x^2 - x + 3}{x^2(x - 3)}$

SUPPLEMENTAL EXERCISES

In Exercises 37 to 42, find the partial fraction decomposition of the given rational expression.

37. $\dfrac{x^2 - 1}{(x - 1)(x + 2)(x - 3)}$

38. $\dfrac{x^2 + x}{x^2(x - 4)}$

39. $\dfrac{-x^4 - 4x^2 + 3x - 6}{x^4(x - 2)}$

40. $\dfrac{3x^2 - 2x - 1}{(x^2 - 1)^2}$

41. $\dfrac{2x^2 + 3x - 1}{x^3 - 1}$

42. $\dfrac{x^3 - 2x^2 + x - 2}{x^4 - x^3 + x - 1}$

There is a short-cut for finding *some* partial fraction decompositions of quadratic polynomials that do not factor over the real numbers. Exercises 43 and 44 give one method and some examples.

43. Show that for real numbers a and b with $a \neq b$,

$$\frac{1}{(b - a)[p(x) + a]} + \frac{1}{(a - b)[p(x) + b]} = \frac{1}{[p(x) + a][p(x) + b]}$$

44. Use the result of Exercise 43 to find the partial fraction decomposition of

a. $\dfrac{1}{(x^2 + 4)(x^2 + 1)}$

b. $\dfrac{1}{(x^2 + 1)(x^2 + 9)}$

c. $\dfrac{1}{(x^2 + x + 1)(x^2 + x + 2)}$

d. $\dfrac{1}{(x^2 + 2x + 4)(x^2 + 2x + 9)}$

PROJECTS

1. Computer algebra systems (CAS) such as *Mathematica* and *Derive* provide computer assistance for partial fraction decompositions. The command in *Mathematica* is **Apart** and the command in *Derive* is **Expand**. Here is an example of using each of these programs to find the partial fraction decomposition of $\dfrac{x^3 - 4x^2 - 19x - 35}{x^2 - 7x}$.

Derive

Start up the *Derive* program. The menu bar on the bottom of the screen should begin with **Author**. If it does not, press the ⎡Esc⎤ key until it does. Now type

$$\boxed{\text{A((x^3-4x^2-19x-35)/(x^2-7x))}}\ \boxed{\text{Enter}}\ \text{E}\ \boxed{\text{Enter}}$$

The **A** allows you to input the expression into the computer. The **E** after the first ⎡Enter⎤ is the command **Expand**, which performs the partial fraction decomposition. The result is displayed as $\dfrac{3}{7 - x} + x + \dfrac{5}{x} + 3$.

Mathematica

Start up the *Mathematica* program. Now type

$$\boxed{\text{Apart[(x^3-4x^2-19x-35)/(x^2-7x)]}}\ \boxed{\text{Enter}}$$

The brackets, **[** and **]**, are not interchangeable with parentheses, **(** and **)**. The result is displayed as $3 - \dfrac{3}{-7 + x} + \dfrac{5}{x} + x$.

Use a CAS program to find partial fraction decompositions for some of the exercises in this section. Include a printout of your work.

SECTION

7.5 INEQUALITIES IN TWO VARIABLES AND SYSTEMS OF INEQUALITIES

Two examples of inequalities in two variables are

$$2x + 3y > 6 \quad \text{and} \quad xy \le 1$$

A solution of an inequality in two variables is an ordered pair (x, y) that satisfies the inequality. For example, $(-2, 4)$ is a solution of the first inequality because $2(-2) + 3(4) > 6$. The ordered pair $(2, 1)$ is not a solution of the second inequality because $(2)(1) \not\le 1$.

The **solution set of an inequality** in two variables is the set of all ordered pairs that satisfy the inequality. The **graph** of an inequality is the graph of the solution set.

To sketch the graph of an inequality, first replace the inequality symbol by an equality sign and sketch the graph of the equation. Use a dashed graph for $<$ or $>$ to indicate that the curve is not part of the solution set. Use a solid graph for $\le$ or $\ge$ to show that the graph *is* part of the solution set.

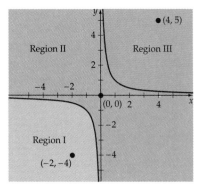

$$xy \geq 1$$

Figure 7.26

It is important to test an ordered pair in each region of the plane defined by the graph. If the ordered pair satisfies the inequality, shade that entire region. Do this for each region into which the graph divides the plane. For example, consider the inequality $xy \geq 1$. **Figure 7.26** shows the three regions of the plane defined by this inequality. Because the inequality is $\geq$, a solid graph is used.

Choose an ordered pair in each of the three regions and determine whether that ordered pair satisfies the inequality. In Region I, choose a point, say $(-2, -4)$. Because $(-2)(-4) \geq 1$, Region I is part of the solution set. In Region II, choose a point, say $(0, 0)$. Because $0 \cdot 0 \ngeq 1$, Region II is not part of the solution set. In Region III, choose $(4, 5)$. Because $4 \cdot 5 \geq 1$, Region III is part of the solution set.

You may choose the coordinates of any point not on the graph of the equation as a test ordered pair; $(0, 0)$ is usually a good choice.

EXAMPLE 1 *Graph a Linear Inequality*

Graph: $3x + 4y > 12$

Solution

Graph the line $3x + 4y = 12$ using a dashed line.

Test the ordered pair $(0, 0)$: $3(0) + 4(0) = 0 \ngtr 12$

Because $(0, 0)$ does not satisfy the inequality, do not shade this region.

Test the ordered pair $(2, 3)$: $3(2) + 4(3) = 18 > 12$

Because $(2, 3)$ satisfies the inequality, the half-plane that includes $(2, 3)$ is the solution set. See **Figure 7.27**.

TRY EXERCISE 6, EXERCISE SET 7.5

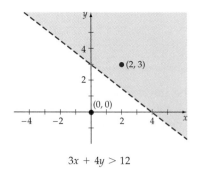

$$3x + 4y > 12$$

Figure 7.27

In general, the solution set of a *linear inequality in two variables* will be one of the regions of the plane separated by a line. Each region is called a **half-plane.**

EXAMPLE 2 *Graph a Nonlinear Inequality*

Graph: $y \leq x^2 + 2x - 3$

Solution

Graph the parabola $y = x^2 + 2x - 3$ using a solid curve.

Test the ordered pair $(0, 0)$: $0 \nleq 0^2 + 2(0) - 3$

Because $(0, 0)$ does not satisfy the inequality, do not shade this region.

Test the ordered pair $(3, 2)$: $2 \leq (3)^2 + 2(3) - 3$

Because $(3, 2)$ satisfies the inequality, shade this region of the plane. See **Figure 7.28.**

TRY EXERCISE 12, EXERCISE SET 7.5

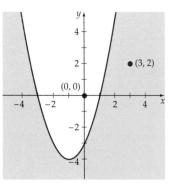

$$y \leq x^2 + 2x - 3$$

Figure 7.28

EXAMPLE 3   Graph an Absolute Value Inequality

Graph: $y \geq |x| + 1$

Solution

Graph the equation $y = |x| + 1$ using a solid graph.

$$\text{Test the ordered pair } (0,0): \quad 0 \not\geq |0| + 1$$

Because $0 \not\geq 1$, $(0,0)$ does not belong to the solution set. Do not shade that portion of the plane that contains $(0,0)$.

$$\text{Test the ordered pair } (0,4): \quad 4 \geq |0| + 1$$

Because $(0,4)$ satisfies the inequality, shade this region. See **Figure 7.29**.

TRY EXERCISE 20, EXERCISE SET 7.5

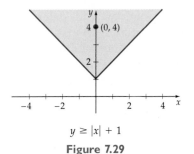

$y \geq |x| + 1$

Figure 7.29

SYSTEM OF INEQUALITIES IN TWO VARIABLES

The **solution set of a system of inequalities** is the intersection of the solution sets of the individual inequalities. To graph the solution set of a system of inequalities, first graph the solution set of each inequality. The solution set of the system of inequalities is the region of the plane represented by the intersection of the shaded regions.

EXAMPLE 4   Graph a System of Linear Inequalities

Graph the solution set of the system of inequalities

$$\begin{cases} 3x - 2y > 6 \\ 2x - 5y \leq 10 \end{cases}$$

Solution

Graph the line $3x - 2y = 6$ using a dashed line. Test the ordered pair $(0,0)$. Because $3(0) - 2(0) \not> 6$, $(0,0)$ does not belong to the solution set. Do not shade the region that contains $(0,0)$. Instead, shade the region below and to the right of the graph of $3x - 2y = 6$; because any ordered pair from this region satisfies $3x - 2y > 6$.

Graph the line $2x - 5y = 10$ using a solid line. Test the ordered pair $(0,0)$. Because $2(0) - 5(0) \leq 10$, shade the region that contains $(0,0)$.

The solution set is the region of the plane represented by the intersection of the solution sets of the individual inequalities. See **Figure 7.30**.

TRY EXERCISE 26, EXERCISE SET 7.5

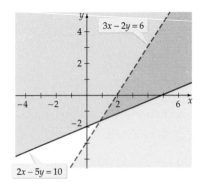

Figure 7.30

EXAMPLE 5 *Graph a Nonlinear System of Inequalities*

Graph the solution set of the system of inequalities

$$\begin{cases} x^2 - y^2 \leq 9 \\ 2x + 3y > 12 \end{cases}$$

Solution

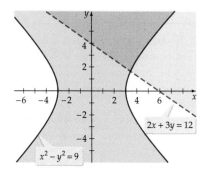

Figure 7.31

Graph the hyperbola $x^2 - y^2 = 9$ by using a solid graph. Test the ordered pair $(0, 0)$. Because $0^2 - 0^2 \leq 9$, shade the region containing the origin. By choosing points in the other regions, you should show that those regions are not part of the solution set.

Graph the line $2x + 3y = 12$ by using a dashed graph. Test the ordered pair $(0, 0)$. Because $2(0) + 3(0) \not> 12$, do not shade the half-plane below the line. Testing the ordered pair $(4, 4)$ will show that we need to shade the half-plane above the line $2x + 3y = 12$.

The solution set is the region of the plane represented by the intersection of the solution sets of the individual inequalities. This intersection is shown by the dark color in **Figure 7.31**.

> **TRY EXERCISE 36, EXERCISE SET 7.5**

EXAMPLE 6 *Identify a System of Inequalities with No Solution*

Graph the solution set of the system of inequalities

$$\begin{cases} x^2 + y^2 \leq 16 \\ x^2 - y^2 \geq 36 \end{cases}$$

Solution

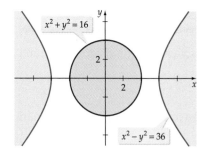

Figure 7.32

Graph the circle $x^2 + y^2 = 16$ by using a solid graph. Test the ordered pair $(0, 0)$. Because $0^2 + 0^2 \leq 16$, shade the inside of the circle.

Graph the hyperbola $x^2 - y^2 = 36$ by using a solid graph. Use ordered pairs from each of the regions defined by the hyperbola to determine that the solution of $x^2 - y^2 > 36$ consists of the region to the right of the right branch of the hyperbola and the region to the left of the left branch.

Because the solution sets of the inequalities do not intersect, the system has no solution. The solution set is the empty set. See **Figure 7.32**.

> **TRY EXERCISE 40, EXERCISE SET 7.5**

EXAMPLE 7   Graph a System of Four Inequalities

Graph the solution set of the system of inequalities

$$\begin{cases} 2x - 3y \le 2 \\ 3x + 4y \ge 12 \\ x \ge -1, \, y \ge 2 \end{cases}$$

Solution

First graph the inequalities $x \ge -1$ and $y \ge 2$. Because $x \ge -1$ and $y \ge 2$, the solution set for this system will be on or above the line $y = 2$ and on or to the right of the line $x = -1$. See **Figure 7.33.**

Graph the solution set of $2x - 3y = 2$ by using a solid graph. Because $2(0) - 3(0) \le 2$, shade the region above the line.

Graph the solution set of $3x + 4y = 12$ by using a solid graph. Test an ordered pair, say $(3, 3)$, to determine that we need to shade above the line $3x + 4y = 12$.

The solution set of the system of equations is the region where the graphs of the solution sets of all four inequalities intersect. This intersection is indicated by the dark color in **Figure 7.34.**

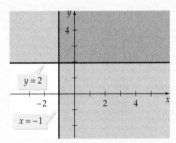

Figure 7.33

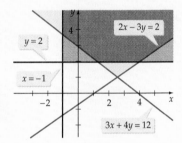

Figure 7.34

TRY EXERCISE 42, EXERCISE SET 7.5

TOPICS FOR DISCUSSION

1. Does the graph of a linear inequality in two variables represent the graph of a function? Why or why not?

2. What is a half-plane?

3. Is it possible for a system of inequalities to have no solution? If so, give an example. If not, explain why not.

4. Let $A = \{(x, y) \mid x + y > 5\}$ and let $B = \{(x, y) \mid x - y < 3\}$. What is the significance of $A \cap B$?

5. Suppose a company makes two types of frying pans: regular and nonstick. Each week the company plans on making at least twice as many nonstick

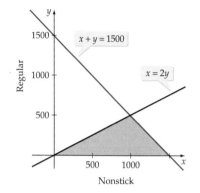

pans as regular. Production facilities are such that the company can make a a maximum of 1500 pans per week. Letting x represent the number of nonstick pans and y represent the number of regular pans, the system of inequalities

$$\begin{cases} x \geq 2y \\ x + y \leq 1500 \\ x \geq 0, y \geq 0 \end{cases}$$

represents this situation. The graph of the solution set is shown at the left. Explain the meaning of the solution set (shown shaded) in the context of this problem.

EXERCISE SET 7.5

In Exercises 1 to 22, sketch the graph of each inequality.

1. $y \leq -2$

2. $x + y > -2$

3. $y \geq 2x + 3$

4. $y < -2x + 1$

5. $2x - 3y < 6$

6. $3x + 4y \leq 4$

7. $4x + 3y \leq 12$

8. $5x - 2y < 8$

9. $y < x^2$

10. $x > y^2$

11. $y \geq x^2 - 2x - 3$

12. $y < 2x^2 - x - 3$

13. $(x - 2)^2 + (y - 1)^2 < 16$

14. $(x + 2)^2 + (y - 3)^2 > 25$

15. $\dfrac{(x - 3)^2}{9} - \dfrac{(y + 1)^2}{16} > 1$ **16.** $\dfrac{(x + 1)^2}{25} - \dfrac{(y - 3)^2}{16} \leq 1$

17. $4x^2 + 9y^2 - 8x + 18y \geq 23$

18. $25x^2 - 16y^2 - 100x - 64y < 64$

19. $y < 2^{x-1}$ **20.** $y > \log_3 x$

21. $y \leq \log_2 (x - 1)$ **22.** $y > 3^x + 1$

In Exercises 23 to 46, sketch the graph of the solution set of each system of inequalities.

23. $\begin{cases} 1 \leq x < 3 \\ -2 < y \leq 4 \end{cases}$ **24.** $\begin{cases} -2 < x < 4 \\ \quad\quad y \geq -1 \end{cases}$

25. $\begin{cases} 3x + 2y \geq 1 \\ x + 2y < -1 \end{cases}$ **26.** $\begin{cases} 2x - 5y < -6 \\ 3x + y < 8 \end{cases}$

27. $\begin{cases} 2x - y \geq -4 \\ 4x - 2y \leq -17 \end{cases}$ **28.** $\begin{cases} 4x + 2y > 5 \\ 6x + 3y > 10 \end{cases}$

29. $\begin{cases} 4x - 3y < 14 \\ 2x + 5y \leq -6 \end{cases}$ **30.** $\begin{cases} 3x + 5y \geq -8 \\ 2x - 3y \geq 1 \end{cases}$

31. $\begin{cases} y < 2x + 3 \\ y > 2x - 2 \end{cases}$ **32.** $\begin{cases} y > 3x + 1 \\ y < 3x - 2 \end{cases}$

33. $\begin{cases} y < 2x - 1 \\ y \geq x^2 + 3x - 7 \end{cases}$ **34.** $\begin{cases} y \leq 2x + 7 \\ y > x^2 + 3x + 1 \end{cases}$

35. $\begin{cases} x^2 + y^2 \leq 49 \\ 9x^2 + 4y^2 \geq 36 \end{cases}$ **36.** $\begin{cases} y < 2x - 1 \\ y > x^2 - 2x + 2 \end{cases}$

37. $\begin{cases} (x - 1)^2 + (y + 1)^2 \leq 16 \\ (x - 1)^2 + (y + 1)^2 \geq 4 \end{cases}$

38. $\begin{cases} (x + 2)^2 + (y - 3)^2 > 25 \\ (x + 2)^2 + (y - 3)^2 < 16 \end{cases}$

39. $\begin{cases} \dfrac{(x - 4)^2}{16} - \dfrac{(y + 2)^2}{9} > 1 \\ \dfrac{(x - 4)^2}{25} + \dfrac{(y + 2)^2}{9} < 1 \end{cases}$

40. $\begin{cases} \dfrac{(x + 1)^2}{36} + \dfrac{(y - 2)^2}{25} < 1 \\ \dfrac{(x + 1)^2}{25} + \dfrac{(y - 2)^2}{36} < 1 \end{cases}$

41. $\begin{cases} 2x - 3y \geq -5 \\ x + 2y \leq 7 \\ x \geq -1, y \geq 0 \end{cases}$ **42.** $\begin{cases} 5x + y \leq 9 \\ 2x + 3y \leq 14 \\ x \geq -2, y \geq 2 \end{cases}$

43. $\begin{cases} 3x + 2y \geq 14 \\ x + 3y \geq 14 \\ x \leq 10, y \leq 8 \end{cases}$ **44.** $\begin{cases} 4x + y \geq 13 \\ 3x + 2y \geq 16 \\ x \leq 15, y \leq 12 \end{cases}$

45. $\begin{cases} 3x + 4y \leq 12 \\ 2x + 5y \leq 10 \\ x \geq 0, y \geq 0 \end{cases}$ **46.** $\begin{cases} 5x + 3y \leq 15 \\ x + 4y \leq 8 \\ x \geq 0, y \geq 0 \end{cases}$

SUPPLEMENTAL EXERCISES

In Exercises 47 to 58, sketch the graph of the inequality.

47. $y < |x|$

48. $y \geq |2x - 4|$

49. $|y| \geq |x|$

50. $|y| \leq |x - 1|$

51. $|x + y| \leq 1$

52. $|x - y| > 1$

53. $|x| + |y| \leq 1$

54. $|x| - |y| > 1$

55. $y > [\![x]\!]$, where $[\![x]\!]$ is the greatest integer function

56. $y > x - [\![x]\!]$, where $[\![x]\!]$ is the greatest integer function

57. Sketch the graphs of $xy > 1$ and $y > 1/x$. Note that the two graphs are not the same, yet the second inequality can be derived from the first by dividing each side by x. Explain.

58. Sketch the graph of $x/y < 1$ and the graph of $x < y$. Note that the two graphs are not the same, yet the second inequality can be derived from the first by multiplying each side by y. Explain.

PROJECTS

1. **A PARALLELOGRAM COORDINATE SYSTEM** The xy-coordinate system described in this chapter consisted of two coordinate lines that intersected at right angles. It is not necessary that coordinate lines intersect at right angles for a coordinate system to exist. Draw two coordinate lines that intersect at 0 for each line but for which the angle between the two axes is 45°. You now have a *parallelogram* coordinate system rather than a *rectangular* coordinate system. Explain the last sentence. Now experiment in this system. For example, is the graph of $3x + 4y = 12$ a straight line in the *parallelogram* coordinate system? In a parallelogram coordinate system, is the graph of $y = x^2$ a parabola?

SECTION 7.6 LINEAR PROGRAMMING

Consider a business analyst who is trying to maximize the profit from the production of a product or an engineer who is trying to minimize the amount of energy an electrical circuit needs to operate. Generally, problems that seek to maximize or minimize a situation are called **optimization problems.** One strategy for solving certain of these problems was developed in the 1940s and is called **linear programming.**

A linear programming problem involves a **linear objective function,** which is the function that must be maximized or minimized. This objective function is subject to some **constraints,** which are inequalities, or equations that restrict the values of the variables. To illustrate these concepts, suppose a manufacturer produces two types of computer monitors: monochrome and color. Past sales experience shows that at least twice as many monochrome monitors are sold as color monitors. Suppose further that the manufacturing plant is capable of producing 12 monitors per day. Let x represent the number of monochrome monitors produced, and let y represent the number of color monitors produced. Then

$$\begin{cases} x \geq 2y \\ x + y \leq 12 \end{cases}$$ • **These are the constraints.**

These two inequalities place a constraint, or restriction, on the manufacturer. For example, the manufacturer cannot produce 5 color monitors, because that would require producing at least 10 monochrome monitors, and $5 + 10 \nleq 12$.

Suppose a profit of \$50 is earned on each monochrome monitor sold and \$75 is earned on each color monitor sold. Then the manufacturer's profit, P, is given by the equation

$$P = 50x + 75y \qquad \bullet \textbf{ Objective function}$$

The equation $P = 50x + 75y$ defines the objective function. The goal of this linear programming problem is to determine how many of each monitor should be produced to maximize the manufacturer's profit and at the same time satisfy the constraints.

Because the manufacturer cannot produce fewer than zero units of either monitor, there are two other implied constraints, $x \geq 0$ and $y \geq 0$. Our linear programming problem now looks like

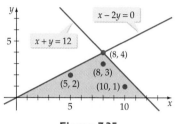

Figure 7.35

Objective function: $P = 50x + 75y$

Constraints: $\begin{cases} x - 2y \geq 0 \\ x + y \leq 12 \\ x \geq 0, y \geq 0 \end{cases}$

To solve this problem, graph the solution set of the constraints. The solution set of the constraints is called the **set of feasible solutions.** Ordered pairs in this set are used to evaluate the objective function to determine which ordered pair maximizes the profit. For example, $(5, 2)$, $(8, 3)$, and $(10, 1)$ are three ordered pairs in the set. See **Figure 7.35.** For these ordered pairs, the profit would be

$$P = 50(5) + 75(2) \ = 400 \qquad \bullet \textbf{ x = 5, y = 2}$$
$$P = 50(8) + 75(3) \ = 625 \qquad \bullet \textbf{ x = 8, y = 3}$$
$$P = 50(10) + 75(1) = 575 \qquad \bullet \textbf{ x = 10, y = 1}$$

It would be impossible to check every ordered pair in the set of feasible solutions to find which maximizes profit. Fortunately, we can find that ordered pair by solving the objective function $P = 50x + 75y$ for y.

$$y = -\frac{2}{3}x + \frac{P}{75}$$

In this form, the objective function is a linear equation whose graph has slope $-2/3$ and y-intercept $P/75$. If P is as large as possible (P a maximum), then the y-intercept will be as large as possible. Thus the maximum profit will occur on the line that has a slope of $-2/3$ and has the largest possible y-intercept and intersects the set of feasible solutions.

From **Figure 7.36,** the largest possible y-intercept occurs when the line passes through the point with coordinates $(8, 4)$. At this point, the profit is

$$P = 50(8) + 75(4) = 700$$

The manufacturer will maximize profit by producing 8 monochrome monitors and 4 color monitors each day. The profit will be \$700 per day.

In general, the goal of any linear programming problem is to maximize or minimize the objective function, subject to the constraints. Minimization problems occur, for example, when a manufacturer wants to minimize the cost of operations.

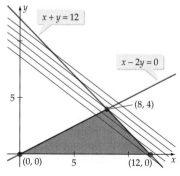

Figure 7.36

Suppose that a cost minimization problem results in the following objective function and constraints.

$$\text{Objective function:}\quad C = 3x + 4y$$

$$\text{Constraints:}\quad \begin{cases} x + y \geq 1 \\ 2x - y \leq 5 \\ x + 2y \leq 10 \\ x \geq 0,\, y \geq 0 \end{cases}$$

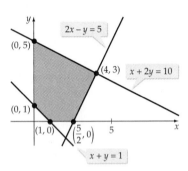

Figure 7.37

Figure 7.37 is the graph of the solution set of the constraints. The task is to find the ordered pair that satisfies all the constraints and that will give the smallest value of C. We again could solve the objective function for y and, because we want to minimize C, find the smallest y-intercept. However, a theorem from linear programming simplifies our task even more. The proof of this theorem, omitted here, is based on the techniques we used to solve our examples.

FUNDAMENTAL LINEAR PROGRAMMING THEOREM

If an objective function has an optimal solution, then that solution will be at a vertex of the set of feasible solutions.

Following is a list of the values of C at the vertices. The minimum value of the objective function occurs at the point whose coordinates are $(1, 0)$.

(x, y)	$C = 3x + 4y$	
$(1, 0)$	$C = 3(1) + 4(0) = 3$	• **Minimum**
$\left(\dfrac{5}{2}, 0\right)$	$C = 3\left(\dfrac{5}{2}\right) + 4(0) = 7.5$	
$(4, 3)$	$C = 3(4) + 4(3) = 24$	• **Maximum**
$(0, 5)$	$C = 3(0) + 4(5) = 20$	
$(0, 1)$	$C = 3(0) + 4(1) = 4$	

The maximum value of the objective function can also be determined from the list. It occurs at $(4, 3)$.

It is important to realize that the maximum or minimum value of an objective function depends on the objective function and on the set of feasible solutions. For example, using the same set of feasible solutions as in **Figure 7.37** but changing the objective function to $C = 2x + 5y$ changes the maximum value of C to 25 at the ordered pair $(0, 5)$. You should verify this result by making a list similar to the one shown above.

EXAMPLE 1 *Solve a Minimization Problem*

Minimize the objective function $C = 4x + 7y$ with the constraints

$$\begin{cases} 3x + y \geq 6 \\ x + y \geq 4 \\ x + 3y \geq 6 \\ x \geq 0,\, y \geq 0 \end{cases}$$

Solution

Determine the set of feasible solutions by graphing the solution set of the inequalities. See **Figure 7.38.** Note that in this instance the set of feasible solutions is an unbounded set.

Find the vertices of the region by solving the following systems of equations. These systems are formed by the equations of the lines that intersect to form a vertex of the set of feasible solutions.

$$\begin{cases} 3x + y = 6 \\ x + y = 4 \end{cases} \quad \begin{cases} x + 3y = 6 \\ x + y = 4 \end{cases}$$

The solutions of the two systems are $(1, 3)$ and $(3, 1)$, respectively. The points $(0, 6)$ and $(6, 0)$ are the vertices on the y- and x-axes.

Evaluate the objective function at each of the four vertices of the set of feasible solutions.

$$\begin{array}{ll} (x, y) & C = 4x + 7y \\ (0, 6) & C = 4(0) + 7(6) = 42 \\ (1, 3) & C = 4(1) + 7(3) = 25 \\ (3, 1) & C = 4(3) + 7(1) = 19 \quad \bullet \textbf{ Minimum} \\ (6, 0) & C = 4(6) + 7(0) = 24 \end{array}$$

The minimum value of the objective function is 19 at $(3, 1)$.

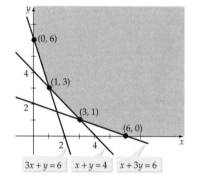

Figure 7.38

$3x + y = 6$ $\quad$ $x + y = 4$ $\quad$ $x + 3y = 6$

TRY EXERCISE 12, EXERCISE SET 7.6

Linear programming can be used to determine the best allocation of the resources available to a company. In fact, the word *programming* refers to a "program to allocate resources."

EXAMPLE 2   *Solve an Applied Minimization Problem*

A manufacturer of animal food makes two grain mixtures, G_1 and G_2. Each kilogram of G_1 contains 300 grams of vitamins, 400 grams of protein, and 100 grams of carbohydrate. Each kilogram of G_2 contains 100 grams of vitamins, 300 grams of protein, and 200 grams of carbohydrate. Minimum nutritional guidelines require that a feed mixture made from these grains contain at least 900 grams of vitamins, 2200 grams of protein, and 800 grams of carbohydrate. G_1 costs $2.00 per kilogram to produce, and G_2 costs $1.25 per kilogram to produce. Find the number of kilograms of each grain mixture that should be produced to minimize cost.

Solution

Let

$$x = \text{the number of kilograms of } G_1$$
$$y = \text{the number of kilograms of } G_2$$

The objective function is the cost function $C = 2x + 1.25y$.

Continued •➤

Because x kilograms of G_1 contain $300x$ grams of vitamins and y kilograms of G_2 contain $100y$ grams of vitamins, the total amount of vitamins contained in x kilograms of G_1 and y kilograms of G_2 is $300x + 100y$. At least 900 grams of vitamins are necessary, so $300x + 100y \geq 900$. Following similar reasoning, we have the constraints

$$\begin{cases} 300x + 100y \geq 900 \\ 400x + 300y \geq 2200 \\ 100x + 200y \geq 800 \\ x \geq 0, y \geq 0 \end{cases}$$

Two of the vertices of the set of feasible solutions (see **Figure 7.39**) can be found by solving two systems of equations. These systems are formed by the equations of the lines that intersect to form a vertex of the set of feasible solutions.

$$\begin{cases} 300x + 100y = 900 \\ 400x + 300y = 2200 \end{cases}$$ • **The vertex is (1, 6).**

$$\begin{cases} 100x + 200y = 800 \\ 400x + 300y = 2200 \end{cases}$$ • **The vertex is (4, 2).**

The vertices on the x- and y-axes are the x- and y-intercepts $(8, 0)$ and $(0, 9)$.

Substitute the coordinates of the vertices into the objective function.

$$
\begin{aligned}
&(x, y) \quad C = 2x + 1.25y \\
&(0, 9) \quad C = 2(0) + 1.25(9) = 11.25 \\
&(1, 6) \quad C = 2(1) + 1.25(6) = 9.50 \qquad \bullet \text{ Minimum} \\
&(4, 2) \quad C = 2(4) + 1.25(2) = 10.50 \\
&(8, 0) \quad C = 2(8) + 1.25(0) = 16.00
\end{aligned}
$$

The minimum value of the objective function is \$9.50. It occurs when the company produces a feed mixture that contains 1 kilogram of G_1 and 6 kilograms of G_2.

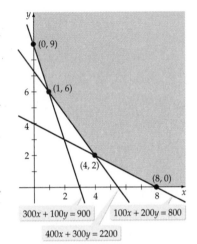

Figure 7.39

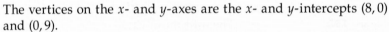

TRY EXERCISE 22, EXERCISE SET 7.6

EXAMPLE 3   *Solve an Applied Maximization Problem*

A chemical firm produces two types of industrial solvents, S_1 and S_2. Each solvent is a mixture of three chemicals. Each kiloliter of S_1 requires 12 liters of chemical 1, 9 liters of chemical 2, and 30 liters of chemical 3. Each kiloliter of S_2 requires 24 liters of chemical 1, 5 liters of chemical 2, and 30 liters of chemical 3. The profit per kiloliter of S_1 is \$100, and the profit per kiloliter of S_2 is \$85. The inventory of the company shows 480 liters of chemical 1, 180 liters of chemical 2, and 720 liters of chemical 3. Assuming the company can sell all the solvent it makes, find the number of kiloliters of each solvent that the company should make to maximize profit.

Solution

Let

$$x = \text{the number of kiloliters of } S_1$$
$$y = \text{the number of kiloliters of } S_2$$

The objective function is the profit function $P = 100x + 85y$.

Because x kiloliters of S_1 require $12x$ liters of chemical 1, and y kiloliters of S_2 require $24y$ liters of chemical 1, the total amount of chemical 1 needed is $12x + 24y$. There are 480 liters of chemical 1 in inventory, so $12x + 24y \leq 480$. Following similar reasoning, we have the constraints

$$\begin{cases} 12x + 24y \leq 480 \\ 9x + 5y \leq 180 \\ 30x + 30y \leq 720 \\ x \geq 0, \, y \geq 0 \end{cases}$$

Two of the vertices of the set of feasible solutions (see **Figure 7.40**) can be found by solving two systems of equations. These systems are formed by the equations of the lines that intersect to form a vertex of the set of feasible solutions.

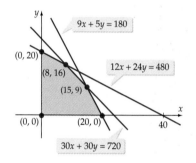

Figure 7.40

$$\begin{cases} 12x + 24y = 480 \\ 30x + 30y = 720 \end{cases} \qquad \bullet \text{ The vertex is } (8, 16).$$

$$\begin{cases} 9x + 5y = 180 \\ 30x + 30y = 720 \end{cases} \qquad \bullet \text{ The vertex is } (15, 9).$$

The vertices on the x- and y-axes are the x- and y-intercepts, $(20, 0)$ and $(0, 20)$.

Substitute the coordinates of the vertices into the objective function.

(x, y)	$P = 100x + 85y$	
$(0, 20)$	$P = 100(0) + 85(20) = 1700$	
$(8, 16)$	$P = 100(8) + 85(16) = 2160$	
$(15, 9)$	$P = 100(15) + 85(9) = 2265$	$\bullet$ **Maximum**
$(20, 0)$	$P = 100(20) + 85(0) = 2000$	

The maximum value of the objective function is \$2265 when the company produces 15 kiloliters of S_1 and 9 kiloliters of S_2.

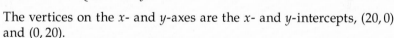

Topics for Discussion

1. What is an optimization problem? Give an example of a situation in which optimization may be the goal.

2. What is a constraint for a linear programming problem? Explain what type of condition might be a constraint for the situation you gave in Exercise 1.

3. What is the objective function for a linear programming problem? Explain what the objective function might be for the situation you gave in Exercise 1.

4. What is the set of feasible solutions for a linear programming problem?

5. If a linear programming problem has an optimal solution, where in the set of feasible solutions will that optimal solution occur?

EXERCISE SET 7.6

In Exercises 1 to 20, solve the linear programming problem. Assume $x \geq 0$ and $y \geq 0$.

1. Minimize $C = 4x + 2y$ with the constraints
$$\begin{cases} x + y \geq 7 \\ 4x + 3y \geq 24 \\ x \leq 10, \ y \leq 10 \end{cases}$$

2. Minimize $C = 5x + 4y$ with the constraints
$$\begin{cases} 3x + 4y \geq 32 \\ x + 4y \geq 24 \\ x \leq 12, \ y \leq 15 \end{cases}$$

3. Maximize $C = 6x + 7y$ with the constraints
$$\begin{cases} x + 2y \leq 16 \\ 5x + 3y \leq 45 \end{cases}$$

4. Maximize $C = 6x + 5y$ with the constraints
$$\begin{cases} 2x + 3y \leq 27 \\ 7x + 3y \leq 42 \end{cases}$$

5. Minimize $C = 5x + 6y$ with the constraints
$$\begin{cases} 4x - 3y \leq 2 \\ 2x + 3y \geq 10 \end{cases}$$

6. Maximize $C = 4x + 5y$ with the constraints
$$\begin{cases} 2x - y \leq 0 \\ 0 \leq y \leq 10 \\ 0 \leq x \leq 10 \end{cases}$$

7. Maximize $C = x + 6y$ with the constraints
$$\begin{cases} 5x + 8y \leq 120 \\ 7x + 16y \leq 192 \end{cases}$$

8. Minimize $C = 4x + 5y$ with the constraints
$$\begin{cases} x + 3y \geq 30 \\ 3x + 4y \geq 60 \end{cases}$$

9. Minimize $C = 4x + y$ with the constraints
$$\begin{cases} 3x + 5y \geq 120 \\ x + y \geq 32 \end{cases}$$

10. Maximize $C = 7x + 2y$ with the constraints
$$\begin{cases} x + 3y \leq 108 \\ 7x + 4y \leq 280 \end{cases}$$

11. Maximize $C = 2x + 7y$ with the constraints
$$\begin{cases} x + y \leq 10 \\ x + 2y \leq 16 \\ 2x + y \leq 16 \end{cases}$$

12. Minimize $C = 4x + 3y$ with the constraints
$$\begin{cases} 2x + y \geq 8 \\ 2x + 3y \geq 16 \\ x + 3y \geq 11 \\ x \leq 20, \ y \leq 20 \end{cases}$$

13. Minimize $C = 3x + 2y$ with the constraints
$$\begin{cases} 3x + y \geq 12 \\ 2x + 7y \geq 21 \\ x + y \geq 8 \end{cases}$$

14. Maximize $C = 2x + 6y$ with the constraints
$$\begin{cases} x + y \leq 12 \\ 3x + 4y \leq 40 \\ x + 2y \leq 18 \end{cases}$$

15. Maximize $C = 3x + 4y$ with the constraints
$$\begin{cases} 2x + y \leq 10 \\ 2x + 3y \leq 18 \\ x - y \leq 2 \end{cases}$$

16. Minimize $C = 3x + 7y$ with the constraints
$$\begin{cases} x + y \geq 9 \\ 3x + 4y \geq 32 \\ x + 2y \geq 12 \end{cases}$$

17. Minimize $C = 3x + 2y$ with the constraints
$$\begin{cases} x + 2y \geq 8 \\ 3x + y \geq 9 \\ x + 4y \geq 12 \end{cases}$$

18. Maximize $C = 4x + 5y$ with the constraints

$$\begin{cases} 3x + 4y \le 250 \\ x + y \le 75 \\ 2x + 3y \le 180 \end{cases}$$

19. Maximize $C = 6x + 7y$ with the constraints

$$\begin{cases} x + 2y \le 900 \\ x + y \le 500 \\ 3x + 2y \le 1200 \end{cases}$$

20. Minimize $C = 11x + 16y$ with the constraints

$$\begin{cases} x + 2y \ge 45 \\ x + y \ge 40 \\ 2x + y \ge 45 \end{cases}$$

21. MAXIMIZE PROFIT A farmer is planning to raise wheat and barley. Each acre of wheat yields a profit of $50, and each acre of barley yields a profit of $70. To sow the crop, two machines, a tractor and tiller, are rented. The tractor is available for 200 hours, and the tiller is available for 100 hours. Sowing an acre of barley requires 3 hours of tractor time and 2 hours of tilling. Sowing an acre of wheat requires 4 hours of tractor time and 1 hour of tilling. How many acres of each crop should be planted to maximize the farmer's profit?

22. MINIMIZE COST An ice cream supplier has two machines that produce vanilla and chocolate ice cream. To meet one of its contractual obligations, the company must produce at least 60 gallons of vanilla ice cream and 100 gallons of chocolate ice cream per hour. One machine makes 4 gallons of vanilla and 5 gallons of chocolate ice cream per hour. The second machine makes 3 gallons of vanilla and 10 gallons of chocolate ice cream per hour. It costs $28 per hour to run machine 1 and $25 per hour to run machine 2. How many hours should each machine be operated to fulfill the contract at the least expense?

23. MAXIMIZE PROFIT A manufacturer makes two types of golf clubs: a starter model and a professional model. The starter model requires 4 hours in the assembly room and 1 hour in the finishing room. The professional model requires 6 hours in the assembly room and 1 hour in the finishing room. The total number of hours available in the assembly room is 108. There are 24 hours available in the finishing room. The profit for each starter model is $35, and the profit for each professional model is $55. Assuming all the sets produced can be sold, find how many of each set should be manufactured to maximize profit.

24. MAXIMIZE PROFIT A company makes two types of telephone answering machines: the standard model and the deluxe model. Each machine passes through three processes: P_1, P_2, and P_3. One standard answering machine requires 1 hour in P_1, 1 hour in P_2, and 2 hours in P_3. One deluxe answering machine requires 3 hours in P_1, 1 hour in P_2, and 1 hour in P_3. Because of employee work schedules, P_1 is available for 24 hours, P_2 is available for 10 hours, and P_3 is available for 16 hours. If the profit is $25 for each standard model and $35 for each deluxe model, how many units of each type should the company produce to maximize profit?

SUPPLEMENTAL EXERCISES

25. MINIMIZE COST A dietitian formulates a special diet from two food groups: A and B. Each ounce of food group A contains 3 units of vitamin A, 1 unit of vitamin C, and 1 unit of vitamin D. Each unit of food group B contains 1 unit of vitamin A, 1 unit of vitamin C, and 3 units of vitamin D. Each ounce of food group A costs 40 cents, and each ounce of food group B costs 10 cents. The dietary constraints are such that at least 24 units of vitamin A, 16 units of vitamin C, and 30 units of vitamin D are required. Find the amount of each food group that should be used to minimize the cost. What is the minimum cost?

26. MAXIMIZE PROFIT Among the many products it produces, an oil refinery makes two specialized petroleum distillates: Pymex A and Pymex B. Each distillate passes through three stages: S_1, S_2, and S_3. Each liter of Pymex A requires 1 hour in S_1, 3 hours in S_2, and 3 hours in S_3. Each liter of Pymex B requires 1 hour in S_1, 4 hours in S_2, and 2 hours in S_3. There are 10 hours available for S_1, 36 hours available for S_2, and 27 hours available for S_3. The profit per liter of Pymex A is $12, and the profit per liter of Pymex B is $9. How many liters of each distillate should be produced to maximize profit? What is the maximum profit?

27. MAXIMIZE PROFIT An engine reconditioning company works on 4- and 6-cylinder engines. Each 4-cylinder engine requires 1 hour for cleaning, 5 hours for overhauling, and 3 hours for testing. Each 6-cylinder engine requires 1 hour for cleaning, 10 hours for overhauling, and 2 hours for testing. The cleaning station is available for at most 9 hours, the overhauling equipment is available for at most 80 hours, and the testing equipment is available for at most 24 hours. For each reconditioned 4-cylinder engine, the company makes a profit of $150. A reconditioned 6-cylinder engine yields a profit of $250. The company can sell all the reconditioned engines it produces. How many of each type should be produced to maximize profit? What is the maximum profit?

28. **MINIMIZE COST** A producer of animal feed makes two food products: F_1 and F_2. The products contain three major ingredients: M_1, M_2, and M_3. Each ton of F_1 requires 200 pounds of M_1, 100 pounds of M_2, and 100 pounds of M_3. Each ton of F_2 requires 100 pounds of M_1, 200 pounds of M_2, and 400 pounds of M_3. There are at least 5000 pounds of M_1 available, at least 7000 pounds of M_2 available, and at least 10,000 pounds of M_3 available. Each ton of F_1 costs \$450 to make, and each ton of F_2 costs \$300 to make. How many tons of each food product should the feed producer make to minimize cost? What is the minimum cost?

PROJECTS

1. **HISTORY OF LINEAR PROGRAMMING** Linear programming has been used successfully to solve a wide range of problems in fields as diverse as providing health care and hardening nuclear silos. Write an essay on linear programming and some of the applications of this procedure in solving practical problems. Include in your essay the contributions of George Danzig, Narendra Karmarkar, and L. G. Khachian.

EXPLORING CONCEPTS WITH TECHNOLOGY

Ill-Conditioned Systems of Equations

Solving systems of equations algebraically as we did in this chapter is not practical for systems of equations that contain a large number of variables. In those cases, a computer solution is the only hope. Computer solutions are not without some problems, however.

Consider the system of equations

$$\begin{cases} 0.24567x + 0.49133y = 0.73700 \\ 0.84312x + 1.68623y = 2.52935 \end{cases}$$

It is easy to verify that the solution of this system of equations is $(1, 1)$. However, change the constant 0.73700 to 0.73701 (add 0.00001) and the constant 2.52935 to 2.52936 (add 0.00001), and the solution is now $(3, 0)$. Thus a very small change in the constant terms produced a dramatic change in the solution. A system of equations of this sort is said to be *ill-conditioned.*

These types of systems are important because computers generally cannot store numbers beyond a certain number of significant digits. Your calculator, for example, probably allows you to enter no more than 10 significant digits. If an exact number cannot be entered, then an approximation to that number is necessary. When a computer is solving an equation or system of equations, the hope is that approximations of the coefficients it uses will give reasonable approximations to the solutions. For ill-conditioned systems of equations, this is not always true.

In the system of equations above, small changes in the constant terms caused a large change in the solution. It is possible that small changes in the coefficients of the variables will also cause large changes in the solution.

In the two systems of equations that follow, examine the effects of approximating the fractional coefficients on the solutions. Try approximating each fraction to the nearest hundredth, to the nearest thousandth, to the nearest ten-thousandth, and then to the limits of your calculator. The exact solution of the first system of equations is $(27, -192, 210)$. The exact solution of the second system of equations is $(-64, 900, -2520, 1820)$.

$$\begin{cases} x + \dfrac{1}{2}y + \dfrac{1}{3}z = 1 \\ \dfrac{1}{2}x + \dfrac{1}{3}y + \dfrac{1}{4}z = 2 \\ \dfrac{1}{3}x + \dfrac{1}{4}y + \dfrac{1}{5}z = 3 \end{cases} \qquad \begin{cases} x + \dfrac{1}{2}y + \dfrac{1}{3}z + \dfrac{1}{4}w = 1 \\ \dfrac{1}{2}x + \dfrac{1}{3}y + \dfrac{1}{4}z + \dfrac{1}{5}w = 2 \\ \dfrac{1}{3}x + \dfrac{1}{4}y + \dfrac{1}{5}z + \dfrac{1}{6}w = 3 \\ \dfrac{1}{4}x + \dfrac{1}{5}y + \dfrac{1}{6}z + \dfrac{1}{7}w = 4 \end{cases}$$

Note how the solutions change as the approximations change and thus how important it is to know whether a system of equations is ill-conditioned. For systems that are not ill-conditioned, approximations of the coefficients yield reasonable approximations of the solution. For ill-conditioned systems of equations, that is not always true.

CHAPTER 7 REVIEW

7.1 Systems of Linear Equations in Two Variables

- A system of equations is two or more equations considered together. A solution of a system of equations in two variables is an ordered pair that satisfies each equation of the system. Equivalent systems of equations have the same solution set.

- A system of equations is consistent if it has one or more solutions. A system of linear equations is independent if it has exactly one solution. A system is dependent if it has infinitely many solutions. An inconsistent system of equations has no solution.

- **Operations That Produce Equivalent Systems of Equations**

  1. Interchange any two equations.
  2. Replace an equation with a nonzero multiple of that equation.
  3. Replace an equation with the sum of an equation and a nonzero constant multiple of another equation in the system.

7.2 Systems of Linear Equations in More Than Two Variables

- An equation of the form $ax + by + cz = d$, with a, b, and c not all zero, is a linear equation in three variables. A solution of a system of equations in three variables is an ordered triple that satisfies each equation of the system.

- The graph of a linear equation in three variables is a plane.

- A linear system of equations for which the constant term is zero for all equations of the system is called a homogeneous system of equations.

7.3 Nonlinear Systems of Equations

- A nonlinear system of equations is a system in which one or more equations of the system are nonlinear.

7.4 Partial Fractions

- A rational expression can be written as the sum of terms whose denominators are factors of the denominator of the rational expression. This is called a partial fraction decomposition.

7.5 Inequalities in Two Variables and Systems of Inequalities

- The graph of an inequality in two variables frequently separates the plane into two or more regions.

- The solution set of a system of inequalities is the intersection of the solution sets of the individual inequalities.

7.6   Linear Programming

- A linear programming problem consists of a linear objective function and a number of constraints, which are inequalities, or equations that restrict the values of the variables.

- The Fundamental Linear Programming Theorem states that if an objective function has an optimal solution, then that solution will be at a vertex of the set of feasible solutions.

CHAPTER 7 TRUE/FALSE EXERCISES

In Exercises 1 to 10, answer true or false. If the statement is false, give an example to show that the statement is false.

1. A system of equations will always have a solution as long as the number of equations is equal to the number of variables.

2. A system of two different quadratic equations can have at most four solutions.

3. A homogeneous system of equations is one in which all the variables have the same exponent.

4. In an xyz-coordinate system, the graph of the set of points formed by the intersection of two different planes is a straight line.

5. It is possible to find a partial fraction decomposition of a rational expression if the degree of the numerator is greater than the degree of the denominator.

6. Two systems of equations with the same solution set have the same equations in their respective systems.

7. The systems of equations

$$\begin{cases} x = 0 \\ y = 0 \end{cases} \quad \text{and} \quad \begin{cases} y = x \\ y = -x \end{cases}$$

are equivalent systems of equations.

8. For a linear programming problem, one or more constraints are used to define the set of feasible solutions.

9. A system of three linear equations in three variables for which two of the planes are parallel and the third plane intersects the first two is a dependent system of equations.

10. The inequality $xy < 1$ and the inequality $y < 1/x$ are equivalent inequalities.

CHAPTER 7 REVIEW EXERCISES

In Exercises 1 to 30, solve each system of equations.

1. $\begin{cases} 2x - 4y = -3 \\ 3x + 8y = -12 \end{cases}$

2. $\begin{cases} 4x - 3y = 15 \\ 2x + 5y = -12 \end{cases}$

3. $\begin{cases} 3x - 4y = -5 \\ y = \dfrac{2}{3}x + 1 \end{cases}$

4. $\begin{cases} 7x + 2y = -14 \\ y = -\dfrac{5}{2}x - 3 \end{cases}$

5. $\begin{cases} y = 2x - 5 \\ x = 4y - 1 \end{cases}$

6. $\begin{cases} y = 3x + 4 \\ x = 4y - 5 \end{cases}$

7. $\begin{cases} 6x + 9y = 15 \\ 10x + 15y = 25 \end{cases}$

8. $\begin{cases} 4x - 8y = 9 \\ 2x - 4y = 5 \end{cases}$

9. $\begin{cases} 2x - 3y + z = -9 \\ 2x + 5y - 2z = 18 \\ 4x - y + 3z = -4 \end{cases}$

10. $\begin{cases} x - 3y + 5z = 1 \\ 2x + 3y - 5z = 15 \\ 3x + 6y + 5z = 15 \end{cases}$

11. $\begin{cases} x + 3y - 5z = -12 \\ 3x - 2y + z = 7 \\ 5x + 4y - 9z = -17 \end{cases}$

12. $\begin{cases} 2x - y + 2z = 5 \\ x + 3y - 3z = 2 \\ 5x - 9y + 8z = 13 \end{cases}$

13. $\begin{cases} 3x + 4y - 6z = 10 \\ 2x + 2y - 3z = 6 \\ x - 6y + 9z = -4 \end{cases}$

14. $\begin{cases} x - 6y + 4z = 6 \\ 4x + 3y - 4z = 1 \\ 5x - 9y + 8z = 13 \end{cases}$

15. $\begin{cases} 2x + 3y - 2z = 0 \\ 3x - y - 4z = 0 \\ 5x + 13y - 4z = 0 \end{cases}$

16. $\begin{cases} 3x - 5y + z = 0 \\ x + 4y - 3z = 0 \\ 2x + y - 2z = 0 \end{cases}$

17. $\begin{cases} x - 2y + z = 1 \\ 3x + 2y - 3z = 1 \end{cases}$

18. $\begin{cases} 2x - 3y + z = 1 \\ 4x + 2y + 3z = 21 \end{cases}$

19. $\begin{cases} y = x^2 - 2x - 3 \\ y = 2x - 7 \end{cases}$

20. $\begin{cases} y = 2x^2 + x \\ y = 2x + 1 \end{cases}$

21. $\begin{cases} y = 3x^2 - x + 1 \\ y = x^2 + 2x - 1 \end{cases}$

22. $\begin{cases} y = 4x^2 - 2x - 3 \\ y = 2x^2 + 3x - 6 \end{cases}$

23. $\begin{cases} (x + 1)^2 + (y - 2)^2 = 4 \\ 2x + y = 4 \end{cases}$

24. $\begin{cases} (x - 1)^2 + (y + 1)^2 = 5 \\ y = 2x - 3 \end{cases}$

25. $\begin{cases} (x - 2)^2 + (y + 2)^2 = 4 \\ (x + 2)^2 + (y + 1)^2 = 17 \end{cases}$

26. $\begin{cases} (x + 1)^2 + (y - 2)^2 = 1 \\ (x - 2)^2 + (y + 2)^2 = 20 \end{cases}$

27. $\begin{cases} x^2 - 3xy + y^2 = -1 \\ 3x^2 - 5xy - 2y^2 = 0 \end{cases}$

28. $\begin{cases} 2x^2 + 2xy - y^2 = -1 \\ 6x^2 + xy - y^2 = 0 \end{cases}$

29. $\begin{cases} 2x^2 - 5xy + 2y^2 = 56 \\ 14x^2 - 3xy - 2y^2 = 56 \end{cases}$

30. $\begin{cases} 2x^2 + 7xy + 6y^2 = 1 \\ 6x^2 + 7xy + 2y^2 = 1 \end{cases}$

In Exercises 31 to 36, find the partial fraction decomposition.

31. $\dfrac{7x - 5}{x^2 - x - 2}$

32. $\dfrac{x + 1}{(x - 1)^2}$

33. $\dfrac{2x - 2}{(x^2 + 1)(x + 2)}$

34. $\dfrac{5x^2 - 10x + 9}{(x - 2)^2(x + 1)}$

35. $\dfrac{11x^2 - x - 2}{x^3 - x}$

36. $\dfrac{x^4 + x^3 + 4x^2 + x + 3}{(x^2 + 1)^2}$

In Exercises 37 to 48, graph the solution set of each inequality.

37. $4x - 5y < 20$

38. $2x + 7y \geq -14$

39. $y \geq 2x^2 - x - 1$

40. $y < x^2 - 5x - 6$

41. $(x - 2)^2 + (y - 1)^2 > 4$

42. $(x + 3)^2 + (y + 1)^2 \leq 9$

43. $\dfrac{(x - 3)^2}{16} - \dfrac{(y + 2)^2}{25} \leq 1$

44. $\dfrac{(x + 1)^2}{9} - \dfrac{(y - 3)^2}{4} < -1$

45. $(2x - y + 1)(x - 2y - 2) > 0$

46. $(2x - 3y - 6)(x + 2y - 4) < 0$

47. $x^2y^2 < 1$

48. $xy \geq 0$

In Exercises 49 to 60, graph the solution set of each system of inequalities.

49. $\begin{cases} 2x - 5y < 9 \\ 3x + 4y \geq 2 \end{cases}$

50. $\begin{cases} 3x + y > 7 \\ 2x + 5y < 9 \end{cases}$

51. $\begin{cases} 2x + 3y > 6 \\ 2x - y > -2 \\ x < 3 \end{cases}$

52. $\begin{cases} 2x + 5y > 10 \\ x - y > -2 \\ x \leq 4 \end{cases}$

53. $\begin{cases} 2x + 3y \leq 18 \\ x + y \leq 7 \\ x \geq 0, y \geq 0 \end{cases}$

54. $\begin{cases} 3x + 5y \geq 25 \\ 2x + 3y \geq 16 \\ x \geq 0, y \geq 0 \end{cases}$

55. $\begin{cases} 3x + y \geq 6 \\ x + 4y \geq 14 \\ 2x + 3y \geq 16 \\ x \geq 0, y \geq 0 \end{cases}$

56. $\begin{cases} 3x + 2y \geq 14 \\ x + y \geq 6 \\ 11x + 4y \leq 48 \\ x \geq 0, y \geq 0 \end{cases}$

57. $\begin{cases} y < x^2 - x - 2 \\ y \geq 2x - 4 \end{cases}$

58. $\begin{cases} y > 2x^2 + x - 1 \\ y > x + 3 \end{cases}$

59. $\begin{cases} x^2 + y^2 - 2x + 4y > 4 \\ y < 2x^2 - 1 \end{cases}$

60. $\begin{cases} x^2 - y^2 - 4x - 2y < -4 \\ x^2 + y^2 - 4x + 4y > 8 \end{cases}$

In Exercises 61 to 66, solve the linear programming problem. In each problem, assume $x \geq 0$ and $y \geq 0$.

61. Objective function: $P = 2x + 2y$
    Constraints: $\begin{cases} x + 2y \leq 14 \\ 5x + 2y \leq 30 \end{cases}$
    Maximize the objective function.

62. Objective function: $P = 4x + 5y$
    Constraints: $\begin{cases} 2x + 3y \leq 24 \\ 4x + 3y \leq 36 \end{cases}$
    Maximize the objective function.

63. Objective function: $P = 4x + y$
    Constraints: $\begin{cases} 5x + 2y \geq 16 \\ x + 2y \geq 8 \\ x \leq 20, y \leq 20 \end{cases}$
    Minimize the objective function.

64. Objective function: $P = 2x + 7y$
    Constraints: $\begin{cases} 4x + 3y \geq 24 \\ 4x + 7y \geq 40 \\ x \leq 10, y \leq 10 \end{cases}$
    Minimize the objective function.

65. Objective function: $P = 6x + 3y$
    Constraints: $\begin{cases} 5x + 2y \geq 20 \\ x + y \geq 7 \\ x + 2y \geq 10 \\ x \leq 15, y \leq 15 \end{cases}$
    Minimize the objective function.

66. Objective function: $P = 5x + 4y$
    Constraints: $\begin{cases} x + y \leq 10 \\ 2x + y \leq 13 \\ 3x + y \leq 18 \end{cases}$
    Maximize the objective function.

67. Find an equation of the form $y = ax^2 + bx + c$ whose graph passes through the points $(1, 0)$, $(-1, 5)$, and $(2, 3)$.

68. Find an equation of the circle that passes through the points $(4, 2)$, $(0, 1)$, and $(3, -1)$.

69. Find an equation of a plane that passes through the points $(2, 1, 2)$, $(3, 1, 0)$, and $(-2, -3, -2)$. Use the equation $z = ax + by + c$.

70. How many liters of a 20% acid solution should be mixed with 10 liters of a 10% acid solution so that the resulting solution is a 16% acid solution?

71. Flying with the wind, a small plane traveled 855 miles in 5 hours. Flying against the same wind, the plane traveled 575 miles in the same time. Find the rate of the wind and the rate of the plane in calm air.

72. A collection of ten coins has a value of $1.25. The collection consists of only nickels, dimes, and quarters. How many of each coin are in the collection? (*Hint:* There is more than one solution.)

73. Consider the ordered triple (a, b, c). Find all real number values for $a, b,$ and c so that the product of any two numbers equals the remaining number.

CHAPTER 7 TEST

In Exercises 1 to 8, solve each system of equations. If a system of equations is inconsistent, so state.

1. $\begin{cases} 3x + 2y = -5 \\ 2x - 5y = -16 \end{cases}$

2. $\begin{cases} x - \dfrac{1}{2}y = 3 \\ 2x - y = 6 \end{cases}$

3. $\begin{cases} x + 3y - z = 8 \\ 2x - 7y + 2z = 1 \\ 4x - y + 3z = 13 \end{cases}$

4. $\begin{cases} 3x - 2y + z = 2 \\ x + 2y - 2z = 1 \\ 4x \quad\ - z = 3 \end{cases}$

5. $\begin{cases} 2x - 3y + z = -1 \\ x + 5y - 2z = 5 \end{cases}$

6. $\begin{cases} 4x + 2y + z = 0 \\ x - 3y - 2z = 0 \\ 3x + 5y + 3z = 0 \end{cases}$

7. $\begin{cases} y = x + 3 \\ y = x^2 + x - 1 \end{cases}$

8. $\begin{cases} y = x^2 - x - 3 \\ y = 2x^2 + 2x - 1 \end{cases}$

In Exercises 9 to 12, graph each inequality.

9. $3x - 4y > 8$

10. $y \le x^2 - 2x - 3$

11. $x^2 + 4y^2 \ge 16$

12. $x + y^2 < 0$

In Exercises 13 to 16, graph each system of inequalities. If the solution set is empty, so state.

13. $\begin{cases} 2x - 5y \le 16 \\ x + 3y \ge -3 \end{cases}$

14. $\begin{cases} x^2 + y^2 > 9 \\ x^2 + y^2 < 4 \end{cases}$

15. $\begin{cases} x + y \ge 8 \\ 2x + y \ge 11 \\ x \ge 0, y \ge 0 \end{cases}$

16. $\begin{cases} 2x + 3y \le 12 \\ x + y \le 5 \\ 3x + 2y \le 11 \\ x \ge 0, y \ge 0 \end{cases}$

In Exercises 17 and 18, find the partial fraction decomposition.

17. $\dfrac{3x - 5}{x^2 - 3x - 4}$

18. $\dfrac{2x + 1}{x(x^2 + 1)}$

19. A farmer has 160 acres available on which to plant oats and barley. It costs $15 per acre for oat seed and $13 per acre for barley seed. The labor cost is $15 per acre for oats and $20 per acre for barley. The farmer has $2200 available to purchase seed and has set aside $2600 for labor. The profit per acre for oats is $120, and the profit per acre for barley is $150. How many acres of oats should the farmer plant to maximize profit?

20. Find an equation of the circle that passes through the points $(3, 5)$, $(-3, -3)$ and $(4, 4)$. (*Hint:* Use $x^2 + y^2 + ax + by + c = 0$.)

MATRICES

Matrices and Error-Correcting Codes

A matrix is a rectangular array of elements, an example of which is shown below. This particular matrix is a modification of what is called a *Hadamard matrix*. One application of Hadamard matrices is the enhancement of images sent to Earth by satellites or space probes such as the Mariner or Voyager.

$$\begin{bmatrix} 1 & 1 & 1 & 1 & 1 & 1 & 1 & 1 \\ 1 & 0 & 1 & 0 & 1 & 0 & 1 & 0 \\ 1 & 1 & 0 & 0 & 1 & 1 & 0 & 0 \\ 1 & 0 & 0 & 1 & 1 & 0 & 0 & 1 \\ 1 & 1 & 1 & 1 & 0 & 0 & 0 & 0 \\ 1 & 0 & 1 & 0 & 0 & 1 & 0 & 1 \\ 1 & 1 & 0 & 0 & 0 & 0 & 1 & 1 \\ 1 & 0 & 0 & 1 & 0 & 1 & 1 & 0 \end{bmatrix}$$

A satellite orbits Earth, ready to transmit images from space.

An image to be transmitted from a space probe to a receiving station is first divided into very small squares called *pixels*. Each pixel is then assigned a number, the magnitude of which is a measure of the darkness of the square. For example, if pixels are assigned numbers from 0 to 63, a value of 0 would correspond to white and a value of 63 would correspond to black. These numbers are then represented as binary numbers, with $0 = 000000_{two}$ and $63 = 111111_{two}$. Using this conversion, the image is represented by a series of 0s and 1s (the 0s and 1s are called *bits*) that are sent to Earth.

Matrices make possible this detailed image of Mars' surface.

To produce an accurate image on Earth, the bits transmitted by a space probe must be received accurately. However, because of what engineers call *noise* (which is similar to static on a radio), some of the 0s are changed to 1s and some of the 1s are changed to 0s. As a result, the image is blurred.

To minimize the effect of noise, *error-correcting codes* are used to help determine whether a bit has been changed. One way of establishing these codes is to use a Hadamard matrix.

How a satellite "sees" the Yolga River in Russia.

8.1 GAUSSIAN ELIMINATION METHOD

A **matrix** is a rectangular array of numbers. Each number in a matrix is called an **element** of the matrix. The matrix below, with three rows and four columns, is called a 3×4 (read "3 by 4") matrix.

$$\begin{bmatrix} 2 & 5 & -2 & 5 \\ -3 & 6 & 4 & 0 \\ 1 & 3 & 7 & 2 \end{bmatrix}$$

A matrix of m rows and n columns is said to be of **order** $m \times n$ or **dimension** $m \times n$. A **square matrix of order** n is a matrix with n rows and n columns. The matrix above has order 3×4. We will use the notation a_{ij} to refer to the element of a matrix in the ith row and jth column. For the matrix given above, $a_{23} = 4$, $a_{31} = 1$, and $a_{13} = -2$.

The elements $a_{11}, a_{22}, a_{33}, \ldots, a_{mm}$ form the **main diagonal** of a matrix. The elements 2, 6, and 7 form the main diagonal of the matrix shown above.

A matrix can be created from a system of linear equations. Consider the system of linear equations

$$\begin{cases} 2x - 3y + z = 2 \\ x \quad\quad - 3z = 4 \\ 4x - y + 4z = 3 \end{cases}$$

Using only the coefficients and constants of this system, we can write the 3×4 matrix

$$\begin{bmatrix} 2 & -3 & 1 & 2 \\ 1 & 0 & -3 & 4 \\ 4 & -1 & 4 & 3 \end{bmatrix}$$

This matrix is called the **augmented matrix** of the system of equations. The matrix formed by the coefficients of the system is the **coefficient matrix**. The matrix formed from the constants is the **constant matrix** for the system. The coefficient matrix and constant matrix for the given system are

Coefficient matrix: $\begin{bmatrix} 2 & -3 & 1 \\ 1 & 0 & -3 \\ 4 & -1 & 4 \end{bmatrix}$ Constant matrix: $\begin{bmatrix} 2 \\ 4 \\ 3 \end{bmatrix}$

We can write a system of equations from an augmented matrix.

Augmented matrix: $\begin{bmatrix} 2 & -1 & 4 & 3 \\ 1 & 1 & 0 & 2 \\ 3 & -2 & -1 & 2 \end{bmatrix}$ $\xrightarrow{\text{System:}}$ $\begin{cases} 2x - y + 4z = 3 \\ x + y \quad = 2 \\ 3x - 2y - z = 2 \end{cases}$

In certain cases, an augmented matrix represents a system of equations that we can solve by back substitution. Consider the following augmented matrix and the equivalent system of equations.

$$\begin{bmatrix} 1 & -3 & 4 & 5 \\ 0 & 1 & 2 & -4 \\ 0 & 0 & 1 & -1 \end{bmatrix} \xrightarrow{\text{equivalent system}} \begin{cases} x - 3y + 4z = 5 \\ y + 2z = -4 \\ z = -1 \end{cases}$$

Solving this system by using back substitution, we find that the solution is $(3, -2, -1)$. The matrix above is in *echelon form.*

ECHELON FORM

A matrix is in **echelon form** if all the following conditions are satisfied.

1. The first nonzero number in any row is a 1.

2. Rows are arranged so that the column containing the first nonzero number in any row is to the left of the column containing the first nonzero number of the next row.

3. All rows consisting entirely of zeros appear at the bottom of the matrix.

Following are three examples of matrices in echelon form.

$$\begin{bmatrix} 1 & -3 & 4 & 2 \\ 0 & 1 & -2 & -1 \\ 0 & 0 & 0 & 0 \end{bmatrix} \quad \begin{bmatrix} 1 & 2 & -1 & 3 \\ 0 & 1 & 2 & -1 \end{bmatrix} \quad \begin{bmatrix} 1 & -1 & 3 & 2 \\ 0 & 1 & 2 & 5 \\ 0 & 0 & 1 & -2 \end{bmatrix}$$

We can write an augmented matrix in echelon form by using **elementary row operations.** These operations are a rewording, in matrix terminology, of the operations that produce equivalent equations.

ELEMENTARY ROW OPERATIONS

Given the augmented matrix for a system of linear equations, each of the following elementary row operations produces a matrix of an equivalent system of equations.

1. Interchanging any two rows

2. Multiplying all the elements in a row by the same nonzero number

3. Replacing a row by the sum of that row and a nonzero multiple of any other row

It is convenient to specify each operation symbolically as follows:

1. Interchanging the ith and jth rows: $R_i \longleftrightarrow R_j$

2. Multiplying the ith row by k, a nonzero constant: kR_i

3. Replacing the jth row by the sum of that row and a nonzero multiple of the ith row: $kR_i + R_j$

To demonstrate these operations, we will use the 3×3 matrix

$$\begin{bmatrix} 2 & 1 & -2 \\ 3 & -2 & 2 \\ 1 & -2 & 3 \end{bmatrix}$$

$$\begin{bmatrix} 2 & 1 & -2 \\ 3 & -2 & 2 \\ 1 & -2 & 3 \end{bmatrix} \xrightarrow{R_1 \longleftrightarrow R_3} \begin{bmatrix} 1 & -2 & 3 \\ 3 & -2 & 2 \\ 2 & 1 & -2 \end{bmatrix}$$

- Interchange row 1 and row 3.

$$\begin{bmatrix} 2 & 1 & -2 \\ 3 & -2 & 2 \\ 1 & -2 & 3 \end{bmatrix} \xrightarrow{-3R_2} \begin{bmatrix} 2 & 1 & -2 \\ -9 & 6 & -6 \\ 1 & -2 & 3 \end{bmatrix}$$

- Multiply row 2 by -3.

$$\begin{bmatrix} 2 & 1 & -2 \\ 3 & -2 & 2 \\ 1 & -2 & 3 \end{bmatrix} \xrightarrow{-2R_3 + R_1} \begin{bmatrix} 0 & 5 & -8 \\ 3 & -2 & 2 \\ 1 & -2 & 3 \end{bmatrix}$$

- Multiply row 3 by -2 and add to row 1. Replace row 1 by the sum.

The **Gaussian elimination method** is an algorithm[1] that uses elementary row operations to solve a system of linear equations. The goal of this method is to rewrite an augmented matrix in echelon form.

We will now demonstrate how to solve a system of two equations in two variables by the Gaussian elimination method. Consider the system of equations

$$\begin{cases} 2x + 5y = -1 \\ 3x - 2y = 8 \end{cases} \qquad (1)$$

The augmented matrix for this system is

$$\begin{bmatrix} 2 & 5 & -1 \\ 3 & -2 & 8 \end{bmatrix}$$

The goal of the Gaussian elimination method is to rewrite the augmented matrix in echelon form by using elementary row operations. The row operations are chosen so that first, there is a 1 as a_{11}; second, there is a 0 as a_{21}; and third, there is a 1 as a_{22}.

Begin by multiplying row 1 by 1/2. The result is a 1 as a_{11}.

$$\begin{bmatrix} 2 & 5 & -1 \\ 3 & -2 & 8 \end{bmatrix} \xrightarrow{\frac{1}{2}R_1} \begin{bmatrix} 1 & \frac{5}{2} & -\frac{1}{2} \\ 3 & -2 & 8 \end{bmatrix}$$

Now multiply row 1 by -3 and add the result to row 2. Replace row 2. The result is a 0 as a_{21}.

$$\begin{bmatrix} 1 & \frac{5}{2} & -\frac{1}{2} \\ 3 & -2 & 8 \end{bmatrix} \xrightarrow{-3R_1 + R_2} \begin{bmatrix} 1 & \frac{5}{2} & -\frac{1}{2} \\ 0 & -\frac{19}{2} & \frac{19}{2} \end{bmatrix}$$

[1] An algorithm is a procedure used in calculations. The word is derived from Al-Khwarizmi, the name of the author of an Arabic algebra book written around A.D. 825.

Now multiply row 2 by $-2/19$. The result is a 1 as a_{22}. The matrix is now in row echelon form.

$$\begin{bmatrix} 1 & \frac{5}{2} & -\frac{1}{2} \\ 0 & -\frac{19}{2} & \frac{19}{2} \end{bmatrix} \xrightarrow{\;-\frac{2}{19}R_2\;} \begin{bmatrix} 1 & \frac{5}{2} & -\frac{1}{2} \\ 0 & 1 & -1 \end{bmatrix}$$

The system of equations written from the echelon form of the matrix is

$$\begin{cases} x + \dfrac{5}{2}y = -\dfrac{1}{2} \\ \qquad\quad y = -1 \end{cases} \qquad (2)$$

To solve by back substitution, replace y in the first equation by -1 and solve for x.

$$x + \left(\frac{5}{2}\right)(-1) = -\frac{1}{2}$$
$$x = 2$$

The solution of System (1) is $(2, -1)$.

To conserve space, we will occasionally perform more than one elementary row operation in one step. For example, the notation

$$\xrightarrow[\;-2R_1 + R_3\;]{\;3R_1 + R_2\;}$$

means that two elementary row operations were performed. First multiply row 1 by 3 and add it to row 2. Replace row 2. Now multiply row 1 by -2 and add it to row 3. Replace row 3.

EXAMPLE 1	*Solve a System of Equations by the Gaussian Elimination Method*

Solve by using the Gaussian elimination method.

$$\begin{cases} 3t - 8u + 8v + 7w = 41 \\ t - 2u + 2v + w = 9 \\ 2t - 2u + 6v - 4w = -1 \\ 2t - 2u + 3v - 3w = 3 \end{cases}$$

Solution

Write the augmented matrix and then use elementary row operations to rewrite the matrix in echelon form.

$$\begin{bmatrix} 3 & -8 & 8 & 7 & 41 \\ 1 & -2 & 2 & 1 & 9 \\ 2 & -2 & 6 & -4 & -1 \\ 2 & -2 & 3 & -3 & 3 \end{bmatrix} \xrightarrow{\;R_1 \longleftrightarrow R_2\;} \begin{bmatrix} 1 & -2 & 2 & 1 & 9 \\ 3 & -8 & 8 & 7 & 41 \\ 2 & -2 & 6 & -4 & -1 \\ 2 & -2 & 3 & -3 & 3 \end{bmatrix}$$

Continued ⬥➤

TAKE NOTE

Following a systematic procedure will help you reduce a matrix to echelon form. Using elementary row operations, change a_{11} to **1** and change the remaining elements in the first column to **0**. Now change a_{22} to **1** and change the remaining elements *below* a_{22} to zero. Now move to a_{33} and repeat the procedure. Continue moving down the main diagonal until you reach a_{nn} or until all remaining elements on the main diagonal are zero.

$$
\begin{array}{l}
-3R_1 + R_2 \\
-2R_1 + R_3 \\
-2R_1 + R_4 \\
\longrightarrow
\end{array}
\begin{bmatrix}
1 & -2 & 2 & 1 & 9 \\
0 & -2 & 2 & 4 & 14 \\
0 & 2 & 2 & -6 & -19 \\
0 & 2 & -1 & -5 & -15
\end{bmatrix}
\begin{array}{l}
-\dfrac{1}{2}R_2 \\
\longrightarrow
\end{array}
\begin{bmatrix}
1 & -2 & 2 & 1 & 9 \\
0 & 1 & -1 & -2 & -7 \\
0 & 2 & 2 & -6 & -19 \\
0 & 2 & -1 & -5 & -15
\end{bmatrix}
$$

$$
\begin{array}{l}
-2R_2 + R_3 \\
-2R_2 + R_4 \\
\longrightarrow
\end{array}
\begin{bmatrix}
1 & -2 & 2 & 1 & 9 \\
0 & 1 & -1 & -2 & -7 \\
0 & 0 & 4 & -2 & -5 \\
0 & 0 & 1 & -1 & -1
\end{bmatrix}
\begin{array}{l}
R_4 \longleftrightarrow R_3 \\
\end{array}
\begin{bmatrix}
1 & -2 & 2 & 1 & 9 \\
0 & 1 & -1 & -2 & -7 \\
0 & 0 & 1 & -1 & -1 \\
0 & 0 & 4 & -2 & -5
\end{bmatrix}
$$

$$
\begin{array}{l}
-4R_3 + R_4 \\
\longrightarrow
\end{array}
\begin{bmatrix}
1 & -2 & 2 & 1 & 9 \\
0 & 1 & -1 & -2 & -7 \\
0 & 0 & 1 & -1 & -1 \\
0 & 0 & 0 & 2 & -1
\end{bmatrix}
\begin{array}{l}
\dfrac{1}{2}R_4 \\
\longrightarrow
\end{array}
\begin{bmatrix}
1 & -2 & 2 & 1 & 9 \\
0 & 1 & -1 & -2 & -7 \\
0 & 0 & 1 & -1 & -1 \\
0 & 0 & 0 & 1 & -\frac{1}{2}
\end{bmatrix}
$$

The last matrix is in echelon form. The system of equations written from the matrix is

$$
\begin{cases}
t - 2u + 2v + w = 9 \\
\quad\ u - v - 2w = -7 \\
\qquad\quad v - w = -1 \\
\qquad\qquad\ w = -\dfrac{1}{2}
\end{cases}
$$

Solve by back substitution. The solution is $(-13/2, -19/2, -3/2, -1/2)$.

TRY EXERCISE 14, EXERCISE SET 8.1

EXAMPLE 2 *Solve a Dependent System of Equations*

Solve using the Gaussian elimination method.

$$
\begin{cases}
x - 3y + 4z = 1 \\
2x - 5y + 3z = 6 \\
x - 2y - z = 5
\end{cases}
$$

Solution

Write the augmented matrix and then use elementary row operations to rewrite the matrix in echelon form.

$$
\begin{bmatrix}
1 & -3 & 4 & 1 \\
2 & -5 & 3 & 6 \\
1 & -2 & -1 & 5
\end{bmatrix}
\begin{array}{l}
-2R_1 + R_2 \\
-R_1 + R_3 \\
\longrightarrow
\end{array}
\begin{bmatrix}
1 & -3 & 4 & 1 \\
0 & 1 & -5 & 4 \\
0 & 1 & -5 & 4
\end{bmatrix}
$$

$$
\begin{array}{l}
-R_2 + R_3 \\
\longrightarrow
\end{array}
\begin{bmatrix}
1 & -3 & 4 & 1 \\
0 & 1 & -5 & 4 \\
0 & 0 & 0 & 0
\end{bmatrix}
$$

$$\begin{cases} x - 3y + 4z = 1 \\ \quad\quad y - 5z = 4 \end{cases}$$ • **Equivalent system**

Any solution of the system of equations is a solution of $y - 5z = 4$. Solving this equation for y, we have $y = 5z + 4$.

$$x - 3y + 4z = 1$$
$$x - 3(5z + 4) + 4z = 1 \qquad \bullet\ \textbf{y = 5z + 4}$$
$$x = 11z + 13 \qquad \bullet\ \textbf{Solve for x.}$$

Both x and y are expressed in terms of z. Let z be any real number c. The solutions of the system of equations are $(11c + 13, 5c + 4, c)$.

TRY EXERCISE 18, EXERCISE SET 8.1

EXAMPLE 3 *Identify an Inconsistent System of Equations*

Solve using the Gaussian elimination method.

$$\begin{cases} x - 3y + z = 5 \\ 3x - 7y + 2z = 12 \\ 2x - 4y + z = 3 \end{cases}$$

Solution

Write the augmented matrix and then use elementary row operations to rewrite the matrix in echelon form.

$$\begin{bmatrix} 1 & -3 & 1 & 5 \\ 3 & -7 & 2 & 12 \\ 2 & -4 & 1 & 3 \end{bmatrix} \begin{array}{c} -3R_1 + R_2 \\ -2R_1 + R_3 \\ \longrightarrow \end{array} \begin{bmatrix} 1 & -3 & 1 & 5 \\ 0 & 2 & -1 & -3 \\ 0 & 2 & -1 & -7 \end{bmatrix}$$

$$\begin{array}{c} \frac{1}{2}R_2 \\ \longrightarrow \end{array} \begin{bmatrix} 1 & -3 & 1 & 5 \\ 0 & 1 & -\frac{1}{2} & -\frac{3}{2} \\ 0 & 2 & -1 & -7 \end{bmatrix} \begin{array}{c} -2R_2 + R_3 \\ \longrightarrow \end{array} \begin{bmatrix} 1 & -3 & 1 & 5 \\ 0 & 1 & -\frac{1}{2} & -\frac{3}{2} \\ 0 & 0 & 0 & -4 \end{bmatrix}$$

$$\begin{cases} x - 3y + z = 5 \\ \quad\quad y - \dfrac{1}{2}z = -\dfrac{3}{2} \qquad \bullet\ \textbf{Equivalent system} \\ \quad\quad\quad 0z = -4 \end{cases}$$

Because the equation $0z = -4$ has no solution, the system of equations has no solution.

TRY EXERCISE 20, EXERCISE SET 8.1

EXAMPLE 4 Solve a Nonsquare System of Equations

Solve the system of equations using the Gaussian elimination method.

$$\begin{cases} x_1 - 2x_2 - 3x_3 - 2x_4 = 1 \\ 2x_1 - 3x_2 - 4x_3 - 2x_4 = 3 \\ x_1 + x_2 + x_3 - 7x_4 = -7 \end{cases}$$

TAKE NOTE

When there are fewer equations than variables (as in Example 4), the system of equations has either no solution or an infinite number of solutions. See the Project at the end of this section.

Solution

Write the augmented matrix and then use elementary row operations to rewrite the matrix in echelon form.

$$\begin{bmatrix} 1 & -2 & -3 & -2 & 1 \\ 2 & -3 & -4 & -2 & 3 \\ 1 & 1 & 1 & -7 & -7 \end{bmatrix} \begin{matrix} -2R_1 + R_2 \\ -1R_1 + R_3 \\ \longrightarrow \end{matrix} \begin{bmatrix} 1 & -2 & -3 & -2 & 1 \\ 0 & 1 & 2 & 2 & 1 \\ 0 & 3 & 4 & -5 & -8 \end{bmatrix}$$

$$\begin{matrix} -3R_2 + R_3 \\ -\dfrac{1}{2}R_3 \\ \longrightarrow \end{matrix} \begin{bmatrix} 1 & -2 & -3 & -2 & 1 \\ 0 & 1 & 2 & 2 & 1 \\ 0 & 0 & 1 & \frac{11}{2} & \frac{11}{2} \end{bmatrix}$$

$$\begin{cases} x_1 - 2x_2 - 3x_3 - 2x_4 = 1 \\ x_2 + 2x_3 + 2x_4 = 1 \qquad \text{• Equivalent system} \\ x_3 + \frac{11}{2}x_4 = \frac{11}{2} \end{cases}$$

Now express each of the variables in terms of x_4. Solve the third equation for x_3.

$$x_3 = -\frac{11}{2}x_4 + \frac{11}{2}$$

Substitute this value into the second equation and solve for x_2.

$$x_2 + 2\left(-\frac{11}{2}x_4 + \frac{11}{2}\right) + 2x_4 = 1$$

$$x_2 = 9x_4 - 10 \qquad \text{• Simplify}$$

Substitute the values for x_2 and x_3 into the first equation and solve for x_1.

$$x_1 - 2(9x_4 - 10) - 3\left(-\frac{11}{2}x_4 + \frac{11}{2}\right) - 2x_4 = 1$$

$$x_1 = \frac{7}{2}x_4 - \frac{5}{2} \qquad \text{• Simplify}$$

If x_4 is any real number c, the solution is of the form

$$\left(\frac{7}{2}c - \frac{5}{2},\ 9c - 10,\ -\frac{11}{2}c + \frac{11}{2},\ c\right)$$

TRY EXERCISE 36, EXERCISE SET 8.1

TOPICS FOR DISCUSSION

1. What is a matrix? What is the order of a matrix?

2. Explain how an augmented matrix differs from the coefficient matrix for a system of equations.

3. Give examples of matrices that are in echelon form and of matrices that are not in echelon form.

4. What are the elementary row operations? Give examples of each one.

5. After elementary row operations have been correctly performed on an augmented matrix, the result is $\begin{bmatrix} 1 & -2 & 3 & 0 \\ 0 & 1 & 2 & -1 \\ 0 & 0 & 0 & 3 \end{bmatrix}$. Does this result indicate that the system of equations has a unique solution, an infinite number of solutions, or no solution?

EXERCISE SET 8.1

In Exercises 1 to 4, write the augmented matrix, the coefficient matrix, and the constant matrix.

1. $\begin{cases} 2x - 3y + z = 1 \\ 3x - 2y + 3z = 0 \\ x \quad\quad + 5z = 4 \end{cases}$

2. $\begin{cases} -3y + 2z = 3 \\ 2x - y \quad\quad = -1 \\ 3x - 2y + 3z = 4 \end{cases}$

3. $\begin{cases} 2x - 3y - 4z + w = 2 \\ 2y + z = 2 \\ x - y + 2z = 4 \\ 3x - 3y - 2z = 1 \end{cases}$

4. $\begin{cases} x - y + 2z + 3w = -2 \\ 2x \quad\quad + z - 2w = 1 \\ 3x \quad\quad\quad - 2w = 3 \\ -x + 3y - z \quad\quad = 3 \end{cases}$

In Exercises 5 to 12, use elementary row operations to write each matrix in echelon form.

5. $\begin{bmatrix} 2 & -1 & 3 & -2 \\ 1 & -1 & 2 & 2 \\ 3 & 2 & -1 & 3 \end{bmatrix}$

6. $\begin{bmatrix} 1 & 2 & 4 & 1 \\ 2 & 2 & 7 & 3 \\ 3 & 6 & 8 & -1 \end{bmatrix}$

7. $\begin{bmatrix} 4 & -5 & -1 & 2 \\ 3 & -4 & 1 & -2 \\ 1 & -2 & -1 & 3 \end{bmatrix}$

8. $\begin{bmatrix} -2 & 1 & -1 & 3 \\ 2 & 2 & 4 & 6 \\ 3 & 1 & -1 & 2 \end{bmatrix}$

9. $\begin{bmatrix} 1 & -2 & 3 & -4 \\ 3 & -6 & 10 & -14 \\ 5 & -8 & 19 & -21 \\ 2 & -4 & 7 & -10 \end{bmatrix}$

10. $\begin{bmatrix} 2 & -1 & 3 & 2 \\ 1 & 2 & -1 & 3 \\ 3 & 5 & -2 & 2 \\ 4 & 3 & 1 & 8 \end{bmatrix}$

11. $\begin{bmatrix} 1 & -3 & 4 & 2 & 1 \\ 2 & -3 & 5 & -2 & -1 \\ -1 & 2 & -3 & 1 & 3 \end{bmatrix}$

12. $\begin{bmatrix} 2 & -1 & 3 & 2 & 2 \\ 1 & -2 & 2 & 1 & -1 \\ 3 & -5 & -1 & -2 & 3 \end{bmatrix}$

In Exercises 13 to 38, solve each system of equations by the Gaussian elimination method.

13. $\begin{cases} x + 2y - 2z = -2 \\ 5x + 9y - 4z = -3 \\ 3x + 4y - 5z = -3 \end{cases}$

14. $\begin{cases} x - 3y + z = 8 \\ 2x - 5y - 3z = 2 \\ x + 4y + z = 1 \end{cases}$

15. $\begin{cases} 3x + 7y - 7z = -4 \\ x + 2y - 3z = 0 \\ 5x + 6y + z = -8 \end{cases}$

16. $\begin{cases} 2x - 3y + 2z = 13 \\ 3x - 4y - 3z = 1 \\ 3x + y - z = 2 \end{cases}$

17. $\begin{cases} x + 2y - 2z = 3 \\ 5x + 8y - 6z = 14 \\ 3x + 4y - 2z = 8 \end{cases}$

18. $\begin{cases} 3x - 5y + 2z = 4 \\ x - 3y + 2z = 4 \\ 5x - 11y + 6z = 12 \end{cases}$

19. $\begin{cases} 3x + 2y - z = 1 \\ 2x + 3y - z = 1 \\ x - y + 2z = 3 \end{cases}$

20. $\begin{cases} 2x + 5y + 2z = -1 \\ x + 2y - 3z = 5 \\ 5x + 12y + z = 10 \end{cases}$

21. $\begin{cases} x - 3y + 2z = 0 \\ 2x - 5y - 2z = 0 \\ 4x - 11y + 2z = 0 \end{cases}$

22. $\begin{cases} x + y - 2z = 0 \\ 3x + 4y - z = 0 \\ 5x + 6y - 5z = 0 \end{cases}$

23. $\begin{cases} 2x + y - 3z = 4 \\ 3x + 2y + z = 2 \end{cases}$

24. $\begin{cases} 3x - 6y + 2z = 2 \\ 2x + 5y - 3z = 2 \end{cases}$

25. $\begin{cases} 2x + 2y - 4z = 4 \\ 2x + 3y - 5z = 4 \\ 4x + 5y - 9z = 8 \end{cases}$

26. $\begin{cases} 3x - 10y + 2z = 34 \\ x - 4y + z = 13 \\ 5x - 2y + 7z = 31 \end{cases}$

27. $\begin{cases} x + 3y + 4z = 11 \\ 2x + 3y + 2z = 7 \\ 4x + 9y + 10z = 20 \\ 3x - 2y + z = 1 \end{cases}$

28. $\begin{cases} x - 4y + 3z = 4 \\ 3x - 10y + 3z = 4 \\ 5x - 18y + 9z = 10 \\ 2x + 2y - 3z = -11 \end{cases}$

29. $\begin{cases} t + 2u - 3v + w = -7 \\ 3t + 5u - 8v + 5w = -8 \\ 2t + 3u - 7v + 3w = -11 \\ 4t + 8u - 10v + 7w = -10 \end{cases}$

30. $\begin{cases} t + 4u + 2v - 3w = 11 \\ 2t + 10u + 3v - 5w = 17 \\ 4t + 16u + 7v - 9w = 34 \\ t + 4u + v - w = 4 \end{cases}$

31. $\begin{cases} 2t - u + 3v + 2w = 2 \\ t - u + 2v + w = 2 \\ 3t - 2v - 3w = 13 \\ 2t + 2u - 2w = 6 \end{cases}$

32. $\begin{cases} 4t + 7u - 10v + 3w = -29 \\ 3t + 5u - 7v + 2w = -20 \\ t + 2u - 3v + w = -9 \\ 2t - u + 2v - 4w = 15 \end{cases}$

33. $\begin{cases} 3t + 10u + 7v - 6w = 7 \\ 2t + 8u + 6v - 5w = 5 \\ t + 4u + 2v - 3w = 2 \\ 4t + 14u + 9v - 8w = 8 \end{cases}$

34. $\begin{cases} t - 3u + 2v + 4w = 13 \\ 3t - 8u + 4v + 13w = 35 \\ 2t - 7u + 8v + 5w = 28 \\ 4t - 11u + 6v + 17w = 56 \end{cases}$

35. $\begin{cases} t - u + 2v - 3w = 9 \\ 4t + 11v - 10w = 46 \\ 3t - u + 8v - 6w = 27 \end{cases}$

36. $\begin{cases} t - u + 3v - 5w = 10 \\ 2t - 3u + 4v + w = 7 \\ 3t + u - 2v - 2w = 6 \end{cases}$

37. $\begin{cases} 3t - 4u + v = 2 \\ t + u - 2v + 3w = 1 \end{cases}$

38. $\begin{cases} 2t + 3v - 4w = 2 \\ t + 2u - 4v + w = -3 \end{cases}$

Some graphing calculators and computer programs contain a program that will assist you in solving a system of linear equations by rewriting the system in echelon form. Try one of these programs for Exercises 39 to 44.

39. $\begin{cases} x_1 + 2x_2 - x_3 + 2x_4 + 3x_5 = 11 \\ x_1 - x_2 + 2x_3 - x_4 + 2x_5 = 0 \\ 2x_1 + x_2 - x_3 + 2x_4 - x_5 = 4 \\ 3x_1 + 2x_2 - x_3 + x_4 - 2x_5 = 2 \\ 2x_1 + x_2 - x_3 - 2x_4 + x_5 = 4 \end{cases}$

40. $\begin{cases} x_1 - 2x_2 + 2x_3 - 3x_4 + 2x_5 = 5 \\ x_1 - 3x_2 - x_3 + 2x_4 - x_5 = -4 \\ 3x_1 + x_2 - 2x_3 + x_4 + 3x_5 = 9 \\ 2x_1 - x_2 + 3x_3 - x_4 - 2x_5 = 2 \\ -x_1 + 2x_2 - 2x_3 + 3x_4 - x_5 = -4 \end{cases}$

41. $\begin{cases} x_1 + 2x_2 - 3x_3 - x_4 + 2x_5 = -10 \\ -x_1 - 3x_2 + x_3 + x_4 - x_5 = 4 \\ 2x_1 + 3x_2 - 5x_3 + 2x_4 + 3x_5 = -20 \\ 3x_1 + 4x_2 - 7x_3 + 3x_4 - 2x_5 = -16 \\ 2x_1 + x_2 - 6x_3 + 4x_4 - 3x_5 = -12 \end{cases}$

42. $\begin{cases} x_1 - 2x_2 + 2x_3 - 3x_4 + x_5 = 5 \\ 2x_1 - 3x_2 + 4x_3 - 5x_4 - x_5 = 13 \\ x_1 + x_2 - 2x_3 + 2x_4 + 2x_5 = -11 \\ 3x_1 - 2x_2 + 2x_3 - 2x_4 - 2x_5 = 7 \\ 4x_1 - 4x_2 + 4x_3 - 5x_4 - x_5 = 12 \end{cases}$

43. **CURVE FITTING** Find a cubic function whose graph passes through the points $(0, 2)$, $(1, 0)$, $(-2, -12)$, and $(3, 8)$. (*Hint:* Use the equation $y = ax^3 + bx^2 + cx + d$.)

44. **CURVE FITTING** Find a cubic function whose graph passes through the points $(0, 0)$, $(1, 1)$, $(2, 6)$, and $(-1, 0)$. (*Hint:* Use the equation $y = ax^3 + bx^2 + cx + d$.)

Supplemental Exercises

In Exercises 45 to 47, use the system of equations

$$\begin{cases} x + 3y - a^2z = a^2 \\ 2x + 3y + az = 2 \\ 3x + 4y + 2z = 3 \end{cases}$$

45. Find all values of a for which the system of equations has a unique solution.

46. Find all values of a for which the system of equations has infinitely many solutions.

47. Find all values of a for which the system of equations has no solutions.

48. Find an equation of the plane that passes through the points $(1, 2, 6)$, $(-1, 1, 7)$, and $(4, 2, 0)$. Use the equation $z = ax + by + c$.

49. Find an equation of the plane that passes through the points $(-1, 0, -4)$, $(2, 1, 5)$, and $(-1, 1, -1)$. Use the equation $z = ax + by + c$.

Projects

1. **Echelon Form by Using a Graphing Calculator** Many graphing calculators have the elementary row operations as built-in functions. Complete this project using one of those calculators.

a. Enter into your calculator the augmented matrix for

$$\begin{cases} 2x - 3y + z = 4 \\ x + 2y - 2z = -2 \\ 3x + y - 3z = 4 \end{cases}$$

b. Complete the following steps to write the augmented matrix in echelon form. *Suggestion:* Suppose that you enter the augmented matrix as A. If you perform an elementary row operation on A, the new matrix will be displayed. However, matrix A has *not* been changed. The new matrix must be saved as another matrix, say B. Now perform the elementary row operations on B and save the result in B. When you have finished, the original matrix will still be in A and the echelon form of matrix A will be in B.

1. $R_1 \leftrightarrow R_2$ **2.** $-2R_1 + R_2 \rightarrow R_2$ **3.** $-3R_1 + R_3 \rightarrow R_3$

4. $-\dfrac{1}{7}R_2$ **5.** $5R_2 + R_3 \rightarrow R_3$ **6.** $-\dfrac{7}{4}R_3$

Section 8.2   THE ALGEBRA OF MATRICES

Besides being convenient for solving systems of equations, matrices are useful tools to model problems in business and science. One very prevalent application of matrices is to spreadsheet programs.

The typical method used in spreadsheets is to number the rows $1, 2, 3, \ldots$ and to identify the columns as $A, B, C, \ldots$. The partial spreadsheet (p. 446) shows how a consumer's car loan is being repaid over a 5-year period. The elements in column A represent the loan amount at the beginning of a year; column B represents the amount owed after a year; and column C represents the amount of interest paid during the year.

$$\begin{array}{c c c c} & A & B & C \\ 1 & 10,000.00 & 8,305.60 & 738.77 \\ 2 & 8,305.60 & 6,470.56 & 598.13 \\ 3 & 6,470.56 & 4,483.22 & 445.82 \\ 4 & 4,483.22 & 2,330.93 & 280.88 \\ 5 & 2,330.93 & 0.00 & 102.24 \end{array}$$

For instance, the element in 3C means that the consumer paid $445.82 in interest during the third year of the loan.

QUESTION   What is the meaning of the element in 3A?

Matrices are effective for situations in which there are a number of items to be classified. For instance, suppose a music store has sales for January as shown in the following matrix.

	Rock	R&B	Rap	Classical	Other
CDs	455	135	65	87	236
Tapes	252	68	32	40	101
Videos	36	4	5	2	28

This matrix indicates, for instance, that the music store sold 40 classical tapes in January.

Now consider a similar matrix for February.

	Rock	R&B	Rap	Classical	Other
CDs	402	128	68	101	255
Tapes	259	35	28	51	115
Videos	28	7	3	5	33

Looking at this matrix and the one for January reveals that the number of R&B tapes sold for the two months is 68 + 35 = 103. By adding the elements in corresponding cells, we obtain the total sales for the two months. In matrix notation, this would be shown as

$$\begin{bmatrix} 455 & 135 & 65 & 87 & 236 \\ 252 & 68 & 32 & 40 & 101 \\ 36 & 4 & 5 & 2 & 28 \end{bmatrix} + \begin{bmatrix} 402 & 128 & 68 & 101 & 255 \\ 259 & 35 & 28 & 51 & 115 \\ 28 & 7 & 3 & 5 & 33 \end{bmatrix} = \begin{bmatrix} 857 & 263 & 133 & 188 & 491 \\ 511 & 103 & 60 & 91 & 216 \\ 64 & 11 & 8 & 7 & 61 \end{bmatrix}$$

In the matrix that represents the sum, 857 (in row 1, column 1) indicates that a total of 857 rock music CDs were sold in January and February. Similarly, a total of 91 (row 2, column 4) classical tapes were sold for the two months.

This example suggests that the addition of two matrices should be performed by adding the corresponding elements. Before we actually state this definition, we first introduce some notation and a definition of equality.

ANSWER   At the beginning of the third year of the loan, the consumer owed $6,470.56.

Throughout this book a matrix will be indicated by using a capital letter or by surrounding a lower-case letter with brackets. For instance, a matrix can be denoted as

$$A \quad \text{or} \quad [a_{ij}]$$

An important concept involving matrices is the principle of equality.

DEFINITION OF EQUALITY OF TWO MATRICES

Two matrices $A = [a_{ij}]$ and $B = [b_{ij}]$ are equal if and only if

$$a_{ij} = b_{ij}$$

for every i and j.

For example, if $A = \begin{bmatrix} a & -2 & b \\ 3 & c & 1 \end{bmatrix}$ and $B = \begin{bmatrix} 3 & x & -4 \\ 3 & -1 & y \end{bmatrix}$, then $A = B$ if and only if $a = 3$, $x = -2$, $b = -4$, $c = -1$, and $y = 1$.

QUESTION   If two matrices A and B are equal, do they have the same order?

DEFINITION OF ADDITION OF MATRICES

If A and B are matrices of order $m \times n$, then the sum of the matrices is the $m \times n$ matrix given by

$$A + B = [a_{ij} + b_{ij}]$$

Here is an example. Let $A = \begin{bmatrix} 2 & -2 & 3 \\ 1 & 3 & -4 \end{bmatrix}$ and $B = \begin{bmatrix} 5 & -2 & 6 \\ -2 & 3 & 5 \end{bmatrix}$. Then

$$A + B = \begin{bmatrix} 2 & -2 & 3 \\ 1 & 3 & -4 \end{bmatrix} + \begin{bmatrix} 5 & -2 & 6 \\ -2 & 3 & 5 \end{bmatrix} = \begin{bmatrix} 2 + 5 & (-2) + (-2) & 3 + 6 \\ 1 + (-2) & 3 + 3 & (-4) + 5 \end{bmatrix}$$

$$= \begin{bmatrix} 7 & -4 & 9 \\ -1 & 6 & 1 \end{bmatrix}$$

Now let $C = \begin{bmatrix} 2 & -3 \\ 4 & 1 \end{bmatrix}$ and $D = \begin{bmatrix} 3 & 2 & 0 \\ 1 & -5 & 3 \end{bmatrix}$. Here $C + D$ is not defined because the matrices do not have the same order.

To define the subtraction of two matrices, we first define the additive inverse of a matrix.

ADDITIVE INVERSE OF A MATRIX

Given the matrix $A = [a_{ij}]$, the additive inverse of A is $-A = [-a_{ij}]$.

ANSWER   Yes. If they were of different order, there would be an element in one matrix for which there was no corresponding element in the second matrix.

For example, if $A = \begin{bmatrix} -2 & 3 & -1 \\ 0 & -1 & 4 \end{bmatrix}$, then the additive inverse of A is

$$-A = -\begin{bmatrix} -2 & 3 & -1 \\ 0 & -1 & 4 \end{bmatrix} = \begin{bmatrix} 2 & -3 & 1 \\ 0 & 1 & -4 \end{bmatrix}$$

Subtraction of two matrices is defined in terms of the additive inverse of a matrix.

DEFINITION OF SUBTRACTION OF MATRICES

Given two matrices A and B of order $m \times n$, then $A - B$ is the sum of A and the additive inverse of B.

$$A - B = A + (-B)$$

As an example, let $A = \begin{bmatrix} 2 & -3 \\ -1 & 2 \\ 2 & 4 \end{bmatrix}$ and $B = \begin{bmatrix} -1 & 2 \\ -4 & 1 \\ 3 & -2 \end{bmatrix}$. Then

$$A - B = \begin{bmatrix} 2 & -3 \\ -1 & 2 \\ 2 & 4 \end{bmatrix} - \begin{bmatrix} -1 & 2 \\ -4 & 1 \\ 3 & -2 \end{bmatrix} = \begin{bmatrix} 2 & -3 \\ -1 & 2 \\ 2 & 4 \end{bmatrix} + \begin{bmatrix} 1 & -2 \\ 4 & -1 \\ -3 & 2 \end{bmatrix} = \begin{bmatrix} 3 & -5 \\ 3 & 1 \\ -1 & 6 \end{bmatrix}$$

Of special importance is the *zero matrix*, which is the matrix that consists of all zeros. The zero matrix is the additive identity for matrices.

DEFINITION OF THE ZERO MATRIX

The $m \times n$ **zero matrix**, denoted by O, is the matrix whose elements are all zeros.

Three examples of zero matrices are

$$\begin{bmatrix} 0 & 0 & 0 \\ 0 & 0 & 0 \end{bmatrix} \qquad \begin{bmatrix} 0 & 0 & 0 & 0 \\ 0 & 0 & 0 & 0 \\ 0 & 0 & 0 & 0 \end{bmatrix} \qquad \begin{bmatrix} 0 & 0 \\ 0 & 0 \end{bmatrix}$$

PROPERTIES OF MATRIX ADDITION

Given matrices A, B, C and the zero matrix O, each of order $m \times n$, then the following properties hold.

Commutative	$A + B = B + A$
Associative	$A + (B + C) = (A + B) + C$
Additive inverse	$A + (-A) = O$
Additive identity	$A + O = O + A = A$

Two types of products involve matrices. The first product we will discuss is the product of a real number and a matrix. Consider the matrix below, which shows the hourly wages for various job classifications in a construction firm before a 6% pay increase.

$$
\begin{array}{cccc}
& \text{Carpenter} & \text{Welder} & \text{Plumber} & \text{Electrician}
\end{array}
$$

$$
\begin{array}{l}
\text{Apprentice} \\
\text{Journeyman}
\end{array}
\begin{bmatrix}
12.75 & 15.86 & 14.76 & 16.87 \\
15.60 & 18.07 & 16.89 & 19.05
\end{bmatrix}
$$

After the pay increase, the pay in each job category will increase by 6%. This can be shown in matrix form as

$$
1.06 \begin{bmatrix} 12.75 & 15.86 & 14.76 & 16.87 \\ 15.60 & 18.07 & 16.89 & 19.05 \end{bmatrix} = \begin{bmatrix} 1.06 \cdot 12.75 & 1.06 \cdot 15.86 & 1.06 \cdot 14.76 & 1.06 \cdot 16.87 \\ 1.06 \cdot 15.60 & 1.06 \cdot 18.07 & 1.06 \cdot 16.89 & 1.06 \cdot 19.05 \end{bmatrix}
$$

$$
= \begin{bmatrix} 13.52 & 16.81 & 15.65 & 17.88 \\ 16.54 & 19.15 & 17.90 & 20.19 \end{bmatrix}
$$

The element in row 1, column 4 indicates that an apprentice electrician will earn \$17.88 per hour after the pay increase.

This example suggests that to multiply a matrix by a constant, we multiply each entry in the matrix by the constant.

DEFINITION OF THE PRODUCT OF A REAL NUMBER AND A MATRIX

Given the $m \times n$ matrix $A = [a_{ij}]$ and the real number c, then $cA = [ca_{ij}]$.

Finding the product of a real number and a matrix is called **scalar multiplication**. As an example of this definition, consider the matrix

$$
A = \begin{bmatrix} 2 & -3 & 1 \\ 3 & 1 & -2 \\ 1 & -1 & 4 \end{bmatrix}
$$

and the constant $c = -2$. Then

$$
-2A = -2 \begin{bmatrix} 2 & -3 & 1 \\ 3 & 1 & -2 \\ 1 & -1 & 4 \end{bmatrix} = \begin{bmatrix} -2(2) & -2(-3) & -2(1) \\ -2(3) & -2(1) & -2(-2) \\ -2(1) & -2(-1) & -2(4) \end{bmatrix} = \begin{bmatrix} -4 & 6 & -2 \\ -6 & -2 & 4 \\ -2 & 2 & -8 \end{bmatrix}
$$

This definition is also used to factor a constant from a matrix.

$$
\begin{bmatrix} \frac{3}{2} & -\frac{5}{4} & \frac{1}{4} \\ \frac{3}{4} & \frac{1}{2} & \frac{5}{2} \end{bmatrix} = \frac{1}{4} \begin{bmatrix} 6 & -5 & 1 \\ 3 & 2 & 10 \end{bmatrix}
$$

PROPERTIES OF SCALAR MULTIPLICATION

Given real numbers a, b, and c and matrices $A = [a_{ij}]$ and $B = [b_{ij}]$ each of order $m \times n$, then

$$
(b + c)A = bA + cA
$$

$$
c(A + B) = cA + cB
$$

$$
a(bA) = (ab)A
$$

| EXAMPLE 1 | Find the Sum of Two Scalar Products |

Given $A = \begin{bmatrix} -2 & 3 \\ 4 & -2 \\ 0 & 4 \end{bmatrix}$ and $B = \begin{bmatrix} 8 & -2 \\ -3 & 2 \\ -4 & 7 \end{bmatrix}$, find $2A + 5B$.

Solution

$$2A + 5B = 2\begin{bmatrix} -2 & 3 \\ 4 & -2 \\ 0 & 4 \end{bmatrix} + 5\begin{bmatrix} 8 & -2 \\ -3 & 2 \\ -4 & 7 \end{bmatrix}$$

$$= \begin{bmatrix} -4 & 6 \\ 8 & -4 \\ 0 & 8 \end{bmatrix} + \begin{bmatrix} 40 & -10 \\ -15 & 10 \\ -20 & 35 \end{bmatrix} = \begin{bmatrix} 36 & -4 \\ -7 & 6 \\ -20 & 43 \end{bmatrix}$$

TRY EXERCISE 6, EXERCISE SET 8.2

Day	5:00 A.M.–5:00 P.M. $.23 per minute
Evening	5:00 P.M.–11:00 P.M. $.17 per minute
Night	11:00 P.M.–5:00 A.M. $.08 per minute

Now we turn to the product of two matrices. This product can be developed by considering long-distance telephone rates. The rates charged by a telephone company depend on the time of day a call is made. For this particular company, the schedule is shown in the table at the left. During one month, the number of long-distance minutes used by a customer of this telephone company was

Day	33 minutes
Evening	48 minutes
Night	15 minutes

The total cost for long-distance telephone service for that month is the sum of the products of the cost per minute and the number of minutes.

$$\text{Total cost} = 0.23(33) + 0.17(48) + 0.08(15) = \$16.95$$

In matrix terms, the cost per minute can be written as the *row* matrix [0.23 0.17 0.08]. The number of minutes of long-distance service used can be written as the *column* matrix $\begin{bmatrix} 33 \\ 48 \\ 15 \end{bmatrix}$. The product of the row matrix and the column matrix is

$$[0.23 \quad 0.17 \quad 0.08]\begin{bmatrix} 33 \\ 48 \\ 15 \end{bmatrix} = 0.23(33) + 0.17(48) + 0.08(15) = 16.95$$

In general, if A is a row matrix of order $1 \times n$,

$$A = [a_1 \quad a_2 \quad \cdots \quad a_n]$$

and B is a column matrix of order $n \times 1$,

$$B = \begin{bmatrix} b_1 \\ b_2 \\ \vdots \\ b_n \end{bmatrix}$$

then the product of A and B, written AB, is

$$AB = [a_1 \quad a_2 \quad a_3 \quad \cdots \quad a_n] \begin{bmatrix} b_1 \\ b_2 \\ \vdots \\ b_n \end{bmatrix} = a_1 b_1 + a_2 b_2 + a_3 b_3 + \cdots + a_n b_n$$

TAKE NOTE

For the example at the right, the number of elements in the row matrix *A* equals the number of elements in the column matrix *B*. If this is not the case, *AB* is not defined. For instance, if

$A = [-2 \quad 3 \quad 1]$ and $B = \begin{bmatrix} 3 \\ 1 \end{bmatrix}$,

then *AB* is not defined.

For example, if $A = [2 \quad -1 \quad 4]$ and $B = \begin{bmatrix} -3 \\ 2 \\ 6 \end{bmatrix}$, then

$$AB = 2(-3) + (-1)(2) + 4(6) = 16$$

Now consider three phone companies ($T_1, T_2,$ and T_3) with different rate structures and two customers (C_1 and C_2).

Telephone Company Rates (cost per minute)

	Day	Night	Evening
T_1	$.23	$.17	$.08
T_2	$.27	$.12	$.10
T_3	$.26	$.15	$.09

Customer Time Chart (minutes)

	C_1	C_2
Day	45	52
Night	73	60
Evening	21	8

In terms of matrices, let the telephone companies' rate structure be denoted by T and the customers' time usage by C. Then

$$T = \begin{bmatrix} 0.23 & 0.17 & 0.08 \\ 0.27 & 0.12 & 0.10 \\ 0.26 & 0.15 & 0.09 \end{bmatrix} \quad \text{and} \quad C = \begin{bmatrix} 45 & 52 \\ 73 & 60 \\ 21 & 8 \end{bmatrix}$$

Let P denote the product TC. This product is determined by extending the concept of the product of a row-and-column matrix. Multiply each row of T and each column of C.

$$P = \begin{bmatrix} 0.23 & 0.17 & 0.08 \\ 0.27 & 0.12 & 0.10 \\ 0.26 & 0.15 & 0.09 \end{bmatrix} \begin{bmatrix} 45 & 52 \\ 73 & 60 \\ 21 & 8 \end{bmatrix}$$

$$= \begin{bmatrix} [0.23 \quad 0.17 \quad 0.08] \begin{bmatrix} 45 \\ 73 \\ 21 \end{bmatrix} & [0.23 \quad 0.17 \quad 0.08] \begin{bmatrix} 52 \\ 60 \\ 8 \end{bmatrix} \\ [0.27 \quad 0.12 \quad 0.10] \begin{bmatrix} 45 \\ 73 \\ 21 \end{bmatrix} & [0.27 \quad 0.12 \quad 0.10] \begin{bmatrix} 52 \\ 60 \\ 8 \end{bmatrix} \\ [0.26 \quad 0.15 \quad 0.09] \begin{bmatrix} 45 \\ 73 \\ 21 \end{bmatrix} & [0.26 \quad 0.15 \quad 0.09] \begin{bmatrix} 52 \\ 60 \\ 8 \end{bmatrix} \end{bmatrix}$$

$$= \begin{bmatrix} 0.23(45) + 0.17(73) + 0.08(21) & 0.23(52) + 0.17(60) + 0.08(8) \\ 0.27(45) + 0.12(73) + 0.10(21) & 0.27(52) + 0.12(60) + 0.10(8) \\ 0.26(45) + 0.15(73) + 0.09(21) & 0.26(52) + 0.15(60) + 0.09(8) \end{bmatrix}$$

$$= \begin{bmatrix} 24.44 & 22.80 \\ 23.01 & 22.04 \\ 24.54 & 23.24 \end{bmatrix}$$

Each entry in P is the total cost for long-distance service that each customer would incur for each of the three telephone companies. For example, $p_{11} = 24.44$ represents the amount company T_1 would charge customer C_1. The entry in row 3, column 2 ($p_{32} = 23.24$) represents the amount company T_3 would charge customer C_2. In each case, the subscripts on an element of P denote the company and the customer, respectively.

Using this application as a model, we now define the product of two matrices. The definition is an extension of the definition of the product of a row matrix and a column matrix.

DEFINITION OF THE PRODUCT OF TWO MATRICES

Let $A = [a_{ij}]$ be a matrix of order $m \times n$, and let $B = [b_{ij}]$ be a matrix of order $n \times p$. Then the product AB is the matrix of order $m \times p$ given by $AB = [c_{ij}]$, where each element c_{ij} is

$$c_{ij} = \begin{bmatrix} a_{i1} & a_{i2} & a_{i3} \cdots a_{in} \end{bmatrix} \begin{bmatrix} b_{1j} \\ b_{2j} \\ b_{3j} \\ . \\ . \\ . \\ b_{nj} \end{bmatrix} = a_{i1}b_{1j} + a_{i2}b_{2j} + a_{i3}b_{3j} + \cdots + a_{in}b_{nj}$$

For the product of two matrices to be possible, the number of columns of the first matrix must equal the number of rows of the second matrix.

$$\underset{m \times n}{A} \quad \cdot \quad \underset{n \times p}{B} \quad = \quad \underset{m \times p}{C}$$

⌞ Must be equal ⌟
Order of product matrix

The product matrix has as many rows as the first matrix and as many columns as the second matrix. For example, let

$$A = \begin{bmatrix} 2 & -3 & 0 \\ 1 & 4 & -1 \end{bmatrix} \quad \text{and} \quad B = \begin{bmatrix} 1 & 0 \\ 4 & -2 \\ 3 & 5 \end{bmatrix}$$

Then A has order 2×3 and B has order 3×2. Thus the order of AB is 2×2.

$$\begin{bmatrix} 2 & -3 & 0 \\ 1 & 4 & -1 \end{bmatrix}_{2\times3} \begin{bmatrix} 1 & 0 \\ 4 & -2 \\ 3 & 5 \end{bmatrix}_{3\times2} = \begin{bmatrix} [2 \ -3 \ 0]\begin{bmatrix}1\\4\\3\end{bmatrix} & [2 \ -3 \ 0]\begin{bmatrix}0\\-2\\5\end{bmatrix} \\ [1 \ 4 \ -1]\begin{bmatrix}1\\4\\3\end{bmatrix} & [1 \ 4 \ -1]\begin{bmatrix}0\\-2\\5\end{bmatrix} \end{bmatrix}_{2\times2}$$

$$= \begin{bmatrix} 2(1) + (-3)(4) + 0(3) & 2(0) + (-3)(-2) + 0(5) \\ 1(1) + 4(4) + (-1)(3) & 1(0) + 4(-2) + (-1)(5) \end{bmatrix}_{2\times2} = \begin{bmatrix} -10 & 6 \\ 14 & -13 \end{bmatrix}_{2\times2}$$

EXAMPLE 2   *Find the Product of Two Matrices*

Find each product.

a. $\begin{bmatrix} 2 & 3 \\ -3 & 1 \\ 1 & -3 \end{bmatrix}\begin{bmatrix} 1 & 2 & -2 & 3 \\ -1 & 0 & 3 & -4 \end{bmatrix}$ **b.** $\begin{bmatrix} 1 & -1 & 3 \\ 2 & 2 & -1 \\ 0 & -2 & 3 \end{bmatrix}\begin{bmatrix} 4 & -2 & 0 \\ -1 & 3 & 1 \\ 2 & -3 & 1 \end{bmatrix}$

Solution

a. $\begin{bmatrix} 2 & 3 \\ -3 & 1 \\ 1 & -3 \end{bmatrix}\begin{bmatrix} 1 & 2 & -2 & 3 \\ -1 & 0 & 3 & -4 \end{bmatrix}$

$$= \begin{bmatrix} 2(1) + 3(-1) & 2(2) + 3(0) & 2(-2) + 3(3) & 2(3) + 3(-4) \\ (-3)(1) + 1(-1) & (-3)(2) + 1(0) & (-3)(-2) + 1(3) & (-3)3 + 1(-4) \\ 1(1) + (-3)(-1) & 1(2) + (-3)(0) & 1(-2) + (-3)(3) & 1(3) + (-3)(-4) \end{bmatrix}$$

$$= \begin{bmatrix} -1 & 4 & 5 & -6 \\ -4 & -6 & 9 & -13 \\ 4 & 2 & -11 & 15 \end{bmatrix}$$

b. $\begin{bmatrix} 1 & -1 & 3 \\ 2 & 2 & -1 \\ 0 & -2 & 3 \end{bmatrix}\begin{bmatrix} 4 & -2 & 0 \\ -1 & 3 & 1 \\ 2 & -3 & 1 \end{bmatrix}$

$$= \begin{bmatrix} 4 + 1 + 6 & -2 + (-3) + (-9) & 0 + (-1) + 3 \\ 8 + (-2) + (-2) & -4 + 6 + 3 & 0 + 2 + (-1) \\ 0 + 2 + 6 & 0 + (-6) + (-9) & 0 + (-2) + 3 \end{bmatrix}$$

$$= \begin{bmatrix} 11 & -14 & 2 \\ 4 & 5 & 1 \\ 8 & -15 & 1 \end{bmatrix}$$

TRY EXERCISE 16, EXERCISE SET 8.2

Graphing calculators can be used to perform matrix operations. You first enter the dimension of the matrix and then its elements. The matrix is stored (usually) as an upper-case letter. Once you have entered the matrices, you can use the regular arithmetic operation keys and the variable names for the matrices to perform many operations. For instance, let

$$A = \begin{bmatrix} -2 & 3 & 4 \\ 1 & 0 & -2 \\ 3 & 1 & -1 \end{bmatrix} \text{ and } B = \begin{bmatrix} 1 & 2 & -1 \\ 3 & 1 & 0 \\ -1 & 1 & 2 \end{bmatrix}.$$ Typical calculator displays for

addition and multiplication of matrices A and B are shown in **Figure 8.1.**

```
[A] + [B]
              [[ -1  5   3 ]
               [ 4   1  -2 ]
               [ 2   2   1 ]]
```

```
[A] * [B]
              [[ 3  3  10 ]
               [ 3  0  -5 ]
               [ 7  6  -5 ]]
```

Figure 8.1

Generally, matrix multiplication is not commutative. That is, given two matrices A and B, $AB \neq BA$. In some cases, as in Example 2a, if the matrices were reversed, the product would not be defined.

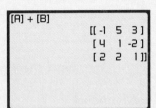

$$\begin{bmatrix} 1 & 2 & -2 & 3 \\ -1 & 0 & 3 & -4 \end{bmatrix}_{2 \times 4} \begin{bmatrix} 2 & 3 \\ -3 & 1 \\ 1 & -3 \end{bmatrix}_{3 \times 2}$$

columns $\neq$ rows

Even in those cases where multiplication is defined, the products AB and BA may not be equal. Finding the product of Example 2b with the matrices reversed illustrates this point.

$$\begin{bmatrix} 4 & -2 & 0 \\ -1 & 3 & 1 \\ 2 & -3 & 1 \end{bmatrix}\begin{bmatrix} 1 & -1 & 3 \\ 2 & 2 & -1 \\ 0 & -2 & 3 \end{bmatrix} = \begin{bmatrix} 0 & -8 & 14 \\ 5 & 5 & -3 \\ -4 & -10 & 12 \end{bmatrix} \neq \begin{bmatrix} 11 & -14 & 2 \\ 4 & 5 & 1 \\ 8 & -15 & 1 \end{bmatrix}$$

Although matrix multiplication is not a commutative operation, the associative property of multiplication and the distributive property do hold for matrices.

PROPERTIES OF MATRIX MULTIPLICATION

Associative property   Given matrices A, B, and C of orders $m \times n$, $n \times p$, and $p \times q$, respectively, then

$$A(BC) = (AB)C$$

Distributive property   Given matrices A_1 and A_2 of order $m \times n$ and matrices B_1 and B_2 of order $n \times p$, then

$$A_1(B_1 + B_2) = A_1B_1 + A_1B_2 \qquad \bullet \text{ Left distributive property}$$

$$(A_1 + A_2)B_1 = A_1B_1 + A_2B_1 \qquad \bullet \text{ Right distributive property}$$

A square matrix that has a 1 for each element on the main diagonal and zeros elsewhere is called an *identity matrix.*

IDENTITY MATRIX

The **identity matrix** of order n, denoted I_n, is the $n \times n$ matrix

$$I_n = \begin{bmatrix} 1 & 0 & 0 & \cdots & 0 \\ 0 & 1 & 0 & \cdots & 0 \\ 0 & 0 & 1 & \cdots & 0 \\ \vdots & \vdots & \vdots & \cdots & \vdots \\ 0 & 0 & 0 & \cdots & 1 \end{bmatrix}_{n \times n}$$

The identity matrix has properties similar to those of the real number 1. For example, the product of a matrix A and I is A.

$$\begin{bmatrix} 2 & -3 & 0 \\ 4 & 7 & -5 \\ 9 & 8 & -6 \end{bmatrix} \begin{bmatrix} 1 & 0 & 0 \\ 0 & 1 & 0 \\ 0 & 0 & 1 \end{bmatrix} = \begin{bmatrix} 2 & -3 & 0 \\ 4 & 7 & -5 \\ 9 & 8 & -6 \end{bmatrix}$$

MULTIPLICATIVE IDENTITY PROPERTY FOR MATRICES

If A is a square matrix of order n, and I_n is the identity matrix of order n, then $AI_n = I_n A = A$.

MATRIX PRODUCTS AND SYSTEMS OF EQUATIONS

Consider the system of equations

$$\begin{cases} 2x + 3y - z = 5 \\ x - 2y + 2z = 6 \\ 4x + y - 3z = 5 \end{cases}$$

This system can be expressed as a product of matrices, as shown below.

$$\begin{bmatrix} 2x + 3y - z \\ x - 2y + 2z \\ 4x + y - 3z \end{bmatrix} = \begin{bmatrix} 5 \\ 6 \\ 5 \end{bmatrix}$$ • **Equality of matrices**

$$\begin{bmatrix} 2 & 3 & -1 \\ 1 & -2 & 2 \\ 4 & 1 & -3 \end{bmatrix} \begin{bmatrix} x \\ y \\ z \end{bmatrix} = \begin{bmatrix} 5 \\ 6 \\ 5 \end{bmatrix}$$ • **Definition of matrix multiplication**

Reversing this procedure, certain matrix products can represent systems of equations. Consider the matrix equation

$$\begin{bmatrix} 4 & 3 & -2 \\ 1 & -2 & 3 \\ 1 & 0 & 5 \end{bmatrix}_{3\times3} \begin{bmatrix} x \\ y \\ z \end{bmatrix}_{3\times1} = \begin{bmatrix} 2 \\ -1 \\ 3 \end{bmatrix}_{3\times1}$$

$$\begin{bmatrix} 4x + 3y - 2z \\ x - 2y + 3z \\ x + 5z \end{bmatrix}_{3\times1} = \begin{bmatrix} 2 \\ -1 \\ 3 \end{bmatrix}_{3\times1}$$ • **Definition of matrix multiplication**

$$\begin{cases} 4x + 3y - 2z = 2 \\ x - 2y + 3z = -1 \\ x + 5z = 3 \end{cases}$$ • **Equality of matrices**

Performing operations on matrices that represent a system of equations is another method of solving systems of equations. This is discussed in the next section.

TOPICS FOR DISCUSSION

1. How are matrices related to spreadsheet programs?

2. Is it always possible to add two matrices? If so, explain why. If not, discuss what conditions must be met for two matrices to be added. Is matrix addition a commutative operation?

3. Is it always possible to multiply two matrices? If so, explain why. If not, discuss what conditions must be met for two matrices to be multiplied. Is matrix multiplication a commutative operation?

4. How does scalar multiplication differ from matrix multiplication?

5. Is it possible that two matrices could be added but not multiplied? Can two matrices be multiplied but not added? Discuss what type of conditions must be met for two matrices to be both added and multiplied.

EXERCISE SET 8.2

In Exercises 1 to 8, find a. **A + B**, b. **A − B**, c. **2B**, and d. **2A − 3B**.

1. $A = \begin{bmatrix} 2 & -1 \\ 3 & 3 \end{bmatrix}$ $B = \begin{bmatrix} -1 & 3 \\ 2 & 1 \end{bmatrix}$

2. $A = \begin{bmatrix} 0 & -2 \\ 2 & 3 \end{bmatrix}$ $B = \begin{bmatrix} 5 & -1 \\ 3 & 0 \end{bmatrix}$

3. $A = \begin{bmatrix} 0 & -1 & 3 \\ 1 & 0 & -2 \end{bmatrix}$ $B = \begin{bmatrix} -3 & 1 & 2 \\ 2 & 5 & -3 \end{bmatrix}$

4. $A = \begin{bmatrix} 2 & -2 & 4 \\ 0 & -3 & -4 \end{bmatrix}$ $B = \begin{bmatrix} 1 & -5 & 6 \\ 4 & -2 & -3 \end{bmatrix}$

5. $A = \begin{bmatrix} -3 & 4 \\ 2 & -3 \\ -1 & 0 \end{bmatrix}$ $B = \begin{bmatrix} 4 & 1 \\ 1 & -2 \\ 3 & -4 \end{bmatrix}$

6. $A = \begin{bmatrix} 2 & -2 \\ 3 & 4 \\ 1 & 0 \end{bmatrix}$ $B = \begin{bmatrix} -1 & 8 \\ 2 & -2 \\ -4 & 3 \end{bmatrix}$

7. $A = \begin{bmatrix} -2 & 3 & -1 \\ 0 & -1 & 2 \\ -4 & 3 & 3 \end{bmatrix}$ $B = \begin{bmatrix} 1 & -2 & 0 \\ 2 & 3 & -1 \\ 3 & -1 & 2 \end{bmatrix}$

8. $A = \begin{bmatrix} 0 & 2 & 0 \\ 1 & -3 & 3 \\ 5 & 4 & -2 \end{bmatrix}$ $B = \begin{bmatrix} -1 & 2 & 4 \\ 3 & 3 & -2 \\ -4 & 4 & 3 \end{bmatrix}$

In Exercises 9 to 16, find AB and BA if possible.

9. $A = \begin{bmatrix} 2 & -3 \\ 1 & 4 \end{bmatrix}$ $B = \begin{bmatrix} -2 & 4 \\ 2 & -3 \end{bmatrix}$

10. $A = \begin{bmatrix} 3 & -2 \\ 4 & 1 \end{bmatrix}$ $B = \begin{bmatrix} -1 & -1 \\ 0 & 4 \end{bmatrix}$

11. $A = \begin{bmatrix} 3 & -1 \\ 2 & 3 \end{bmatrix}$ $B = \begin{bmatrix} 4 & 1 \\ 2 & -3 \end{bmatrix}$

12. $A = \begin{bmatrix} -3 & 2 \\ 2 & -2 \end{bmatrix}$ $B = \begin{bmatrix} 0 & 2 \\ -2 & 4 \end{bmatrix}$

13. $A = \begin{bmatrix} 2 & -1 \\ 0 & 3 \\ 1 & -2 \end{bmatrix}$ $B = \begin{bmatrix} 1 & -2 & 3 \\ 2 & 0 & 1 \end{bmatrix}$

14. $A = \begin{bmatrix} -1 & 3 \\ 2 & 1 \\ -3 & -2 \end{bmatrix}$ $B = \begin{bmatrix} 0 & -1 & 2 \\ 1 & 2 & -4 \end{bmatrix}$

15. $A = \begin{bmatrix} 2 & -1 & 3 \\ 0 & 2 & -1 \\ 0 & 0 & 2 \end{bmatrix}$ $B = \begin{bmatrix} 2 & 0 & 0 \\ 1 & -1 & 0 \\ 2 & -1 & -2 \end{bmatrix}$

16. $A = \begin{bmatrix} -1 & 2 & 0 \\ 2 & -1 & 1 \\ -2 & 2 & -1 \end{bmatrix}$ $B = \begin{bmatrix} 2 & -1 & 0 \\ 1 & 5 & -1 \\ 0 & -1 & 3 \end{bmatrix}$

In Exercises 17 to 24, find AB if possible.

17. $A = \begin{bmatrix} 1 & -2 & 3 \end{bmatrix}$ $B = \begin{bmatrix} 1 & 0 \\ 2 & -1 \\ 1 & 2 \end{bmatrix}$

18. $A = \begin{bmatrix} -2 & 3 \\ 1 & -2 \\ 0 & 2 \end{bmatrix}$ $B = \begin{bmatrix} 3 \\ -2 \end{bmatrix}$

19. $A = \begin{bmatrix} 2 & -1 \\ 3 & 3 \end{bmatrix}$ $B = \begin{bmatrix} 1 & -2 \\ 3 & 1 \\ 0 & -2 \end{bmatrix}$

20. $A = \begin{bmatrix} 2 & 0 & -1 \\ 3 & 4 & -3 \end{bmatrix}$ $B = \begin{bmatrix} 3 & -1 & 0 \\ 2 & 4 & 5 \end{bmatrix}$

21. $A = \begin{bmatrix} 2 & 3 \\ -4 & -6 \end{bmatrix}$ $B = \begin{bmatrix} 3 & 6 \\ -2 & -4 \end{bmatrix}$

22. $A = \begin{bmatrix} 2 & -1 & 3 \\ -1 & 2 & 1 \end{bmatrix}$ $B = \begin{bmatrix} 1 & 3 & 2 \\ 2 & -1 & 0 \\ 3 & 1 & 2 \end{bmatrix}$

23. $A = \begin{bmatrix} 1 & 2 & -2 & 3 \\ 0 & -2 & 1 & -3 \end{bmatrix}$ $B = \begin{bmatrix} -2 & 0 \\ 4 & -2 \end{bmatrix}$

24. $A = \begin{bmatrix} 2 & -2 & 4 \\ 1 & 0 & -1 \\ 2 & 1 & 3 \end{bmatrix}$ $B = \begin{bmatrix} 2 & 1 & -3 & 0 \\ 0 & -2 & 1 & -2 \\ 1 & -1 & 0 & 2 \end{bmatrix}$

In Exercises 25 to 28, given the matrices

$$A = \begin{bmatrix} -1 & 3 \\ 2 & -1 \\ 3 & 1 \end{bmatrix} \text{ and } B = \begin{bmatrix} 0 & -2 \\ 1 & 3 \\ 4 & -3 \end{bmatrix}$$

find the 3 × 2 matrix X that is a solution of the equation.

25. $3X + A = B$ 26. $2A - 3X = 5B$

27. $2X - A = X + B$ 28. $3X + 2B = X - 2A$

In Exercises 29 to 32, use the matrices

$$A = \begin{bmatrix} 2 & -3 \\ 1 & -1 \end{bmatrix} \text{ and } B = \begin{bmatrix} 3 & -1 & 0 \\ 2 & -2 & -1 \\ 1 & 0 & 2 \end{bmatrix}$$

If A is a square matrix, then $A^n = A \cdot A \cdot A \cdots A$, where the matrix A is repeated n times.

29. Find A^2.   30. Find A^3.

31. Find B^2.   32. Find B^3.

In Exercises 33 to 38, find the system of equations that is equivalent to the given matrix equation.

33. $\begin{bmatrix} 3 & -8 \\ 4 & 3 \end{bmatrix} \begin{bmatrix} x \\ y \end{bmatrix} = \begin{bmatrix} 11 \\ 1 \end{bmatrix}$ 34. $\begin{bmatrix} 2 & 7 \\ 3 & -4 \end{bmatrix} \begin{bmatrix} x \\ y \end{bmatrix} = \begin{bmatrix} 1 \\ 16 \end{bmatrix}$

35. $\begin{bmatrix} 1 & -3 & -2 \\ 3 & 1 & 0 \\ 2 & -4 & 5 \end{bmatrix} \begin{bmatrix} x \\ y \\ z \end{bmatrix} = \begin{bmatrix} 6 \\ 2 \\ 1 \end{bmatrix}$

36. $\begin{bmatrix} 2 & 0 & 5 \\ 3 & -5 & 1 \\ 4 & -7 & 6 \end{bmatrix} \begin{bmatrix} x \\ y \\ z \end{bmatrix} = \begin{bmatrix} 9 \\ 7 \\ 14 \end{bmatrix}$

37. $\begin{bmatrix} 2 & -1 & 0 & 2 \\ 4 & 1 & 2 & -3 \\ 6 & 0 & 1 & -2 \\ 5 & 2 & -1 & -4 \end{bmatrix} \begin{bmatrix} x_1 \\ x_2 \\ x_3 \\ x_4 \end{bmatrix} = \begin{bmatrix} 5 \\ 6 \\ 10 \\ 8 \end{bmatrix}$

38. $\begin{bmatrix} 5 & -1 & 2 & -3 \\ 4 & 0 & 2 & 0 \\ 2 & -2 & 5 & -4 \\ 3 & 1 & -3 & 4 \end{bmatrix} \begin{bmatrix} x_1 \\ x_2 \\ x_3 \\ x_4 \end{bmatrix} = \begin{bmatrix} -2 \\ 2 \\ -1 \\ 2 \end{bmatrix}$

39. **LIFE SCIENCES** Biologists use capture-recapture models to estimate how many animals live in a certain area. A sample of, say, fish are caught and tagged. When sub-

sequent samples of fish are caught, a biologist can use a capture history matrix to record (with a 1) which, if any, of the fish in the original sample are caught again. The rows of this matrix represent the particular fish (each has its own identification number), and the columns represent the number of the sample in which the fish was caught. Here is a small capture history matrix.

Samples

$$\begin{array}{c} & \begin{array}{cccc} 1 & 2 & 3 & 4 \end{array} \\ \begin{array}{c} \text{Fish A} \\ \text{Fish B} \\ \text{Fish C} \end{array} & \left[\begin{array}{cccc} 1 & 0 & 0 & 1 \\ 0 & 1 & 1 & 1 \\ 0 & 0 & 1 & 1 \end{array}\right] \end{array}$$

a. What is the dimension of this matrix? Write a sentence that explains the meaning of dimension in this case.

b. What is the meaning of the 1 in row A, column 4?

c. Which fish was captured the most times?

40. LIFE SCIENCE Biologists can use a predator-prey matrix to study the relationships among animals in an ecosystem. Each row and each column represents an animal in the system. A 1 as an element in the matrix indicates that the animal represented by that row preys on the animal represented by that column. A 0 indicates that the animal in that row does not prey on the animal in that column. A simple predator-prey matrix is shown below. The abbreviations are H = hawk, R = rabbit, S = snake, C = coyote.

$$\begin{array}{c} & \begin{array}{cccc} H & R & S & C \end{array} \\ \begin{array}{c} H \\ R \\ S \\ C \end{array} & \left[\begin{array}{cccc} 0 & 1 & 1 & 0 \\ 0 & 0 & 0 & 0 \\ 1 & 1 & 0 & 0 \\ 0 & 1 & 1 & 0 \end{array}\right] \end{array}$$

a. What is the dimension of this matrix? Write a sentence that explains the meaning of dimension in this case.

b. What is the meaning of the 0 in row 2, column 1?

c. What is the meaning of there being all zeros in column C?

d. What is the meaning of all zeros in row R?

41. BUSINESS The matrix below shows the sales revenues, in millions of dollars, that a pharmaceutical company received from various divisions in different parts of the country. The abbreviations are W = western states, N = northern states, S = southern states, and E = eastern states.

$$\begin{array}{c} & \begin{array}{cccc} W & N & S & E \end{array} \\ \begin{array}{c} \text{Patented drugs} \\ \text{Generic drugs} \\ \text{Nonprescription drugs} \end{array} & \left[\begin{array}{cccc} 2.0 & 1.4 & 3.0 & 1.4 \\ 0.8 & 1.1 & 2.0 & 0.9 \\ 3.6 & 1.2 & 4.5 & 1.5 \end{array}\right] \end{array}$$

Suppose the business plan for this company indicates that it anticipates a 2% decrease in sales (because of competition) for each of its drug divisions for each region of the country. Express, to the nearest ten thousand dollars, this matrix as a scalar product and compute the anticipated sales matrix.

42. SALARY SCHEDULES The partial current-year salary matrix for an elementary school district is given below. Column A indicates a B.A. degree, column B a B.A. degree plus 15 graduate units, column C an M.A. degree, and column D an M.A. degree plus 30 additional graduate units. The rows give the numbers of years of teaching experience. Each entry is the annual salary in thousands of dollars.

$$\begin{array}{c} & \begin{array}{cccc} A & B & C & D \end{array} \\ \begin{array}{c} \\ \text{Years} \end{array} \begin{array}{c} \text{0 to 5} \\ \text{5 to 9} \\ \text{10 to 15} \end{array} & \left[\begin{array}{cccc} 18.0 & 18.9 & 20.0 & 21.5 \\ 19.0 & 20.3 & 22.5 & 24.5 \\ 20.0 & 21.4 & 24.0 & 27.0 \end{array}\right] \end{array}$$

Express, as matrix scalar multiplication to the nearest hundred dollars, the result of the school board's approving a 6% salary increase for all teachers in this district, and compute the scalar product.

43. SPORTS The matrices for the number of wins and losses at home, H, and away, A, are shown for the top 3 finishers of the 1995 American League East division baseball teams.

$$H = \begin{array}{c} & \begin{array}{cc} W & L \end{array} \\ & \left[\begin{array}{cc} 42 & 30 \\ 46 & 26 \\ 36 & 36 \end{array}\right] \begin{array}{c} \text{Boston} \\ \text{New York} \\ \text{Baltimore} \end{array} \end{array} \qquad A = \begin{array}{c} & \begin{array}{cc} W & L \end{array} \\ & \left[\begin{array}{cc} 44 & 28 \\ 33 & 39 \\ 35 & 37 \end{array}\right] \begin{array}{c} \text{Boston} \\ \text{New York} \\ \text{Baltimore} \end{array} \end{array}$$

a. Find $H + A$.

b. Write a sentence that explains the meaning of the sum of the two matrices.

c. Find $H - A$.

d. Write a sentence that explains the meaning of the difference of the two matrices.

44. BUSINESS Let A represent the number of televisions of various sizes in two stores of a company in one city, and let B represent the same situation for the company in a second city.

$$A = \begin{array}{c} & \begin{array}{ccc} \text{19-inch} & \text{25-inch} & \text{40-inch} \end{array} \\ & \left[\begin{array}{ccc} 23 & 35 & 49 \\ 32 & 41 & 24 \end{array}\right] \begin{array}{c} \text{Store 1} \\ \text{Store 2} \end{array} \end{array}$$

$$B = \begin{array}{c} & \begin{array}{ccc} \text{19-inch} & \text{25-inch} & \text{40-inch} \end{array} \\ & \left[\begin{array}{ccc} 19 & 28 & 36 \\ 25 & 38 & 26 \end{array}\right] \begin{array}{c} \text{Store 1} \\ \text{Store 2} \end{array} \end{array}$$

a. Find $A + B$.

b. Write a sentence that explains the meaning of the sum of the two matrices.

45. GEOMETRIC TRANSFORMATION Consider the rectangle shown below. Each pair of *x*- and *y*-coordinates of the points shown appears as a column of a matrix of dimension 2×4. This is matrix *A* below. Matrix *T* is called a *translation matrix*.

$$A = \begin{bmatrix} -2 & 4 & 2 & -4 \\ 5 & 2 & -2 & 1 \end{bmatrix}$$

$$T = \begin{bmatrix} 2 & 2 & 2 & 2 \\ -1 & -1 & -1 & -1 \end{bmatrix}$$

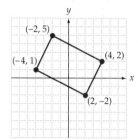

a. Find $A + T$.

b. Using the columns of $A + T$ as the *x*- and *y*-coordinates of four points, plot the points on the same coordinate grid as the original rectangle. Construct a polygon by connecting the points in the order of the columns. What polygon have you constructed?

c. ✏ Write a sentence that explains the relationship between the original polygon and the new one.

46. GEOMETRIC TRANSFORMATION Use the information in Exercise 45.

a. Find $A - T$.

b. Using the columns of $A - T$ as the *x*- and *y*-coordinates of four points, plot the points on the same coordinate grid as the original rectangle. Construct a polygon by connecting the points in the order of the columns. What polygon have you constructed?

c. ✏ Write a sentence that explains the relationship between the original polygon and the new one.

47. GEOMETRIC TRANSFORMATION Consider the information in Exercise 45 and the matrix $R = \begin{bmatrix} 0 & -1 \\ 1 & 0 \end{bmatrix}$.

a. Find $R \cdot A$.

b. Using the columns of $R \cdot A$ as the *x*- and *y*-coordinates of four points, plot the points on the same coordinate grid as the original rectangle. Construct a polygon by connecting the points in the order of the columns. What polygon have you constructed?

c. ✏ Write a sentence that explains the meaning of the product of these two matrices.

48. GEOMETRIC TRANSFORMATION Consider the information in Exercise 45 and the matrix $R = \begin{bmatrix} 0 & 1 \\ 1 & 0 \end{bmatrix}$.

a. Find $R \cdot A$.

b. Using the columns of $R \cdot A$ as the *x*- and *y*-coordinates of four points, plot the points on the same coordinate grid as the original rectangle. Construct a polygon by connecting the points in the order of the columns. What polygon have you constructed?

c. ✏ Write a sentence that explains the meaning of the product of these two matrices.

49. BUSINESS INVENTORY Matrix *A* gives the stock on hand of four products in a warehouse at the beginning of the week, and matrix *B* gives the stock on hand for the same four items at the end of the week. Find and interpret $A - B$.

	Blue	Green	Red	
	530	650	815	Pens
$A =$	190	385	715	Pencils
	485	600	610	Ink
	150	210	305	Colored Lead

	Blue	Green	Red	
	480	500	675	Pens
$B =$	175	215	345	Pencils
	400	350	480	Ink
	70	95	280	Colored Lead

50. BUSINESS SERVICES Matrix *A* gives the number of employees in the divisions of a company in the west coast branch, and matrix *B* gives the same information for the east coast branch. Find and interpret $A + B$.

	Engineering	Admini- stration	Data Processing	
	315	200	415	Division I
$A =$	285	175	300	Division II
	275	195	250	Division III

	Engineering	Admini- stration	Data Processing	
	200	175	350	Division I
$B =$	150	90	180	Division II
	105	50	175	Division III

51. BUSINESS INVENTORY The total unit sales matrix for three computer stores is given by

	Monitors	Printers	Computers	Drives	
	25	31	35	12	Store A
$S =$	20	12	30	15	Store B
	16	19	25	18	Store C

The unit pricing matrix in dollars for the three stores is given by

	Store A	Store B	Store C	
	250	225	315	Monitor
$P =$	180	210	225	Printer
	400	425	450	Computer
	89	95	78	Drive

Find the gross income matrix.

52. YOUTH SPORTS The total unit sales matrix at three soccer games in a summer league for children is given by

	Soft Drinks	Hot Dogs	Candy	Popcorn	
	52	50	75	20	Game 1
$S =$	45	48	80	20	Game 2
	62	70	78	25	Game 3

The unit pricing matrix in dollars for the wholesale cost of each item and the retail price of each item is given by

	Wholesale	Retail	
	0.25	0.50	Soft Drinks
$P =$	0.30	0.75	Hot Dogs
	0.15	0.45	Candy
	0.10	0.50	Popcorn

Use matrix multiplication to find the total cost and total revenue at each game.

In Exercises 53 to 58, use a graphing calculator to perform the indicated operations on matrices A and B.

$$A = \begin{bmatrix} 2 & -1 & 3 & 5 & -1 \\ 2 & 0 & 2 & -1 & 1 \\ -1 & -3 & 2 & 3 & 3 \\ 5 & -4 & 1 & 0 & 3 \\ 0 & 2 & -1 & 4 & 3 \end{bmatrix}$$

$$B = \begin{bmatrix} 0 & -2 & 1 & 7 & 2 \\ -3 & 0 & 2 & 3 & 1 \\ -2 & 1 & 1 & 4 & 5 \\ 6 & 4 & -4 & 2 & -3 \\ 3 & -2 & -5 & 1 & 3 \end{bmatrix}$$

53. AB **54.** BA **55.** A^3

56. B^3 **57.** $A^2 + B^2$ **58.** $AB - BA$

SUPPLEMENTAL EXERCISES

The elements of a matrix can be complex numbers. In Exercises 59 to 68, let

$$A = \begin{bmatrix} 2 + 3i & 1 - 2i \\ 1 + i & 2 - i \end{bmatrix} \quad \text{and} \quad B = \begin{bmatrix} 1 - i & 2 + 3i \\ 3 + 2i & 4 - i \end{bmatrix}$$

Perform the indicated operations.

59. $3A$ **60.** $-2B$ **61.** $2iB$ **62.** $3iA$

63. $A + B$ **64.** $A - B$ **65.** AB **66.** BA

67. A^2 **68.** B^2

Matrices with complex number elements play a role in the theory of the atom. The following three matrices, called Pauli spin matrices, were used by **Wolfgang Pauli** in his early study of the electron. Use these matrices in Exercises 69 to 71.

$$\sigma_1 = \begin{bmatrix} 0 & 1 \\ 1 & 0 \end{bmatrix} \quad \sigma_2 = \begin{bmatrix} 0 & -i \\ i & 0 \end{bmatrix} \quad \sigma_3 = \begin{bmatrix} 1 & 0 \\ 0 & -1 \end{bmatrix}$$

69. Show that $(\sigma_i)^2 = I_2$ for $i = 1, 2,$ and 3.

70. Show that $\sigma_1 \cdot \sigma_2 = i\sigma_3$.

71. Show that $\sigma_1 \cdot \sigma_2 + \sigma_2 \cdot \sigma_1 = O$.

72. Given two real numbers a and b and a matrix A of order 2×2, prove that $(a + b)A = aA + bA$.

73. Given two real numbers a and b and a matrix A of order 2×2, prove that $a(bA) = (ab)A$.

PROJECTS

1. **MATRICES IN GRAPHICS ART** Matrices can be used to translate and dilate geometric figures in the plane. These concepts are important in graphics design and art. Consider the kite shown at the right.

a. Find the coordinates of each vertex of the kite, and prepare a matrix K of dimension 2×4, where the columns are the x- and y-coordinates, respectively, of the vertices of the kite.

b. Find the product $\frac{1}{2}K = M$. Plot the points of M, using the columns as the x- and y-coordinates, respectively, of the vertices of a new kite.

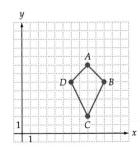

c. Find the length of any two line segments of the original kite and the corresponding lengths for the new kite. Show that the **scale factor,** which is the ratio $\dfrac{\text{length of new}}{\text{length of original}}$, is $\dfrac{1}{2}$ for each pair of line segments. Observe that this is the coefficient of K in the product in **b**.

d. Find the product $2K = N$. Plot the points of N again, using the columns as the x- and y-coordinates, respectively, of the vertices of a new kite. Show that the new kite has a scale factor of 2.

e. Consider the vertices A, B, C, and D of the original kite and the corresponding vertices A', B', C', and D' of the kite from **c**. Show that the lines through AA', BB', CC', and DD' pass through the origin of the coordinate system. This point of intersection is called the **center of dilation.**

SECTION

8.3 THE INVERSE OF A MATRIX

Recall that the multiplicative inverse of a nonzero real number c is $1/c$, the number whose product with c is 1. For example, the multiplicative inverse of $\dfrac{2}{3}$ is $\dfrac{3}{2}$ because $\dfrac{2}{3} \cdot \dfrac{3}{2} = 1$.

For some square matrices we can define a multiplicative inverse.

MULTIPLICATIVE INVERSE OF A MATRIX

If A is a square matrix of order n, then the **inverse** of matrix A, denoted by A^{-1}, has the property that

$$A \cdot A^{-1} = A^{-1} \cdot A = I_n$$

where I_n is the identity matrix of order n.

As we will see shortly, not all square matrices have a multiplicative inverse.

QUESTION   Are there any real numbers that do not have a multiplicative inverse?

A procedure for finding the inverse (we will simply say *inverse* for *multiplicative inverse*) uses elementary row operations. The procedure will be illustrated by finding the inverse of a 2 × 2 matrix.

ANSWER   The real number zero does not have a multiplicative inverse.

Let $A = \begin{bmatrix} 2 & 7 \\ 1 & 4 \end{bmatrix}$. To the matrix A we will merge the identity matrix I_2 to the right of A and denote this new matrix by $[A:I_2]$.

$$[A:I_2] = \begin{bmatrix} 2 & 7 & 1 & 0 \\ 1 & 4 & 0 & 1 \end{bmatrix}$$

$$A \longrightarrow \uparrow \qquad \uparrow \!\!\!\!\!\!\! \underline{\quad\quad} \; I_2$$

Now we use elementary row operations in a manner similar to that of the Gaussian elimination method. The goal is to produce

$$[I_2:A^{-1}] = \begin{bmatrix} 1 & 0 & b_{11} & b_{12} \\ 0 & 1 & b_{21} & b_{22} \end{bmatrix}$$

$$I_2 \longrightarrow \uparrow \qquad \uparrow \!\!\!\!\!\!\! \underline{\quad\quad} \; A^{-1}$$

In this form, the inverse matrix is the matrix that is to the right of the identity matrix. That is,

$$A^{-1} = \begin{bmatrix} b_{11} & b_{12} \\ b_{21} & b_{22} \end{bmatrix}$$

To find A^{-1}, we first use a series of elementary row operations that will result in a 1 in the first row and the first column.

$$\begin{bmatrix} 2 & 7 & 1 & 0 \\ 1 & 4 & 0 & 1 \end{bmatrix} \xrightarrow{\frac{1}{2} R_1} \begin{bmatrix} 1 & \frac{7}{2} & \frac{1}{2} & 0 \\ 1 & 4 & 0 & 1 \end{bmatrix} \xrightarrow{-1R_1 + R_2} \begin{bmatrix} 1 & \frac{7}{2} & \frac{1}{2} & 0 \\ 0 & \frac{1}{2} & -\frac{1}{2} & 1 \end{bmatrix}$$

$$\xrightarrow{2R_2} \begin{bmatrix} 1 & \frac{7}{2} & \frac{1}{2} & 0 \\ 0 & 1 & -1 & 2 \end{bmatrix} \xrightarrow{-\frac{7}{2}R_2 + R_1} \begin{bmatrix} 1 & 0 & 4 & -7 \\ 0 & 1 & -1 & 2 \end{bmatrix}$$

The inverse matrix is the matrix to the right of the identity matrix. Therefore,

$$A^{-1} = \begin{bmatrix} 4 & -7 \\ -1 & 2 \end{bmatrix}$$

Each elementary row operation is chosen to advance the process of transforming the original matrix into the identity matrix.

EXAMPLE 1   *Find the Inverse of a 3 × 3 Matrix*

Find the inverse of the matrix $A = \begin{bmatrix} 1 & -1 & 2 \\ 2 & 0 & 6 \\ 3 & -5 & 7 \end{bmatrix}$.

Solution

$$\begin{bmatrix} 1 & -1 & 2 & 1 & 0 & 0 \\ 2 & 0 & 6 & 0 & 1 & 0 \\ 3 & -5 & 7 & 0 & 0 & 1 \end{bmatrix}$$

• **Merge the given matrix with the identity matrix I_3.**

$$-2R_1 + R_2 \atop -3R_1 + R_3 \longrightarrow \begin{bmatrix} 1 & -1 & 2 & 1 & 0 & 0 \\ 0 & 2 & 2 & -2 & 1 & 0 \\ 0 & -2 & 1 & -3 & 0 & 1 \end{bmatrix}$$

• Because a_{11} is already 1, we next produce zeros in a_{21} and a_{31}.

$$\frac{1}{2}R_2 \longrightarrow \begin{bmatrix} 1 & -1 & 2 & 1 & 0 & 0 \\ 0 & 1 & 1 & -1 & \frac{1}{2} & 0 \\ 0 & -2 & 1 & -3 & 0 & 1 \end{bmatrix}$$

• Produce a 1 in a_{22}.

$$2R_2 + R_3 \longrightarrow \begin{bmatrix} 1 & -1 & 2 & 1 & 0 & 0 \\ 0 & 1 & 1 & -1 & \frac{1}{2} & 0 \\ 0 & 0 & 3 & -5 & 1 & 1 \end{bmatrix}$$

• Produce a 0 in a_{32}.

$$\frac{1}{3}R_3 \longrightarrow \begin{bmatrix} 1 & -1 & 2 & 1 & 0 & 0 \\ 0 & 1 & 1 & -1 & \frac{1}{2} & 0 \\ 0 & 0 & 1 & -\frac{5}{3} & \frac{1}{3} & \frac{1}{3} \end{bmatrix}$$

• Produce a 1 in a_{33}.

$$-1R_3 + R_2 \atop -2R_3 + R_1 \longrightarrow \begin{bmatrix} 1 & -1 & 0 & \frac{13}{3} & -\frac{2}{3} & -\frac{2}{3} \\ 0 & 1 & 0 & \frac{2}{3} & \frac{1}{6} & -\frac{1}{3} \\ 0 & 0 & 1 & -\frac{5}{3} & \frac{1}{3} & \frac{1}{3} \end{bmatrix}$$

• Now work upward. Produce a 0 in a_{23} and a_{13}.

$$R_2 + R_1 \longrightarrow \begin{bmatrix} 1 & 0 & 0 & 5 & -\frac{1}{2} & -1 \\ 0 & 1 & 0 & \frac{2}{3} & \frac{1}{6} & -\frac{1}{3} \\ 0 & 0 & 1 & -\frac{5}{3} & \frac{1}{3} & \frac{1}{3} \end{bmatrix}$$

• Produce a 0 in a_{12}.

The inverse matrix is $A^{-1} = \begin{bmatrix} 5 & -\frac{1}{2} & -1 \\ \frac{2}{3} & \frac{1}{6} & -\frac{1}{3} \\ -\frac{5}{3} & \frac{1}{3} & \frac{1}{3} \end{bmatrix}$.

You should verify that this matrix satisfies the condition of an inverse matrix. That is, show that $A^{-1} \cdot A = A \cdot A^{-1} = I_3$.

TRY EXERCISE 6, EXERCISE SET 8.3

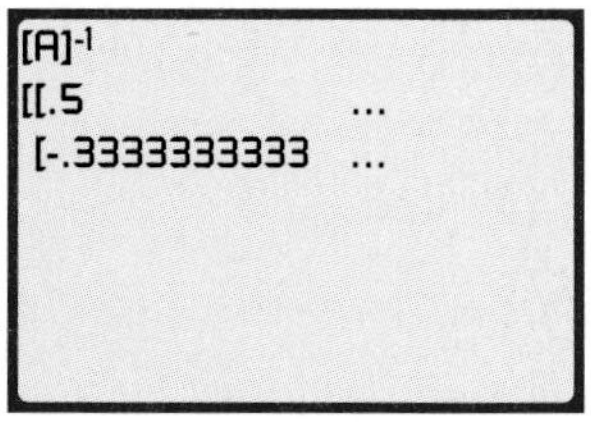

The inverse of a matrix can be found by using a graphing calculator. Enter and store the matrix in some variable, say A. To compute the inverse of A, use the $\boxed{x^{-1}}$ key to calculate the inverse. For instance, let $A = \begin{bmatrix} 4 & 3 \\ 2 & 3 \end{bmatrix}$. A typical calculator display of the inverse of A is shown in **Figure 8.3.** Because the elements of the matrix are decimals, it is possible to see only the first column of the inverse matrix. Use the arrow keys to see the remaining columns.

Another possibility for viewing the inverse of A is to use the function on your calculator that converts a decimal to a fraction. This will change the decimals to fractions, and you will be able to see more columns of the matrix.

```
[A]⁻¹
[[.5          ...
 [-.3333333333 ...
```

Figure 8.3

A **singular matrix** is a matrix that does not have a multiplicative inverse. A matrix that has a multiplicative inverse is a **nonsingular matrix.** As you apply the procedure beginning on page 462 to a singular matrix, there will come a point where there are all zeros in a row of the *original* matrix. When that condition exists, the original matrix does not have an inverse.

EXAMPLE 2 *Identify a Singular Matrix*

Show that the matrix $\begin{bmatrix} 1 & -1 & -1 \\ 2 & -3 & 0 \\ 1 & -2 & 1 \end{bmatrix}$ is a singular matrix.

Solution

$$\begin{bmatrix} 1 & -1 & -1 & 1 & 0 & 0 \\ 2 & -3 & 0 & 0 & 1 & 0 \\ 1 & -2 & 1 & 0 & 0 & 1 \end{bmatrix} \xrightarrow[\substack{-2R_1 + R_2 \\ -1R_1 + R_3}]{} \begin{bmatrix} 1 & -1 & -1 & 1 & 0 & 0 \\ 0 & -1 & 2 & -2 & 1 & 0 \\ 0 & -1 & 2 & -1 & 0 & 1 \end{bmatrix}$$

$$\xrightarrow{-1 \cdot R_2} \begin{bmatrix} 1 & -1 & -1 & 1 & 0 & 0 \\ 0 & 1 & -2 & 2 & -1 & 0 \\ 0 & -1 & 2 & -1 & 0 & 1 \end{bmatrix} \xrightarrow{R_2 + R_3} \begin{bmatrix} 1 & -1 & -1 & -1 & 0 & 0 \\ 0 & 1 & -2 & 2 & -1 & 0 \\ 0 & 0 & 0 & 1 & -1 & 1 \end{bmatrix}$$

There are zeros in a row of the original matrix. The original matrix does not have an inverse.

TRY EXERCISE 10, EXERCISE SET 8.3

SOLVING SYSTEMS OF EQUATIONS USING INVERSE MATRICES

Systems of linear equations can be solved by finding the inverse of the coefficient matrix. Consider the system of equations

$$\begin{cases} 3x_1 + 4x_2 = -1 \\ 3x_1 + 5x_2 = 1 \end{cases} \qquad (1)$$

Using matrix multiplication and the concept of equality of matrices, we can write this system as a matrix equation.

$$\begin{bmatrix} 3 & 4 \\ 3 & 5 \end{bmatrix} \begin{bmatrix} x_1 \\ x_2 \end{bmatrix} = \begin{bmatrix} -1 \\ 1 \end{bmatrix} \qquad (2)$$

If we let

$$A = \begin{bmatrix} 3 & 4 \\ 3 & 5 \end{bmatrix} \qquad X = \begin{bmatrix} x_1 \\ x_2 \end{bmatrix} \qquad B = \begin{bmatrix} -1 \\ 1 \end{bmatrix}$$

then Equation (2) can be written as $AX = B$. The inverse of the coefficient matrix A is $A^{-1} = \begin{bmatrix} \frac{5}{3} & -\frac{4}{3} \\ -1 & 1 \end{bmatrix}$.

To solve the system of equations, multiply each side of the equation $AX = B$ by the inverse A^{-1}.

$$\begin{bmatrix} \frac{5}{3} & -\frac{4}{3} \\ -1 & 1 \end{bmatrix}\begin{bmatrix} 3 & 4 \\ 3 & 5 \end{bmatrix}\begin{bmatrix} x_1 \\ x_2 \end{bmatrix} = \begin{bmatrix} \frac{5}{3} & -\frac{4}{3} \\ -1 & 1 \end{bmatrix}\begin{bmatrix} -1 \\ 1 \end{bmatrix}$$

$$\begin{bmatrix} x_1 \\ x_2 \end{bmatrix} = \begin{bmatrix} -3 \\ 2 \end{bmatrix}$$

Thus $x_1 = -3$ and $x_2 = 2$. The solution to System (1) is $(-3, 2)$.

TAKE NOTE

The disadvantage of using the inverse matrix method to solve a system of equations is that this method will not work if the system is dependent or inconsistent. In addition, this method cannot distinguish between inconsistent and dependent systems. However, in some applications this method is very efficient. See the material on input-output analysis later in this section.

EXAMPLE 3   *Solve a System of Equations by Using the Inverse of the Coefficient Matrix*

Find the solution of the system of equations by using the inverse of the coefficient matrix.

$$\begin{cases} x_1 + 7x_3 = 20 \\ 2x_1 + x_2 - x_3 = -3 \\ 7x_1 + 3x_2 + x_3 = 2 \end{cases} \quad (1)$$

Solution

Write the system as a matrix equation.

$$\begin{bmatrix} 1 & 0 & 7 \\ 2 & 1 & -1 \\ 7 & 3 & 1 \end{bmatrix}\begin{bmatrix} x_1 \\ x_2 \\ x_3 \end{bmatrix} = \begin{bmatrix} 20 \\ -3 \\ 2 \end{bmatrix} \quad (2)$$

The inverse of the coefficient matrix is $\begin{bmatrix} -\frac{4}{3} & -7 & \frac{7}{3} \\ 3 & 16 & -5 \\ \frac{1}{3} & 1 & -\frac{1}{3} \end{bmatrix}$.

Multiply each side of the matrix equation (2) by the inverse.

$$\begin{bmatrix} -\frac{4}{3} & -7 & \frac{7}{3} \\ 3 & 16 & -5 \\ \frac{1}{3} & 1 & -\frac{1}{3} \end{bmatrix}\begin{bmatrix} 1 & 0 & 7 \\ 2 & 1 & -1 \\ 7 & 3 & 1 \end{bmatrix}\begin{bmatrix} x_1 \\ x_2 \\ x_3 \end{bmatrix} = \begin{bmatrix} -\frac{4}{3} & -7 & \frac{7}{3} \\ 3 & 16 & -5 \\ \frac{1}{3} & 1 & -\frac{1}{3} \end{bmatrix}\begin{bmatrix} 20 \\ -3 \\ 2 \end{bmatrix}$$

$$\begin{bmatrix} x_1 \\ x_2 \\ x_3 \end{bmatrix} = \begin{bmatrix} -1 \\ 2 \\ 3 \end{bmatrix}$$

Thus $x_1 = -1$, $x_2 = 2$, and $x_3 = 3$. The solution to System (1) is $(-1, 2, 3)$.

TRY EXERCISE 20, EXERCISE SET 8.3

The advantage of using the inverse matrix to solve a system of equations is not apparent unless it is necessary to solve repeatedly a system of equations with the same coefficient matrix but different constant matrices. *Input-output analysis* is one such application of this method.

INPUT-OUTPUT ANALYSIS

In an economy, some of the output of an industry is used by the industry to produce its own product. For example, an electric company uses water and electricity to produce electricity, and a water company uses water and electricity to produce drinking water. **Input-output analysis** attempts to determine the necessary output of industries to satisfy each other's demands plus the demands of consumers. Wassily Leontief, a Harvard economist, was awarded the Nobel prize for his work in this field.

An **input-output matrix** is used to express the interdependence among industries in an economy. Each column of this matrix gives the dollar values of the inputs an industry needs to produce $1 worth of output.

To illustrate the concepts, we will assume an economy with only three industries: agriculture, transportation, and oil. Suppose that to produce $1 worth of agricultural products requires $.05 worth of agriculture, $.02 worth of transportation, and $.05 worth of oil. To produce $1 worth of transportation requires $.10 worth of agriculture, $.08 worth of transportation, and $.10 worth of oil. To produce $1 worth of oil requires $.10 worth of agriculture, $.15 worth of transportation, and $.13 worth of oil. The input-output matrix A is

$$
\begin{array}{c}
\textbf{Input requirements of} \\
\begin{array}{ccc}
\text{Agriculture} & \text{Transportation} & \text{Oil}
\end{array}
\end{array}
$$

$$
\textbf{from}\quad
\begin{array}{l}
\text{Agriculture} \\
\text{Transportation} \\
\text{Oil}
\end{array}
\begin{bmatrix}
0.05 & 0.10 & 0.10 \\
0.02 & 0.08 & 0.15 \\
0.05 & 0.10 & 0.13
\end{bmatrix}
$$

Consumers (other than the industries themselves) want to purchase some of the output from these industries. The amount of output that the consumer will want is called the **final demand** on the economy. This is represented by a column matrix.

Suppose in our example that the final demand is $3 billion worth of agriculture, $1 billion worth of transportation, and $2 billion worth of oil. The final demand matrix is

$$
\begin{bmatrix} 3 \\ 1 \\ 2 \end{bmatrix} = D
$$

We represent the total output of each industry (in billions of dollars) as follows:

$$x = \text{total output of agriculture}$$

$$y = \text{total output of transportation}$$

$$z = \text{total output of oil}$$

The object of input-output analysis is to determine the values of x, y, and z that will satisfy the amount the consumer demands. To find these values, consider agriculture. The amount of agriculture left for the consumer (demand d) is

$$d = x - (\text{amount of agriculture used by industries}) \qquad (1)$$

To find the amount of agriculture used by the three industries in our economy, refer to the input-output matrix. Production of x billion dollars worth of agriculture takes $0.05x$ of agriculture, production of y billion dollars worth of transportation takes $0.10y$ of agriculture, and production of z billion dollars worth of oil takes $0.10z$ of agriculture. Thus,

Amount of agriculture used by industries $= 0.05x + 0.10y + 0.10z$ (2)

Combining Equations (1) and (2), we have

$$d = x - (0.05x + 0.10y + 0.10z)$$
$$3 = 0.95x - 0.10y - 0.10z \qquad \bullet \textbf{ d is \$3 billion for agriculture.}$$

We could continue this way for each of the other industries. The result would be a system of equations. Instead, however, we will use a matrix approach.

If $X =$ total output of the three industries of the economy, then

$$X = \begin{bmatrix} x \\ y \\ z \end{bmatrix}$$

The product of A, the input-output matrix, and X is

$$AX = \begin{bmatrix} 0.05 & 0.10 & 0.10 \\ 0.02 & 0.08 & 0.15 \\ 0.05 & 0.10 & 0.13 \end{bmatrix} \begin{bmatrix} x \\ y \\ z \end{bmatrix}$$

This matrix represents the dollar amount of products used in production for all three industries. Thus the amount available for consumer demand is $X - AX$. As a matrix equation, we can write

$$X - AX = D$$

Solving this equation for X, we determine the output necessary to meet the needs of our industries and the consumer.

$$IX - AX = D \qquad \bullet \textbf{\textit{I} is the identity matrix. Thus \textit{IX} = X.}$$
$$(I - A)X = D \qquad \bullet \textbf{ Right distributive property}$$
$$X = (I - A)^{-1}D \qquad \bullet \textbf{ Assuming the inverse of (\textit{I} $-$ A) exists}$$

The last equation states that the solution to an input-output problem can be found by multiplying the demand matrix D by the inverse of $(I - A)$. In our example, we have

$$I - A = \begin{bmatrix} 1 & 0 & 0 \\ 0 & 1 & 0 \\ 0 & 0 & 1 \end{bmatrix} - \begin{bmatrix} 0.05 & 0.10 & 0.10 \\ 0.02 & 0.08 & 0.15 \\ 0.05 & 0.10 & 0.13 \end{bmatrix} = \begin{bmatrix} 0.95 & -0.10 & -0.10 \\ -0.02 & 0.92 & -0.15 \\ -0.05 & -0.10 & 0.87 \end{bmatrix}$$

$$(I - A)^{-1} \approx \begin{bmatrix} 1.063 & 0.131 & 0.145 \\ 0.034 & 1.112 & 0.196 \\ 0.065 & 0.135 & 1.180 \end{bmatrix}$$

The consumer demand is

$$X = (I - A)^{-1}D$$
$$X \approx \begin{bmatrix} 1.063 & 0.131 & 0.145 \\ 0.034 & 1.112 & 0.196 \\ 0.065 & 0.135 & 1.180 \end{bmatrix} \begin{bmatrix} 3 \\ 1 \\ 2 \end{bmatrix} = \begin{bmatrix} 3.61 \\ 1.61 \\ 2.69 \end{bmatrix}$$

This matrix indicates that $3.61 billion worth of agriculture, $1.61 billion worth of transportation, and $2.69 billion worth of oil must be produced by the industries to satisfy consumers' demands and the industries' internal requirements.

If we change the final demand matrix,

$$D = \begin{bmatrix} 2 \\ 2 \\ 3 \end{bmatrix}$$

then the total output of the economy can be found as

$$X \approx \begin{bmatrix} 1.063 & 0.131 & 0.145 \\ 0.034 & 1.112 & 0.196 \\ 0.065 & 0.135 & 1.180 \end{bmatrix} \begin{bmatrix} 2 \\ 2 \\ 3 \end{bmatrix} = \begin{bmatrix} 2.82 \\ 2.88 \\ 3.94 \end{bmatrix}$$

Thus agriculture must produce output worth $2.82 billion, transportation must produce output worth $2.88 billion, and oil must produce output worth $3.94 billion to satisfy the given consumer demand and the industries' internal requirements.

TOPICS FOR DISCUSSION

1. Explain how to find the inverse of a matrix.

2. Explain the difference between a singular matrix and a nonsingular matrix.

3. Discuss the advantages and disadvantages of solving a system of equations by using an inverse matrix.

4. Do all square matrices have an inverse? If not, give an example of a square matrix that does not have an inverse.

EXERCISE SET 8.3

In Exercises 1 to 10, find the inverse of the given matrix.

1. $\begin{bmatrix} 1 & -3 \\ -2 & 5 \end{bmatrix}$

2. $\begin{bmatrix} 1 & 2 \\ -2 & -3 \end{bmatrix}$

3. $\begin{bmatrix} 1 & 4 \\ 2 & 10 \end{bmatrix}$

4. $\begin{bmatrix} -2 & 3 \\ -6 & -8 \end{bmatrix}$

5. $\begin{bmatrix} 1 & 2 & -1 \\ 2 & 5 & 1 \\ 3 & 6 & -2 \end{bmatrix}$

6. $\begin{bmatrix} 1 & 3 & -2 \\ -1 & -5 & 6 \\ 2 & 6 & -3 \end{bmatrix}$

7. $\begin{bmatrix} 1 & 2 & -1 \\ 2 & 6 & 1 \\ 3 & 6 & -4 \end{bmatrix}$

8. $\begin{bmatrix} 2 & 1 & -1 \\ 6 & 4 & -1 \\ 4 & 2 & -3 \end{bmatrix}$

9. $\begin{bmatrix} 2 & 4 & -4 \\ 1 & 3 & -4 \\ 2 & 4 & -3 \end{bmatrix}$

10. $\begin{bmatrix} 1 & -2 & 2 \\ 2 & -3 & 1 \\ 3 & -6 & 6 \end{bmatrix}$

In Exercises 11 to 14, use a graphing calculator to find the inverse of the given matrix.

11. $\begin{bmatrix} 1 & -1 & 2 & 1 \\ 2 & -1 & 5 & 1 \\ 3 & -3 & 7 & 5 \\ -2 & 3 & -4 & -1 \end{bmatrix}$

12. $\begin{bmatrix} 1 & 1 & -1 & 2 \\ 3 & 2 & -1 & 5 \\ 2 & 2 & -1 & 5 \\ 4 & 4 & -4 & 7 \end{bmatrix}$

13. $\begin{bmatrix} 1 & -1 & 1 & 3 \\ 2 & -1 & 4 & 8 \\ 1 & 1 & 6 & 10 \\ -1 & 5 & 5 & 4 \end{bmatrix}$

14. $\begin{bmatrix} 1 & -1 & 1 & 2 \\ 2 & -1 & 6 & 6 \\ 3 & -1 & 12 & 12 \\ -2 & -1 & -14 & -10 \end{bmatrix}$

In Exercises 15 to 24, solve each system of equations by using inverse matrix methods.

15. $\begin{cases} x + 4y = 6 \\ 2x + 7y = 11 \end{cases}$

16. $\begin{cases} 2x + 3y = 5 \\ x + 2y = 4 \end{cases}$

17. $\begin{cases} x - 2y = 8 \\ 3x + 2y = -1 \end{cases}$

18. $\begin{cases} 3x - 5y = -18 \\ 2x - 3y = -11 \end{cases}$

19. $\begin{cases} x + y + 2z = 4 \\ 2x + 3y + 3z = 5 \\ 3x + 3y + 7z = 14 \end{cases}$

20. $\begin{cases} x + 2y - z = 5 \\ 2x + 3y - z = 8 \\ 3x + 6y - 2z = 14 \end{cases}$

21. $\begin{cases} x + 2y + 2z = 5 \\ -2x - 5y - 2z = 8 \\ 2x + 4y + 7z = 19 \end{cases}$

22. $\begin{cases} x - y + 3z = 5 \\ 3x - y + 10z = 16 \\ 2x - 2y + 5z = 9 \end{cases}$

23. $\begin{cases} w + 2x + z = 6 \\ 2w + 5x + y + 2z = 10 \\ 2w + 4x + y + z = 8 \\ 3w + 6x + 4z = 16 \end{cases}$

24. $\begin{cases} w - x + 2y = 5 \\ 2w - x + 6y + 2z = 16 \\ 3w - 2x + 9y + 4z = 28 \\ w - 2x - z = 2 \end{cases}$

In Exercises 25–28, solve each application by writing a system of equations that models the conditions and then applying inverse matrix methods.

25. BUSINESS REVENUE A vacation resort offers a helicopter tour of an island. The price for an adult ticket is $20; the price for a children's ticket is $15. The records of the tour operator show that 100 people took the tour on Saturday and 120 people took the tour on Sunday. The total receipts for Saturday were $1900, and on Sunday the receipts were $2275. Find the number of adults and the number of children who took the tour on Saturday and on Sunday.

26. BUSINESS REVENUE A company sells a standard and a deluxe model tape recorder. Each standard tape recorder costs $45 to manufacture, and each deluxe model costs $60 to manufacture. The January manufacturing budget for 90 of these recorders was $4650; the February budget for 100 recorders was $5250. Find the number of each type of recorder manufactured in January and in February.

27. SOIL SCIENCE The following table shows the active chemical content of three different soil additives.

Additive	Grams per 100 Grams		
	Ammonium Nitrate	Phosphorus	Iron
1	30	10	10
2	40	15	10
3	50	5	5

A soil chemist wants to prepare two chemical samples. The first sample contains 380 grams of ammonium nitrate, 95 grams of phosphorus, and 85 grams of iron. The second sample requires 380 grams of ammonium nitrate, 110 grams of phosphorus, and 90 grams of iron. How many grams of each additive are required for sample 1, and how many grams of each additive are required for sample 2?

28. NUTRITION The following table shows the carbohydrate, fat, and protein content of three food types.

Food Type	Grams per 100 Grams		
	Carbohydrate	Fat	Protein
I	13	10	13
II	4	4	3
III	1	0	10

A nutritionist must prepare two diets from these three food groups. The first diet must contain 23 grams of carbohydrate, 18 grams of fat, and 39 grams of protein. The second diet must contain 35 grams of carbohydrate, 28 grams of fat, and 42 grams of protein. How many grams of each food type are required for the first diet, and how many grams of each food type are required for the second diet?

In Exercises 29 to 32, use a graphing calculator to find the inverse of each matrix.

29. $\begin{bmatrix} 2 & -2 & 3 & 1 \\ 5 & 2 & -2 & 3 \\ 6 & -1 & 2 & 3 \\ 2 & 3 & -1 & 5 \end{bmatrix}$

30. $\begin{bmatrix} 3 & -1 & 0 & 1 \\ 2 & -2 & 3 & 0 \\ -1 & -3 & 5 & 3 \\ 5 & 3 & -2 & 1 \end{bmatrix}$

31. $\begin{bmatrix} -\frac{2}{7} & 4 & -\frac{1}{6} \\ -2 & \sqrt{2} & -3 \\ \sqrt{3} & 3 & -\sqrt{5} \end{bmatrix}$

32. $\begin{bmatrix} 6 & \pi & -\frac{4}{7} \\ -5 & \sqrt{7} & 2 \\ \frac{5}{6} & -\sqrt{3} & \sqrt{10} \end{bmatrix}$

33. INPUT-OUTPUT ANALYSIS A simplified economy has three industries: manufacturing, transportation, and service. The input-output matrix for this economy is

$$\begin{bmatrix} 0.20 & 0.15 & 0.10 \\ 0.10 & 0.30 & 0.25 \\ 0.20 & 0.10 & 0.10 \end{bmatrix}$$

Find the gross output needed to satisfy the consumer demand of $120 million worth of manufacturing, $60 million worth of transportation, and $55 million worth of service.

34. INPUT-OUTPUT ANALYSIS A four-sector economy consists of manufacturing, agriculture, service, and transportation. The input-output matrix for this economy is

$$\begin{bmatrix} 0.10 & 0.05 & 0.20 & 0.15 \\ 0.20 & 0.10 & 0.30 & 0.10 \\ 0.05 & 0.30 & 0.20 & 0.40 \\ 0.10 & 0.20 & 0.15 & 0.20 \end{bmatrix}$$

Find the gross output needed to satisfy a consumer demand of $80 million worth of manufacturing, $100 million worth of agriculture, $50 million worth of service, and $80 million worth of transportation.

35. INPUT-OUTPUT ANALYSIS A conglomerate is composed of three industries: coal, iron, and steel. To produce $1 worth of coal requires $.05 worth of coal, $.02 worth of iron, and $.10 worth of steel. To produce $1 worth of iron requires $.20 worth of coal, $.03 worth of iron, and $.12 worth of steel. To produce $1 worth of steel requires $.15 worth of coal, $.25 worth of iron, and $.05 worth of steel. How much should each industry produce to allow for a consumer demand of $30 million worth of coal, $5 million worth of iron, and $25 million worth of steel?

36. INPUT-OUTPUT ANALYSIS A conglomerate has three divisions: plastics, semiconductors, and computers. For each $1 worth of output, the plastics division needs $.01 worth of plastics, $.03 worth of semiconductors, and $.10 worth of computers. Each $1 worth of output from the semiconductor division requires $.08 worth from plastics, $.05 worth from semiconductors, and $.15 worth from computers. For each $1 worth of output, the computer division needs $.20 worth from plastics, $.20 worth from semiconductors, and $.10 worth from computers. The conglomerate estimates consumer demand of $100 million worth from the plastics division, $75 million worth from the semiconductor division, and $150 million worth from the computer division. At what level should each division produce to satisfy this demand?

SUPPLEMENTAL EXERCISES

37. Let $A = \begin{bmatrix} 2 & -3 \\ -6 & 9 \end{bmatrix}$ and $B = \begin{bmatrix} -3 & 15 \\ -2 & 10 \end{bmatrix}$. Show that $AB = O$, the 2×2 zero matrix. This illustrates that for matrices, if $AB = O$, it is not necessarily so that $A = O$ or $B = O$.

38. Show that if a matrix A has an inverse and $AB = O$, then $B = O$.

39. Let $A = \begin{bmatrix} 2 & -1 \\ -4 & 2 \end{bmatrix}$, $B = \begin{bmatrix} 3 & 4 \\ 1 & 5 \end{bmatrix}$, and $C = \begin{bmatrix} 4 & 7 \\ 3 & 11 \end{bmatrix}$. Show that $AB = AC$ but that $B \neq C$. This illustrates that the cancellation rule of real numbers may not apply to matrices.

40. (Continuation of Exercise 39.) Show that if A is a matrix that has an inverse and $AB = AC$, then $B = C$.

41. Show that if $A = \begin{bmatrix} a & b \\ c & d \end{bmatrix}$ and $ad - bc \neq 0$ then

$$A^{-1} = \frac{1}{ad - bc} \begin{bmatrix} d & -b \\ -c & a \end{bmatrix}$$

42. Use the result of Exercise 41 to show that a square matrix of order 2 has an inverse if and only if $ad - bc \neq 0$.

43. Use the result of Exercise 41 to find the inverse of each matrix.

a. $\begin{bmatrix} 2 & -3 \\ 4 & -5 \end{bmatrix}$ b. $\begin{bmatrix} 5 & 6 \\ 3 & 4 \end{bmatrix}$ c. $\begin{bmatrix} 0 & -1 \\ 4 & 4 \end{bmatrix}$

44. Let $A = \begin{bmatrix} 3 & -2 \\ 1 & 1 \end{bmatrix}$ and $B = \begin{bmatrix} 2 & -1 \\ 2 & 3 \end{bmatrix}$. Use Exercise 41 to show that

$$A^{-1} = \frac{1}{5} \begin{bmatrix} 1 & 2 \\ -1 & 3 \end{bmatrix} \quad \text{and} \quad B^{-1} = \frac{1}{8} \begin{bmatrix} 3 & 1 \\ -2 & 2 \end{bmatrix}$$

Now show that $(AB)^{-1} = B^{-1} \cdot A^{-1}$.

45. Generalize the last result in Exercise 44. That is, show that if A and B are square matrices of order n and each has an inverse matrix, then $(AB)^{-1} = B^{-1} \cdot A^{-1}$. (*Hint:* Begin with the equation $(AB)(AB)^{-1} = I$, where I is the identity matrix. Now multiply each side of the equation by A^{-1} and then by B^{-1}.)

PROJECTS

1. Cryptography is the study of the techniques of concealing the meaning of a message. The message that is to be concealed is called **plaintext.** The concealed message is called **ciphertext.** One way to change plaintext to ciphertext is to give each letter of the alphabet a numerical equivalent. Then matrices are used to scramble the numbers so that it is difficult to determine which number is associated with which letter.

   a. One way to assign each letter a number is to use the ASCII coding system. Determine how this system assigns a number to each letter and punctuation mark.

   b. Now write a short sentence, such as "THE BUCK STOPS HERE." Group the letters of the sentence into packets of, say, 3, using 0 (zero) for a space. Our sentence would look like

(THE)(0BU)(CK0)(STO)(PS0)(HER)(E.0)

Replace each letter and punctuation mark by its numerical ASCII equivalent. For our message, the first three groups would be

$$(84 \quad 72 \quad 69)(48 \quad 66 \quad 85)(67 \quad 75 \quad 48)\dots$$

Place these numbers in a matrix, using the set of three numbers as a column. For our example, the first three columns are

$$W = \begin{bmatrix} 84 & 48 & 67 & \dots \\ 72 & 66 & 75 & \dots \\ 69 & 85 & 48 & \dots \end{bmatrix}$$

c. Now construct a 3×3 matrix E that has an inverse. You can use any 3×3 matrix as long as you can find the inverse. (A graphing calculator may be useful here.)

d. Find the product $E \cdot W = M$. The numbers in the matrix M would be sent as the coded message. Do this for your message.

e. The person who receives this message would multiply the matrix M by E^{-1} to restore the message to its original form. Do this for your message.

SECTION

8.4 DETERMINANTS

Associated with each square matrix A is a number called the *determinant* of A. We will denote the determinant of the matrix A by $\det(A)$ or by $|A|$. For the remainder of this chapter, we assume that all matrices are square matrices.

THE DETERMINANT OF A 2 × 2 MATRIX

The **determinant** of the matrix $A = [a_{ij}]$ of order 2 is

$$|A| = \begin{vmatrix} a_{11} & a_{12} \\ a_{21} & a_{22} \end{vmatrix} = a_{11}a_{22} - a_{21}a_{12}$$

Caution   Be careful not to confuse the notation for a matrix and that for a determinant. The symbol [ ] (brackets) is used for a matrix; the symbol | | (vertical bars) is used for the determinant of a matrix.

An easy way to remember the formula for the determinant of a 2×2 matrix is to recognize that the determinant is the difference between the products of the diagonal elements. That is,

$$\begin{vmatrix} a_{11} & a_{12} \\ a_{21} & a_{22} \end{vmatrix} = a_{11}a_{22} - a_{21}a_{12}$$

EXAMPLE 1   *Find the Value of a Determinant*

Find the value of the determinant of the matrix $A = \begin{bmatrix} 5 & 3 \\ 2 & -3 \end{bmatrix}$.

Solution

$$|A| = \begin{vmatrix} 5 & 3 \\ 2 & -3 \end{vmatrix} = 5(-3) - 2(3) = -15 - 6 = -21$$

TRY EXERCISE 2, EXERCISE SET 8.4

MINORS AND COEFFICIENTS

To define the determinant of a matrix of order greater than 2, we first need two other definitions.

THE MINOR OF A MATRIX

The **minor** M_{ij} of the element a_{ij} of a square matrix A of order $n \geq 3$ is the determinant of the matrix of order $n - 1$ obtained by deleting the ith row and the jth column of A.

Consider the matrix $A = \begin{bmatrix} 2 & -1 & 5 \\ 4 & 3 & -7 \\ 8 & -7 & 6 \end{bmatrix}$. The minor M_{23} is the determinant of the matrix A formed by deleting row 2 and column 3 from A.

$$M_{23} = \begin{vmatrix} 2 & -1 \\ 8 & -7 \end{vmatrix} \qquad \bullet \begin{vmatrix} 2 & -1 & 5 \\ 4 & 3 & -7 \\ 8 & -7 & 6 \end{vmatrix}$$

$$= 2(-7) - 8(-1) = -14 + 8 = -6$$

The minor M_{31} is the determinant of the matrix A formed by deleting row 3 and column 1 from A.

$$M_{31} = \begin{vmatrix} -1 & 5 \\ 3 & -7 \end{vmatrix} \qquad \bullet \begin{vmatrix} 2 & -1 & 5 \\ 4 & 3 & -7 \\ 8 & -7 & 6 \end{vmatrix}$$

$$= (-1)(-7) - 3(5) = 7 - 15 = -8$$

The second definition we need is that of the *cofactor* of a matrix.

COFACTOR OF A MATRIX

The **cofactor** C_{ij} of the element a_{ij} of a square matrix A is given by $C_{ij} = (-1)^{i+j}M_{ij}$, where M_{ij} is the minor of a_{ij}.

When $i + j$ is an even integer, $(-1)^{i+j} = 1$. When $i + j$ is an odd integer, $(-1)^{i+j} = -1$. Thus

$$C_{ij} = \begin{cases} M_{ij}, & i + j \text{ is an even integer} \\ -M_{ij}, & i + j \text{ is an odd integer} \end{cases}$$

EXAMPLE 2   *Find the Minor and Cofactor of a Matrix*

Given $A = \begin{bmatrix} 4 & 3 & -2 \\ 5 & -2 & 4 \\ 3 & -2 & -6 \end{bmatrix}$, find M_{32} and C_{12}.

Solution

$$M_{32} = \begin{vmatrix} 4 & -2 \\ 5 & 4 \end{vmatrix} = 4(4) - 5(-2) = 16 + 10 = 26$$

$$C_{12} = (-1)^{1+2}M_{12} = -M_{12} = -\begin{vmatrix} 5 & 4 \\ 3 & -6 \end{vmatrix} = -(-30 - 12) = 42$$

TRY EXERCISE 14, EXERCISE SET 8.4

Cofactors are used to evaluate the determinant of a matrix of order 3 or greater. The technique used to evaluate a determinant by using cofactors is called *expanding by cofactors.*

DETERMINANTS BY EXPANDING BY COFACTORS

Given the square matrix A of order 3 or greater, the value of the determinant of A is the sum of the products of the elements of any row or column and their cofactors. For the rth row of A, the value of the determinant of A is

$$|A| = a_{r1}C_{r1} + a_{r2}C_{r2} + a_{r3}C_{r3} + \cdots + a_{rn}C_{rn}$$

For the cth column of A, the determinant of A is

$$|A| = a_{1c}C_{1c} + a_{2c}C_{2c} + a_{3c}C_{3c} + \cdots + a_{nc}C_{nc}$$

This theorem states that the value of a determinant can be found by expanding by cofactors of *any* row or column. The value of the determinant is the same in each case. To illustrate the method, consider the matrix $A = \begin{bmatrix} 2 & 3 & -1 \\ 4 & -2 & 3 \\ 1 & -3 & 4 \end{bmatrix}$. Expanding the determinant of A by some row, say row 2, gives

$$|A| = \begin{vmatrix} 2 & 3 & -1 \\ 4 & -2 & 3 \\ 1 & -3 & 4 \end{vmatrix} = 4C_{21} + (-2)C_{22} + 3C_{23}$$

$$= 4(-1)^{2+1}M_{21} + (-2)(-1)^{2+2}M_{22} + 3(-1)^{2+3}M_{23}$$

$$= (-4)\begin{vmatrix} 3 & -1 \\ -3 & 4 \end{vmatrix} + (-2)\begin{vmatrix} 2 & -1 \\ 1 & 4 \end{vmatrix} + (-3)\begin{vmatrix} 2 & 3 \\ 1 & -3 \end{vmatrix}$$

$$= (-4)9 + (-2)9 + (-3)(-9) = -27$$

Expanding the determinant of A by some column, say column 3, gives

$$|A| = \begin{vmatrix} 2 & 3 & -1 \\ 4 & -2 & 3 \\ 1 & -3 & 4 \end{vmatrix} = (-1)C_{13} + 3C_{23} + 4C_{33}$$

$$= (-1)(-1)^{1+3}M_{13} + 3(-1)^{2+3}M_{23} + 4(-1)^{3+3}M_{33}$$

$$= (-1)\begin{vmatrix} 4 & -2 \\ 1 & -3 \end{vmatrix} + (-3)\begin{vmatrix} 2 & 3 \\ 1 & -3 \end{vmatrix} + 4\begin{vmatrix} 2 & 3 \\ 4 & -2 \end{vmatrix}$$

$$= (-1)(-10) + (-3)(-9) + 4(-16) = -27$$

The value of the determinant of A is the same whether we expanded by cofactors of the elements of a row or by cofactors of the elements of a column. When evaluating a determinant, choose the most convenient row or column, which usually is the row or column containing the most zeros.

TAKE NOTE

Example 3 illustrates that choosing a row or column with the most zeros and then expanding about that row or column will reduce the number of calculations you must perform. For Example 3 we have $0 \cdot C_{32} = 0$, and it is not necessary to compute C_{32}.

EXAMPLE 3 *Evaluate a Determinant by Cofactors*

Evaluate the determinant of $A = \begin{bmatrix} 5 & -3 & -1 \\ -2 & 1 & -1 \\ 1 & 0 & 2 \end{bmatrix}$ by expanding by cofactors.

Solution

Because $a_{32} = 0$, expand using row 3 or column 2. Row 3 will be used here.

$$|A| = 1C_{31} + 0C_{32} + 2C_{33} = 1(-1)^{3+1}M_{31} + 0(-1)^{3+2}M_{32} + 2(-1)^{3+3}M_{33}$$

$$= 1\begin{vmatrix} -3 & -1 \\ 1 & -1 \end{vmatrix} + 0 + 2\begin{vmatrix} 5 & -3 \\ -2 & 1 \end{vmatrix} = 1[3 - (-1)] + 0 + 2[5 - 6]$$

$$= 4 - 2 = 2$$

TRY EXERCISE 20, EXERCISE SET 8.4

The determinant of a matrix can be found by using a graphing calculator. Many of these calculators use *det* as the operation that produces the value of the determinant. For instance, if $A = \begin{bmatrix} -2 & 3 & 4 \\ 1 & 0 & -2 \\ 3 & 1 & -1 \end{bmatrix}$, then a typical calculator display of the determinant of A is shown in **Figure 8.4**.

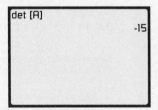

Figure 8.4

EFFECTS OF ELEMENTARY ROW OPERATIONS ON THE VALUE OF A DETERMINANT OF A MATRIX

If A is a square matrix of order n, then the following elementary row operations produce the indicated changes in the determinant of A.

1. Interchanging any two rows of A changes the sign of $|A|$.

2. Multiplying a row of A by a constant k multiplies the determinant of A by k.

3. Adding a multiple of a row of A to another row does not change the value of the determinant of A.

To illustrate these properties, consider the matrix $A = \begin{bmatrix} 2 & 3 \\ 1 & -2 \end{bmatrix}$. The determinant of A is $|A| = 2(-2) - 1(3) = -7$. Now consider each of the elementary row operations.

Interchange the rows of A and evaluate the determinant.

$$\begin{vmatrix} 1 & -2 \\ 2 & 3 \end{vmatrix} = 1(3) - 2(-2) = 3 + 4 = 7 = -|A|$$

Multiply row 2 of A by -3 and evaluate the determinant.

$$\begin{vmatrix} 2 & 3 \\ -3 & 6 \end{vmatrix} = 2(6) - (-3)3 = 12 + 9 = 21 = -3|A|$$

Multiply row 1 of A by -2 and add it to row 2. Then evaluate the determinant.

$$\begin{vmatrix} 2 & 3 \\ -3 & -8 \end{vmatrix} = 2(-8) - (-3)(3) = -16 + 9 = -7 = |A|.$$

These elementary row operations are often used to rewrite a matrix in *triangular form*. A matrix is in **triangular form** if all elements below or above the main diagonal are zero. The matrices

$$A = \begin{bmatrix} 2 & -2 & 3 & 1 \\ 0 & -2 & 4 & 2 \\ 0 & 0 & 6 & 9 \\ 0 & 0 & 0 & -5 \end{bmatrix} \quad \text{and} \quad B = \begin{bmatrix} 3 & 0 & 0 & 0 \\ 2 & -3 & 0 & 0 \\ 6 & 4 & -2 & 0 \\ 8 & 3 & 4 & 2 \end{bmatrix}$$

are in triangular form.

DETERMINANT OF A MATRIX IN TRIANGULAR FORM

Let A be a square matrix of order n in triangular form. The determinant of A is the product of the elements on the main diagonal.

$$|A| = a_{11}a_{22}a_{33} \cdots a_{nn}$$

For the matrices A and B given above,

$$|A| = 2(-2)(6)(-5) = 120$$
$$|B| = 3(-3)(-2)(2) = 36$$

EXAMPLE 4   *Evaluate a Determinant by Elementary Row Operations*

Evaluate the determinant by rewriting in triangular form.

$$\begin{vmatrix} 2 & 1 & -1 & 3 \\ 2 & 2 & 0 & 1 \\ 4 & 5 & 4 & -3 \\ 2 & 2 & 7 & -3 \end{vmatrix}$$

Solution

Rewrite the matrix in triangular form by using elementary row operations.

$$\begin{vmatrix} 2 & 1 & -1 & 3 \\ 2 & 2 & 0 & 1 \\ 4 & 5 & 4 & -3 \\ 2 & 2 & 7 & -3 \end{vmatrix} \begin{matrix} -1R_1 + R_2 \\ -2R_1 + R_3 \\ -1R_1 + R_4 \\ = \end{matrix} \begin{vmatrix} 2 & 1 & -1 & 3 \\ 0 & 1 & 1 & -2 \\ 0 & 3 & 6 & -9 \\ 0 & 1 & 8 & -6 \end{vmatrix}$$

$$\begin{matrix} \text{Factor 3,} \\ \text{from row 3.} \\ = \end{matrix} \; 3\begin{vmatrix} 2 & 1 & -1 & 3 \\ 0 & 1 & 1 & -2 \\ 0 & 1 & 2 & -3 \\ 0 & 1 & 8 & -6 \end{vmatrix} \begin{matrix} -1R_2 + R_3 \\ -1R_2 + R_4 \\ = \end{matrix} \; 3\begin{vmatrix} 2 & 1 & -1 & 3 \\ 0 & 1 & 1 & -2 \\ 0 & 0 & 1 & -1 \\ 0 & 0 & 7 & -4 \end{vmatrix}$$

$$-7R_3 + R_4 = 3\begin{vmatrix} 2 & 1 & -1 & 3 \\ 0 & 1 & 1 & -2 \\ 0 & 0 & 1 & -1 \\ 0 & 0 & 0 & 3 \end{vmatrix} = 3(2)(1)(1)(3) = 18$$

TRY EXERCISE 42, EXERCISE SET 8.4

In some cases it is possible to recognize when the determinant of a matrix is zero.

CONDITIONS FOR A ZERO DETERMINANT

If A is a square matrix, then $|A| = 0$ when any one of the following is true.

1. A row (column) consists entirely of zeros.

2. Two rows (columns) are identical.

3. One row (column) is a constant multiple of a second row (column).

Proof To prove part 2 of this theorem, let A be the given matrix and let $D = |A|$. Now interchange the two identical rows. Then $|A| = -D$. Thus

$$D = -D$$

Zero is the only real number that is its own additive inverse, and hence $D = |A| = 0$. ◆

The proofs of the other two properties are left as exercises.

QUESTION   If I is the identity matrix of order n, what is the value of $|I|$?

The last property of determinants that we will discuss is a product property.

PRODUCT PROPERTY OF DETERMINANTS

If A and B are square matrices of order n, then

$$|AB| = |A||B|$$

Recall that a singular matrix is one that does not have a multiplicative inverse. The Product Property of Determinants can be used to determine whether a matrix has an inverse.

ANSWER   The identity matrix is in diagonal form with 1s on the main diagonal. Thus $|I|$ is a product of 1s, or $|I| = 1$.

Consider a matrix A with an inverse A^{-1}. Then, by the last theorem,

$$|A \cdot A^{-1}| = |A||A^{-1}|$$

But $A \cdot A^{-1} = I$, the identity matrix, and $|I| = 1$. Therefore,

$$1 = |A||A^{-1}|$$

From the last equation, $|A| \neq 0$. And, in particular,

$$|A^{-1}| = \frac{1}{|A|}$$

These results are summarized in the following theorem.

EXISTENCE OF THE INVERSE OF A SQUARE MATRIX

If A is a square matrix of order n, then A has a multiplicative inverse if and only if $|A| \neq 0$. Furthermore,

$$|A^{-1}| = \frac{1}{|A|}$$

We proved only part of this theorem. It remains to show that given $|A| \neq 0$, then A has an inverse. This proof can be found in most texts on linear algebra.

TOPICS FOR DISCUSSION

1. Discuss the difference between a matrix and a determinant.

2. Explain the difference between the minor and the cofactor of an element of a matrix.

3. Discuss how determinants are used to discover whether a matrix has an inverse.

4. Explain how to calculate the value of a determinant by expanding by cofactors.

5. Discuss how elementary row operations are used to find the determinant of a matrix.

EXERCISE SET 8.4

In Exercises 1 to 8, evaluate the determinants.

1. $\begin{vmatrix} 2 & -1 \\ 3 & 5 \end{vmatrix}$

2. $\begin{vmatrix} 2 & 9 \\ -6 & 2 \end{vmatrix}$

3. $\begin{vmatrix} 5 & 0 \\ 2 & -3 \end{vmatrix}$

4. $\begin{vmatrix} 0 & -8 \\ 3 & 4 \end{vmatrix}$

5. $\begin{vmatrix} 4 & 6 \\ 2 & 3 \end{vmatrix}$

6. $\begin{vmatrix} -3 & 6 \\ 4 & -8 \end{vmatrix}$

7. $\begin{vmatrix} 0 & 9 \\ 0 & -2 \end{vmatrix}$

8. $\begin{vmatrix} -3 & 9 \\ 0 & 0 \end{vmatrix}$

In Exercises 9 to 12, evaluate the indicated minor and cofactor for the determinant

$$\begin{vmatrix} 5 & -2 & -3 \\ 2 & 4 & -1 \\ 4 & -5 & 6 \end{vmatrix}$$

9. M_{11}, C_{11}

10. M_{21}, C_{21}

11. M_{32}, C_{32}

12. M_{33}, C_{33}

In Exercises 13 to 16, evaluate the indicated minor and co-factor for the determinant

$$\begin{vmatrix} 3 & -2 & 3 \\ 1 & 3 & 0 \\ 6 & -2 & 3 \end{vmatrix}$$

13. M_{22}, C_{22} 14. M_{13}, C_{13} 15. M_{31}, C_{31} 16. M_{23}, C_{23}

In Exercises 17 to 26, evaluate the determinant by expanding by cofactors.

17. $\begin{vmatrix} 2 & -3 & 1 \\ 2 & 0 & 2 \\ 3 & -2 & 4 \end{vmatrix}$

18. $\begin{vmatrix} 3 & 1 & -2 \\ 2 & -5 & 4 \\ 3 & 2 & 1 \end{vmatrix}$

19. $\begin{vmatrix} -2 & 3 & 2 \\ 1 & 2 & -3 \\ -4 & -2 & 1 \end{vmatrix}$

20. $\begin{vmatrix} 3 & -2 & 0 \\ 2 & -3 & 2 \\ 8 & -2 & 5 \end{vmatrix}$

21. $\begin{vmatrix} 2 & -3 & 10 \\ 0 & 2 & -3 \\ 0 & 0 & 5 \end{vmatrix}$

22. $\begin{vmatrix} 6 & 0 & 0 \\ 2 & -3 & 0 \\ 7 & -8 & 2 \end{vmatrix}$

23. $\begin{vmatrix} 0 & -2 & 4 \\ 1 & 0 & -7 \\ 5 & -6 & 0 \end{vmatrix}$

24. $\begin{vmatrix} 5 & -8 & 0 \\ 2 & 0 & -7 \\ 0 & -2 & -1 \end{vmatrix}$

25. $\begin{vmatrix} 4 & -3 & 3 \\ 2 & 1 & -4 \\ 6 & -2 & -1 \end{vmatrix}$

26. $\begin{vmatrix} -2 & 3 & 9 \\ 4 & -2 & -6 \\ 0 & -8 & -24 \end{vmatrix}$

In Exercises 27 to 40, without expanding, give a reason for each equality.

27. $\begin{vmatrix} 2 & -1 & 3 \\ 0 & 0 & 0 \\ 3 & 4 & 1 \end{vmatrix} = 0$

28. $\begin{vmatrix} 2 & 3 & 0 \\ 1 & -2 & 0 \\ 4 & 1 & 0 \end{vmatrix} = 0$

29. $\begin{vmatrix} 1 & 4 & -1 \\ 2 & 4 & 12 \\ 3 & 1 & 4 \end{vmatrix} = 2\begin{vmatrix} 1 & 4 & -1 \\ 1 & 2 & 6 \\ 3 & 1 & 4 \end{vmatrix}$

30. $\begin{vmatrix} 1 & -3 & 4 \\ 4 & 6 & 1 \\ 0 & -9 & 3 \end{vmatrix} = -3\begin{vmatrix} 1 & 1 & 4 \\ 4 & -2 & 1 \\ 0 & 3 & 3 \end{vmatrix}$

31. $\begin{vmatrix} 1 & 5 & -2 \\ 2 & -1 & 4 \\ 3 & 0 & -2 \end{vmatrix} = \begin{vmatrix} 1 & 5 & -2 \\ 0 & -11 & 8 \\ 3 & 0 & -2 \end{vmatrix}$

32. $\begin{vmatrix} 1 & 1 & -3 \\ 2 & 2 & 5 \\ 1 & -2 & 4 \end{vmatrix} = \begin{vmatrix} 1 & 1 & -3 \\ 2 & 2 & 5 \\ 0 & -3 & 7 \end{vmatrix}$

33. $\begin{vmatrix} 4 & -3 & 2 \\ 6 & 2 & 1 \\ -2 & 2 & 4 \end{vmatrix} = 2\begin{vmatrix} 2 & -3 & 2 \\ 3 & 2 & 1 \\ -1 & 2 & 4 \end{vmatrix}$

34. $\begin{vmatrix} 2 & -1 & 3 \\ 3 & 0 & 1 \\ -4 & 2 & -6 \end{vmatrix} = 0$

35. $\begin{vmatrix} 2 & -4 & 5 \\ 0 & 3 & 4 \\ 0 & 0 & -2 \end{vmatrix} = -12$

36. $\begin{vmatrix} 3 & 0 & 0 \\ 2 & -1 & 0 \\ 3 & 4 & 5 \end{vmatrix} = -15$

37. $\begin{vmatrix} 3 & 5 & -2 \\ 2 & 1 & 0 \\ 9 & -2 & -3 \end{vmatrix} = -\begin{vmatrix} 9 & -2 & -3 \\ 2 & 1 & 0 \\ 3 & 5 & -2 \end{vmatrix}$

38. $\begin{vmatrix} 6 & 0 & -2 \\ 2 & -1 & -3 \\ 1 & 5 & -7 \end{vmatrix} = -\begin{vmatrix} 0 & 6 & -2 \\ -1 & 2 & -3 \\ 5 & 1 & -7 \end{vmatrix}$

39. $a^3\begin{vmatrix} 1 & 1 & 1 \\ a & a & a \\ a^2 & a^2 & a^2 \end{vmatrix} = \begin{vmatrix} a & a & a \\ a^2 & a^2 & a^2 \\ a^3 & a^3 & a^3 \end{vmatrix}$

40. $\begin{vmatrix} 1 & 1 & 1 \\ 2 & 2 & 2 \\ 3 & 3 & 3 \end{vmatrix} = 0$

In Exercises 41 to 50, evaluate the determinant by first rewriting the determinant in triangular form.

41. $\begin{vmatrix} 2 & 4 & 1 \\ 1 & 2 & -1 \\ 1 & 2 & 2 \end{vmatrix}$

42. $\begin{vmatrix} 3 & -2 & -1 \\ 1 & 2 & 4 \\ 2 & -2 & 3 \end{vmatrix}$

43. $\begin{vmatrix} 1 & 2 & -1 \\ 2 & 3 & 1 \\ 3 & 4 & 3 \end{vmatrix}$

44. $\begin{vmatrix} 1 & 2 & 5 \\ -1 & 1 & -2 \\ 3 & 1 & 10 \end{vmatrix}$

45. $\begin{vmatrix} 0 & -1 & 1 \\ 1 & 0 & -2 \\ 2 & 2 & 0 \end{vmatrix}$

46. $\begin{vmatrix} 2 & -1 & 3 \\ 1 & 1 & 1 \\ 3 & -4 & 5 \end{vmatrix}$

47. $\begin{vmatrix} 1 & 2 & -1 & 2 \\ 1 & -2 & 0 & 3 \\ 3 & 0 & 1 & 5 \\ -2 & -4 & 1 & 6 \end{vmatrix}$

48. $\begin{vmatrix} 1 & -1 & -1 & 2 \\ 0 & 2 & 4 & 6 \\ 1 & 1 & 4 & 12 \\ 1 & -1 & 0 & 8 \end{vmatrix}$

49. $\begin{vmatrix} 1 & 2 & 3 & -1 \\ 6 & 5 & 9 & 8 \\ 2 & 4 & 12 & -1 \\ 1 & 2 & 6 & -1 \end{vmatrix}$

50. $\begin{vmatrix} 1 & 2 & 0 & -2 \\ -1 & 1 & 3 & 5 \\ 2 & 1 & 4 & 0 \\ -2 & 5 & 2 & 6 \end{vmatrix}$

In Exercises 51 to 54, use a graphing calculator to find the value of the determinant of the matrix.

51. $\begin{vmatrix} 2 & -2 & 3 & 1 \\ 5 & 2 & -2 & 3 \\ 6 & -1 & 2 & 3 \\ 2 & 3 & -1 & 5 \end{vmatrix}$

52. $\begin{vmatrix} 3 & -1 & 0 & 1 \\ 2 & -2 & 3 & 0 \\ -1 & -3 & 5 & 3 \\ 5 & 3 & -2 & 1 \end{vmatrix}$

53. $\begin{bmatrix} -\frac{2}{7} & 4 & -\frac{1}{6} \\ -2 & \sqrt{2} & -3 \\ \sqrt{3} & 3 & -\sqrt{5} \end{bmatrix}$

54. $\begin{bmatrix} 6 & \pi & -\frac{4}{7} \\ -5 & \sqrt{7} & 2 \\ \frac{5}{6} & -\sqrt{3} & \sqrt{10} \end{bmatrix}$

SUPPLEMENTAL EXERCISES

The area of a triangle with vertices (x_1, y_1), (x_2, y_2), and (x_3, y_3) can be given as the absolute value of the determinant

$$\frac{1}{2} \begin{vmatrix} x_1 & y_1 & 1 \\ x_2 & y_2 & 1 \\ x_3 & y_3 & 1 \end{vmatrix}$$

Use this formula to find the area of each triangle whose coordinates are given in Exercises 55 to 58.

55. $(2, 3)$, $(-1, 0)$, $(4, 8)$

56. $(-3, 4)$, $(1, 5)$, $(5, -2)$

57. $(4, 9)$, $(8, 2)$, $(-3, -2)$

58. $(0, 4)$, $(-5, 7)$, $(2, 9)$

59. Given a square matrix of order 3 where one row is a constant multiple of a second row, show that the determinant of the matrix is zero. (*Hint:* Use an elementary

row operation and part 2 of the theorem for conditions for a zero determinant.)

60. Given a square matrix of order 3 with a zero as every element in a column, show that the determinant of the matrix is zero. (*Hint:* Expand the determinant by cofactors using the column of zeros.)

61. Show that the determinant $\begin{vmatrix} x & y & 1 \\ x_1 & y_1 & 1 \\ x_2 & y_2 & 1 \end{vmatrix} = 0$ is the equation of a line through the points (x_1, y_1) and (x_2, y_2).

62. Use Exercise 61 to find the equation of the line passing through the points $(2, 3)$ and $(-1, 4)$.

63. Use Exercise 61 to find the equation of the line passing through the points $(-3, 4)$ and $(2, -3)$.

64. Show that $\begin{vmatrix} a_1 & b_1 \\ a_2 & b_2 \end{vmatrix} = \begin{vmatrix} a_1 & b_1 \\ ka_1 + a_2 & kb_1 + b_2 \end{vmatrix}$.

What property of determinants does this illustrate?

65. Surveyors use a formula to find the area of a plot of land. *Surveyor's Area Formula:* If the vertices (x_1, y_1), (x_2, y_2), (x_3, y_3), ..., (x_n, y_n) of a simple polygon are listed counterclockwise around the perimeter, the area of the polygon is

$$A = \frac{1}{2} \left\{ \begin{vmatrix} x_1 & x_2 \\ y_1 & y_2 \end{vmatrix} + \begin{vmatrix} x_2 & x_3 \\ y_2 & y_3 \end{vmatrix} + \begin{vmatrix} x_3 & x_4 \\ y_3 & y_4 \end{vmatrix} + \cdots + \begin{vmatrix} x_n & x_1 \\ y_n & y_1 \end{vmatrix} \right\}$$

Use the Surveyor's Area Formula to find the area of the polygon with vertices $(8, -4)$, $(25, 5)$, $(15, 9)$, $(17, 20)$, and $(0, 10)$.

PROJECTS

1. Consider the rectangle in the accompanying diagram. Construct a matrix M of dimension 2×4 where the columns are the x- and y-coordinates, respectively, of successive vertices of the rectangle.

   a. Consider the matrix $A = \begin{bmatrix} 2 & 1 \\ 3 & 2 \end{bmatrix}$. Find the product AM. Plot the points of AM again, using the columns as x- and y-coordinates, respectively, of successive vertices of a new figure. Show that the area of the new figure is the same as the area of the original rectangle and that $\det(A) = 1$.

   b. Proceed as in **a.**, but use the matrix $A = \begin{bmatrix} 3 & 1 \\ 1 & 1 \end{bmatrix}$ and show that the area of the new figure is twice the area of the original rectangle. Show that $\det(A) = 2$.

   c. Proceed as in **a.**, but use the matrix $A = \begin{bmatrix} 1 & 2 \\ 0.5 & 1 \end{bmatrix}$ and show that the new figure is a line segment and that, therefore, the figure has no area. Show that $\det(A) = 0$.

   d. Make a conjecture as to how the value of the determinant of A influences the area of the figure represented by AM.

   e. Let $A = \begin{bmatrix} 2 & 1 \\ 5 & 2 \end{bmatrix}$ and repeat **a.** Does your conjecture from **d.** still hold for this matrix? If not, refine your original conjecture.

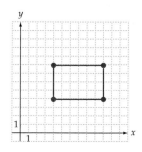

8.5 CRAMER'S RULE

An application of determinants is to solve a system of linear equations. Consider the system

$$\begin{cases} a_{11}x_1 + a_{12}x_2 = b_1 \\ a_{21}x_1 + a_{22}x_2 = b_2 \end{cases}$$

To eliminate x_2 from this system, we first multiply the top equation by a_{22} and the bottom equation by a_{12}. Then we subtract.

$$\begin{aligned}
a_{22}a_{11}x_1 + a_{22}a_{12}x_2 &= a_{22}b_1 \\
a_{12}a_{21}x_1 + a_{12}a_{22}x_2 &= a_{12}b_2 \\
\hline
(a_{22}a_{11} - a_{12}a_{21})x_1 \qquad\quad &= a_{22}b_1 - a_{12}b_2 \\
x_1 &= \frac{a_{22}b_1 - a_{12}b_2}{a_{22}a_{11} - a_{12}a_{21}}
\end{aligned}$$

$$\text{or} \qquad x_1 = \frac{\begin{vmatrix} b_1 & a_{12} \\ b_2 & a_{22} \end{vmatrix}}{\begin{vmatrix} a_{11} & a_{12} \\ a_{21} & a_{22} \end{vmatrix}}, \qquad \begin{vmatrix} a_{11} & a_{12} \\ a_{21} & a_{22} \end{vmatrix} \neq 0$$

We can find x_2 in a similar manner. The results are given in Cramer's Rule for a System of Two Linear Equations.

CRAMER'S RULE FOR A SYSTEM OF TWO LINEAR EQUATIONS

Let
$$\begin{cases} a_{11}x_1 + a_{12}x_2 = b_1 \\ a_{21}x_1 + a_{22}x_2 = b_2 \end{cases}$$

be the system of equations for which the determinant of the coefficient matrix is not zero. The solution of the system of equations is the ordered pair whose coordinates are

$$x_1 = \frac{\begin{vmatrix} b_1 & a_{12} \\ b_2 & a_{22} \end{vmatrix}}{\begin{vmatrix} a_{11} & a_{12} \\ a_{21} & a_{22} \end{vmatrix}} \qquad \text{and} \qquad x_2 = \frac{\begin{vmatrix} a_{11} & b_1 \\ a_{21} & b_2 \end{vmatrix}}{\begin{vmatrix} a_{11} & a_{12} \\ a_{21} & a_{22} \end{vmatrix}}$$

Note that the denominator is the determinant of the coefficient matrix of the variables. The numerator of x_1 is formed by replacing column 1 of the coefficient determinant with the constants b_1 and b_2. The numerator of x_2 is formed by replacing column 2 of the coefficient determinant with the constants b_1 and b_2.

EXAMPLE 1	*Solve a System of Equations by Using Cramer's Rule*

Solve the following system of equations using Cramer's Rule.

$$\begin{cases} 5x_1 - 3x_2 = 6 \\ 2x_1 + 4x_2 = -7 \end{cases}$$

Solution

$$x_1 = \frac{\begin{vmatrix} 6 & -3 \\ -7 & 4 \end{vmatrix}}{\begin{vmatrix} 5 & -3 \\ 2 & 4 \end{vmatrix}} = \frac{3}{26} \qquad x_2 = \frac{\begin{vmatrix} 5 & 6 \\ 2 & -7 \end{vmatrix}}{\begin{vmatrix} 5 & -3 \\ 2 & 4 \end{vmatrix}} = -\frac{47}{26}$$

The solution is $(3/26, -47/26)$.

TRY EXERCISE 4, EXERCISE SET 8.5

Cramer's Rule can be used for a system of three linear equations in three variables. For example, consider the system of equations

$$\begin{cases} 2x - 3y + z = 2 \\ 4x + 2z = -3 \\ 3x + y - 2z = 1 \end{cases} \qquad (1)$$

To solve this system of equations, we extend the concepts behind the solution for a system of two linear equations. The solution of the system has the form (x, y, z), where

$$x = \frac{D_x}{D} \qquad y = \frac{D_y}{D} \qquad z = \frac{D_z}{D}$$

The determinant D is the determinant of the coefficient matrix. The determinants D_x, D_y, and D_z are the determinants of the matrices formed by replacing the first, second, and third columns, respectively, by the constants. For System (1),

$$x = \frac{D_x}{D} \qquad y = \frac{D_y}{D} \qquad z = \frac{D_z}{D}$$

where $\quad D = \begin{vmatrix} 2 & -3 & 1 \\ 4 & 0 & 2 \\ 3 & 1 & -2 \end{vmatrix} = -42 \qquad D_x = \begin{vmatrix} 2 & -3 & 1 \\ -3 & 0 & 2 \\ 1 & 1 & -2 \end{vmatrix} = 5$

$$D_y = \begin{vmatrix} 2 & 2 & 1 \\ 4 & -3 & 2 \\ 3 & 1 & -2 \end{vmatrix} = 49 \qquad D_z = \begin{vmatrix} 2 & -3 & 2 \\ 4 & 0 & -3 \\ 3 & 1 & 1 \end{vmatrix} = 53$$

Thus

$$x = -\frac{5}{42} \qquad y = -\frac{7}{6} \qquad z = -\frac{53}{42}$$

The solution of System (1) is

$$\left(-\frac{5}{42}, -\frac{7}{6}, -\frac{53}{42} \right)$$

Cramer's Rule can be extended to a system of n linear equations in n variables.

CRAMER'S RULE

Let

$$\begin{cases} a_{11}x_1 + a_{12}x_2 + a_{13}x_3 + \cdots + a_{1n}x_n = b_1 \\ a_{21}x_1 + a_{22}x_2 + a_{23}x_3 + \cdots + a_{2n}x_n = b_2 \\ a_{31}x_1 + a_{32}x_2 + a_{33}x_3 + \cdots + a_{3n}x_n = b_3 \\ \qquad \vdots \qquad \vdots \qquad \vdots \qquad\qquad \vdots \qquad \vdots \\ a_{n1}x_1 + a_{n2}x_2 + a_{n3}x_3 + \cdots + a_{nn}x_n = b_n \end{cases}$$

be a system of n equations in n variables. The solution of the system is given by $(x_1, x_2, x_3, \ldots, x_n)$, where

$$x_1 = \frac{D_1}{D} \qquad x_2 = \frac{D_2}{D} \qquad \cdots \qquad x_i = \frac{D_i}{D} \qquad \cdots \qquad x_n = \frac{D_n}{D}$$

and D is the determinant of the coefficient matrix, $D \neq 0$. D_i is the determinant formed by replacing the ith column of the coefficient matrix with the column of constants $b_1, b_2, b_3, \ldots, b_n$.

Because the determinant of the coefficient matrix must be nonzero for us to use Cramer's Rule, this method is not appropriate for systems of linear equations with no solution or infinitely many solutions. In fact, the only time a system of linear equations has a unique solution is when the coefficient determinant is not zero, a fact summarized in the following theorem.

SYSTEMS OF LINEAR EQUATIONS WITH UNIQUE SOLUTIONS

A system of n linear equations in n variables has a unique solution if and only if the determinant of the coefficient matrix is not zero.

Cramer's Rule is also useful when we want to determine the value of only a single variable in a system of equations.

EXAMPLE 2   *Determine the Value of a Single Variable in a System of Linear Equations*

Find x_3 for the system of equations

$$\begin{cases} 4x_1 \qquad\quad + 3x_3 - 2x_4 = \;\;\; 2 \\ 3x_1 + \;\; x_2 + 2x_3 - \;\; x_4 = \;\;\; 4 \\ \;\; x_1 - 6x_2 - 2x_3 + 2x_4 = \;\;\; 0 \\ 2x_1 + 2x_2 \qquad\quad - \;\; x_4 = -1 \end{cases}$$

Continued •▶

Solution

Find D and D_3.

$$D = \begin{vmatrix} 4 & 0 & 3 & -2 \\ 3 & 1 & 2 & -1 \\ 1 & -6 & -2 & 2 \\ 2 & 2 & 0 & -1 \end{vmatrix} = 39 \qquad D_3 = \begin{vmatrix} 4 & 0 & 2 & -2 \\ 3 & 1 & 4 & -1 \\ 1 & -6 & 0 & 2 \\ 2 & 2 & -1 & -1 \end{vmatrix} = 96$$

Thus $x_3 = 96/39 = 32/13$.

> **TRY EXERCISE 24, EXERCISE SET 8.5**

TOPICS FOR DISCUSSION

1. Discuss the advantages and disadvantages of using Cramer's Rule to solve a system of equations.

2. Can Cramer's Rule be used to solve any system of linear equations? If not, explain when Cramer's Rule will lead to a solution and when it will not.

EXERCISE SET 8.5

In Exercises 1 to 20, solve each system of equations by using Cramer's Rule.

1. $\begin{cases} 3x_1 + 4x_2 = 8 \\ 4x_1 - 5x_2 = 1 \end{cases}$

2. $\begin{cases} x_1 - 3x_2 = 9 \\ 2x_1 - 4x_2 = -3 \end{cases}$

3. $\begin{cases} 5x_1 + 4x_2 = -1 \\ 3x_1 - 6x_2 = 5 \end{cases}$

4. $\begin{cases} 2x_1 + 5x_2 = 9 \\ 5x_1 + 7x_2 = 8 \end{cases}$

5. $\begin{cases} 7x_1 + 2x_2 = 0 \\ 2x_1 + x_2 = -3 \end{cases}$

6. $\begin{cases} 3x_1 - 8x_2 = 1 \\ 4x_1 + 5x_2 = -2 \end{cases}$

7. $\begin{cases} 3x_1 - 7x_2 = 0 \\ 2x_1 + 4x_2 = 0 \end{cases}$

8. $\begin{cases} 5x_1 + 4x_2 = -3 \\ 2x_1 - x_2 = 0 \end{cases}$

9. $\begin{cases} 1.2x_1 + 0.3x_2 = 2.1 \\ 0.8x_1 - 1.4x_2 = -1.6 \end{cases}$

10. $\begin{cases} 3.2x_1 - 4.2x_2 = 1.1 \\ 0.7x_1 + 3.2x_2 = -3.4 \end{cases}$

11. $\begin{cases} 3x_1 - 4x_2 + 2x_3 = 1 \\ x_1 - x_2 + 2x_3 = -2 \\ 2x_1 + 2x_2 + 3x_3 = -3 \end{cases}$

12. $\begin{cases} 5x_1 - 2x_2 + 3x_3 = -2 \\ 3x_1 + x_2 - 2x_3 = 3 \\ x_1 - 2x_2 + 3x_3 = -1 \end{cases}$

13. $\begin{cases} x_1 + 4x_2 - 2x_3 = 0 \\ 3x_1 - 2x_2 + 3x_3 = 4 \\ 2x_1 + x_2 - 3x_3 = -1 \end{cases}$

14. $\begin{cases} 4x_1 - x_2 + 2x_3 = 6 \\ x_1 + 3x_2 - x_3 = -1 \\ 2x_1 + 3x_2 - 2x_3 = 5 \end{cases}$

15. $\begin{cases} 2x_2 - 3x_3 = 1 \\ 3x_1 - 5x_2 + x_3 = 0 \\ 4x_1 + 2x_3 = -3 \end{cases}$

16. $\begin{cases} 2x_1 + 5x_2 = 1 \\ x_1 - 3x_3 = -2 \\ 2x_1 - x_2 + 2x_3 = 4 \end{cases}$

17. $\begin{cases} 4x_1 - 5x_2 + x_3 = -2 \\ 3x_1 + x_2 = 4 \\ x_1 - x_2 + 3x_3 = 0 \end{cases}$

18. $\begin{cases} 3x_1 - x_2 + x_3 = 5 \\ x_1 + 3x_3 = -2 \\ 2x_1 + 2x_2 - 5x_3 = 0 \end{cases}$

19. $\begin{cases} 2x_1 + 2x_2 - 3x_3 = 0 \\ x_1 - 3x_2 + 2x_3 = 0 \\ 4x_1 - x_2 + 3x_3 = 0 \end{cases}$

20. $\begin{cases} x_1 + 3x_2 = -2 \\ 2x_1 - 3x_2 + x_3 = 1 \\ 4x_1 + 5x_2 - 2x_3 = 0 \end{cases}$

In Exercises 21 to 26, solve for the indicated variable.

21. Solve for x_2: $\begin{cases} 2x_1 - 3x_2 + 4x_3 - x_4 = 1 \\ x_1 + 2x_2 + 2x_4 = -1 \\ 3x_1 + x_2 - 2x_4 = 2 \\ x_1 - 3x_2 + 2x_3 - x_4 = 3 \end{cases}$

22. Solve for x_4: $\begin{cases} 3x_1 + x_2 - 2x_3 + 3x_4 = 4 \\ 2x_1 - 3x_2 + 2x_3 = -2 \\ x_1 + x_2 - 2x_3 + 2x_4 = 3 \\ 2x_1 + 3x_3 - 2x_4 = 4 \end{cases}$

23. Solve for x_1: $\begin{cases} x_1 - 3x_2 + 2x_3 + 4x_4 = 0 \\ 3x_1 + 5x_2 - 6x_3 + 2x_4 = -2 \\ 2x_1 - x_2 + 9x_3 + 8x_4 = 0 \\ x_1 + x_2 + x_3 - 8x_4 = -3 \end{cases}$

24. Solve for x_3: $\begin{cases} 2x_1 + 5x_2 - 5x_3 - 3x_4 = -3 \\ x_1 + 7x_2 + 8x_3 - x_4 = 4 \\ 4x_1 + x_3 + x_4 = 3 \\ 3x_1 + 2x_2 - x_3 = 0 \end{cases}$

25. Solve for x_4: $\begin{cases} 3x_2 - x_3 + 2x_4 = 1 \\ 5x_1 + x_2 + 3x_3 - x_4 = -4 \\ x_1 - 2x_2 + 9x_4 = 5 \\ 2x_1 + 2x_3 = 3 \end{cases}$

26. Solve for x_1: $\begin{cases} 4x_1 + x_2 - 3x_4 = 4 \\ 5x_1 + 2x_2 - 2x_3 + x_4 = 7 \\ x_1 - 3x_2 + 2x_3 - 2x_4 = -6 \\ 3x_3 + 4x_4 = -7 \end{cases}$

SUPPLEMENTAL EXERCISES

27. A solution of the system of equations
$$\begin{cases} 2x_1 - 3x_2 + x_3 = 9 \\ x_1 + x_2 - 2x_3 = -3 \\ 4x_1 - x_2 - 3x_3 = 3 \end{cases}$$
is $(1, -2, 1)$. However, this solution cannot be found by using Cramer's Rule. Explain.

28. Verify the solution for x_2 given in Cramer's Rule for a System of Two Equations by solving the system of equations
$$\begin{cases} a_{11}x_1 + a_{12}x_2 = b_1 \\ a_{21}x_1 + a_{22}x_2 = b_2 \end{cases}$$
for x_2 by using the elimination method.

29. For what values of k does the system of equations
$$\begin{cases} kx + 3y = 7 \\ kx - 2y = 5 \end{cases}$$
have a unique solution?

30. For what values of k does the system of equations
$$\begin{cases} kx + 4y = 5 \\ 9x - ky = 2 \end{cases}$$
have a unique solution?

31. For what values of k does the system of equations
$$\begin{cases} x + 2y - 3z = 4 \\ 2x + ky - 4z = 5 \\ x - 2y + z = 6 \end{cases}$$
have a unique solution?

32. For what values of k does the system of equations
$$\begin{cases} kx_1 + x_2 = 1 \\ x_2 - 4x_3 = 1 \\ x_1 + kx_3 = 1 \end{cases}$$
have a unique solution?

33. Find real values for r and s so that $ru + sv = w$, where u, v, and w are complex numbers and $u = 2 + 3i$, $v = 4 - 2i$, and $w = -6 + 15i$.

34. Find real values for r and s such that $ru + sv = w$, where $u = 3 - 4i$, $v = 1 + 2i$, and $w = 4 - 22i$.

PROJECTS

1. Prove Cramer's Rule for a system of three linear equations in three variables.

EXPLORING CONCEPTS WITH TECHNOLOGY

Stochastic Matrices

Matrices can be used to predict how percents of populations will change over time. Consider two neighborhood supermarkets, Super A and Super B. Each week Super A loses 5% of its customers to Super B, and each week Super B loses 8% of its customers to Super A. If this trend continues and if Super A currently has 40% of the neighborhood customers and Super B the remaining 60% of the neighborhood customers, what percent of the neighborhood will each have after n weeks?

We will approach this problem by examining the changes on a week-by-week basis. Because Super A loses 5% of its customers each week, it retains 95% of its customers. It has 40% of the neighborhood customers now, so after 1 week it will have 95% of its 40% share, or 38% ($0.95 \cdot 0.40$) of the customers. In that same week, it gains 8% of the customers of Super B. Because Super B has 60% of the neighborhood customers, Super A's gain is 4.8% ($0.08 \cdot 0.60$). After 1 week, Super A has 38% + 4.8% = 42.8% of the neighborhood customers. Super B has the remaining 57.2% of the customers.

The changes for the second week are calculated similarly. Super A retains 95% of its 42.8% and gains 8% of Super B's 57.2%. After week 2, Super A has

$$0.95 \cdot 0.428 + 0.08 \cdot 0.572 \approx 0.452$$

or approximately 45.2%, of the neighborhood customers. Super B has the remaining 54.8%.

We could continue in this way, but using matrices is a more convenient way to proceed. Let $T = \begin{bmatrix} 0.95 & 0.05 \\ 0.08 & 0.92 \end{bmatrix}$, where column 1 represents the percent retained by Super A and column 2 represents the percent retained by Super B. Let $X = [0.40 \quad 0.60]$ be the current market shares of Super A and Super B, respectively. Now form the product XT.

$$[0.40 \quad 0.60] \begin{bmatrix} 0.95 & 0.05 \\ 0.08 & 0.92 \end{bmatrix} = [0.428 \quad 0.572]$$

For the second week, multiply the market share after week 1 by T.

$$[0.428 \quad 0.572] \begin{bmatrix} 0.95 & 0.05 \\ 0.08 & 0.92 \end{bmatrix} \approx [0.452 \quad 0.548]$$

The last product can also be expressed as

$$[0.452 \quad 0.548] = [0.428 \quad 0.572] \begin{bmatrix} 0.95 & 0.05 \\ 0.08 & 0.92 \end{bmatrix} = [0.40 \quad 0.60] \overbrace{\begin{bmatrix} 0.95 & 0.05 \\ 0.08 & 0.92 \end{bmatrix} \begin{bmatrix} 0.95 & 0.05 \\ 0.08 & 0.92 \end{bmatrix}}^{[0.428 \quad 0.572]}$$

$$= [0.40 \quad 0.60] \begin{bmatrix} 0.95 & 0.05 \\ 0.08 & 0.92 \end{bmatrix}^2 = XT^2$$

Note that the exponent on T corresponds to the fact that 2 weeks have passed. In general, the market share after n weeks is XT^n. The matrix T is called a **stochastic matrix.** A stochastic matrix is characterized by the fact that each element of the matrix is nonnegative and the sum of the elements in each row is 1.

Use a calculator to calculate the market share of Super A and Super B after 20 weeks, 40 weeks, 60 weeks, and 100 weeks. What observations do you draw from your calculations? We started this problem with the assumption that Super A had 40% of the market and Super B had 60% of the market. Suppose, however, that originally Super A had 99% of the market and Super B 1%. Does this affect the market share each will have after 100 weeks? If Super A had 1%

of the market and Super B had 99% of the market, what will the market share of each be after 100 weeks?

As another example, suppose each of three department stores is vying for the business of the other two stores. In one month, Store A loses 15% of its customers to Store B and 8% of its customers to Store C. Store B loses 10% of its customers to Store A and 12% to Store C. Store C loses 5% to Store A and 9% to Store B. Assuming these three stores have 100% of the market and the trend continues, determine what market share each will have after 100 months.

CHAPTER 8 REVIEW

8.1 Gaussian Elimination Method

- A matrix is a rectangular array of numbers. A matrix with m rows and n columns is of order $m \times n$ or dimension $m \times n$.

- For a system of equations, it is possible to form a coefficient matrix, an augmented matrix, and a constant matrix.

- A matrix is in echelon form if all the following conditions are satisfied:

  1. The first nonzero number in any row is a 1.

  2. Rows are arranged so that the column containing the first nonzero number is to the left of the column containing the first nonzero number of the next row.

  3. All rows consisting entirely of zeros appear at the bottom of the matrix.

- The Gaussian elimination method uses elementary row operations to solve a system of linear equations.

- **Elementary Row Operations**
  The elementary row operations for a matrix are

  1. Interchanging two rows

  2. Multiplying all the elements in a row by the same nonzero number

  3. Replacing a row by the sum of that row and a nonzero multiple of any other row

8.2 The Algebra of Matrices

- Two matrices $A = [a_{ij}]$ and $B = [b_{ij}]$ are equal if and only if $a_{ij} = b_{ij}$ for every i and j.

- The sum of two matrices of the same order is the matrix whose elements are the sum of the corresponding elements of the two matrices.

- The $m \times n$ zero matrix is the matrix whose elements are all zeros.

- Taking the product of a real number and a matrix is called scalar multiplication.

- In order for us to multiply two matrices, the number of columns of the first matrix must equal the number of rows of the second matrix.

- In general, matrix multiplication is not commutative.

- The multiplicative identity matrix is the matrix with 1s on the main diagonal and zeros everywhere else.

8.3 The Inverse of a Matrix

- The multiplicative inverse of a square matrix A, denoted by A^{-1}, has the property that

$$A \cdot A^{-1} = A^{-1} \cdot A = I_n$$

where I_n is the multiplicative identity matrix.

- A singular matrix is one that does not have a multiplicative inverse.

- Input-output analysis attempts to determine the necessary output of industries to satisfy each other's demands plus the demands of consumers.

8.4 Determinants

- Associated with each square matrix is a number called the determinant of the matrix.

- The minor of the element a_{ij} of a square matrix A is the determinant of the matrix obtained by deleting the ith row and the jth column of A.

- The cofactor of the element a_{ij} of a square matrix A is $(-1)^{i+j}M_{ij}$, where M_{ij} is the minor of a_{ij}.

- The value of a determinant can be found by multiplying the elements of any row or column by their respective cofactors and then adding the results. This is called expanding by cofactors.

8.5 Cramer's Rule

- Cramer's Rule is a method of solving a system of n equations in n variables by using determinants.

CHAPTER 8 TRUE/FALSE EXERCISES

In Exercises 1 to 15, answer true or false. If the statement is false, give an example to show that the statement is false.

1. If $A = \begin{bmatrix} 2 & 3 \\ 1 & 4 \end{bmatrix}$, then $A^2 = \begin{bmatrix} 4 & 9 \\ 1 & 16 \end{bmatrix}$.

2. Every matrix has an additive inverse.

3. Every square matrix has a multiplicative inverse.

4. Let the matrices A, B, and C be square matrices of order n. If $AB = AC$, then $B = C$.

5. It is possible to find the determinant of every square matrix.

6. If A and B are square matrices of order n, then

$$\det(A + B) = \det(A) + \det(B)$$

7. Cramer's Rule can be used to solve any system of three equations in three variables.

8. If A and B are matrices of order n, then $AB - BA = O$.

9. A nonsingular matrix has a multiplicative inverse.

10. If A, B, and C are square matrices of order n, then the product ABC depends on which two matrices are multiplied first. That is, $(AB)C$ produces a different result from $A(BC)$.

11. The Gaussian elimination method for solving a system of linear equations can be applied only to systems of equations that have the same number of variables as equations.

12. If A is a square matrix of order n, then $\det(2A) = 2\det(A)$.

13. If A and B are matrices, then the product AB is defined when the number of columns of A equals the number of rows of B.

14. If A and B are square matrices of order n and $AB = O$ (the zero matrix), then $A = O$ or $B = O$.

15. If $A = \begin{bmatrix} 3 & 6 \\ -1 & -2 \end{bmatrix}$, then $A^5 = A$.

CHAPTER 8 REVIEW EXERCISES

In Exercises 1 to 18, perform the indicated operations. Let

$$A = \begin{bmatrix} 2 & -1 & 3 \\ 3 & 2 & -1 \end{bmatrix}, B = \begin{bmatrix} 0 & -2 \\ 4 & 2 \\ 1 & -3 \end{bmatrix}, C = \begin{bmatrix} 2 & 6 & 1 \\ 1 & 2 & -1 \\ 2 & 4 & -1 \end{bmatrix},$$

and $D = \begin{bmatrix} -3 & 4 & 2 \\ 4 & -2 & 5 \end{bmatrix}$.

1. $3A$
2. $-2B$
3. $-A + D$
4. $2A - 3D$
5. AB
6. DB
7. BA
8. BD
9. C^2
10. C^3
11. BAC
12. ADB
13. $AB - BA$
14. $DB - BD$
15. $(A - D)C$
16. $AC - DC$
17. C^{-1}
18. $|C|$

In Exercises 19 to 34, solve the system of equations by using the Gaussian elimination method.

19. $\begin{cases} 2x - 3y = 7 \\ 3x - 4y = 10 \end{cases}$

20. $\begin{cases} 3x + 4y = -9 \\ 2x + 3y = -7 \end{cases}$

21. $\begin{cases} 4x - 5y = 12 \\ 3x + y = 9 \end{cases}$

22. $\begin{cases} 2x - 5y = 10 \\ 5x + 2y = 4 \end{cases}$

23. $\begin{cases} x + 2y + 3z = 5 \\ 3x + 8y + 11z = 17 \\ 2x + 6y + 7z = 12 \end{cases}$

24. $\begin{cases} x - y + 3z = 10 \\ 2x - y + 7z = 24 \\ 3x - 6y + 7z = 21 \end{cases}$

25. $\begin{cases} 2x - y - z = 4 \\ x - 2y - 2z = 5 \\ 3x - 3y - 8z = 19 \end{cases}$

26. $\begin{cases} 3x - 7y + 8z = 10 \\ x - 3y + 2z = 0 \\ 2x - 8y + 7z = 5 \end{cases}$

27. $\begin{cases} 4x - 9y + 6z = 54 \\ 3x - 8y + 8z = 49 \\ x - 3y + 2z = 17 \end{cases}$

28. $\begin{cases} 3x + 8y - 5z = 6 \\ 2x + 9y - z = -8 \\ x - 4y - 2z = 16 \end{cases}$

29. $\begin{cases} x + y + 2z = -5 \\ 2x + 3y + 5z = -13 \\ 2x + 5y + 7z = -19 \end{cases}$

30. $\begin{cases} x - 2y + 3z = 9 \\ 3x - 5y + 8z = 25 \\ x \quad\quad - z = 5 \end{cases}$

31. $\begin{cases} w + 2x - y + 2z = 1 \\ 3w + 8x + y + 4z = 1 \\ 2w + 7x + 3y + 2z = 0 \\ w + 3x - 2y + 5z = 6 \end{cases}$

32. $\begin{cases} w - 3x - 2y + z = -1 \\ 2w - 5x \quad\quad + 3z = 1 \\ 3w - 7x + 3y \quad\quad = -18 \\ 2w - 3x - 5y - 2z = -8 \end{cases}$

33. $\begin{cases} w + 3x + y - 4z = 3 \\ w + 4x + 3y - 6z = 5 \\ 2w + 8x + 7y - 5z = 11 \\ 2w + 5x \quad\quad - 6z = 4 \end{cases}$

34. $\begin{cases} w + 4x - 2y + 3z = 6 \\ 2w + 9x - y + 5z = 13 \\ w + 7x + 6y + 5z = 9 \\ 3w + 14x \qquad + 7z = 20 \end{cases}$

In Exercises 35 to 46, find the inverse, if it exists, of the given matrix.

35. $\begin{bmatrix} 2 & -2 \\ 3 & -2 \end{bmatrix}$
36. $\begin{bmatrix} 3 & 4 \\ 2 & 3 \end{bmatrix}$
37. $\begin{bmatrix} -2 & 3 \\ 2 & 4 \end{bmatrix}$

38. $\begin{bmatrix} 5 & -4 \\ 3 & 2 \end{bmatrix}$
39. $\begin{bmatrix} 1 & 2 & 1 \\ 2 & 6 & 4 \\ 3 & 8 & 6 \end{bmatrix}$
40. $\begin{bmatrix} 1 & -3 & 2 \\ 3 & -8 & 7 \\ 2 & -3 & 6 \end{bmatrix}$

41. $\begin{bmatrix} 3 & -2 & 7 \\ 2 & -1 & 5 \\ 3 & 0 & 10 \end{bmatrix}$
42. $\begin{bmatrix} 4 & 9 & -11 \\ 3 & 7 & -8 \\ 2 & 6 & -3 \end{bmatrix}$

43. $\begin{bmatrix} 1 & -1 & 2 & 3 \\ 2 & -1 & 6 & 5 \\ 3 & -1 & 9 & 6 \\ 2 & -2 & 4 & 7 \end{bmatrix}$
44. $\begin{bmatrix} 1 & 2 & -2 & 1 \\ 3 & 7 & -3 & 1 \\ 2 & 7 & 4 & 3 \\ 1 & 4 & 2 & 4 \end{bmatrix}$

45. $\begin{bmatrix} 3 & 7 & -1 & 8 \\ 2 & 5 & 0 & 5 \\ 3 & 6 & -4 & 8 \\ 2 & 4 & -4 & 4 \end{bmatrix}$
46. $\begin{bmatrix} 3 & 1 & 5 & -5 \\ 2 & 1 & 4 & -3 \\ 3 & 0 & 4 & -3 \\ 4 & 1 & 8 & 1 \end{bmatrix}$

In Exercises 47 to 50, solve the given system of equations for each set of constants. Use the inverse matrix method.

47. $\begin{cases} 3x + 4y = b_1 \\ 2x + 3y = b_2 \end{cases}$
a. $b_1 = 2, b_2 = -3$
b. $b_1 = -2, b_2 = 4$

48. $\begin{cases} 2x - 5y = b_1 \\ 3x - 7y = b_2 \end{cases}$
a. $b_1 = -3, b_2 = 4$
b. $b_1 = 2, b_2 = -5$

49. $\begin{cases} 2x + y - z = b_1 \\ 4x + 4y + z = b_2 \\ 2x + 2y - 3z = b_3 \end{cases}$
a. $b_1 = -1, b_2 = 2, b_3 = 4$
b. $b_1 = -2, b_2 = 3, b_3 = 0$

50. $\begin{cases} 3x - 2y + z = b_1 \\ 3x - y + 3z = b_2 \\ 6x - 4y + z = b_3 \end{cases}$
a. $b_1 = 0, b_2 = 3, b_3 = -2$
b. $b_1 = 1, b_2 = 2, b_3 = -4$

In Exercises 51 to 58, evaluate each determinant by using elementary row or column operations.

51. $\begin{vmatrix} 2 & 6 & 4 \\ 1 & 2 & 1 \\ 3 & 8 & 6 \end{vmatrix}$
52. $\begin{vmatrix} 3 & 0 & 10 \\ 3 & -2 & 7 \\ 2 & -1 & 5 \end{vmatrix}$

53. $\begin{vmatrix} 3 & -8 & 7 \\ 2 & -3 & 6 \\ 1 & -3 & 2 \end{vmatrix}$
54. $\begin{vmatrix} 4 & 9 & -11 \\ 2 & 6 & -3 \\ 3 & 7 & -8 \end{vmatrix}$

55. $\begin{vmatrix} 1 & -1 & 2 & 1 \\ 2 & -1 & 6 & 3 \\ 3 & -1 & 8 & 7 \\ 3 & 0 & 9 & 9 \end{vmatrix}$
56. $\begin{vmatrix} 1 & 2 & -2 & 3 \\ 3 & 7 & -3 & 11 \\ 2 & 3 & -5 & 11 \\ 2 & 6 & 1 & 8 \end{vmatrix}$

57. $\begin{vmatrix} 1 & 2 & -2 & 1 \\ 2 & 5 & -3 & 1 \\ 2 & 0 & -10 & 1 \\ 3 & 8 & -4 & 1 \end{vmatrix}$
58. $\begin{vmatrix} 1 & 3 & -2 & 0 \\ 3 & 11 & -4 & 4 \\ 2 & 9 & -8 & 2 \\ 3 & 12 & -10 & 2 \end{vmatrix}$

In Exercises 59 to 64, solve each system of equations by using Cramer's Rule.

59. $\begin{cases} 2x_1 - 3x_2 = 2 \\ 3x_1 + 5x_2 = 2 \end{cases}$
60. $\begin{cases} 3x_1 + 4x_2 = -3 \\ 5x_1 - 2x_2 = 2 \end{cases}$

61. $\begin{cases} 2x_1 + x_2 - 3x_3 = 2 \\ 3x_1 + 2x_2 + x_3 = 1 \\ x_1 - 3x_2 + 4x_3 = -2 \end{cases}$
62. $\begin{cases} 3x_1 + 2x_2 - x_3 = 0 \\ x_1 + 3x_2 - 2x_3 = 3 \\ 4x_1 - x_2 - 5x_3 = -1 \end{cases}$

63. $\begin{cases} 2x_2 + 5x_3 = 2 \\ 2x_1 - 5x_2 + x_3 = 4 \\ 4x_1 + 3x_2 = 2 \end{cases}$
64. $\begin{cases} 2x_1 - 3x_2 - 4x_3 = 2 \\ x_1 - 2x_2 + 2x_3 = -1 \\ 2x_1 + 7x_2 - x_3 = 2 \end{cases}$

In Exercises 65 and 66, use Cramer's Rule to solve for the indicated variable.

65. Solve for x_3: $\begin{cases} x_1 - 3x_2 + x_3 + 2x_4 = 3 \\ 2x_1 + 7x_2 - 3x_3 + x_4 = 2 \\ -x_1 + 4x_2 + 2x_3 - 3x_4 = -1 \\ 3x_1 + x_2 - x_3 - 2x_4 = 0 \end{cases}$

66. Solve for x_2: $\begin{cases} 2x_1 + 3x_2 - 2x_3 + x_4 = -2 \\ x_1 - x_2 - 3x_3 + 2x_4 = 2 \\ 3x_1 + 3x_2 - 4x_3 - x_4 = 4 \\ 5x_1 - 5x_2 - x_3 + 2x_4 = 7 \end{cases}$

In Exercises 67 and 68, solve the input-output problem.

67. **BUSINESS RESOURCE ALLOCATION** An electronics conglomerate has three divisions, which produce computers, monitors, and disk drives. For each $1 worth of output, the computer division needs $.05 worth of computers, $.02 worth of monitors, and $.03 worth of disk drives. For each $1 worth of output, the monitor division needs $.06 worth of computers, $.04 worth of monitors, and $.03 worth of disk drives. For each $1 worth of output, the disk drive division requires $.08 worth of computers, $.04 worth of monitors, and $.05 worth of disk drives. Sales estimates are $30 million for the computer division, $12 million for the monitor division, and $21 million for the disk drive

division. At what level should each division produce to satisfy this demand?

68. BUSINESS RESOURCE ALLOCATION A manufacturing conglomerate has three divisions, which produce paper, lumber, and prefabricated walls. For each \$1 worth of output, the lumber division needs \$.07 worth of lumber, \$.03 worth of paper, and \$.03 worth of prefabricated walls. For each \$1 worth of output, the paper division needs \$.04 worth of lumber, \$.07 worth of paper, and \$.03 worth of prefabricated walls. For each \$1 worth of output, the prefabricated walls division requires \$.07 worth of lumber, \$.04 worth of paper, and \$.02 worth of prefabricated walls. Sales estimates are \$27 million for the lumber division, \$18 million for the paper division, and \$10 million for the prefabricated walls division. At what level should each division produce to satisfy this demand?

CHAPTER 8 TEST

1. Write the augmented matrix, the coefficient matrix, and the constant matrix for the system of equations

$$\begin{cases} 2x + 3y - 3z = 4 \\ 3x \quad\quad + 2z = -1 \\ 4x - 4y + 2z = 3 \end{cases}$$

2. Write a system of equations that is equivalent to the

augmented matrix $\begin{bmatrix} 3 & -2 & 5 & -1 & 9 \\ 2 & 3 & -1 & 4 & 8 \\ 1 & 0 & 3 & 2 & -1 \end{bmatrix}$.

In Exercises 3 to 5, solve the system of equations by using the Gaussian elimination method.

3. $\begin{cases} x - 2y + 3z = 10 \\ 2x - 3y + 8z = 23 \\ -x + 3y - 2z = -9 \end{cases}$

4. $\begin{cases} 2x + 6y - z = 1 \\ x + 3y - z = 1 \\ 3x + 10y - 2z = 1 \end{cases}$

5. $\begin{cases} w + 2x - 3y + 2z = 11 \\ 2w + 5x - 8y + 5z = 28 \\ -2w - 4x + 7y - z = -18 \end{cases}$

In Exercises 6 to 18, let $A = \begin{bmatrix} -1 & 3 & 2 \\ 1 & 4 & -1 \end{bmatrix}$,

$B = \begin{bmatrix} 2 & -1 & 3 \\ 4 & -2 & -1 \\ 3 & 2 & 2 \end{bmatrix}$, **and** $C = \begin{bmatrix} 1 & -2 & 3 \\ 2 & -3 & 8 \\ -1 & 3 & -2 \end{bmatrix}$. **Perform each possible operation. If an operation is not possible, so state.**

6. $-3A$

7. $A + B$

8. $3B - 2C$

9. AB

10. $AB - A$

11. CA

12. $BC - CB$

13. A^2

14. B^2

15. C^{-1}

16. Find the minor and cofactor of b_{21} for matrix B.

17. Find the determinant of B by expanding by cofactors of row 3.

18. Find the determinant of C by using elementary row operations.

19. Find the value of z for the following system of equations by using Cramer's Rule.

$$\begin{cases} 3x + 2y - z = 12 \\ 2x - 3y + 2z = -1 \\ 5x + 6y + 3z = 4 \end{cases}$$

20. A simplified economy has three major industries: mining, manufacturing, and transportation. The input-output matrix for this economy is

$$\begin{bmatrix} 0.15 & 0.23 & 0.11 \\ 0.08 & 0.10 & 0.05 \\ 0.16 & 0.11 & 0.07 \end{bmatrix}$$

Set up, but do not solve, a matrix equation that, when solved, will determine the gross output needed to satisfy consumer demand of \$50 million worth of mining, \$32 million worth of manufacturing, and \$8 million worth of transportation.

SEQUENCES, SERIES, AND PROBABILITY

A fractal is a geometric figure consisting of a basic pattern which keeps repeating on a smaller and smaller scale. Here are three other examples.

A Snowflake with Infinite Perimeter

We started this book with a discussion of infinities. Now that we have reached the last chapter, it seems appropriate to end with a discussion of infinity.

We begin with an equilateral triangle each side of which is 1 unit long. The perimeter, then, is 3 units. Now construct an identical but smaller triangle onto the middle third of each side. The snowflake now has 12 line segments each of length 1/3 unit. The perimeter of the snowflake is 12(1/3) or 4 units. Repeat the procedure and construct identical but smaller triangles on the middle third of each of the 12 sides of the snowflake. The result is 48 line segments each of length 1/9 unit. The perimeter is 48(1/9) = 16/3 units. Continuing this procedure, we can show that the perimeter of each succeeding snowflake is 4/3 the perimeter of the preceding one. Thus the perimeter continues to grow without bound and becomes infinite. See figures below.

Now examine the area of each snowflake. Let A be the area of the original triangle. In the second stage, each new triangle has an area that is 1/9 of A. Because there are 3 new triangles, the area of the snowflake is the original area plus the area of these 3 triangles: $A + 3(1/9)A = (4/3)A$.

For the next snowflake, each new triangle has an area equal to $(1/81)A$. The total area of the 12 new triangles is $12[(1/81)A] = (4/27)A$. The area of the snowflake is now $(4/3)A + (4/27)A = (40/27)A$.

At each stage after the first, the area added is 4/9 the preceding area. Continuing in this way, it is possible to show that the area of each succeeding snowflake approaches $(8/5)A$. That is, the area is finite.

Thus we have a snowflake with infinite perimeter and finite area. Creating figures such as this snowflake is part of the realm of fractals.

9.1 INFINITE SEQUENCES AND SUMMATION NOTATION

The *ordered* list of numbers 2, 4, 8, 16, 32, ... is called an infinite sequence. The list is ordered simply because order makes a difference. The sequence 2, 8, 4, 16, 32, ... contains the same numbers, but in a different order. Therefore, it is a different infinite sequence.

An infinite sequence can be thought of as a pairing between positive numbers and real numbers. For example, 1, 4, 9, 16, 25, 36, ..., n^2, ... pairs a natural number with its square.

1	2	3	4	5	6	...	n	...
↓	↓	↓	↓	↓	↓		↓	
1	4	9	16	25	36	...	n^2	...

This pairing of numbers enables us to define an infinite sequence as a function with domain the positive integers.

INFINITE SEQUENCE

An **infinite sequence** is a function whose domain is the positive integers and whose range is a set of real numbers.

Although the positive integers do not include zero, it is occasionally convenient to include zero in the domain of an infinite sequence. Also, we will frequently use the word *sequence* instead of the phrase *infinite sequence*.

As an example of a sequence, let $f(n) = 2n - 1$. The range of this function is

$$f(1), f(2), f(3), f(4), \ldots, \quad f(n), \quad \ldots$$
$$1, \quad 3, \quad 5, \quad 7, \quad \ldots, \quad 2n - 1, \ldots$$

The elements in the range of a sequence are called the **terms** of the sequence. For our example, the terms are 1, 3, 5, 7, ..., $2n - 1$, ... The **first term** of the sequence is 1, the **second term** is 3, and so on. The **nth term**, or the **general term**, is $2n - 1$.

Rather than use functional notation for sequences, it is customary to use a subscript notation. Thus a_n represents the nth term of a sequence. Using this notation, we would write

$$a_n = 2n - 1$$

Thus $a_1 = 1$, $a_2 = 3$, $a_3 = 5$, $a_4 = 7$.

EXAMPLE 1   Find the Terms of a Sequence

a. Find the first three terms of the sequence $a_n = \dfrac{1}{n(n + 1)}$.

b. Find the eighth term of the sequence $a_n = \dfrac{2^n}{n^2}$.

Solution

a. $a_1 = \dfrac{1}{1(1 + 1)} = \dfrac{1}{2}, a_2 = \dfrac{1}{2(2 + 1)} = \dfrac{1}{6}, a_3 = \dfrac{1}{3(3 + 1)} = \dfrac{1}{12}$

b. $a_8 = \dfrac{2^8}{8^2} = \dfrac{256}{64} = 4$

TRY EXERCISE 6, EXERCISE SET 9.1

An **alternating sequence** is one in which the signs of the terms *alternate* between positive and negative values. The sequence defined by $a_n = (-1)^{n+1} \cdot 1/n$ is an alternating sequence.

$$a_1 = (-1)^{1+1} \cdot \frac{1}{1} = 1 \qquad a_2 = (-1)^{2+1} \cdot \frac{1}{2} = -\frac{1}{2} \qquad a_3 = (-1)^{3+1} \cdot \frac{1}{3} = \frac{1}{3}$$

The first six terms of the sequence are

$$1, -\frac{1}{2}, \frac{1}{3}, -\frac{1}{4}, \frac{1}{5}, -\frac{1}{6}$$

A **recursively defined sequence** is one in which each succeeding term of the sequence is defined by using some of the preceding terms. For example, let $a_1 = 1$, $a_2 = 1$, and $a_{n+1} = a_{n-1} + a_n$.

$$a_3 = a_1 + a_2 = 1 + 1 = 2 \qquad \bullet\, n = 2$$
$$a_4 = a_2 + a_3 = 1 + 2 = 3 \qquad \bullet\, n = 3$$
$$a_5 = a_3 + a_4 = 2 + 3 = 5 \qquad \bullet\, n = 4$$
$$a_6 = a_4 + a_5 = 3 + 5 = 8 \qquad \bullet\, n = 5$$

This recursive sequence 1, 1, 2, 3, 5, 8, ... is called the Fibonacci sequence, named after Leonardo Fibonacci (1180?–?1250), an Italian mathematician.

EXAMPLE 2   Find Terms of a Sequence Defined Recursively

Let $a_1 = 1$ and $a_n = na_{n-1}$. Find a_2, a_3, and a_4.

Solution

$$a_2 = 2a_1 = 2 \cdot 1 = 2 \qquad a_3 = 3a_2 = 3 \cdot 2 = 6 \qquad a_4 = 4a_3 = 4 \cdot 6 = 24$$

TRY EXERCISE 28, EXERCISE SET 9.1

FACTORIALS

It is possible to find an nth term formula for the sequence defined recursively in Example 2 by

$$a_1 = 1 \qquad a_n = na_{n-1}$$

Consider the term a_5 of that sequence.

$$
\begin{aligned}
a_5 &= 5a_4 \\
&= 5 \cdot 4a_3 & \bullet\ a_4 &= 4a_3 \\
&= 5 \cdot 4 \cdot 3a_2 & \bullet\ a_3 &= 3a_2 \\
&= 5 \cdot 4 \cdot 3 \cdot 2a_1 & \bullet\ a_2 &= 2a_1 \\
&= 5 \cdot 4 \cdot 3 \cdot 2 \cdot 1 & \bullet\ a_1 &= 1
\end{aligned}
$$

Continuing in this manner for a_n, we have

$$
\begin{aligned}
a_n &= na_{n-1} \\
&= n(n-1)a_{n-2} \\
&= n(n-1)(n-2)a_{n-3} \\
&\ \ \vdots \\
&= n(n-1)(n-2)(n-3) \cdots 2 \cdot 1
\end{aligned}
$$

The number $n \cdot (n-1) \cdots 3 \cdot 2 \cdot 1$ is called **n factorial** and is written $n!$.

THE FACTORIAL OF A NUMBER

If n is a positive integer, then $n!$, which is read "n factorial," is

$$n! = n \cdot (n-1) \cdots 3 \cdot 2 \cdot 1$$

We also define

$$0! = 1$$

It may seem strange to define $0! = 1$, but we shall see later that it is a reasonable definition.

Examples of factorials include

$$5! = 5 \cdot 4 \cdot 3 \cdot 2 \cdot 1 = 120$$

$$10! = 10 \cdot 9 \cdot 8 \cdot 7 \cdot 6 \cdot 5 \cdot 4 \cdot 3 \cdot 2 \cdot 1 = 3{,}628{,}800$$

Note that we can write $12!$ as

$$12! = 12 \cdot 11! = 12 \cdot 11 \cdot 10! = 12 \cdot 11 \cdot 10 \cdot 9!$$

In general,

$$n! = n \cdot (n-1)!$$

EXAMPLE 3   *Evaluate Factorial Expressions*

Evaluate each factorial expression.   **a.** $\dfrac{8!}{5!}$ **b.** $6! - 4!$

Solution

a. $\dfrac{8!}{5!} = \dfrac{8 \cdot 7 \cdot 6 \cdot 5!}{5!} = 8 \cdot 7 \cdot 6 = 336$

b. $6! - 4! = (6 \cdot 5 \cdot 4 \cdot 3 \cdot 2 \cdot 1) - (4 \cdot 3 \cdot 2 \cdot 1) = 720 - 24 = 696$

TRY EXERCISE 42, EXERCISE SET 9.1

PARTIAL SUMS AND SUMMATION NOTATION

Another important way of obtaining a sequence is by adding the terms of a given sequence. For example, consider the sequence whose general term is given by $a_n = 1/2^n$. The terms of this sequence are

$$\frac{1}{2}, \frac{1}{4}, \frac{1}{8}, \frac{1}{16}, \frac{1}{32}, \ldots, \frac{1}{2^n}, \ldots$$

From this sequence we can generate a new sequence that is the sum of the terms of $1/2^n$.

$$S_1 = \frac{1}{2}$$

$$S_2 = \frac{1}{2} + \frac{1}{4} = \frac{3}{4}$$

$$S_3 = \frac{1}{2} + \frac{1}{4} + \frac{1}{8} = \frac{7}{8}$$

$$S_4 = \frac{1}{2} + \frac{1}{4} + \frac{1}{8} + \frac{1}{16} = \frac{15}{16}$$

and, in general, $S_n = \dfrac{1}{2} + \dfrac{1}{4} + \dfrac{1}{8} + \dfrac{1}{16} + \cdots + \dfrac{1}{2^n}$

The term S_n is called the **nth partial sum** of the infinite sequence, and the sequence $S_1, S_2, S_3, \ldots, S_n$ is called the **sequence of partial sums.**

A convenient notation used for partial sums is called **summation notation.** The sum of the first n terms of a sequence a_n is represented by using the Greek letter Σ (sigma).

$$\sum_{i=1}^{n} a_i = a_1 + a_2 + a_3 + \cdots + a_n$$

This sum is called a **series.** The letter i is called the **index of the summation;** n is the **upper limit** of the summation; 1 is the **lower limit** of the summation.

POINT OF INTEREST

Leonard Euler (1707–1783) found that the sequence of terms given by

$$S_n = 1 - \frac{1}{3} + \frac{1}{5} - \frac{1}{7} + \cdots$$
$$+ \frac{(-1)^{n-1}}{2n - 1}$$

became closer and closer to $\pi/4$ as n increased. In summation notation, we would write each term as

$$S_n = \sum_{k=1}^{n} \frac{(-1)^{k-1}}{2k - 1}$$

EXAMPLE 4 *Evaluating Series*

Evaluate each series.   **a.** $\displaystyle\sum_{i=1}^{4} \frac{i}{i+1}$ **b.** $\displaystyle\sum_{j=2}^{5} (-1)^j j^2$

Solution

a. $\displaystyle\sum_{i=1}^{4} \frac{i}{i+1} = \frac{1}{2} + \frac{2}{3} + \frac{3}{4} + \frac{4}{5} = \frac{163}{60}$

b. $\displaystyle\sum_{j=2}^{5} (-1)^j j^2 = (-1)^2 2^2 + (-1)^3 3^2 + (-1)^4 4^2 + (-1)^5 5^2$

$$= 4 - 9 + 16 - 25 = -14$$

TRY EXERCISE 52, EXERCISE SET 9.1

TAKE NOTE

Example 4b illustrates that it is not necessary for a summation to begin at 1. The index of the summation can be any letter.

PROPERTIES OF SUMMATION NOTATION

If a_n and b_n are sequences and c is a real number, then

1. $\displaystyle\sum_{i=1}^{n} (a_i \pm b_i) = \sum_{i=1}^{n} a_i \pm \sum_{i=1}^{n} b_i$

2. $\displaystyle\sum_{i=1}^{n} c a_i = c \sum_{i=1}^{n} a_i$

3. $\displaystyle\sum_{i=1}^{n} c = nc$

The proof of property (1) depends on the commutative and associative properties of real numbers.

$$\sum_{i=1}^{n} (a_i \pm b_i) = (a_1 \pm b_1) + (a_2 \pm b_2) + \cdots + (a_n \pm b_n)$$

$$= (a_1 + a_2 + \cdots + a_n) \pm (b_1 + b_2 + \cdots + b_n)$$

$$= \sum_{i=1}^{n} a_i \pm \sum_{i=1}^{n} b_i$$

Property (2) is proved by using the distributive property; this is left as an exercise.

To prove property (3), let $a_n = c$. That is, each a_n is equal to the same constant c. (This is called a **constant sequence**.) Then

$$\sum_{i=1}^{n} a_n = a_1 + a_2 + \cdots + a_n = \underbrace{c + c + \cdots + c}_{\textbf{\textit{n} terms}} = nc$$

TOPICS FOR DISCUSSION

1. Discuss the difference between a finite sequence and an infinite sequence. Give an example of each type.

2. Discuss the difference between a sequence and a series.

3. What is a recursive sequence? Give an example of a recursive sequence.

4. What is an alternating sequence? Give an example of an alternating sequence.

EXERCISE SET 9.1

In Exercises 1 to 24, find the first three terms and the eighth term of the sequence that has the given nth term.

1. $a_n = n(n - 1)$

2. $a_n = 2n$

3. $a_n = 1 - \dfrac{1}{n}$

4. $a_n = \dfrac{n + 1}{n}$

5. $a_n = \dfrac{(-1)^{n+1}}{n^2}$

6. $a_n = \dfrac{(-1)^{n+1}}{n(n + 1)}$

7. $a_n = \dfrac{(-1)^{2n-1}}{3n}$

8. $a_n = \dfrac{(-1)^n}{2n - 1}$

9. $a_n = \left(\dfrac{2}{3}\right)^n$

10. $a_n = \left(\dfrac{-1}{2}\right)^n$

11. $a_n = 1 + (-1)^n$

12. $a_n = 1 + (-0.1)^n$

13. $a_n = (1.1)^n$

14. $a_n = \dfrac{n}{n^2 + 1}$

15. $a_n = \dfrac{(-1)^{n+1}}{\sqrt{n}}$

16. $a_n = \dfrac{3^{n-1}}{2^n}$

17. $a_n = n!$

18. $a_n = \dfrac{n!}{(n - 1)!}$

19. $a_n = \log n$

20. $a_n = \ln n$ (natural logarithm)

21. a_n is the digit in the nth place in the decimal expansion of 1/7.

22. a_n is the digit in the nth place in the decimal expansion of 1/13.

23. $a_n = 3$

24. $a_n = -2$

In Exercises 25 to 34, find the first three terms of each recursively defined sequence.

25. $a_1 = 5, a_n = 2a_{n-1}$

26. $a_1 = 2, a_n = 3a_{n-1}$

27. $a_1 = 2, a_n = na_{n-1}$

28. $a_1 = 1, a_n = n^2 a_{n-1}$

29. $a_1 = 2, a_n = (a_{n-1})^2$

30. $a_1 = 4, a_n = \dfrac{1}{a_{n-1}}$

31. $a_1 = 2, a_n = 2na_{n-1}$

32. $a_1 = 2, a_n = (-3)na_{n-1}$

33. $a_1 = 3, a_n = (a_{n-1})^{1/n}$

34. $a_1 = 2, a_n = (a_{n-1})^n$

35. $a_1 = 1, a_2 = 3, a_n = \dfrac{1}{2}(a_{n-1} + a_{n-2})$. Find $a_3, a_4,$ and a_5.

36. $a_1 = 1, a_2 = 4, a_n = (a_{n-1})(a_{n-2})$. Find $a_3, a_4,$ and a_5.

In Exercises 37 to 44, evaluate the factorial expression.

37. $7! - 6!$

38. $(4!)^2$

39. $\dfrac{9!}{7!}$

40. $\dfrac{10!}{5!}$

41. $\dfrac{8!}{3!\,5!}$

42. $\dfrac{12!}{4!\,8!}$

43. $\dfrac{100!}{99!}$

44. $\dfrac{100!}{98!\,2!}$

In Exercises 45 to 58, evaluate the series.

45. $\displaystyle\sum_{i=1}^{5} i$

46. $\displaystyle\sum_{i=1}^{4} i^2$

47. $\displaystyle\sum_{i=1}^{5} i(i - 1)$

48. $\displaystyle\sum_{i=1}^{7} (2i + 1)$

49. $\displaystyle\sum_{k=1}^{4} \dfrac{1}{k}$

50. $\displaystyle\sum_{k=1}^{6} \dfrac{1}{k(k + 1)}$

51. $\displaystyle\sum_{j=1}^{8} 2j$

52. $\displaystyle\sum_{i=1}^{6} (2i + 1)(2i - 1)$

53. $\displaystyle\sum_{i=3}^{5} (-1)^i 2^i$

54. $\displaystyle\sum_{i=3}^{5} \dfrac{(-1)^i}{2^i}$

55. $\displaystyle\sum_{n=1}^{7} \log \dfrac{n + 1}{n}$

56. $\displaystyle\sum_{n=2}^{8} \ln \dfrac{n}{n + 1}$

57. $\displaystyle\sum_{k=0}^{8} \frac{8!}{k!\,(8-k)!}$ **58.** $\displaystyle\sum_{k=0}^{7} \frac{1}{k!}$

In Exercises 59 to 66, write the given series in summation notation.

59. $\dfrac{1}{1} + \dfrac{1}{4} + \dfrac{1}{9} + \dfrac{1}{16} + \dfrac{1}{25} + \dfrac{1}{36}$

60. $2 + 4 + 6 + 8 + 10 + 12 + 14$

61. $2 - 4 + 8 - 16 + 32 - 64 + 128$

62. $1 - 8 + 27 - 64 + 125$

63. $7 + 10 + 13 + 16 + 19$

64. $30 + 26 + 22 + 18 + 14 + 10$

65. $\dfrac{1}{2} + \dfrac{1}{4} + \dfrac{1}{8} + \dfrac{1}{16}$

66. $1 - \dfrac{2}{3} + \dfrac{4}{9} - \dfrac{8}{27} + \dfrac{16}{81} - \dfrac{32}{243}$

SUPPLEMENTAL EXERCISES

67. **NEWTON'S METHOD** Newton's approximation to the square root of a number is given by the recursive sequence

$$a_1 = \frac{N}{2} \qquad a_n = \frac{1}{2}\left(a_{n-1} + \frac{N}{a_{n-1}}\right)$$

Approximate $\sqrt{7}$ by computing a_4. Compare this result with the calculator value of $\sqrt{7} \approx 2.6457513$.

68. Use the formula in Exercise 67 to approximate $\sqrt{10}$ by finding a_5.

69. Let $a_1 = N$ and $a_n = \sqrt{a_{n-1}}$. Find a_{20} when $N = 7$. (*Hint:* Enter 7 into your calculator and then press the $\boxed{\sqrt{}}$ key nineteen times. Make a conjecture as to the value of a_{100}.)

70. Let $a_n = i^n$, where i is the imaginary unit. Find the first eight terms of the sequence defined by a_n. Find a_{237}.

71. Let $a_n = \left[\dfrac{1}{2}(-1 + i\sqrt{3})\right]^n$. Find the first six terms of the sequence defined by a_n. Find a_{99}.

72. **STIRLING'S FORMULA** By using a calculator, evaluate $\sqrt{2\pi n}\,(n/e)^n$, where e is the base of the natural logarithms for $n = 10$, 20, and 30. This formula is called Stirling's formula and is used as an approximation for $n!$. For $n > 20$, the error in the approximation is less than 0.1%.

73. Prove that $\displaystyle\sum_{i=1}^{n} ca_i = c \sum_{i=1}^{n} a_i$, where c is a constant.

PROJECTS

1. **FORMULAS FOR INFINITE SEQUENCES** It is not possible to define an infinite sequence by giving a finite number of terms of the sequence. For instance, the question "What is the next term in the sequence 2, 4, 6, 8, ...?" does not have a unique answer.

a. Verify this statement by finding a formula for a_n such that the first four terms of the sequence are 2, 4, 6, 8 and the next term is 43. *Suggestion:* The formula

$$a_n = \frac{n(n-1)(n-2)(n-3)(n-4)}{4!} + 2n$$

generates the sequence 2, 4, 6, 8, 15 for $n = 1, 2, 3, 4, 5$.

b. Extend the result in **a.** by finding a formula for a_n that will give the first four terms as 2, 4, 6, 8 and the fifth term as x, where x is any real number.

SECTION

9.2 ARITHMETIC SEQUENCES AND SERIES

ARITHMETIC SEQUENCES

Note that in the sequence

$$2, 5, 8, 11, 14, \ldots, 3n - 1, \ldots$$

the difference between successive terms is always 3. Such a sequence is an *arithmetic sequence* or an *arithmetic progression.* These sequences have the following property: The difference between successive terms is the same constant. This constant is called the *common difference.* For the sequence above, the common difference is 3.

In general, an arithmetic sequence can be defined as follows:

ARITHMETIC SEQUENCE

Let d be a real number. A sequence a_n is an **arithmetic sequence** if

$$a_{i+1} - a_i = d \quad \text{for all } i$$

The number d is the **common difference** for the sequence.

Further examples of arithmetic sequences include

$$3, 8, 13, 18, \ldots, 5n - 2, \ldots$$

$$11, 7, 3, -1, \ldots, -4n + 15, \ldots$$

$$1, 2, 3, 4, \ldots, n, \ldots$$

Consider an arithmetic sequence in which the first term is a_1 and the common difference is d. By adding the common difference to each successive term of the arithmetic sequence, we can find a formula for the nth term.

$$a_1 = a_1$$
$$a_2 = a_1 + d$$
$$a_3 = a_2 + d = a_1 + d + d = a_1 + 2d$$
$$a_4 = a_3 + d = a_1 + 2d + d = a_1 + 3d$$

Note the relationship between the term number and the multiplier of d. The multiplier is 1 less than the term number.

FORMULA FOR THE nTH TERM OF AN ARITHMETIC SEQUENCE

The **nth term of an arithmetic sequence** with common difference of d is given by

$$a_n = a_1 + (n - 1)d$$

EXAMPLE 1 *Find the nth Term of an Arithmetic Sequence*

a. Find the twenty-fifth term of the arithmetic sequence whose first three terms are $-12, -6, 0$.

b. The fifteenth term of an arithmetic sequence is -3 and the first term is 25. Find the tenth term.

Solution

a. Find the common difference: $d = a_2 - a_1 = -6 - (-12) = 6$. Use the formula $a_n = a_1 + (n-1)d$ with $n = 25$.

$$a_{25} = -12 + (25-1)(6) = -12 + 24(6) = -12 + 144 = 132$$

b. Solve the equation $a_n = a_1 + (n-1)d$ for d, given that $n = 15$, $a_1 = 25$, and $a_{15} = -3$.

$$-3 = 25 + (14)d$$
$$d = -2$$

Now find the tenth term.

$$a_n = a_1 + (n-1)d$$
$$a_{10} = 25 + (9)(-2) = 7 \qquad \bullet \, n = 10, \, a_1 = 25, \, d = -2$$

TRY EXERCISE 16, EXERCISE SET 9.2

ARITHMETIC SERIES

Consider the arithmetic sequence given by

$$1, 3, 5, \ldots, 2n - 1, \ldots$$

Adding successive terms of this sequence, we generate a sequence of partial sums. The sum of the terms of an arithmetic sequence is called an **arithmetic series.**

$$S_1 = 1$$
$$S_2 = 1 + 3 = 4$$
$$S_3 = 1 + 3 + 5 = 9$$
$$S_4 = 1 + 3 + 5 + 7 = 16$$
$$S_5 = 1 + 3 + 5 + 7 + 9 = 25$$
$$\vdots \qquad \vdots$$
$$S_n = 1 + 3 + \cdots + (2n - 1) \stackrel{?}{=} n^2$$

The first five terms of this sequence are 1, 4, 9, 16, 25. It appears from this example that the sum of the first n odd integers is n^2. Shortly, we will be able to prove this result by using the following formula.

FORMULA FOR THE *n*TH PARTIAL SUM OF AN ARITHMETIC SEQUENCE

The **nth partial sum S_n of an arithmetic sequence a_n** with common difference d is

$$S_n = \frac{n}{2}(a_1 + a_n)$$

Proof We write S_n in both forward and reverse order.

$$S_n = a_1 + a_2 + a_3 + \cdots + a_{n-2} + a_{n-1} + a_n$$

$$S_n = a_n + a_{n-1} + a_{n-2} + \cdots + a_3 + a_2 + a_1$$

Add the two partial sums.

$$2S_n = (a_1 + a_n) + (a_2 + a_{n-1}) + (a_3 + a_{n-2}) + \cdots \qquad (1)$$
$$+ (a_{n-2} + a_3) + (a_{n-1} + a_2) + (a_n + a_1)$$

Consider the term $(a_3 + a_{n-2})$. Using the formula for the nth term of an arithmetic sequence, we have

$$a_3 \quad\quad = a_1 + (3-1)d = a_1 + 2d$$

$$a_{n-2} \quad\quad = a_1 + [(n-2)-1]d = a_1 + nd - 3d$$

Thus $\quad a_3 + a_{n-2} = (a_1 + 2d) + (a_1 + nd - 3d)$
$$= a_1 + (a_1 + nd - d) = a_1 + [a_1 + (n-1)d]$$
$$= a_1 + a_n$$

In a similar manner, we can show that each term in parentheses in the Equation (1) equals $(a_1 + a_n)$. Because there are n such terms, we have

$$2S_n = n(a_1 + a_n)$$

$$S_n = \frac{n}{2}(a_1 + a_n) \qquad \blacklozenge$$

There is an alternative form of the formula for the sum of n terms of an arithmetic sequence.

ALTERNATIVE FORMULA FOR THE SUM OF AN ARITHMETIC SERIES

The nth partial sum S_n of an arithmetic sequence with common difference d is

$$S_n = \frac{n[2a_1 + (n-1)d]}{2}$$

The proof of this theorem is left as an exercise.

EXAMPLE 2 *Find a Partial Sum of an Arithmetic Sequence*

a. Find the sum of the first 100 positive odd integers.

b. Find the sum of the first 50 terms of the arithmetic sequence whose first three terms are 2, 13/4, and 9/2.

Solution

Use the formula $S_n = \dfrac{n}{2}[2a_1 + (n-1)d]$.

a. We have $a_1 = 1$, $d = 2$, and $n = 100$. Thus

$$S_{100} = \frac{100}{2}[2(1) + (100-1)2] = 10{,}000$$

b. We have $a_1 = 2$, $d = \dfrac{5}{4}$, and $n = 50$. Thus

$$S_{50} = \frac{50}{2}\left[2(2) + (50-1)\frac{5}{4}\right] = \frac{6525}{4}$$

TRY EXERCISE 22, EXERCISE SET 9.2

The first n positive integers 1, 2, 3, 4, ..., n are part of an arithmetic sequence with a common difference of 1, $a_1 = 1$, and $a_n = n$. A formula for the sum of the first n positive integers can be found by using the formula for the nth partial sum of an arithmetic sequence.

$$S_n = \frac{n}{2}(a_1 + a_n)$$

Replacing a_1 by 1 and a_n by n yields

$$S_n = \frac{n}{2}(1 + n) = \frac{n(n+1)}{2}$$

This proves the following theorem.

SUM OF THE FIRST n POSITIVE INTEGERS

The sum of the first n positive integers is given by

$$S_n = \frac{n(n+1)}{2}$$

To find the sum of the first 85 positive integers, use $n = 85$.

$$S_{85} = \frac{85(85+1)}{2} = 3655$$

ARITHMETIC MEANS

The **arithmetic mean** of two numbers a and b is $(a + b)/2$. The three numbers a, $(a + b)/2$, b form an arithmetic sequence. In general, given two numbers a and b, it is possible to insert k numbers $c_1, c_2, \ldots, c_k$ in such a way that the sequence

$$a, c_1, c_2, \ldots, c_k, b$$

is an arithmetic sequence. This is called *inserting k arithmetic means between a and b.*

EXAMPLE 3 *Insert Arithmetic Means*

Insert three arithmetic means between 3 and 13.

Solution

After we insert the three terms, the sequence will be

$$a = 3, c_1, c_2, c_3, b = 13$$

The first term of the sequence is 3, the fifth term is 13, and n is 5. Thus

$$a_n = a_1 + (n - 1)d$$
$$13 = 3 + 4d$$
$$d = \frac{5}{2}$$

The three arithmetic means are

$$c_1 = a + d = 3 + \frac{5}{2} = \frac{11}{2}$$

$$c_2 = a + 2d = 3 + 2\left(\frac{5}{2}\right) = 8$$

$$c_3 = a + 3d = 3 + 3\left(\frac{5}{2}\right) = \frac{21}{2}$$

TRY EXERCISE 34, EXERCISE SET 9.2

TOPICS FOR DISCUSSION

1. Discuss what distinguishes an arithmetic sequence from all other types of sequences.

2. Is $a_n = 2/n$ a possible formula for the nth term of an arithmetic sequence? Why or why not?

3. Discuss the characteristics of an arithmetic series. Give an example of an arithmetic series.

4. Consider the series $\sum_{k=1}^{n} f(k)$. Discuss how you can determine whether this is an arithmetic series.

EXERCISE SET 9.2

In Exercises 1 to 14, find the ninth, twenty-fourth, and *n*th terms of the arithmetic sequence.

1. 6, 10, 14, ...

2. 7, 12, 17, ...

3. 6, 4, 2, ...

4. 11, 4, −3, ...

5. −8, −5, −2, ...

6. −15, −9, −3, ...

7. 1, 4, 7, ...

8. −4, 1, 6, ...

9. $a, a + 2, a + 4, \ldots$

10. $a − 3, a + 1, a + 5, \ldots$

11. $\log 7, \log 14, \log 28, \ldots$

12. $\ln 4, \ln 16, \ln 64, \ldots$

13. $\log a, \log a^2, \log a^3, \ldots$

14. $\log_2 5, \log_2 5a, \log_2 5a^2, \ldots$

15. The fourth and fifth terms of an arithmetic sequence are 13 and 15. Find the twentieth term.

16. The sixth and eighth terms of an arithmetic sequence are −14 and −20. Find the fifteenth term.

17. The fifth and seventh terms of an arithmetic sequence are −19 and −29. Find the seventeenth term.

18. The fourth and seventh terms of an arithmetic sequence are 22 and 34. Find the twenty-third term.

In Exercises 19 to 32, find the *n*th partial sum of the arithmetic sequence.

19. $a_n = 3n + 2; n = 10$

20. $a_n = 4n − 3; n = 12$

21. $a_n = 3 − 5n; n = 15$

22. $a_n = 1 − 2n; n = 20$

23. $a_n = 6n; n = 12$

24. $a_n = 7n; n = 14$

25. $a_n = n + 8; n = 25$

26. $a_n = n − 4; n = 25$

27. $a_n = −n; n = 30$

28. $a_n = 4 − n; n = 40$

29. $a_n = n + x; n = 12$

30. $a_n = 2n − x; n = 15$

31. $a_n = nx; n = 20$

32. $a_n = −nx; n = 14$

In Exercises 33 to 36, insert *k* arithmetic means between the given numbers.

33. −1 and 23; $k = 5$

34. 7 and 19; $k = 5$

35. 3 and $\dfrac{1}{2}$; $k = 4$

36. $\dfrac{11}{3}$ and 6; $k = 4$

37. Show that the sum of the first *n* positive odd integers is n^2.

38. Show that the sum of the first *n* positive even integers is $n^2 + n$.

39. STACKING LOGS Logs are stacked so that there are 25 logs in the bottom row, 24 logs in the second row, and so on, decreasing by 1 log each row. How many logs are stacked in the sixth row? How many logs are there in all six rows?

40. THEATER SEATING The seating section in a theater has 27 seats in the first row, 29 seats in the second row, and so on, increasing by 2 seats each row for a total of 10 rows. How many seats are in the tenth row, and how many seats are there in the section?

41. CONTEST PRIZES A contest offers 15 prizes. The first prize is $5000, and each successive prize is $250 less than the preceding prize. What is the value of the fifteenth prize? What is the total amount of money distributed in prizes?

42. PHYSICAL FITNESS An exercise program calls for walking 15 minutes each day for a week. Each week thereafter, the amount of time spent walking increases by 5 minutes per day. In how many weeks will a person be walking 60 minutes each day?

43. PHYSICS An object dropped from a cliff will fall 16 feet the first second, 48 feet the second second, 80 feet the third second, and so on, increasing by 32 feet each second. What is the total distance the object will fall in 7 seconds?

44. PHYSICS The distance a ball rolls down a ramp each second is given by the arithmetic sequence whose *n*th term is $2n − 1$ feet. Find the distance the ball rolls during the tenth second and the total distance the ball travels in 10 seconds.

SUPPLEMENTAL EXERCISES

45. If $f(x)$ is a linear function, show that $f(n)$, where *n* is a positive integer, is an arithmetic sequence.

46. Find the formula for a_n in terms of a_1 and *n* for the sequence that is defined recursively by $a_1 = 3$, $a_n = a_{n-1} + 5$.

47. Find a formula for a_n in terms of a_1 and *n* for the sequence that is defined recursively by $a_1 = 4$, $a_n = a_{n-1} − 3$.

48. Suppose a_n and b_n are two sequences such that $a_1 = 4$, $a_n = b_{n-1} + 5$ and $b_1 = 2$, $b_n = a_{n-1} + 1$. Show that a_n and b_n are arithmetic sequences. Find a_{100}.

49. Suppose a_n and b_n are two sequences such that $a_1 = 1$, $a_n = b_{n-1} + 7$ and $b_1 = -2$, $b_n = a_{n-1} + 1$. Show that a_n and b_n are arithmetic sequences. Find a_{50}.

PROJECTS

1. **ANGLES OF A TRIANGLE** The sum of the interior angles of a triangle is 180°.

    a. Using this fact, what is the sum of the interior angles of a quadrilateral?

    b. What is the sum of the interior angles of a pentagon?

    c. What is the sum of the interior angles of a hexagon?

    d. On the basis of your previous results, what is the apparent formula for the sum of the interior angles of a polygon of n sides?

2. **PROVE A FORMULA** Prove the Alternative Formula for the Sum of an Arithmetic Series.

SECTION

9.3 GEOMETRIC SEQUENCES AND SERIES

GEOMETRIC SEQUENCES

Arithmetic sequences are characterized by a common *difference* between successive terms. A *geometric sequence* is characterized by a common *ratio* between successive terms.

The sequence

$$3, 6, 12, 24, \ldots, 3(2^{n-1}), \ldots$$

is a geometric sequence. Note that the ratio of any two successive terms is 2.

$$\frac{6}{3} = 2 \qquad \frac{12}{6} = 2 \qquad \frac{24}{12} = 2$$

GEOMETRIC SEQUENCE

Let r be a nonzero constant real number. A sequence is a **geometric sequence** if

$$\frac{a_{i+1}}{a_i} = r \quad \text{for all positive integers } i.$$

> **EXAMPLE 1** *Determine Whether a Sequence Is a Geometric Sequence*

Which of the following are geometric sequences?

a. $4, -2, 1, \ldots, 4\left(-\dfrac{1}{2}\right)^{n-1}, \ldots$ **b.** $1, 4, 9, \ldots, n^2, \ldots$

Solution

To determine whether the sequence is a geometric sequence, calculate the ratio of successive terms.

a. $\dfrac{a_{i+1}}{a_i} = \dfrac{4(-1/2)^i}{4(-1/2)^{i-1}} = -\dfrac{1}{2}.$ • Because the ratio of successive terms is a constant, the sequence is a geometric sequence.

b. $\dfrac{a_{i+1}}{a_i} = \dfrac{(i+1)^2}{i^2} = \left(1 + \dfrac{1}{i}\right)^2$ • Because the ratio of successive terms is not a constant, the sequence is not a geometric sequence.

> **TRY EXERCISE 6, EXERCISE SET 9.3**

Consider a geometric sequence in which the first term is a_1 and the common ratio is r. By multiplying each successive term of the geometric sequence by the common ratio, we can derive a formula for the nth term.

$$a_1 = a_1$$
$$a_2 = a_1 r$$
$$a_3 = a_2 r = (a_1 r)r = a_1 r^2$$
$$a_4 = a_3 r = (a_1 r^2)r = a_1 r^3$$

Note the relationship between the number of the term and the number that is the exponent on r. The exponent on r is 1 less than the number of the term. With this observation, we can write a formula for the nth term of a geometric sequence.

> **nTH TERM OF A GEOMETRIC SEQUENCE**
>
> The **nth term of a geometric sequence** with first term a_1 and common ratio r is
>
> $$a_n = a_1 r^{n-1}$$

> **EXAMPLE 2** *Find the nth Term of a Geometric Sequence*

Find the nth term of the geometric sequence whose first three terms are

a. $4, 8/3, 16/9, \ldots$ **b.** $5, -10, 20, \ldots$

Solution

a. $r = \dfrac{8/3}{4} = \dfrac{2}{3}$ and $a_1 = 4$. Thus $a_n = 4\left(\dfrac{2}{3}\right)^{n-1}$.

b. $r = \dfrac{-10}{5} = -2$ and $a_1 = 5$. Thus $a_n = 5(-2)^{n-1}$.

TRY EXERCISE 18, EXERCISE SET 9.3

GEOMETRIC SEQUENCES

Adding the terms of a geometric sequence, we can define the nth partial sum of a geometric sequence in a manner similar to that of an arithmetic sequence. Consider the geometric sequence $1, 2, 4, 8, \ldots, 2^{n-1}, \ldots$

$$S_1 = 1$$

$$S_2 = 1 + 2 = 3$$

$$S_3 = 1 + 2 + 4 = 7$$

$$S_4 = 1 + 2 + 4 + 8 = 15$$

$$\vdots \qquad \vdots$$

$$S_n = 1 + 2 + 4 + 8 + \cdots + 2^{n-1}$$

The first four terms of the sequence of partial sums are 1, 3, 7, 15.

To find a general formula for S_n, the nth term of the sequence of partial sums of a geometric sequence, let

$$S_n = a_1 + a_1 r + a_1 r^2 + \cdots + a_1 r^{n-1}$$

Multiply each side of this equation by r.

$$S_n = a_1 + a_1 r + a_1 r^2 + \cdots + a_1 r^{n-2} + a_1 r^{n-1}$$

$$rS_n = \qquad a_1 r + a_1 r^2 + \cdots + a_1 r^{n-2} + a_1 r^{n-1} + a_1 r^{n}$$

Subtract the two equations.

$$S_n - rS_n = a_1 - a_1 r^{n}$$

$$S_n(1 - r) = a_1(1 - r^{n}) \qquad \bullet \text{ Factor out the common factors.}$$

$$S_n = \frac{a_1(1 - r^{n})}{1 - r} \qquad \bullet \, r \neq 1$$

This proves the following theorem.

> **THE nTH PARTIAL SUM OF A GEOMETRIC SEQUENCE**
>
> The **nth partial sum of a geometric sequence** with first term a_1 and common ratio r is
>
> $$S_n = \frac{a_1(1 - r^{n})}{1 - r} \quad r \neq 1$$

QUESTION   If $r = 1$, what is the nth partial sum of a geometric sequence?

| EXAMPLE 3 | **Find the nth Partial Sum of a Geometric Sequence** |

Find the partial sum of each geometric sequence.

a. $5, 15, 45, \ldots, 5(3)^{n-1}, \ldots$; $n = 4$ **b.** $\displaystyle\sum_{n=1}^{17} 3\left(\frac{3}{4}\right)^{n-1}$

Solution

a. We have $a_1 = 5$, $r = 3$, and $n = 4$. Thus

$$S_4 = \frac{5[1 - 3^4]}{1 - 3} = \frac{5(-80)}{-2} = 200$$

b. When $n = 1$, $a_1 = 3$. The first term is 3. The second term is 9/4. Therefore, the common ratio is $r = 3/4$. Thus

$$S_{17} = \frac{3[1 - (3/4)^{17}]}{1 - (3/4)} \approx 11.909797.$$

TRY EXERCISE 40, EXERCISE SET 9.3

INFINITE GEOMETRIC SERIES

Following are two examples of geometric sequences for which $|r| < 1$.

$$3, \frac{3}{4}, \frac{3}{16}, \frac{3}{64}, \frac{3}{256}, \frac{3}{1024}, \ldots \qquad \bullet\, r = \frac{1}{4}$$

$$2, -1, \frac{1}{2}, -\frac{1}{4}, \frac{1}{8}, -\frac{1}{16}, \frac{1}{32}, \ldots \qquad \bullet\, r = -\frac{1}{2}$$

Note that when the absolute value of the common ratio of a geometric sequence is less than 1, the terms of the geometric sequence approach zero as n increases. We write, for $|r| < 1$, $|r|^n \to 0$ as $n \to \infty$.

Consider again the geometric sequence

$$3, \frac{3}{4}, \frac{3}{16}, \frac{3}{64}, \frac{3}{256}, \frac{3}{1024}, \ldots$$

Table 9.1

n	S_n	r^n
3	3.93750000	0.01562500
6	3.99902344	0.00024414
9	3.99998474	0.00000381
12	3.99999976	0.00000006

The nth partial sums for $n = 3$, 6, 9, and 12 are given in Table 9.1, along with the value of r^n. As n increases, S_n is closer to 4 and r^n is closer to zero. By finding more values of S_n for larger values of n, we would find that $S_n \to 4$ as $n \to \infty$. As n becomes larger and larger, S_n is the nth partial sum of more and more terms of the sequence. The sum of *all* the terms of a sequence is called an **infinite series**. If the sequence is a geometric sequence, we have an **infinite geometric series**.

ANSWER   When $r = 1$, the sequence is the constant sequence a_1. The nth partial sum of a constant sequence is na_1.

SUM OF AN INFINITE GEOMETRIC SEQUENCE

If a_n is a geometric sequence with $|r| < 1$ and first term a_1, then the sum of the infinite geometric series is

$$S = \frac{a_1}{1 - r}$$

A formal proof of this formula requires topics that typically are studied in calculus. We can, however, give an intuitive argument. Start with the formula for the nth partial sum of a geometric sequence.

$$S_n = \frac{a_1(1 - r^n)}{1 - r}$$

When $|r| < 1$, $|r|^n \approx 0$ when n is large. Thus

$$S_n = \frac{a_1(1 - r^n)}{1 - r} \approx \frac{a_1(1 - 0)}{1 - r} = \frac{a_1}{1 - r}$$

An infinite series is represented by $\sum_{n=1}^{\infty} a_n$. One application of infinite geometric series concerns repeating decimals. Consider the repeating decimal

$$0.\overline{6} = \frac{6}{10} + \frac{6}{100} + \frac{6}{1000} + \frac{6}{10,000} + \cdots$$

The right-hand side is a geometric series with $a_1 = 6/10$ and common ratio $r = 1/10$. Thus

$$S = \frac{6/10}{1 - (1/10)} = \frac{6/10}{9/10} = \frac{2}{3}$$

The repeating decimal $0.\overline{6} = 2/3$. We can write any repeating decimal as a ratio of two integers by using the formula for the sum of an infinite geometric series.

> **TAKE NOTE**
>
> The sum of an infinite geometric series is not defined when $|r| \geq 1$. For instance, the infinite geometric series
>
> $2 + 4 + 8 + \cdots + 2^n + \cdots$
>
> with $r = 2$ increases without bound. However, applying the formula $S = \dfrac{a_1}{1 - r}$ with $r = 2$ and $a_1 = 2$ gives $S = -2$, which is not correct.

EXAMPLE 4   *Find the Value of an Infinite Geometric Series*

a.   Evaluate the infinite geometric series $\sum_{n=1}^{\infty} \left(-\frac{2}{3}\right)^{n-1}$.

b.   Write $0.3\overline{45}$ as the ratio of two integers in lowest terms.

Solution

a.   To find the first term, we let $n = 1$. Then $a_1 = (-2/3)^{1-1} = (-2/3)^0 = 1$. The common ratio $r = -2/3$. Thus

$$S = \frac{1}{1 - (-2/3)} = \frac{1}{(5/3)} = \frac{3}{5}$$

Continued ▸➤

b. $0.3\overline{45} = \dfrac{3}{10} + \left[\dfrac{45}{1000} + \dfrac{45}{100,000} + \dfrac{45}{10,000,000} + \cdots \right]$

The terms in the brackets form an infinite geometric series. Evaluate that series with $a_1 = 45/1000$ and $r = 1/100$, and then add the term $3/10$.

$$\frac{45}{1000} + \frac{45}{100,000} + \frac{45}{10,000,000} + \cdots = \frac{45/1000}{1 - (1/100)} = \frac{1}{22}$$

Thus $0.3\overline{45} = \dfrac{3}{10} + \dfrac{1}{22} = \dfrac{19}{55}$

TRY EXERCISE 62, EXERCISE SET 9.3

FUTURE VALUE OF AN ANNUITY

In an earlier chapter we discussed compound interest by using exponential functions. As an extension of this idea, suppose that for each of the next 5 years, A dollars are deposited on December 31 into an account earning $i\%$ annual interest compounded annually. Using the compound interest formula, we can find the total value of all the deposits. Table 9.2 shows the growth of the investment.

Table 9.2

Deposit number	Value of each deposit	
1	$A(1 + i)^4$	Value of first deposit after 4 years
2	$A(1 + i)^3$	Value of second deposit after 3 years
3	$A(1 + i)^2$	Value of third deposit after 2 years
4	$A(1 + i)$	Value of fourth deposit after 1 year
5	A	Value of fifth deposit

The total value of the investment after the last deposit, called the **future value** of the investment, is the sum of the values of all the deposits.

$$P_5 = A + A(1 + i) + A(1 + i)^2 + A(1 + i)^3 + A(1 + i)^4$$

This is a geometric series with first term A and common ratio $1 + i$. Thus, using the formula for the nth partial sum of a geometric sequence

$$S = \frac{a_1(1 - r^n)}{1 - r}$$

we have

$$P_5 = \frac{A[1 - (1 + i)^5]}{1 - (1 + i)} = \frac{A[(1 + i)^5 - 1]}{i}$$

Deposits of equal amounts at equal intervals of time are called **annuities.** When the amounts are deposited at the end of a compounding period (as in our example), we have an **ordinary annuity.**

FUTURE VALUE OF AN ORDINARY ANNUITY

Let $r = i/n$ and $m = nt$, where i is the annual interest rate, n is the number of compounding periods per year, and t is the number of years. Then the future value of an ordinary annuity is given by

$$P = \frac{A[(1 + r)^m - 1]}{r}$$

EXAMPLE 5 *Find the Future Value of an Ordinary Annuity*

An employee savings plan allows any employee to deposit $25 at the end of each month into a savings account earning 6% annual interest compounded monthly. Find the future value of this savings plan if an employee makes the deposits for 10 years.

Solution

We are given $A = 25$, $i = 0.06$, $n = 12$, and $t = 10$. Thus,

$$r = \frac{i}{n} = \frac{0.06}{12} = 0.005 \quad \text{and} \quad m = nt = 12(10) = 120$$

$$P = \frac{25[(1 + 0.005)^{120} - 1]}{0.005} \approx 4096.9837$$

The future value is $4096.98.

TRY EXERCISE 70, EXERCISE SET 9.3

TOPICS FOR DISCUSSION

1. Discuss what distinguishes a geometric sequence from all other types of sequences.

2. Is $a_n = n^2$ a possible formula for the nth term of a geometric sequence? Why or why not?

3. Discuss the characteristics of a geometric series.

4. Consider the series $\sum\limits_{k=1}^{n} f(k)$. Explain how you can determine whether this is a geometric series.

EXERCISE SET 9.3

In Exercises 1 to 12, determine which sequences are geometric. For geometric sequences, find the common ratio.

1. $4, 16, 64, \ldots, 4^n, \ldots$

2. $1, 6, 36, \ldots, 6^{n-1}, \ldots$

3. $1, \dfrac{1}{2}, \dfrac{1}{3}, \ldots, \dfrac{1}{n}, \ldots$

4. $\dfrac{1}{2}, \dfrac{1}{4}, \dfrac{1}{8}, \ldots, \dfrac{1}{2^n}, \ldots$

5. $2^x, 2^{2x}, 2^{3x}, \ldots, 2^{nx}, \ldots$

6. $e^x, -e^{2x}, e^{3x}, \ldots, (-1)^{n-1}e^{nx}, \ldots$

7. $3, 6, 12, \ldots, 3(2^{n-1}), \ldots$

8. $5, -10, 20, \ldots, 5(-2)^{n-1}, \ldots$

9. $x^2, x^4, x^6, \ldots, x^{2n}, \ldots$

10. $3x, 6x^2, 9x^3, \ldots, 3nx^n, \ldots$

11. $\ln 5, \ln 10, \ln 15, \ldots, \ln 5n, \ldots$

12. $\log x, \log x^2, \log x^4, \ldots, \log x^{2^{n-1}}, \ldots$

In Exercises 13 to 32, find the nth term of the geometric sequence.

13. $2, 8, 32, \ldots$

14. $1, 5, 25, \ldots$

15. $-4, 12, -36, \ldots$

16. $-3, 6, -12, \ldots$

17. $6, 4, \dfrac{8}{3}, \ldots$

18. $8, 6, \dfrac{9}{2}, \ldots$

19. $-6, 5, -\dfrac{25}{6}, \ldots$

20. $-2, \dfrac{4}{3}, -\dfrac{8}{9}, \ldots$

21. $9, -3, 1, \ldots$

22. $8, -\dfrac{4}{3}, \dfrac{2}{9}, \ldots$

23. $1, -x, x^2, \ldots$

24. $2, 2a, 2a^2, \ldots$

25. $c^2, c^5, c^8, \ldots$

26. $-x^2, x^4, -x^6, \ldots$

27. $\dfrac{3}{100}, \dfrac{3}{10,000}, \dfrac{3}{1,000,000}, \ldots$

28. $\dfrac{7}{10}, \dfrac{7}{10,000}, \dfrac{7}{10,000,000}, \ldots$

29. $0.5, 0.05, 0.005, \ldots$

30. $0.4, 0.004, 0.00004, \ldots$

31. $0.45, 0.0045, 0.000045, \ldots$

32. $0.234, 0.000234, 0.000000234, \ldots$

33. Find the third term of a geometric sequence whose first term is 2 and whose fifth term is 162.

34. Find the fourth term of a geometric sequence whose third term is 1 and whose eighth term is 1/32.

35. Find the second term of a geometric sequence whose third term is 4/3 and whose sixth term is $-32/81$.

36. Find the fifth term of a geometric sequence whose fourth term is 8/9 and whose seventh term is 64/243.

In Exercises 37 to 46, find the sum of the geometric series.

37. $\displaystyle\sum_{n=1}^{5} 3^n$

38. $\displaystyle\sum_{n=1}^{7} 2^n$

39. $\displaystyle\sum_{n=1}^{6} \left(\dfrac{2}{3}\right)^n$

40. $\displaystyle\sum_{n=1}^{14} \left(\dfrac{4}{3}\right)^n$

41. $\displaystyle\sum_{n=0}^{8} \left(-\dfrac{2}{5}\right)^n$

42. $\displaystyle\sum_{n=0}^{7} \left(-\dfrac{1}{3}\right)^n$

43. $\displaystyle\sum_{n=1}^{10} (-2)^{n-1}$

44. $\displaystyle\sum_{n=0}^{7} 2(5)^n$

45. $\displaystyle\sum_{n=0}^{9} 5(3)^n$

46. $\displaystyle\sum_{n=0}^{10} 2(-4)^n$

In Exercises 47 to 56, find the sum of the infinite geometric series.

47. $\displaystyle\sum_{n=1}^{\infty} \left(\dfrac{1}{3}\right)^n$

48. $\displaystyle\sum_{n=1}^{\infty} \left(\dfrac{3}{4}\right)^n$

49. $\displaystyle\sum_{n=1}^{\infty} \left(-\dfrac{2}{3}\right)^n$

50. $\displaystyle\sum_{n=1}^{\infty} \left(-\dfrac{3}{5}\right)^n$

51. $\displaystyle\sum_{n=1}^{\infty} \left(\dfrac{9}{100}\right)^n$

52. $\displaystyle\sum_{n=1}^{\infty} \left(\dfrac{7}{10}\right)^n$

53. $\displaystyle\sum_{n=1}^{\infty} (0.1)^n$

54. $\displaystyle\sum_{n=1}^{\infty} (0.5)^n$

55. $\displaystyle\sum_{n=0}^{\infty} (-0.4)^n$

56. $\displaystyle\sum_{n=0}^{\infty} (-0.8)^n$

In Exercises 57 to 68, write each rational number as the quotient of two integers in simplest form.

57. $0.\overline{3}$

58. $0.\overline{5}$

59. $0.\overline{45}$

60. $0.\overline{63}$

61. $0.\overline{123}$

62. $0.\overline{395}$

63. $0.4\overline{22}$

64. $0.3\overline{55}$

65. $0.25\overline{4}$

66. $0.37\overline{2}$

67. $1.20\overline{84}$

68. $2.25\overline{90}$

69. TIME VALUE OF MONEY Find the future value of an ordinary annuity that calls for depositing $100 at the end of every 6 months for 8 years into an account that earns 9% interest compounded semiannually.

70. TIME VALUE OF MONEY To save for the replacement of a computer, a business deposits $250 at the end of each month into an account that earns 8% annual interest compounded monthly. Find the future value of the ordinary annuity in 4 years.

SUPPLEMENTAL EXERCISES

71. If the sequence a_n is a geometric sequence, make a conjecture about the sequence $\log a_n$ and give a proof.

72. If the sequence a_n is an arithmetic sequence, make a conjecture about the sequence 2^{a_n} and give a proof.

73. Does $\displaystyle\sum_{i=0}^{\infty} x^i$ $(x \neq 0)$ represent an infinite geometric series? Why or why not?

74. Consider a square with a side of length 1. Construct another square inside the first one by connecting the midpoints of the first square. What is the area of the inscribed square? Continue constructing squares in the same way. Find the area of the nth inscribed square.

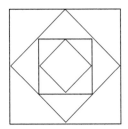

75. The product $P_n = a_1 \cdot a_2 \cdot a_3 \cdots a_n$ is called the nth partial product of a sequence. Find a formula for the nth partial product of the geometric sequence whose nth term is ar^{n-1}.

76. Let $f(x) = ab^x$, $a, b > 0$. Show that if x is restricted to positive integers n, then $f(n)$ is a geometric sequence.

77. PHYSICS A ball is dropped from a height of 5 feet. The ball rebounds 80% of the distance after each fall. Use an infinite geometric series to find the total distance the ball will travel.

78. PENDULUM The bob of a pendulum swings through an arc of 30 inches on its first swing. Each successive swing is 90% of the length of the previous swing. Find the total distance the bob will travel.

79. GENEALOGY Some people can trace their ancestry back ten generations, which means two parents, four grandparents, eight great-grandparents, and so on. How many grandparents does such a family tree include?

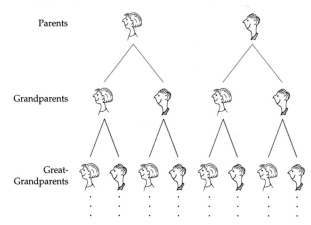

PROJECTS

1. **FRACTALS** The snowflake that was created at the beginning of this chapter is an example of a fractal. Here is another example of a fractal. Begin with a square with each side 1 unit long. Construct another, smaller square onto the middle third of each side. Continue this procedure of constructing similar but smaller squares on each of the line segments. The figure at the right shows the result after the process has been completed twice.

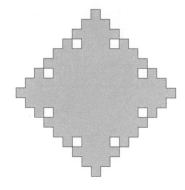

a. What is the perimeter of the figure after the process has been completed n times?

b. As n approaches infinity, what value does the perimeter approach?

c. What is the area of the figure after the process has been completed n times?

d. As n approaches infinity, what value does the area approach? *Suggestion:* The series in **c.** is a geometric series after the first term.

Consider the sequence

$$\frac{1}{1 \cdot 2}, \frac{1}{2 \cdot 3}, \frac{1}{3 \cdot 4}, \cdots, \frac{1}{n(n + 1)}, \cdots$$

and the sequence of partial sums for this sequence:

$$S_1 = \frac{1}{1 \cdot 2} = \frac{1}{2}$$

$$S_2 = \frac{1}{1 \cdot 2} + \frac{1}{2 \cdot 3} = \frac{2}{3}$$

$$S_3 = \frac{1}{1 \cdot 2} + \frac{1}{2 \cdot 3} + \frac{1}{3 \cdot 4} = \frac{3}{4}$$

$$S_4 = \frac{1}{1 \cdot 2} + \frac{1}{2 \cdot 3} + \frac{1}{3 \cdot 4} + \frac{1}{4 \cdot 5} = \frac{4}{5}$$

This pattern suggests the conjecture that

$$S_n = \frac{1}{1 \cdot 2} + \frac{1}{2 \cdot 3} + \frac{1}{3 \cdot 4} + \cdots + \frac{1}{n(n + 1)} = \frac{n}{n + 1}$$

How can we be sure that the pattern does not break down when $n = 50$ or maybe $n = 2000$ or some other large number? As we will show, this conjecture is true for all values of n.

As a second example, consider the conjecture that the expression $n^2 - n + 41$ is a prime number for all positive integers. To test this conjecture, we will try various values of n. See Table 9.3. The results suggest that the conjecture is true. But again, how can we be sure? In fact, this conjecture is false when $n = 41$. In that case we have

$$n^2 - n + 41 = (41)^2 - 41 + 41 = (41)^2$$

and $(41)^2$ is not a prime.

The last example illustrates that just verifying a conjecture for a few values of n does not constitute a proof of the conjecture. To prove theorems about statements involving positive integers, a process called *mathematical induction* is used. This process is based on an axiom called the *induction axiom*.

Table 9.3

n	$n^2 - n + 41$	
1	41	Prime
2	43	Prime
3	47	Prime
4	53	Prime
5	61	Prime

INDUCTION AXIOM

Suppose S is a set of positive integers with the following two properties:

1. 1 is an element of S.

2. If the positive integer k is in S, then $k + 1$ is in S.

Then S contains all the positive integers.

Part 2 of this axiom states that if some positive integer, say 8, is in S, then $8 + 1$, or 9, is in S. But because 9 is in S, part 2 says that $9 + 1$, or 10, is in S, and so on. Part 1 states that 1 is in S. Thus 2 is in S; thus 3 is in S; thus 4 is in S; .... Therefore all the positive integers are in S.

The induction axiom is used to prove the *Principle of Mathematical Induction.*

PRINCIPLE OF MATHEMATICAL INDUCTION

Let P_n be a statement about a positive integer n. If

1. P_1 is true, and

2. The truth of P_k implies the truth of P_{k+1}

then P_n is true for all positive integers.

Part 2 of the Principle of Mathematical Induction is referred to as the **induction hypothesis.** When applying this step, we assume the statement P_k is true and then try to prove that P_{k+1} is also true.

As an example, we will prove that the first conjecture we made in this section is true for all positive integers. Every induction proof has the two distinct parts stated in the theorem. First we must show that the result is true for $n = 1$. Second, we assume the statement is true for some positive integer k and, using that assumption, prove the statement is true for $n = k + 1$.

Prove that

$$S_n = \frac{1}{1 \cdot 2} + \frac{1}{2 \cdot 3} + \frac{1}{3 \cdot 4} + \cdots + \frac{1}{n(n+1)} = \frac{n}{n+1}$$

for all positive integers n.

Proof

1. For $n = 1$,

$$S_1 = \frac{1}{1(1+1)} = \frac{1}{2}, \text{ and } \frac{n}{n+1} = \frac{1}{1+1} = \frac{1}{2}$$

The statement is true for $n = 1$.

2. Assume the statement is true for some positive integer k.

$$S_k = \frac{1}{1 \cdot 2} + \frac{1}{2 \cdot 3} + \frac{1}{3 \cdot 4} + \cdots + \frac{1}{k(k+1)} = \frac{k}{k+1} \qquad \bullet \text{ Induction hypothesis}$$

Now verify that the formula is true when $n = k + 1$. That is, verify that

$$S_{k+1} = \frac{k+1}{(k+1)+1} = \frac{k+1}{k+2} \qquad \bullet \text{ This is the goal of the induction proof.}$$

It is helpful, when proving a theorem about sums, to note that

$$S_{k+1} = S_k + a_{k+1}$$

Begin by noting that $a_k = \dfrac{1}{k(k+1)}$; thus, $a_{k+1} = \dfrac{1}{(k+1)(k+2)}$.

$$S_{k+1} = S_k + a_{k+1}$$

$$= \frac{k}{k+1} + \frac{1}{(k+1)(k+2)}$$

• **By the induction hypothesis and substituting for a_{k+1}**

$$= \frac{k(k+2)}{(k+1)(k+2)} + \frac{1}{(k+1)(k+2)}$$

$$= \frac{k(k+2)+1}{(k+1)(k+2)} = \frac{k^2+2k+1}{(k+1)(k+2)} = \frac{(k+1)^2}{(k+1)(k+2)}$$

$$S_{k+1} = \frac{k+1}{k+2}$$

Because we have verified the two parts of the Principle of Mathematical Induction, we can conclude that the statement is true for all positive integers. ◆

EXAMPLE 1   *Prove by Mathematical Induction*

Prove that $1^2 + 2^2 + 3^2 + \cdots + n^2 = \dfrac{n(n+1)(2n+1)}{6}$.

Solution

Verify the two parts of the Principle of Mathematical Induction.

1. Let $n = 1$.

$$S_1 = 1^2 = 1 = \frac{1(1+1)(2 \cdot 1 + 1)}{6}$$

2. Assume the statement is true for some positive integer k.

$$S_k = 1^2 + 2^2 + 3^2 + \cdots + k^2 = \frac{k(k+1)(2k+1)}{6}$$ • **Induction hypothesis**

Verify that the statement is true when $n = k + 1$. Show that

$$S_{k+1} = \frac{(k+1)(k+2)(2k+3)}{6}$$

Because $a_k = k^2$, $a_{k+1} = (k+1)^2$.

$$S_{k+1} = S_k + a_{k+1}$$

$$= \frac{k(k+1)(2k+1)}{6} + (k+1)^2$$

$$= \frac{k(k+1)(2k+1)}{6} + \frac{6(k+1)^2}{6} = \frac{k(k+1)(2k+1) + 6(k+1)^2}{6}$$

$$= \frac{(k+1)[k(2k+1) + 6(k+1)]}{6} = \frac{(k+1)(2k^2 + 7k + 6)}{6}$$

$$S_{k+1} = \frac{(k+1)(k+2)(2k+3)}{6}$$

By the Principle of Mathematical Induction, the statement is true for all positive integers.

TRY EXERCISE 8, EXERCISE SET 9.4

Mathematical induction can also be used to prove statements about sequences, products, and inequalities.

| EXAMPLE 2 | **Prove a Product Formula by Mathematical Induction** |

Prove that

$$\left(1 + \frac{1}{1}\right)\left(1 + \frac{1}{2}\right)\left(1 + \frac{1}{3}\right)\cdots\left(1 + \frac{1}{n}\right) = n + 1$$

Solution

1. Verify for $n = 1$.

$$\left(1 + \frac{1}{1}\right) = 2, \text{ and } 1 + 1 = 2$$

2. Assume the statement is true for some positive integer k.

$$P_k = \left(1 + \frac{1}{1}\right)\left(1 + \frac{1}{2}\right)\left(1 + \frac{1}{3}\right)\cdots\left(1 + \frac{1}{k}\right) = k + 1 \qquad \bullet \text{ **Induction hypothesis**}$$

Verify that the statement is true when $n = k + 1$. That is, prove $P_{k+1} = k + 2$.

$$P_{k+1} = \left(1 + \frac{1}{1}\right)\left(1 + \frac{1}{2}\right)\left(1 + \frac{1}{3}\right)\cdots\left(1 + \frac{1}{k}\right)\left(1 + \frac{1}{k + 1}\right)$$

$$= P_k\left(1 + \frac{1}{k + 1}\right) = (k + 1)\left(1 + \frac{1}{k + 1}\right) = k + 1 + 1$$

$$P_{k+1} = k + 2$$

By the Principle of Mathematical Induction the statement is true for all positive integers.

| **TRY EXERCISE 12, EXERCISE SET 9.4** |

MATHEMATICAL INDUCTION AND INEQUALITIES

| EXAMPLE 3 | **Prove an Inequality by Mathematical Induction** |

Prove that $1 + 2n \leq 3^n$ for all positive integers.

Solution

1. Let $n = 1$. Then $1 + 2(1) = 3 \leq 3^1$. The statement is true when n is 1.

2. Assume the statement is true for some positive integer k.

$$1 + 2k \leq 3^k \qquad \bullet \text{ **Induction hypothesis**}$$

Continued •➤

Now prove the statement is true for $n = k + 1$. That is, prove that $1 + 2(k + 1) \le 3^{k+1}$.

$$3^{k+1} = 3^k(3)$$
$$\ge (1 + 2k)(3) \quad \bullet \textbf{ Because by the induction hypothesis, } \mathbf{1 + 2k \le 3^k.}$$
$$= 6k + 3$$
$$> 2k + 2 + 1 \quad \bullet \mathbf{6k > 2k, \text{ and } 3 = 2 + 1.}$$
$$= 2(k + 1) + 1$$

Thus $1 + 2(k + 1) \le 3^{k+1}$.

By the Principle of Mathematical Induction, $1 + 2k \le 3^k$ for all positive integers.

TRY EXERCISE 16, EXERCISE SET 9.4

The Principle of Mathematical Induction can be extended to cases where the beginning index is greater than 1.

EXTENDED PRINCIPLE OF MATHEMATICAL INDUCTION

Let P_n be a statement about a positive integer n. If

1. P_j is true for some positive integer j, and

2. For $k \ge j$ the truth of P_k implies the truth of P_{k+1}

then P_n is true for all positive integers $n \ge j$.

EXAMPLE 4   *Prove an Inequality by Mathematical Induction*

For $n \ge 3$, prove that $n^2 > 2n + 1$.

Solution

1. Let $n = 3$. Then $3^2 = 9$; $2(3) + 1 = 7$. Thus $n^2 > 2n + 1$ for $n = 3$.

2. Assume the statement is true for some positive integer $k \ge 3$.

$$k^2 > 2k + 1 \quad \bullet \textbf{ Induction hypothesis}$$

Verify that the statement is true when $n = k + 1$. That is, show that

$$(k + 1)^2 > 2(k + 1) + 1 = 2k + 3$$

$$(k + 1)^2 = k^2 + 2k + 1$$
$$> (2k + 1) + 2k + 1 \quad \bullet \textbf{ Induction hypothesis}$$
$$> 2k + 1 + 1 + 1 \quad \bullet \mathbf{2k > 1}$$
$$= 2k + 3$$

Thus $(k + 1)^2 > 2k + 3$.

By the Extended Principle of Mathematical Induction, $n^2 > 2n + 1$ for all $n \ge 3$.

TRY EXERCISE 20, EXERCISE SET 9.4

TOPICS FOR DISCUSSION

1. Discuss the purpose of mathematical induction.

2. What is an induction hypothesis?

3. What is the Extended Principle of Mathematical Induction and how is it used?

4. Mathematical induction can be used to prove that $x^m \cdot x^n = x^{m+n}$ when m and n are natural numbers. Explain why mathematical induction cannot be used if we are attempting to prove the result when m and n are real numbers.

EXERCISE SET 9.4

In Exercises 1 to 12, use mathematical induction to prove each statement.

1. $\displaystyle\sum_{i=1}^{n} (3i - 2) = 1 + 4 + 7 + \cdots + 3n - 2 = \dfrac{n(3n-1)}{2}$

2. $\displaystyle\sum_{i=1}^{n} 2i = 2 + 4 + 6 + \cdots + 2n = n(n+1)$

3. $\displaystyle\sum_{i=1}^{n} i^3 = 1 + 8 + 27 + \cdots + n^3 = \dfrac{n^2(n+1)^2}{4}$

4. $\displaystyle\sum_{i=1}^{n} 2^i = 2 + 4 + 8 + \cdots + 2^n = 2(2^n - 1)$

5. $\displaystyle\sum_{i=1}^{n} (4i - 1) = 3 + 7 + 11 + \cdots + 4n - 1 = n(2n+1)$

6. $\displaystyle\sum_{i=1}^{n} 3^i = 3 + 9 + 27 + \cdots + 3^n = \dfrac{3(3^n - 1)}{2}$

7. $\displaystyle\sum_{i=1}^{n} (2i - 1)^3 = 1 + 27 + 125 + \cdots + (2n-1)^3$

$$= n^2(2n^2 - 1)$$

8. $\displaystyle\sum_{i=1}^{n} i(i+1) = 2 + 6 + 12 + \cdots + n(n+1)$

$$= \dfrac{n(n+1)(n+2)}{3}$$

9. $\displaystyle\sum_{i=1}^{n} \dfrac{1}{(2i-1)(2i+1)} = \dfrac{1}{1 \cdot 3} + \dfrac{1}{3 \cdot 5} + \dfrac{1}{5 \cdot 7} + \cdots$

$$+ \dfrac{1}{(2n-1)(2n+1)} = \dfrac{n}{2n+1}$$

10. $\displaystyle\sum_{i=1}^{n} \dfrac{1}{2i(2i+2)} = \dfrac{1}{2 \cdot 4} + \dfrac{1}{4 \cdot 6} + \dfrac{1}{6 \cdot 8} + \cdots$

$$+ \dfrac{1}{2n(2n+2)} = \dfrac{n}{4(n+1)}$$

11. $\displaystyle\sum_{i=1}^{n} i^4 = 1 + 16 + 81 + \cdots + n^4$

$$= \dfrac{n(n+1)(2n+1)(3n^2 + 3n - 1)}{30}$$

12. $P_n = \left(1 - \dfrac{1}{2}\right)\left(1 - \dfrac{1}{3}\right)\left(1 - \dfrac{1}{4}\right)\cdots\left(1 - \dfrac{1}{n+1}\right) = \dfrac{1}{n+1}$

In Exercises 13 to 20, use mathematical induction to prove each inequality.

13. $\left(\dfrac{3}{2}\right)^n > n + 1, \; n \geq 4$ **14.** $\left(\dfrac{4}{3}\right)^n > n, \; n \geq 7$

15. If $0 < a < 1$, show that $a^{n+1} < a^n$ for all positive integers n.

16. If $a > 1$, show that $a^{n+1} > a^n$ for all positive integers n.

17. $1 \cdot 2 \cdot 3 \cdot \cdots \cdot n > 2^n, \; n \geq 4$

18. $\dfrac{1}{\sqrt{1}} + \dfrac{1}{\sqrt{2}} + \dfrac{1}{\sqrt{3}} + \cdots + \dfrac{1}{\sqrt{n}} \geq \sqrt{n}$

19. For $a > 0$, show that $(1 + a)^n \geq 1 + na$ for all positive integers.

20. $\log_{10} n < n$ for all positive integers. (*Hint:* Because $\log_{10} x$ is an increasing function, $\log_{10}(n + 1) \leq \log_{10}(n + n)$.)

In Exercises 21 to 30, use mathematical induction to prove each statement.

21. 2 is a factor of $n^2 + n$ for all positive integers n.

22. 3 is a factor of $n^3 - n$ for all positive integers n.

23. 4 is a factor of $5^n - 1$ for all positive integers n. (*Hint:* $5^{k+1} - 1 = 5 \cdot 5^k - 5 + 4$.)

24. 5 is a factor of $6^n - 1$ for all positive integers n.

25. $(xy)^n = x^n y^n$ for all positive integers n.

26. $\left(\dfrac{x}{y}\right)^n = \dfrac{x^n}{y^n}$ for all positive integers n.

27. For $a \neq b$, show that $(a - b)$ is a factor of $a^n - b^n$, where n is a positive integer. *Hint:*

$$a^{k+1} - b^{k+1} = (a \cdot a^k - ab^k) + (ab^k - b \cdot b^k)$$

28. For $a \neq -b$, show that $(a + b)$ is a factor of $a^{2n+1} + b^{2n+1}$, where n is a positive integer. *Hint:*

$$a^{2k+3} + b^{2k+3} = (a^{2k+2} + b^{2k+2})(a + b) - ab(a^{2k+1} + b^{2k+1})$$

29. $\displaystyle\sum_{k=1}^{n} ar^{k-1} = \dfrac{a(1 - r^n)}{1 - r}$ for $r \neq 1$

30. $\displaystyle\sum_{k=1}^{n} (ak + b) = \dfrac{n[(n + 1)a + 2b]}{2}$

SUPPLEMENTAL EXERCISES

In Exercises 31 to 35, use mathematical induction to prove each statement.

31. Using a calculator, find the smallest integer N for which $\log N! > N$. Now prove that $\log n! > n$ for all $n > N$.

32. Let a_n be a sequence for which there is a number r and an integer N for which $a_{n+1}/a_n < r$ for $n \geq N$. Show that $a_{N+k} < a_N r^k$ for each positive integer k.

33. For constant positive integers m and n, show that $(x^m)^n = x^{mn}$.

34. Prove that $\displaystyle\sum_{i=0}^{n} \dfrac{1}{i!} \leq 3 - \dfrac{1}{n}$ for all positive integers n.

35. Prove that $\left(\dfrac{n + 1}{n}\right)^n < n$ for all integers $n \geq 3$.

PROJECTS

1. STEPS IN A MATHEMATICAL INDUCTION PROOF In every proof by mathematical induction, it is important that both parts of the Principle of Mathematical Induction be verified. For instance, consider the formula

$$2 + 4 + 8 + \cdots + 2^n \overset{?}{=} 2^{n+1} + 1$$

a. Show that if we assume the formula is true for some positive integer k, then the formula is true for $k + 1$.

b. Show that the formula is not true for $n = 1$.

c. Show that the formula is not valid for any value of n by showing that the left side is always an even number and the right side is always an odd number.

d. Explain how this shows that both steps of the Principle of Mathematical Induction must be verified.

2. THE TOWER OF HANOI The Tower of Hanoi is a game that consists of three pegs and n disks of distinct diameter arranged on one of the pegs such that the largest disk is on the bottom, then the next largest, and so on. The object of the game is to move all the disks from one peg to a second peg. The rules require that only one disk be moved at a time and that a larger disk may not be placed on a smaller disk. All pegs may be used.

a. Show that it is possible to complete the game in $2^n - 1$ moves.

b. A legend says that in the center of the universe, high priests have the task of moving 64 golden disks from one of three diamond needles by using the rules of the Tower of Hanoi game. When they have completed the transfer, the universe will cease to exist. If one move is made every second, and the priests started 5 billion years ago (the approximate age of Earth), how many more years does the legend predict the universe will continue to exist?

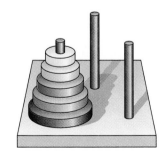

SECTION 9.5 THE BINOMIAL THEOREM

In certain situations in mathematics, it is necessary to write $(a + b)^n$ as the sum of its terms. Because $(a + b)$ is a binomial, this process is called **expanding the binomial.** For small values of n, it is relatively easy to write the expansion by using multiplication.

Earlier in the text, we found

$$(a + b)^1 = a + b$$

$$(a + b)^2 = a^2 + 2ab + b^2$$

$$(a + b)^3 = a^3 + 3a^2b + 3ab^2 + b^3$$

Building on these expansions, we can write a few more.

$$(a + b)^4 = a^4 + 4a^3b + 6a^2b^2 + 4ab^3 + b^4$$

$$(a + b)^5 = a^5 + 5a^4b + 10a^3b^2 + 10a^2b^3 + 5ab^4 + b^5$$

We could continue to build on previous expansions and eventually have quite a comprehensive list of binomial expansions. Instead, however, we will look for a theorem that will enable us to expand $(a + b)^n$ directly without multiplying.

Look at the variable parts of each expansion above. Note that for each $n = 1, 2, 3, 4, 5$

- The first term is a^n. The exponent on a decreases by 1 for each successive term.

- The exponent on b increases by 1 for each successive term. The last term is b^n.

- The degree of each term is n.

To find a pattern for the coefficients in each expansion, first note that there are $n + 1$ terms and that the coefficient of the first and last term is 1. To find the remaining coefficients, consider the expansion of $(a + b)^5$.

$$(a + b)^5 = a^5 + 5a^4b + 10a^3b^2 + 10a^2b^3 + 5ab^4 + b^5$$

$$\frac{5}{1} = 5 \qquad \frac{5 \cdot 4}{2 \cdot 1} = 10 \qquad \frac{5 \cdot 4 \cdot 3}{3 \cdot 2 \cdot 1} = 10 \qquad \frac{5 \cdot 4 \cdot 3 \cdot 2}{4 \cdot 3 \cdot 2 \cdot 1} = 5$$

Observe from these patterns that there is a strong relationship to factorials. In fact, we can express each coefficient by using factorial notation.

$$\frac{5!}{1! \, 4!} = 5 \qquad \frac{5!}{2! \, 3!} = 10 \qquad \frac{5!}{3! \, 2!} = 10 \qquad \frac{5!}{4! \, 1!} = 5$$

In each denominator, the first factorial is the exponent of b and the second factorial is the exponent of a.

In general, we will conjecture that the coefficient of the term $a^{n-k}b^k$ in the expansion of $(a + b)^n$ is $\dfrac{n!}{k!\,(n-k)!}$. Each coefficient of a term of a binomial expansion is called a **binomial coefficient** and is denoted by $\dbinom{n}{k}$.

FORMULA FOR A BINOMIAL COEFFICIENT

The coefficient of the term whose variable part is $a^{n-k}b^k$ in the expansion of $(a + b)^n$ is

$$\binom{n}{k} = \frac{n!}{k!\,(n-k)!}$$

The first term of the expansion of $(a + b)^n$ can be thought of as $a^n b^0$. In that case, we can calculate the coefficient of that term as

$$\binom{n}{0} = \frac{n!}{0!\,(n-0)!} = \frac{n!}{1 \cdot n!} = 1$$

EXAMPLE 1 Evaluate a Binomial Coefficient

Evaluate each binomial coefficient. **a.** $\dbinom{9}{6}$ **b.** $\dbinom{10}{10}$

Solution

a. $\dbinom{9}{6} = \dfrac{9!}{6!\,(9-6)!} = \dfrac{9!}{6!\,3!} = \dfrac{9 \cdot 8 \cdot 7 \cdot 6!}{6! \cdot 3 \cdot 2 \cdot 1} = 84$

b. $\dbinom{10}{10} = \dfrac{10!}{10!\,(10-10)!} = \dfrac{10!}{10!\,0!} = 1.$ • Remember that $0! = 1$.

TRY EXERCISE 4, EXERCISE SET 9.5

We are now ready to state the Binomial Theorem for positive integers.

BINOMIAL THEOREM FOR POSITIVE INTEGERS

If n is a positive integer, then

$$(a + b)^n = \sum_{i=0}^{n} \binom{n}{i} a^{n-i} b^i$$

$$= \binom{n}{0} a^n + \binom{n}{1} a^{n-1}b + \binom{n}{2} a^{n-2}b^2 + \cdots + \binom{n}{n} b^n$$

| EXAMPLE 2 | *Expand the Sum of Two Terms* |

Expand: $(2x^2 + 3)^4$

Solution

$$(2x^2 + 3)^4 = \binom{4}{0}(2x^2)^4 + \binom{4}{1}(2x^2)^3(3) + \binom{4}{2}(2x^2)^2(3)^2$$
$$+ \binom{4}{3}(2x^2)(3)^3 + \binom{4}{4}(3)^4$$
$$= 16x^8 + 96x^6 + 216x^4 + 216x^2 + 81$$

TRY EXERCISE 18, EXERCISE SET 9.5

| EXAMPLE 3 | *Expand a Difference of Two Terms* |

Expand: $(\sqrt{x} - 2y)^5$

Solution

TAKE NOTE

If exactly one of the terms *a* and *b* is negative, the terms of the expansion alternate in sign.

$$(\sqrt{x} - 2y)^5 = \binom{5}{0}(\sqrt{x})^5 + \binom{5}{1}(\sqrt{x})^4(-2y) + \binom{5}{2}(\sqrt{x})^3(-2y)^2$$
$$+ \binom{5}{3}(\sqrt{x})^2(-2y)^3 + \binom{5}{4}(\sqrt{x})(-2y)^4 + \binom{5}{5}(-2y)^5$$
$$= x^{5/2} - 10x^2y + 40x^{3/2}y^2 - 80xy^3 + 80x^{1/2}y^4 - 32y^5$$

TRY EXERCISE 20, EXERCISE SET 9.5

THE *i*TH TERM OF A BINOMIAL EXPANSION

The Binomial Theorem can also be used to find a specific term in the expansion of $(a + b)^n$.

FORMULA FOR THE *i*TH TERM OF A BINOMIAL EXPANSION

TAKE NOTE

The exponent on *b* is 1 *less* than the term number.

The *i*th term of the expansion of $(a + b)^n$ is given by

$$\binom{n}{i-1}a^{n-i+1}b^{i-1}$$

| EXAMPLE 4 | *Find the ith Term of a Binomial Expansion* |

Find the fourth term in the expansion of $(2x^3 - 3y^2)^5$.

Continued • ➤

Solution

With $a = 2x^3$ and $b = -3y^2$, and using the last theorem with $i = 4$ and $n = 5$, we have

$$\binom{5}{3}(2x^3)^2(-3y^2)^3 = -1080x^6y^6$$

The fourth term is $-1080x^6y^6$.

TRY EXERCISE 34, EXERCISE SET 9.5

A pattern for the coefficients of the terms of an expanded binomial can be found by writing the coefficients in a triangular array known as **Pascal's Triangle.** See Figure 9.1.

Each row begins and ends with the number 1. Any other number in a row is the sum of the two closest numbers above it. For example, $4 + 6 = 10$. Thus each succeeding row can be found from the preceding row.

$(a + b)^1$:					1		1				
$(a + b)^2$:				1		2		1			
$(a + b)^3$:			1		3		3		1		
$(a + b)^4$:		1		4		6		4		1	
$(a + b)^5$:	1		5		10		10		5		1
$(a + b)^6$:	1	6		15		20		15		6	1

Figure 9.1

Pascal's triangle can be used to expand a binomial for small values of n. For instance, the seventh row of Pascal's Triangle is

$$1 \quad 7 \quad 21 \quad 35 \quad 35 \quad 21 \quad 7 \quad 1$$

Therefore,

$$(a + b)^7 = a^7 + 7a^6b + 21a^5b^2 + 35a^4b^3 + 35a^3b^4 + 21a^2b^5 + 7ab^6 + b^7$$

TOPICS FOR DISCUSSION

1. Discuss the use of the Binomial Theorem.

2. Can the Binomial Theorem be used to expand $(a + b)^n$, n a natural number, for any expressions for a and b? Why or why not?

3. What is Pascal's Triangle and how is it related to expanding a binomial?

4. Explain how Pascal's Triangle suggests that $\binom{n - 1}{k - 1} + \binom{n - 1}{k} = \binom{n}{k}$.

EXERCISE SET 9.5

In Exercises 1 to 8, evaluate the binomial coefficients.

1. $\dbinom{7}{4}$ **2.** $\dbinom{8}{6}$ **3.** $\dbinom{9}{2}$ **4.** $\dbinom{10}{5}$

5. $\dbinom{12}{9}$ **6.** $\dbinom{6}{5}$ **7.** $\dbinom{11}{0}$ **8.** $\dbinom{14}{14}$

In Exercises 9 to 28, expand the binomial.

9. $(x - y)^6$ **10.** $(a - b)^5$ **11.** $(x + 3)^5$

12. $(x - 5)^4$ **13.** $(2x - 1)^7$ **14.** $(2x + y)^6$

15. $(x + 3y)^6$ **16.** $(x - 4y)^5$ **17.** $(2x - 5y)^4$

18. $(3x + 2y)^4$ **19.** $\left(x + \dfrac{1}{x}\right)^6$ **20.** $(2x - \sqrt{y})^7$

21. $(x^2 - 4)^7$ **22.** $(x - y^3)^6$ **23.** $(2x^2 + y^3)^5$

24. $(2x - y^3)^6$ **25.** $\left(\dfrac{2}{x} - \dfrac{x}{2}\right)^4$ **26.** $\left(\dfrac{a}{b} + \dfrac{b}{a}\right)^3$

27. $(s^{-2} + s^2)^6$ **28.** $(2r^{-1} + s^{-1})^5$

In Exercises 29 to 36, find the indicated term without expanding.

29. $(3x - y)^{10}$; eighth term

30. $(x + 2y)^{12}$; fourth term

31. $(x + 4y)^{12}$; third term

32. $(2x - 1)^{14}$; thirteenth term

33. $(\sqrt{x} - \sqrt{y})^9$; fifth term

34. $(x^{-1/2} + x^{1/2})^{10}$; sixth term

35. $\left(\dfrac{a}{b} + \dfrac{b}{a}\right)^{11}$; ninth term

36. $\left(\dfrac{3}{x} - \dfrac{x}{3}\right)^{13}$; seventh term

37. Find the term that contains b^8 in the expansion of $(2a - b)^{10}$.

38. Find the term that contains s^7 in the expansion of $(3r + 2s)^9$.

39. Find the term that contains y^8 in the expansion of $(2x + y^2)^6$.

40. Find the term that contains b^9 in the expansion of $(a - b^3)^8$.

41. Find the middle term of $(3a - b)^{10}$.

42. Find the middle term of $(a + b^2)^8$.

43. Find the two middle terms of $(s^{-1} + s)^9$.

44. Find the two middle terms of $(x^{1/2} - y^{1/2})^7$.

In Exercises 45 to 50, use the Binomial Theorem to simplify the powers of the complex numbers.

45. $(2 - i)^4$ **46.** $(3 + 2i)^3$

47. $(1 + 2i)^5$ **48.** $(1 - 3i)^5$

49. $\left(\dfrac{\sqrt{2}}{2} + i\dfrac{\sqrt{2}}{2}\right)^8$ **50.** $\left(\dfrac{1}{2} + i\dfrac{\sqrt{3}}{2}\right)^6$

SUPPLEMENTAL EXERCISES

51. Let n be a positive integer. Expand and simplify $\dfrac{(x + h)^n - x^n}{h}$, where x is any real number and $h \neq 0$.

52. Show that $\dbinom{n}{k} = \dbinom{n}{n - k}$ for all positive integers n and k with $0 \leq k \leq n$.

53. Show that $\displaystyle\sum_{k=0}^{n} \dbinom{n}{k} = 2^n$. (*Hint:* Use the Binomial Theorem with $x = 1$, $y = 1$.)

54. Prove that $\dbinom{n}{k} + \dbinom{n}{k + 1} = \dbinom{n + 1}{k + 1}$, n and k integers, $0 \leq k \leq n$.

55. Prove that $\displaystyle\sum_{i=0}^{n} (-1)^i \dbinom{n}{i} = 0$.

56. Approximate $(0.98)^8$ by evaluating the first three terms of $(1 - 0.02)^8$.

57. Approximate $(1.02)^8$ by evaluating the first three terms of $(1 + 0.02)^8$.

There is an extension of the Binomial Theorem called the *Multinomial Theorem*. This theorem is used in determining probabilities. *Multinomial Theorem:*   If n, r, and k are positive integers, then the coefficient of $a^r b^k c^{n-r-k}$ in the expansion of $(a + b + c)^n$ is

$$\frac{n!}{r!\, k!\,(n - r - k)!}$$

In Exercises 58 to 61, use the Multinomial Theorem to find the indicated coefficient.

58. Find the coefficient of $a^2 b^3 c^5$ in the expansion of $(a + b + c)^{10}$.

59. Find the coefficient of $a^5 b^2 c^2$ in the expansion of $(a + b + c)^9$.

60. Find the coefficient of $a^4 b^5$ in the expansion of $(a + b + c)^9$.

61. Find the coefficient of $a^3 c^5$ in the expansion of $(a + b + c)^8$.

PROJECTS

1. PASCAL'S TRIANGLE Write an essay on Pascal's Triangle. Include some of the earliest known examples of the triangle and some of its applications.

2. SOME OTHER FUNCTIONS Do some research and determine a definition for positive integers for each of the following types of numbers. Give examples of calculations using each type of number.

   a. Pochammer (m, n) **b.** double factorial $(n!!)$

SECTION

9.6 PERMUTATIONS AND COMBINATIONS

FUNDAMENTAL COUNTING PRINCIPLE

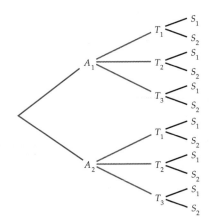

Figure 9.2

Suppose an electronics store offers a three-component stereo system for $250. A buyer must choose one amplifier, one tuner, and one pair of speakers. If the store has two models of amplifiers, three models of tuners, and two speaker models, how many different stereo systems could a consumer purchase?

This problem belongs to a class of problems called *counting problems*. The problem is to determine the number of ways in which the conditions of the problem can be satisfied. One way to do this is to make a tree diagram and then count the items on the list. We will organize the list in a table using A_1 and A_2 for the amplifiers; T_1, T_2, and T_3 for the tuners; and S_1 and S_2 for the speakers. See **Figure 9.2.**

By counting the possible systems that can be purchased, we find there are 12 different systems. Another way to arrive at this result is to find the product of the number of options available.

$$\begin{array}{ccccccc} \text{Number of} & \times & \text{number of} & \times & \text{number of} & = & \text{number of} \\ \text{amplifiers} & & \text{tuners} & & \text{speakers} & & \text{systems} \\ 2 & \times & 3 & \times & 2 & = & 12 \end{array}$$

In some states, a standard car license plate consists of a nonzero digit, followed by three letters, followed by three more digits. What is the maximum number of car license plates of this type that could be issued? If we begin a list of the possible license plates, it soon becomes apparent that listing them all would be very time-consuming and impractical.

1AAA000,   1AAA001,   1AAA002,   1AAA003, . . .

Instead, the following counting principle is used. This principle forms the basis for all counting problems.

FUNDAMENTAL COUNTING PRINCIPLE

Let T_1, T_2, T_3, ..., T_n be a sequence of n conditions. Suppose that T_1 can occur in m_1 ways, T_2 can occur in m_2 ways, T_3 can occur in m_3 ways, and so on until finally T_n can occur in m_n ways. Then the number of ways of satisfying the conditions T_1, T_2, T_3, ..., T_n in succession is given by the product

$$m_1 m_2 m_3 \cdots m_n$$

Table 9.4

Condition	Number of ways
T_1: a nonzero digit	$m_1 = 9$
T_2: a letter	$m_2 = 26$
T_3: a letter	$m_3 = 26$
T_4: a letter	$m_4 = 26$
T_5: a digit	$m_5 = 10$
T_6: a digit	$m_6 = 10$
T_7: a digit	$m_7 = 10$

To apply the counting principle to the license plate problem first find the number of ways each condition can be satisfied, as shown in Table 9.4. Thus, we have

$$\begin{array}{l}\text{Number of car} \\ \text{license plates}\end{array} = 9 \cdot 26 \cdot 26 \cdot 26 \cdot 10 \cdot 10 \cdot 10 = 158{,}184{,}000$$

EXAMPLE 1   *Apply the Fundamental Counting Principle*

An automobile dealer offers three mid-size cars. A customer selecting one of these cars must choose one of three different engines, one of five different colors, and one of four different interior packages. How many different selections can the customer make?

Solution

T_1: mid-size car $m_1 = 3$

T_2: engine $m_2 = 3$

T_3: color $m_3 = 5$

T_4: interior $m_4 = 4$

Number of different selections $= 3 \cdot 3 \cdot 5 \cdot 4 = 180$.

TRY EXERCISE 12, EXERCISE SET 9.6

PERMUTATION

An application of the Fundamental Counting Principle is to determine the number of arrangements of distinct elements in a definite order.

PERMUTATION

A **permutation** is an arrangement of distinct objects in a definite order.

For example, *abc* and *bca* are two of the possible permutations of the three elements *a*, *b*, *c*.

Consider a race with 10 runners. In how many different orders can the runners finish first, second, and third (assuming no ties)?

Any one of the 10 runners could finish first: $m_1 = 10$

Any one of the remaining 9 runners could be second: $m_2 = 9$

Any one of the remaining 8 runners could be third: $m_3 = 8$

By the Fundamental Counting Principle, there are $10 \cdot 9 \cdot 8 = 720$ possible first-, second-, and third-place finishes for the 10 runners. Using the language of permutations, we would say, "There are 720 permutations of 10 objects (the runners) taken 3 (the possible finishes) at a time."

Permutations occur so frequently in counting problems that a formula, rather than the counting principle, is often used.

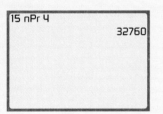

```
15 nPr 4
              32760
```

FORMULA FOR A PERMUTATION OF *n* DISTINCT OBJECTS TAKEN *r* AT A TIME

The number of permutations of n distinct objects taken r at a time is

$$P(n,r) = \frac{n!}{(n - r)!}$$

EXAMPLE 2   Find the Number of Permutations

In how many ways can a president, vice president, secretary, and treasurer be selected from a committee of fifteen people?

Solution

There are fifteen distinct people to place in four positions. Thus $n = 15$ and $r = 4$.

$$P(15, 4) = \frac{15!}{(15 - 4)!} = \frac{15!}{11!} = \frac{15 \cdot 14 \cdot 13 \cdot 12 \cdot 11!}{11!} = 32{,}760$$

TRY EXERCISE 16, EXERCISE SET 9.6

EXAMPLE 3   Find the Number of Seating Permutations

Six people attend a movie and all sit in the same row with six seats.

a. Find the number of ways the group can sit together.

b. Find the number of ways the group can sit together if two people in the group must sit side-by-side.

c. Find the number of ways the group can sit together if two people in the group refuse to sit side-by-side.

Solution

a. There are six distinct people to place in six distinct positions. Thus $n = 6$ and $r = 6$.

$$P(6,6) = \frac{6!}{(6-6)!} = \frac{6!}{0!} = \frac{6!}{1} = 720$$

b. Think of the two people who must sit together as a single object and count the number of arrangements of the *five* objects (AB), C, D, E, F. Thus $n = 5$ and $r = 5$.

$$P(5,5) = \frac{5!}{(5-5)!} = \frac{5!}{0!} = \frac{5!}{1} = 120$$

There are also 120 arrangements with A and B reversed (BA), C, D, E, F. Thus the total number of arrangements is $120 + 120 = 240$.

c. From **a.**, there are 720 possible seating arrangements. From **b.**, there are 240 arrangements with two specific people next to each other. Thus there are $720 - 240 = 480$ arrangements where two specific people are not seated together.

TRY EXERCISE 22, EXERCISE SET 9.6

COMBINATIONS

Up to this point, we have been counting the number of distinct arrangements of objects. In some cases we may be interested in determining the number of ways of selecting objects without regard to the order of the selection. For example, suppose we want to select a committee of three people from five candidates denoted by A, B, C, D, and E. One possible committee is A, C, D. If we select D, C, A, we still have the same committee because the order of the selection is not important. An arrangement of objects for which the order of the selection is not important is a **combination**.

TAKE NOTE

Recall that a binomial coefficient is given by $\binom{n}{r} = \frac{n!}{r!(n-r)!}$ which is the same as $C(n, r)$.

FORMULA FOR THE COMBINATION OF n OBJECTS TAKEN r AT A TIME

The number of combinations of n objects taken r at a time is

$$C(n,r) = \frac{n!}{r!\,(n-r)!}$$

TAKE NOTE

Some calculators use the notation nCr to represent a combination of n objects taken r at a time.

EXAMPLE 4   *Find the Number of Combinations*

A standard deck of playing cards consists of fifty-two cards. How many five-card hands can be chosen from this deck?

Continued •➤

Solution

We have $n = 52$ and $r = 5$. Thus

$$C(52, 5) = \frac{52!}{5!(52 - 5)!} = \frac{52!}{5! \, 47!} = \frac{52 \cdot 51 \cdot 50 \cdot 49 \cdot 48 \cdot 47!}{5 \cdot 4 \cdot 3 \cdot 2 \cdot 1 \cdot 47!} = 2{,}598{,}960$$

TRY EXERCISE 20, EXERCISE SET 9.6

EXAMPLE 5 *Find the Number of Combinations*

A chemist has nine samples of a solution, of which four are type A and five are type B. If the chemist chooses three of the solutions at random, determine in how many ways the chemist can have exactly one type A solution.

Solution

The chemist has chosen three solutions, one of which is type A. If one is type A, then two are type B. The number of ways of choosing one type A solution from four type A solutions is $C(4, 1)$.

$$C(4, 1) = \frac{4!}{1! \, (4 - 1)!} = \frac{4!}{1! \, 3!} = 4$$

The number of ways of choosing two type B solutions from five type B solutions is $C(5, 2)$.

$$C(5, 2) = \frac{5!}{2! \, (5 - 2)!} = \frac{5!}{2! \, 3!} = 10$$

By the counting principle, there are

$$C(4, 1) \cdot C(5, 2) = 4 \cdot 10 = 40$$

ways to have one type A and two type B solutions.

TRY EXERCISE 30, EXERCISE SET 9.6

The difficult part of counting is determining whether to use the counting principle, the permutation formula, or the combination formula. Following is a summary of guidelines.

GUIDELINES FOR SOLVING COUNTING PROBLEMS

1. The counting principle will always work but is not always the easiest method to apply.

2. When reading a problem, ask yourself, "Is the order of the selection process important?" If the answer is yes, the arrangements are permutations. If the answer is no, the arrangements are combinations.

TOPICS FOR DISCUSSION

1. Discuss the Fundamental Counting Principle and how it is used.

2. Discuss the difference between a permutation and a combination.

3. Explain why $\binom{n}{k}$ occurs as a coefficient in the binomial formula.

4. Discuss the difference between counting with replacement and counting without replacement.

EXERCISE SET 9.6

In Exercises 1 to 10, evaluate each quantity.

1. $P(6,2)$ 2. $P(8,7)$ 3. $C(8,4)$ 4. $C(9,2)$

5. $P(8,0)$ 6. $P(9,9)$ 7. $C(7,7)$ 8. $C(6,0)$

9. $C(10,4)$ 10. $P(10,4)$

11. **COMPUTER SYSTEMS** A computer manufacturer offers a computer system with three different disk drives, two different monitors, and two different keyboards. How many different computer systems could a consumer purchase from this manufacturer?

12. **COLOR MONITORS** A computer monitor produces color by blending colors on *palettes*. If a computer monitor has four palettes and each palette has four colors, how many blended colors can be formed? Assume each palette must be used each time.

13. **LIGHT SWITCHES** A large conference room has four doors. At the entrance to each door there is a single light switch. How many different configurations of "on" and "off" are possible for the light switches?

14. **COMPUTER MEMORY** An integer is stored in a computer's memory as a series of zeros and ones. Each memory unit contains 8 spaces for a zero or a one. The first space is used for the sign of the number, and the remaining 7 spaces are used for the integer. How many positive integers can be stored in one memory unit of this computer?

15. **SCHEDULING** In how many different ways can six employees be assigned to six different jobs?

16. **CONTEST WINNERS** First-, second-, and third-place prizes are to be awarded in a dance contest in which twelve contestants are entered. In how many ways can the prizes be awarded?

17. **MAIL BOXES** There are five mailboxes outside a post office. In how many ways can three letters be deposited into the five boxes?

18. **COMMITTEE MEMBERSHIP** How many different committees of three people can be selected from nine people?

19. **TEST QUESTIONS** A professor provides to a class 25 possible essay questions for an upcoming test. Of the 25 questions, the professor will ask 5 of the questions on the exam. How many different tests can the professor prepare?

20. **TENNIS MATCHES** Twenty-six people enter a tennis tournament. How many different first-round matches are possible if each player can be matched with any other player?

21. **EMPLOYEE INITIALS** A company has more than 676 employees. Explain why there must be at least 2 employees who have the same first and last initials.

22. **SEATING ARRANGEMENTS** A car holds six passengers, three in the front seat and three in the back seat. How many different seating arrangements of six people are possible if one person refuses to sit in front and one person refuses to sit in back?

23. **COMMITTEE MEMBERSHIP** A committee of six people is chosen from six senators and eight representatives. How many committees are possible if there are to be three senators and three representatives on the committee?

24. **ARRANGING NUMBERS** The numbers 1, 2, 3, 4, 5, 6 are to be arranged. How many different arrangements are possible under each of the following conditions?

    a. All the even numbers come first.

    b. The arrangements are such that the numbers alternate between even and odd.

25. **TEST QUESTIONS** A true-false examination contains ten questions. In how many ways can a person answer the questions on this test by just guessing? Assume that all questions are answered.

26. **TEST QUESTIONS** A twenty-question, four-option multiple-choice examination is given as a pre-employment test. In how many ways could a prospective employee answer the questions on this test by just guessing? Assume that all questions are answered.

27. **STATE LOTTERY** A state lottery game requires a person to select six different numbers from forty numbers. The order of the selection is not important. In how many ways can this be done?

28. **TEST QUESTIONS** A student must answer eight of ten questions on an exam. How many different choices can the student make?

29. **ACCEPTANCE SAMPLING** A warehouse receives a shipment of ten computers, of which three are defective. Five computers are then randomly selected from the ten and delivered to a store.

   a. In how many ways can the store receive no defective computers?

   b. In how many ways can the store receive one defective computer?

   c. In how many ways can the store receive all three defective computers?

30. **CONTEST** Fifteen students, of whom seven are seniors, are selected as semifinalists for a literary award. Of the fifteen students, ten finalists will be selected.

   a. In how many ways can ten finalists be selected from the fifteen students?

   b. In how many ways can the ten finalists contain three seniors?

   c. In how many ways can the ten finalists contain at least five seniors?

31. **SERIAL NUMBERS** A television manufacturer uses a code for the serial number of a television. The first symbol is the letter A, B, or C and represents the location of the manufacturing plant. The next two symbols are 01, 02, ..., 12 and represent the month in which the set was manufactured. The next symbol is a 5, 6, 7, 8, or 9 and represents the year the set was manufactured. The last seven symbols are digits. How many serial numbers are possible?

32. **CARD GAMES** Five cards are chosen at random from an ordinary deck of playing cards. In how many ways can the cards be chosen under each of the following conditions?

   a. All are hearts.   **b.** All are the same suit.

   c. Exactly three are kings.

   d. Two or more are aces.

33. **ACCEPTANCE SAMPLING** A quality control inspector receives a shipment of ten computer disk drives and randomly selects three of the drives for testing. If two of the disk drives in the shipment are defective, find the number of ways in which the inspector could select at most one defective drive.

34. **BASKETBALL TEAMS** A basketball team has twelve members. In how many ways can five players be chosen under each of the following conditions?

   a. The selection is random.

   b. The two tallest players are always among the five selected.

35. **ARRANGING NUMBERS** The numbers 1, 2, 3, 4, 5, 6 are arranged in random order. In how many ways can the numbers 1 and 2 appear next to one another and in the order 1, 2?

36. **OCCUPANCY PROBLEM** Seven identical balls are randomly placed in seven available containers in such a way that two balls are in one container. Of the remaining six containers, each receives at most one ball. Find the number of ways in which this can be accomplished.

37. **LINES IN A PLANE** Seven points lie in a plane in such a way that no three points lie on the same line. How many lines are determined by seven points?

38. **CHESS MATCHES** A chess tournament has twelve participants. How many games must be scheduled if every player must play every other player exactly once?

39. **CONTEST WINNERS** Eight couples attend a benefit at which two prizes are given. In how many ways can two names be randomly drawn so that the prizes are not awarded to the same couple?

40. **GEOMETRY** Suppose there are twelve distinct points on a circle. How many different triangles can be formed with vertices at the given points?

41. **TEST QUESTIONS** In how many ways can a student answer a twenty-question true-false test if the student marks ten of the questions true and ten of the questions false?

42. **COMMITTEE MEMBERSHIP** From a group of fifteen people a committee of eight is formed. From the committee a president, secretary, and treasurer are selected. Find the number of ways in which the two consecutive operations can be carried out.

43. **COMMITTEE MEMBERSHIP** From a group of twenty people a committee of twelve is formed. From the committee of twelve, a subcommittee of four people is chosen. Find the number of ways in which the two consecutive operations can be carried out.

SUPPLEMENTAL EXERCISES

44. **LINES IN A PLANE** Generalize Exercise 37. That is, given n points in a plane, no three of which lie on the same line, how many lines are determined by n points?

45. **BIRTHDAYS** Seven people are asked the month of their birth. In how many ways can each of the following conditions exist?

   a. No two people have a birthday in the same month.

   b. At least two people have a birthday in the same month.

46. **SUMS OF COINS** From a penny, nickel, dime, and quarter, how many different sums of money can be formed using one or more of the coins?

47. **BIOLOGY** Five sticks of equal length are broken into a short piece and a long piece. The ten pieces are randomly arranged in five pairs. In how many ways will each pair consist of a long stick and a short stick? (This exercise actually has a practical side. When cells are exposed to harmful radiation, some chromosomes break. If two long sides unite or two short sides unite, the cell dies.)

48. **ARRANGING NUMBERS** Four random digits are drawn (repetitions are allowed). Among the four digits, in how many ways can two or more repetitions occur?

49. **RANDOM WALK** An aimless tourist, standing on a street corner, tosses a coin. If the result is heads, the tourist walks one block north. If the result is tails, the tourist walks one block south. At the new corner, the coin is tossed again and the same rule applied. If the coin is tossed ten times, in how many ways will the tourist be back at the original corner? This problem is an elementary example of what is called a *random walk*. Random walk problems have many applications in physics, chemistry, and economics.

PROJECTS

1. **EXPLAIN PERMUTATIONS AND COMBINATIONS** Write an outline of a lesson that you could use to teach permutations and combinations. Include at least five examples of permutations and five examples of combinations.

2. **APPLICATION OF COUNTING** Calculating the number of ways in which balls can be distributed in boxes has a variety of applications. For instance, a traffic engineer may want to know how traffic accidents (the balls) are distributed throughout the days of the week (the boxes). Or a physicist may want to know how electrons (the balls) can be distributed in the energy orbits (the boxes) of an atom. The formula for counting the number of ways n in which distinguishable balls can be placed in k distinguishable boxes, where each box must have at least 1 ball, is given by

$$\binom{k}{0}k^n - \binom{k}{1}(k-1)^n + \binom{k}{2}(k-2)^n + \cdots + (-1)^{k-1}\binom{k}{k-1}$$

Box 1   Box 2   Box 3

Box 1   Box 2   Box 3

The word *distinguishable* is important. This formula refers to counting under circumstances similar to the first figure at the right, where the boxes are numbered and the balls are numbered. The second figure shows a situation that is not covered by this formula. Although the boxes are numbered, the balls are not and are therefore *indistinguishable*.

   a. A computer network consists of five computers and three printers. How many possible connections can be made if each computer must be hooked to a printer and all printers are used?

   b. A supermarket has four checkout lanes. Assuming shoppers are efficient and will not leave a checkout lane empty, in how many ways can ten shoppers line up for the checkout lanes?

9.7 INTRODUCTION TO PROBABILITY

Many events in the world around us have random character, such as the chances of an accident occurring on a certain freeway, the chances of winning a state lottery, and the chances that the nucleus of an atom will undergo fission. By repeatedly observing such events, it is often possible to recognize certain patterns. **Probability** is the mathematical study of random patterns.

When a weather reporter predicts a 30% chance of rain, the forecaster is saying that similar weather conditions have led to rain 30 times out of 100. When a fair coin is tossed, we expect heads to occur 1/2, or 50%, of the time. The numbers 30% (or 0.3) and 1/2 are the probabilities of the events.

SAMPLE SPACES

An activity with an observable outcome is called an **experiment**. Examples of experiments include

1. Flipping a coin and observing the side facing upward

2. Observing the incidence of a disease in a certain population

3. Observing the length of time a person waits in a checkout line in a grocery store

The **sample space** of an experiment is the set of *all possible* outcomes of that experiment.

Consider the experiment of tossing one coin three times and recording the number of occurences of the upward side of the coin. The sample space is

$$S = \{HHH, HHT, HTH, THH, HTT, THT, TTH, TTT\}$$

EXAMPLE 1 *List the Elements of a Sample Space*

Suppose that among five batteries, two are defective. Two batteries are randomly drawn from the five and tested for defects. List the elements in the sample space.

Solution

Label the nondefective batteries N_1, N_2, N_3 and the defective batteries D_1, D_2. The sample space is

$$S = \{N_1 D_1, N_2 D_1, N_3 D_1, N_1 D_2, N_2 D_2, N_3 D_2, N_1 N_2, N_1 N_3, N_2 N_3, D_1 D_2\}$$

TRY EXERCISE 6, EXERCISE SET 9.7

EVENTS IN A SAMPLE SPACE

An **event** E is any subset of a sample space. For the sample space defined in Example 1, several of the events we could define are

E_1: There are no defective batteries.

E_2: At least one battery is defective.

E_3: Both batteries are defective.

Because an event is a subset of the sample space, each of these events can be expressed as a set.

$$E_1 = \{N_1 N_2, N_1 N_3, N_2 N_3\}$$

$$E_2 = \{N_1 D_1, N_2 D_1, N_3 D_1, N_1 D_2, N_2 D_2, N_3 D_2, D_1 D_2\}$$

$$E_3 = \{D_1 D_2\}$$

There are two methods by which elements are drawn from a sample space: with replacement and without replacement. *With replacement* means that after the element is drawn, it is returned to the sample space. The same element could be selected on the next drawing. When elements are drawn *without replacement,* an element drawn is not returned to the sample space and therefore is not available for any subsequent drawing.

EXAMPLE 2   *List the Elements of an Event*

A two-digit number is formed by choosing from the digits 1, 2, 3, 4 both with replacement and without replacement. Express each event as a set.

a. E_1: The second digit is greater than or equal to the first digit.

b. E_2: Both digits are less than zero.

Solution

a.   With replacement: $E_1 = \{11, 12, 13, 14, 22, 23, 24, 33, 34, 44\}$

Without replacement: $E_1 = \{12, 13, 14, 23, 24, 34\}$

b. $E_3 = \varnothing$
Choosing from the digits 1, 2, 3, 4, this event is impossible. The impossible event is denoted by the empty set or null set.

TRY EXERCISE 14, EXERCISE SET 9.7

The probability of an event is defined in terms of the concepts of sample space and event.

PROBABILITY OF AN EVENT

Let $n(S)$ and $n(E)$ represent the number of elements in the sample space S and the event E, respectively. The probability of event E, $P(E)$, is

$$P(E) = \frac{n(E)}{n(S)}$$

Because E is a subset of S, $n(E) \leq n(S)$. Thus $P(E) \leq 1$. If E is an impossible event, then $E = \varnothing$ and $n(E) = 0$. Thus $P(E) = 0$. If E is the event that *always* occurs, then $E = S$ and $n(E) = n(S)$. Thus $P(E) = 1$. Combining these elements, we have, for any event E,

$$0 \leq P(E) \leq 1$$

EXAMPLE 3   *Calculate the Probability of an Event*

A coin is tossed three times. What is the probability of each outcome?

a. E_1: Two or more heads will appear.

b. E_2: At least one tail will appear.

Solution

First determine the number of elements in the sample space. The sample space for this experiment is

$$S = \{HHH, HHT, HTH, THH, HTT, THT, TTH, TTT\}$$

Therefore $n(S) = 8$. Now determine the number of elements in each event. Then calculate the probability of the event by using $P(E) = n(E)/n(S)$.

a. $E_1 = \{HHH, HHT, HTH, THH\}$

$$P(E_1) = \frac{n(E_1)}{n(S)} = \frac{4}{8} = \frac{1}{2}$$

b. $E_2 = \{HHT, HTH, THH, HTT, THT, TTH, TTT\}$

$$P(E_2) = \frac{n(E_2)}{n(S)} = \frac{7}{8}$$

TRY EXERCISE 22, EXERCISE SET 9.7

Calculating probabilities by listing and then counting the elements of a sample space is not always practical. Instead, we will use the counting principles developed in the last section to determine the number of elements in the sample space and in an event.

EXAMPLE 4   *Use the Counting Principles to Calculate a Probability*

A state lottery game allows a person to choose five numbers from the integers 1 to 40. Repetitions of numbers are not allowed. If three or more numbers match the numbers chosen by the lottery, the player wins a prize. Find the probability that a player will match

a. exactly three numbers

b. exactly four numbers

Solution

The sample space S is the number of ways in which five numbers can be chosen from forty numbers. This is a combination because the order of the drawing is not important.

$$n(S) = C(40, 5) = \frac{40!}{5! \, 35!} = 658{,}008$$

We will call the five numbers chosen by the state lottery "lucky" and the remaining thirty-five numbers "unlucky."

a. Let E_1 be the event a player has three lucky and therefore two unlucky numbers. The three lucky numbers are chosen from the five lucky numbers. There are $C(5, 3)$ ways to do this. The two unlucky numbers are chosen from the thirty-five unlucky numbers. There are $C(35, 2)$ ways to do this. By the counting principle, the number of ways the event E_1 can occur is

$$n(E_1) = C(5, 3) \cdot C(35, 2) = 10 \cdot 595 = 5950$$

$$P(E_1) = \frac{n(E_1)}{n(S)} = \frac{C(5, 3) \cdot C(35, 2)}{C(40, 5)} = \frac{5950}{658{,}008} \approx 0.009042$$

b. Let E_2 be the event a player has four lucky numbers and one unlucky number. The number of ways a person can select four lucky numbers and one unlucky number is $C(5, 4) \cdot C(35, 1)$.

$$P(E_2) = \frac{n(E_2)}{n(S)} = \frac{C(5, 4) \cdot C(35, 1)}{C(40, 5)} = \frac{175}{658{,}008} \approx 0.000266$$

TRY EXERCISE 32, EXERCISE SET 9.7

The expression "one or the other of two events occurs" is written as the union of the two sets. For example, suppose an experiment leads to the sample space $S = \{1, 2, 3, 4, 5, 6\}$ and the events are

Draw a number less than four, $E_1 = \{1, 2, 3\}$

Draw an even number, $E_2 = \{2, 4, 6\}$

Then the event $E_1 \cup E_2$ is described by drawing a number less than four *or* an even number. Thus

$$E_1 \cup E_2 = \{1, 2, 3\} \cup \{2, 4, 6\} = \{1, 2, 3, 4, 6\}$$

Two events E_1 and E_2 that cannot occur at the same time are **mutually exclusive** events. Using set notation, if $E_1 \cap E_2 = \varnothing$, then E_1 and E_2 are mutually exclusive.

For example, using the same sample space $\{1, 2, 3, 4, 5, 6\}$, a third event is

$$\text{Draw an odd number, } E_3 = \{1, 3, 5\}$$

Then $E_2 \cap E_3 = \varnothing$ and the events E_2 and E_3 are mutually exclusive. On the other hand,

$$E_1 \cap E_2 = \{2\}$$

so the events E_1 and E_2 are not mutually exclusive.

One of the axioms of probability involves the union of mutually exclusive events.

A PROBABILITY AXIOM

If E_1 and E_2 are mutually exclusive events, then

$$P(E_1 \cup E_2) = P(E_1) + P(E_2)$$

If the events are not mutually exclusive, the addition rule for probabilities can be used.

ADDITION RULE FOR PROBABILITIES

If E_1 and E_2 are two events, then

$$P(E_1 \cup E_2) = P(E_1) + P(E_2) - P(E_1 \cap E_2)$$

The probability axiom and addition rule are useful when calculating probabilities of events connected by the word *or*.

Using the calculations of Example 4, we can find the probability that a player will have three or four lucky numbers in the lottery. Because the events E_1 and E_2 as defined in Example 4 are mutually exclusive,

$$P(E_1 \cup E_2) = P(E_1) + P(E_2) = 0.009042 + 0.000266 = 0.009308$$

As an example of nonmutually exclusive events, draw a card at random from a deck of ordinary playing cards. Find the probability of drawing an ace or a heart.

$$S = \{52 \text{ ordinary playing cards}\}$$

Let $E_1 = \{$an ace$\}$ and $E_2 = \{$a heart$\}$. Then

$$P(E_1) = \frac{n(E_1)}{n(S)} = \frac{4}{52} = \frac{1}{13} \qquad P(E_2) = \frac{n(E_2)}{n(S)} = \frac{13}{52} = \frac{1}{4}$$

We have $E_1 \cup E_2 = \{$an ace *or* a heart$\}$ and $E_1 \cap E_2 = \{$ace of hearts$\}$. First, we find $P(E_1 \cap E_2)$.

$$P(E_1 \cap E_2) = \frac{n(E_1 \cap E_2)}{n(S)} = \frac{1}{52}$$

Now we can find $P(E_1 \cup E_2)$.

$$P(E_1 \cup E_2) = P(E_1) + P(E_2) - P(E_1 \cap E_2) = \frac{1}{13} + \frac{1}{4} - \frac{1}{52} = \frac{16}{52} = \frac{4}{13}$$

Two events are **independent** if the outcome of the first event does not influence the outcome of the second event. As an example, consider tossing a fair coin twice. The outcome of the first toss has no bearing on the outcome of the second toss. The two events are independent.

Now consider drawing two cards in succession, without replacement, from a regular deck of playing cards. The probability that the second card drawn will be an ace depends on the card drawn first.

PROBABILITY RULE FOR INDEPENDENT EVENTS

If E_1 and E_2 are two independent events, then the probability that both E_1 *and* E_2 will occur is

$$P(E_1) \cdot P(E_2)$$

EXAMPLE 5   *Calculate a Probability for Independent Events*

A coin is tossed and then a die is rolled. What is the probability that the coin will show a head and that the die will show a six?

Solution

The events are independent because the result of one does not influence the probability of the other. $P(\text{head}) = 1/6$ and $P(\text{six}) = 1/12$. Thus the probability of tossing a head and rolling a six is

$$P(\text{head}) \cdot P(\text{six}) = \frac{1}{2} \cdot \frac{1}{6} = \frac{1}{12}$$

TRY EXERCISE 34, EXERCISE SET 9.7

Some probabilities can be calculated from formulas. One of the most important of those formulas is the *Binomial Probability Formula*. This formula is used to calculate probabilities for *independent* events.

Airlines "overbook" flights. That is, they sell more tickets than there are seats on the plane. An airline company can determine the probability that some number of passengers will be "bumped" on a certain flight by using the Binomial Probability Formula. For instance, suppose a plane has 200 seats and the airline sells 240 tickets. If the probability that a person will show up for this flight is 0.8, then the probability that one or more people will have to be bumped is given by

$$\sum_{k=1}^{40} \binom{240}{200+k} 0.8^{200+k} 0.2^{40-k}$$

Using a computer reveals that this sum is approximately 0.08. That is, there is only an 8% chance that someone will have to be bumped, even though the number of tickets sold exceeded the plane's capacity by 40.

BINOMIAL PROBABILITY FORMULA

Let an experiment consist of n trials for which the probability of success on a single trial is p and the probability of failure is $q = 1 - p$. Then the probability of k successes in n trials is given by

$$\binom{n}{k} p^k q^{n-k}$$

EXAMPLE 6   *Use the Binomial Formula*

A multiple-choice exam consists of ten questions. For each question there are four possible choices, of which only one is correct. If someone randomly guesses at the answers, what is the probability of guessing six answers correctly?

Solution

Selecting an answer is one trial of the experiment. Because there are ten questions, $n = 10$. There are four possible choices for each question, of which only one is correct. Therefore,

$$p = \frac{1}{4} \quad \text{and} \quad q = 1 - p = 1 - \frac{1}{4} = \frac{3}{4}$$

A success for this experiment occurs each time a correct answer is guessed. Thus $k = 6$. By the Binomial Probability Formula,

$$P = \binom{10}{6}\left(\frac{1}{4}\right)^6\left(\frac{3}{4}\right)^4 \approx 0.016222$$

The probability of guessing six answers correctly is approximately 0.0162.

TRY EXERCISE 40, EXERCISE SET 9.7

Following are five guidelines for calculating probabilities.

GUIDELINES FOR CALCULATING A PROBABILITY

1. The word "or" usually means to add the probabilities of each event.

2. The word "and" usually means to multiply the probabilities of each event.

3. The phrase "at least n" means n or more. At least 5 is 5 or more.

4. The phrase "at most n" means n or less. At most 5 is 5 or less.

5. "Exactly n" means just that. Exactly 5 heads in 7 tosses of a coin means 5 heads *and therefore* 2 tails.

TOPICS FOR DISCUSSION

1. What is the meaning of probability and what are the possible values of a probability?

2. What is the sample space of an experiment? What are events and how are they related to the sample space?

3. Discuss the difference between mutually exclusive events and events that are not mutually exclusive. Give examples of each type.

4. Discuss the Addition Rule for Probabilities and how it is used.

5. What is the Binomial Probability Formula and how is it used?

EXERCISE SET 9.7

In Exercises 1 to 10, list the elements in the sample space defined by the given experiment.

1. Two people are selected from two senators and three representatives.

2. A letter is chosen at random from the word "Tennessee."

3. A fair coin is tossed and then a random integer between 1 and 4, inclusive, is selected.

4. A fair coin is tossed four times.

5. Two identical tennis balls are randomly placed in three tennis ball cans.

6. Two people are selected from among one Republican, one Democrat, and one Independent.

7. Three cards are randomly chosen from the ace of hearts, ace of spades, ace of clubs, and ace of diamonds.

8. Three letters addressed to *A*, *B*, and *C*, respectively, are randomly put into three envelopes addressed to *A*, *B*, and *C*, respectively.

9. Two vowels are randomly chosen from a, e, i, o, and u.

10. Three computer disks are randomly chosen from one defective disk and three nondefective disks.

In Exercises 11 to 15, use the sample space defined by the experiment of tossing a fair coin four times. Express each event as a subset of the sample space.

11. There are no tails.

12. There are exactly two heads.

13. There are at most two heads.

14. There are more than two heads.

15. There are twelve tails.

In Exercises 16 to 20, use the sample space defined by the experiment of choosing two random numbers, in succession, from the integers 1, 2, 3, 4, 5, and 6. The numbers are chosen with replacement. Express each event as a subset of the sample space.

16. The sum of the numbers is 7.

17. The two numbers are the same.

18. The first number is greater than the second number.

19. The second number is a 4.

20. The sum of the two numbers is greater than 1.

In Exercises 21 through 44, calculate the probabilities of the events.

21. **CARD GAMES** From a deck of regular playing cards, one card is chosen at random. Find the probability of each event.

  a. The card is a king.   b. The card is a spade.

22. **NUMBER THEORY** A single number is chosen from the digits 1, 2, 3, 4, 5, and 6. Find the probability that the number is an even number or a number divisible by 3.

23. **ECONOMICS** An economist predicts that the probability of an increase in gross domestic product (GDP) is 0.64 and that the probability of an increase in inflation is 0.55. The economist also predicts that the probability of an increase in GDP *and* inflation is 0.22. Find the probability of an increase in GDP *or* an increase in inflation.

24. **NUMBER THEORY** Four digits are selected from the digits 1, 2, 3, and 4, and a number is formed. Find the probability that the number is greater than 3000, assuming digits can be repeated.

25. **Building Industry** An owner of a construction company has bid for the contracts on two buildings. If the contractor estimates that the probability of getting the first contract is 1/2, that of getting the second contract is 1/5, and that of getting both contracts is 1/10, find the probability that the contractor will get at least one of the two building contracts.

26. **Acceptance Sampling** A shipment of ten calculators contains two defective calculators. Two calculators are chosen from the shipment. Find the probability of each event.

    a. Both are defective.

    b. At least one is defective.

27. **Number Theory** Five random digits are selected from 0 to 9 with replacement. What is the probability (to the nearest hundredth) that 0 does not occur?

28. **Queuing Theory** Six persons are arranged in a line. What is the probability that two specific people, say A and B, are standing next to each other?

29. **Lottery** A box contains 500 envelopes, of which 50 have $100 in cash, 75 have $50 in cash, and 125 have $25 in cash. If an envelope is selected at random from this box, what is the probability that it will contain at least $50?

30. **Jury Selection** A jury of twelve people is selected from thirty people: fifteen women and fifteen men. What is the probability that the jury will have six men and six women?

31. **Queuing Theory** Three girls and three boys are randomly placed in six adjacent seats. What is the probability that the boys and girls will be in alternating seats?

32. **Committee Membership** A committee of four is chosen from three accountants and five actuaries. Find the probability that the committee consists of two accountants and two actuaries.

33. **Extra-Sensory Perception** A magician claims to be able to read minds. To test this claim, five cards numbered 1 to 5 are used. A subject selects two cards from the five and concentrates on the numbers. What is the probability that the magician can correctly identify the two cards by just guessing?

34. **Card Games** One card is randomly drawn from a regular deck of playing cards. The card is replaced and another card is drawn. Are the events independent? What is the probability that both cards drawn are aces?

35. **Scheduling** A meeting is scheduled by randomly choosing a weekday and then randomly choosing an hour between 8:00 A.M. and 4:00 P.M. What is the probability that Monday at 8:00 A.M. is chosen?

36. **National Defense** A missile radar detection system consists of two radar screens. The probability that any one of the radar screens will detect an incoming missile is 0.95. If radar detections are assumed to be independent events, what is the probability that a missile that enters the detection space of the radar will be detected?

37. **Oil Industry** An oil drilling venture involves drilling four wells in different parts of the country. For each well, the probability that it will be profitable is 0.10, and the probability that it will be unprofitable is 0.90. If these events are independent, what is the probability of drilling at least one unprofitable well?

38. **Manufacturing** A manufacturer of CD-ROMs claims that only 1 of every 1000 CD-ROMs manufactured is defective. If this claim is correct and if defective CD-ROMs are independent events, what is the probability that of the next three CD-ROMs produced, all are not defective?

39. **Preference Testing** A software firm is considering marketing two newly designed spreadsheet programs, A and B. To test the appeal of the programs, the firm installs them in four corporations. After 2 weeks, the firm asks each corporation to evaluate each program. If the corporations have no preference, what is the probability that all four will choose product A?

40. **Agriculture** A fruit grower claims that one-fourth of the orange trees in a grove crop have suffered frost damage. Find the probability that among eight orange trees, exactly three have frost damage.

41. **Quality Control** A quality control inspector receives a shipment of 20 computer monitors. From the 20 monitors, the inspector randomly chooses 5 for inspection. If the probability of a monitor being defective is 0.05, what is the probability that at least one of the monitors chosen by the inspector is defective?

42. **Lottery** Consider a lottery that sells 1000 tickets and awards two prizes. If you purchase 10 tickets, what is the probability that you will win a prize?

43. **Airline Scheduling** An airline estimates that 75% of the people who make a reservation for a certain flight will actually show up for the flight. Suppose the airline sells 25 tickets on this flight and the plane has room for 20 passengers. What is the probability that 21 or more people with tickets will show up for the flight?

44. **Airline Scheduling** Suppose that an airplane's engines operate independently and that the probability that any one engine will fail is 0.03. A plane can make a safe landing if at least one-half of its engines operate. Is a safe flight more likely to occur in a two-engine or a four-engine plane? Why?

SUPPLEMENTAL EXERCISES

45. **SPREAD OF A RUMOR** A club has nine members. One member starts a rumor by telling it to a second club member, who repeats the rumor to a third person, and so on. At each stage, the recipient of the rumor is chosen at random from the nine club members. What is the probability that the rumor will be told three times without returning to the originator?

46. **EXTRA-SENSORY PERCEPTION** As a test for extra-sensory perception (ESP), ten cards, five black and five white, are shuffled, and then a person looks at each card. In another room, the ESP subject attempts to guess whether the card is black or white. The ESP subject must guess black five times and white five times. If the ESP subject has no extra-sensory perception, what is the probability that the subject will correctly name eight of the ten cards?

47. **TELEPHONE NUMBER EXTENSIONS** The telephone extensions at a university are four-digit numbers chosen from the digits 1 through 9. If two telephone numbers are randomly chosen from the telephone book, what is the probability that the last two digits (and no others) match?

48. **ARRANGING LETTERS OF A WORD** Each arrangement of the letters of the word "Tennessee" is written on a piece of paper, and all the pieces of paper are placed in a bowl. One piece of paper is selected at random. What is the probability that the first letter in the arrangement is a T?

PROJECTS

1. **MONTE HALL PROBLEM** The grand prize in a game show is behind one of three curtains. A contestant selects one of the three curtains, say curtain A. To add drama to the show, the game host reveals a prize behind one of the other curtains. This prize is not the grand prize. Now the contestant has an opportunity to cancel the original choice (curtain A) and choose the remaining closed curtain or stay with the original choice. What is the probability that the contestant will now choose the grand prize? *Note:* This problem is sometimes referred to as the *Monte Hall* problem after the game show *Let's Make a Deal.* For more information on this problem, read the "Ask Marilyn" columns (by syndicated columnist Marilyn vos Savant) in which this problem was discussed (see *Parade* magazine, September 9, 1990, p. 15).

2. **PROBABILITY AND AUTOMATIC GARAGE DOOR OPENERS** A home construction company builds 500 homes in a planned community, and each home has one garage with an automatic garage door opener. The garage door opener has six switches that can be set to either 0 or 1. Suppose a homeowner chooses some sequence, say 011101. Assuming that all the homes in the development are sold, what is the probability that at least two of the homeowners have chosen the same sequence and could therefore open their neighbor's garage?

EXPLORING CONCEPTS WITH TECHNOLOGY

Mathematical Expectation

Expectation E is a number used to determine the fairness of a gambling game. It is defined as the probability P of winning a bet times the amount A available to win.

$$E = P \cdot A$$

A game is called fair if the expectation of the game equals the amount bet. For example, if you and a friend each bet $1 on who can guess the side facing up on the flip of a coin, then the expectation is $E = \dfrac{1}{2} \cdot \$2 = \$1$. Because the amount of your bet equals the expectation, the game is fair.

When a game is unfair, it benefits one of the players. If you bet $1 and your friend bets $2 on who can guess the flip of a coin, your expectation is $E = \frac{1}{2} \cdot \$3 = \$1.50$. Because your expectation is greater than the amount you bet, the game is advantageous to you. Your friend's expectation is also $1.50, which is less than the amount your friend bet. This is a disadvantage to your friend.

Keno is a game of chance played in many casinos. In this game, a large basket contains 80 balls numbered from 1 to 80. From these balls, the casino randomly chooses 20 balls. The number of ways in which the dealer can choose 20 balls from 80 is the number of combinations of 80 things chosen 20 at a time, or $C(80, 20)$.

In one particular game, a gambler can bet $1 and mark five numbers. The gambler will win a prize if three of the five numbers marked are included in the 20 numbered balls chosen by the dealer. By the counting principle, there are $C(20, 3) \cdot C(60, 2) = 2{,}017{,}800$ ways the gambler can do this. The probability of this event is $\frac{C(20, 3) \cdot C(60, 2)}{C(80, 5)} \approx 0.0839$. The amount the gambler wins for this event is $2 (the $1 bet plus $1 from the casino), so the expectation of the gambler is approximately $.17 (0.0839 · $2).

Each casino has different rules and different methods of awarding prizes. The tables below give the prizes for a $2 bet for some of the possible choices a gambler can make at four casinos. Complete the Expectation column. In each case, the Mark column indicates how many numbers the gambler marked, and the Catch column shows how many of the numbers marked by the gambler were also chosen by the dealer.

Casino 1

Mark	Catch	Win	Expectation
6	4	$8	
6	5	$176	
6	6	$2960	

Casino 2

Mark	Catch	Win	Expectation
6	4	$6	
6	5	$160	
6	6	$3900	

Casino 3

Mark	Catch	Win	Expectation
6	4	$8	
6	5	$180	
6	6	$3000	

Casino 4

Mark	Catch	Win	Expectation
6	4	$6	
6	5	$176	
6	6	$3000	

Adding the expectations in each column gives you the total expectation for marking six numbers. Find the total expectation for each casino. Which casino offers the gambler the greatest expectation?

CHAPTER 9 REVIEW

9.1 Infinite Sequences and Summation Notation

- An infinite sequence is a function whose domain is the positive integers and whose range is a set of real numbers.

- An alternating sequence is one in which the signs of the terms alternate between positive and negative values.

- A recursively defined sequence is one in which each succeeding term of the sequence is defined by using some of the preceding terms.

- If n is a positive integer, then n factorial, $n!$, is the product of the first n positive integers.

$$n! = n(n-1)(n-2) \cdots 3 \cdot 2 \cdot 1$$

- If a_n is a sequence, then $S_n = \sum_{i=1}^{n} a_i$ is the nth partial sum of the sequence.

9.2 Arithmetic Sequences and Series

- Given that d is a real number, the sequence a_n is an arithmetic sequence if $a_{i+1} - a_i = d$ for all i. The number d is called the common difference for the sequence.

- The nth term of an arithmetic sequence with common difference of d is $a_n = a_1 + (n-1)d$.

- If a_n is an arithmetic sequence, then the nth partial sum S_n of the sequence is given by

$$S_n = \frac{n}{2}(a_1 + a_n)$$

9.3 Geometric Sequences and Series

- Given that $r \neq 0$ is a constant real number, the sequence a_n is a geometric sequence if $a_{i+1}/a_i = r$ for all positive integers i. The ratio r is called the common ratio for the geometric sequence.

- The nth term of the geometric sequence is $a_n = a_1 r^{n-1}$, where a_1 is the first term of the sequence and r is the common ratio.

- If a_n is a geometric sequence, then the nth partial sum of the sequence is given by

$$S_n = \frac{a_1(1 - r^n)}{1 - r} \quad r \neq 1$$

- If $|r| < 1$, then the sum of an infinite geometric series is given by

$$S = \frac{a_1}{1 - r}$$

9.4 Mathematical Induction

- *Principle of Mathematical Induction*
Let P_n be a statement that involves positive integers. If

1. P_1 is true, and

2. The truth of P_k implies the truth of P_{k+1}

then P_n is true for all positive integers.

9.5 The Binomial Theorem

- *Binomial Theorem for Positive Integers*
If n is a positive integer, then

$$(a+b)^n = \sum_{i=0}^{n} \binom{n}{i} a^{n-i} b^i$$

- The ith term of the expansion of $(a+b)^n$ is

$$\binom{n}{i-1} a^{n-i+1} b^{i-1}$$

9.6 Permutations and Combinations

- The Fundamental Counting Principle is used to count the number of ways in which a sequence of n conditions can occur.

- A permutation is an arrangement of distinct objects in a definite order. The formula for the permutations of n distinct objects taken r at a time is

$$P(n, r) = \frac{n!}{(n-r)!}$$

- A combination is an arrangement of objects for which the order of the selection is not important. The formula for the number of combinations of n objects taken r at a time is

$$C(n, r) = \frac{n!}{r!(n-r)!}$$

9.7 Introduction to Probability

- Probability is the mathematical study of random patterns. The sample space of an experiment is the set of all possible outcomes of that experiment. An event is any subset of a sample space.

- If S is the sample space of an experiment and E is an event in the sample space, then the probability of the event is given by

$$P(E) = \frac{n(E)}{n(S)}$$

where $n(E)$ and $n(S)$ are the number of elements in E and S, respectively.

- *Addition Rule for Probabilities*
If E_1 and E_2 are two events, then

$$P(E_1 \cup E_2) = P(E_1) + P(E_2) - P(E_1 \cap E_2)$$

- *Probability Rule for Independent Events*
If E_1 and E_2 are two independent events, then the probability that both E_1 and E_2 will occur is

$$P(E_1) \cdot P(E_2)$$

- **Binomial Probability Formula**
  Let an experiment consist of n trials for which the probability of success on a single trial is p and the probability of failure is $q = 1 - p$. Then the probability of k successes in n trials is given by

$$\binom{n}{k} p^k q^{n-k}$$

CHAPTER 9 TRUE/FALSE EXERCISES

In Exercises 1 to 15, answer true or false. If the statement is false, give an example to show that the statement is false.

1. $0! \cdot 4! = 0$

2. $\left(\sum_{i=1}^{3} a_i\right)\left(\sum_{i=1}^{3} b_i\right) = \sum_{i=1}^{3} a_i b_i$

3. $\dfrac{n(n-1)(n-2)\cdots(n-k+1)}{k!} = C(n, k)$

4. No two terms of a sequence can be equal.

5. $1, 8, 27, 64, \ldots, k^3, \ldots$ is a geometric sequence.

6. $a_1 = 2$, $a_{n+1} = a_n - 3$ defines an arithmetic sequence.

7. $0.\overline{9} = 1$

8. Adding all the terms of an infinite sequence produces an infinite sum.

9. Because the first step of an induction proof is normally easy, this step can be omitted.

10. In the expansion of $(a + b)^8$, the exponent on a for the fifth term is 5.

11. The counting principle states that if there are n ways to satisfy one condition and m ways to satisfy a second condition, then there are $n + m$ ways to satisfy both conditions.

12. The number of permutations of n things taken r at a time is given by $n!/r!$.

13. If E is an event in a sample space, then $0 \leq P(E) \leq 1$, where $P(E)$ is the probability of E.

14. If A and B are mutually exclusive events, then $P(A \cap B) = 1$.

15. If a coin is tossed five times, then the probability of observing HHHHH is the same as the probability of observing HTHHT.

CHAPTER 9 REVIEW EXERCISES

In Exercises 1 to 20, find the third and seventh terms of the sequence defined by a_n.

1. $a_n = n^2$

2. $a_n = n!$

3. $a_n = 3n + 2$

4. $a_n = 1 - 2n$

5. $a_n = 2^{-n}$

6. $a_n = 3^n$

7. $a_n = \dfrac{1}{n!}$

8. $a_n = \dfrac{1}{n}$

9. $a_n = \left(\dfrac{2}{3}\right)^n$

10. $a_n = \left(-\dfrac{4}{3}\right)^n$

11. $a_1 = 2$, $a_n = 3a_{n-1}$

12. $a_1 = -1$, $a_n = 2a_{n-1}$

13. $a_1 = 1$, $a_n = -na_{n-1}$

14. $a_1 = 2$, $a_n = n^2 a_{n-1}$

15. $a_1 = 4$, $a_n = a_{n-1} + 2$

16. $a_1 = 3$, $a_n = a_{n-1} - 3$

17. $a_1 = 1$, $a_2 = 2$, $a_n = a_{n-1}a_{n-2}$

18. $a_1 = 1$, $a_2 = 2$, $a_n = a_{n-1}/a_{n-2}$

19. $a_1 = -1$, $a_n = 3na_{n-1}$

20. $a_1 = 2$, $a_n = -2na_{n-1}$

21–40. Classify each sequence defined in Exercises 1 to 20 as arithmetic, geometric, or neither.

In Exercises 41 to 56, find the indicated sum of the series.

41. $\displaystyle\sum_{n=1}^{9} (2n - 3)$

42. $\displaystyle\sum_{i=1}^{11} (1 - 3i)$

43. $\displaystyle\sum_{k=1}^{8} (4k + 1)$

44. $\displaystyle\sum_{i=1}^{10} (i^2 + 3)$

45. $\displaystyle\sum_{n=1}^{6} 3 \cdot 2^n$

46. $\displaystyle\sum_{i=1}^{5} 2 \cdot 4^{i-1}$

47. $\displaystyle\sum_{k=1}^{9} (-1)^k 3^k$

48. $\displaystyle\sum_{i=1}^{8} (-1)^{i+1} 2^i$

49. $\displaystyle\sum_{i=1}^{10} \left(\dfrac{2}{3}\right)^i$

50. $\displaystyle\sum_{i=1}^{11} \left(\dfrac{3}{2}\right)^i$

51. $\displaystyle\sum_{n=1}^{9} \dfrac{(-1)^{n+1}}{n^2}$

52. $\displaystyle\sum_{k=1}^{5} \dfrac{(-1)^{k+1}}{k!}$

53. $\displaystyle\sum_{n=1}^{\infty} \left(\dfrac{1}{4}\right)^n$

54. $\displaystyle\sum_{i=1}^{\infty} \left(-\dfrac{5}{6}\right)^i$

55. $\displaystyle\sum_{k=1}^{\infty} \left(-\dfrac{4}{5}\right)^k$

56. $\displaystyle\sum_{j=0}^{\infty} \left(\dfrac{1}{5}\right)^j$

In Exercises 57 to 64, prove each statement by mathematical induction.

57. $\sum_{i=1}^{n} (5i + 1) = \dfrac{n(5n + 7)}{2}$ **58.** $\sum_{i=1}^{n} (3 - 4i) = n(1 - 2n)$

59. $\sum_{i=0}^{n} \left(-\dfrac{1}{2} \right)^i = \dfrac{2[1 - (-1/2)^{n+1}]}{3}$

60. $\sum_{i=0}^{n} (-1)^i = \dfrac{1 - (-1)^{n+1}}{2}$

61. $n^n \geq n!$ **62.** $n! > 4^n, \quad n \geq 9$

63. 3 is a factor of $n^3 + 2n$ for all positive integers n.

64. Let $a_1 = \sqrt{2}$ and $a_n = (\sqrt{2})^{a_{n-1}}$. Prove that $a_n < 2$ for all positive integers n.

In Exercises 65 to 68, use the Binomial Theorem to expand each binomial.

65. $(4a - b)^5$ **66.** $(x + 3y)^6$

67. $(\sqrt{a} + 2\sqrt{b})^8$ **68.** $\left(2x - \dfrac{1}{2x} \right)^7$

69. Find the fifth term in the expansion of $(3x - 4y)^7$.

70. Find the eighth term in the expansion of $(1 - 3x)^9$.

71. COMPUTER PASSWORDS A computer password consists of eight letters. How many passwords are possible? Assume there is no difference between lower-case and upper-case letters.

72. SERIAL NUMBERS The serial number on an airplane consists of the letter N, followed by six numerals, followed by one letter. How many serial numbers are possible?

73. COMMITTEE MEMBERSHIP From a committee of fifteen members, a president, vice president, and treasurer are elected. In how many ways can this be accomplished?

74. SCHEDULING The emergency staff for a hospital consists of four supervisors and twelve regular employees. How many shifts of four people can be formed if each shift must contain exactly one supervisor?

75. COMMITTEE MEMBERSHIP From twelve people, a committee of five people is formed. In how many ways can this

be accomplished if there are two people among the twelve who refuse to serve together on the committee?

76. ACCEPTANCE SAMPLING A shipment of ten calculators contains two defective ones. A quality control inspector randomly chooses four of the calculators for testing. What is the probability that the inspector will choose one defective calculator?

77. SUMS OF COINS A nickel, dime, and quarter are tossed. What is the probability that the nickel and dime will show heads and the quarter will show tails? What is the probability that only one of the coins will show tails?

78. ARRANGEMENTS OF CARDS A deck of ten cards contains five red and five black cards. If four cards are drawn from the deck, what is the probability that two are red and two are black?

79. NUMBER THEORY For the 1000 numbers 000 to 999, what is the probability that the middle digit is greater than the other two digits?

80. NUMBER THEORY Two numbers are chosen, with replacement, from the digits 1, 2, 3, 4, 5, and 6, and their sum is recorded. Now two more digits are selected and their sum noted. This process continues until the sum is 7 or the original sum is obtained. If the original sum was 9, what is the probability of having another sum of 9 before having a sum of 7? (*Hint:* Assume the events are independent. The probability can be found by summing an infinite geometric series.)

81. CARD GAMES Which of the following has the greater probability: drawing an ace and a ten-card (ten, jack, queen, or king) from a regular deck of fifty-two playing cards or drawing an ace and a ten-card from two decks of regular playing cards?

82. NUMBER THEORY From the digits 1, 2, 3, 4, and 5, two numbers are chosen without replacement. What is the probability that the second number is greater than the first number?

83. EMPLOYEE BADGES A room contains twelve people who are wearing badges numbered 1 to 12. If three people are randomly selected, what is the probability that the person wearing badge 6 will be included?

CHAPTER 9 TEST

In Exercises 1 to 3, find the third and fifth terms of the sequence defined by a_n.

1. $a_n = \dfrac{2^n}{n!}$ **2.** $a_n = \dfrac{(-1)^{n+1}}{2n}$

3. $a_1 = 3, \; a_n = 2a_{n-1}$

In Exercises 4 to 6, classify each sequence as an arithmetic sequence, a geometric sequence, or neither.

4. $a_n = -2n + 3$ **5.** $a_n = 2n^2$

6. $a_n = \dfrac{(-1)^{n-1}}{3^n}$

In Exercises 7 to 9, find the indicated sum of the series.

7. $\sum\limits_{i=1}^{6} \dfrac{1}{i}$

8. $\sum\limits_{j=1}^{10} \dfrac{1}{2^j}$

9. $\sum\limits_{k=1}^{20} (3k - 2)$

10. The third term of an arithmetic sequence is 7 and the eighth term is 22. Find the twentieth term.

11. Find the sum of the infinite geometric series given by $\sum\limits_{k=1}^{\infty} \left(\dfrac{3}{8}\right)^k$.

12. Write $0.\overline{15}$ as the quotient of integers in simplest form.

In Exercises 13 and 14, prove the statement by mathematical induction.

13. $\sum\limits_{i=1}^{n} (2 - 3i) = \dfrac{n(1 - 3n)}{2}$

14. $n! > 3^n, \quad n \geq 7$

15. Write the binomial expansion of $(x - 2y)^5$.

16. Write the binomial expansion of $\left(x + \dfrac{1}{x}\right)^6$.

17. Find the sixth term in the expansion of $(3x + 2y)^8$.

18. Three cards are randomly chosen from a regular deck of playing cards. In how many ways can the cards be chosen so that the three cards are different?

19. A serial number consists of seven characters. The first three characters are upper-case letters of the alphabet. The next two characters are selected from the digits 1 through 9. The last two characters are upper-case letters of the alphabet. How many serial numbers are possible if no letter or number can be used twice in the same serial number?

20. Five cards are randomly selected from a deck of cards containing eight black cards and ten red cards. What is the probability that three black cards and two red cards are selected?

SOLUTIONS TO SELECTED EXERCISES

Exercise Set 1.1, page 8

2. a. Integers: 21, 53

   b. Rational numbers: $5.\overline{17}$, -4.25, $1/4$, 21, 53, $0.45454545\ldots$

   c. Irrational numbers: π

   d. Real numbers: All of the given numbers are real numbers.

   e. Prime numbers: 53

   f. Composite numbers: 21

4. $\{0, 1, 2, 3, 4\} \cap \{1, 3, 6, 10\} = \{1, 3\}$

16. Commutative property of addition

18. Symmetric property of equality

30. $\dfrac{2a}{5} + \dfrac{3a}{7} = \dfrac{2a \cdot 7}{5 \cdot 7} + \dfrac{3a \cdot 5}{7 \cdot 5} = \dfrac{29a}{35}$

Exercise Set 1.2, page 15

16. $\xleftarrow{\ \ \ \ \ \ }[\ \ \ \ )\xrightarrow{\ \ \ \ \ \ }$ $[-2, 1)$
    $\quad\quad$ -2 0 1

56. $|x + 6| + |x - 2| = x + 6 - (x - 2)$
    $$= x + 6 - x + 2$$
    $$= 8$$

64. $|-5 - 8| = 13$

80. $|y + 3| > 6$

Exercise Set 1.3, page 27

24. $\left(\dfrac{2ab^2c^3}{5ab^2}\right)^3 = \left(\dfrac{2c^3}{5}\right)^3 = \dfrac{8c^9}{125}$

36. $\dfrac{x^{1/3}y^{5/6}}{x^{3/2}y^{1/6}} = x^{1/3-3/2}y^{5/6-1/6} = x^{2/6-9/6}y^{4/6} = \dfrac{y^{2/3}}{x^{7/6}}$

76. $\sqrt{18x^2y^5} = \sqrt{9x^2y^4}\,\sqrt{2y} = 3|x|y^2\sqrt{2y}$

84. $5\sqrt[3]{3} + 2\sqrt[3]{81} = 5\sqrt[3]{3} + 2\sqrt[3]{27 \cdot 3} = 5\sqrt[3]{3} + 2 \cdot 3\sqrt[3]{3}$
    $$= 5\sqrt[3]{3} + 6\sqrt[3]{3} = 11\sqrt[3]{3}$$

92. $(3\sqrt{5y} - 4)^2 = (3\sqrt{5y} - 4)(3\sqrt{5y} - 4)$
    $$= 9 \cdot 5y - 12\sqrt{5y} - 12\sqrt{5y} + 16$$
    $$= 45y - 24\sqrt{5y} + 16$$

104. $\dfrac{2}{\sqrt[4]{4y}} = \dfrac{2}{\sqrt[4]{4y}} \cdot \dfrac{\sqrt[4]{4y^3}}{\sqrt[4]{4y^3}} = \dfrac{2\sqrt[4]{4y^3}}{2y} = \dfrac{\sqrt[4]{4y^3}}{y}$

110. $\dfrac{5}{\sqrt{y} - \sqrt{3}} \cdot \dfrac{(\sqrt{y} + \sqrt{3})}{\sqrt{y} + \sqrt{3}} = \dfrac{5\sqrt{y} + 5\sqrt{3}}{y - 3}$

Exercise Set 1.4, page 35

2. $3 + \sqrt{-25} = 3 + i\sqrt{25} = 3 + 5i$

16. $(-3 + i) - (-8 + 2i) = -3 + i + 8 - 2i = 5 - i$

24. $(5 - 3i)(-2 - 4i) = -10 - 20i + 6i + 12i^2$
    $$= -10 - 14i - 12 = -22 - 14i$$

34. $\dfrac{5 - 7i}{5 + 7i} = \dfrac{5 - 7i}{5 + 7i} \cdot \dfrac{5 - 7i}{5 - 7i} = \dfrac{25 - 35i - 35i + 49i^2}{25 - 49i^2}$
    $$= \dfrac{25 - 70i - 49}{25 + 49} = \dfrac{-24 - 70i}{74} = -\dfrac{12}{37} - \dfrac{35}{37}i$$

54. $i^{28} = (i^4)^7 = 1$

68. $\sqrt{-3}\,\sqrt{-121} = i\sqrt{3} \cdot i\sqrt{121} = i\sqrt{3} \cdot 11i$
    $$= 11i^2\sqrt{3} = -11\sqrt{3}$$

Exercise Set 1.5, page 41

24. $(5y^2 - 7y + 3) + (2y^2 + 8y + 1) = 7y^2 + y + 4$

32. $(5x - 7)(3x^2 - 8x - 5)$
    $$= 15x^3 - 40x^2 - 25x - 21x^2 + 56x + 35$$
    $$= 15x^3 - 61x^2 + 31x + 35$$

60. $(4x^2 - 3y)(4x^2 + 3y) = (4x^2)^2 - (3y)^2 = 16x^4 - 9y^2$

72. $-x^2 - 5x + 4 = -(-5)^2 - 5(-5) + 4$
    $$= -25 + 25 + 4 = 4$$

82. $\dfrac{1}{6}n^3 - \dfrac{1}{2}n^2 + \dfrac{1}{3}n = \dfrac{1}{6}(21)^3 - \dfrac{1}{2}(21)^2 + \dfrac{1}{3}(21)$
    $$= 1330 \text{ committees}$$

84. a. $4.3 \times 10^{-6}(1000)^2 - 2.1 \times 10^{-4}(1000)$
    $$= 4.09 \text{ seconds}$$

    b. $4.3 \times 10^{-6}(5000)^2 - 2.1 \times 10^{-4}(5000)$
    $$= 106.45 \text{ seconds}$$

    c. $4.3 \times 10^{-6}(10{,}000)^2 - 2.1 \times 10^{-4}(10{,}000)$
    $$= 427.9 \text{ seconds}$$

Exercise Set 1.6, page 52

6. $6a^3b^2 - 12a^2b + 72ab^3 = 6ab(a^2b - 2a + 12b^2)$

18. $b^2 + 12b - 28 = (b + 14)(b - 2)$

22. $57y^2 + y - 6 = (19y - 6)(3y + 1)$

30. $b^2 - 4ac = 8^2 - 4(16)(-35) = 2304 = 48^2$
    The trinomial is factorable over the integers.

38. $81b^2 - 16c^2 = (9b - 4c)(9b + 4c)$

48. $b^2 - 24b + 144 = (b - 12)^2$

54. $b^3 + 64 = (b + 4)(b^2 - 4b + 16)$

64. $81y^4 - 16 = (9y^2 - 4)(9y^2 + 4)$
$$= (3y - 2)(3y + 2)(9y^2 + 4)$$

74. $4y^2 - 4yz + z^2 - 9 = (2y - z)^2 - 9 = (2y - z)^2 - 3^2$
$$= (2y - z - 3)(2y - z + 3)$$

Exercise Set 1.7, page 60

2. $\dfrac{2x^2 - 5x - 12}{2x^2 + 5x + 3} = \dfrac{(2x + 3)(x - 4)}{(2x + 3)(x + 1)} = \dfrac{x - 4}{x + 1}$

16. $\dfrac{x^2 - 16}{x^2 + 7x + 12} \cdot \dfrac{x^2 - 4x - 21}{x^2 - 4x}$
$$= \dfrac{(x - 4)(x + 4)(x + 3)(x - 7)}{(x + 3)(x + 4)x(x - 4)} = \dfrac{x - 7}{x}$$

30. $\dfrac{3y - 1}{3y + 1} - \dfrac{2y - 5}{y - 3} = \dfrac{(3y - 1)(y - 3)}{(3y + 1)(y - 3)} - \dfrac{(2y - 5)(3y + 1)}{(y - 3)(3y + 1)}$
$$= \dfrac{(3y^2 - 10y + 3) - (6y^2 - 13y - 5)}{(3y + 1)(y - 3)}$$
$$= \dfrac{-3y^2 + 3y + 8}{(3y + 1)(y - 3)}$$

42. $\dfrac{3 - \dfrac{2}{a}}{5 + \dfrac{3}{a}} = \dfrac{\left(3 - \dfrac{2}{a}\right)a}{\left(5 + \dfrac{3}{a}\right)a} = \dfrac{3a - 2}{5a + 3}$

60. $\dfrac{e^{-2} - f^{-1}}{ef} = \dfrac{\dfrac{1}{e^2} - \dfrac{1}{f}}{ef} = \dfrac{f - e^2}{e^2 f} \div \dfrac{ef}{1}$
$$= \dfrac{f - e^2}{e^2 f} \cdot \dfrac{1}{ef} = \dfrac{f - e^2}{e^3 f^2}$$

64. a. $\dfrac{v_1 + v_2}{1 + \dfrac{v_1 v_2}{c^2}} = \dfrac{1.2 \times 10^8 + 2.4 \times 10^8}{1 + \dfrac{(1.2 \times 10^8)(2.4 \times 10^8)}{(6.7 \times 10^8)^2}} \approx 3.4 \times 10^8$

b. $\dfrac{v_1 + v_2}{1 + \dfrac{v_1 \cdot v_2}{c^2}} = \dfrac{c^2(v_1 + v_2)}{c^2\left(1 + \dfrac{v_1 \cdot v_2}{c^2}\right)} = \dfrac{c^2(v_1 + v_2)}{c^2 + v_1 \cdot v_2}$

Exercise Set 2.1, page 75

2. $-3y + 20 = 2$
$$-3y = -18$$
$$y = 6$$

12. $\dfrac{1}{2}x + 7 - \dfrac{1}{4}x = \dfrac{19}{2}$
$$4\left(\dfrac{1}{2}x + 7 - \dfrac{1}{4}x\right) = 4\left(\dfrac{19}{2}\right)$$
$$2x + 28 - x = 38$$
$$x = 38 - 28$$
$$x = 10$$

20. $5(x + 4)(x - 4) = (x - 3)(5x + 4)$
$$5(x^2 - 16) = 5x^2 - 11x - 12$$
$$5x^2 - 80 = 5x^2 - 11x - 12$$
$$-80 + 12 = -11x$$
$$-68 = -11x$$
$$\dfrac{68}{11} = x$$

30. $2x + \dfrac{1}{3} = \dfrac{6x + 1}{3}$ • **Rewrite the left side.**
$$\dfrac{6x + 1}{3} = \dfrac{6x + 1}{3}$$

This equation is an identity.

38. $\dfrac{4}{y + 2} = \dfrac{7}{y - 4}$ $y \neq 2, y \neq 4$
$$4(y - 4) = 7(y + 2)$$
$$4y - 16 = 7y + 14$$
$$4y - 7y = 14 + 16$$
$$-3y = 30$$
$$y = -10$$

58. Earnings $= 990a - 9000$
$$19,000 = 990a - 9000$$ • **Substitute 19,000 for "Earnings."**
$$28,000 = 990a$$
$$a = \dfrac{28,000}{990} \approx 28.28$$

The average male high school graduate can expect to earn an annual income of $19,000 at age 28 (to the nearest year).

Exercise Set 2.2, page 85

4. $A = P + Prt$
$$A = P(1 + rt)$$ • **Factor.**
$$P = \dfrac{A}{(1 + rt)}$$

22. $P = 2l + 2w$, $w = \dfrac{1}{2}l + 1$
$$110 = 2l + 2\left(\dfrac{1}{2}l + 1\right)$$ • **Substitute for w.**
$$110 = 2l + l + 2$$ • **Simplify.**
$$108 = 3l$$
$$36 = l$$
$$l = 36 \text{ meters}$$
$$w = \dfrac{1}{2}l + 1 = \dfrac{1}{2}(36) + 1 = 19 \text{ meters}$$

30. Let t_1 = the time it takes to travel to the island.
Let t_2 = the time it takes to make the return trip.

$$t_1 + t_2 = 7.5$$
$$t_2 = 7.5 - t_1$$
$$15t_1 = 10t_2$$
$$15t_1 = 10(7.5 - t_1) \quad \bullet \textbf{ Substitute for } t_2.$$
$$15t_1 = 75 - 10t_1$$
$$25t_1 = 75$$
$$t_1 = 3 \text{ hours}$$
$$D = 15t_1 = 15(3) = 45 \text{ nautical miles}$$

36. Let x = the number of glasses of orange juice.

Profit = revenue − cost
$$\$2337 = 0.75x - 0.18x$$
$$2337 = 0.57x$$
$$x = \frac{2337}{0.57}$$
$$x = 4100$$

40. Let x = the amount of money invested at 5%.

5%	x
7%	$7500 - x$

$$0.05x + 0.07(7500 - x) = 405$$
$$0.05x + 525 - 0.07x = 405$$
$$-0.02x = -120$$
$$x = 6000$$
$$7500 - x = 1500$$

$6000 was invested at 5%. $1500 was invested at 7%.

44. Let x = the number of liters of the 40% solution to be mixed with the 24% solution.

0.40	x
0.24	4
0.30	$4 + x$

$$0.40x + 0.24(4) = 0.30(4 + x)$$
$$0.40x + 0.96 = 1.2 + 0.30x$$
$$0.10x = 0.24$$
$$x = 2.4$$

Thus 2.4 liters of 40% sulfuric acid should be mixed with 4 liters of a 24% sulfuric acid solution, to produce the 30% solution.

54. Let x = the number of hours needed to print the report if both the printers are used.
Printer A prints 1/3 of the report every hour.
Printer B prints 1/4 of the report every hour.
Thus

$$\frac{1}{3}x + \frac{1}{4}x = 1$$
$$4x + 3x = 12 \cdot 1$$
$$7x = 12$$
$$x = \frac{12}{7} \approx 1.71$$

It would take approximately 1.71 hours to print the report.

Exercise Set 2.3, page 96

4. $12w^2 - 41w + 24 = 0$
$$(4w - 3)(3w - 8) = 0$$
$$4w - 3 = 0 \quad \text{or} \quad 3w - 8 = 0$$
$$w = \frac{3}{4} \qquad\qquad w = \frac{8}{3}$$

18. $y^2 = 225$
$$y = \pm\sqrt{225}$$
$$y = \pm 15$$

34. $\qquad x^2 - 6x = 0$
$$x^2 - 6x + 9 = 9$$
$$(x - 3)^2 = 9$$
$$x - 3 = \pm\sqrt{9}$$
$$x = 3 \pm 3$$
$$x = 3 + 3 \quad \text{or} \quad x = 3 - 3$$
$$x = 6 \qquad\qquad x = 0$$

38. $2x^2 + 10x - 3 = 0$
$$2x^2 + 10x = 3$$
$$2(x^2 + 5x) = 3$$
$$x^2 + 5x = \frac{3}{2}$$
$$x^2 + 5x + \frac{25}{4} = \frac{3}{2} + \frac{25}{4}$$
$$\left(x + \frac{5}{2}\right)^2 = \frac{31}{4}$$
$$x + \frac{5}{2} = \pm\sqrt{\frac{31}{4}}$$
$$x = -\frac{5}{2} \pm \frac{\sqrt{31}}{2}$$
$$x = \frac{-5 + \sqrt{31}}{2} \quad \text{or} \quad x = \frac{-5 - \sqrt{31}}{2}$$

48. $2x^2 + 4x - 1 = 0$
$$x = \frac{-4 \pm \sqrt{4^2 - 4(2)(-1)}}{4}$$
$$x = \frac{-4 \pm \sqrt{16 + 8}}{4} = \frac{-4 \pm \sqrt{24}}{4}$$
$$x = \frac{-4 \pm 2\sqrt{6}}{4} = \frac{-2 \pm \sqrt{6}}{2}$$
$$x = \frac{-2 + \sqrt{6}}{2} \quad \text{or} \quad x = \frac{-2 - \sqrt{6}}{2}$$

58. $x^2 + 3x - 11 = 0$
$$b^2 - 4ac = 3^2 - 4(1)(-11) = 9 + 44 = 53 > 0$$

Thus the equation has two distinct real roots.

68. $\left(10 \text{ feet} + \frac{1}{4} \text{ inch}\right)^2 = (10 \text{ feet})^2 + x^2$

$\sqrt{(120.25)^2 - (120)^2} = x$ • **Change feet to inches.**

$7.75 = x$

To the nearest inch, the concrete will rise 8 inches.

74. Let P = perimeter and A = area.

$A = 4800 = lw$

$l = \dfrac{4800}{w}$

$P = 4w + 2l = 400$

$2w + l = 200$

$2w + \dfrac{4800}{w} = 200$ • **Substitute for l.**

$2w^2 + 4800 = 200w$

$w^2 - 100w + 2400 = 0$

$(w - 60)(w - 40) = 0$

$w = 60$ or $w = 40$

$l = \dfrac{4800}{60} = 80$ or $l = \dfrac{4800}{40} = 120$

There are two solutions: 60 yards × 80 yards or 40 yards × 120 yards.

Exercise Set 2.4, page 105

6. $x^4 - 36x^2 = 0$

$x^2(x^2 - 36) = 0$

$x^2(x - 6)(x + 6) = 0$

$x = 0, x = 6, x = -6$

14. $\sqrt{10 - x} = 4$ Check: $\sqrt{10 - (-6)} = 4$

$10 - x = 16$ $\sqrt{16} = 4$

$-x = 6$ $4 = 4$

$x = -6$

The solution is -6.

16. $x = \sqrt{5 - x} + 5$

$(x - 5)^2 = (\sqrt{5 - x})^2$

$x^2 - 10x + 25 = 5 - x$

$x^2 - 9x + 20 = 0$

$(x - 5)(x - 4) = 0$

$x = 5$ or $x = 4$

Check: $5 = \sqrt{5 - 5} + 5$ $4 = \sqrt{5 - 4} + 5$

$5 = 0 + 5$ $4 = 1 + 5$

$5 = 5$ $4 = 6$ False

The solution is 5.

20. $\sqrt{x + 7} - 2 = \sqrt{x - 9}$

$(\sqrt{x + 7} - 2)^2 = (\sqrt{x - 9})^2$

$x + 7 - 4\sqrt{x + 7} + 4 = x - 9$

$-4\sqrt{x + 7} = -20$

$(\sqrt{x + 7})^2 = (5)^2$

$x + 7 = 25$

$x = 18$

Check: $\sqrt{18 + 7} - 2 = \sqrt{18 - 9}$

$\sqrt{25} - 2 = \sqrt{9}$

$5 - 2 = 3$

$3 = 3$

The solution is 18.

32. $(4z + 7)^{1/3} = 2$

$[(4z + 7)^{1/3}]^3 = 2^3$

$4z + 7 = 8$

$4z = 1$

$z = \dfrac{1}{4}$

Check: $\left[4\left(\dfrac{1}{4}\right) + 7\right]^{1/3} = 2$

$8^{1/3} = 2$

$2 = 2$

The solution is $\dfrac{1}{4}$.

42. $x^4 - 10x^2 + 9 = 0$ • **Let $u = x^2$.**

$u^2 - 10u + 9 = 0$

$(u - 9)(u - 1) = 0$

$u = 9$ or $u = 1$

$x^2 = 9$ $x^2 = 1$

$x = \pm 3$ $x = \pm 1$

The solutions are 3, -3, 1, and -1.

52. $6x^{2/3} - 7x^{1/3} - 20 = 0$ • **Let $u = x^{1/3}$.**

$6u^2 - 7u - 20 = 0$

$(3u + 4)(2u - 5) = 0$

$u = -\dfrac{4}{3}$ or $u = \dfrac{5}{2}$

$x^{1/3} = -\dfrac{4}{3}$ $x^{1/3} = \dfrac{5}{2}$

$(x^{1/3})^3 = \left(-\dfrac{4}{3}\right)^3$ $(x^{1/3})^3 = \left(\dfrac{5}{2}\right)^3$

$x = -\dfrac{64}{27}$ $x = \dfrac{125}{8}$

The solutions are $-64/27$ and $125/8$.

Exercise Set 2.5, page 118

8. $-4(x - 5) \geq 2x + 15$

$\quad -4x + 20 \geq 2x + 15$

$\quad\quad -6x \geq -5$

$\quad\quad\quad x \leq \dfrac{5}{6}$

The solution set is $\{x \mid x \leq 5/6\}$.

12. Let m = the number of miles driven.
Company A: $29 + 0.12m$
Company B: $22 + 0.21m$

$$29 + 0.12m < 22 + 0.21m$$
$$77.\overline{7} < m$$

Company A is less expensive if you drive at least 78 miles.

16. $2x + 5 > -16$ and $2x + 5 < 9$

$\quad\quad 2x > -21$ and $\quad\quad 2x < 4$

$\quad\quad x > -\dfrac{21}{2}$ and $\quad\quad x < 2$

$\left\{x \mid x > -\dfrac{21}{2}\right\} \cup \{x \mid x < 2\} = \left\{x \mid -\dfrac{21}{2} < x < 2\right\}$

The solution set is $\{x \mid -21/2 < x < 2\}$.

28. $\quad 41 \leq \quad F \quad \leq 68$

$\quad 41 \leq \dfrac{9}{5}C + 32 \leq 68$

$\quad\quad 9 \leq \quad \dfrac{9}{5}C \quad \leq 36$

$\dfrac{5}{9}(9) \leq \left(\dfrac{5}{9}\right)\left(\dfrac{9}{5}\right)C \leq \dfrac{5}{9}(36)$

$\quad\quad 5 \leq \quad C \quad \leq 20$

36. $\quad x^2 + 5x + 6 < 0$

$(x + 2)(x + 3) = 0$

$x = -2$ and $x = -3$ • **Critical values**

Use a test number from each of the intervals $(-\infty, -3)$, $(-3, -2)$, and $(-2, \infty)$ to determine where $x^2 + 5x + 6$ is negative.

$$+ + + + + + + | - | + + + + +$$
$$\quad\quad\quad\quad -3\ -2 \quad 0$$

In interval notation the solution set is $(-3, -2)$.

44. $R = 312x - 3x^2$

$\quad\quad 312x - 3x^2 \geq 5925$

$\quad -3x^2 + 312x - 5925 \geq 0$

$\quad 3(-x^2 + 104x - 1975) \geq 0$

$\quad 3(-x + 25)(x - 79) \geq 0$

$x = 79$ and $x = 25$ • **Critical values**

Use a test number from each of the intervals $(-\infty, 25)$, $(25, 79)$, and $(79, \infty)$ to determine where $3(-x + 25)(x - 79)$ is positive.

$$- - - | + + + + + | - - -$$
$$\quad\quad 25 \quad\quad 79$$

In interval notation the solution set is $[25, 79]$.

46. $\dfrac{x - 2}{x + 3} > 0$

$x = 2$ and $x = -3$ • **Critical values**

Use a test number from each of the intervals $(-\infty, -3)$, $(-3, 2)$, and $(2, \infty)$ to determine where $(x - 2)/(x + 3)$ is positive. The solution set is $(-\infty, -3) \cup (2, \infty)$.

54. $\quad\quad \dfrac{3x + 1}{x - 2} \geq 4$

$\quad\quad \dfrac{3x + 1}{x - 2} - 4 \geq 0$

$\quad \dfrac{3x + 1 - 4(x - 2)}{x - 2} \geq 0$

$\quad\quad\quad \dfrac{-x + 9}{x - 2} \geq 0$

$x = 2$ and $x = 9$ • **Critical values**

Use a test number from each of the intervals $(-\infty, 2)$, $(2, 9)$, and $(9, \infty)$ to determine where $(-x + 9)/(x - 2)$ is positive. The solution set is $(2, 9]$.

Exercise Set 2.6, page 124

10. $|2x - 3| = 21$

$2x - 3 = 21$ or $2x - 3 = -21$

$\quad\quad 2x = 24 \quad\quad\quad 2x = -18$

$\quad\quad\quad x = 12 \quad\quad\quad\quad x = -9$

The solutions are 12 and -9.

30. $|2x - 9| < 7$

$\quad -7 < 2x - 9 < 7$

$\quad\quad 2 < \quad 2x \quad < 16$

$\quad\quad 1 < \quad x \quad < 8$

In interval notation, the solution set is $(1, 8)$.

34. $|2x - 5| \geq 1$

$2x - 5 \leq -1$ or $2x - 5 \geq 1$

$2x \leq 4$ $2x \geq 6$

$x \leq 2$ $x \geq 3$

In interval notation, the solution set is $(-\infty, 2] \cup [3, \infty)$.

48. The volume of the beaker is given by

$$V = \pi(2^2)h = 4\pi h$$

The tolerance is 15 cubic centimeters, and the standard is 750 cubic centimeters, so we have

$|V - 750| \leq 15$

$|4\pi h - 750| \leq 15$

$-15 \leq 4\pi h - 750 \leq 15$

$735 \leq \quad 4\pi h \quad \leq 765$

$\dfrac{735}{4\pi} \leq \quad h \quad \leq \dfrac{765}{4\pi}$

$58.5 \leq \quad h \quad \leq 60.9$ • **Round to the nearest 0.1.**

To ensure that we have 750 cubic centimeters of solution within a tolerance of 15 cubic centimeters, we should fill the beaker to a height of at least 58.5 centimeters but no higher than 60.9 centimeters.

50. $|x^2 - 2| > 1$

$x^2 - 2 = 1$ or $x^2 - 2 = -1$

$x^2 = 3$ $x^2 = 1$

$x = \pm\sqrt{3}$ $x = \pm 1$ • **Critical values**

Use a test number from each of the intervals $(-\infty, -\sqrt{3})$, $(-\sqrt{3}, -1)$, $(-1, 1)$, $(1, \sqrt{3})$, and $(\sqrt{3}, \infty)$ to determine where $|x^2 - 2| > 1$. The solution set is $(-\infty, -\sqrt{3}) \cup (-1, 1) \cup (\sqrt{3}, \infty)$.

Exercise Set 3.1, page 141

26.

30.

32.

40. y-intercept: $(0, -15/4)$
x-intercept: $(5, 0)$

66. $r = \sqrt{(1 - (-2))^2 + (7 - 5)^2}$

$\quad = \sqrt{9 + 4} = \sqrt{13}$

Using the standard form

$$(x - h)^2 + (y - k)^2 = r^2$$

with $h = -2$, $k = 5$, and $r = \sqrt{13}$ yields

$$(x + 2)^2 + (y - 5)^2 = (\sqrt{13})^2$$

68. $x^2 + y^2 - 6x - 4y + 12 = 0$

$x^2 - 6x + y^2 - 4y = -12$

$x^2 - 6x + 9 + y^2 - 4y + 4 = -12 + 9 + 4$

$(x - 3)^2 + (y - 2)^2 = 1^2$

center $(3, 2)$, radius 1

Exercise Set 3.2, page 156

2. Given $g(x) = 2x^2 + 3$

a. $g(3) = 2(3)^2 + 3 = 18 + 3 = 21$

b. $g(-1) = 2(-1)^2 + 3 = 2 + 3 = 5$

c. $g(0) = 2(0)^2 + 3 = 0 + 3 = 3$

d. $g(1/2) = 2(1/2)^2 + 3 = 1/2 + 3 = 7/2$

e. $g(c) = 2(c)^2 + 3 = 2c^2 + 3$

f. $g(c + 5) = 2(c + 5)^2 + 3 = 2c^2 + 20c + 50 + 3$

$\quad = 2c^2 + 20c + 53$

10. a. Because $0 \leq 0 \leq 5$, $Q(0) = 4$.

b. Because $6 < e < 7$, $Q(e) = -e + 9$.

c. Because $1 < n < 2$, $Q(n) = 4$.

d. Because $1 < m \leq 2$, $8 < m^2 + 7 \leq 11$. Thus

$$Q(m^2 + 7) = \sqrt{(m^2 + 7) - 7} = \sqrt{m^2} = m$$

14. $x^2 - 2y = 2$ • **Solve for y.**

$-2y = -x^2 + 2$

$y = \dfrac{1}{2}x^2 - 1$

y is a function of x because each x value will yield one and only one y value.

28. The domain is the set of all real numbers.

40. Domain all real numbers

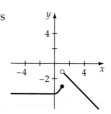

48. $C(4.75) = 0.85 - 0.50 \text{ int}(1 - 4.75)$

$\qquad = 0.85 - 0.50(-4) = 2.85$

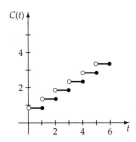

50. a. This is the graph of a function. Every vertical line intersects the graph in at most one point.

b. This is not the graph of a function. Some vertical lines intersect the graph at two points.

c. This is not the graph of a function. The vertical line at $x = -2$ intersects the graph at more than one point.

d. This is the graph of a function. Every vertical line intersects the graph at exactly one point.

66. $V(t) = 44{,}000 - 4200t$

68. a. $V(x) = (30 - 2x)^2 x$

$\qquad = (900 - 120x + 4x^2)x$

$\qquad = 900x - 120x^2 + 4x^3$

b. Domain $\{x \mid 0 < x < 15\}$

72. $d(A, B) = \sqrt{1 + x^2}$. The time required to swim from A to B at 2 mph is $\sqrt{1 + x^2}/2$.

$d(B, C) = 3 - x$. The time required to run from B to C at 8 mph is $(3 - x)/8$.

Thus the total time to reach point C is

$$t = \frac{\sqrt{1 + x^2}}{2} + \frac{3 - x}{8} \text{ hours}$$

Exercise Set 3.3, page 172

2. $m = \dfrac{1 - 4}{5 - (-2)} = -\dfrac{3}{7}$

16. $m = -1$
$\quad b = 1$

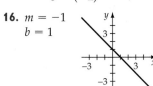

28. $y - 5 = -2(x - 0)$

$\qquad y = -2x + 5$

40. $\qquad f(x) = 0$

$\qquad -2x + 4 = 0$

$\qquad\qquad -2x = -4$

$\qquad\qquad\quad x = -2$

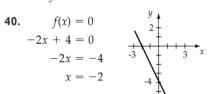

44. $\qquad f_1(x) = f_2(x)$

$\qquad -2x - 11 = 3x + 7$

$\qquad\quad -5x - 11 = 7$

$\qquad\qquad -5x = 18$

$\qquad\qquad\quad x = -\dfrac{18}{5} = -3.6$

48. a. Determine the slope of the line between $(0, 11\,000)$ and $(8, 52\,000)$.

$$m = \frac{52{,}000 - 11{,}000}{8 - 0} = 5125$$

Using the slope-intercept formula, the equation of the line is

$$N(t) = 5125t + 11{,}000$$

where $N(t)$ is the number of reported cases of AIDS for men in year t.

b. The year 1989 corresponds to $t = 3$. Evaluate the function at 3.

$$N(3) = 5125(3) + 11{,}000 = 26{,}375$$

c. The year 2000 corresponds to $t = 14$. Evaluate the function at 14.

$$N(14) = 5125(14) + 11{,}000 = 82{,}750$$

d. The number of reported cases of AIDS in men is increasing at a rate of approximately 5125 cases per year.

50. $P(x) = R(x) - C(x)$

$\quad P(x) = 124x - (78.5x + 5005)$

$\quad P(x) = 45.5x - 5005$

$\quad 45.5x - 5005 = 0$

$\qquad\quad 45.5x = 5005$

$\qquad\qquad x = 110 \qquad$ • **The break-even point**

60. a. and b.

Winning time, in seconds

Graph with T(t), values 20, 21, 22, 23, 24, t-axis 8 16 24 32 40
Year (t = 0 represents 1948)

Answers will vary. However, $T(t) = -0.071t + 24.3$ is one possibility.

c. Answers will vary. However, if you use $T(t)$ as above, then

$$T(52) = -0.071(52) + 24.3 \approx 20.6 \text{ seconds}$$

66. a. The slope of the radius from $(0,0)$ to $(\sqrt{15}, 1)$ is $1/\sqrt{15}$. The slope of the linear path of the rock is $-\sqrt{15}$. The path of the rock is given by

$$y - 1 = -\sqrt{15}(x - \sqrt{15})$$
$$y - 1 = -\sqrt{15}x + 15$$
$$y = -\sqrt{15}x + 16$$

Every point on the wall has a y value of 14. Thus

$$14 = -\sqrt{15}x + 16$$
$$-2 = -\sqrt{15}x$$
$$x = \frac{2}{\sqrt{15}} \approx 0.52$$

The rock hits the wall at $(0.52, 14)$.

Exercise Set 3.4, page 184

10. $f(x) = x^2 + 6x - 1$
$\quad = x^2 + 6x + 9 + (-1 - 9)$
$\quad = (x + 3)^2 - 10$

vertex $(-3, -10)$
axis of symmetry $x = -3$

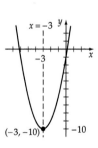

20. $h = -\dfrac{b}{2a} = -\dfrac{-6}{2(1)} = 3$

$k = f(3) = 3^2 - 6(3) = -9$

vertex $(3, -9)$

$f(x) = (x - 3)^2 - 9$

30. $f(x) = -x^2 - 6x$
$\quad = -(x^2 + 6x)$
$\quad = -(x^2 + 6x + 9) + 9$
$\quad = -(x + 3)^2 + 9$

Maximum value of f is 9 when $x = -3$.

40. a. $l + w = 240$, so $w = 240 - l$.

b. $A = lw = l(240 - l) = 240l - l^2$.

c. The l value of the vertex point of the graph of $A = 240l - l^2$ is

$$\frac{-b}{2a} = \frac{-240}{2(-1)} = 120$$

Thus $l = 120$ meters and $w = 240 - 120 = 120$ meters are the dimensions that produce the greatest area.

54. Let $x = $ the number of parcels.

a. $R(x) = xp = x(22 - 0.01x) = -0.01x^2 + 22x$

b. $P(x) = R(x) - C(x)$
$\quad = (-0.01x^2 + 22x) - (2025 + 7x)$
$\quad = -0.01x^2 + 15x - 2025$

c. $-\dfrac{b}{2a} = -\dfrac{15}{2(-0.01)} = 750$

The maximum profit is

$$P(750) = -0.01(750)^2 + 15(750) - 2025 = \$3600$$

d. The price per parcel that yields the maximum profit is

$$p(750) = 22 - 0.01(750) = \$14.50$$

e. The break-even point(s) occur when $R(x) = C(x)$.

$$-0.01x^2 + 22x = 2025 + 7x$$
$$0 = 0.01x^2 - 15x + 2025$$
$$x = \frac{-(-15) \pm \sqrt{(-15)^2 - 4(0.01)(2025)}}{2(0.01)}$$

$x = 150$ and $x = 1350$ are the break-even points.

Thus the minimum number of parcels the air freight company must ship to break even is 150.

56. $h(t) = -16t^2 + 64t + 80$

$$t = \frac{-b}{2a} = \frac{-64}{2(-16)} = 2$$

$$h(2) = -16(2)^2 + 64(2) + 80$$
$$= -64 + 128 + 80 = 144$$

a. The vertex $(2, 144)$ gives us the maximum height of 144 feet.

b. The vertex of the graph of h is $(2, 144)$, so the time when it achieves this maximum height is at time $t = 2$ seconds.

c. $-16t^2 + 64t + 80 = 0$ • **Solve for t with h = 0.**
$\quad -16(t^2 - 4t - 5) = 0$
$\quad -16(t + 1)(t - 5) = 0$
$\quad t = -1 \qquad t - 5 = 0$
$\quad$ no $\qquad\qquad t = 5$

The projectile will have a height of 0 feet at time $t = 5$ seconds.

Exercise Set 3.5, page 195

14. Symmetrical with respect to the x-axis, because replacing y with $-y$ leaves the equation unaltered. The graph is not symmetric with respect to the y-axis, because replacing x with $-x$ alters the equation.

24. The graph is symmetric with respect to the origin because $(-y) = (-x)^3 - (-x)$ simplifies to $-y = -x^3 + x$, which is equivalent to the original equation $y = x^3 - x$.

44. Even, because $h(-x) = (-x)^2 + 1 = x^2 + 1 = h(x)$.

58.

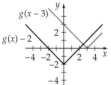

60.

62.

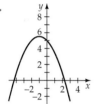

64. a.

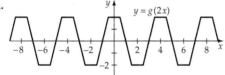

b.

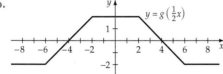

Exercise Set 3.6, page 206

10. $f(x) + g(x) = \sqrt{x-4} - x$ domain $\{x \mid x \geq 4\}$
$f(x) - g(x) = \sqrt{x-4} + x$ domain $\{x \mid x \geq 4\}$
$f(x)g(x) = -x\sqrt{x-4}$ domain $\{x \mid x \geq 4\}$
$f(x)/g(x) = -\dfrac{\sqrt{x-4}}{x}$ domain $\{x \mid x \geq 4\}$

14. $(f + g)(x) = (x^2 - 3x + 2) + (2x - 4) = x^2 - x - 2$
$(f + g)(-7) = (-7)^2 - (-7) - 2 = 49 + 7 - 2 = 54$

30. $\dfrac{f(x + h) - f(x)}{h} = \dfrac{[4(x + h) - 5] - (4x - 5)}{h}$

$= \dfrac{4x + 4(h) - 5 - 4x + 5}{h}$

$= \dfrac{4(h)}{h} = 4$

38. $(g \circ f)(x) = g[f(x)] = g[2x - 7]$
$= 3[2x - 7] + 2 = 6x - 19$
$(f \circ g)(x) = f[g(x)] = f[3x + 2]$
$= 2[3x + 2] - 7 = 6x - 3$

50. $(f \circ g)(4) = f[g(4)]$
$= f[4^2 - 5(4)]$
$= f[-4] = 2(-4) + 3 = -5$

66. a. $l = 3 - 0.5t$ for $0 \leq t \leq 6$. $l = -3 + 0.5t$ for $t > 6$. In either case, $l = |3 - 0.5t|$. $w = |2 - 0.2t|$ as in Example 7.

b. $A(t) = |3 - 0.5t| \, |2 - 0.2t|$

c. A is decreasing on $[0, 6]$ and on $[8, 10]$.
A is increasing on $[6, 8]$ and on $[10, 14]$.

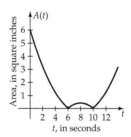

d. The highest point on the graph of A for $0 \leq t \leq 14$ occurs when $t = 0$ seconds.

72. a. On $[2, 3]$,

$a = 2$

$\Delta t = 3 - 2 = 1$

$f(a + \Delta t) = f(3) = 6 \cdot 3^2 = 54$

$f(a) = f(2) = 6 \cdot 2^2 = 24$

Average velocity $= \dfrac{f(a + \Delta t) - f(a)}{\Delta t}$

$= \dfrac{f(3) - f(2)}{1}$

$= 54 - 24 = 30$ feet per second

This is identical to the slope of the line through $(2, f(2))$ and $(3, f(3))$ because

$m = \dfrac{f(3) - f(2)}{3 - 2} = f(3) - f(2)$

b. On $[2, 2.5]$,

$a = 2$

$\Delta t = 2.5 - 2 = 0.5$

$f(a + \Delta t) = f(2.5) = 6(2.5)^2 = 37.5$

Average velocity $= \dfrac{f(2.5) - f(2)}{0.5}$

$= \dfrac{37.5 - 24}{0.5}$

$= \dfrac{13.5}{0.5} = 27$ feet per second

c. On $[2, 2.1]$,

$a = 2$

$\Delta t = 2.1 - 2 = 0.1$

$f(a + \Delta t) = f(2.1) = 6(2.1)^2 = 26.46$

Average velocity $= \dfrac{f(2.1) - f(2)}{0.1}$

$= \dfrac{26.46 - 24}{0.1}$

$= \dfrac{2.46}{0.1} = 24.6$ feet per second

d. On $[2, 2.01]$,

$a = 2$

$\Delta t = 2.01 - 2 = 0.01$

$f(a + \Delta t) = f(2.01) = 6(2.01)^2 = 24.2406$

Average velocity $= \dfrac{f(2.01) - f(2)}{0.01}$

$= \dfrac{24.2406 - 24}{0.01}$

$= \dfrac{0.2406}{0.01} = 24.06$ feet per second

e. On $[2, 2.001]$,

$a = 2$

$\Delta t = 2.001 - 2 = 0.001$

$f(a + \Delta t) = f(2.001) = 6(2.001)^2 = 24.024006$

Average velocity $= \dfrac{f(2.001) - f(2)}{0.001}$

$= \dfrac{24.024006 - 24}{0.001}$

$= \dfrac{0.024006}{0.001} = 24.006$ feet per second

f. On $[2, 2 + \Delta t]$,

$\dfrac{f(2 + \Delta t) - f(2)}{\Delta t} = \dfrac{6(2 + \Delta t)^2 - 24}{\Delta t}$

$= \dfrac{6(4 + 4(\Delta t) + (\Delta t)^2) - 24}{\Delta t}$

$= \dfrac{24 + 24(\Delta t) + 6(\Delta t)^2 - 24}{\Delta t}$

$= \dfrac{24\Delta t + 6(\Delta t)^2}{\Delta t} = 24 + 6(\Delta t)$

As Δt approaches zero, the average velocity seems to approach 24 feet per second.

Exercise Set 3.7, page 215

2. $(f \circ g)(x) = f[g(x)] = f[2x + 6]$

$= \dfrac{1}{2}[2x + 6] - 3 = x + 3 - 3 = x$

$(g \circ f)(x) = g[f(x)] = g\left[\dfrac{1}{2}x - 3\right]$

$= 2\left[\dfrac{1}{2}x - 3\right] + 6 = x - 6 + 6 = x$

10. $g(x) = \dfrac{2}{3}x + 4$

$y = \dfrac{2}{3}x + 4$

$x = \dfrac{2}{3}y + 4$ • **Interchange x and y.**

$x - 4 = \dfrac{2}{3}y$ • **Solve for y.**

$\dfrac{3}{2}x - 6 = y$

Thus $g^{-1}(x) = \dfrac{3}{2}x - 6.$

18. $G(x) = \dfrac{3x}{x - 5}, \quad x \neq 5$

$y = \dfrac{3x}{x - 5}$

$x = \dfrac{3y}{y - 5}$ • **Interchange x and y.**

$xy - 5x = 3y$ • **Solve for y.**

$xy - 3y = 5x$

$y = \dfrac{5x}{x - 3}$

Thus $G^{-1}(x) = \dfrac{5x}{x - 3}, \quad x \neq 3.$

32. $f(x) = x^2 + 6x - 6, x \geq -3$

Domain f is $\{x \mid x \geq -3\}$, range f is $\{y \mid y \geq -15\}$.

$$y = x^2 + 6x - 6$$
$$x = y^2 + 6y - 6 \quad \bullet \text{ Interchange } x \text{ and } y.$$
$$x + 6 = y^2 + 6y$$
$$x + 15 = y^2 + 6y + 9 \quad \bullet \text{ Complete the square.}$$
$$x + 15 = (y + 3)^2$$

Choose the positive root, because the range of f^{-1} is $\{y \mid y \geq -3\}$.

$$\sqrt{x + 15} = y + 3$$
$$-3 + \sqrt{x + 15} = y$$

Thus $f^{-1}(x) = -3 + \sqrt{x + 15}$, domain f^{-1} is $\{x \mid x \geq -15\}$, and range f^{-1} is $\{y \mid y \geq -3\}$.

36.

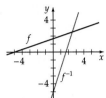

Exercise Set 3.8, page 221

22. $d = kw$

$$6 = k \cdot 80$$
$$\frac{6}{80} = k$$
$$k = \frac{3}{40}$$

Thus $d = \frac{3}{40} \cdot 100 = 7.5$ inches.

24. $r = kv^2$

$$140 = k \cdot 60^2$$
$$\frac{140}{60^2} = k$$
$$\frac{7}{180} = k$$

Thus $r = \frac{7}{180} \cdot 65^2 \approx 164$ feet.

28. $I = \dfrac{k}{d^2}$

$$50 = \frac{k}{10^2}$$
$$5000 = k$$

Thus $I = \dfrac{5000}{d^2} = \dfrac{5000}{15^2} = \dfrac{5000}{225} \approx 22.2$ footcandles.

30. $L = kwd^2$

$$200 = k \cdot 2 \cdot 6^2$$
$$k = \frac{200}{2 \cdot 6^2} = \frac{25}{9}$$

Thus $L = \dfrac{25}{9} \cdot 4 \cdot 4^2 = \dfrac{1600}{9} \approx 177.8$ pounds.

34. $L = k\dfrac{wd^2}{l}$

$$800 = k\frac{4 \cdot 8^2}{12}$$
$$\frac{12 \cdot 800}{4 \cdot 8^2} = k$$
$$37.5 = k$$

Thus $L = 37.5\dfrac{3.5 \cdot 6^2}{16} = 295.3125 \approx 295$ pounds.

Exercise Set 4.1, page 241

6.
$$
\begin{array}{r}
x^2 - 9 \\
2x^2 - x - 5 \overline{)2x^4 - x^3 - 23x^2 + 9x + 45} \\
\underline{2x^4 - x^3 - 5x^2} \\
-18x^2 + 9x + 45 \\
\underline{-18x^2 + 9x + 45} \\
0
\end{array}
$$

12.
$$
\begin{array}{r|rrrr}
5 & 5 & 6 & -8 & 1 \\
& & 25 & 155 & 735 \\
\hline
& 5 & 31 & 147 & 736
\end{array}
$$

$$\frac{5x^3 + 6x^2 - 8x + 1}{x - 5} = 5x^2 + 31x + 147 + \frac{736}{x - 5}$$

32.
$$
\begin{array}{r|rrrr}
3 & 2 & -1 & 3 & -1 \\
& & 6 & 15 & 54 \\
\hline
& 2 & 5 & 18 & 53
\end{array}
$$

$$P(c) = P(3) = 53$$

42.
$$
\begin{array}{r|rrrr}
-6 & 1 & 4 & -27 & -90 \\
& & -6 & 12 & 90 \\
\hline
& 1 & -2 & -15 & 0
\end{array}
$$

A remainder of 0 implies that $x + 6$ is a factor of $P(x)$.

64. Synthetic division by $n = 8$ and $n = 9$ shows that at least 9 buttons are required.

$$
\begin{array}{r|rrrrrr}
8 & 1 & -10 & 35 & -50 & 24 & 0 \\
& & 8 & -16 & 152 & 816 & 6720 \\
\hline
& 1 & -2 & 19 & 102 & 840 & 6720
\end{array}
$$

$$
\begin{array}{r|rrrrrr}
9 & 1 & -10 & 35 & -50 & 24 & 0 \\
& & 9 & -9 & 234 & 1656 & 15120 \\
\hline
& 1 & -1 & 26 & 184 & 1680 & 15120
\end{array}
$$

When $n = 8$, $P(n) < 15{,}000$. When $n = 9$, $P(n) \geq 15{,}000$. Therefore, 9 buttons are needed.

Exercise Set 4.2, page 249

2. Because $a_n = -2$ is negative and $n = 3$ is odd, the graph of P goes up to the far left and down to the far right.

26. The volume of the box is $V = lwh$, with $h = x$, $l = 18 - 2x$, and $w = \dfrac{42 - 3x}{2}$. Therefore, the volume is

$$
\begin{aligned}
V(x) &= (18 - 2x)\left(\frac{42 - 3x}{2}\right)x \\
&= 3x^3 - 69x^2 + 378x
\end{aligned}
$$

Use a graphing utility to graph $y = V(x)$. The graph is shown below. The value of x that produces the maximum volume is 3.571 inches (to the nearest 0.001 inch). The maximum volume is approximately 607 cubic inches.

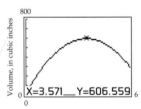

34.
$$
\begin{array}{r|rrrr}
0 & 4 & -1 & -6 & 1 \\
& & 0 & 0 & 0 \\
\hline
& 4 & -1 & -6 & 1 \quad \bullet\, P(0) = 1
\end{array}
$$

$$
\begin{array}{r|rrrr}
1 & 4 & -1 & -6 & 1 \\
& & 4 & 3 & -3 \\
\hline
& 4 & 3 & -3 & -2 \quad \bullet\, P(1) = -2
\end{array}
$$

Because $P(0)$ and $P(1)$ have opposite signs, P must have a real zero between 0 and 1.

Exercise Set 4.3, page 259

12. $p = \pm 1,\ \pm 2,\ \pm 3,\ \pm 5,\ \pm 6,\ \pm 10,\ \pm 15,\ \pm 30$

$q = \pm 1$

$\dfrac{p}{q} = \pm 1,\ \pm 2,\ \pm 3,\ \pm 5,\ \pm 6,\ \pm 10,\ \pm 15,\ \pm 30$

24.
$$
\begin{array}{r|rrrr}
3 & 1 & 0 & -19 & -28 \\
& & 3 & 9 & -30 \\
\hline
& 1 & 3 & -10 & -58
\end{array}
\qquad
\begin{array}{r|rrrr}
-2 & 1 & 0 & -19 & -28 \\
& & -2 & 4 & 30 \\
\hline
& 1 & -2 & -15 & 2
\end{array}
$$

$$
\begin{array}{r|rrrr}
4 & 1 & 0 & -19 & -28 \\
& & 4 & 16 & -12 \\
\hline
& 1 & 4 & -3 & -40
\end{array}
\qquad
\begin{array}{r|rrrr}
-3 & 1 & 0 & -19 & -28 \\
& & -3 & 9 & 30 \\
\hline
& 1 & -3 & -10 & 2
\end{array}
$$

$$
\begin{array}{r|rrrr}
5 & 1 & 0 & -19 & -28 \\
& & 5 & 25 & 30 \\
\hline
& 1 & 5 & 6 & 2
\end{array}
\qquad
\begin{array}{r|rrrr}
-4 & 1 & 0 & -19 & -28 \\
& & -4 & 16 & 12 \\
\hline
& 1 & -4 & -3 & -16
\end{array}
$$

$$
\begin{array}{r|rrrr}
-5 & 1 & 0 & -19 & -28 \\
& & -5 & 25 & -30 \\
\hline
& 1 & -5 & 6 & -58
\end{array}
$$

5 is an upper bound and -5 is a lower bound.

36. One positive real zero because the polynomial P has one variation in sign.

$$
P(-x) = (-x)^3 - 19(-x) - 30 = -x^3 + 19x - 30
$$

2 or no negative real zeros because $-x^3 + 19x - 30$ has two variations in sign.

48. One positive and two or no negative real zeros (see Exercise 36).

$$
\begin{array}{r|rrrr}
5 & 1 & 0 & -19 & -30 \\
& & 5 & 25 & 30 \\
\hline
& 1 & 5 & 6 & 0
\end{array}
$$

The reduced polynomial is

$$
x^2 + 5x + 6 = (x + 3)(x + 2)
$$

which has -3 and -2 as zeros. The zeros of $x^3 - 19x - 30$ are 5, -2, and -3.

Exercise Set 4.4, page 265

2.
$$
\begin{array}{r|rrrr}
5 + 3i & 3 & -29 & 92 & 34 \\
& & 15 + 9i & -97 + 3i & -34 \\
\hline
& 3 & -14 + 9i & -5 + 3i & 0
\end{array}
$$

$$
\begin{array}{r|rrr}
5 - 3i & 3 & -14 + 9i & -5 + 3i \\
& & 15 - 9i & 5 - 3i \\
\hline
& 3 & 1 & 0
\end{array}
$$

The reduced polynomial $3x + 1$ has $-1/3$ as a zero. The zeros of $3x^3 - 29x^2 + 92x + 34$ are $5 + 3i$, $5 - 3i$, and $-1/3$.

20. The graph of $P(x) = 4x^3 + 3x^2 + 16x + 12$ is shown at top of page S13. Applying Descartes' Rule of Signs, we find that the real zeros are all negative numbers. From the Upper- and Lower-Bound Theorem there is no real zero less than -1, and from the Rational Zero Theorem the possible rational zeros (that are negative and greater than -1) are $p/q = -1/2, -1/4, -3/4$. From the graph, it appears that $-3/4$ is a zero.

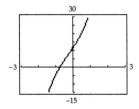

Using synthetic division, we have

$$
-\frac{3}{4} \begin{array}{|rrrr} 4 & 3 & 16 & 12 \\ & -3 & 0 & -12 \\ \hline 4 & 0 & 16 & 0 \end{array}
$$

Thus $-3/4$ is a zero, and by the Factor Theorem,

$$4x^3 + 3x^2 + 16x + 12 = \left(x + \frac{3}{4}\right)(4x^2 + 16) = 0$$

Solving $4x^2 + 16 = 0$, we have $x = -2i$ and $x = 2i$. The solutions of the original equation are $-3/4$, $-2i$, and $2i$.

28. $6x^3 - 23x^2 - 4x = x(6x^2 - 23x - 4)$
$$= x(6x + 1)(x - 4)$$

48. Because P has real coefficients, use the Conjugate Pair Theorem.

$P = (x - [3 + 2i])(x - [3 - 2i])(x - 7)$
$\quad = (x - 3 - 2i)(x - 3 + 2i)(x - 7)$
$\quad = (x^2 - 6x + 13)(x - 7)$
$\quad = x^3 - 13x^2 + 55x - 91$

Exercise Set 4.5, page 280

2. $x^2 - 4 = 0$
$(x - 2)(x + 2) = 0$
$x = 2$ or $x = -2$

The vertical asymptotes are $x = 2$ and $x = -2$.

6. The horizontal asymptote is $y = 0$ (x-axis) because the degree of the denominator is larger than the degree of the numerator.

10. Vertical asymptote: $x - 2 = 0$
$$x = 2$$

Horizontal asymptote: $y = 0$

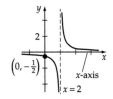

26. Vertical asymptote: $x^2 - 6x + 9 = 0$
$(x - 3)(x - 3) = 0$
$x = 3$

The horizontal asymptote is $y = 1/1 = 1$ (the Theorem on Horizontal Asymptotes) because numerator and denominator both have degree 2.

34.

$$
\begin{array}{r}
x + 1 \\
x^2 - 3x + 5 \overline{\big)\, x^3 - 2x^2 + 3x + 4} \\
\underline{x^3 - 3x^2 + 5x} \\
x^2 - 2x + 4 \\
\underline{x^2 - 3x + 5} \\
x - 1
\end{array}
$$

$$F(x) = x + 1 + \frac{x - 1}{x^2 - 3x + 5}$$

Slant asymptote: $y = x + 1$

40. Vertical asymptote: $2x + 5 = 0$
$$2x = -5$$
$$x = -\frac{5}{2}$$

The vertical asymptote is $x = -5/2$.

Slant asymptote:

$$
\begin{array}{r}
\frac{1}{2}x - \frac{13}{4} \\
2x + 5 \overline{\big)\, x^2 - 4x - 5} \\
\underline{x^2 + \frac{5}{2}x} \\
-\frac{13}{2}x - 5 \\
\underline{-\frac{13}{2}x - \frac{65}{4}} \\
\frac{45}{4}
\end{array}
$$

$$F(x) = \frac{1}{2}x - \frac{13}{4} + \frac{45/4}{2x + 5}$$

The slant asymptote is $y = \dfrac{1}{2}x - \dfrac{13}{4}$.

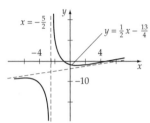

50. $F(x) = \dfrac{x^2 - x - 12}{x^2 - 2x - 8} = \dfrac{(x - 4)(x + 3)}{(x - 4)(x + 2)} = \dfrac{x + 3}{x + 2}, \; x \neq 4$

The function F is undefined at $x = 4$. Thus the graph of F is the graph of $y = \dfrac{x + 3}{x + 2}$ with an open circle at $(4, 7/6)$.

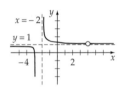

68. $A(x) = \dfrac{40{,}000 + 20x + 0.0001x^2}{x}$

a. $A(5000) = \dfrac{40{,}000 + 20(5000) + 0.0001(5000)^2}{500}$

$\qquad\quad = \$28.50$

b. $A(10{,}000) = \$25$

c. $y = 0.0001x + 20$

d. Use a graphing utility to graph $y = A(x)$.

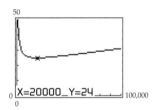

To minimize cost, 20,000 books should be published.

Exercise Set 5.1, page 296

36.    **40.**

44.   **48.**

50. Graph $f(x) = 3^{-x} - 4$. Then use the features of a graphing utility to locate the zero. The graph is shown below. To the nearest hundredth, the zero of f is -1.26.

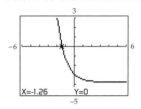

Exercise Set 5.2, page 304

2. $\log_{10} 1000 = 3$

$\qquad\quad 1000 = 10^3$

12. $\qquad 3^5 = 243$

$\qquad \log_3 243 = 5$

22. $\log_{10} \dfrac{1}{1000} = n$

$\qquad\quad 10^n = \dfrac{1}{1000}$

$\qquad\quad 10^n = 10^{-3}$

$\qquad\qquad n = -3$

32. $\log_b (x^2 y^3) = \log_b x^2 + \log_b y^3 = 2 \log_b x + 3 \log_b y$

42. $5 \log_3 x - 4 \log_3 y + 2 \log_3 z$

$\qquad = \log_3 x^5 - \log_3 y^4 + \log_3 z^2$

$\qquad = \log_3 \dfrac{x^5 z^2}{y^4}$

52. $\log_5 37 = \dfrac{\log 37}{\log 5} \approx 2.2436$

Exercise Set 5.3, page 312

12.   **14.**

20.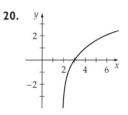

44. With $y = M$ and $x = r$, graph $M = 5 \log r - 5$ and then use the features of your graphing utility to determine the value of r when $M = 1.5$. The star *Achernar* is approximately 20.0 parsecs from Earth.

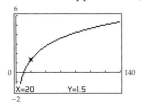

Exercise Set 5.4, page 320

2. $3^x = 243$

$3^x = 3^5$

$x = 5$

10. $\qquad 6^x = 50$

$\log (6^x) = \log 50$

$x \log 6 = \log 50$

$x = \dfrac{\log 50}{\log 6} \approx 2.18$

22. $\log_3 x + \log_3 (x + 6) = 3$

$\log_3 [x(x + 6)] = 3$

$3^3 = x(x + 6)$

$27 = x^2 + 6x$

$x^2 + 6x - 27 = 0$

$(x + 9)(x - 3) = 0$

$x = -9 \quad \text{or} \quad x = 3$

Because $\log_3 x$ is defined only for $x > 0$, the only solution is $x = 3$.

30. $\ln x = \dfrac{1}{2} \ln \left(2x + \dfrac{5}{2} \right) + \dfrac{1}{2} \ln 2$

$\qquad = \dfrac{1}{2} \left[\ln \left(2x + \dfrac{5}{2} \right) + \ln 2 \right]$

$\ln x = \dfrac{1}{2} \ln \left[2\left(2x + \dfrac{5}{2} \right) \right]$

$\ln x = \dfrac{1}{2} \ln (4x + 5)$

$\ln x = \ln (4x + 5)^{1/2}$

$x = \sqrt{4x + 5}$

$x^2 = 4x + 5$

$0 = x^2 - 4x - 5$

$0 = (x - 5)(x + 1)$

$x = 5 \quad \text{or} \quad x = -1$

Check: $\ln 5 = \dfrac{1}{2} \ln \left(10 + \dfrac{5}{2} \right) + \dfrac{1}{2} \ln 2$

$\qquad 1.6094 \approx 1.2629 + 0.3466$

Because $\ln (-1)$ is not defined, -1 is not a solution. Thus the only solution is $x = 5$.

34. $\qquad \dfrac{10^x + 10^{-x}}{2} = 8$

$10^x + 10^{-x} = 16$

$10^x(10^x + 10^{-x}) = (16)10^x$ • **Multiply each side by 10^x.**

$10^{2x} + 1 = 16(10^x)$

$10^{2x} - 16(10^x) + 1 = 0$

$u^2 - 16u + 1 = 0$ • **Let $u = 10^x$.**

$u = \dfrac{16 \pm \sqrt{16^2 - 4(1)(1)}}{2} = 8 \pm 3\sqrt{7}$

$10^x = 8 \pm 3\sqrt{7}$ • **Replace u with 10^x.**

$\log 10^x = \log (8 \pm 3\sqrt{7})$

$x = \log (8 \pm 3\sqrt{7}) \approx \pm 1.20241$

52. $\text{pH} = -\log (1.26 \times 10^{-12}) \approx 11.9$

Thus the ammonia is a basic solution.

54. $\qquad \text{pH} = 3.1$

$-\log [H_3O^+] = 3.1$

$H_3O^+ = 10^{-3.1} \approx 7.9 \times 10^{-4}$

68. a. $\qquad t = \dfrac{9}{24} \ln \dfrac{24 + v}{24 - v}$

$1.5 = \dfrac{9}{24} \ln \dfrac{24 + v}{24 - v}$

$4 = \ln \dfrac{24 + v}{24 - v}$

$e^4 = \dfrac{24 + v}{24 - v}$ • **$N = \ln M$ means $e^N = M$.**

$(24 - v)e^4 = 24 + v$

$-v - ve^4 = 24 - 24e^4$

$v(-1 - e^4) = 24 - 24e^4$

$v = \dfrac{24 - 24e^4}{-1 - e^4} \approx 23.14$

The velocity is 23.14 feet per second.

b. The vertical asymptote is $v = 24$.

c. Because of the air resistance, the object cannot exceed a velocity of 24 feet per second.

Exercise Set 5.5, page 333

4. a. $P = 12{,}500, r = 0.08, t = 10, n = 1.$

$B = 12{,}500\left(1 + \dfrac{0.08}{1} \right)^{10} \approx \$26{,}986.56$

b. $n = 365$

$$B = 12{,}500\left(1 + \frac{0.08}{365}\right)^{3650} \approx \$27{,}816.82$$

c. $n = 8760$

$$B = 12{,}500\left(1 + \frac{0.08}{8760}\right)^{87600} \approx \$27{,}819.16$$

6. $P = 32{,}000,\ r = 0.08,\ t = 3.$

$$B = Pe^{rt} = 32{,}000e^{3(0.08)} \approx \$40{,}679.97$$

10. $t = \dfrac{\ln 3}{r} \qquad r = 0.055$

$$t = \frac{\ln 3}{0.055}$$

$t \approx 20$ years (to the nearest year)

18. $P(x) = 20{,}899(1.027)^x$

a. $P(9) = 20{,}899(1.027)^9 \approx 26{,}562$

The population will be 26,562,000 people in the year 2000.

b. Because $P(x)$ is in thousands, 25,000,000 is 25,000 thousands. Replace $P(x)$ by 25,000.

$$25{,}000 = 20{,}899(1.027)^x$$

$$\frac{25{,}000}{20{,}899} = (1.027)^x$$

$$x = \frac{\log\,(25{,}000/20{,}899)}{\log\,(1.027)}$$

$$x \approx 6.7$$

The population will first exceed 25,000,000 in approximately 7 years, or in 1998.

20.
$$N(t) = N_0 e^{kt}$$
$$N(138) = N_0 e^{138k}$$
$$0.5N_0 = N_0 e^{138k}$$
$$0.5 = e^{138k}$$
$$\ln 0.5 = 138k$$
$$k = \frac{\ln 0.5}{138} \approx -0.005023$$
$$N(t) = N_0(0.5)^{t/138} \approx N_0 e^{-0.005023t}$$

24.
$$N(t) = N_0(0.5)^{t/5730}$$
$$0.65N_0 = N_0(0.5)^{t/5730}$$
$$\ln 0.65 = \ln\,(0.5)^{t/5730}$$
$$0.65 = (0.5)^{t/5700}$$
$$t = 5730\,\frac{\ln 0.65}{\ln 0.5} \approx 3561$$

The bone is approximately 3561 years old.

28. $N(3.4 \times 10^{-5}) = 10 \log\left(\dfrac{3.4 \times 10^{-5}}{10^{-16}}\right)$

$$= 10 \log\,(3.4 \times 10^{11})$$
$$= 10(\log 3.4 + \log 10^{11})$$
$$= 10 \log 3.4 + 110 \approx 115.3 \text{ decibels}$$

36. a.

b. Here is an algebraic solution. An approximate solution can be obtained from the graph.

$$v = 64(1 - e^{-t/2})$$
$$50 = 64(1 - e^{-t/2})$$
$$\frac{50}{64} = (1 - e^{-t/2})$$
$$1 - \frac{50}{64} = e^{-t/2}$$
$$\ln\left(1 - \frac{50}{64}\right) = -\frac{t}{2}$$
$$t = -2\ln\left(1 - \frac{50}{64}\right) \approx 3.0$$

c. As $t \to \infty$, $e^{-t/2} \to 0$. Therefore, $64(1 - e^{-t/2}) \to 64$. The horizontal asymptote is $v = 64$.

d. Because of the air resistance, the velocity of the object will never exceed 64 feet per second.

40. a.

$$P(t) = \frac{mP_0}{P_0 + (m - P_0)e^{-kt}}$$

$$900 = \frac{800(5500)}{800 + 4700e^{-k}} \qquad \bullet\ t = 1,\ P(1) = 900,\ m = 5500,\ P_0 = 800$$

$$800 + 4700e^{-k} = \frac{800(5500)}{900}$$

$$4700e^{-k} = \frac{800(5500)}{900} - 800$$

$$e^{-k} = \frac{\dfrac{800(5500)}{900} - 800}{4700}$$

$$k = -\ln\left(\frac{\dfrac{800(5500)}{900} - 800}{4700}\right)$$

$$k \approx 0.14$$

b.
$$P(t) = \frac{800(5500)}{800 + 4700e^{-0.14t}}$$

$$2000 = \frac{800(5500)}{800 + 4700e^{-0.14t}}$$

$$800 + 4700e^{-0.14t} = \frac{800(5500)}{2000}$$

$$e^{-0.14t} = \frac{1400}{4700}$$

$$-0.14t = \ln\frac{14}{47}$$

$$t \approx 8.6506448$$

The population will first exceed 2000 in the eighth year after 1990, or in 1998.

30. vertex $(2, -3)$, focus $(0, -3)$

$(h, k) = (2, -3)$, so $h = 2$ and $k = -3$.

Focus is $(h + p, k) = (2 + p, -3) = (0, -3)$.

Therefore, $2 + p = 0$ and $p = -2$.

$$(y - k)^2 = 4p(x - h)$$
$$(y + 3)^2 = 4(-2)(x - 2)$$
$$(y + 3)^2 = -8(x - 2)$$

36. The mirror surface is given by

$$4py = x^2$$

The point $(100, 3.75375)$ is on the surface, so

$$4p(3.75375) = 100^2$$

$$p = \frac{100^2}{4(3.75375)}$$

$$p \approx 666 \text{ inches}$$

Exercise Set 6.1, page 352

4. Comparing $x^2 = 4py$ with $x^2 = -\dfrac{1}{4}y$, we have

$$4p = -\frac{1}{4} \quad \text{or} \quad p = -\frac{1}{16}$$

vertex $(0, 0)$

focus $(0, -1/16)$

directrix $y = 1/16$

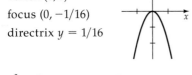

20. $x^2 + 5x - 4y - 1 = 0$

$$x^2 + 5x = 4y + 1$$

$$x^2 + 5x + \frac{25}{4} = 4y + 1 + \frac{25}{4} \quad \bullet \text{ Complete the square.}$$

$$\left(x + \frac{5}{2}\right)^2 = 4\left(y + \frac{29}{16}\right) \quad \bullet h = -\frac{5}{2}, k = -\frac{29}{16}$$

$$4p = 4 \qquad \bullet \text{ Compare to}$$
$$p = 1 \qquad (x - h)^2 = 4p(y - k)^2.$$

vertex $(-5/2, -29/16)$

focus $(h, k + p) = (-5/2, -13/16)$

directrix $y = k - p = -45/16$

28. vertex $(0, 0)$, focus $(5, 0)$, $p = 5$ because focus is $(p, 0)$

$$y^2 = 4px$$
$$y^2 = 4(5)x$$
$$y^2 = 20x$$

Exercise Set 6.2, page 362

20. $25x^2 + 12y^2 = 300$

$$\frac{x^2}{12} + \frac{y^2}{25} = 1 \qquad \bullet a^2 = 25, b^2 = 12, \quad c^2 = 25 - 12$$
$$a = 5, \quad b = 2\sqrt{3}, \quad c = \sqrt{13}$$

center $(0, 0)$

vertices $(0, 5)$ and $(0, -5)$

foci $(0, \sqrt{13})$ and $(0, -\sqrt{13})$

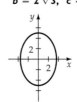

26.
$$9x^2 + 16y^2 + 36x - 16y - 104 = 0$$
$$9x^2 + 36x + 16y^2 - 16y - 104 = 0$$
$$9(x^2 + 4x) + 16(y^2 - y) = 104$$
$$9(x^2 + 4x + 4) + 16\left(y^2 - y + \frac{1}{4}\right) = 104 + 36 + 4$$
$$9(x + 2)^2 + 16\left(y - \frac{1}{2}\right)^2 = 144$$
$$\frac{(x + 2)^2}{16} + \frac{\left(y - \frac{1}{2}\right)^2}{9} = 1$$

center $(-2, 1/2)$

$a = 4, b = 3,$

$c = \sqrt{4^2 - 3^2} = \sqrt{7}$

vertices $(2, 1/2)$ and $(-6, 1/2)$,

foci $(-2 + \sqrt{7}, 1/2)$ and

$(-2 - \sqrt{7}, 1/2)$

42. center $(-4, 1) = (h, k)$. Therefore, $h = -4$ and $k = 1$.
Length of minor axis is 8, so $2b = 8$ or $b = 4$.
The equation of the ellipse is of the form

$$\frac{(x - h)^2}{a^2} + \frac{(y - k)^2}{b^2} = 1$$

$$\frac{(x + 4)^2}{a^2} + \frac{(y - 1)^2}{16} = 1 \quad \bullet\, h = -4, k = 1, b = 4$$

$$\frac{(0 + 4)^2}{a^2} + \frac{(4 - 1)^2}{16} = 1 \quad \bullet\, \textbf{The point } (0, 4) \textbf{ is on the}$$
$$\textbf{graph. Thus } x = 0 \textbf{ and } y = 4$$
$$\textbf{satisfy the equation.}$$

$$\frac{16}{a^2} + \frac{9}{16} = 1 \quad \bullet\, \textbf{Solve for } a^2.$$

$$\frac{16}{a^2} = \frac{7}{16}$$

$$a^2 = \frac{256}{7}$$

$$\frac{(x + 4)^2}{256/7} + \frac{(y - 1)^2}{16} = 1$$

48. Because the foci are $(0, -3)$ and $(0, 3)$, $c = 3$ and center is $(0, 0)$, the midpoint of the line segment between $(0, -3)$ and $(0, 3)$.

$$e = \frac{c}{a}$$

$$\frac{1}{4} = \frac{3}{a} \quad \bullet\, \textbf{e} = \frac{\textbf{1}}{\textbf{4}}$$

$$a = 12$$

$$3^2 = 12^2 - b^2 \quad \bullet\, \textbf{c}^2 = \textbf{a}^2 - \textbf{b}^2$$

$$b^2 = 144 - 9 = 135 \quad \bullet\, \textbf{Solve for } \textbf{b}^2.$$

The equation of the ellipse is $\dfrac{x^2}{135} + \dfrac{y^2}{144} = 1$.

54. The mean distance is $a = 67.08$ million miles.

Aphelion $= a + c = 67.58$ million miles

Thus $c = 67.58 - a = 0.50$ million miles.

$b = \sqrt{a^2 - c^2} = \sqrt{67.08^2 - 0.50^2} \approx 67.078$

An equation of the orbit of Venus is

$$\frac{x^2}{67.08^2} + \frac{y^2}{67.078^2} = 1$$

56. The length of the semimajor axis is 50 feet. Thus

$$c^2 = a^2 - b^2$$

$$32^2 = 50^2 - b^2$$

$$b^2 = 50^2 - 32^2$$

$$b = \sqrt{50^2 - 32^2}$$

$$b \approx 38.4 \text{ feet}$$

Exercise Set 6.3, page 374

4. $\dfrac{y^2}{25} - \dfrac{x^2}{36} = 1$

$a^2 = 25 \qquad b^2 = 36 \qquad c^2 = a^2 + b^2 = 25 + 36 = 61$

$a = 5 \qquad\quad b = 6 \qquad\quad c = \sqrt{61}$

Transverse axis is on y-axis because y^2 term is positive.

center $(0, 0)$

foci $(0, \sqrt{61})$ and $(0, -\sqrt{61})$

asymptotes $y = \dfrac{5}{6}x$ and $y = -\dfrac{5}{6}x$

vertices $(0, 5)$ and $(0, -5)$

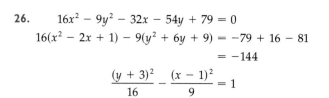

26. $\qquad 16x^2 - 9y^2 - 32x - 54y + 79 = 0$

$16(x^2 - 2x + 1) - 9(y^2 + 6y + 9) = -79 + 16 - 81$

$$= -144$$

$$\frac{(y + 3)^2}{16} - \frac{(x - 1)^2}{9} = 1$$

Transverse axis is parallel to y-axis because y^2 term is positive. Center is at $(1, -3)$; $a^2 = 16$ so $a = 4$.

vertices $(h, k + a) = (1, 1)$

$\qquad\qquad (h, k - a) = (1, -7)$

$c^2 = a^2 + b^2 = 16 + 9 = 25$

$\quad c = \sqrt{25} = 5$

foci $(h, k + c) = (1, 2)$

$\qquad (h, k - c) = (1, -8)$

Because $b^2 = 9$ and $b = 3$, the asymptotes are

$y + 3 = \dfrac{4}{3}(x - 1)$ and

$y + 3 = -\dfrac{4}{3}(x - 1)$.

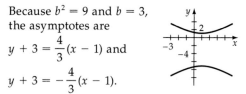

48. Because the vertices are $(2, 3)$ and $(-2, 3)$, $a = 2$ and center is $(0, 3)$.

$e = \dfrac{c}{a} \qquad\qquad c^2 = a^2 + b^2$

$\qquad\qquad\qquad\quad 5^2 = 2^2 + b^2$

$\dfrac{5}{2} = \dfrac{c}{2} \qquad\qquad b^2 = 25 - 4 = 21$

$c = 5$

Substituting into the standard equation yields

$$\frac{x^2}{4} - \frac{(y - 3)^2}{21} = 1.$$

54. a. Because the transmitters are 300 miles apart,
$2c = 300$ and $c = 150$.

$2a = \text{rate} \times \text{time}$

$2a = 0.186 \times 800 = 148.8$ miles

Thus $a = 74.4$ miles.

$b = \sqrt{c^2 - a^2}$

$\quad = \sqrt{150^2 - 74.4^2} \approx 130.25$ miles

The ship is located on the hyperbola given by

$$\frac{x^2}{74.4^2} - \frac{y^2}{130.25^2} = 1$$

b. The ship will reach the coastline when $y = 0$. Thus

$$\frac{x^2}{74.4^2} - \frac{0^2}{130.25^2} = 1$$

$$\frac{x^2}{74.4^2} = 1$$

$$x^2 = 74.4^2$$

$$x = 74.4 \text{ miles}$$

The ship reaches the coastline 74.4 miles to the right of transmitter T_2.

Exercise Set 7.1, page 389

6. $\begin{cases} 8x + 3y = -7 & (1) \\ x = 3y + 15 & (2) \end{cases}$

$8(3y + 15) + 3y = -7$ • **Replace x in Eq. (1).**

$24y + 120 + 3y = -7$ • **Simplify.**

$27y = -127$

$y = -\dfrac{127}{27}$

$x = 3\left(-\dfrac{127}{27}\right) + 15 = \dfrac{8}{9}$ • **Substitute** $-\dfrac{127}{27}$ **for y in Eq. (2).**

The solution is $(8/9, -127/27)$.

18. $\begin{cases} 3x - 4y = 8 & (1) \\ 6x - 8y = 9 & (2) \end{cases}$

$8y = 6x - 9$ • **Solve Eq. (2) for y.**

$y = \dfrac{3}{4}x - \dfrac{9}{8}$

$3x - 4\left(\dfrac{3}{4}x - \dfrac{9}{8}\right) = 8$ • **Replace y in Eq. (1).**

$3x - 3x + \dfrac{9}{2} = 8$ • **Simplify.**

$\dfrac{9}{2} = 8$

This is a false equation. Therefore, the system of equations is inconsistent and has no solution.

20. $\begin{cases} 5x + 2y = 2 & (1) \\ y = -\dfrac{5}{2}x + 1 & (2) \end{cases}$

$5x + 2\left(-\dfrac{5}{2}x + 1\right) = 2$ • **Replace y in Eq. (1).**

$5x - 5x + 2 = 2$ • **Simplify.**

$2 = 2$

This is a true statement, therefore the system of equations is dependent. Let $x = c$. Then $y = -\dfrac{5}{2}c + 1$.

Thus the solutions are $\left(c, -\dfrac{5}{2}c + 1\right)$.

24. $\begin{cases} 3x - 8y = -6 & (1) \\ -5x + 4y = 10 & (2) \end{cases}$

$\begin{array}{r} 3x - 8y = -6 \\ -10x + 8y = 20 \\ \hline -7x \quad\quad = 14 \end{array}$ • **2 times Eq. (2)**

$x = -2$

$3(-2) - 8y = -6$ • **Substitute −2 for x in Eq. (1). Solve for y.**

$-8y = 0$

$y = 0$

The solution is $(-2, 0)$.

28. $\begin{cases} 4x + 5y = 2 & (1) \\ 8x - 15y = 9 & (2) \end{cases}$

$\begin{array}{r} 12x + 15y = 6 \\ 8x - 15y = 9 \\ \hline 20x \quad\quad = 15 \end{array}$ • **3 times Eq. (1)**

$x = \dfrac{3}{4}$

$4\left(\dfrac{3}{4}\right) + 5y = 2$ • **Substitute** $\dfrac{3}{4}$ **for x in Eq. (1).**

$3 + 5y = 2$ • **Solve for y.**

$y = -\dfrac{1}{5}$

The solution is $(3/4, -1/5)$.

44. Let r = the rate of the canoeist.
Let w = the rate of the current.
Rate of canoeist with the current: $r + w$
Rate of canoeist against the current: $r - w$

$$r \cdot t = d$$
$$(r + w) \cdot 2 = 12 \quad (1)$$
$$(r - w) \cdot 4 = 12 \quad (2)$$

$r + w = 6$	• **Divide Eq. (1) by 2.**
$\underline{r - w = 3}$	• **Divide Eq. (2) by 4.**
$2r \quad\quad = 9$	
$r = 4.5$	

$$4.5 + w = 6$$
$$w = 1.5$$

Rate of canoeist = 4.5 mph
Rate of current = 1.5 mph

Exercise Set 7.2, page 401

12. $\begin{cases} 3x + 2y - 5z = 6 \quad (1) \\ 5x - 4y + 3z = -12 \quad (2) \\ 4x + 5y - 2z = 15 \quad (3) \end{cases}$

$15x + 10y - 25z = 30$	• **5 times Eq. (1)**
$\underline{-15x + 12y - 9z = 36}$	• **−3 times Eq. (2)**
$22y - 34z = 66$	• **Divide by 2.**
$11y - 17z = 33 \quad (4)$	

$12x + 8y - 20z = 24$	• **4 times Eq. (1)**
$\underline{-12x - 15y + 6z = -45}$	• **−3 times Eq. (3)**
$-7y - 14z = -21$	• **Divide by −7.**
$y + 2z = 3 \quad (5)$	

$11y - 17z = 33 \quad (4)$	
$\underline{-11y - 22z = -33}$	• **−11 times Eq. (5)**
$-39z = 0$	
$z = 0 \quad (6)$	

$$11y - 17(0) = 33$$
$$y = 3$$
$$3x + 2(3) - 5(0) = 6$$
$$x = 0$$

The solution is $(0, 3, 0)$.

16. $\begin{cases} 2x + 3y + 2z = 14 \quad (1) \\ x - 3y + 4z = 4 \quad (2) \\ -x + 12y - 6z = 2 \quad (3) \end{cases}$

$2x + 3y + 2z = 14 \quad (1)$	
$\underline{-2x + 6y - 8z = -8}$	• **−2 times Eq. (2)**
$9y - 6z = 6$	• **Divide by 3.**
$3y - 2z = 2 \quad (4)$	

$2x + 3y + 2z = 14 \quad (1)$	
$\underline{-2x + 24y - 12z = 4}$	• **2 times Eq. (3)**
$27y - 10z = 18 \quad (5)$	

$-27y + 18z = -18$	• **−9 times Eq. (4)**
$\underline{27y - 10z = 18 \quad (5)}$	
$8z = 0$	
$z = 0 \quad (6)$	

$3y - 2(0) = 2$	• **Substitute $z = 0$ in**
$y = \dfrac{2}{3}$	**Eq. (4).**

$2x + 3\left(\dfrac{2}{3}\right) + 2(0) = 14$	• **Substitute $y = 2/3$ and**
$x = 6$	$z = 0$ **in Eq. (1).**

The solution is $(6, 2/3, 0)$.

18. $\begin{cases} 2x + 3y - 6z = 4 \quad (1) \\ 3x - 2y - 9z = -7 \quad (2) \\ 2x + 5y - 6z = 8 \quad (3) \end{cases}$

$6x + 9y - 18z = 12$	• **3 times Eq. (1)**
$\underline{-6x + 4y + 18z = 14}$	• **−2 times Eq. (2)**
$13y = 26$	
$y = 2 \quad (4)$	

$2x + 3y - 6z = 4 \quad (1)$	
$\underline{-2x - 5y + 6z = -8}$	• **−1 times Eq. (3)**
$-2y = -4$	
$y = 2 \quad (5)$	

$y = 2 \quad (4)$	
$\underline{-y = -2}$	• **−1 times Eq. (5)**
$0 = 0 \quad (6)$	

The equations are dependent. Let $z = c$.

$2x + 3(2) - 6c = 4$	• **Substitute $y = 2$ and**
$x = 3c - 1$	$z = c$ **in Eq. (1).**

The solutions are $(3c - 1, 2, c)$.

20. $\begin{cases} x - 3y + 4z = 9 \quad (1) \\ 3x - 8y - 2z = 4 \quad (2) \end{cases}$

$-3x + 9y - 12z = -27$	• **−3 times Eq. (1)**
$\underline{3x - 8y - 2z = 4} \quad (2)$	
$y - 14z = -23 \quad (3)$	
$y = 14z - 23$	• **Solve Eq. (3) for y.**

$$x - 3(14z - 23) + 4z = 9$$ • Substitute $14z - 23$ for y in Eq. (1).

$$x = 38z - 60$$ • Solve for x.

Let $z = c$. The solutions are $(38c - 60, 14c - 23, c)$.

32. $\begin{cases} 5x + 2y + 3z = 0 & (1) \\ 3x + y - 2z = 0 & (2) \\ 4x - 7y + 5z = 0 & (3) \end{cases}$

$\begin{array}{lr} 15x + 6y + 9z = 0 & \text{• 3 times Eq. (1)} \\ -15x - 5y + 10z = 0 & \text{• } -5 \text{ times Eq. (2)} \\ \hline y + 19z = 0 \quad (4) & \end{array}$

$\begin{array}{lr} 20x + 8y + 12z = 0 & \text{• 4 times Eq. (1)} \\ -20x + 35y - 25z = 0 & \text{• } -5 \text{ times Eq. (3)} \\ \hline 43y - 13z = 0 \quad (5) & \end{array}$

$\begin{array}{lr} -43y - 817z = 0 & \text{• } -43 \text{ times Eq. (4)} \\ 43y - 13z = 0 \quad (5) & \\ \hline -830z = 0 & \\ z = 0 \quad (6) & \end{array}$

Solving by back substitution, the only solution is $(0, 0, 0)$.

36. $x^2 + y^2 + ax + by + c = 0$

$\begin{cases} 0 + 36 + a(0) + b(6) + c = 0 & \text{• Let } x = 0, y = 6. \\ 1 + 25 + a(1) + b(5) + c = 0 & \text{• Let } x = 1, y = 5. \\ 49 + 1 + a(-7) + b(-1) + c = 0 & \text{• Let } x = -7, y = -1. \end{cases}$

$\begin{cases} 6b + c = -36 & (1) \\ a + 5b + c = -26 & (2) \\ -7a - b + c = -50 & (3) \end{cases}$

$\begin{array}{lr} 7a + 35b + 7c = -182 & \text{• 7 times Eq. (2)} \\ -7a - b + c = -50 \quad (3) & \\ \hline 34b + 8c = -232 & \\ 17b + 4c = -116 \quad (4) & \end{array}$

$\begin{array}{lr} -24b - 4c = 144 & \text{• } -4 \text{ times Eq. (1)} \\ 17b + 4c = -116 \quad (4) & \\ \hline -7b = 28 & \\ b = -4 & \end{array}$

$\begin{array}{lr} 17(-4) + 4c = -116 & \text{• Substitute } -4 \text{ for} \\ c = -12 & \quad b \text{ in Eq. (4).} \end{array}$

$\begin{array}{lr} -7a - (-4) - 12 = -50 & \text{• Substitute } -4 \text{ for} \\ a = 6 & \quad b \text{ and } -12 \text{ for } c \text{ in} \\ & \quad \text{Eq. (3).} \end{array}$

An equation of a circle whose graph passes through the three given points is $x^2 + y^2 + 6x - 4y - 12 = 0$.

Exercise Set 7.3, page 407

8. $\begin{cases} x - 2y = 3 & (1) \\ xy = -1 & (2) \end{cases}$

$$x = 2y + 3$$ • Solve Eq. (1) for x.

$$(2y + 3)y = -1$$ • Replace x by $2y + 3$ in Eq. (2).

$$2y^2 + 3y + 1 = 0$$ • Solve for y.

$$(2y + 1)(y + 1) = 0$$

$$y = -\frac{1}{2} \quad \text{or} \quad y = -1$$

$\begin{array}{ll} x - 2\left(-\dfrac{1}{2}\right) = 3 & x - 2(-1) = 3 \\ x = 2 & x = 1 \end{array}$ • Substitute for y in Eq. (1).

The solutions are $(2, -1/2)$ and $(1, -1)$.

16. $\begin{cases} 3x^2 - 2y^2 = 19 & (1) \\ x^2 - y^2 = 5 & (2) \end{cases}$

$\begin{array}{lr} 3x^2 - 2y^2 = 19 \quad (1) & \\ -3x^2 + 3y^2 = -15 & \text{• Multiply Eq. (2) by } -3. \\ \hline y^2 = 4 & \text{• Add the equations.} \\ y = \pm 2 & \text{• Solve for } y. \end{array}$

$\begin{array}{lr} x^2 - (-2)^2 = 5 & \text{• Substitute } -2 \text{ for } y \text{ in Eq. (2).} \\ x^2 - 4 = 5 & \\ x^2 = 9 & \\ x = \pm 3 & \end{array}$

$\begin{array}{lr} x^2 - 2^2 = 5 & \text{• Substitute 2 for } y \text{ in Eq. (2).} \\ x^2 - 4 = 5 & \\ x^2 = 9 & \\ x = \pm 3 & \end{array}$

The solutions are $(-2, -3)$, $(-2, 3)$, $(2, -3)$, and $(2, 3)$.

20. $\begin{cases} 2x^2 + 3y^2 = 11 & (1) \\ 3x^2 + 2y^2 = 19 & (2) \end{cases}$

Use the elimination method to eliminate y^2.

$\begin{array}{lr} 4x^2 + 6y^2 = 22 & \text{• 2 times Eq. (1)} \\ -9x^2 - 6y^2 = -57 & \text{• } -3 \text{ times Eq. (2)} \\ \hline -5x^2 = -35 & \\ x^2 = 7 & \end{array}$

$\begin{array}{lr} 2(7) + 3y^2 = 11 & \text{• Substitute for } x \text{ in Eq. (1).} \\ 3y^2 = -3 & \\ y^2 = -1 & \end{array}$

$y^2 = -1$ has no real number solutions. The graphs of the equations do not intersect. The system is inconsistent and has no solution.

28. $\begin{cases} (x + 2)^2 + (y - 3)^2 = 10 \\ (x - 3)^2 + (y + 1)^2 = 13 \end{cases}$

$$x^2 + 4x + 4 + y^2 - 6y + 9 = 10 \quad (1)$$
$$\underline{x^2 - 6x + 9 + y^2 + 2y + 1 = 13} \quad (2)$$
$$10x - 5 \qquad -8y + 8 = -3 \qquad \text{• Subtract.}$$

$$10x - 8y = -6$$

$$y = \frac{5x + 3}{4} \quad (3) \quad \begin{array}{l}\text{• Solve} \\ \text{for } y.\end{array}$$

$$(x + 2)^2 + \left(\frac{5x - 9}{4}\right)^2 = 10 \quad \begin{array}{l}\text{• Substitute} \\ \text{for } y.\end{array}$$

$$x^2 + 4x + 4 + \frac{25x^2 - 90x + 81}{16} = 10 \quad \begin{array}{l}\text{• Solve} \\ \text{for } x.\end{array}$$

$$16x^2 + 64x + 64 + 25x^2 - 90x + 81 = 160$$

$$41x^2 - 26x - 15 = 0$$

$$(41x + 15)(x - 1) = 0$$

$$x = -\frac{15}{41} \quad \text{or} \quad x = 1$$

$$y = \frac{5}{4}\left(-\frac{15}{41}\right) + \frac{3}{4} \quad \text{or} \quad y = \frac{5(1) + 3}{4} \qquad \begin{array}{l}\text{• Substitute} \\ \text{for } x \text{ into} \\ \text{Eq. (3).}\end{array}$$

$$y = \frac{12}{41} \qquad\qquad y = 2$$

The solutions are $(-15/41, 12/41)$ and $(1, 2)$.

Exercise Set 7.4, page 415

14. $\dfrac{7x + 44}{x^2 + 10x + 24} = \dfrac{7x + 44}{(x + 4)(x + 6)} = \dfrac{A}{x + 4} + \dfrac{B}{x + 6}$

$$7x + 44 = A(x + 6) + B(x + 4)$$

$$7x + 44 = (A + B)x + (6A + 4B)$$

$$\begin{cases} 7 = A + B \\ 44 = 6A + 4B \end{cases}$$

The solution is $A = 8, B = -1$.

$$\frac{7x + 44}{x^2 + 10x + 24} = \frac{8}{x + 4} + \frac{-1}{x + 6}$$

22. $\dfrac{x - 18}{x(x - 3)^2} = \dfrac{A}{x} + \dfrac{B}{x - 3} + \dfrac{C}{(x - 3)^2}$

$$x - 18 = A(x - 3)^2 + Bx(x - 3) + Cx$$

$$x - 18 = Ax^2 - 6Ax + 9A + Bx^2 - 3Bx + Cx$$

$$x - 18 = (A + B)x^2 + (-6A - 3B + C)x + 9A$$

$$\begin{cases} 0 = A + B \\ 1 = -6A - 3B + C \\ -18 = 9A \end{cases}$$

The solution is $A = -2, B = 2, C = -5$.

$$\frac{x - 18}{x(x - 3)^2} = \frac{-2}{x} + \frac{2}{x - 3} + \frac{-5}{(x - 3)^2}$$

24. $x^3 - x^2 + 10x - 10 = (x - 1)(x^2 + 10)$

$$\frac{9x^2 - 3x + 49}{(x - 1)(x^2 + 10)} = \frac{A}{x - 1} + \frac{Bx + C}{x^2 + 10}$$

$$9x^2 - 3x + 49 = A(x^2 + 10) + (Bx + C)(x - 1)$$

$$9x^2 - 3x + 49 = (A + B)x^2 + (-B + C)x + (10A - C)$$

$$\begin{cases} 9 = A + B \\ -3 = -B + C \\ 49 = 10A - C \end{cases}$$

The solution is $A = 5, B = 4, C = 1$.

$$\frac{9x^2 - 3x + 49}{x^3 - x^2 + 10x - 10} = \frac{5}{x - 1} + \frac{4x + 1}{x^2 + 10}$$

30. $\dfrac{2x^3 + 9x + 1}{(x^2 + 7)^2} = \dfrac{Ax + B}{x^2 + 7} + \dfrac{Cx + D}{(x^2 + 7)^2}$

$$2x^3 + 9x + 1 = (Ax + B)(x^2 + 7) + Cx + D$$

$$2x^3 + 9x + 1 = Ax^3 + Bx^2 + (7A + C)x + (7B + D)$$

$$\begin{cases} 2 = A \\ 0 = B \\ 9 = 7A + C \\ 1 = 7B + D \end{cases}$$

The solutions are $A = 2, B = 0, C = -5, D = 1$.

$$\frac{2x^3 + 9x + 1}{x^4 + 14x^2 + 49} = \frac{2x}{x^2 + 7} + \frac{-5x + 1}{(x^2 + 7)^2}$$

34. $\begin{array}{r} x + 1 \\ 2x^2 + 3x - 2 \overline{\smash{\big)}\ 2x^3 + 5x^2 + 3x - 8} \\ \underline{2x^3 + 3x^2 - 2x} \\ 2x^2 + 5x - 8 \\ \underline{2x^2 + 3x - 2} \\ 2x - 6 \end{array}$

$$\frac{2x^3 + 5x^2 + 3x - 8}{2x^2 + 3x - 2} = x + 1 + \frac{2x - 6}{2x^2 + 3x - 2}$$

$$\frac{2x - 6}{(2x - 1)(x + 2)} = \frac{A}{2x - 1} + \frac{B}{x + 2}$$

$$2x - 6 = A(x + 2) + B(2x - 1)$$

$$2x - 6 = Ax + 2A + 2Bx - B$$

$$2x - 6 = (A + 2B)x + (2A - B)$$

$$\begin{cases} 2 = A + 2B \\ -6 = 2A - B \end{cases}$$

The solutions are $A = -2, B = 2$.

$$\frac{2x^3 + 5x^2 + 3x - 8}{2x^2 + 3x - 2} = x + 1 + \frac{-2}{2x - 1} + \frac{2}{x + 2}$$

Exercise Set 7.5, page 421

6.

12.

20.

26.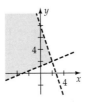

36.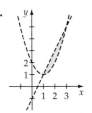

40.

42.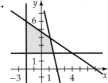

(x, y)	Cost	
$(0, 20)$	500	
$(12, 4)$	436	• **Minimum**
$(20, 0)$	560	

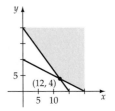

To achieve the minimum cost, use machine 1 for 12 hours and machine 2 for 4 hours.

24. Let $x =$ number of standard models.
Let $y =$ number of deluxe models.
Profit $= 25x + 35y$

Constraints: $\begin{cases} x + 3y \leq 24 \\ x + y \leq 10 \\ 2x + y \leq 16 \\ x \geq 0, y \geq 0 \end{cases}$

(x, y)	Profit	
$(0, 0)$	0	
$(0, 8)$	280	
$(6, 4)$	290	
$(3, 7)$	320	• **Maximum**
$(8, 0)$	200	

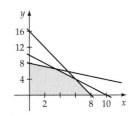

To maximize profits, produce 3 standard models and 7 deluxe models.

Exercise Set 7.6, page 428

12. $C = 4x + 3y$

(x, y)	C	
$(0, 8)$	24	
$(2, 4)$	20	• **Minimum**
$(5, 2)$	26	
$(11, 0)$	44	
$(20, 0)$	80	
$(20, 20)$	140	
$(0, 20)$	60	

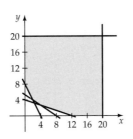

22. $x =$ hours of machine 1 use
$y =$ hours of machine 2 use
Cost $= 28x + 25y$

Constraints: $\begin{cases} 4x + 3y \geq 60 \\ 5x + 10y \geq 100 \\ x \geq 0, y \geq 0 \end{cases}$

Exercise Set 8.1, page 443

14. $\begin{bmatrix} 1 & -3 & 1 & 8 \\ 2 & -5 & -3 & 2 \\ 1 & 4 & 1 & 1 \end{bmatrix} \xrightarrow[\substack{-2R_1 + R_2 \\ -1R_1 + R_3}]{} \begin{bmatrix} 1 & -3 & 1 & 8 \\ 0 & 1 & -5 & -14 \\ 0 & 7 & 0 & -7 \end{bmatrix}$

$\xrightarrow{-7R_2 + R_3} \begin{bmatrix} 1 & -3 & 1 & 8 \\ 0 & 1 & -5 & -14 \\ 0 & 0 & 35 & 91 \end{bmatrix}$

$\xrightarrow{\frac{1}{35}R_3} \begin{bmatrix} 1 & -3 & 1 & 8 \\ 0 & 1 & -5 & -14 \\ 0 & 0 & 1 & \frac{13}{5} \end{bmatrix}$

$\begin{cases} x - 3y + z = 8 \\ y - 5z = -14 \\ z = \dfrac{13}{5} \end{cases}$

By back substitution, the solution is $(12/5, -1, 13/5)$.

18. $\begin{bmatrix} 3 & -5 & 2 & 4 \\ 1 & -3 & 2 & 4 \\ 5 & -11 & 6 & 12 \end{bmatrix} \xrightarrow{R_2 \longleftrightarrow R_1} \begin{bmatrix} 1 & -3 & 2 & 4 \\ 3 & -5 & 2 & 4 \\ 5 & -11 & 6 & 12 \end{bmatrix}$

$\xrightarrow[-5R_1 + R_3]{-3R_1 + R_2} \begin{bmatrix} 1 & -3 & 2 & 4 \\ 0 & 4 & -4 & -8 \\ 0 & 4 & -4 & -8 \end{bmatrix} \xrightarrow{\frac{1}{4}R_2} \begin{bmatrix} 1 & -3 & 2 & 4 \\ 0 & 1 & -1 & -2 \\ 0 & 4 & -4 & -8 \end{bmatrix}$

$\xrightarrow{-4R_2 + R_3} \begin{bmatrix} 1 & -3 & 2 & 4 \\ 0 & 1 & -1 & -2 \\ 0 & 0 & 0 & 0 \end{bmatrix}$

$\begin{cases} x - 3y + 2z = 4 & (1) \\ \quad\quad y - z = -2 & (2) \\ \quad\quad\quad 0 = 0 & (3) \end{cases}$

$y - z = -2 \quad$ or $\quad y = z - 2$

$x - 3(z - 2) + 2z = 4 \qquad$ • Substitute $z - 2$
for y in Eq. (1).

$x - 3z + 6 + 2z = 4$

$\qquad\qquad x = z - 2$

Let $z = c$. The solutions are $(c - 2, c - 2, c)$.

20. $\begin{bmatrix} 2 & 5 & 2 & -1 \\ 1 & 2 & -3 & 5 \\ 5 & 12 & 1 & 10 \end{bmatrix} \xrightarrow{R_2 \longleftrightarrow R_1} \begin{bmatrix} 1 & 2 & -3 & 5 \\ 2 & 5 & 2 & -1 \\ 5 & 12 & 1 & 10 \end{bmatrix}$

$\xrightarrow[-5R_1 + R_3]{-2R_1 + R_2} \begin{bmatrix} 1 & 2 & -3 & 5 \\ 0 & 1 & 8 & -11 \\ 0 & 2 & 16 & -15 \end{bmatrix}$

$\xrightarrow{-2R_2 + R_3} \begin{bmatrix} 1 & 2 & -3 & 5 \\ 0 & 1 & 8 & -11 \\ 0 & 0 & 0 & 7 \end{bmatrix}$

$\begin{cases} x + 2y - 3z = 5 \\ \quad\quad y + 8z = -11 \\ \quad\quad\quad 0 = 7 \end{cases}$

Because $0 = 7$ is a false equation, the system of equations has no solution.

36. $\begin{bmatrix} 1 & -1 & 3 & -5 & 10 \\ 2 & -3 & 4 & 1 & 7 \\ 3 & 1 & -2 & -2 & 6 \end{bmatrix}$

$\xrightarrow[-3R_1 + R_3]{-2R_1 + R_2} \begin{bmatrix} 1 & -1 & 3 & -5 & 10 \\ 0 & -1 & -2 & 11 & -13 \\ 0 & 4 & -11 & 13 & -24 \end{bmatrix}$

$\xrightarrow{-1R_2} \begin{bmatrix} 1 & -1 & 3 & -5 & 10 \\ 0 & 1 & 2 & -11 & 13 \\ 0 & 4 & -11 & 13 & -24 \end{bmatrix}$

$\xrightarrow{-4R_2 + R_3} \begin{bmatrix} 1 & -1 & 3 & -5 & 10 \\ 0 & 1 & 2 & -11 & 13 \\ 0 & 0 & -19 & 57 & -76 \end{bmatrix}$

$\xrightarrow{-\frac{1}{19}R_3} \begin{bmatrix} 1 & -1 & 3 & -5 & 10 \\ 0 & 1 & 2 & -11 & 13 \\ 0 & 0 & 1 & -3 & 4 \end{bmatrix}$

$\begin{cases} t - u + 3v - 5w = 10 & (1) \\ \quad\quad u + 2v - 11w = 13 & (2) \\ \quad\quad\quad v - 3w = 4 & (3) \end{cases}$

$v = 3w + 4$

$u + 2(3w + 4) - 11w = 13 \qquad$ • Substitute $3w + 4$
for v in Eq. (2).

$\qquad\qquad u = 5w + 5$

$t - (5w + 5) + 3(3w + 4) - 5w = 10 \qquad$ • Substitute $5w + 5$
for u and $3w + 4$
for v in Eq. (1).

$\qquad\qquad t = w + 3$

Let w be any real number c. The solution of the system of equations is $(c + 3, 5c + 5, 3c + 4, c)$.

Exercise Set 8.2, page 456

6. a. $A + B = \begin{bmatrix} 2 & -2 \\ 3 & 4 \\ 1 & 0 \end{bmatrix} + \begin{bmatrix} -1 & 8 \\ 2 & -2 \\ -4 & 3 \end{bmatrix} = \begin{bmatrix} 1 & 6 \\ 5 & 2 \\ -3 & 3 \end{bmatrix}$

b. $A - B = \begin{bmatrix} 2 & -2 \\ 3 & 4 \\ 1 & 0 \end{bmatrix} - \begin{bmatrix} -1 & 8 \\ 2 & -2 \\ -4 & 3 \end{bmatrix} = \begin{bmatrix} 3 & -10 \\ 1 & 6 \\ 5 & -3 \end{bmatrix}$

c. $2B = 2\begin{bmatrix} -1 & 8 \\ 2 & -2 \\ -4 & 3 \end{bmatrix} = \begin{bmatrix} -2 & 16 \\ 4 & -4 \\ -8 & 6 \end{bmatrix}$

d. $2A - 3B = 2\begin{bmatrix} 2 & -2 \\ 3 & 4 \\ 1 & 0 \end{bmatrix} - 3\begin{bmatrix} -1 & 8 \\ 2 & -2 \\ -4 & 3 \end{bmatrix} = \begin{bmatrix} 7 & -28 \\ 0 & 14 \\ 14 & -9 \end{bmatrix}$

16. $AB = \begin{bmatrix} -1 & 2 & 0 \\ 2 & -1 & 1 \\ -2 & 2 & -1 \end{bmatrix}\begin{bmatrix} 2 & -1 & 0 \\ 1 & 5 & -1 \\ 0 & -1 & 3 \end{bmatrix}$

$= \begin{bmatrix} (-1)(2) + (2)(1) + (0)(0) \\ (2)(2) + (-1)(1) + (1)(0) \\ (-2)(2) + (2)(1) + (-1)(0) \end{bmatrix}$

$$(-1)(-1) + (2)(5) \quad + (0)(-1)$$
$$(2)(-1) \quad + (-1)(5) + (1)(-1)$$
$$(-2)(-1) + (2)(5) \quad + (-1)(-1)$$

$$\begin{bmatrix} (-1)(0) + (2)(-1) \quad + (0)(3) \\ (2)(0) \quad + (-1)(-1) + (1)(3) \\ (-2)(0) + (2)(-1) \quad + (-1)(3) \end{bmatrix}$$

$$= \begin{bmatrix} 0 & 11 & -2 \\ 3 & -8 & 4 \\ -2 & 13 & -5 \end{bmatrix}$$

$$BA = \begin{bmatrix} 2 & -1 & 0 \\ 1 & 5 & -1 \\ 0 & -1 & 3 \end{bmatrix} \begin{bmatrix} -1 & 2 & 0 \\ 2 & -1 & 1 \\ -2 & 2 & -1 \end{bmatrix}$$

$$= \begin{bmatrix} (2)(-1) + (-1)(2) + (0)(-2) \\ (1)(-1) + (5)(2) \quad + (-1)(-2) \\ (0)(-1) + (-1)(2) + (3)(-2) \end{bmatrix}$$

$$\begin{matrix} (2)(2) + (-1)(-1) + (0)(2) \\ (1)(2) + (5)(-1) \quad + (-1)(2) \\ (0)(2) + (-1)(-1) + (3)(2) \end{matrix}$$

$$\begin{bmatrix} (2)(0) + (-1)(1) + (0)(-1) \\ (1)(0) + (5)(1) \quad + (-1)(-1) \\ (0)(0) + (-1)(1) + (3)(-1) \end{bmatrix}$$

$$= \begin{bmatrix} -4 & 5 & -1 \\ 11 & -5 & 6 \\ -8 & 7 & -4 \end{bmatrix}$$

Exercise Set 8.3, page 468

6. $\begin{bmatrix} 1 & 3 & -2 & 1 & 0 & 0 \\ -1 & -5 & 6 & 0 & 1 & 0 \\ 2 & 6 & -3 & 0 & 0 & 1 \end{bmatrix}$

$\xrightarrow[\substack{R_1 + R_2 \\ -2R_1 + R_3}]{} \begin{bmatrix} 1 & 3 & -2 & 1 & 0 & 0 \\ 0 & -2 & 4 & 1 & 1 & 0 \\ 0 & 0 & 1 & -2 & 0 & 1 \end{bmatrix}$

$\xrightarrow[\substack{-\frac{1}{2}R_2}]{} \begin{bmatrix} 1 & 3 & -2 & 1 & 0 & 0 \\ 0 & 1 & -2 & -\frac{1}{2} & -\frac{1}{2} & 0 \\ 0 & 0 & 1 & -2 & 0 & 1 \end{bmatrix}$

$\xrightarrow[\substack{2R_3 + R_2 \\ 2R_3 + R_1}]{} \begin{bmatrix} 1 & 3 & 0 & -3 & 0 & 2 \\ 0 & 1 & 0 & -\frac{9}{2} & -\frac{1}{2} & 2 \\ 0 & 0 & 1 & -2 & 0 & 1 \end{bmatrix}$

$\xrightarrow[\substack{-3R_2 + R_1}]{} \begin{bmatrix} 1 & 0 & 0 & \frac{21}{2} & \frac{3}{2} & -4 \\ 0 & 1 & 0 & -\frac{9}{2} & -\frac{1}{2} & 2 \\ 0 & 0 & 1 & -2 & 0 & 1 \end{bmatrix}$

The inverse matrix is $\begin{bmatrix} \frac{21}{2} & \frac{3}{2} & -4 \\ -\frac{9}{2} & -\frac{1}{2} & 2 \\ -2 & 0 & 1 \end{bmatrix}$.

10. $\begin{bmatrix} 1 & -2 & 2 & 1 & 0 & 0 \\ 2 & -3 & 1 & 0 & 1 & 0 \\ 3 & -6 & 6 & 0 & 0 & 1 \end{bmatrix}$

$\xrightarrow[\substack{-2R_1 + R_2 \\ -3R_1 + R_3}]{} \begin{bmatrix} 1 & -2 & 2 & 1 & 0 & 0 \\ 0 & 1 & -3 & -2 & 1 & 0 \\ 0 & 0 & 0 & -3 & 0 & 1 \end{bmatrix}$

Because there are zeros below the rightmost 1 along the main diagonal, the matrix does not have an inverse.

20. $\begin{bmatrix} 1 & 2 & -1 \\ 2 & 3 & -1 \\ 3 & 6 & -2 \end{bmatrix} \begin{bmatrix} x \\ y \\ z \end{bmatrix} = \begin{bmatrix} 5 \\ 8 \\ 14 \end{bmatrix}$

The inverse of the coefficient matrix is

$$\begin{bmatrix} 0 & 2 & -1 \\ -1 & -1 & 1 \\ -3 & 0 & 1 \end{bmatrix}$$

Multiplying each side of the equation by the inverse, we have

$$\begin{bmatrix} x \\ y \\ z \end{bmatrix} = \begin{bmatrix} 0 & 2 & -1 \\ -1 & -1 & 1 \\ -3 & 0 & 1 \end{bmatrix} \begin{bmatrix} 5 \\ 8 \\ 14 \end{bmatrix} = \begin{bmatrix} 2 \\ 1 \\ -1 \end{bmatrix}$$

The solution is $(2, 1, -1)$.

Exercise Set 8.4, page 478

2. $\begin{vmatrix} 2 & 9 \\ -6 & 2 \end{vmatrix} = 2 \cdot 2 - (-6)(9) = 4 + 54 = 58$

14. $M_{13} = \begin{vmatrix} 1 & 3 \\ 6 & -2 \end{vmatrix} = 1(-2) - 6(3) = -2 - 18 = -20$

$C_{13} = (-1)^{1+3} \cdot M_{13} = 1 \cdot M_{13} = 1(-20) = -20$

20. Expanding with cofactors of row 1 yields

$\begin{vmatrix} 3 & -2 & 0 \\ 2 & -3 & 2 \\ 8 & -2 & 5 \end{vmatrix} = 3C_{11} + (-2)C_{12} + 0 \cdot C_{13}$

$= 3 \begin{vmatrix} -3 & 2 \\ -2 & 5 \end{vmatrix} + 2 \begin{vmatrix} 2 & 2 \\ 8 & 5 \end{vmatrix} + 0 \begin{vmatrix} 2 & -3 \\ 8 & -2 \end{vmatrix}$

$= 3(-15 + 4) + 2(10 - 16) + 0$

$= 3(-11) + 2(-6) = -33 + (-12)$

$= -45$

42. Let $D = \begin{vmatrix} 3 & -2 & -1 \\ 1 & 2 & 4 \\ 2 & -2 & 3 \end{vmatrix}$. Then

$$D \overset{R_1 \leftrightarrow R_2}{=} - \begin{vmatrix} 1 & 2 & 4 \\ 3 & -2 & -1 \\ 2 & -2 & 3 \end{vmatrix} \overset{\substack{-3R_1 + R_2 \\ -2R_1 + R_3}}{=} - \begin{vmatrix} 1 & 2 & 4 \\ 0 & -8 & -13 \\ 0 & -6 & -5 \end{vmatrix}$$

$$\overset{-\frac{1}{8}R_2}{=} 8 \begin{vmatrix} 1 & 2 & 4 \\ 0 & 1 & \frac{13}{8} \\ 0 & -6 & -5 \end{vmatrix} \overset{6R_2 + R_3}{=} 8 \begin{vmatrix} 1 & 2 & 4 \\ 0 & 1 & \frac{13}{8} \\ 0 & 0 & \frac{19}{4} \end{vmatrix}$$

$$= 8(1)(1)(\tfrac{19}{4}) = 38$$

Exercise Set 8.5, page 484

4. $x_1 = \dfrac{\begin{vmatrix} 9 & 5 \\ 8 & 7 \end{vmatrix}}{\begin{vmatrix} 2 & 5 \\ 5 & 7 \end{vmatrix}} = \dfrac{63 - 40}{14 - 25} = \dfrac{23}{-11} = -\dfrac{23}{11}$

$x_2 = \dfrac{\begin{vmatrix} 2 & 9 \\ 5 & 8 \end{vmatrix}}{\begin{vmatrix} 2 & 5 \\ 5 & 7 \end{vmatrix}} = \dfrac{16 - 45}{14 - 25} = \dfrac{-29}{-11} = \dfrac{29}{11}$

The solution is $(-23/11, 29/11)$.

24. $x_3 = \dfrac{\begin{vmatrix} 2 & 5 & -3 & -3 \\ 1 & 7 & 4 & -1 \\ 4 & 0 & 3 & 1 \\ 3 & 2 & 0 & 0 \end{vmatrix}}{\begin{vmatrix} 2 & 5 & -5 & -3 \\ 1 & 7 & 8 & -1 \\ 4 & 0 & 1 & 1 \\ 3 & 2 & -1 & 0 \end{vmatrix}} = \dfrac{157}{168}$

Exercise Set 9.1, page 497

6. $a_n = \dfrac{(-1)^{n+1}}{n(n+1)}$, $a_1 = \dfrac{(-1)^{1+1}}{1(1+1)} = \dfrac{1}{2}$,

$a_2 = \dfrac{(-1)^{2+1}}{2(2+1)} = -\dfrac{1}{6}$, $a_3 = \dfrac{(-1)^{3+1}}{3(3+1)} = \dfrac{1}{12}$,

$a_8 = \dfrac{(-1)^{8+1}}{8(8+1)} = -\dfrac{1}{72}$

28. $a_1 = 1$, $a_2 = 2^2 \cdot a_1 = 4 \cdot 1 = 4$, $a_3 = 3^2 \cdot a_2 = 9 \cdot 4 = 36$

42. $\dfrac{12!}{4!\,8!} = \dfrac{12 \cdot 11 \cdot 10 \cdot 9 \cdot 8!}{4!\,8!} = \dfrac{12 \cdot 11 \cdot 10 \cdot 9}{4 \cdot 3 \cdot 2 \cdot 1} = 495$

52. $\displaystyle\sum_{i=1}^{6}(2i+1)(2i-1) = \sum_{i=1}^{6}(4i^2 - 1)$

$$= (4 \cdot 1^2 - 1) + (4 \cdot 2^2 - 1)$$
$$+ (4 \cdot 3^2 - 1) + (4 \cdot 4^2 - 1)$$
$$+ (4 \cdot 5^2 - 1) + (4 \cdot 6^2 - 1)$$
$$= 3 + 15 + 35 + 63 + 99 + 143$$
$$= 358$$

Exercise Set 9.2, page 504

16. $a_6 = -14$, $a_8 = -20$

$$a_8 = a_6 + 2d$$

$$\frac{a_8 - a_6}{2} = d \quad \bullet \textbf{ Solve for } \textbf{d.}$$

$$\frac{-20 - (-14)}{2} = d$$

$$-3 = d$$

$$a_n = a_1 + (n-1)d$$

$$a_6 = a_1 + (6-1)(-3)$$

$$-14 = a_1 + (-15)$$

$$a_1 = 1$$

$$a_{15} = 1 + (15-1)(-3) = 1 + (14)(-3) = -41$$

22. $S_{20} = \dfrac{20}{2}(a_1 + a_{20})$

$$a_1 = 1 - 2(1) = -1$$

$$a_{20} = 1 - 2(20) = -39$$

$$S_{20} = 10[-1 + (-39)] = 10(-40) = -400$$

34. $a = 7$, c_1, c_2, c_3, c_4, c_5, $b = 19$

$$a_n = a_1 + (n-1)d$$

$$19 = 7 + (7-1)d \quad \bullet \textbf{ There are 7 terms, so } \textit{n} = \textbf{7.}$$

$$19 = 7 + 6d$$

$$d = 2$$

$$c_1 = a_1 + d = 7 + 2 = 9$$

$$c_2 = a_1 + 2d = 7 + 4 = 11$$

$$c_3 = a_1 + 3d = 7 + 6 = 13$$

$$c_4 = a_1 + 4d = 7 + 8 = 15$$

$$c_5 = a_1 + 5d = 7 + 10 = 17$$

Exercise Set 9.3, page 511

6. $\dfrac{a_{i+1}}{a_i} = \dfrac{(-1)^i e^{(i+1)x}}{(-1)^{i-1} e^{ix}} = -e^{(i+1)x - ix} = -e^x$

Because x is a constant, $-e^x$ is a constant and the sequence is a geometric sequence.

18. $\dfrac{a_2}{a_1} = \dfrac{6}{8} = \dfrac{3}{4} = r$

$a_n = a_1 r^{n-1}$

$a_n = 8\left(\dfrac{3}{4}\right)^{n-1}$

40. $r = \dfrac{4}{3},\ a_1 = \dfrac{4}{3},\ n = 14$

$S_n = \dfrac{a_1(1 - r^n)}{1 - r}$

$S_{14} = \dfrac{\dfrac{4}{3}\left[1 - \left(\dfrac{4}{3}\right)^{14}\right]}{1 - \dfrac{4}{3}} = \dfrac{\dfrac{4}{3}\left[\dfrac{-263{,}652{,}487}{4{,}782{,}969}\right]}{-\dfrac{1}{3}} \approx 220.49$

62. $0.3\overline{95} = \dfrac{3}{10} + \dfrac{95}{1000} + \dfrac{95}{100{,}000} + \cdots = \dfrac{3}{10} + \dfrac{95/1000}{1 - 1/100}$

$= \dfrac{3}{10} + \dfrac{95}{990} = \dfrac{392}{990} = \dfrac{196}{495}$

70. $P = \dfrac{A[(1 + r)^m - 1]}{r};\quad A = 250,\ r = \dfrac{0.08}{12},\ m = 12(4)$

$P = \dfrac{250[(1 + 0.08/12)^{48} - 1]}{0.08/12} \approx 14087.48$

Exercise Set 9.4, page 519

8. $S_n = 2 + 6 + 12 + \cdots + n(n + 1) = \dfrac{n(n + 1)(n + 2)}{3}$

1. When $n = 1$, $S_1 = 1(1 + 1) = 2$; $\dfrac{1(1 + 1)(1 + 2)}{3} = 2$

   Therefore, the statement is true for $n = 1$.

2. Assume the statement is true for $n = k$.

   $S_k = 2 + 6 + 12 + \cdots + k(k + 1)$

   $= \dfrac{k(k + 1)(k + 2)}{3}$ • **Induction hypothesis**

   Prove the statement is true for $n = k + 1$. That is, prove

   $S_{k+1} = \dfrac{(k + 1)(k + 2)(k + 3)}{3}.$

   Because $a_k = k(k + 1)$ and $a_{k+1} = (k + 1)(k + 2)$,

   $S_{k+1} = S_k + a_{k+1} = \dfrac{k(k + 1)(k + 2)}{3} + (k + 1)(k + 2)$

   $= \dfrac{k(k + 1)(k + 2) + 3(k + 1)(k + 2)}{3}$

   $= \dfrac{(k + 1)(k + 2)(k + 3)}{3}$ • **Factor out (k + 1) and (k + 2) from each term.**

   By the Principle of Mathematical Induction, the statement is true for all positive integers n.

12. $P_n = \left(1 - \dfrac{1}{2}\right)\left(1 - \dfrac{1}{3}\right)\cdots\left(1 - \dfrac{1}{n + 1}\right) = \dfrac{1}{n + 1}$

1. Let $n = 1$; then $P_1 = \left(1 - \dfrac{1}{2}\right) = \dfrac{1}{2}$; $\dfrac{1}{1 + 1} = \dfrac{1}{2}$

   The statement is true for $n = 1$.

2. Assume the statement is true for $n = k$.

   $P_k = \left(1 - \dfrac{1}{2}\right)\left(1 - \dfrac{1}{3}\right)\cdots\left(1 - \dfrac{1}{k + 1}\right) = \dfrac{1}{k + 1}$

   Prove the statement is true for $n = k + 1$. That is, prove

   $P_{k+1} = \left(1 - \dfrac{1}{2}\right)\left(1 - \dfrac{1}{3}\right)\cdots\left(1 - \dfrac{1}{k + 1}\right)\left(1 - \dfrac{1}{k + 2}\right)$

   $= \dfrac{1}{k + 2}$

   Because $a_k = \left(1 - \dfrac{1}{k + 1}\right)$ and $a_{k+1} = \left(1 - \dfrac{1}{k + 2}\right)$,

   $P_{k+1} = P_k \cdot a_{k+1} = \dfrac{1}{k + 1} \cdot \left(1 - \dfrac{1}{k + 2}\right)$

   $= \dfrac{1}{k + 1} \cdot \dfrac{k + 1}{k + 2} = \dfrac{1}{k + 2}$

   By the Principle of Mathematical Induction, the statement is true for all positive integers n.

16. If $a > 1$, show that $a^{n+1} > a^n$ for all positive integers n.

1. Because $a > 1$, $a \cdot a > a \cdot 1$ or $a^2 > a$. Thus the statement is true when $n = 1$.

2. Assume the statement is true for $n = k$.

   $a^{k+1} > a^k$ • **Induction hypothesis**

   Prove the statement is true for $n = k + 1$. That is, prove

   $a^{k+2} > a^{k+1}$

   Because $a^{k+1} > a^k$ and $a > 0$,

   $a(a^{k+1}) > a(a^k)$

   $a^{k+2} > a^{k+1}$

   By the Principle of Mathematical Induction, the statement is true for all positive integers n.

20. 1. Let $n = 1$. Because $\log_{10} 1 = 0$,

   $\log_{10} 1 < 1$

   The inequality is true for $n = 1$.

2. Assume $\log_{10} k < k$ is true for some positive integer k (induction hypothesis). Prove the inequality is true for $n = k + 1$. That is, prove $\log_{10}(k + 1) < k + 1$ is true when $n = k + 1$.

   $\log_{10}(k + 1) \le \log_{10}(k + k)$

   $= \log_{10} 2k = \log_{10} 2 + \log_{10} k < 1 + k$

   Thus $\log_{10}(k + 1) < k + 1$. By the Principle of Mathematical Induction, $\log_{10} n < n$ for all positive integers n.

Exercise Set 9.5, page 525

4. $\dbinom{10}{5} = \dfrac{10!}{5!\,5!} = \dfrac{10 \cdot 9 \cdot 8 \cdot 7 \cdot 6 \cdot 5!}{5!\,5!} = \dfrac{10 \cdot 9 \cdot 8 \cdot 7 \cdot 6}{5 \cdot 4 \cdot 3 \cdot 2 \cdot 1}$

$= 252$

18. $(3x + 2y)^4$

$= (3x)^4 + 4(3x)^3(2y) + 6(3x)^2(2y)^2 + 4(3x)(2y)^3 + (2y)^4$

$= 81x^4 + 216x^3y + 216x^2y^2 + 96xy^3 + 16y^4$

20. $(2x - \sqrt{y})^7 = \dbinom{7}{0}(2x)^7 + \dbinom{7}{1}(2x)^6(-\sqrt{y})$

$+ \dbinom{7}{2}(2x)^5(-\sqrt{y})^2 + \dbinom{7}{3}(2x)^4(-\sqrt{y})^3$

$+ \dbinom{7}{4}(2x)^3(-\sqrt{y})^4 + \dbinom{7}{5}(2x)^2(-\sqrt{y})^5$

$+ \dbinom{7}{6}(2x)(-\sqrt{y})^6 + \dbinom{7}{7}(-\sqrt{y})^7$

$= 128x^7 - 448x^6\sqrt{y} + 672x^5y - 560x^4y\sqrt{y}$

$+ 280x^3y^2 - 84x^2y^2\sqrt{y}$

$+ 14xy^3 - y^3\sqrt{y}$

34. $\dbinom{10}{6 - 1}(x^{-1/2})^{10-6+1}(x^{1/2})^{6-1} = \dbinom{10}{5}(x^{-1/2})^5(x^{1/2})^5 = 252$

Exercise Set 9.6, page 531

12. Because there are 4 palettes and each palette contains 4 colors, by the counting principle there are $4 \cdot 4 \cdot 4 \cdot 4 = 256$ possible colors.

16. There are three possible finishes (first, second, and third) for the 12 contestants. Because the order of finish is important, these are the permutations of the 12 contestants selected 3 at a time.

$P(12, 3) = \dfrac{12!}{(12 - 3)!} = \dfrac{12!}{9!} = 12 \cdot 11 \cdot 10 = 1320$

There are 1320 possible finishes.

20. Player A matched against Player B is the same tennis match as Player B matched against Player A. Therefore, this is a combination of 26 players selected 2 at a time.

$C(26, 2) = \dfrac{26!}{2!(26 - 2)!} = \dfrac{26!}{2!\,24!} = \dfrac{26 \cdot 25 \cdot 24!}{2 \cdot 1 \cdot 24!} = 325$

There are 325 possible first-round matches.

22. The person who refuses to sit in the back seat can be placed in any one of the 3 front seats. Similarly, the person who refuses to sit in the front can be placed in any of the 3 back seats. The remaining 4 people can sit in any of the remaining seats. The number of seating arrangements is

$3 \cdot 3 \cdot 4 \cdot 3 \cdot 2 \cdot 1 = 216$

30. a. The number of ways in which 10 finalists can be selected from 15 semifinalists is the combination of 15 students selected 10 at a time.

$C(15, 10) = 3003$

There are 3003 ways in which the finalists can be chosen.

b. The number of ways in which the 10 finalists can include 3 seniors is the product of the combination of 7 seniors selected 3 at a time and the combination of 8 remaining students selected 7 at a time.

$C(7, 3)C(8, 7) = 35 \cdot 8 = 280$

There are 280 ways in which the finalists can include 3 seniors.

c. "At least five seniors" means 5 or 6 or 7 seniors are finalists (there are only 7 seniors). Because the events are related by "or," sum the number of ways each event can occur.

$C(7, 5)C(8, 5) + C(7, 6)C(8, 4) + C(7, 7)C(8, 3)$

$= 21 \cdot 56 + 7 \cdot 70 + 1 \cdot 56 = 1176 + 490 + 56 = 1722$

There are 1722 ways in which the finalists can include at least 5 seniors.

Exercise Set 9.7, page 541

6. Let R represent the Republican, D the Democrat, and I the Independent. The sample space is

$\{(R, D), (R, I), (D, I)\}$

14. $\{HHHT, HHTH, HTHH, THHH, HHHH\}$

22. Let $E = \{2, 4, 6\}$, $T = \{3, 6\}$, $S = \{1, 2, 3, 4, 5, 6\}$.

$E \cup T = \{2, 3, 4, 6\}$

$P(E \cup T) = \dfrac{N(E \cup T)}{N(S)} = \dfrac{4}{6} = \dfrac{2}{3}$

32. $\dfrac{C(3, 2) \cdot C(5, 2)}{C(8, 4)} = \dfrac{3 \cdot 10}{70} = \dfrac{3}{7}$

34. Yes, because the probability of an ace on each draw is $4/52 = 1/13$.

$P(2 \text{ aces}) = \dfrac{1}{13} \cdot \dfrac{1}{13} = \dfrac{1}{169}$

40. This is a binomial experiment; $p = 1/4$; $q = 3/4$, $n = 8$, $k = 3$.

$\dbinom{8}{3}\left(\dfrac{1}{4}\right)^3\left(\dfrac{3}{4}\right)^5 = 56\left(\dfrac{1}{64}\right)\left(\dfrac{243}{1024}\right) \approx 0.2076$

ANSWERS TO ODD-NUMBERED EXERCISES

Exercise Set 1.1, page 8

1. a. $-3, 4, 11, 57$ are integers. **b.** $-3, 4, 1/5, 11, 3.14, 57$ are rational numbers. **c.** $0.25225222522225\ldots$ is an irrational number. **d.** All are real numbers. **e.** 11 is a prime number. **f.** 4, 57 are composite numbers. **3.** $\{1, 3\}$ **5.** $\{1, 3\}$ **7.** $\{0, 2, 4\}$
9. $\{0, 1, 2, 3, 4, 5, 11\}$ **11.** $\{1, 3, 5, 6, 10, 11\}$ **13.** $\{0, 1, 2, 3, 4, 6, 8, 10\}$ **15.** associative property of addition
17. identity property of multiplication **19.** reflexive property of equality **21.** transitive property of equality
23. inverse property of addition **25.** commutative property of addition **27.** transitive property of equality **29.** $-3a/7$
31. $-7a/20$ **33.** $-10/21$ **35.** $-18/5$ **37.** $-2a/15$ **39. a.** $26/55$ of the pool **b.** $26x/165$ of the pool **41.** $8/59$
43. $2 - 1 \neq 1 - 2$ **45.** $(3 - 1) - 5 = -3; 3 - (1 - 5) = 7$ **47.** False **49.** True **51.** False **53.** True **55.** True **57. a.** $0.\overline{72}$
b. 0.825 **c.** $0.\overline{285714}$ **d.** $0.1\overline{35}$ **59.** $0.16620626, 0.16662040, 0.16666204, 0.16667$; $\dfrac{\sqrt{x + 9} - 3}{x}$ is approaching $0.1\overline{6} = 1/6$.

61. $A = \{4, 6, 8, 9, 10\}$ **63.** $C = \{53, 59\}$ **65.** all the properties except for the identity property of addition and the inverse properties
67. all the properties **69. a.** $5 + 7$ **b.** $23 + 7$

Exercise Set 1.2, page 15

1.
3. $5/2 < 4$ **5.** $2/3 > 0.6666$ **7.** $1.75 < 2.23$ **9.** $0.\overline{36} = 4/11$ **11.** $10/5 = 2$ **13.** $\pi > 3.14159$

15. $(3, 5)$ **17.** $(-\infty, 3)$ **19.** $[0, 3)$ **21.** $(-\infty, -3) \cup [2, \infty)$

23. $(3, 4)$ **25.** $(-1, 3]$ **27.** $-4 \leq x \leq 1$ **29.** $1 < x < 5$

31. $x \geq 2.5$ **33.** $x < 2$ **35.** $x \leq 2$ or $x > 3$ **37.** $x < 3$ or $x > 3$

39. $(-\infty, 3]$ **41.** $-1 < x \leq 2, (-1, 2]$ **43.** $1 \leq x \leq 4$ **45.** $-2 \leq x < \pi$ **47.** 4

49. 27.4 **51.** -11 **53.** $y^2 + 10$ **55.** $1 + \pi$ **57.** 9 **59.** $x + 7$ **61.** 7 **63.** 8 **65.** 50 **67.** 33 **69.** $\pi + 3$ **71.** $5/12$
73. $|a - 2|$ **75.** $|m - n|$ or $|n - m|$ **77.** $|a - 4| < z$ **79.** $|x + 2| < 7$ **81.** $(-\infty, 3) \cup (3, \infty)$ **83.** $(-3, 3)$ **85.** False **87.** True
91. $I \leq 120$ **93.** $2 \leq A < 3$ **95.** $|x - 2| < |x - 6|$ **97.** $|x - 3| > |x + 7|$ **99.** $2 < |x - 4| < 7$ **101.** $|x - a| < \delta$

Exercise Set 1.3, page 27

1. -256 **3.** 1 **5.** $81/16$ **7.** 8 **9.** -3 **11.** -16 **13.** 16 **15.** $1/3$ **17.** $1/9$ **19.** $3/4$ **21.** $9/4$ **23.** $6x^7 y^4$ **25.** $\dfrac{36x^4}{y^4}$

27. $\dfrac{y}{3x}$ **29.** $\dfrac{a + b}{ab}$ **31.** $\dfrac{16}{b^2}$ **33.** $3|xy^3|$ **35.** $a^{1/2} b^{3/10}$ **37.** $a^2 + 7a$ **39.** $p - q$ **41.** $m^2 n^{3/2}$ **43.** 1 **45.** $r^{(m-n)/(mn)}$ **47.** 2.1×10^7

49. 9.5×10^{-4} **51.** 6500 **53.** 0.000217 **55.** $\sqrt{3x}$ **57.** $5\sqrt[4]{xy}$ **59.** $\sqrt[3]{(5w)^2}$ **61.** $(17k)^{1/3}$ **63.** $a^{2/5}$ **65.** $\left(\dfrac{7a}{3}\right)^{1/2}$ **67.** $3\sqrt{5}$

69. $2\sqrt[3]{3}$ **71.** $-3\sqrt[3]{3}$ **73.** $-2\sqrt[4]{4}$ **75.** $2x|y|\sqrt{6x}$ **77.** $-2ay^2\sqrt[3]{2y}$ **79.** $6\sqrt{3}$ **81.** $-3\sqrt{2}$ **83.** 0 **85.** $22x^2 y\sqrt[3]{y}$
87. $29 + 11\sqrt{5}$ **89.** $2x - 9$ **91.** $50y + 10\sqrt{6yz} + 3z$ **93.** $x + 10\sqrt{x - 3} + 22$ **95.** $2x + 14\sqrt{2x + 5} + 54$ **97.** $\sqrt{2}$ **99.** $\dfrac{\sqrt{10}}{6}$

101. $\dfrac{3\sqrt[3]{4}}{2}$ **103.** $\dfrac{2\sqrt[3]{x}}{x}$ **105.** $\dfrac{\sqrt{5}}{3}$ **107.** $\dfrac{\sqrt{6xy}}{9y}$ **109.** $\dfrac{3(\sqrt{5} - \sqrt{x})}{5 - x}$ **111.** $-\dfrac{2\sqrt{7} + 7}{3}$ **113. a.** $2^8 = 256$ **b.** $2^{16} = 65,536$

115. 1.97×10^4 seconds **117.** $\$4873.50$ **119. a.** 81% **b.** 66% **121.** $x^3 y^{n+1}$ **123.** $\dfrac{y^{n+1}}{x^{5n}}$ **125.** $8/5$ **127.** $-19/12$ **129.** $3^{(3^3)}$

133. $\dfrac{1}{\sqrt{4 + h} + 2}$ **135.** $\dfrac{1}{\sqrt{a + h} + a}$ **137.** $\dfrac{1}{\sqrt{n^2 + 1} + n}$

Exercise Set 1.4, page 35

1. $2 + 3i$ **3.** $4 - 11i$ **5.** $8 + i\sqrt{3}$ **7.** $7 + 4i$ **9.** $9i$ **11.** $5 + 12i$ **13.** $4 - 3i$ **15.** $-2 - 5i$ **17.** $12 - 2i$ **19.** $-2 + 11i$

21. $16 + 16i$ **23.** $23 + 2i$ **25.** 74 **27.** $-117 - i$ **29.** $\dfrac{1}{2} - \dfrac{1}{2}i$ **31.** $\dfrac{7}{58} + \dfrac{3}{58}i$ **33.** $\dfrac{5}{13} + \dfrac{12}{13}i$ **35.** $\dfrac{1}{61} + \dfrac{11}{61}i$ **37.** $1 - 6i$

39. $-16 - 30i$ **41.** $-29 - 17i$ **43.** $-2 - 2i$ **45.** -16 **47.** $75i$ **49.** $-i$ **51.** i **53.** -1 **55.** -1 **57.** $-i$ **59.** i **61.** -1

63. $-i$ **65.** -2 **67.** $-8\sqrt{5}$ **69.** 11 **71.** $9 + 40i$ **73.** $\frac{1}{2} \pm \frac{\sqrt{3}}{2}i$ **75.** $-1 \pm i$ **77.** $-\frac{3}{2} \pm \frac{\sqrt{3}}{2}i$ **79.** $-\frac{1}{4} \pm \frac{\sqrt{23}}{4}i$ **81.** 5
83. $\sqrt{29}$ **85.** $\sqrt{65}$ **87.** 3 **95.** no **97.** $66 + 6\sqrt{5} - 14\sqrt{3}$ **99.** 1 **101.** 0

Exercise Set 1.5, page 41

1. D **3.** H **5.** G **7.** B **9.** J **11. a.** $x^2 + 2x - 7$ **b.** 2 **c.** 1, 2, -7 **d.** 1 **e.** $x^2, 2x, -7$ **13. a.** $x^3 - 1$ **b.** 3 **c.** 1, -1
d. 1 **e.** $x^3, -1$ **15. a.** $2x^4 + 3x^3 + 4x^2 + 5$ **b.** 4 **c.** 2, 3, 4, 5 **d.** 2 **e.** $2x^4, 3x^3, 4x^2, 5$ **17.** 3 **19.** 5 **21.** 2
23. $5x^2 + 11x + 3$ **25.** $9w^3 + 8w^2 - 2w + 6$ **27.** $-2r^2 + 3r - 12$ **29.** $-3u^2 - 2u + 4$ **31.** $8x^3 + 18x^2 - 67x + 40$
33. $6x^4 - 19x^3 + 26x^2 - 29x + 10$ **35.** $10x^2 + 22x + 4$ **37.** $y^2 + 3y + 2$ **39.** $4z^2 - 19z + 12$ **41.** $a^2 + 3a - 18$
43. $b^2 + 2b - 24$ **45.** $10x^2 - 57xy + 77y^2$ **47.** $18x^2 + 55xy + 25y^2$ **49.** $12w^2 - 40wx + 33x^2$ **51.** $6p^2 - 11pq - 35q^2$
53. $12d^2 + 4d - 8$ **55.** $r^3 + s^3$ **57.** $60c^3 - 49c^2 + 4$ **59.** $9x^2 - 25$ **61.** $9x^4 - 6x^2y + y^2$ **63.** $16w^2 + 8wz + z^2$
65. $x^2 - 4x + 4 + 2xy - 4y + y^2$ **67.** $x^2 + 10x + 25 - y^2$ **69.** 29 **71.** -17 **73.** -1 **75.** 33 **77. a.** 48.46 **b.** 51.44
79. a. 10.994998 **b.** 10.99949998 **81.** 11,175 matches **83.** 14.8, 90.4 **85.** $a^3 + 3a^2b + 3ab^2 + b^3$ **87.** $x^3 - 3x^2 + 3x - 1$
89. $8x^3 - 36x^2y + 54xy^2 - 27y^3$

Exercise Set 1.6, page 52

1. $5(x + 4)$ **3.** $-3x(5x + 4)$ **5.** $2xy(5x + 3 - 7y)$ **7.** $(x - 3)(2a + 4b)$ **9.** $(3x + 1)(x^2 + 2)$ **11.** $(x - 1)(ax + b)$
13. $(3w + 2)(2w^2 - 5)$ **15.** $(x + 3)(x + 4)$ **17.** $(a - 12)(a + 2)$ **19.** $(6x + 1)(x + 4)$ **21.** $(17x + 4)(3x - 1)$ **23.** $(3x + 8y)(2x - 5y)$
25. $(x^2 + 5)(x^2 + 1)$ **27.** $(6x^2 + 5)(x^2 + 3)$ **29.** factorable over the integers **31.** not factorable over the integers **33.** not factorable over the integers **35.** $(x - 3)(x + 3)$ **37.** $(2a - 7)(2a + 7)$ **39.** $(1 - 10x)(1 + 10x)$ **41.** $(x^2 - 3)(x^2 + 3)$ **43.** $(x + 3)(x + 7)$
45. $(x + 5)^2$ **47.** $(a - 7)^2$ **49.** $(2x + 3)^2$ **51.** $(z^2 + 2w^2)^2$ **53.** $(x - 2)(x^2 + 2x + 4)$ **55.** $(2x - 3y)(4x^2 + 6xy + 9y^2)$
57. $(2 - x^2)(4 + 2x^2 + x^4)$ **59.** $(x - 3)(x^2 - 3x + 3)$ **61.** $2(3x - 1)(3x + 1)$ **63.** $(2x - 1)(2x + 1)(4x^2 + 1)$ **65.** $a(3x - 2y)(4x - 5y)$
67. $b(3x + 4)(x - 1)(x + 1)$ **69.** $2b(6x + y)^2$ **71.** $(w - 3)(w^2 - 12w + 39)$ **73.** $(x + 3y - 1)(x + 3y + 1)$ **75.** not factorable over the
integers **77.** $(2x - 5)^2(3x + 5)$ **79.** $(2x - y)(2x + y + 1)$ **81.** 8 **83.** 64 **85.** $(x^n - 1)(x^n + 1)(x^{2n} + 1)$ **87.** $\pi(R - r)(R + r)$
89. $r^2(4 - \pi)$ **91.** $(x + 3i)(x - 3i)$ **93.** $(2x + 9i)(2x - 9i)$ **95.** 0 **97.** 0

Exercise Set 1.7, page 60

1. $\dfrac{x + 4}{3}$ **3.** $\dfrac{x - 3}{x - 2}$ **5.** $\dfrac{a^2 - 2a + 4}{a - 2}$ **7.** $-\dfrac{x + 8}{x + 2}$ **9.** $-\dfrac{4y^2 + 7}{y + 7}$ **11.** $-\dfrac{8}{a^3b}$ **13.** $\dfrac{10}{27q^2}$ **15.** $\dfrac{x(3x + 7)}{2x + 3}$ **17.** $\dfrac{x + 3}{2x + 3}$
19. $\dfrac{(2y + 3)(3y - 4)}{(2y - 3)(y + 1)}$ **21.** $\dfrac{1}{a - 8}$ **23.** $\dfrac{3p - 2}{r}$ **25.** $\dfrac{8x(x - 4)}{(x - 5)(x + 3)}$ **27.** $\dfrac{3y - 4}{y + 4}$ **29.** $\dfrac{7z(2z - 5)}{(2z - 3)(z - 5)}$ **31.** $\dfrac{-2x^2 + 14x - 3}{(x - 3)(x + 3)(x + 4)}$
33. $\dfrac{(2x - 1)(x + 5)}{x(x - 5)}$ **35.** $\dfrac{-q^2 + 12q + 5}{(q - 3)(q + 5)}$ **37.** $\dfrac{3x^2 - 7x - 13}{(x + 3)(x + 4)(x - 3)(x - 4)}$ **39.** $\dfrac{(x + 2)(3x - 1)}{x^2}$ **41.** $\dfrac{4x + 1}{x - 1}$ **43.** $\dfrac{x - 2y}{y(y - x)}$
45. $\dfrac{(5x + 9)(x + 3)}{(x + 2)(4x + 3)}$ **47.** $\dfrac{(b + 3)(b - 1)}{(b - 2)(b + 2)}$ **49.** $\dfrac{x - 1}{x}$ **51.** $2 - m^2$ **53.** $\dfrac{-x^2 + 5x + 1}{x^2}$ **55.** $\dfrac{-x - 7}{x^2 + 6x - 3}$ **57.** $\dfrac{2x - 3}{x + 3}$ **59.** $\dfrac{a + b}{ab(a - b)}$
61. $\dfrac{(b - a)(b + a)}{ab(a^2 + b^2)}$ **63. a.** 136.55 mph **b.** $\dfrac{2v_1v_2}{v_1 + v_2}$ **65.** $\dfrac{2x + 1}{x(x + 1)}$ **67.** $\dfrac{3x^2 - 4}{x(x - 2)(x + 2)}$ **69.** $\dfrac{x^2 + 9x + 25}{(x + 5)^2}$ **71.** $\dfrac{x(1 - 4xy)}{(1 - 2xy)(1 + 2xy)}$
73. $R\left[\dfrac{(1 + i)^n - 1}{i(1 + i)^n}\right]$

Chapter 1 True/False Exercises, page 65

1. True **2.** False; if $a = 1/2$, then $(1/2)^2 = 1/4 < 1/2$. **3.** True **4.** False; $\sqrt{2} + (-\sqrt{2}) = 0$, which is a rational number.
5. False; $(2 \oplus 4) \oplus 6 \neq 2 \oplus (4 \oplus 6)$. **6.** False; $x > a$ is written as (a, ∞). **7.** False; $\sqrt{(-2)^2} \neq -2$. **8.** True **9.** True **10.** True

Chapter 1 Review Exercises, page 66

1. integer, rational number, real number, prime number **3.** rational number, real number **5.** $\{1, 2, 3, 5, 7, 11\}$ **7.** distributive
property **9.** associative property of multiplication **11.** identity property of addition **13.** symmetric property of equality
15. $(-4, 2]$ **17.** $-3 \le x < 2$ **19.** 7 **21.** $4 - \pi$ **23.** 17 **25.** -36 **27.** $12x^9y^3$ **29.** 6.2×10^5
31. 35,000 **33.** $-a^2 - 2a - 1$ **35.** $6x^4 + 5x^3 - 13x^2 + 22x - 20$ **37.** $3(x + 5)^2$ **39.** $4(5a^2 - b^2)$ **41.** $\dfrac{3x - 2}{x + 4}$ **43.** $\dfrac{2x + 3}{2x - 5}$

45. $\dfrac{x(3x + 10)}{(x + 3)(x - 3)(x + 4)}$ **47.** $\dfrac{2x - 9}{3x - 17}$ **49.** 5   **51.** $x^{17/12}$ **53.** $x^{3/4}y^2$ **55.** $4|a|b^3\sqrt{3b}$ **57.** $6|x|\sqrt{2y}$ **59.** $\dfrac{3|y|\sqrt{15y}}{5}$ **61.** $\dfrac{7\sqrt[3]{4x}}{2}$

63. $-3y^2\sqrt[3]{5x^2y}$ **65.** $3 - 8i$, conjugate $3 + 8i$ **67.** $5 + 2i$ **69.** $25 - 19i$ **71.** 1   **73.** 1

Chapter 1 Test, page 67

1. distributive property   **2.** $\{0, 1, 2, 3, 4, 5, 6, 7, 8, 9\}$ **3.** -3 **4.** 7   **5.** $\dfrac{4}{9x^4y^2}$ **6.** $\dfrac{96bc^2}{a^5}$ **7.** 1.37×10^{-3}

8. $x^3 - 2x^2 + 5xy - 2x^2y - 2y^2$ **9.** -94 **10.** $(7x - 1)(x + 5)$ **11.** $(a - 4b)(3x - 2)$ **12.** $2x(2x - y)(4x^2 + 2xy + y^2)$

13. $(x + y)(x - 1)(x^2 + x + 1)$ **14.** $\dfrac{(x - 2)(x^2 + x + 1)}{(x + 1)(x - 1)}$ **15.** $\dfrac{(x - 6)(x + 1)}{(x + 3)(x - 2)(x - 3)}$ **16.** $\dfrac{x(x + 2)}{x - 3}$ **17.** $\dfrac{3a^2 - 3ab - 10a + 5b}{a(2a - b)}$

18. $\dfrac{x(2x - 1)}{2x + 1}$ **19.** $\dfrac{x^{5/6}}{y^{9/4}}$ **20.** $7xy\sqrt[3]{3xy}$ **21.** $\dfrac{\sqrt[4]{8x}}{2}$ **22.** $\dfrac{3\sqrt{x} - 6}{x - 4}$ **23.** $2 + 11i$ **24.** $11 - 10i$ **25.** $\dfrac{6}{13} - \dfrac{4}{13}i$

Exercise Set 2.1, page 75

1. 15   **3.** -4 **5.** 9/2   **7.** 108/23   **9.** 2/9   **11.** 12   **13.** 16   **15.** 9   **17.** 75   **19.** 1/2   **21.** 22/13   **23.** 95/18   **25.** 1200
27. 2500   **29.** identity   **31.** conditional equation   **33.** contradiction   **35.** contradiction   **37.** 31   **39.** 2   **41.** no solution
43. no solution   **45.** 7/2   **47.** 6   **49.** -4 **51.** -12 **53.** 1   **55.** no solution   **57.** $a = 33$ **59.** maximum 166 beats per
minute, minimum 127 beats per minute   **61.** no   **63.** yes   **67.** $10\sqrt{7}/7$ **69.** $-7\sqrt{3}/6$ **71.** $1/(a - b)$

Exercise Set 2.2, page 85

1. $h = \dfrac{3V}{\pi r^2}$ **3.** $t = \dfrac{I}{Pr}$ **5.** $m_1 = \dfrac{Fd^2}{Gm_2}$ **7.** $v_0 = \dfrac{s + 16t^2}{t}$ **9.** $T_w = \dfrac{-Q_w + m_wc_wT_f}{m_wc_w}$ **11.** $d = \dfrac{a_n - a_1}{n - 1}$ **13.** $r = \dfrac{S - a_1}{S}$

15. $f_1 = \dfrac{w_1f - w_2f_2 + w_2f}{w_1}$ **17.** $v_{LC} = \dfrac{f_{LC}v - f_vv}{f_v}$ **19.** 100   **21.** 30 feet by 57 feet   **23.** 12 centimeters, 36 centimeters, 36 centimeters

25. 872, 873   **27.** 18, 20   **29.** 240 meters   **31.** 2 hours   **33.** 98   **35.** 850   **37.** \$937.50   **39.** \$7600 invested at 8%, \$6400 invested
at 6.5%   **41.** \$3750   **43.** $18\frac{2}{11}$ grams   **45.** 64 liters   **47.** 1200 at \$14 and 1800 at \$25   **49.** $6\frac{2}{3}$ pounds of the \$12 coffee and
$13\frac{1}{3}$ pounds of the \$9 coffee   **51.** 10 grams   **53.** 7.875 hours   **55.** $13\frac{1}{3}$ hours   **57.** \$10.05 for book, \$0.05 for bookmark
59. 6.25 feet   **61.** 40 pounds   **63.** 1384 feet   **65.** 84 years old

Exercise Set 2.3, page 96

1. $5, -3$ **3.** $-24, 3/8$ **5.** $0, 7/3$ **7.** $4/3, -2/5$ **9.** $1/2, -4$ **11.** 8, 2   **13.** 3, 8/3   **15.** 7   **17.** ± 9 **19.** $\pm 2\sqrt{6}$ **21.** $\pm 2i$

23. $11, -1$ **25.** 7/2   **27.** $-1/2$ **29.** $-3 \pm 2\sqrt{2}$ **31.** $5, -3$ **33.** $0, -10$ **35.** $\dfrac{-3 \pm \sqrt{13}}{2}$ **37.** $\dfrac{-2 \pm \sqrt{6}}{2}$ **39.** $\dfrac{4 \pm \sqrt{13}}{3}$

41. $\dfrac{-3 \pm 2\sqrt{6}}{3}$ **43.** $-3, 5$ **45.** $\dfrac{-1 \pm \sqrt{5}}{2}$ **47.** $\dfrac{-2 \pm \sqrt{2}}{2}$ **49.** $\dfrac{5 \pm i\sqrt{11}}{6}$ **51.** $\dfrac{-3 \pm \sqrt{41}}{4}$ **53.** $-\dfrac{\sqrt{2}}{2}, -\sqrt{2}$ **55.** $\dfrac{3 \pm i\sqrt{11}}{2}$

57. 81, two distinct real numbers   **59.** -116, two distinct nonreal complex numbers   **61.** 0, one real number   **63.** ± 24 **65.** 49/4

67. 76.4 inches   **69.** 26.8 centimeters   **71.** 10 centimeters by 3.5 centimeters   **73.** 100 feet by 150 feet   **75.** 10, 12   **77.** $\dfrac{5 + \sqrt{29}}{2}$

79. 35 mph for the first part and 45 mph for the last part   **81.** 12 hours   **83.** 2.02 seconds   **85.** yes   **87.** yes   **89.** yes

91. $\dfrac{v_0 \pm \sqrt{v_0^2 - 2g(s - s_0)}}{g}$ **93.** $\dfrac{2 \pm \sqrt{4 + 3x}}{x}$ **95.** $-\dfrac{y}{6} \pm \dfrac{yi}{6}\sqrt{47}$ **97.** $\dfrac{-1 \pm \sqrt{33 + 4x}}{2}$ **99.** 22

Exercise Set 2.4, page 105

1. $0, \pm 5$ **3.** $2, \pm 1$ **5.** $0, \pm 3$ **7.** $0, -5, 8$ **9.** $0, \pm 4$ **11.** $2, -1 \pm i\sqrt{3}$ **13.** 40   **15.** 3   **17.** 7   **19.** 7   **21.** 9   **23.** 5/2
25. $1, -6$ **27.** 4   **29.** 1, 5   **31.** 2   **33.** $23, -31$ **35.** $2, -1/8$ **37.** $0, 1/256$ **39.** $-1, -59/3$ **41.** $\pm\sqrt{7}, \pm\sqrt{2}$ **43.** $\pm 2, \pm\sqrt{6}/2$
45. $\sqrt[3]{2}, -\sqrt[3]{3}$ **47.** $-\sqrt[3]{36}/3, \sqrt[3]{98}/7$ **49.** 1, 16   **51.** $-1/27, 64$ **53.** $\pm\sqrt{15}/3$ **55.** ± 1 **57.** $1/2, -1/5$ **59.** 256/81, 16
61. $\pm 0.62, \pm 1.62$ **63.** $\pm 0.34, \pm 2.98$ **65.** $x = \pm\sqrt{9 - y^2}$ **67.** $x = y + 2\sqrt{yz} + z$ **69.** $x = (7 - 2y^2)/(2y)$ **71.** 9, 36   **73.** 3 inches

75. 10.5 mm   **77.** 87 feet   **79. a.** 8.93 inches   **b.** $5\sqrt{3}$ inches   **81.** $s = \left(\dfrac{-275 + 5\sqrt{3025 + 176T}}{2}\right)^2$

Exercise Set 2.5, page 118

1. $\{x \mid x < 4\}$ **3.** $\{x \mid x < -6\}$ **5.** $\{x \mid x \le -3\}$ **7.** $\{x \mid x \ge -13/8\}$ **9.** $\{x \mid x < 2\}$ **11.** if you write more than 57 checks a month
13. at least 34 sales   **15.** $\{x \mid -3/4 < x \le 4\}$ **17.** $\{x \mid 1/3 \le x \le 11/3\}$ **19.** $\{x \mid -3/8 \le x < 11/4\}$ **21.** $\{x \mid x < 1\}$ **23.** $\{x \mid x > -1\}$
25. the set of all real numbers   **27.** $20 \le C \le 40$ **29.** $\{12, 14, 16\}$, $\{14, 16, 18\}$ **31.** $(-\infty, -7) \cup (0, \infty)$ **33.** $[-4, 4]$ **35.** $(-5, -2)$
37. $(-\infty, -4] \cup [7, \infty)$ **39.** $[-1/2, 4/3]$ **41.** $(-\infty, -5/4] \cup [3/2, \infty)$ **43.** $(0, 210)$ **45.** $(-4, 1)$ **47.** $[-29/2, -8)$ **49.** $[-4, -7/2)$
51. $(-\infty, -1) \cup (2, 4)$ **53.** $(-\infty, 5) \cup [12, \infty)$ **55.** $(-2/3, 0) \cup (5/2, \infty)$ **57.** $(-\infty, 5)$ **59.** $[4, \infty)$ **61.** $[-9, \infty)$ **63.** $[-3, 3]$
65. $(-\infty, -4] \cup [4, -\infty)$ **67.** $(-\infty, -3] \cup [5, \infty)$ **69.** $(-\infty, \infty)$ **71.** $(-\infty, 3) \cup (3, 6) \cup (6, \infty)$ **73.** $(-3, \infty)$ **75.** $[-\sqrt{3}, 0] \cup [\sqrt{3}, \infty)$
77. $(-\infty, \infty)$ **79.** $[-2, 0] \cup [5, \infty)$ **81.** $(-\infty, -2\sqrt{6}] \cup [2\sqrt{6}, \infty)$ **83.** $(-\infty, -2\sqrt{14}] \cup [2\sqrt{14}, \infty)$ **85.** 1 second $< t <$ 3 seconds

Exercise Set 2.6, page 124

1. $-4, 4$ **3.** $7, 3$ **5.** $-5, -7$ **7.** $-34, 6$ **9.** $8, -3$ **11.** $2, -8$ **13.** $20, -12$ **15.** no solution   **17.** $12, -18$
19. $(a + b)/2, (a - b)/2$ **21.** $(a + \delta, a - \delta)$ **23.** $(-4, 4)$ **25.** $(-8, 10)$ **27.** $(-\infty, -33) \cup (27, \infty)$ **29.** $(-\infty, -3/2) \cup (5/2, \infty)$
31. $(-\infty, -8] \cup [2, \infty)$ **33.** $[-4/3, 8]$ **35.** $(-\infty, -4] \cup [28/5, \infty)$ **37.** $(-\infty, \infty)$ **39.** $\{4\}$ **41.** no solution   **43.** $(-\infty, \infty)$
45. $(3 - b, 3 + b)$ **47.** maximum radius 4.480 inches, minimum radius 4.432 inches   **49.** $(-\sqrt{2}, 0) \cup (0, \sqrt{2})$ **51.** $(-4, -2) \cup (2, 4)$
53. $(-\infty, -5] \cup [-4, -3] \cup [-2, \infty)$ **55.** $(-\infty, -\sqrt{26}] \cup [-\sqrt{17}, \sqrt{17}] \cup [\sqrt{26}, \infty)$ **57.** False; not true for $x < 0$.   **59.** True   **61.** True
63. $\{x \mid x \ge -4\}$ **65.** $\{x \mid x \le -7\}$ **67.** $\{x \mid x \ge -7/2\}$ **69.** $\{x \mid x = 3, -5\}$ **71.** $(-5, -1) \cup (1, 5)$ **73.** $(-7, -3] \cup [3, 7)$
75. $(a - \delta, a) \cup (a, a + \delta)$ **77.** $(2, 4) \cup (8, 10)$ **79.** $(-\infty, -20/3] \cup [28/3, \infty)$ **81.** $(1/2, \infty)$ **83. a.** $|x - 3| < 8$ **b.** $|x - j| < k$
85. a. $|s - 4.25| \le 0.01$ **b.** $4.24 \le s \le 4.26$

Chapter 2 True/False Exercises, page 127

1. False; $(-3)^2 = 9$.   **2.** False; one has solution set $\{3\}$, and the other has solution set $\{3, -4\}$.   **3.** True   **4.** True   **5.** False; $100 > 1$
but $1/100 \not> 1/1$.   **6.** False; the discriminant is $b^2 - 4ac$.   **7.** False; $\sqrt{1} + \sqrt{1} = 1 + 1 = 2$ but $1 + 1 = 2 \ne 2^2$.   **8.** True   **9.** False;
$3x^2 - 48 = 0$ has roots of 4 and -4.   **10.** True

Chapter 2 Review Exercises, page 128

1. $3/2$ **3.** $1/2$ **5.** $-38/15$ **7.** $3, 2$ **9.** $(1 \pm \sqrt{13})/6$ **11.** $0, 5/3$ **13.** $\pm 2\sqrt{3}/3, \pm \sqrt{10}/2$ **15.** -5 **17.** 4 **19.** -4
21. $-2, -4$ **23.** $5, 1$ **25.** $2, -3$ **27.** $-2, -1$ **29.** $14, -31\frac{1}{2}$ **31.** $(-\infty, 2]$ **33.** $[-5, 2]$ **35.** $145/9 \le C \le 35$
37. $(-\infty, 0] \cup [3, 4]$ **39.** $(-\infty, -3) \cup (4, \infty)$ **41.** $(-\infty, 5/2] \cup (3, \infty)$ **43.** $(2/3, 2)$ **45.** $(-2, 0) \cup (0, 2)$ **47.** $(1, 2) \cup (2, 3)$
49. $h = \dfrac{V}{\pi r^2}$ **51.** $b_1 = \dfrac{2A - hb_2}{h}$ **53.** $m = \dfrac{e}{c^2}$ **55.** 80 **57.** 24 nautical miles   **59.** $1750 in the 4% account, $3750 in the
6% account   **61.** $864   **63.** 18 hours   **65.** 13 feet

Chapter 2 Test, page 129

1. 9.6 or 48/5   **2.** $-14/3$ **3.** $x = \dfrac{c - cd}{a - c}, a \ne c$ **4.** $-2 \pm \sqrt{5}$ **5.** $\dfrac{-1 \pm 2\sqrt{7}}{3}$ **6.** $-4, 1, -\dfrac{1}{2} - \dfrac{\sqrt{3}}{2}i, -\dfrac{1}{2} + \dfrac{\sqrt{3}}{2}i$ **7.** $2, 3$
8. $8/27, -64$ **9.** $67, -61$ **10.** $x \le 5/2$ **11.** $[-4, -1) \cup [3, \infty)$ **12.** $2 \le x \le 13/3$ **13.** $-1, -6$ **14.** $(-7, -1)$
15. $(-\infty, -5/3] \cup [3, \infty)$ **16.** 2 mph   **17.** $6500 at 8.2%; $2500 at 6.5%   **18.** 2.25 liters   **19.** 15 hours   **20.** more than 100 miles

Exercise Set 3.1, page 141

1.

3. a.

b.

c. $t \approx 11.0$ means that $10.95 \le t < 11.05$.

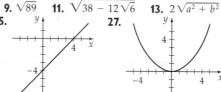

5. $7\sqrt{5}$ **7.** $\sqrt{1261}$ **9.** $\sqrt{89}$ **11.** $\sqrt{38 - 12\sqrt{6}}$ **13.** $2\sqrt{a^2 + b^2}$ **15.** $-x\sqrt{10}$ **17.** $(12, 0), (-4, 0)$ **19.** $(3, 2)$ **21.** $(6, 4)$
23. $(-0.875, 3.91)$ **25.**   **27.**   **29.**   **31.**   **33.**

35.

37.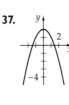

39. $(0, 12/5)$, $(6, 0)$

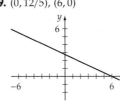

41. $(0, \sqrt{5})$, $(0, -\sqrt{5})$, $(5, 0)$

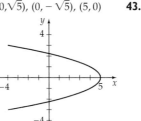

43. $(0, 4)$, $(0, -4)$, $(-4, 0)$

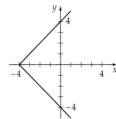

45. $(0, \pm 2)$, $(\pm 2, 0)$

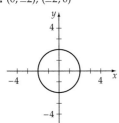

47. $(0, \pm 4)$, $(\pm 4, 0)$

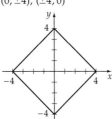

49. center $(0, 0)$, radius 6
51. center $(0, 0)$, radius 10
53. center $(1, 3)$, radius 7
55. center $(-2, -5)$, radius 5
57. center $(8, 0)$, radius 1/2
59. $(x - 4)^2 + (y - 1)^2 = 2^2$
61. $(x - 1/2)^2 + (y - 1/4)^2 = (\sqrt{5})^2$
63. $(x - 0)^2 + (y - 0)^2 = 5^2$
65. $(x - 1)^2 + (y - 3)^2 = 5^2$
67. center $(3, 0)$, radius 2
69. center $(2, 5)$, radius 3
71. center $(7, -4)$, radius 3
73. center $(-1/2, 0)$, radius 4
75. center $(1/2, -1/3)$, radius 1/6

77.   **79.**   **81.** **83.** **85.** **87.**

89. $(13, 5)$ **91.** $(7, -6)$ **93.** $(5/2, 71/4)$ **95.** $(5/2, 17/8)$ **97.** yes   **99.** $x^2 - 6x + y^2 - 8y = 0$ **101.** $9x^2 + 25y^2 = 225$
103. $(x + 1)^2 + (y - 7)^2 = 5^2$ **105.** $(x - 7)^2 + (y - 11)^2 = 11^2$ **107.** $(x + 3)^2 + (y - 3)^2 = 3^2$

Exercise Set 3.2, page 156

1. a. 5   **b.** -4 **c.** -1 **d.** 1   **e.** $3k - 1$ **f.** $3k + 5$ **3. a.** $\sqrt{5}$ **b.** 3   **c.** 3   **d.** $\sqrt{21}$ **e.** $\sqrt{r^2 + 2r + 6}$ **f.** $\sqrt{c^2 + 5}$
5. a. 1/2   **b.** 1/2   **c.** 5/3   **d.** 1   **e.** $\dfrac{1}{c^2 + 4}$ **f.** $\dfrac{1}{|2 + h|}$ **7. a.** 1   **b.** 1   **c.** -1 **d.** -1 **e.** 1   **f.** -1 **9. a.** -11 **b.** 6
c. $3c + 1$ **d.** $-k^2 - 2k + 10$ **11.** yes   **13.** no   **15.** no   **17.** yes   **19.** no   **21.** yes   **23.** yes   **25.** yes   **27.** all real numbers
29. all real numbers   **31.** $\{x \mid x \neq -2\}$ **33.** $\{x \mid x \geq -7\}$ **35.** $\{x \mid -2 \leq x \leq 2\}$ **37.** $\{x \mid x > -4\}$
39.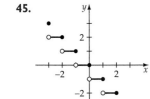

domain: all real numbers

41.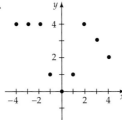

domain: $\{-4, -3, -2, -1, 0, 1, 2, 3, 4\}$

43.

domain: $\{x \mid -6 \leq x \leq 6\}$

45.

domain: $\{x \mid -3 \leq x \leq 3\}$

47. a. $C(3.97) = 1.16$
b. $c(w)$

49. a, b, and **d.**
51. decreasing on $(-\infty, 0]$; increasing on $[0, \infty)$
53. increasing on $(-\infty, \infty)$
55. decreasing on $(-\infty, -3]$; increasing on $[-3, 0]$; decreasing on $[0, 3]$; increasing on $[3, \infty)$
57. constant on $(-\infty, 0]$; increasing on $[0, \infty)$
59. decreasing on $(-\infty, 0]$; constant on $[0, 1]$; increasing on $[1, \infty)$
61. g and F
63. a. $w = 25 - l$ **b.** $A = 25l - l^2$
65. $v(t) = 80,000 - 6500t$, $0 \leq t \leq 10$
67. a. $C(x) = 2000 + 22.80x$ **b.** $R(x) = 37.00x$ **c.** $P(x) = 14.20x - 2000$ **69.** $h = 15 - 5r$ **71.** $d = \sqrt{(3t)^2 + 50^2}$
73. $d = \sqrt{(45 - 8t)^2 + (6t)^2}$ **75.** 275, 375, 385, 390, 394   **77.** $c = -2$ or $c = 3$ **79.** 1 is not in the range of f.

81.

Dot MODE

83. 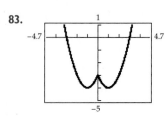 **85.**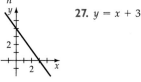

87. 4
89. 2
91. a. 36
b. 13
c. 12
d. 30
e. $13k - 2$
f. $8k - 11$

93. $4\sqrt{21}$ **95.** $1, -3$ **97.**

Exercise Set 3.3, page 172

1. $-3/2$ **3.** $-1/2$ **5.** The line does not have slope.   **7.** 6   **9.** 9/19   **11.** $\dfrac{f(3 + h) - f(3)}{h}$ **13.** $\dfrac{f(h) - f(0)}{h}$

15.   **17.**   **19.**   **21.**   **23.**   **25.**   **27.** $y = x + 3$

29. $y = \dfrac{3}{4}x + \dfrac{1}{2}$ **31.** $y = (0)x + 4 = 4$ **33.** $y = -4x - 10$ **35.** $y = -\dfrac{3}{4}x + \dfrac{13}{4}$ **37.** $y = \dfrac{12}{5}x - \dfrac{29}{5}$ **43.** 1/3   **45.** 16/3

47. a. $N(t) = 1200t + 900$ **b.** $N(6) = 8100$ **c.** $N(14) = 17,700$ **d.** 1200 cases per year   **49.** $P(x) = 40.50x - 1782$, $x = 44$, the break-even point   **51.** $P(x) = 79x - 10,270$, $x = 130$, the break-even point   **53. a.** \$275   **b.** \$283   **c.** \$355   **d.** \$8
55. a. $C(t) = 19,500.00 + 6.75t$ **b.** $R(t) = 55.00t$ **c.** $P(t) = 48.25t - 19,500.00$ **d.** approximately 405 days   **57.** $T \approx 19.62$ seconds
59. a.

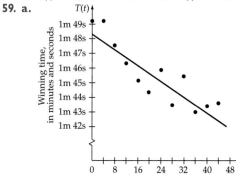

Year ($t = 0$ represents the year 1948)

b. Answers will vary. However, if you use the line through $(0, 108.2)$ and $(40, 103.2)$ then the equation of the line is $T(t) = -0.125t + 108.2$.
c. Using the equation in **b.** with $t = 52$ yields 101.7 seconds.

61. $y = -\dfrac{3}{4}x + \dfrac{15}{4}$ **63.** $y = x + 1$ **65.** $x = -5$ ft

67. a. $Q = (3, 10)$, $m = 5$ **b.** $Q = (2.1, 5.41)$, $m = 4.1$
c. $Q = (2.01, 5.0401)$, $m = 4.01$ **d.** 4
73. $y = -2x + 11$ **75.** $5x + 3y = 15$ **77.** $3x + y = 17$

81. $(9/2, 81/4)$ **83. a.** $P \approx \begin{cases} 0.00875t + 0.1, & 0 \le t \le 8 \\ 0.014t + 0.058, & 8 < t \le 13 \end{cases}$ **b.** $\approx 28\%$

Exercise Set 3.4, page 184

1. d   **3.** b   **5.** g   **7.** c   **9.** $f(x) = (x + 2)^2 - 3$
vertex: $(-2, -3)$
axis of symmetry: $x = -2$

11. $f(x) = (x - 4)^2 - 11$
vertex: $(4, -11)$
axis of symmetry: $x = 4$

13. $f(x) = (x - (-3/2))^2 - 5/4$
vertex: $(-3/2, -5/4)$
axis of symmetry: $x = -3/2$

15. $f(x) = -(x - 2)^2 + 6$
vertex: $(2, 6)$
axis of symmetry: $x = 2$

17. $f(x) = -3(x - 1/2)^2 + 31/4$
vertex: $(1/2, 31/4)$
axis of symmetry: $x = 1/2$

19. vertex: $(5, -25)$, $f(x) = (x - 5)^2 - 25$
21. vertex: $(0, -10)$, $f(x) = x^2 - 10$
23. vertex: $(3, 10)$, $f(x) = -(x - 3)^2 + 10$
25. vertex: $(3/4, 47/8)$, $f(x) = 2(x - 3/4)^2 + 47/8$
27. vertex: $(1/8, 17/16)$, $f(x) = -4(x - 1/8)^2 + 17/16$
29. -16, minimum
31. 11, maximum
33. $-1/8$, minimum
35. -11, minimum
37. 35, maximum

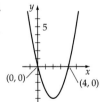

39. a. 27 feet **b.** $22\frac{5}{16}$ feet **c.** 20.1 feet from the center **41. a.** $w = \dfrac{600 - 2l}{3}$ **b.** $A = 200l - \dfrac{2}{3}l^2$ **c.** $w = 100$ feet, $l = 150$ feet

43. y-intercept $(0, 0)$; x-intercepts $(0, 0)$ and $(-6, 0)$ **45.** y-intercept $(0, -6)$; no x-intercepts **47.** 740 units yield a maximum revenue of \$109,520. **49.** 85 units yield a maximum profit of \$24.25. **51.** $P(x) = -0.1x^2 + 50x - 1840$, break-even points: $x = 40$ and $x = 460$ **53. a.** $R(x) = -0.25x^2 + 30.00x$ **b.** $P(x) = -0.25x^2 + 27.50x - 180$ **c.** \$576.25 **d.** 55 **55. a.** $t = 4$ seconds **b.** 256 feet

c. $t = 8$ seconds **57.** $r = \dfrac{48}{4 + \pi} \approx 6.72$ feet, $h = r \approx 6.72$ feet **59.** $f(x) = \dfrac{3}{4}x^2 - 3x + 4$ **61. a.** $w = 16 - x$ **b.** $A = 16x - x^2$

63. The discriminant is $b^2 - 4(1)(-1) = b^2 + 4$, which is positive for all b. **65.** increases the height of each point on the graph by c units. **67.** 4, 4

Exercise Set 3.5, page 195

1. 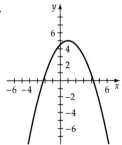 **3.**   **5.** 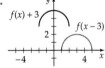 **7.** 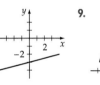 **9.** 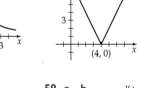 **11.**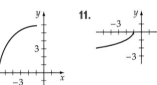

13. a. no **b.** yes **15. a.** no **b.** no **17. a.** yes **b.** yes **19. a.** yes **b.** yes **21. a.** yes **b.** yes **23.** no **25.** yes
27. yes **29.** yes **31.** 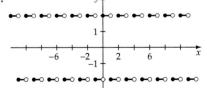 **33.** **35.** **37.**

39. 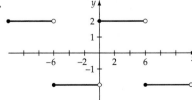 **41.** **43.** even **45.** odd **47.** even **49.** even **51.** even **53.** even **55.** neither **57. a., b.** 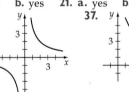 **59. a., b.**

61. **63. a.** **b.**

65. a.

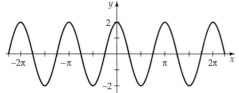

b.

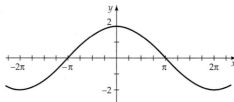

67.

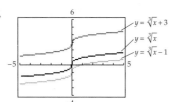

69.

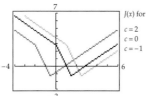

71.

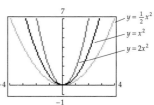

73.

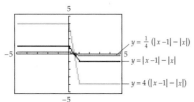

75. a.

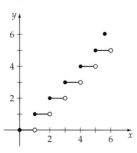

c.

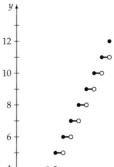

77. a. $f(x) = \dfrac{2}{(x+1)^2 + 1} + 1$

b. $f(x) = -\dfrac{2}{(x-2)^2 + 1}$

b.

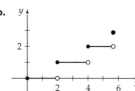

Exercise Set 3.6, page 206

1. $f(x) + g(x) = x^2 - x - 12$, domain all real numbers
$f(x) - g(x) = x^2 - 3x - 18$, domain all real numbers
$f(x) \cdot g(x) = x^3 + x^2 - 21x - 45$, domain all real numbers
$f(x)/g(x) = x - 5$, domain $\{x \mid x \neq -3\}$

3. $f(x) + g(x) = 3x + 12$, domain all real numbers
$f(x) - g(x) = x + 4$, domain all real numbers
$f(x) \cdot g(x) = 2x^2 + 16x + 32$, domain all real numbers
$f(x)/g(x) = 2$, domain $\{x \mid x \neq -4\}$

5. $f(x) + g(x) = x^3 + 2x^2 + 8x$, domain all real numbers
$f(x) - g(x) = x^3 + 2x^2 + 6x$, domain all real numbers
$f(x) \cdot g(x) = x^4 + 2x^3 + 7x^2$, domain all real numbers
$f(x)/g(x) = x^2 + 2x + 7$, domain $\{x \mid x \neq 0\}$

7. $f(x) + g(x) = 4x^2 + 7x - 12$, domain all real numbers
$f(x) - g(x) = x - 2$, domain all real numbers
$f(x) \cdot g(x) = 4x^4 + 14x^3 - 12x^2 - 41x + 35$, domain all real numbers
$f(x)/g(x) = 1 + \dfrac{x - 2}{2x^2 + 3x - 5}$, domain $\{x \mid x \neq 1, x \neq -5/2\}$

9. $f(x) + g(x) = \sqrt{x - 3} + x$, domain $\{x \mid x \geq 3\}$
$f(x) - g(x) = \sqrt{x - 3} - x$, domain $\{x \mid x \geq 3\}$
$f(x) \cdot g(x) = x\sqrt{x - 3}$, domain $\{x \mid x \geq 3\}$
$f(x)/g(x) = \dfrac{\sqrt{x - 3}}{x}$, domain $\{x \mid x \geq 3\}$

11. $f(x) + g(x) = \sqrt{4 - x^2} + 2 + x$, domain $\{x \mid -2 \leq x \leq 2\}$
$f(x) - g(x) = \sqrt{4 - x^2} - 2 - x$, domain $\{x \mid -2 \leq x \leq 2\}$
$f(x) \cdot g(x) = (\sqrt{4 - x^2})(2 + x)$, domain $\{x \mid -2 \leq x \leq 2\}$
$f(x)/g(x) = \dfrac{\sqrt{4 - x^2}}{2 + x}$, domain $\{x \mid -2 < x \leq 2\}$

13. 18 **15.** $-9/4$ **17.** 30 **19.** 12 **21.** 300 **23.** $-384/125$ **25.** $-5/2$ **27.** $-1/4$ **29.** 2 **31.** $2x + h$ **33.** $4x + 2h + 4$

35. $-8x - 4h$ **37.** $(g \circ f)(x) = 6x + 3$
$(f \circ g)(x) = 6x - 16$

39. $(g \circ f)(x) = x^2 + 4x + 1$
$(f \circ g)(x) = x^2 + 8x + 11$

41. $(g \circ f)(x) = -5x^3 - 10x$
$(f \circ g)(x) = -125x^3 - 10x$

43. $(g \circ f)(x) = \dfrac{1 - 5x}{x + 1}$
$(f \circ g)(x) = \dfrac{2}{3x - 4}$

45. $(g \circ f)(x) = \dfrac{\sqrt{1 - x^2}}{x}$
$(f \circ g)(x) = \dfrac{1}{x - 1}$

47. $(g \circ f)(x) = -\dfrac{2|5 - x|}{3}$
$(f \circ g)(x) = \dfrac{3|x|}{|5x + 2|}$

49. 66 **51.** 51 **53.** -4 **55.** 41 **57.** $-3848/625$

59. $6 + 2\sqrt{3}$　**61.** $16c^2 + 4c - 6$　**63.** $9k^4 + 36k^3 + 45k^2 + 18k - 4$　**65. a.** $A(t) = \pi(1.5t)^2$, $A(2) = 9\pi$ square feet ≈ 28.27 square feet
b. $V(t) = 2.25\pi t^3$, $V(3) = 60.75\pi$ cubic feet ≈ 190.85 cubic feet　**67. a.** $d(t) = \sqrt{(48 - t)^2 - 4^2}$　**b.** $s(35) = 13$ feet, $d(35) \approx 12.37$ feet
69. $(Y \circ F)(x)$ converts x inches to yards.　**71. a.** 99.8; this is identical to the slope of the line through $(0, C(0))$ and $(1, C(1))$.
b. 156.2　**c.** -49.7　**d.** -30.8　**e.** -16.4　**f.** 0　**79. a.** $(s \circ m)(x) = 87 + 49{,}300/x$　**b.** \$89

Exercise Set 3.7, page 215

9. $f^{-1}(x) = \dfrac{x - 1}{4}$　**11.** $F^{-1}(x) = \dfrac{1 - x}{6}$　**13.** $j^{-1}(t) = \dfrac{t - 1}{2}$　**15.** $f^{-1}(v) = \sqrt[3]{1 - v}$　**17.** $f^{-1}(x) = -\dfrac{4x}{x + 3}$, $x \neq -3$

19. $M^{-1}(t) = \dfrac{5}{1 - t}$, $t \neq 1$　**21.** $r^{-1}(t) = -\sqrt{\dfrac{1}{t}}$, $t > 0$　**23.** $J^{-1}(x) = \sqrt{x - 4}$, $x \geq 4$

25. $f^{-1}(x) = \sqrt{x - 3}$, domain $\{x \mid x \geq 3\}$, range $\{y \mid y \geq 0\}$　**27.** $f^{-1}(x) = x^2$, domain $\{x \mid x \geq 0\}$, range $\{y \mid y \geq 0\}$
　　 f has domain $\{x \mid x \geq 0\}$, range $\{y \mid y \geq 3\}$.　　　　　　 f has domain $\{x \mid x \geq 0\}$, range $\{y \mid y \geq 0\}$.
29. $f^{-1}(x) = \sqrt{9 - x^2}$, domain $\{x \mid 0 \leq x \leq 3\}$, range $\{y \mid 0 \leq y \leq 3\}$　**31.** $f^{-1}(x) = 2 + \sqrt{x + 3}$, domain $\{x \mid x \geq -3\}$, range $\{y \mid y \geq 2\}$
　　 f has domain $\{x \mid 0 \leq x \leq 3\}$, range $\{y \mid 0 \leq y \leq 3\}$.　　　 f has domain $\{x \mid x \geq 2\}$, range $\{y \mid y \geq -3\}$.
33. $f^{-1}(x) = -4 - \sqrt{x + 25}$, domain $\{x \mid x \geq -25\}$, range $\{y \mid y \leq -4\}$　**35.**　　　　　**37.**
　　 f has domain $\{x \mid x \leq -4\}$, range $\{y \mid y \geq -25\}$.

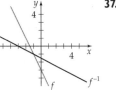

39. 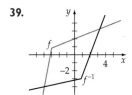　**41.** 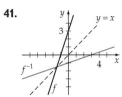　**43.** 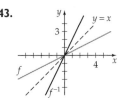　**45.** 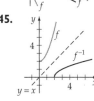　**47.**

49. $f^{-1}(x) = \dfrac{x - b}{a}$, $a \neq 0$　**51.** $f^{-1}(x) = \dfrac{x + 1}{1 - x}$, $x \neq 1$　**53.** no　**55.** yes　**57.** yes　**59.** no　**61.** 5　**63.** 4

65. The reflection of f about the line $y = x$ yields f. Thus f is its own inverse.

Exercise Set 3.8, page 221

1. $d = kt$　**3.** $y = k/x$　**5.** $m = knp$　**7.** $V = klwh$　**9.** $A = ks^2$　**11.** $F = km_1m_2/d^2$　**13.** $y = kx$, $k = 4/3$　**15.** $r = kt^2$, $k = 1/81$
17. $T = krs^2$, $k = 7/25$　**19.** $V = klwh$, $k = 1$　**21.** 1.02 liters　**23.** 437.5 pounds per square foot　**25. a.** approximately 3.3 seconds
b. approximately 3.7 feet　**27.** 112 decibels　**29. a.** 9 times larger　**b.** 3 times larger　**c.** 27 times larger　**31.** 6 times larger
33. approximately 3.2 miles per second　**35.** approximately 3950 pounds　**37.** $d \approx 142$ million miles

Chapter 3 True/False Exercises, page 228

1. False. Let $f(x) = x^2$. Then $f(3) = f(-3) = 9$, but $3 \neq -3$.　**2.** False. $f(x) = x^2$ does not have an inverse function.　**3.** False. Let
$f(x) = 2x$, $g(x) = 3x$. Then $f(g(0)) = 0$ and $g(f(0)) = 0$, but f and g are not inverse functions.　**4.** True　**5.** False. Let $f(x) = 3x$.
$[f(x)]^2 = 9x^2$, whereas $f[f(x)] = f(3x) = 3(3x) = 9x$.　**6.** False. Let $f(x) = x^2$. Then $f(1) = 1$, $f(2) = 4$. Thus $f(2)/f(1) = 4 \neq 2/1$.
7. True　**8.** False. $f(-1 + 3) = f(2) = 2$. $f(-1) + f(3) = 1 + 3 = 4$.　**9.** True　**10.** True　**11.** True　**12.** True

Chapter 3 Review Exercises, page 229

1. $\sqrt{181}$　**3.** $(-1/2, 10)$　**5.** center $(3, -4)$, radius 9　**7.** $(x - 2)^2 + (y + 3)^2 = 5^2$　**9. a.** 2　**b.** 10　**c.** $3t^2 + 4t - 5$
d. $3x^2 + 6xh + 3h^2 + 4x + 4h - 5$　**e.** $9t^2 + 12t - 15$　**f.** $27t^2 + 12t - 5$　**11. a.** 5　**b.** -11　**c.** $x^2 - 12x + 32$　**d.** $x^2 + 4x - 8$
13. $8x + 4h - 3$　**15.** 　**17.** 　**19.**

increasing on $[3, \infty)$
decreasing on $(-\infty, 3]$

increasing on $[-2, 2]$
constant on $(-\infty, -2] \cup [2, \infty)$

increasing on $(-\infty, \infty)$

21. domain {x | x is a real number}   **23.** domain {x | −5 ≤ x ≤ 5}   **25.** $y = -2x + 1$ **27.** $y = \dfrac{3}{4}x + \dfrac{19}{2}$ **29.** $f(x) = (x + 3)^2 + 1$

31. $f(x) = -(x + 4)^2 + 19$ **33.** $f(x) = -3(x - 2/3)^2 - 11/3$ **35.** $(1, 8)$ **37.** $(5, 161)$ **39.** $4\sqrt{5}/5$ **41.**

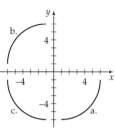

43. symmetric to the y-axis   **45.** symmetric to the origin   **47.** symmetric to the x-axis, the y-axis, and the origin
49. symmetric to the x-axis, the y-axis, and the origin
51.

53.

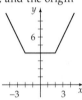

55.

   a. domain all real numbers
      range {y | y ≤ 4}
   b. even

   a. domain all real numbers
      range {y | y ≥ 4}
   b. even

   a. domain all real numbers
      range all real numbers
   b. odd

57. $F(x) = (x + 2)^2 - 11$ **59.** $P(x) = 3(x - 0)^2 - 4$ **61.** $W(x) = -4(x + 3/4)^2 + 33/4$ **63.**

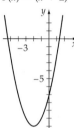

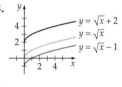

65.

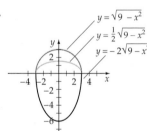

67.

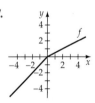

69. $f(x) + g(x) = x^2 + x - 6$, domain all real numbers
    $f(x) - g(x) = x^2 - x - 12$, domain all real numbers
    $f(x) \cdot g(x) = x^3 + 3x^2 - 9x - 27$, domain all real numbers
    $f(x)/g(x) = x - 3$, domain {x | x ≠ −3}
71. yes
73. yes

75. $f^{-1}(x) = \dfrac{x + 4}{3}$ **77.** $h^{-1}(x) = -2x - 4$ **79.** 25, 25   **81. a.** 18 **b.** 15 **c.** 13.5 **d.** 12.03 **e.** 12

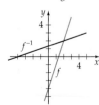

Chapter 3 Test, page 231

1. midpoint $(1,1)$; length $2\sqrt{13}$ **2.** $(0,\sqrt{2})$, $(0,-\sqrt{2})$, $(-4,0)$

3.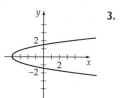

4. center $(2,-1)$; radius 3

5. domain $\{x \mid x \geq 4 \text{ or } x \leq -4\}$ **6.** $3\sqrt{10}/5$ **7.** 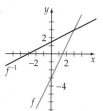   **8.**

domain all real numbers; range $\{y \mid y \geq 2\}$

increasing on $(-\infty, 2]$
decreasing on $[2, \infty)$

9. a. $R = 12.00x$ **b.** $P = 11.25x - 875$ **c.** $x = 78$ **10.**

11. a. even   **b.** odd   **c.** neither

12. $y = -\dfrac{2}{3}x + \dfrac{2}{3}$

13. -12 minimum

14. $x^2 + x - 3$; $\dfrac{x^2 - 1}{x - 2}$, $x \neq 2$

15. $2x + h$

16. $x - 2\sqrt{x - 2} - 1$ **17.** $f^{-1}(x) = \sqrt{x + 9}$
domain of f: $\{x \mid x \geq 0\}$
range of f: $\{y \mid y \geq -9\}$
domain of f^{-1}: $\{x \mid x \geq -9\}$
range of f^{-1}: $\{y \mid y \geq 0\}$

18. $f^{-1}(x) = \dfrac{1}{2}x + \dfrac{3}{2}$

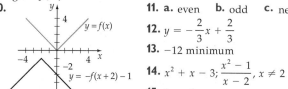

19. a. 25 feet per second   **b.** 22.5 feet per second   **c.** 20.05 feet per second   **20.** 5.69 lumens

Exercise Set 4.1, page 241

1. $5x^2 - 9x + 10 + \dfrac{-10}{x + 3}$ **3.** $x^3 + 5x^2 - 9x - 45$ **5.** $x^2 - \dfrac{100}{3} + \dfrac{100x/3 + 10/3}{3x^2 + x + 1}$ **7.** $4x^2 + 1 + \dfrac{11}{5x^2 - 2}$

9. $x + 4 + \dfrac{6x - 3}{x^2 + x - 4}$ **11.** $4x^2 + 3x + 12 + \dfrac{17}{x - 2}$ **13.** $4x^2 - 4x + 2 + \dfrac{1}{x + 1}$ **15.** $x^4 + 4x^3 + 6x^2 + 24x + 101 + \dfrac{403}{x - 4}$

17. $x^4 + x^3 + x^2 + x + 1$ **19.** $8x^2 + 6$ **21.** $x^7 + 2x^6 + 5x^5 + 10x^4 + 21x^3 + 42x^2 + 85x + 170 + \dfrac{344}{x - 2}$

23. $x^5 - 3x^4 + 9x^3 - 27x^2 + 81x - 242 + \dfrac{716}{x + 3}$ **25.** $3x - 3.1 + \dfrac{4.07}{x - 0.3}$ **27.** $2x^2 - 11x - 17 + \dfrac{3}{x}$ **29.** $1 + \dfrac{6}{x + 2}$ **31.** 25

33. 45 **35.** -2230 **37.** -80 **39.** -187 **41.** yes   **43.** yes   **45.** yes   **47.** yes   **49.** yes   **51.** yes   **63.** 13
65. $(x + 3)(x - 1)(x - 2)$ **67.** $(x + 3)(x + 2)(x + 1)(x - 4)$ **71.** 13 **73.** By the Factor Theorem, $P(x)$ has a factor of $x - c$ if and only
if $P(c) = 0$. However, $P(c) = 4c^4 + 7c^2 + 12$, which is greater than 0 for any real number c.

Exercise Set 4.2, page 249

1. up to far left, up to far right   **3.** down to far left, up to far right   **5.** down to far left, down to far right
7. down to far left, up to far right   **9.** up to far left, down to far right   **11.** $(-2, -5)$, minimum   **13.** $(-4, 17)$, maximum
15. length 62.5 feet by width 125 feet   **17.** 6

19. $(-2.1, 5.0)$, maximum
$(1.4, -16.9)$, minimum

21. $(4, -77)$, minimum
$(-2, 31)$, maximum

23. $(-1, -14)$ and $(3, -14)$, minima
$(1, 2)$, maximum

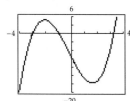

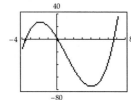

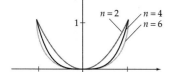

25. 2.137 inches, 337.1 cubic inches   **27.** 103.3 kilometers/hour; 1:0511 P.M.   **29.** 55°, 1:44 P.M.   **31.** 0.4   **39.** $-7/2, -2, 3$

41. $0, 2/5, 1$ **43.** $-5, -7/3, 11/2$ **45.** $-3, 0, 2$ **47.** 1   **49.** $-2.4, 0.4$ **51.** $-6.2, 0.2$

53.

55. When n is an odd number, the graph passes through the x-axis. When n is an even number, the graph touches the x-axis.

57. 3 and 4

59. False. As one example, let $P(x) = x^3 - 5x^2 + 6x$, $a = 1$, and $b = 4$. In this case, $P(a) > 0$ and $P(b) > 0$; however, $P(x)$ has a zero when $x = 2$.

61.

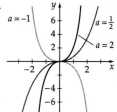

Exercise Set 4.3, page 259

1. 3 (multiplicity 2), -5 (multiplicity 1)   **3.** 0 (multiplicity 2), $-5/3$ (multiplicity 2)

5. 2 (multiplicity 1), -2 (multiplicity 1), -3 (multiplicity 2)   **7.** 5 (multiplicity 2), -2 (multiplicity 2)

9. -3 (multiplicity 1), 3 (multiplicity 1), -1 (multiplicity 1), 1 (multiplicity 1)   **11.** $\pm 1, \pm 2, \pm 4, \pm 8$

13. $\pm 1, \pm 2, \pm 3, \pm 4, \pm 6, \pm 12, \pm 1/2, \pm 3/2$ **15.** $\pm 1, \pm 2, \pm 4, \pm 1/2, \pm 1/3, \pm 2/3, \pm 4/3, \pm 1/6$ **17.** $\pm 1, \pm 3, \pm 9, \pm 1/2, \pm 3/2, \pm 9/2$

19. $\pm 1, \pm 7, \pm 1/2, \pm 7/2, \pm 1/4, \pm 7/4$ **21.** $\pm 1, \pm 2, \pm 4, \pm 8, \pm 16, \pm 32$ **23.** upper bound 2, lower bound -5

25. upper bound 4, lower bound -4 **27.** upper bound 1, lower bound -4 **29.** upper bound 1, lower bound -5

31. upper bound 4, lower bound -2 **33.** upper bound 2, lower bound -1 **35.** one positive, two or no negative

37. two or no positive, one negative   **39.** one positive, three or one negative   **41.** one positive, two or no negative

43. three or one positive, one negative   **45.** one positive, no negative   **47.** $2, -1, -4$ **49.** $3, -4, 1/2$

51. $1/2, -1/3, -2$ (multiplicity 2)   **53.** $1, -1, -9/2$ **55.** $1/2, 4, \sqrt{3}, -\sqrt{3}$ **57.** $2, -1$ (multiplicity 2)

59. $0, -2, 1 + \sqrt{2}, 1 - \sqrt{2}$ **61.** -1 (multiplicity 3), 2   **63.** $-3/2, 1$ (multiplicity 2), 8   **65.** $-5, -1, 1, 4$ **67.** $-2.5, -1, 2, 3.5$

69. $-3, -0.5, 2, 3.5, 5$ **71.** $-0.5, 2$ **79.** yes   **81.** yes

Exercise Set 4.4, page 265

1. $1 - i, 1/2$ **3.** $i, -3$ **5.** $-i\sqrt{2}, 1, \sqrt{5}, -\sqrt{5}$ **7.** $1 + \dfrac{1}{2}i, 1 - \dfrac{1}{2}i$ **9.** $1 - 3i, 1 + 2i, 1 - 2i$ **11.** $-i, 3, -1$ (multiplicity 2)

13. $2, -3, 2i, -2i$ **15.** $1/2, -3, 1 + 5i, 1 - 5i$ **17.** 1 (multiplicity 3), $3 + 2i, 3 - 2i$ **19.** $\dfrac{3}{2}, \dfrac{-1 + i\sqrt{7}}{2}, \dfrac{-1 - i\sqrt{7}}{2}$ **21.** $-2/3, 3/4, 5/2$

23. $-i, i, 3$ (multiplicity 2)   **25.** -3 (multiplicity 2), 1 (multiplicity 2)   **27.** $x(x - 2)(x + 1)$ **29.** $x(x^2 + 9)$ **31.** $(x^2 + 6)(x + 2)(x - 2)$

33. $(x^2 + 2)(x^2 + 1)$ **35.** $(x + 1)(x - 3)(x^2 + 4)$ **37.** $x^3 - 3x^2 - 10x + 24$ **39.** $x^3 - 3x^2 + 4x - 12$

41. $x^4 - 10x^3 + 63x^2 - 214x + 290$ **43.** $x^5 - 22x^4 + 212x^3 - 1012x^2 + 2251x - 1830$ **45.** $4x^3 - 19x^2 + 224x - 159$

47. $x^3 + 13x + 116$ **49.** $x^4 - 18x^3 + 131x^2 - 458x + 650$ **51.** $3x^3 - 12x^2 + 3x + 18$ **53.** $-2x^4 + 4x^3 + 36x^2 - 140x + 150$

55. Because $x^3 - x^2 - ix^2 - 9x + 9 + 9i$ does not have real coefficients, the theorem does not apply.   **57.** $P(x) = (x - 2)^3(x^2 + 9)$

59. $P(x) = \dfrac{1}{2}x^5 - 4x^4 + \dfrac{25}{2}x^3 - 19x^2 + 14x - 4$

Exercise Set 4.5, page 280

1. vertical asymptotes: $x = 0, x = -3$ **3.** vertical asymptotes: $x = 4/3, x = -1/2$ **5.** horizontal asymptote: $y = 4$

7. horizontal asymptote: $y = 30$ **9.** vertical asymptote: $x = -4$ **11.** vertical asymptote: $x = 3$
horizontal asymptote: $y = 0$ horizontal asymptote: $y = 0$

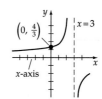

13. vertical asymptote: $x = 0$
horizontal asymptote: $y = 0$

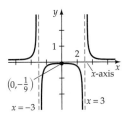

15. vertical asymptote: $x = -4$
horizontal asymptote: $y = 1$

17. vertical asymptote: $x = 2$
horizontal asymptote: $y = -1$

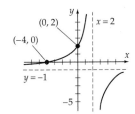

19. vertical asymptotes: $x = 3$, $x = -3$
horizontal asymptote: $y = 0$

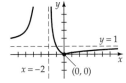

21. vertical asymptotes: $x = -3$, $x = 1$
horizontal asymptote: $y = 0$

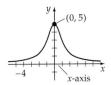

23. vertical asymptotes: $x = 3$, $x = -3$
horizontal asymptote: $y = 0$

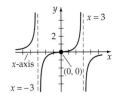

25. vertical asymptote: $x = -2$
horizontal asymptote: $y = 1$

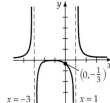

27. vertical asymptote: none
horizontal asymptote: $y = 0$

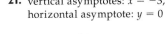

29. vertical asymptotes: $x = 3$, $x = -3$
horizontal asymptote: $y = 2$

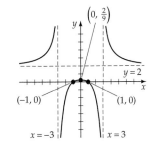

31. vertical asymptotes: $x = -1 + \sqrt{2}$, $x = -1 - \sqrt{2}$
horizontal asymptote: $y = 1$ **33.** $y = 3x - 7$ **35.** $y = x$ **37.** vertical asymptote: $x = 0$
slant asymptote: $y = x$

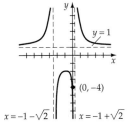

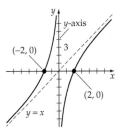

39. vertical asymptote: $x = -3$
slant asymptote: $y = x - 6$

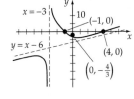

41. vertical asymptote: $x = 4$
slant asymptote: $y = 2x + 13$

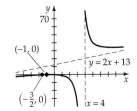

43. vertical asymptote: $x = -2$
slant asymptote: $y = x - 3$

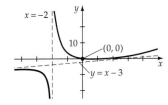

45. vertical asymptotes: $x = 2$, $x = -2$ **47.**
slant asymptote: $y = x$

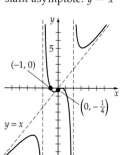

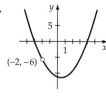

49.

51.

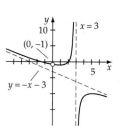

53.

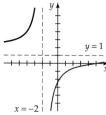

55.

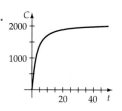

57. $x = -1.3$, $x = 2.3$ **65. a.** \$1333.33 **b.** \$8000 **c.**
59. none
61. $x = -2$, $x = 1$, $x = 3$
63. $x = -3$, $x = 4$

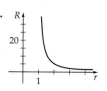

67. a. 611   **d.**
b. 1777
c. $y = 2000$

18 weeks   **69. a.** As the radius of the blood vessel gets   **c.**
smaller, the resistance gets larger.
b. As the radius of the blood vessel gets
larger, the resistance approaches zero.

71. $(-2, 2)$ **73.** $(0, 1), (-4, 1)$

Chapter 4 True/False Exercises, page 286

1. False; $x - i$ has a zero of i, but it does not have a zero of $-i$.   **2.** False; Descartes' Rule of Signs indicates that $x^3 - x^2 + x - 1$ has 3 or 1 positive zeros.   **3.** True   **4.** True   **5.** False; $f(x) = \dfrac{x}{x^2 + 1}$ does not have a vertical asymptote.

6. False; $f(x) = \dfrac{(x-2)^2}{(x-3)(x-2)} = \dfrac{x-2}{x-3}$, $x \neq 2$.   **7.** True   **8.** True   **9.** True   **10.** True   **11.** True   **12.** True   **13.** True

14. False; $x^2 + 1$ does not have a real zero.

Chapter 4 Review Exercises, page 286

1. $x + 4 + \dfrac{-5x - 29}{x^2 + x + 3}$ **3.** $-x + \dfrac{3x^2 - 12x - 3}{x^3 + x}$ **5.** $3x^2 - 5x - 1$ **7.** $4x^2 + x + 8 + \dfrac{22}{x - 3}$ **9.** $3x^2 - 6x + 7 + \dfrac{-13}{x + 2}$

11. $3x^2 + 5x - 11$ **13.** 77   **15.** 33   **21.**   **23.**   **25.**   **27.** $\pm 1, \pm 2, \pm 3, \pm 6$

29. $\pm 1, \pm 2, \pm 3, \pm 4, \pm 6, \pm 12, \pm 1/3, \pm 2/3, \pm 4/3, \pm 1/5, \pm 2/5, \pm 3/5, \pm 4/5, \pm 6/5, \pm 12/5, \pm 1/15, \pm 2/15, \pm 4/15$ **31.** ± 1
33. no positive and three or one negative   **35.** one positive and one negative   **37.** $1, -2, -5$ **39.** -2 (multiplicity 2), $-1/2, -4/3$
41. 1 (multiplicity 4)   **43.** $2x^3 - 3x^2 - 23x + 12$ **45.** $x^4 - 3x^3 + 27x^2 - 75x + 50$ **47.** vertical asymptote: $x = -2$, horizontal asymptote: $y = 3$ **49.** vertical asymptote: $x = -1$, slant asymptote: $y = 2x + 3$

51.   **53.**   **55.** 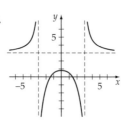   **57.** 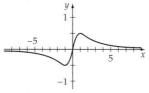   **59.** 1.325   **61.** 0.786

Chapter 4 Test, page 287

1. $3x^2 - x + 6 - \dfrac{13}{x+2}$ **2.** 43   **4.** up to the far left and down to the far right   **5.** 0, 2/3, −3   **6.** $P(1) < 0, P(2) > 0$. Therefore P has a zero between 1 and 2.   **7.** 2 (multiplicity 2), −2 (multiplicity 2), 3/2 (multiplicity 1), −1 (multiplicity 3)   **8.** ±1, ±3, ±1/2, ±3/2, ±1/3, ±1/6   **9.** upper bound 4, lower bound −5   **10.** 4, 2, or 0 positive zeros, no negative zero.   **11.** 1/2, 3, −2   **12.** i, 2, −1/2   **13.** 0, 1 (multiplicity 2), $2 + i$, $2 - i$ **14.** $x^4 - 5x^3 + 8x^2 - 6x$ **15.** vertical asymptotes: $x = 3, x = 2$ **16.** horizontal asymptote: $y = 3/2$ **17.**   **18.**   **19.** 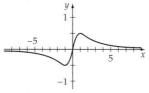   **20.** 1.8

Exercise Set 5.1, page 296

1. 4.72880   **3.** 442.335   **5.** 2.17458   **7.** 164.022   **9.** 5.65223   **11.** 0.969476   **13.** 70.4503   **15.** 19.8130   **17.** 14.0940   **19.** 15.1543   **21.** 3353.33   **23.** 8103.08

25.   **27.**   **29.**   **31.**   **33.**

35.   **37.**   **39.**   **41.**   **43.**   **45.**

47.   **49.** 1.58   **51.** 0.69   **53.** 0.79   **55.** 0.80

57. a. 0.6922   **b.** $f(x)$ approaches $y = 1$
59. a. 20,000   **b.** 40,000   **c.** 320,000
61. a. $228.26   **b.** $10,956.48   **c.** $1956.48
63. a. $12,428.73   **b.** $10,367.67   **c.** 34

65.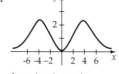

domain: $(-\infty, \infty)$
range: $(-1, 1)$
f is an odd function.

67.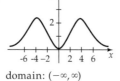

domain: $(-\infty, \infty)$
range: $[0, 2.2)$
f is an even function.

69.

domain: $(-\infty, \infty)$
range: $(0.5, \infty)$
f is neither even nor odd.

71.

domain: $(-\infty, 0]$
range: $[0, 1)$
f is neither even nor odd.

73. a.

domain: $(-\infty, \infty)$
range: $[2, \infty)$

b.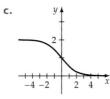

domain: $(-\infty, \infty)$
range: $(-\infty, \infty)$

75. a. -2
  b. 4
  c. does not exist
  d. h is not a real-valued
     exponential function
     because b is not a
     positive constant.

77.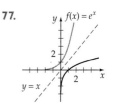

81. e^{π} **83. a.** vertical asymptote: none   **c.**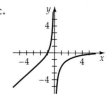
          horizontal asymptote: $y = 0$
                                        $y = 2$
        b. none

85. a. vertical asymptote: $x = 0$ **c.**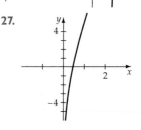
          horizontal asymptote: $y = 0$
        b. (-10)

87. a. vertical asymptote: $x = 0$ **c.**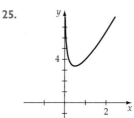
          horizontal asymptote: $y = 0$

  b. $\left(\dfrac{1}{2}, 0\right)$

Exercise Set 5.2, page 304

1. $10^2 = 100$ **3.** $5^3 = 125$ **5.** $3^4 = 81$ **7.** $b^t = r$ **9.** $3^{-3} = 1/27$ **11.** $\log_2 16 = 4$ **13.** $\log_7 343 = 3$ **15.** $\log_{10} 10,000 = 4$
17. $\log_b j = k$ **19.** $\log_b b = 1$ **21.** 6   **23.** 5   **25.** 3   **27.** -2 **29.** 0   **31.** $\log_b x + \log_b y + \log_b z$ **33.** $\log_3 x - 4\log_3 z$
35. $1/2 \log_b x - 3\log_b y$ **37.** $\log_b x + 2/3 \log_b y - 1/3 \log_b z$ **39.** $1/2 \log_7 x + 1/2 \log_7 z - 2\log_7 y$ **41.** $\log_{10} [x^2(x + 5)]$
43. $\log_b \sqrt{\dfrac{(x - y)^3 (x + y)}{z}}$ **45.** $\log_8 (x + y)$ **47.** $\ln [x^2(x - 3)^4]$ **49.** $\ln (xz/y)$ **51.** 1.5395   **53.** 0.86719   **55.** -1.7308

57. -2.3219 **59.** 0.87357   **61.** 3.06   **63.** 0.00334   **65.** 7.40   **67.** 0.300   **69.** $[1, 10^{1000}]$ **71.** $[e^e, e^{e^3}]$ **73.** $(0, 1)$
77. reflexive property of equality, definition of $\log_b x = n$

Exercise Set 5.3, page 312

1.   **3.**   **5.**   **7.**   **9.**

11.   **13.**   **15.**   **17.**   **19.**

21.   **23.**   **25.**   **27.**

29. $f(x) \to 0$　　　**31.** Yes　　　**33.** False　　**35. a.** 49　**b.** 2.2　**37.** $6,774

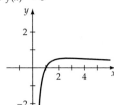

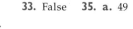

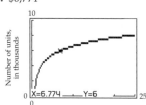

Note: x and y values shown are rounded.

39. 4.0%　　　　　　　　　　　　　　　　　　**41.** 8.6　**43.**　　　　　　　　　　　120.2 parsecs

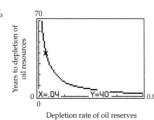

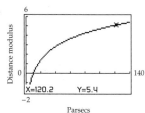

Note: x and y values shown are rounded.　　　Note: x and y values shown are rounded.

45.　　　　　　　　5.0 parsecs　**47.** $f(x)$ and $g(x)$ are inverse functions.　**49.**

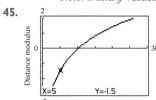

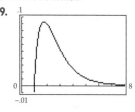

Note: x and y values shown are rounded.
51. domain $\{x \mid x \geq 1\}$, range $\{y \mid y \geq 0\}$　**53.** domain $\{x \mid -1 < x < 1\}$, range $\{y \mid y \geq 100\}$　**55.** domain $\{x \mid x > 1\}$, range all real numbers
57. a. 3　**b.** 1.386　**c.** 3.296　**59.** The domain of F is $\{x \mid x \neq 0\}$, but the domain of G is $\{x \mid x > 0\}$　**61.** 8.8

Exercise Set 5.4, page 320

1. 6　**3.** $-3/2$　**5.** $-6/5$　**7.** 3　**9.** $\dfrac{\log 70}{\log 5}$　**11.** $-\dfrac{\log 120}{\log 3}$　**13.** $\dfrac{\log 315 - 3}{2}$　**15.** $\ln 10$　**17.** $\dfrac{\ln 2 - \ln 3}{\ln 6}$　**19.** $\dfrac{3 \log 2 - \log 5}{2 \log 2 + \log 5}$
21. $2 + 2\sqrt{2}$　**23.** 199/95　**25.** 3　**27.** 10^{10}　**29.** 2　**31.** 5　**33.** $\log (20 + \sqrt{401})$　**35.** $1/2 \log (3/2)$　**37.** $\ln (15 \pm 4\sqrt{14})$
39. $\ln (1 + \sqrt{65}) - \ln 8$　**41.** 1.61　**43.** 0.96　**45.** 2.20　**47.** -1.93　**49.** -1.34　**51.** 2.2　**53.** 3.2×10^{-5}
55. a. 8500, 10,285　**b.** in 6 years　**57. a.** 60°F　**b.** in 27 minutes
59. a.　　　　　　　　　　　　**b.** 48 hours　　**61. a.**　　　　　　　　**b.** 27 years

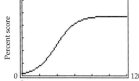　**c.** $P = 100$　　　　　　　　　　　**c.** $B = 1000$
　　　　　　　　　　　　d. As the number of hours　　　　　　　　**d.** As the number of years
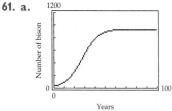　　　of training increases, it　　　　　　　　increases, the bison
　　　　　　　　　　　　is impossible to score over　　　　　　　population will never
　　　　　　　　　　　　100% on the test.　　　　　　　　　　exceed 1000.

63. a.　　　　　　　　　　　　**b.** 77 years　**65. a.**　　　　　　　**b.** 78 years　**67. a.** 146.9 feet per second
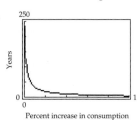　**c.** 1.9%　　　　　　　　　　　　**c.** 1.9%　　**b.** $v = 175$
　　　　　　　　　　　　　　　　　　　　　　　　d. more　　**c.** It is impossible for the
　　　　　　　　　　　　　　　　　　　　　　　　　　　　parachutist to fall faster
　　　　　　　　　　　　　　　　　　　　　　　　　　　　than 175 feet per second.

69. a. 1.72
  b. $v = 100$
  c. The object cannot fall faster than 100 feet per second.

71. a.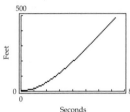

b. 2.6 seconds

73. a. the integers from 0 through 196
  b. 138

75. The second step; because log 0.5 < 0, the inequality sign must be reversed.   **77.** $x = y/(y - 1)$ **79.** $e^{0.336} \approx 1.4$ **81.** $(0, 1/\ln 2)$

Exercise Set 5.5, page 333

1. a. $9724.05   **b.** $11,256.80   **3. a.** $48,885.72   **b.** $49,282.20   **c.** $49,283.29   **5.** $24,730.82   **7.** 8.8 years   **9.** $t = (\ln 3)/r$
11. 14 years   **13. a.** 2200 bacteria   **b.** 17,600 bacteria   **15.** $N(t) = 22{,}600e^{0.01368t}$ **17. a.** 10,543,000   **b.** 2041
19. a.

Microgram of Na vs Time, in hours

  b. 3.18 micrograms   **c.** 15 hours   **d.** 30 hours   **21.** 6601 years ago
  23. 2378 years old
  25. a. 0.056   **b.** 42°   **c.** 54 minutes
    d. will never reach 34°
  27. 10 times
  29. 3.01
  31. a. 211 hours   **b.** 1386 hours
  33. 3.1 years

35. a.

v vs t graph

  b. 0.98 seconds
  c. $v = 32$
  d. As time increases, the velocity never exceeds 32 feet per second.

37. a.

s vs t graph

  b. 2.5 seconds
  c. 24.56
  d. The average speed of the object was 24.56 feet per second between $t = 1$ and $t = 2$.

39. a. 0.29
  b. 1994

41. a.

R vs t graph

  b. 1196 squirrels per year
  c. 7.94 years
  d. The maximum rate of growth of squirrels is 1196 squirrels per year.

43. a.

R vs t graph

  b. 32 bison per year
  c. 26.78 years
  d. The maximum rate of growth of bison is 32 bison per year.

45. 45 hours   **47. a.**

P vs x graph

  b. 0.504
  c. 8.08 million liters
  d. As the number of liters of water becomes very large ($x \to \infty$), the probability of running out of water becomes very small (approaches 0).

49. a. 2.2 seconds   **b.** It is impossible to exceed a velocity of 100 feet per second.   **51. a.** 21.7; 0.87   **b.** 1086; 0.88   **c.** 72,382; 0.92
53. 5 years   **55.** 100 times   **57. a.** 96   **b.** 3385   **c.** 13,395   **d.** 39,751   **59.** 6.93%

Chapter 5 True/False Exercises, page 342

1. True   **2.** True   **3.** True   **4.** False; because f is not defined for negative values of x, and thus $g(f(x))$ is undefined for negative values of x.   **5.** False; $h(x)$ is not an increasing function for $0 < b < 1$.   **6.** False; $j(x)$ is not an increasing function for $0 < b < 1$.
7. True   **8.** True   **9.** True   **10.** True   **11.** False; $\log x + \log y = \log (xy)$.   **12.** True   **13.** True   **14.** True

Chapter 5 Review Exercises, page 342

1. 2   **3.** 3   **5.** -2 **7.** -3 **9.** ± 1000 **11.** 7   **13.** 15.6729   **15.** 5.47395   **17.** 13.6458

19.

21.

23.

25.

27.

29.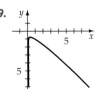

31.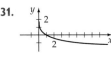

33. $4^3 = 64$ **35.** $(\sqrt{2})^4 = 4$ **37.** $\log_5 125 = 3$ **39.** $\log_{10} 1 = 0$
41. $2 \log_b x + 3 \log_b y - \log_b z$ **43.** $\ln x + 3 \ln y$ **45.** $\log (x^2 \sqrt[3]{x + 1})$
47. $\ln \dfrac{\sqrt{2xy}}{z^3}$ **49.** 2.86754   **51.** −0.117233   **53.** 295   **55.** 1.41×10^{22}
57. $\ln 30/\ln 4$ **59.** 4   **61.** 4   **63.** $\ln 3/(2 \ln 4)$ **65.** 10^{1000} **67.** 1,000,005
69. 81   **71.** 4   **73.** 4.2   **75. a.** \$20,323.79   **b.** \$20,339.99   **77.** \$4,438.10
79. $N(t) = e^{0.8047t}$ **81.** $N(t) = 3.783e^{0.0558t}$

Chapter 5 Test, page 343

1. 4.88936   **2.** 3.85743   **3.**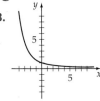

4.

5.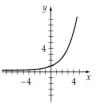

6. $b^c = 5x - 3$ **7.** $\log_3 y = x/2$ **8.** $2 \log_b x + 4 \log_b y - 3 \log_b z$ **9.** $2 \log_b z - 3 \log_b y - 1/2 \log_b x$ **10.** $\log_{10} \dfrac{2x + 3}{(x - 2)^3}$

11. 690 years   **12.** 1.7925   **13.**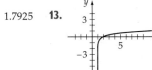

14.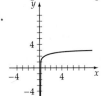

15.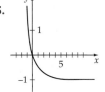

16. 1   **17.** 1.9206   **18.** 1   **19.** \$29,539.62   **20.** \$3109.26

Exercise Set 6.1, page 352

1. vertex: $(0, 0)$
focus: $(0, -1)$
directrix: $y = 1$

3. vertex: $(0, 0)$
focus: $(1/12, 0)$
directrix: $x = -1/12$

5. vertex: $(2, -3)$
focus: $(2, -1)$
directrix: $y = -5$

7. vertex: $(2, -4)$
focus: $(1, -4)$
directrix: $x = 3$

9. vertex: $(-4, 1)$
focus: $(-7/2, 1)$
directrix: $x = -9/2$

11. vertex: $(2, 2)$
focus: $(2, 5/2)$
directrix: $y = 3/2$

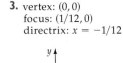

13. vertex: $(-4, -10)$
focus: $(-4, -39/4)$
directrix: $y = -41/4$

15. vertex: $(-7/4, 3/2)$
focus: $(-2, 3/2)$
directrix: $x = -3/2$

17. vertex: $(-5, -3)$
focus: $(-9/2, -3)$
directrix: $x = -11/2$

19. vertex: $(-3/2, 13/12)$
focus: $(-3/2, 1/3)$
directrix: $y = 11/6$

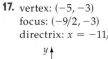 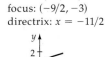

21. vertex: $(2, -5/4)$
focus: $(2, -3/4)$
directrix: $y = -7/4$

23. vertex: $(9/2, -1)$
focus: $(35/8, -1)$
directrix: $x = 37/8$

25. vertex: $(1, 1/9)$
focus: $(1, 31/36)$
directrix: $y = -23/36$

27. $x^2 = -16y$ **29.** $(x + 1)^2 = 4(y - 2)$ **31.** $(x - 3)^2 = 4(y + 4)$ **33.** $(x + 4)^2 = 4(y - 1)$ **35.** on axis 4 feet above vertex
37. $a = 1.5$ inches   **39.** $(-0.3660, -0.3660)$ and $(1.3660, 1.3660)$ **41.** $(-1.5616, 3.8769)$ and $(2.5616, 12.1231)$ **43.** 4
45. $4|p|$ **47.**   **49.**   **51.** $x^2 + y^2 - 8x - 8y - 2xy = 0$

Exercise Set 6.2, page 362

1. vertices: $(0, 5), (0, -5)$
foci: $(0, 3), (0, -3)$

3. vertices: $(3, 0), (-3, 0)$
foci: $(\sqrt{5}, 0), (-\sqrt{5}, 0)$

5. vertices: $(0, 3), (0, -3)$
foci: $(0, \sqrt{2}), (0, -\sqrt{2})$

7. vertices: $(0, 4), (0, -4)$
foci: $(0, \sqrt{55}/2), (0, -\sqrt{55}/2)$

9. vertices: $(8, -2), (-2, -2)$
foci: $(6, -2), (0, -2)$

11. vertices: $(-2, 5), (-2, -5)$
foci: $(-2, 4), (-2, -4)$

13. vertices: $(1 + \sqrt{21}, 3), (1 - \sqrt{21}, 3)$
foci: $(1 + \sqrt{17}, 3), (1 - \sqrt{17}, 3)$

15. vertices: $(1, 2), (1, -4)$
foci: $(1, -1 + \sqrt{65}/3), (1, -1 - \sqrt{65}/3)$

17. vertices: $(2, 0), (-2, 0)$
foci: $(1, 0), (-1, 0)$

19. vertices: $(0, 5), (0, -5)$
foci: $(0, 3), (0, -3)$

21. vertices: $(0, 4), (0, -4)$
foci: $(0, \sqrt{39}/2), (0, -\sqrt{39}/2)$

23. vertices: $(3, 6), (3, 2)$
foci: $(3, 4 + \sqrt{3}), (3, 4 - \sqrt{3})$

25. vertices: $(-1, -3), (5, -3)$
foci: $(0, -3), (4, -3)$

27. vertices: $(2, 4)$, $(2, -4)$
foci: $(2, \sqrt{7})$, $(2, -\sqrt{7})$

29. vertices: $(-1, 6)$, $(-1, -4)$
foci: $(-1, 4)$, $(-1, -2)$

31. vertices: $(11/2, -1)$, $(1/2, -1)$
foci: $(3 + \sqrt{17}/2, -1)$, $(3 - \sqrt{17}/2, -1)$

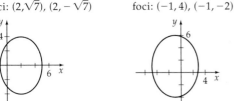

33. $\dfrac{x^2}{25} + \dfrac{y^2}{9} = 1$ **35.** $\dfrac{x^2}{36} + \dfrac{y^2}{16} = 1$ **37.** $\dfrac{x^2}{36} + \dfrac{y^2}{81/8} = 1$ **39.** $\dfrac{(x+2)^2}{16} + \dfrac{(y-4)^2}{7} = 1$ **41.** $\dfrac{(x-2)^2}{25/24} + \dfrac{(y-4)^2}{25} = 1$

43. $\dfrac{(x-5)^2}{16} + \dfrac{(y-1)^2}{25} = 1$ **45.** $\dfrac{x^2}{25} + \dfrac{y^2}{21} = 1$ **47.** $\dfrac{x^2}{20} + \dfrac{y^2}{36} = 1$ **49.** $\dfrac{(x-1)^2}{25} + \dfrac{(y-3)^2}{21} = 1$ **51.** $\dfrac{x^2}{80} + \dfrac{y^2}{144} = 1$

53. $\dfrac{x^2}{884.74^2} + \dfrac{y^2}{883.35^2} = 1$ **55.** 40 feet **57.** $\dfrac{(x - 9\sqrt{15}/2)^2}{324} + \dfrac{y^2}{81/4} = 1$ **59.** $\dfrac{x^2}{36} + \dfrac{y^2}{27} = 1$ **61.** $\dfrac{(x-1)^2}{16} + \dfrac{(y-2)^2}{12} = 1$

63. $9/2$ **67.** $x = \pm \dfrac{9\sqrt{5}}{5}$

Exercise Set 6.3, page 374

1. center: $(0, 0)$
vertices: $(\pm 4, 0)$
foci: $(\pm\sqrt{41}, 0)$
asymptotes: $y = \pm 5x/4$

3. center: $(0, 0)$
vertices: $(0, \pm 2)$
foci: $(0, \pm\sqrt{29})$
asymptotes: $y = \pm 2x/5$

5. center: $(0, 0)$
vertices: $(\pm\sqrt{7}, 0)$
foci: $(\pm 4, 0)$
asymptotes: $y = \pm 3\sqrt{7}x/7$

7. center: $(0, 0)$
vertices: $(\pm 3/2, 0)$
foci: $(\pm\sqrt{73}/2, 0)$
asymptotes: $y = \pm 8x/3$

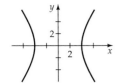
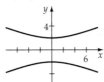
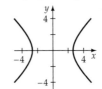

9. center: $(3, -4)$
vertices: $(7, -4)$, $(-1, -4)$
foci: $(8, -4)$, $(-2, -4)$
asymptotes: $y + 4 = \pm 3(x - 3)/4$

11. center: $(1, -2)$
vertices: $(1, 0)$, $(1, -4)$
foci: $(1, -2 \pm 2\sqrt{5})$
asymptotes: $y + 2 = \pm(x - 1)/2$

13. center: $(-2, 0)$
vertices: $(1, 0)$, $(-5, 0)$
foci: $(-2 \pm \sqrt{34}, 0)$
asymptotes: $y = \pm 5(x + 2)/3$

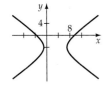
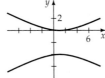

15. center: $(1, -1)$
vertices: $(7/3, -1)$, $(-1/3, -1)$
foci: $(1 \pm \sqrt{97}/3, -1)$
asymptotes: $y + 1 = \pm 9(x - 1)/4$

17. center: $(0, 0)$
vertices: $(\pm 3, 0)$
foci: $(\pm 3\sqrt{2}, 0)$
asymptotes: $y = \pm x$

19. center: $(0, 0)$
vertices: $(0, \pm 3)$
foci: $(0, \pm 5)$
asymptotes: $y = \pm 3x/4$

21. center: $(0, 0)$
vertices: $(0, \pm 2/3)$
foci: $(0, \pm\sqrt{5}/3)$
asymptotes: $y = \pm 2x$

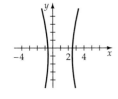

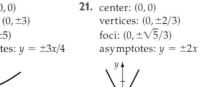
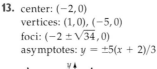

23. center: $(3, 4)$
vertices: $(3, 6)$, $(3, 2)$
foci: $(3, 4 \pm 2\sqrt{2})$
asymptotes: $y - 4 = \pm(x - 3)$

25. center: $(-2, -1)$
vertices: $(-2, 2)$, $(-2, -4)$
foci: $(-2, -1 \pm \sqrt{13})$
asymptotes: $y + 1 = \pm 3(x + 2)/2$

27. $y = \dfrac{-6 \pm \sqrt{36 + 4(4x^2 + 32x + 39)}}{-2}$

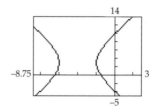

29. $y = \dfrac{64 \pm \sqrt{4096 + 64(9x^2 - 36x + 116)}}{-32}$

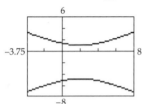

31. $y = \dfrac{18 \pm \sqrt{324 + 36(4x^2 + 8x - 6)}}{-18}$

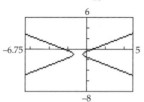

33. $\dfrac{x^2}{9} - \dfrac{y^2}{7} = 1$ **35.** $\dfrac{y^2}{20} - \dfrac{x^2}{5} = 1$

37. $\dfrac{y^2}{9} - \dfrac{x^2}{36/7} = 1$ **39.** $\dfrac{y^2}{16} - \dfrac{x^2}{64} = 1$ **41.** $\dfrac{(x - 4)^2}{4} - \dfrac{(y - 3)^2}{5} = 1$ **43.** $\dfrac{(x - 4)^2}{144/41} - \dfrac{(y + 2)^2}{225/41} = 1$ **45.** $\dfrac{(y - 2)^2}{3} - \dfrac{(x - 7)^2}{12} = 1$

47. $\dfrac{(y - 7)^2}{1} - \dfrac{(x - 1)^2}{3} = 1$ **49.** $\dfrac{x^2}{4} - \dfrac{y^2}{12} = 1$ **51.** $\dfrac{(x - 4)^2}{36/7} - \dfrac{(y - 1)^2}{4} = 1$ and $\dfrac{(y - 1)^2}{36/7} - \dfrac{(x - 4)^2}{4} = 1$

53. a. $\dfrac{x^2}{2162.25} - \dfrac{y^2}{13,462.75} = 1$ **55.** ellipse   **57.** parabola   **59.** parabola   **61.** ellipse
b. 221 miles

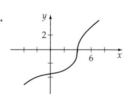

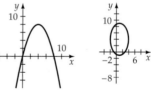

63. $\dfrac{x^2}{1} - \dfrac{y^2}{3} = 1$ **65.** $\dfrac{y^2}{9} - \dfrac{x^2}{7} = 1$ **67.** $x = \pm\dfrac{16\sqrt{41}}{41}$ **73.**

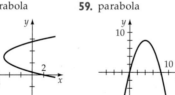

Chapter 6 True/False Exercises, page 379

1. False; a parabola has no asymptotes.   **2.** True   **3.** False; by keeping foci fixed and varying asymptotes, we can make conjugate axis any size needed.   **4.** False; $\dfrac{x^2}{25} + \dfrac{y^2}{9} = 1$ and $\dfrac{x^2}{36} + \dfrac{y^2}{20} = 1$ have the same c's but different a's.   **5.** False; parabolas have no asymptotes.   **6.** True   **7.** False; the graph of a parabola can be a function.   **8.** True   **9.** True

Chapter 6 Review Exercises, page 379

1. vertices: $(\pm 2, 0)$
foci: $(\pm 2\sqrt{2}, 0)$
asymptotes: $y = \pm x$

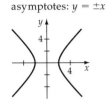

3. vertices: $(-1, -1)$, $(7, -1)$
foci: $(3 \pm 2\sqrt{3}, -1)$

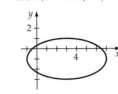

5. vertex: $(-2, 1)$
foci: $(-29/16, 1)$

7. vertices: $(-2, -2)$, $(-2, 4)$
foci: $(-2, 1 \pm \sqrt{5})$

9. vertices: $(-5, 2/3)$, $(7, 2/3)$
foci: $(1 \pm 2\sqrt{13}, 2/3)$
asymptotes: $y - 2/3 = \pm 2(x - 1)/3$

11. vertex: $(-7/2, -1)$
focus: $(-7/2, -3)$

13. $\dfrac{(x - 2)^2}{25} + \dfrac{(y - 3)^2}{16} = 1$

15. $\dfrac{(x + 2)^2}{4} - \dfrac{(y - 2)^2}{5} = 1$

17. $x^2 = 3(y + 2)/2$ or $(y + 2)^2 = 12x$

19. $\dfrac{x^2}{36} - \dfrac{y^2}{4/9} = 1$

21. $(y - 3)^2 = -8x$

23. $\dfrac{(x - 1)^2}{25} + \dfrac{(y - 1)^2}{9} = 1$

Chapter 6 Test, page 380

1. focus: $(0, 2)$
vertex: $(0, 0)$
directrix: $y = -2$

2. focus: $(-2, 4)$, vertex: $(-2, 1)$, directrix: $y = -2$

3. $(y + 2)^2 = -8(x - 1)$

4.

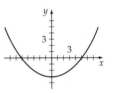

5. vertices: $(0, 8)$, $(0, -8)$
foci: $(0, \sqrt{55})$ $(0, -\sqrt{55})$

6.

7. vertices: $(3, 4)$, $(3, -6)$
foci: $(3, 3)$, $(3, -5)$

8. $\dfrac{x^2}{45} + \dfrac{(y + 3)^2}{9} = 1$

9. $\dfrac{4}{5}$

10.

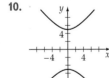

11. vertices: $(6, 0)$, $(-6, 0)$
foci: $(-10, 0)$, $(10, 0)$
asymptotes: $y = \pm 4x/3$

12.

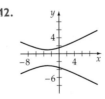

13. vertices: $(-5, 10)$, $(-5, -2)$
foci: $(-5, 4 + 3\sqrt{5})$, $(-5, 4 - 3\sqrt{5})$

14. $\dfrac{(x + 4)^2}{4} - \dfrac{(y + 3)^2}{30} = 1$

15. $(y - 4)^2 = 16(x - 2)$

Exercise Set 7.1, page 389

1. $(2, -4)$ **3.** $(-6/5, 27/5)$ **5.** $(3, 4)$ **7.** $(1, -1)$ **9.** $(3, -4)$ **11.** $(2, 5)$ **13.** $(-1, -1)$ **15.** $(62/25, 34/25)$ **17.** no solution
19. $(c, -4c/3 + 2)$ **21.** $(2, -4)$ **23.** $(0, 3)$ **25.** $(3c/5, c)$ **27.** $(-1/2, 2/3)$ **29.** no solution **31.** $(-6, 3)$ **33.** $(2, -3/2)$
35. $(2\sqrt{3}, 3)$ **37.** $(38/(17\pi), 3/17)$ **39.** $(\sqrt{2}, \sqrt{3})$ **41.** plane: 120 mph, wind: 30 mph **43.** boat: 25 mph, current: 5 mph
45. \$12 per kilogram for iron, \$16 per kilogram for lead **47.** 12 nickels, 7 dimes **49.** 86 **51.** \$14,000 at 6%, \$11,000 at 6.5%
53. 8 gm of 40% gold, 12 gm of 60% gold **55.** 20 ml of 13% solution, 30 ml of 18% solution **57.** $x = -58/17$, $y = 52/17$
59. $x = -2$, $y = -1$ **61.** $x = 153/26$, $y = 151/26$ **63.** $x = 2 + 3i$, $y = 1 - 2i$ **65.** $x = 3 - 5i$, $y = 4i$

Exercise Set 7.2, page 401

1. $(2, -1, 3)$ **3.** $(2, 0, -3)$ **5.** $(2, -3, 1)$ **7.** $(-5, 1, -1)$ **9.** $(3, -5, 0)$ **11.** $(0, 2, 3)$ **13.** $(5c - 25, 48 - 9c, c)$
15. $(3, -1, 0)$ **17.** no solution   **19.** $((50 - 11c)/11, (11c - 18)/11, c)$ **21.** no solution   **23.** $((25 + 4c)/29, (55 - 26c)/29, c)$
25. $(0, 0, 0)$ **27.** $(5c/14, 4c/7, c)$ **29.** $(-11c, -6c, c)$ **31.** $(0, 0, 0)$ **33.** $y = 2x^2 - x - 3$ **35.** $x^2 + y^2 - 4x + 2y - 20 = 0$
37. center $(-7, -2)$, radius 13   **39.** 5 dimes, 10 nickels, 4 quarters   **41.** 685   **43.** $(3, 5, 2, -3)$ **45.** $(1, -2, -1, 3)$
47. $(14a - 7b - 8, -6a + 2b + 5, a, b)$ **49.** $A = -13/2$ **51.** $A \neq -3, A \neq 1$ **53.** $A = -3$ **55.** $3x - 5y - 2z = -2$

Exercise Set 7.3, page 407

1. $(1, 0), (2, 2)$ **3.** $((2 + \sqrt{2})/2, (-6 + \sqrt{2})/2), ((2 - \sqrt{2})/2, (-6 - \sqrt{2})/2)$ **5.** $(5, 18)$ **7.** $(4, 6), (6, 4)$ **9.** $(-3/2, -4), (2, 3)$
11. $(19/29, -11/29), (1, 1)$ **13.** $(-2, 9), (1, -3), (-1, 1)$ **15.** $(-2, 1), (-2, -1), (2, 1), (2, -1)$ **17.** $(4, 2), (-4, 2), (4, -2), (-4, -2)$
19. no solution   **21.** $(12/5, 1/5), (2, 1)$ **23.** $(26/5, -3/5), (1, -2)$ **25.** $(39/10, -7/10), (3, 2)$
27. $((-3 + \sqrt{3})/2, (1 + \sqrt{3})/2), ((-3 - \sqrt{3})/2, (1 - \sqrt{3})/2)$ **29.** $(19/13, 22/13), (1, 4)$ **31.** no solution   **33.** $(0, 1), (1, 2)$
35. $(0.7035, 0.4949)$ **37.** $(1.7549, 1.3247)$ **39.** $(-0.7071, 0.7071), (0.7071, 0.7071)$ **41.** $(1, 5)$ **43.** $(-1, 1), (1, -1)$ **45.** $(1, -2), (-1, 2)$

Exercise Set 7.4, page 415

1. $A = -3, B = 4$ **3.** $A = -2/5, B = 1/5$ **5.** $A = 1, B = -1, C = 4$ **7.** $A = 1, B = 3, C = 2$ **9.** $A = 1, B = 0, C = 1, D = 0$
11. $\dfrac{3}{x} + \dfrac{5}{x + 4}$ **13.** $\dfrac{7}{x - 9} + \dfrac{-4}{x + 2}$ **15.** $\dfrac{5}{2x + 3} + \dfrac{3}{2x + 5}$ **17.** $\dfrac{20}{11(3x + 5)} + \dfrac{-3}{11(x - 2)}$ **19.** $x + 3 + \dfrac{1}{x - 2} + \dfrac{-1}{x + 2}$
21. $\dfrac{1}{x} + \dfrac{2}{x + 7} + \dfrac{-28}{(x + 7)^2}$ **23.** $\dfrac{2}{x} + \dfrac{3x - 1}{x^2 - 3x + 1}$ **25.** $\dfrac{2}{x + 3} + \dfrac{-1}{(x + 3)^2} + \dfrac{4}{x^2 + 1}$ **27.** $\dfrac{3}{x - 4} + \dfrac{5}{(x - 4)^2}$ **29.** $\dfrac{3x - 1}{x^2 + 10} + \dfrac{4x}{(x^2 + 10)^2}$
31. $\dfrac{1}{2k(k - x)} + \dfrac{1}{2k(k + x)}$ **33.** $x + \dfrac{1}{x} + \dfrac{-2}{x - 1}$ **35.** $2x - 2 + \dfrac{3}{x^2 - x - 1}$ **37.** $\dfrac{1}{5(x + 2)} + \dfrac{4}{5(x - 3)}$
39. $\dfrac{1}{x} + \dfrac{2}{x^2} + \dfrac{3}{x^4} + \dfrac{-2}{x - 2}$ **41.** $\dfrac{4}{3(x - 1)} + \dfrac{2x + 7}{3(x^2 + x + 1)}$

Exercise Set 7.5, page 421

1. 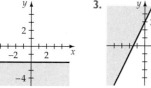   **3.** 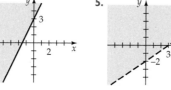   **5.** 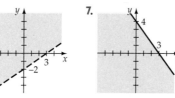   **7.** 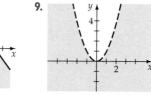   **9.**

11. 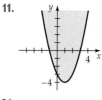   **13.** 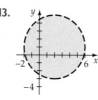   **15.** 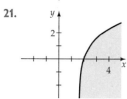   **17.** 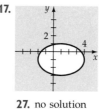   **19.**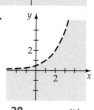

21.   **23.** 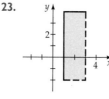   **25.** 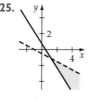   **27.** no solution 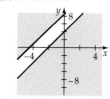   **29.**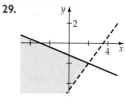

31. 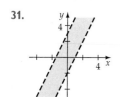   **33.** 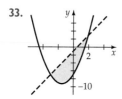   **35.** 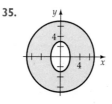   **37.** 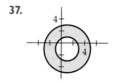   **39.**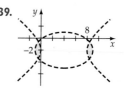

41.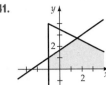

43.

45.

47.

49.

51.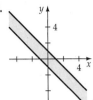

53.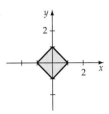

55.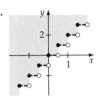

57. a. 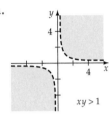 **b.**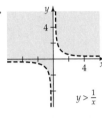

If x is a negative number, then the inequality is reversed when both sides of the inequality are multiplied by a negative number.

Exercise Set 7.6, page 428

1. minimum at $(0, 8)$: 16   **3.** maximum at $(6, 5)$: 71   **5.** minimum at $(0, 10/3)$: 20   **7.** maximum at $(0, 12)$: 72
9. minimum at $(0, 32)$: 32   **11.** maximum at $(0, 8)$: 56   **13.** minimum at $(2, 6)$: 18   **15.** maximum at $(3, 4)$: 25
17. minimum at $(2, 3)$: 12   **19.** maximum at $(100, 400)$: 3400   **21.** 20 acres of wheat and 40 acres of barley
23. 0 starter sets and 18 pro sets   **25.** 24 ounces of group B and 0 ounces of group A; yields a minimum cost of $2.40.
27. two 4-cylinder engines and seven 6-cylinder engines; yields a maximum profit of $2050.

Chapter 7 True/False Exercises, page 432

1. False; $\begin{cases} x + y = 1 \\ x + y = 2 \end{cases}$ has no solution.   **2.** True   **3.** False; a homogeneous system is one where the constant term in each equation is zero.

4. True   **5.** True   **6.** False; $\begin{cases} x + y = 2 \\ x + 2y = 3 \end{cases}$ and $\begin{cases} 2x + 3y = 5 \\ 2x - 2y = 0 \end{cases}$ are two systems with the same solution but no common equations.   **7.** True

8. True   **9.** False; it is inconsistent.   **10.** False; $(-1, 1)$ satisfies the first equation but not the second, and $(-2, -1)$ satisfies the second but not the first.

Chapter 7 Review Exercises, page 432

1. $(-18/7, -15/28)$ **3.** $(-3, -1)$ **5.** $(3, 1)$ **7.** $((5 - 3c)/2, c)$ **9.** $(1/2, 3, -1)$ **11.** $((7c - 3)/11, (16c - 43)/11, c)$
13. $(2, (3c + 2)/2, c)$ **15.** $(14c/11, -2c/11, c)$ **17.** $((c + 1)/2, (3c - 1)/4, c)$ **19.** $(2, -3)$ **21.** no real solution   **23.** $(1/5, 18/5), (1, 2)$
25. $(2, 0), (18/17, -64/17)$ **27.** $(2, 1), (-2, -1)$ **29.** $(2, -3), (-2, 3)$ **31.** $\dfrac{3}{x - 2} + \dfrac{4}{x + 1}$ **33.** $\dfrac{6x - 2}{5(x^2 + 1)} + \dfrac{-6}{5(x + 2)}$

35. $\dfrac{2}{x} + \dfrac{4}{x - 1} + \dfrac{5}{x + 1}$ **37.**   **39.**   **41.**   **43.**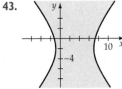

45.   **47.**   **49.** 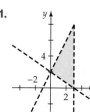   **51.** 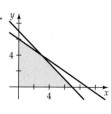   **53.**

55.   **57.** 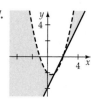   **59.** 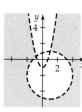   **61.** maximum at $(4,5)$: 18   **63.** minimum at $(0,8)$: 8

65. minimum at $(2,5)$: 27   **67.** $y = \dfrac{11}{6}x^2 - \dfrac{5}{2}x + \dfrac{2}{3}$ **69.** $z = -2x + 3y + 3$ **71.** wind: 28 mph, plane: 143 mph
73. $(0,0,0)$, $(1,1,1)$, $(1,-1,-1)$, $(-1,-1,1)$, $(-1,1,-1)$

Chapter 7 Test, page 434

1. $(-3,2)$ **2.** $((6+c)/2, c)$ **3.** $(173/39, 29/39, -4/3)$ **4.** $((c+3)/4, (7c+1)/8, c)$ **5.** $((c+10)/13, (5c+11)/13, c)$
6. $(c/14, -9c/14, c)$ **7.** $(2,5)$, $(-2,1)$ **8.** $(-2,3)$, $(-1,-1)$ **9.**   **10.** 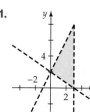   **11.**

12.   **13.**   **14.** no graph; the solution set is the empty set.   **15.**   **16.**

17. $\dfrac{7}{5(x-4)} + \dfrac{8}{5(x+1)}$ **18.** $\dfrac{1}{x} + \dfrac{-x+2}{x^2+1}$ **19.** 680/7 acres of oats and 400/7 acres of barley   **20.** $x^2 + y^2 - 2y - 24 = 0$

Exercise Set 8.1, page 443

1. $\begin{bmatrix} 2 & -3 & 1 & 1 \\ 3 & -2 & 3 & 0 \\ 1 & 0 & 5 & 4 \end{bmatrix} \begin{bmatrix} 2 & -3 & 1 \\ 3 & -2 & 3 \\ 1 & 0 & 5 \end{bmatrix} \begin{bmatrix} 1 \\ 0 \\ 4 \end{bmatrix}$ **3.** $\begin{bmatrix} 2 & -3 & -4 & 1 & 2 \\ 0 & 2 & 1 & 0 & 2 \\ 1 & -1 & 2 & 0 & 4 \\ 3 & -3 & -2 & 0 & 1 \end{bmatrix} \begin{bmatrix} 2 & -3 & -4 & 1 \\ 0 & 2 & 1 & 0 \\ 1 & -1 & 2 & 0 \\ 3 & -3 & -2 & 0 \end{bmatrix} \begin{bmatrix} 2 \\ 2 \\ 4 \\ 1 \end{bmatrix}$ **5.** $\begin{bmatrix} 1 & -1 & 2 & 2 \\ 0 & 1 & -1 & -6 \\ 0 & 0 & 1 & -27/2 \end{bmatrix}$

7. $\begin{bmatrix} 1 & -2 & -1 & 3 \\ 0 & 1 & 2 & -11/2 \\ 0 & 0 & 1 & -13/6 \end{bmatrix}$ **9.** $\begin{bmatrix} 1 & -2 & 3 & -4 \\ 0 & 1 & 2 & -1/2 \\ 0 & 0 & 1 & -2 \\ 0 & 0 & 0 & 0 \end{bmatrix}$ **11.** $\begin{bmatrix} 1 & -3 & 4 & 2 & 1 \\ 0 & 1 & -1 & -2 & -1 \\ 0 & 0 & 0 & 1 & 3 \end{bmatrix}$ **13.** $(2,-1,1)$ **15.** $(1,-2,-1)$

17. $(2-2c, 2c+1/2, c)$ **19.** $(1/2, 1/2, 3/2)$ **21.** $(16c, 6c, c)$ **23.** $(7c+6, -11c-8, c)$ **25.** $(c+2, c, c)$ **27.** no solution
29. $(2,-2,3,4)$ **31.** $(21/10, -8/5, 2/5, -5/2)$ **33.** $(3, -3/2, 1, -1)$ **35.** $(27c/2 + 39, 5c/2 + 10, -4c - 10, c)$
37. $(c_1 - 12c_2/7 + 6/7, c_1 - 9c_2/7 + 1/7, c_1, c_2)$ **39.** $(1, 0, -2, 1, 2)$ **41.** $(77c + 151)/3, (-25c - 50)/3, (14c + 34)/3, -3c - 7, c)$
43. $f(x) = x^3 - 2x^2 - x + 2$ **45.** all values of a except $a \neq 1$ and $a \neq -6$ **47.** $a = -6$ **49.** $z = 2x + 3y - 2$

Exercise Set 8.2, page 456

1. a. $\begin{bmatrix} 1 & 2 \\ 5 & 4 \end{bmatrix}$ **b.** $\begin{bmatrix} 3 & -4 \\ 1 & 2 \end{bmatrix}$ **c.** $\begin{bmatrix} -2 & 6 \\ 4 & 2 \end{bmatrix}$ **d.** $\begin{bmatrix} 7 & -11 \\ 0 & 3 \end{bmatrix}$ **3. a.** $\begin{bmatrix} -3 & 0 & 5 \\ 3 & 5 & -5 \end{bmatrix}$ **b.** $\begin{bmatrix} 3 & -2 & 1 \\ -1 & -5 & 1 \end{bmatrix}$ **c.** $\begin{bmatrix} -6 & 2 & 4 \\ 4 & 10 & -6 \end{bmatrix}$

d. $\begin{bmatrix} 9 & -5 & 0 \\ -4 & -15 & 5 \end{bmatrix}$ **5. a.** $\begin{bmatrix} 1 & 5 \\ 3 & -5 \\ 2 & -4 \end{bmatrix}$ **b.** $\begin{bmatrix} -7 & 3 \\ 1 & -1 \\ -4 & 4 \end{bmatrix}$ **c.** $\begin{bmatrix} 8 & 2 \\ 2 & -4 \\ 6 & -8 \end{bmatrix}$ **d.** $\begin{bmatrix} -18 & 5 \\ 1 & 0 \\ -11 & 12 \end{bmatrix}$ **7. a.** $\begin{bmatrix} -1 & 1 & -1 \\ 2 & 2 & 1 \\ -1 & 2 & 5 \end{bmatrix}$

b. $\begin{bmatrix} -3 & 5 & -1 \\ -2 & -4 & 3 \\ -7 & 4 & 1 \end{bmatrix}$ **c.** $\begin{bmatrix} 2 & -4 & 0 \\ 4 & 6 & -2 \\ 6 & -2 & 4 \end{bmatrix}$ **d.** $\begin{bmatrix} -7 & 12 & -2 \\ -6 & -11 & 7 \\ -17 & 9 & 0 \end{bmatrix}$ **9.** $\begin{bmatrix} -10 & 17 \\ 6 & -8 \end{bmatrix}, \begin{bmatrix} 0 & 22 \\ 1 & -18 \end{bmatrix}$ **11.** $\begin{bmatrix} 10 & 6 \\ 14 & -7 \end{bmatrix}, \begin{bmatrix} 14 & -1 \\ 0 & -11 \end{bmatrix}$

13. $\begin{bmatrix} 0 & -4 & 5 \\ 6 & 0 & 3 \\ -3 & -2 & 1 \end{bmatrix}, \begin{bmatrix} 5 & -13 \\ 5 & -4 \end{bmatrix}$ **15.** $\begin{bmatrix} 9 & -2 & -6 \\ 0 & -1 & 2 \\ 4 & -2 & -4 \end{bmatrix}, \begin{bmatrix} 4 & -2 & 6 \\ 2 & -3 & 4 \\ 4 & -4 & 3 \end{bmatrix}$ **17.** $[0, 8]$ **19.** The product is not possible. **21.** $\begin{bmatrix} 0 & 0 \\ 0 & 0 \end{bmatrix}$

23. The product is not possible. **25.** $\begin{bmatrix} 1/3 & -5/3 \\ -1/3 & 4/3 \\ 1/3 & -4/3 \end{bmatrix}$ **27.** $\begin{bmatrix} -1 & 1 \\ 3 & 2 \\ 7 & -2 \end{bmatrix}$ **29.** $\begin{bmatrix} 1 & -3 \\ 1 & -2 \end{bmatrix}$ **31.** $\begin{bmatrix} 7 & -1 & 1 \\ 1 & 2 & 0 \\ 5 & -1 & 4 \end{bmatrix}$

33. $\begin{cases} 3x - 8y = 11 \\ 4x + 3y = 1 \end{cases}$ **35.** $\begin{cases} x - 3y - 2z = 6 \\ 3x + y = 2 \\ 2x - 4y + 5z = 1 \end{cases}$ **37.** $\begin{cases} 2x_1 - x_2 + 2x_4 = 5 \\ 4x_1 + x_2 + 2x_3 - 3x_4 = 6 \\ 6x_1 + x_3 - 2x_4 = 10 \\ 5x_1 + 2x_2 - x_3 - 4x_4 = 8 \end{cases}$

39. a. 3×4, Three different fish were caught in 4 different samples. **b.** Fish A was caught in sample 4. **c.** Fish B

41. $\begin{bmatrix} 1.96 & 1.37 & 2.94 & 1.37 \\ 0.78 & 1.08 & 1.96 & 0.88 \\ 3.53 & 1.18 & 4.41 & 1.47 \end{bmatrix}$ **43. a.** $\begin{bmatrix} 86 & 58 \\ 79 & 65 \\ 71 & 73 \end{bmatrix}$ **b.** The matrix represents the total number of wins and losses for each team.

c. $\begin{bmatrix} -2 & 2 \\ 13 & -13 \\ 1 & -1 \end{bmatrix}$ **d.** The matrix represents the difference between performance at home and performance away.

45. a. $\begin{bmatrix} 0 & 6 & 4 & -2 \\ 4 & 1 & -3 & 0 \end{bmatrix}$ **b.** rectangle **c.** The new rectangle is shifted 2 units right and 1 unit down from the original rectangle.

47. a. $\begin{bmatrix} -5 & -2 & 2 & -1 \\ -2 & 4 & 2 & -4 \end{bmatrix}$ **b.** rectangle **c.** Second rectangle obtained by rotating first by 90°.

49. $A - B = \begin{bmatrix} 50 & 150 & 140 \\ 15 & 170 & 370 \\ 85 & 250 & 130 \\ 80 & 115 & 25 \end{bmatrix}$ **51.** $\begin{bmatrix} 26,898 & 28,150 & 31,536 \\ 20,495 & 21,195 & 23,670 \\ 19,022 & 19,925 & 21,969 \end{bmatrix}$ **53.** $\begin{bmatrix} 24 & 21 & -12 & 32 & 0 \\ -7 & -8 & 3 & 21 & 20 \\ 32 & 10 & -32 & 1 & 5 \\ 19 & -15 & -17 & 30 & 20 \\ 29 & 9 & -28 & 13 & -6 \end{bmatrix}$ **55.** $\begin{bmatrix} 46 & -100 & 36 & 273 & 93 \\ 82 & -93 & 19 & 27 & 97 \\ 73 & -10 & -23 & 109 & 83 \\ 212 & -189 & 52 & 37 & 156 \\ 68 & -22 & 54 & 221 & 58 \end{bmatrix}$

$A - B$ is the number sold of each item during the week.

57. $\begin{bmatrix} 76 & -8 & -25 & 30 & 6 \\ 14 & 16 & -10 & 14 & 2 \\ 39 & 0 & -45 & 22 & 27 \\ 0 & -4 & 23 & 83 & -16 \\ 56 & -20 & -22 & 7 & 5 \end{bmatrix}$ **59.** $\begin{bmatrix} 6 + 9i & 3 - 6i \\ 3 + 3i & 6 - 3i \end{bmatrix}$ **61.** $\begin{bmatrix} 2 + 2i & -6 + 4i \\ -4 + 6i & 2 + 8i \end{bmatrix}$ **63.** $\begin{bmatrix} 3 + 2i & 3 + i \\ 4 + 3i & 6 - 2i \end{bmatrix}$

65. $\begin{bmatrix} 12 - 3i & -3 + 3i \\ 10 + i & 6 - i \end{bmatrix}$ **67.** $\begin{bmatrix} -2 + 11i & 8 - 6i \\ 2 + 6i & 6 - 5i \end{bmatrix}$

Exercise Set 8.3, page 468

1. $\begin{bmatrix} -5 & -3 \\ -2 & -1 \end{bmatrix}$ **3.** $\begin{bmatrix} 5 & -2 \\ -1 & 1/2 \end{bmatrix}$ **5.** $\begin{bmatrix} -16 & -2 & 7 \\ 7 & 1 & -3 \\ -3 & 0 & 1 \end{bmatrix}$ **7.** $\begin{bmatrix} 15 & -1 & -4 \\ -11/2 & 1/2 & 3/2 \\ 3 & 0 & -1 \end{bmatrix}$ **9.** $\begin{bmatrix} 7/2 & -2 & -2 \\ -5/2 & 1 & 2 \\ -1 & 0 & 1 \end{bmatrix}$

11. $\begin{bmatrix} 19/2 & -1/2 & -3/2 & 3/2 \\ 7/4 & 1/4 & -1/4 & 3/4 \\ -7/2 & 1/2 & 1/2 & -1/2 \\ 1/4 & -1/4 & 1/4 & 1/4 \end{bmatrix}$ **13.** $\begin{bmatrix} 2 & 3/5 & -7/5 & 4/5 \\ 4 & -7/5 & -2/5 & 4/5 \\ -6 & 14/5 & -1/5 & -3/5 \\ 3 & -8/5 & 2/5 & 1/5 \end{bmatrix}$ **15.** $(2, 1)$ **17.** $(7/4, -25/8)$ **19.** $(1, -1, 2)$ **21.** $(23, -12, 3)$

23. $(0, 4, -6, -2)$ **25.** on Saturday 80 adults, 20 children   **27.** Sample 1: 500 g of additive 1, 200 g of additive 2, 300 g of additive 3
on Sunday 95 adults, 25 children   Sample 2: 400 g of additive 1, 400 g of additive 2, 200 g of additive 3

29. $\begin{bmatrix} -5.667 & -3.667 & 5 & 0.333 \\ -27.667 & -18.667 & 24 & 2.333 \\ -19.333 & -13.333 & 17 & 1.667 \\ 15 & 10 & -13 & -1 \end{bmatrix}$ **31.** $\begin{bmatrix} -0.150 & -0.217 & 0.302 \\ 0.248 & -0.024 & 0.013 \\ 0.217 & -0.200 & -0.195 \end{bmatrix}$

33. $194.67 million worth of manufacturing, $156.03 million worth of transportation, $121.82 million worth of services
35. $39.69 million worth of coal, $14.30 million worth of iron, $32.30 million worth of steel

43. a. $\begin{bmatrix} -5/2 & 3/2 \\ -2 & 1 \end{bmatrix}$ **b.** $\begin{bmatrix} 2 & -3 \\ -3/2 & 5/2 \end{bmatrix}$ **c.** $\begin{bmatrix} 1 & 1/4 \\ -1 & 0 \end{bmatrix}$

Exercise Set 8.4, page 478

1. 13   **3.** -15 **5.** 0   **7.** 0   **9.** 19, 19   **11.** 1, -1 **13.** $-9, -9$ **15.** $-9, -9$ **17.** 10   **19.** 53   **21.** 20   **23.** 46
25. 0   **27.** Row 2 consists of zeros, so the determinant is zero.   **29.** 2 was factored from row 2.
31. Row 1 was multiplied by -2 and added to row 2.   **33.** 2 was factored from column 1.
35. The matrix is in diagonal form. The value of the determinant is the product of the terms on the main diagonal.
37. Row 1 and row 3 were interchanged, so the sign of the determinant was changed.
39. Each row of the determinant was multiplied by a.   **41.** 0   **43.** 0   **45.** 6   **47.** -90 **49.** 21   **51.** 3   **53.** -38.933
55. 9/2 square units   **57.** $46\frac{1}{2}$ square units   **63.** $7x + 5y = -1$ **65.** 263.5

Exercise Set 8.5, page 484

1. $x_1 = 44/31, x_2 = 29/31$ **3.** $x_1 = 1/3, x_2 = -2/3$ **5.** $x_1 = 2, x_2 = -7$ **7.** $x_1 = 0, x_2 = 0$ **9.** $x_1 = 1.28125, x_2 = 1.875$
11. $x_1 = 21/17, x_2 = -3/17, x_3 = -29/17$ **13.** $x_1 = 32/49, x_2 = 13/49, x_3 = 6/7$ **15.** $x_1 = -29/64, x_2 = -25/64, x_3 = -19/32$
17. $x_1 = 50/53, x_2 = 62/53, x_3 = 4/53$ **19.** $x_1 = 0, x_2 = 0, x_3 = 0$ **21.** $x_2 = -35/19$ **23.** $x_1 = -121/131$ **25.** $x_4 = 4/3$
27. The determinant of the coefficient matrix is zero, so Cramer's Rule cannot be used. The system of equations has infinitely many
solutions.   **29.** all values of k except $k = 0$ **31.** all values of k except $k = 2$ **33.** $r = 3, s = -3$

Chapter 8 True/False Exercises, page 488

1. False; $A^2 = A \cdot A = \begin{bmatrix} 7 & 18 \\ 6 & 19 \end{bmatrix}$.   **2.** True   **3.** False; a singular matrix does not have a multiplicative inverse.

4. False; as an example, $A = \begin{bmatrix} 2 & -1 \\ -4 & 2 \end{bmatrix}$, $B = \begin{bmatrix} 3 & 4 \\ 1 & 5 \end{bmatrix}$, and $C = \begin{bmatrix} 4 & 7 \\ 3 & 11 \end{bmatrix}$. $AB = AC$ but $B \neq C$.   **5.** True   **6.** False; for example, if

$A = \begin{bmatrix} 1 & 4 \\ -2 & 3 \end{bmatrix}$ and $B = \begin{bmatrix} 2 & 0 \\ -1 & 5 \end{bmatrix}$, then $\det(A) + \det(B) \neq \det(A + B)$.   **7.** False; if the determinant of the coefficient matrix is zero,

Cramer's Rule cannot be used to solve the system of equations.   **8.** False; matrix multiplication is not commutative—that is, $AB \neq BA$,
$AB - BA \neq 0$.   **9.** True   **10.** False; by the Associative Property of Matrix Multiplication, given A, B, and C square matrices of order
n, $(AB)C = A(BC)$.   **11.** False; if the number of equations is less than the number of variables, the Gaussian elimination method can be
used to solve the system of linear equations. If the system of equations has a solution, the solutions will be given in terms of one or
more of the variables.   **12.** False; for example, for a 2×2 matrix, $\det(2A) = 2 \cdot 2 \det(A)$, and for a 3×3 matrix,

$\det(2A) = 4 \cdot 2 \det(A)$.   **13.** True   **14.** False; for example, given $A = \begin{bmatrix} -3 & 2 \\ -6 & 4 \end{bmatrix}$ and $B = \begin{bmatrix} 2 & 4 \\ 3 & 6 \end{bmatrix}$, then $AB = \begin{bmatrix} 0 & 0 \\ 0 & 0 \end{bmatrix} = O$, but $A \neq O$
and $B \neq O$.   **15.** True

Chapter 8 Review Exercises, page 488

1. $\begin{bmatrix} 6 & -3 & 9 \\ 9 & 6 & -3 \end{bmatrix}$ **3.** $\begin{bmatrix} -5 & 5 & -1 \\ 1 & -4 & 6 \end{bmatrix}$ **5.** $\begin{bmatrix} -1 & -15 \\ 7 & 1 \end{bmatrix}$ **7.** $\begin{bmatrix} -6 & -4 & 2 \\ 14 & 0 & 10 \\ -7 & -7 & 6 \end{bmatrix}$ **9.** $\begin{bmatrix} 12 & 28 & -5 \\ 2 & 6 & 0 \\ 6 & 16 & -1 \end{bmatrix}$ **11.** $\begin{bmatrix} -12 & -36 & -4 \\ 48 & 124 & 4 \\ -9 & -32 & -6 \end{bmatrix}$

13. not possible   **15.** $\begin{bmatrix} 7 & 24 & 9 \\ -10 & -22 & 1 \end{bmatrix}$ **17.** $\begin{bmatrix} -1 & -5 & 4 \\ 1/2 & 2 & -3/2 \\ 0 & -2 & 1 \end{bmatrix}$ **19.** $(2,-1)$ **21.** $(3,0)$ **23.** $(3,1,0)$ **25.** $(1,0,-2)$

27. $(3,-4,1)$ **29.** $(-c-2,-c-3,c)$ **31.** $(1,-2,2,3)$ **33.** $(-37c+2,16c,-7c+1,c)$ **35.** $\begin{bmatrix} -1 & 1 \\ -3/2 & 1 \end{bmatrix}$ **37.** $\begin{bmatrix} -2/7 & 3/14 \\ 1/7 & 1/7 \end{bmatrix}$

39. $\begin{bmatrix} 2 & -2 & 1 \\ 0 & 3/2 & -1 \\ -1 & -1 & 1 \end{bmatrix}$ **41.** $\begin{bmatrix} -10 & 20 & -3 \\ -5 & 9 & -1 \\ 3 & -6 & 1 \end{bmatrix}$ **43.** $\begin{bmatrix} -1 & -7 & 4 & 2 \\ -6 & -3 & 2 & 3 \\ 1 & 2 & -1 & -1 \\ -2 & 0 & 0 & 1 \end{bmatrix}$ **45.** The matrix does not have an inverse.

47. a. $(18,-13)$ **b.** $(-22,16)$ **49. a.** $(-18/7,23/7,-6/7)$ **b.** $(-31/14,20/7,3/7)$ **51.** -2 **53.** -1 **55.** 0 **57.** 0
59. $x_1=16/19, x_2=-2/19$ **61.** $x_1=13/44, x_2=1/4, x_3=-17/44$ **63.** $x_1=18/23, x_2=-26/69, x_3=38/69$ **65.** $x_3=115/126$
67. \$34.47 million computer division, \$14.20 million disk drive division, \$23.64 million moniter division.

Chapter 8 Test, page 490

1. $\begin{bmatrix} 2 & 3 & -3 & 4 \\ 3 & 0 & 2 & -1 \\ 4 & -4 & 2 & 3 \end{bmatrix}, \begin{bmatrix} 2 & 3 & -3 \\ 3 & 0 & 2 \\ 4 & -4 & 2 \end{bmatrix}\begin{bmatrix} 4 \\ -1 \\ 3 \end{bmatrix}$ **2.** $\begin{cases} 3x-2y+5z-w=9 \\ 2x+3y-z+4w=8 \\ x+3z+2w=-1 \end{cases}$ **3.** $(2,-1,2)$ **4.** $(3,-1,-1)$

5. $(3c-5,-7c+14,4-3c,c)$ **6.** $\begin{bmatrix} 3 & -9 & -6 \\ -3 & -12 & 3 \end{bmatrix}$ **7.** $A+B$ is not defined.   **8.** $\begin{bmatrix} 4 & 1 & 3 \\ 8 & 0 & -19 \\ 11 & 0 & 10 \end{bmatrix}$ **9.** $\begin{bmatrix} 16 & -1 & -2 \\ 15 & -11 & -3 \end{bmatrix}$

10. $\begin{bmatrix} 17 & -4 & -4 \\ 14 & -15 & -2 \end{bmatrix}$ **11.** CA is not defined.   **12.** $\begin{bmatrix} -6 & -1 & -19 \\ -15 & -25 & -27 \\ 1 & 3 & 31 \end{bmatrix}$ **13.** A^2 is not defined.   **14.** $\begin{bmatrix} 9 & 6 & 13 \\ -3 & -2 & 12 \\ 20 & -3 & 11 \end{bmatrix}$

15. $\begin{bmatrix} 18 & -5 & 7 \\ 4 & -1 & 2 \\ -3 & 1 & -1 \end{bmatrix}$ **16.** $M_{21}=-8, C_{21}=8$ **17.** 49 **18.** -1 **19.** $-140/41$ **20.** $\left(\begin{bmatrix} 1 & 0 & 0 \\ 0 & 1 & 0 \\ 0 & 0 & 1 \end{bmatrix} - \begin{bmatrix} 0.15 & 0.23 & 0.11 \\ 0.08 & 0.10 & 0.05 \\ 0.16 & 0.11 & 0.07 \end{bmatrix}\right)^{-1}\begin{bmatrix} 50 \\ 32 \\ 8 \end{bmatrix}$

Exercise Set 9.1, page 497

1. $0, 2, 6, a_8=56$ **3.** $0, 1/2, 2/3, a_8=7/8$ **5.** $1, -1/4, 1/9, a_8=-1/64$ **7.** $-1/3, -1/6, -1/9, a_8=-1/24$
9. $2/3, 4/9, 8/27, a_8=256/6561$ **11.** $0, 2, 0, a_8=2$ **13.** $1.1, 1.21, 1.331, a_8=2.14358881$ **15.** $1, -\sqrt{2}/2, \sqrt{3}/3, a_8=-\sqrt{2}/4$
17. $1, 2, 6, a_8=40320$ **19.** $0, 0.3010, 0.4771, a_8=0.9031$ **21.** $1, 4, 2, a_8=4$ **23.** $3, 3, 3, a_8=3$ **25.** $5, 10, 20$ **27.** $2, 4, 12$
29. $2, 4, 16$ **31.** $2, 8, 48$ **33.** $3, \sqrt{3}, \sqrt[6]{3}$ **35.** $2, 5/2, 9/4$ **37.** 4320 **39.** 72 **41.** 56 **43.** 100 **45.** 15 **47.** 40 **49.** $25/12$

51. 72 **53.** -24 **55.** $3\log 2$ **57.** 256 **59.** $\sum_{i=1}^{6}\frac{1}{i^2}$ **61.** $\sum_{i=1}^{7}2^i(-1)^{i+1}$ **63.** $\sum_{i=0}^{4}(7+3i)$ **65.** $\sum_{i=1}^{4}\frac{1}{2^i}$ **67.** 2.6457520

69. $a_{20}\approx 1.0000037, a_{100}\approx 1$ **71.** $\frac{1}{2}(-1+i\sqrt{3}), \frac{1}{2}(-1-i\sqrt{3}), 1, \frac{1}{2}(-1+i\sqrt{3}), \frac{1}{2}(-1-i\sqrt{3}), 1, a_{99}=1$

Exercise Set 9.2, page 504

1. $a_9=38, a_{24}=98, a_n=4n+2$ **3.** $a_9=-10, a_{24}=-40, a_n=8-2n$ **5.** $a_9=16, a_{24}=61, a_n=3n-11$
7. $a_9=25, a_{24}=70, a_n=3n-2$ **9.** $a_9=a+16, a_{24}=a+46, a_n=a+2n-2$
11. $a_9=\log 7+8\log 2, a_{24}=\log 7+23\log 2, a_n=\log 7+(n-1)\log 2$ **13.** $a_9=9\log a, a_{24}=24\log a, a_n=n\log a$
15. 45 **17.** -79 **19.** 185 **21.** -555 **23.** 468 **25.** 525 **27.** -465 **29.** $78+12x$ **31.** $210x$ **33.** $3, 7, 11, 15, 19$
35. $5/2, 2, 3/2, 1$ **39.** 20 on 6th row, 135 in the 6 rows   **41.** \$1500, \$48,750   **43.** 784 feet   **47.** $a_n=7-3n$ **49.** $a_{50}=197$

Exercise Set 9.3, page 511

1. geometric, $r=4$ **3.** not geometric   **5.** geometric, $r=2^x$ **7.** geometric, $r=2$ **9.** geometric, $r=x^2$ **11.** not geometric
13. 2^{2n-1} **15.** $-4(-3)^{n-1}$ **17.** $6(2/3)^{n-1}$ **19.** $-6(-5/6)^{n-1}$ **21.** $(-1/3)^{n-3}$ **23.** $(-x)^{n-1}$ **25.** c^{3n-1} **27.** $3(1/100)^n$ **29.** $5(0.1)^n$
31. $45(0.01)^n$ **33.** 18 **35.** -2 **37.** 363 **39.** $1330/729$ **41.** $\dfrac{279{,}091}{390{,}625}$ **43.** -341 **45.** $147{,}620$ **47.** $1/2$ **49.** $-2/5$ **51.** $9/91$
53. $1/9$ **55.** $5/7$ **57.** $1/3$ **59.** $5/11$ **61.** $41/333$ **63.** $422/999$ **65.** $229/900$ **67.** $997/825$ **69.** \$2271.93
71. Because $\log r$ is a constant, the sequence $\log a_n$ is an arithmetic sequence.   **73.** Yes. The common ratio is x.   **75.** $a^n r^{[(n-1)n]/2}$
77. 45 feet   **79.** 2044

Exercise Set 9.4, page 519

No answers are provided because each exercise is a verification.

Exercise Set 9.5, page 525

1. 35   **3.** 36   **5.** 220   **7.** 1   **9.** $x^6 - 6x^5y + 15x^4y^2 - 20x^3y^3 + 15x^2y^4 - 6xy^5 + y^6$ **11.** $x^5 + 15x^4 + 90x^3 + 270x^2 + 405x + 243$
13. $128x^7 - 448x^6 + 672x^5 - 560x^4 + 280x^3 - 84x^2 + 14x - 1$ **15.** $x^6 + 18x^5y + 135x^4y^2 + 540x^3y^3 + 1215x^2y^4 + 1458xy^5 + 729y^6$
17. $16x^4 - 160x^3y + 600x^2y^2 - 1000xy^3 + 625y^4$ **19.** $x^6 + 6x^4 + 15x^2 + 20 + 15/x^2 + 6/x^4 + 1/x^6$
21. $x^{14} - 28x^{12} + 336x^{10} - 2240x^8 + 8960x^6 - 21{,}504x^4 + 28{,}672x^2 - 16{,}384$ **23.** $32x^{10} + 80x^8y^3 + 80x^6y^6 + 40x^4y^9 + 10x^2y^{12} + y^{15}$
25. $16/x^4 - 16/x^2 + 6 - x^2 + x^4/16$ **27.** $s^{-12} + 6s^{-8} + 15s^{-4} + 20 + 15s^4 + 6s^8 + s^{12}$ **29.** $-3240x^3y^7$ **31.** $1056x^{10}y^2$ **33.** $126x^2y^2\sqrt{x}$
35. $165b^5/a^5$ **37.** $180a^2b^8$ **39.** $60x^2y^8$ **41.** $-61{,}236a^5b^5$ **43.** $126s^{-1}, 126s$ **45.** $-7 - 24i$ **47.** $41 - 38i$ **49.** 1

51. $nx^{n-1} + \dfrac{n(n-1)x^{n-2}h}{2} + \dfrac{n(n-1)(n-2)x^{n-3}h^2}{6} + \cdots + h^{n-1}$ **57.** 1.1712   **59.** 756   **61.** 56

Exercise Set 9.6, page 531

1. 30   **3.** 70   **5.** 1   **7.** 1   **9.** 210   **11.** 12   **13.** 16   **15.** 720   **17.** 125   **19.** 53,130
21. There are 676 ways to arrange 26 letters taken 2 at a time. Now if there are more than 676 employees, then at least 2 employees will have the same first and last initials.   **23.** 1120   **25.** 1024   **27.** 3,838,380   **29. a.** 21   **b.** 105   **c.** 21   **31.** 1.8×10^9
33. 112   **35.** 120   **37.** 21   **39.** 112   **41.** 184,756   **43.** 62,355,150   **45. a.** 3,991,680   **b.** 31,840,128   **47.** 120   **49.** 252

Exercise Set 9.7, page 541

1. $\{S_1R_1, S_1R_2, S_1R_3, S_2R_1, S_2R_2, S_2R_3, R_1R_2, R_1R_3, R_2R_3, S_1S_2\}$ **3.** $\{H1, H2, H3, H4, T1, T2, T3, T4\}$ **5.** Let the three cans be represented by A, B, and C and let (x, y) represent the cans that balls 1 and 2 are placed in; e.g., (A, B) means ball 1 in can A and ball 2 in can B. $S = \{(A, A), (A, B), (A, C), (B, B), (B, C), (B, A), (C, C), (C, A), (C, B)\}$ **7.** $\{HSC, HSD, HCD, SCD\}$ **9.** $\{ae, ai, ao, au, ei, eo, eu, io, iu, ou\}$
11. $\{HHHH\}$ **13.** $\{TTTT, HTTT, THTT, TTHT, TTTH, TTHH, THTH, HTHT, THHT, HTTH, HHTT\}$ **15.** $\varnothing$
17. $\{(1,1), (2,2), (3,3), (4,4), (5,5), (6,6)\}$ **19.** $\{(1,4), (2,4), (3,4), (4,4), (5,4), (6,4)\}$ **21. a.** 1/13   **b.** 1/4   **23.** 0.97   **25.** 3/5   **27.** 0.59
29. 0.25   **31.** 0.1   **33.** 0.1   **35.** 0.025   **37.** 0.9999   **39.** 1/16   **41.** 0.2262   **43.** 0.2137   **45.** $(7/8)^2$ **47.** $\dfrac{56}{729}$

Chapter 9 True/False Exercises, page 546

1. False; $0! \cdot 4! = 1 \cdot 4 \cdot 3 \cdot 2 \cdot 1 = 24$.   **2.** False; $\left(\sum\limits_{i=1}^{3} i\right)\left(\sum\limits_{i=1}^{3} i\right) \neq \sum\limits_{i=1}^{3} i^2$.   **3.** True   **4.** False; the constant sequence has all terms equal.

5. False; $\dfrac{(k+1)^3}{k^3} = (1 + 1/k)^3$ is not a constant.   **6.** True   **7.** True   **8.** False; $\sum\limits_{i=1}^{\infty} \dfrac{1}{2^i} = 1$.   **9.** False; see Project 1, Section 9.4.

10. False; the exponent is 4.   **11.** False; there are $m \cdot n$ ways.   **12.** False; $P(n, r) = \dfrac{n!}{(n-r)!}$.   **13.** True   **14.** False; $P(A \cap B) = P(\varnothing) = 0$.
15. True

Chapter 9 Review Exercises, page 546

1. $a_3 = 9, a_7 = 49$ **3.** $a_3 = 11, a_7 = 23$ **5.** $a_3 = 1/8, a_7 = 1/128$ **7.** $a_3 = 1/6, a_7 = 1/5040$ **9.** $a_3 = 8/27, a_7 = 128/2187$
11. $a_3 = 18, a_7 = 1458$ **13.** $a_3 = 6, a_7 = 5040$ **15.** $a_3 = 8, a_7 = 16$ **17.** $a_3 = 2, a_7 = 256$ **19.** $a_3 = -54, a_7 = -3{,}674{,}160$
21. neither   **23.** arithmetic   **25.** geometric   **27.** neither   **29.** geometric   **31.** geometric   **33.** neither   **35.** arithmetic
37. neither   **39.** neither   **41.** 63   **43.** 152   **45.** 378   **47.** $-14{,}763$ **49.** 1.9653   **51.** 0.8280   **53.** 1/3   **55.** $-4/9$
65. $1024a^5 - 1280a^4b + 640a^3b^2 - 160a^2b^3 + 20ab^4 - b^5$
67. $a^4 + 16a^{7/2}b^{1/2} + 112a^3b + 448a^{5/2}b^{3/2} + 1120a^2b^2 + 1792a^{3/2}b^{5/2} + 1792ab^3 + 1024a^{1/2}b^{7/2} + 256b^4$ **69.** $241{,}920x^3y^4$ **71.** 26^8 **73.** 2730
75. 672   **77.** 1/8, 3/8   **79.** 0.285   **81.** drawing an ace and a ten-card from one deck   **83.** 1/4

Chapter 9 Test, page 547

1. $a_3 = 4/3, a_5 = 4/15$ **2.** $a_3 = 1/6, a_5 = 1/10$ **3.** $a_3 = 12, a_5 = 48$ **4.** arithmetic   **5.** neither   **6.** geometric   **7.** 49/20
8. 1023/1024   **9.** 590   **10.** 58   **11.** 3/5   **12.** 5/33   **15.** $x^5 - 10x^4y + 40x^3y^2 - 80x^2y^3 + 80xy^4 - 32y^5$
16. $x^6 + 6x^4 + 15x^2 + 20 + 15/x^2 + 6/x^4 + 1/x^6$ **17.** $48{,}384x^3y^5$ **18.** 132,600   **19.** 568,339,200   **20.** 0.294118

Glossary

abscissa The x-coordinate of an ordered pair. (Section 3.1)

absolute minimum A minimum value of a function f that is also the smallest range value of f. (Section 4.2)

absolute value The absolute value of the real number a, denoted $|a|$, equals a when $a \geq 0$ and equals $-a$ when $a < 0$. (Section 1.2)

addition Addition of the two real numbers a and b is designated by $a + b = c$, where c is the sum and the real numbers a and b are called terms. (Section 1.1)

additive inverse The number $-b$ is called the additive inverse of b. (Section 1.1)

additive inverse of a polynomial If $P(x)$ is a polynomial, then $-P(x)$ is the additive inverse of $P(x)$. (Section 1.5)

alternating sequence A sequence in which the signs of the terms alternate between positive and negative values. (Section 9.1)

annuities Deposits of equal amounts at equal intervals of time. (Section 9.3)

antilogarithm In $\log_a M = N$, the number M. (Section 5.2)

argument The independent variable of a function. (Section 5.2)

arithmetic mean The arithmetic mean of two numbers a and b is $(a + b)/2$. (Section 9.2)

arithmetic sequence A sequence in which the difference between any two successive terms is constant. (Section 9.2)

arithmetic series The sum of the terms of an arithmetic sequence. (Section 9.2)

asymptotes A line (or curve) approached by another curve in the sense that the perpendicular distance from a point on the curve to the asymptote approaches zero as the point moves an infinite distance from the origin of the coordinate system. (Section 4.5)

augmented matrix A matrix consisting of the coefficients and constants of a system of equations. (Section 8.2)

average velocity The ratio of the change in distance to the change in time. (Section 3.6)

axis of symmery of a parabola The line that passes through the focus and is perpendicular to the directrix. (Section 6.1)

base In the expression b^x, b is the base. (Section 1.3)

binomial A simplified polynomial that has two terms. (Section 1.5)

binomial coefficient The coefficient of a term of a binomial expansion. (Section 9.5)

Boyle's Law The volume V of a sample of gas (at a constant temperature) varies inversely as the pressure P. (Section 3.8)

break-even point The value of x for which $R(x) = C(x)$ (revenue equals cost). (Section 3.3)

Cartesian coordinate system A two-dimensional coordinate system formed by the intersection of two number lines. (Section 3.1)

center of a hyperbola The midpoint of the transverse axis. (Section 6.3)

center of an ellipse The midpoint of the major axis. (Section 6.2)

circle The set of points in a plane that are a fixed distance from a specified point. (Section 3.1)

closed interval $[a, b]$ represents all real numbers between a and b, including a and including b. (Section 1.2)

coefficient The constant of a monomial. (Section 1.5)

coefficient matrix The matrix formed by the coefficients of a system of equations. (Section 8.1)

cofactor $(-1)^{i+j}M_{ij}$, where M_{ij} is the minor of the matrix. (Section 8.4)

combination An arrangement of objects for which the order of the selection is not important. (Section 9.6)

combined variation A variation that involves more than one type of variation. (Section 3.8)

common difference In an arithmetic sequence, the difference between any two successive terms. (Section 9.2)

common logarithm A logarithm with a base of 10. (Section 5.2)

complex conjugates The complex numbers $a + bi$ and $a - bi$. (Section 1.4)

complex number A number in the form $a + bi$, where a and b are real numbers and i is the imaginary unit. (Section 1.4)

composite function A function formed from the composition of two functions f and g, given by $(f \circ g)(x) = f(g(x))$. (Section 3.6)

composite number A composite number is an integer greater than 1 that is not a prime number. (Section 1.1)

compound continuously In compounding interest, to increase the number of compounding periods without bound. (Section 5.5)

compound inequality An inequality formed by joining two inequalities with the connective word *and* or *or*. (Section 2.5)

compound interest Interest that is added to principal at regular intervals so that interest is paid on interest as well as on principal. (Section 5.5)

conditional equation Any equation that is true for some values of the variable but not true for other values of the variable. (Section 2.1)

conjugate axis of a hyperbola The axis that passes through the center of the hyperbola and is perpendicular to the transverse axis. (Section 6.3)

conjugates The complex numbers $a + bi$ and $a - bi$. (Section 1.4)

consistent A system of equation for which the graphs intersect at a single point. (Section 7.1)

constant function A function of the form $f(x) = a$, where a is a real number. (Section 3.2)

constant matrix The matrix formed from the constants of a system of equations. (Section 8.1)

constant of proportionality In the direct variation equation $y = kx$, the value of k. (Section 3.8)

constant polynomial A nonzero constant, such as 5. (Section 1.5)

constant sequence A sequence in which each term is the same. (Section 9.1)

constant term A monomial with no variable part. (Section 1.5)

constraints Equations or inequalities that force the solution of a linear programming problem to lie within a particular set. (Section 7.6)

contradiction An equation that has no solutions. (Section 2.1)

coordinate The number associated with a particular point on a real number line. (Section 1.2)

coordinate axis A line on which each real number can be designated by a point. (Section 1.1)

coordinate plane The set of all points on a flat two-dimensional surface. (Section 3.1)

coordinates An ordered pair of numbers. (Section 3.1)

cost function The function, C, that gives a manufacturer's cost to produce x units of a product. (Section 3.3)

critical values of a rational expression The numbers that cause the numerator or the denominator of the rational expression to equal zero. (Section 2.5)

cube The product of the same three factors. (Section 1.6)

cube root One of the three equal factors of a cube. (Section 1.6)

cubic equation An equation of the form $ax^3 + bx^2 + cx + d = 0$, where $a \neq 0$. (Section 2.4)

decreasing function A function f for which, for all x_1 and x_2 in the domain of f, $f(x_1) > f(x_2)$ whenever $x_1 < x_2$. (Section 3.2)

degree of a monomial The sum of the exponents of the variables in the monomial. (Section 1.5)

degree of a polynomial The largest degree of the terms in the polynomial. (Section 1.5)

denominator The nonzero real number b in the fraction a/b. (Section 1.1)

dependent A system of equations which has an infinite number of solutions. (Section 7.1)

dependent variable For a function defined by an equation, the variable that represents elements of the range. (Section 3.2)

Descartes' Rule of Signs A theorem that describes the number of positive or negative zeros a polynomial function may have. (Section 4.3)

determinant A square array of elements having a value determined by a rule involving the sum of the products of certain elements. (Section 8.4)

diagonal form A matrix is in diagonal form if all elements below and above the main diagonal are zero. (Section 8.4)

difference If $a - b = c$ then c is called the difference of a and b. (Section 1.1)

difference of two squares An expression of the form $a^2 - b^2$. (Section 1.6)

difference quotient The quotient defined by $[f(x + h) - f(x)]/h$. (Section 3.6)

dimension of a matrix A matrix of m rows and n columns has dimension $m \times n$ (read "m by n"). (Section 8.1)

directly proportional If $y = kx$, the variable y varies directly as the variable x, or y is directly proportional to x. (Section 3.8)

directrix A line perpendicular to the line containing the foci of a conic section. (Section 6.1)

discriminant For $ax^2 + bx + c$ where $a \neq 0$, the discriminant is $b^2 - 4ac$. (Section 2.3)

dividend A quantity to be divided. (Section 4.1)

division The division of a and b, designated by $a \div b$. (Section 1.1)

divisor The quantity by which another quantity, the dividend, is to be divided. (Section 4.1)

domain The set of all the first coordinates of the ordered pairs of a function. (Section 3.2)

domain of a rational expression The set of all real numbers that can be used as replacements for the variable in the rational expression. (Section 1.7)

double root A root of an equation that is repeated twice. (Section 2.3)

eccentricity A measure used to describe a characteristic of a conic section. The value of the eccentricity is c/a, where c is the distance from the center to a focus and a is the distance from the center to a vertex. (Section 6.2)

echelon form A form of a matrix in which the first nonzero element in any row is a 1, the rows are arranged so that the column containing the first nonzero number in any row is to the left of the column containing the first nonzero number of the next row, and all rows consisting entirely of zeros appear at the bottom of the matrix. (Section 8.1)

element of a set Each member of the set. (Section 1.1)

element of a matrix Each member in the matrix. (Section 8.1)

elementary row operation An operation performed on the rows of a matrix. (Section 8.1)

elimination method A method of solving a system of equations. (Section 7.1)

ellipse The set of all points in a plane, the sum of whose distances from two fixed points (foci) is a positive constant. (Section 6.2)

empty set The set without any elements. (Section 1.1)

equals a equals b (denoted by $a = b$) if $a - b = 0$. (Section 1.2)

equation A statement of equality between two numbers or two expressions. (Sections 1.1, 2.1)

equivalent systems of equations Systems of equations that have exactly the same solution(s). (Section 7.1)

equivalent equations Equations that have exactly the same solution(s). (Section 2.1)

equivalent inequalities Inequalities that have the same solution set. (Section 2.5)

evaluate a polynomial To substitute the given value(s) for uthe variable(s) and then perform the indicated operations using the Order of Operations Agreement. (Section 1.5)

even function A function f for which $f(-x) = f(x)$ for all x in the domain of f. (Section 3.5)

event Any subset of a sample space. (Section 9.7)

experiment An activity with an observable outcome. (Section 9.7)

exponent In the expression b^n, n is the exponent. (Section 1.3)

exponential decay function A function of the form $A(x) = Ae^{kt}$, where $k < 0$ and $t \geq 0$. (Section 5.5)

exponential equation An equation in which a variable appears as an exponent in a term of the equation. (Section 5.4)

exponential function   A function defined by $f(x) = b^x$ where $b > 0$, $b \neq 1$, and x is any real number. (Section 5.1)

exponential growth function   A function of the form $A(x) = Ae^{kt}$, where $k > 0$ and $t \geq 0$. (Section 5.5)

exponential notation   An expression written in the form b^x. (Section 1.3)

extraneous solution   An apparent solution of an equation that is not a solution of the original equation. (Section 2.4)

factor by grouping   To factor by first grouping together pairs of terms that have a common factor. (Section 1.6)

factoring   Writing a polynomial as a product of polynomials of lower degree. (Section 1.6)

factoring over the integers   Factoring by using only polynomial factors that have integer coefficients. (Section 1.6)

factors   If $ab = c$, then a and b are called factors of c. (Section 1.1)

final demand   The amount of output that a consumer will want. (Section 8.3)

formula   An equation that expresses known relationships between two or more variables. (Section 2.2)

function   A set of ordered pairs in which no two ordered pairs that have the same first coordinate have different second coordinates. (Section 3.2)

Fundamental Theorem of Algebra   If $P(x)$ is a polynomial with complex number coefficients and is of degree greater than or equal to 1, then $P(x)$ has at least one real zero. (Section 4.4)

future value   The total value of an investment after the last deposit. (Section 9.3)

Gaussian elimination method   An algorithm that uses elementary row operations to solve a system of linear equations. (Section 8.2)

general form of the equation of a circle   An equation of the form $x^2 + y^2 + ax + by + c = 0$. (Section 3.1)

general form of the equation of a line   An equation of the form $Ax + By + C = 0$, where A, B, and C are real numbers and both A and B are not 0. (Section 3.3)

geometric sequence   A sequence in which the ratio of any two successive terms is a constant. (Section 9.3)

graph of a function   The graph of all the ordered pairs that belong to the function. (Section 3.2)

graph of an equation   The set of all points whose coordinates satisfy the equation. (Section 3.1)

greater than a is greater than b (denoted by $a > b$) if $a - b$ is positive. (Section 1.2)

greatest integer function   The function, denoted by $f(x) = [\![x]\!]$, for which the value of the function is the greatest integer less than or equal to x. (Section 3.2)

half-life   The time required for the disintegration of half of the atoms in a sample of a radioactive substance. (Section 5.5)

half-open interval $(a, b]$ represents all real numbers between a and b, not including a, but including b; $[a, b)$ represents all real numbers between a and b, including a, but not including b. (Section 1.2)

half-plane   Each region in a plane separated by a line. (Section 7.5)

homogeneous system of equations   A linear system of equations for which the constant term of each equation is 0. (Section 7.2)

horizontal asymptote   If $\lim\limits_{x \to \pm\infty} f(x) = A$, then $y = A$ is a horizontal asymptote of the graph of f. (Section 4.5)

hyperbola   The set of all points in a plane, the difference between whose distances from two fixed points (foci) is a positive constant. (Section 6.3)

hypotenuse   In a right triangle, the side opposite the $90°$ angle. (Section 2.3)

identity   An equation that is true for *every* real number for which all terms of the equation are defined. (Section 2.1)

identity matrix   An $n \times n$ matrix that has 1s on the main diagonal and 0s as the remaining elements. (Section 8.2)

imaginary number   A number in the form ai where i is the imaginary unit and a is a real number. (Section 1.4)

imaginary part of a complex number   The real number b for the complex number $a + bi$. (Section 1.4)

imaginary unit   The number i, defined so that $i^2 = -1$. (Section 1.4)

inconsistent system of equations   A system of equations that has no solution. (Section 7.1)

increasing function   A function f for which, for all elements x_1 and x_2 in the domain of f, $f(x_1) < f(x_2)$, whenever $x_1 < x_2$. (Section 3.2)

independent events   Two events for which the outcome of the first event does not influence the outcome of the second event. (Section 9.7)

independent variable   For a function defined by an equation, the variable that represents elements of the domain. (Section 3.2)

index of a radical   In the expression $\sqrt[n]{a}$, the positive integer n is the index of the radical. (Section 1.3)

infinite sequence   A function whose domain is the positive integers and whose range is a set of real numbers. (Section 9.1)

infinite series   The sum of all the terms of an infinite sequence. (Section 9.3)

integers   The numbers $\ldots -4, -3, -2, -1, 0, 1, 2, 3, \ldots$. (Section 1.1)

intercept   Any point on a graph that has an x- or a y-coordinate of 0; a point where the graph intersects the x- or the y-axis. (Section 3.1)

interest   Money paid for the use of money. (Section 5.5)

intersection of sets   The intersection of sets A and B, denoted by $A \cap B$, is the set of all elements that belong to both set A and set B. (Section 1.1)

interval notation   A compact notation used to represent subsets of real numbers. (Section 1.2)

inverse of a matrix   The inverse of matrix A, denoted by A^{-1}, is the matrix with the property that $AA^{-1} = I$, the identity matrix. (Section 8.3)

inverse function   The function, denoted by f^{-1}, that is formed by interchanging the x and y coordinates of a function f. (Section 3.7)

inversely proportional   If $y = k/x$, the variable y varies inversely as the variable x, or y is inversely proportional to x. (Section 3.8)

irrational numbers   The set of all nonterminating, nonrepeating decimals. (Section 1.1)

irreducible over the reals   A quadratic factor with no real zeros. (Section 4.4)

leading coefficient The coefficient a_n of a polynomial of degree n. (Section 1.5)

legs In a right triangle, the two sides other than the hypotenuse. (Section 2.3)

less than a is less than b (denoted by $a < b$) if $b - a$ is positive. (Section 1.2)

like radicals Radicals that have the same radicand and the same index. (Section 1.1)

like terms Terms that have exactly the same variables raised to the same powers. (Section 1.5)

linear equation An equation that can be written in the form $ax + b = 0$ where a and b are real numbers, and $a \neq 0$. (Section 2.1)

linear extrapolation A linear approximation of data beyond the given values. (Section 3.3)

linear function A function of the form $f(x) = ax + b$. (Section 3.3)

linear interpolation A linear approximation of data between the given values. (Section 3.3)

linear objective function The function to be maximized or minimized in a linear programming problem. (Section 7.6)

linear programming A technique of solving some types of maximization or minimization problems. (Section 7.6)

linear system of equations A system of equations in which each equation is a linear equation. (Section 7.1)

logarithmic function f with base b $y = \log_b x$ if and only if $b^x = y$. (Section 5.2)

logarithmic equation An equation that involves logarithms. (Section 5.4)

lower bound A real number a for which no zero of the polynomial function P is less than a. (Section 4.3)

major axis The longer axis of the graph of an ellipse. (Section 6.2)

matrix A rectangular array of numbers. (Section 8.1)

maximum value of a quadratic function If $a < 0$, then the vertex (h, k) is the highest point on the graph of $f(x) = a(x - h)^2 + k$, and k is the maximum value of the function f. (Section 3.4)

maximum value of a function The largest range element of the function. (Section 3.2)

midpoint The point on a line segment that is equidistant from the endpoints of the segment. (Section 3.1)

minimum value of a function The smallest range element of the function. (Section 3.2)

minimum value (of a quadratic function) If $a > 0$, then the vertex (h, k) is the lowest point on the graph of $f(x) = a(x - h)^2 + k$, and k is the minimum value of the function f. (Section 3.4)

minor of a matrix The determinant formed by removing the ith row and jth column of the determinant of the matrix; denoted by M_{ij}. (Section 8.4)

minor axis The shorter axis of the graph of an ellipse. (Section 6.2)

monomial A constant, a variable, or a product of a constant and one or more variables, with the variables having only nonnegative integer exponents. (Section 1.5)

multiplication Multiplication of the real numbers a and b is designated by ab. (Section 1.1)

multiplicative inverse The multiplicative inverse or reciprocal of the nonzero number b is $1/b$. (Section 1.1)

mutually exclusive events Two events A and B for which $A \cap B = \emptyset$. (Section 9.7)

n factorial ($n!$) $n! = n(n - 1)(n - 2) \cdots 3 \cdot 2 \cdot 1$, n a natural number. $0! = 1$. (Section 9.1)

natural exponential function The function defined by $f(x) = e^x$ for all real numbers x. (Section 5.1)

natural logarithm A logarithm with base e. (Section 5.2)

natural number A positive integer. (Section 1.1)

negative integer An integer less than 0. (Section 1.1)

negative real number A real number less than 0. (Section 1.2)

nonfactorable over the integers A polynomial that cannot be factored into the product of two polynomials having integer coefficients. (Section 1.6)

nonlinear system of equations A system of equations in which one or more equations are not linear equations. (Section 7.3)

nonsingular matrix A matrix that has a multiplicative inverse. (Section 8.3)

nth partial sum The sum of the first n terms of a sequence. (Section 9.1)

null set The set without any elements; denoted by $\emptyset$. (Section 1.1)

numerator The real number a in the fraction a/b. (Section 1.1)

numerical coefficient The constant in a monomial. (Section 1.5)

odd function A function for which $f(-x) = -f(x)$. (Section 3.5)

one-to-one function A function that satisfies the additional condition that given any y value, there is only one x value paired with that given y value. (Section 3.2)

open interval (a, b) represents all real numbers between a and b, not including a and not including b. (Section 1.2)

optimization problem A problem that requires a situation to be maximized or minimized. (Section 7.6)

order of a matrix A matrix of m rows and n columns has order $m \times n$ (read "m by n"). (Section 8.1)

ordinary annuity An annuity for which the amounts are deposited at the end of a compounding period. (Section 9.3)

ordinate The y-coordinate of an ordered pair. (Section 3.1)

origin The point, $(0, 0)$, where the x- and y-axes intersect. (Section 3.1)

parabola The set of all points in a plane that are equidistant from a fixed line (directrix) and a fixed point (focus) not on the directrix. (Section 6.1)

parallel lines Two nonintersecting lines in a plane. (Section 3.3)

partial fraction decomposition The method by which a more complicated rational expression is written as a sum of simpler rational expressions. (Section 7.4)

Pascal's Triangle A triangular array of the coefficients of the terms of expanded binomials. (Section 9.5)

perfect-square trinomial A trinomial that is the square of a binomial. (Section 1.6)

permutation An arrangement of distinct objects in a definite order. (Section 9.6)

perpendicular lines Two lines that intersect to form adjacent angles each of which measures 90°. (Section 3.3)

pH The negative of the common logarithm of the molar hydronium-ion concentration, M. (Section 5.4)

piecewise-defined function A function represented by more than one equation. (Section 3.2)

plot a point To draw a dot at the point's location in the coordinate plane. (Section 3.1)

point-slope form The equation of a straight line written in the form $y - y_1 = m(x - x_1)$. (Section 3.3)

polynomial A sum of a finite number of monomials. (Section 1.5)

positive integer An integer greater than zero. (Section 1.1)

positive real number A number to the right of the origin. (Section 1.2)

power The expression b^n is the nth power of b. (Section 1.3)

prime number A positive integer greater than 1 that has no positive-integer factors other than itself and 1. (Section 1.1)

principal An amount of money invested. (Section 5.5)

principal square root The positive square root of a number. (Section 1.3)

probability The mathematical study of random patterns. (Section 9.7)

product If $ab = c$, then c is the product. (Section 1.1)

profit function The function, P, that gives a manufacturer's profit from selling x units of a product. (Section 3.3)

quadrants The four regions formed by the axes. (Section 3.1)

quadratic equation An equation that can be written in the standard quadratic form $ax^2 + bx + c = 0, a \neq 0$. (Section 2.3)

quadratic formula If $ax^2 + bx + c = 0, a \neq 0$, then

$$x = \frac{-b \pm \sqrt{b^2 - 4ac}}{2a}.$$ (Section 2.3)

quadratic function A function that can be represented by an equation of the form $f(x) = ax^2 + bx + c, a \neq 0$. (Section 3.4)

quadratic in form A polynomial that can be written in the form $au^2 + bu + c = 0$, where $a \neq 0$. (Section 2.4)

quotient If $a \div b = c$, then c is the quotient of a and b; the number obtained when dividing one quantity by another. (Sections 1.1, 4.1)

radicals Expressions using the notation $\sqrt[n]{b}$, also used to denote roots. (Section 1.3)

radicand In the expression $\sqrt[n]{b}$, the number b is the radicand. (Section 1.3)

radius The distance from the center of a circle or sphere to a point on the circle or sphere. (Section 3.1)

range The set of all the second coordinates of the ordered pairs of a function. (Section 3.2)

rational expression A fraction in which the numerator and the denominator are polynomials. (Section 1.7)

rational function A function that can be expressed as a quotient of polynomials. (Section 4.5)

rational inequalities An inequality that involves rational expressions. (Section 2.5)

rational numbers The set of all terminating or repeating decimals. (Section 1.1)

rationalize the denominator To write a fraction in an equivalent form that does not involve any radicals in the denominator. (Section 1.1)

real number line A coordinate axis used to represent the real numbers geometrically. (Section 1.1)

real numbers The set of all rational or irrational numbers. (Section 1.1)

real part of a complex number The real number a for the complex number $a + bi$. (Section 1.4)

reciprocal The multiplicative inverse or reciprocal of the nonzero number b is $1/b$. (Section 1.1)

reciprocal function The function denoted by $1/f$. (Section 3.7)

recursively defined sequence A sequence in which each succeeding term of the sequence is defined using one or more of the preceding terms. (Section 9.1)

reduced polynomial The polynomial formed from $P(x)/(x - a)$, where a is a zero of $P(x)$. (Section 4.3)

remainder The number left over when one integer is divided by another. (Section 4.1)

revenue function The function, R, that gives a manufacturer's revenue from the sale of x units of a product. (Section 3.3)

right triangle A triangle that contains one 90° angle. (Section 2.3)

roots of an equation The values of the variable that satisfy an equation. (Section 2.1)

roots of a polynomial The values of x for which a polynomial $P(x)$ is equal to 0. (Section 4.1)

sample space The set of all possible outcomes of an experiment. (Section 9.7)

scalar multiplication The product of a real number and a vector or a matrix. (Section 8.2)

scientific notation A number in the form $a \times 10^n$, where n is an integer and $1 \leq a < 10$. (Section 1.3)

sequence of partial sums A sequence formed from the partials sums of another sequence. (Section 9.1)

series The indicated sum of a sequence. (Section 9.1)

set of feasible solutions The solution set of the constraints of a linear programming problem. (Section 7.6)

set-builder notation Makes use of a variable and a characteristic property of the elements of the set alone possess. (Section 1.1)

simple interest Interest that is a fixed percent r, per time period t, of the amount of money invested. (Section 5.5)

simple zero A zero of multiplicity 1. (Section 4.3)

simplified A rational expression is simplified when 1 is the only common polynomial factor of both the numerator and the denominator. (Section 1.7)

simplify a rational expression To factor the numerator and the denominator of the rational expression. (Section 1.7)

singular matrix A matrix that does not have a multiplicative inverse. (Section 8.3)

slant asymptote A linear asymptote that is not a vertical or horizontal line. (Section 4.5)

slope of a line The ratio of the change in y to the change in x between two points on the line. (Section 3.3)

slope-intercept form The equation of a line written in the form $f(x) = mx + b$; the slope is m and the y-intercept is $(0, b)$. (Section 3.3)

smooth, continuous curve   A curve that does not have sharp corners, a break, or a hole. (Section 4.2)

solution of a system of equations in two variables   An ordered pair that is a solution of both equations of the system. (Section 7.1)

solution set of a system of inequalities   The intersection of the solution sets of the individual inequalities. (Section 7.5)

solution set of an inequality in one variable   The set of all solutions of the inequality. (Section 2.5)

solution set of an inequality in two variables   The set of all ordered pairs that satisfy the inequality. (Section 7.5)

solutions of an equation   The values of the variable that satisfy the equation. (Section 2.1)

solve an equation   To find all values of the variable that satisfy the equation. (Section 2.1)

square matrix of order n   A matrix with n rows and n columns. (Section 8.1)

square root   If the index n equals 2, then the radical $\sqrt[n]{b}$ is written as simply $\sqrt{b}$, and it is referred to as the principal square root of b or simply the square root of b. (Section 1.3)

standard form of a quadratic function   A quadratic function $f(x) = ax^2 + bx + c$ written in the form $f(x) = a(x - h)^2 + k$. (Section 3.4)

standard form of the equation of a circle   An equation of a circle written in the form $(x - h)^2 + (y - k)^2 = r^2$, where (h, k) is the center and r is the radius. (Section 3.1)

standard form of a polynomial   A polynomial in the variable x written with decreasing powers of x. (Section 1.5)

standard quadratic form   A quadratic equation written in the form $ax^2 + bx + c = 0, a \neq 0$. (Section 2.3)

step function   A function of the form $f(x) = [\![x]\!]$, where $[\![x]\!]$ is the greatest integer function. (Section 3.2)

subset   Set A is a subset of set B if every element of set A is also an element of set B. (Section 1.1)

substitution method   A method of solving a system of equations. (Section 7.1)

subtraction   Subtraction of the real numbers a and b is designated by $a - b$. (Section 1.1)

sum   If $a + b = c$, then c is the sum. (Section 1.1)

summation notation   A convenient notation used for partial sums. (Section 9.1)

symmetric with respect to a line   A graph is symmetric with respect to a line L if, for each point P on the graph, there is a point P' on the graph such that the line L is the perpendicular bisector of the line segment PP'. (Section 3.4)

symmetric with respect to a point   A graph is symmetric with respect to a point Q if, for each point P on the graph, there is a point P' on the graph such that Q is the midpoint of the line segment PP'. (Section 3.5)

symmetric with respect to the x-axis   A graph is symmetric with respect to the x-axis if, whenever the point given by (x, y) is on the graph, then $(x, -y)$ is also on the graph. (Section 3.5)

symmetric with respect to the y-axis   A graph is symmetric with respect to the y-axis if, whenever the point given by (x, y) is on the graph, then $(-x, y)$ is also on the graph. (Section 3.5)

synthetic division   A procedure for dividing a polynomial by a binomial of the form $x - c$. (Section 4.1)

system of equations   Two or more equations considered together. (Section 7.1)

term   Each monomial of a polynomial. (Section 1.5)

terms   If $a + b = c$, then the real numbers a and b are called the terms. (Section 1.1)

terms of a sequence   The elements in the range of the sequence. (Section 9.1)

third-degree equation   A cubic equation. (Section 2.4)

tolerance   The acceptable amount by which a dimension may differ from a given standard. (Section 2.6)

transverse axis of a hyperbola   The line segment joining the intercepts. (Section 6.3)

triangular form of a matrix   A matrix in which the entries above or below the main diagonal are zero. (Section 8.2)

trinomial   A simplified polynomial that has three terms. (Section 1.5)

trivial solution   For a homogeneous system of equations, the solution that contains all zeros. (Section 7.2)

turning point   A point where a function changes from an increasing function to a decreasing function or vice versa. (Section 4.2)

union   The union of sets A and B, denoted by $A \cup B$, is the set of all elements belonging to set A, to set B, or both. (Section 1.1)

upper bound   A real number b for which no zero of the polynomal function P is greater than b. (Section 4.3)

variation   Many real-life situations involve variables that are related by a type of function called a variation. (Section 3.8)

variation constant   In the direct variation equation $y = kx$, the value of k. (Section 3.8)

varies directly   If $y = kx$, the variable y varies directly as the variable x. (Section 3.8)

varies directly as the nth power   If y varies directly as the nth power of x, then $y = kx^n$, where k is a constant. (Section 3.8)

varies inversely   If $y = k/x$, the variable y varies inversely as the variable x. (Section 3.8)

varies inversely as the nth power   If y varies inversely as the nth power of x, then $y = k/x^n$, where k is a constant. (Section 3.8)

varies jointly   The variable z varies jointly as the variables x and y if and only if $z = kxy$, where k is a constant. (Section 3.8)

vertex of a parabola   The lowest point of a parabola that opens up or the highest point on a parabola that opens down; the midpoint of the line segment joining the focus and directrix of the parabola. (Sections 3.4, 6.1)

vertical asymptote   If $\lim\limits_{x \to a} f(x) = \pm\infty$, then $x = a$ is a vertical asymptote of the graph of f. (Section 4.5)

vertices of a hyperbola   The points where the hyperbola intersects the transverse axis. (Section 6.3)

vertices of an ellipse   The endpoints of the major axis of the ellipse. (Section 6.2)

x-axis   A horizontal coordinate axis. (Section 3.1)

x-coordinate   In the ordered pair (a, b), the real number a. (Section 3.1)

x-intercept   A point at which a graph crosses the x-axis. (Section 3.1)

y-axis   A vertical coordinate axis. (Section 3.1)

y-coordinate   In the ordered pair (a, b), the real number b. (Section 3.1)

y-intercept   A point at which a graph crosses the y-axis. (Section 3.1)

z-axis   A third coordinate axis perpendicular to the xy plane. (Section 7.2)

zero of a polynomial   Any value of x that causes a polynomial in x to equal 0 is called a zero of the polynomial. (Section 2.5)

zero matrix   A matrix in which all the elements are 0. (Section 8.2)

zero of multiplicity k   If a polynomial $P(x)$ has $(x - r)$ as a factor exactly k times, then r is a zero of multiplicity k of $P(x)$. (Section 4.3)

zero product property   The zero product property states that if the product of two factors equals 0, then at least one of the factors is 0. (Section 2.3)

zeros of a function   For a function f, the values of x for which $f(x) = 0$. (Section 4.1)

Index

direction. Repeat for two rows.

Step 2. Join the rows referring to the Placement Diagram; press seams in one direction.

Step 3. Join one each Half-Log Cabin block and Half-Log Cabin Reversed block as shown in **Figure 5**; repeat. Press seams in one direction.

Step 4. Sew a pieced triangle unit to each short end of the pieced center section to complete the top; press seams toward triangle units.

Completing the Runner

Step 1. Using the completed top as a pattern, cut the backing piece the exact size of the completed top. Layer with wrong sides together.

Step 2. Prepare 4½ yards multicolored silk for binding and bind edges.

Step 3. Hand- or machine-stitch in the ditch of some seams to hold backing and pieced top layers together to finish. 🎁

FIGURE 4 Join 4 Log Cabin blocks to make a row.

FIGURE 5 Join 1 Half-Log Cabin block and 1 reversed block.

Metric Conversion Charts

Metric Conversions

U.S. Measurement		Multiplied by		Metric Measurement
yards	x	.9144	=	meters (m)
yards	x	91.44	=	centimeters (cm)
inches	x	2.54	=	centimeters (cm)
inches	x	25.40	=	millimeters (mm)
inches	x	.0254	=	meters (m)

Metric Measurement		Multiplied by		U.S. Measurement
centimeters	x	.3937	=	inches
meters	x	1.0936	=	yards

Standard Equivalents

U.S. Measurement		Metric Measurement		
1/8 inch	=	3.20 mm	=	0.32 cm
1/4 inch	=	6.35 mm	=	0.635 cm
3/8 inch	=	9.50 mm	=	0.95 cm
1/2 inch	=	12.70 mm	=	1.27 cm
5/8 inch	=	15.90 mm	=	1.59 cm
3/4 inch	=	19.10 mm	=	1.91 cm
7/8 inch	=	22.20 mm	=	2.22 cm
1 inch	=	25.40 mm	=	2.54 cm
1/8 yard	=	11.43 cm	=	0.11 m
1/4 yard	=	22.86 cm	=	0.23 m
3/8 yard	=	34.29 cm	=	0.34 m
1/2 yard	=	45.72 cm	=	0.46 m
5/8 yard	=	57.15 cm	=	0.57 m
3/4 yard	=	68.58 cm	=	0.69 m
7/8 yard	=	80.00 cm	=	0.80 m
1 yard	=	91.44 cm	=	0.91 m

U.S. Measurement		Metric Measurement		
1 1/8 yard	=	102.87 cm	=	1.03 m
1 1/4 yard	=	114.30 cm	=	1.14 m
1 3/8 yard	=	125.73 cm	=	1.26 m
1 1/2 yard	=	137.16 cm	=	1.37 m
1 5/8 yard	=	148.59 cm	=	1.49 m
1 3/4 yard	=	160.02 cm	=	1.60 m
1 7/8 yard	=	171.44 cm	=	1.71 m
2 yards	=	182.88 cm	=	1.83 m
2 1/8 yards	=	194.31 cm	=	1.94 m
2 1/4 yards	=	205.74 cm	=	2.06 m
2 3/8 yards	=	217.17 cm	=	2.17 m
2 1/2 yards	=	228.60 cm	=	2.29 m
2 5/8 yards	=	240.03 cm	=	2.40 m
2 3/4 yards	=	251.46 cm	=	2.51 m
2 7/8 yards	=	262.88 cm	=	2.63 m
3 yards	=	274.32 cm	=	2.74 m
3 1/8 yards	=	285.75 cm	=	2.86 m
3 1/4 yards	=	297.18 cm	=	2.97 m
3 3/8 yards	=	308.61 cm	=	3.09 m
3 1/2 yards	=	320.04 cm	=	3.20 m
3 5/8 yards	=	331.47 cm	=	3.31 m
3 3/4 yards	=	342.90 cm	=	3.43 m
3 7/8 yards	=	354.32 cm	=	3.54 m
4 yards	=	365.76 cm	=	3.66 m
4 1/8 yards	=	377.19 cm	=	3.77 m
4 1/4 yards	=	388.62 cm	=	3.89 m
4 3/8 yards	=	400.05 cm	=	4.00 m
4 1/2 yards	=	411.48 cm	=	4.11 m
4 5/8 yards	=	422.91 cm	=	4.23 m
4 3/4 yards	=	434.34 cm	=	4.34 m
4 7/8 yards	=	445.76 cm	=	4.46 m
5 yards	=	457.20 cm	=	4.57 m

Special Thanks

We would like to thank the talented quilt designers whose work is featured in this collection.

Frieda Anderson
Mini Log Cabin Heart, 22
Shimmering Foliage, 102

Joan Ballew
Baby Steps, 89

Pat Campbell
Hanging Diamonds, 41

Holly Daniels
Bed of Roses, 119
Stained Glass Diamonds, 168

Lucy Fazely
Joined Hearts, 151
Peaceful Journey, 165

**Lucy Fazely &
    Michael L. Burns**
Bear Paw Star, 63

Sandra L. Hatch
Crazy-Patch Log, 19

Connie Kauffman
Sunshine & Shadows, 15

Crazy Logs Kid's Quilt, 27
Plaid Lap Robe & Candle Mat, 45
Batik Bedspread, 83
Images of Africa, 86

Chris Malone
Posies 'Round the Cabin, 123

Dorothy Milligan
Red Starburst, 67

Patsy Moreland
Tulip Quartet, 147

Cynthia Myerberg
Rules of Chaos, 154

Shirley Palmer
Tulips Around the Cabin, 142

Connie Rand
Courthouse Steps Variation, 80
Log Cabin Stable, 126

Judith Sandstrom
Diamonds & Squares, 33
Framed Stars, 99

Christine Schultz
Sylvan Flora, 137

Marian Shenk
Pink Dogwood Trails, 93
Delft Cabins, 106
Stained Glass Star Flower, 111
Cabins Around the Posy, 116

Willow Ann Sirch
Silk Jewels, 171

Ruth Swasey
Lopsided Log Cabin, 72
Overlapping Cabins, 77
Simplified Log Cabin, 134

Jodi Warner
Sweetheart Log Cabin, 36

Julie Weaver
Cabins of the Bear, 50
Carolina Cabins, 55
Patriotic Cabins, 160

Fabrics & Supplies

Page 15: Sunshine & Shadows—Hobbs Heirloom fusible batting.

Page 19: Crazy-Patch Log—Bright Bitz fabrics by Jan Mullen for Marcus Brothers. Machine-quilted by Dianne Hodgkins.

Page 27: Crazy Logs Kid's Quilt—Hobbs Heirloom fusible batting and Sulky threads.

Page 45: Plaid Lap Robe & Candle Mat—Hobbs Heirloom fusible batting.

Page 50: Cabins of the Bear—Signature variegated thread from American & Efird, and Warm & Natural cotton batting from The Warm Co.

Page 55: Carolina Cabins—Lite Steam-A-Seam 2.

Page 63: Bear Paw Star—Classic Cottons/Fabric Country Traditions and Design Palette fabrics, Warm & Natural cotton batting from The Warm Co., 6½″ x 24″ ruler from Quilter's Rule, and Coats Dual Duty all-purpose thread. Quilt pieced on a Pfaff Creative 2140 sewing machine.

Page 83: Batik Bedspread—Hoffman California fabrics. Machine-quilted by Kathy Slater.

Page 99: Framed Stars—Pellon WonderUnder fusible web and Stitch-n-Tear fabric stabilizer, DMC thread, Fiskars rotary-cutting tools, Hobbs Heirloom cotton batting, and Sharpie extra-fine-point marker.

Page 147: Tulip Quartet—Pellon fleece, Olfa rotary cutter, and Omnigrid 18″ ruler and cutting mat.

Page 151: Joined Hearts—Basics and Textures Fabrics from Northcott, Warm & Natural cotton batting from The Warm Co., 6½″ x 24″ ruler from Quilter's Rule and Dual Duty Plus all-purpose thread, and machine-quilting and crafts thread from Coats.

Page 160: Patriotic Cabins—Warm & Natural cotton batting.

Page 165: Peaceful Journey—Designer Essential Scrolls and Kona Multi-Dye fabrics provided by Kaufman Fine Fabrics, Warm & Natural cotton batting from The Warm Co., and Coats Dual Duty Plus all-purpose and Star multicolored quilting and craft thread used to make sample. The quilt was stitched on a Pfaff Creative 2140 sewing machine.

Page 171: Silk Jewels—Dupioni silk available from Jan Banann & Co. at P.O. Box 8486, Warwick, RI 02888.